MONITORING ECOLOGICAL CONDITION AT REGIONAL SCALES

MONITORING ECOLOGICAL CONDITION AT REGIONAL SCALES

Proceedings of the Third Symposium on
the Environmental Monitoring and Assessment Program (EMAP)
Albany, NY, U.S.A., 8–11 April, 1997

Edited by

Shabeg Sandhu

National Health and Environmental Effects Research Laboratory, U.S. Environmental Protection Agency, Research Triangle Park, NC, U.S.A.

Laura Jackson

National Health and Environmental Effects Research Laboratory, U.S. Environmental Protection Agency, Research Triangle Park, NC, U.S.A.

Kay Austin

National Center for Environmental Assessment
U.S. Environmental Protection Agency, Washington, DC, U.S.A.

Jeffrey Hyland

National Oceanographic and Atmospheric Administration
Charleston, SC, U.S.A.

Brian Melzian

Atlantic Ecology Division,
National Health and Environmental Effects Research Laboratory
U.S. Environmental Protection Agency, Narranganestt, RI, U.S.A.

Kevin Summers

Gulf Ecology Division
National Health and Environmental Effects Research Laboratory
U.S. Environmental Protection Agency, Gulf Breeze, FL, U.S.A.

Technical Editors
Frederick de Serres and Paul Celmer
Technology Planning and Management Corporation

Reprinted from *Environmental Monitoring and Assessment*, Volume 51, Nos. 1–2, 1998

Springer-Science+Business Media, B.V.

A C.I.P. Catalogue for this book is available from the Library of Congress

ISBN 978-0-7923-5070-5 ISBN 978-94-011-4976-1 (eBook)
DOI 10.1007/978-94-011-4976-1

Printed on acid-free paper

TABLE OF CONTENTS

MONITORING ECOLOGICAL CONDITION AT REGIONAL SCALES

Proceedings of the Third Symposium on
the Environmental Monitoring and Assessment Program (EMAP)
Albany, NY, U.S.A., April 8–11, 1997

PREFACE

The Environmental Monitoring and Assessment Program was created by EPA to develop the capability for tracking the changing conditions of our natural resources and to give environmental policy the advantages of a sound scientific understanding of trends. Former EPA Administrators recognized early that contemporary monitoring programs could not even quantify simple unknowns like the number of lakes suffering from acid rain, let along determine if national control policies were benefiting these lakes. Today, adding to acidification impacts are truly complex problems such as determining the effects of climate change, of increases in ultraviolet light, toxic chemicals, eutrophication and critical habitat loss. Also today, the Government Performance and Results Act seeks to have agencies develop performance standards based on results rather than simply on levels of programmatic activities. The charge to EMAP with respect to measuring the condition of ecosystems is, therefore, the same today as it was a decade ago. We welcome the increasing urgency for sound scientific monitoring methods and data by efforts to protect and improve the environment.

Systematic nationwide monitoring of natural resources is more than any one program can accomplish, however. In an era of declining budgets, it is crucial that monitoring programs at all levels of government coordinate and share environmental data. EMAP resources are dwarfed by the more than $500 million spent on federal monitoring activities each year. Fortunately, the CENR has taken the initiative to forge a national monitoring framework among federal agencies and to coordinate the advancement of the science of environmental monitoring. The original goals of EMAP are now those of the CENR National Framework and EMAP has pledged its full support of the CENR activities to put the Framework into practice.

EMAP has been the subject of more than 20 peer reviews, a fact which seems to me to be unprecedented for a monitoring program. In 1995, the EPA Office of Research and Development demonstrated the importance of peer review by reshaping the EMAP strategy in accordance with those peer reviews. The EMAP strategy is to develop methods for measuring the integrity of ecosystems as well as methods to detect trends in these measures. Working through the CENR initiatives, EMAP will support the evaluation of large scale, multi-resource assessments such as the Mid-Atlantic region of the U.S. Through the EPA grants programs, almost $12 million each year will be targeted to stimulate the development of new ecological and landscape indicators for monitoring.

This third EMAP Symposium brings together scientists and managers to discuss the changes in EMAP and recent advances in monitoring science during the creation of a national agenda for monitoring. The symposium illustrates that trends in acid rain effects are still an important objective to EMAP and that monitoring designs can now include acidification as one of many stressors on our environment. By the time of the fourth EMAP Symposium, there will be proof-of-concept for regional-scale assessments from the CENR Mid-Atlantic pilot study. I hope that this proceedings demonstrated that EMAP continues to promote new concepts in monitoring and invite you to plan to join us in the next meeting in 1999.

Gilman Veith
Associate Laboratory Director for Ecology
National Health and Environmental Effects Research Laboratory

ENVIRONMENTAL MONITORING AND RESEARCH INITIATIVE: A PRIORITY ACTIVITY FOR THE COMMITTEE ON ENVIRONMENTAL AND NATURAL RESOURCES

DONALD PRYOR, ROSINA BIERBAUM, and JERRY MELILLO

National Science and Technology Council

Abstract: The Committee on Environment and Natural Resources (CENR) has recognized a high priority need to integrate and coordinate federal agencies' efforts in order to enable a comprehensive evaluation of our nation's environmental resources and ecological systems. The federal government spends about $640 million per year collecting data about our forests, agricultural and rangelands, lakes, rivers, estuaries, and coastal marine systems. These efforts have significantly aided the progress in preserving and protecting the environment in recent decades but are not sufficiently coordinated to provide us a truly comprehensive status report or full understanding of the causes and effects of environmental change. This paper describes the Committee on Environment and Natural Resources and its functions, provides a status report on the Environmental Monitoring and Research Initiative, and offers some perspectives on the factors that will make the initiative and its contributing programs a success. In particular, the paper discusses the potential relationship with the Environmental Monitoring and Assessment Program (EMAP).

1. Introduction

Reinventing our national environmental monitoring and research programs is, in the words of the Vice President, "one of the most significant efforts underway in our government today." The federal government conducts monitoring activities at more than 15,000 sites nationwide. The information we have derived from these sites has been a major factor in the environmental progress we have made over the past 25 years. However, the efforts are not sufficiently coordinated to provide us a truly comprehensive status report nor designed to enable full understanding of the causes and effects of environmental change.

The Committee on Environment and Natural Resources (CENR) has made it a high priority to integrate and coordinate agencies' efforts in order to enable a comprehensive evaluation of our nation's environmental resources and ecological systems. It is important to ask how each federal monitoring activity can fit into this integrated picture. A particular question for this meeting is how the Environmental Monitoring and Assessment Program (EMAP) might fit in.

2. CENR and the Environmental Monitoring Team

The Committee on Environment and Natural Resources is one of nine standing committees of the National Science and Technology Council (NSTC), the first cabinet-level science and technology entity. CENR's primary function is to coordinate environmental and

Environmental Monitoring and Assessment **51:** 3–14, 1998.

natural resources research and development across the federal agencies[1]. More specifically, it is chartered to:

- Develop and implement interagency environmental research strategy
- Facilitate planning, coordination, and communication
- Identify and recommend budget priorities
- Ensure a strong link between science and policy
- Provide review, analysis, and recommendations to the NSTC

By working together with a wide range of stakeholders including academia, industry, public interest groups, and state and local governments, the CENR has identified critical shortfalls in our understanding of important environmental issues and has worked to make sure they are addressed. One of these issues is environmental monitoring. In July of 1995 an Environmental Monitoring Team was established and charged "to develop a national framework for integration and coordination of environmental monitoring and related research through collaboration and building upon existing networks and programs."

The Environmental Monitoring Team produced a conceptual framework for how federal environmental monitoring activities can fit together. It described three "tiers" (Figure 1):

- Inventories/remote sensing, including programs such as the National Wetlands Inventory, Coastal Change Analysis Program (C-CAP), Multi-Resolution Land Characterization (MRLC), Gap Analysis Program, etc.
- National and regional surveys, including, among others, the Natural Resources Inventory (NRI), Forest Inventory and Analysis (FIA), Forest Health Monitoring, National and State and Local Air Monitoring Systems (NAMS/SLAMS), National Stream Gauging Network, and, of course, EMAP.
- Intensive monitoring and research sites or "index sites," such as Long Term Ecological Research (LTER) sites, Forest Service Experimental Forests, National Estuarine Research Reserves, and National Water Quality Assessment sites.

Interconnection of these tiers can combine the data collection and the process understanding that, together, will enable evaluation of the status, trends, and even predictions of the state of the environment. The Team's "Framework" report evolved through several drafts since early 1996. A final version is now available and can be accessed through the Environmental Monitoring Initiative's web site at "http://www.epa.gov/monitor."

---

[1]Committee on Environment and Natural Resources (1995), "Preparing for the Future through Science and Technology: An Agenda for Environmental and Natural Resource Research"

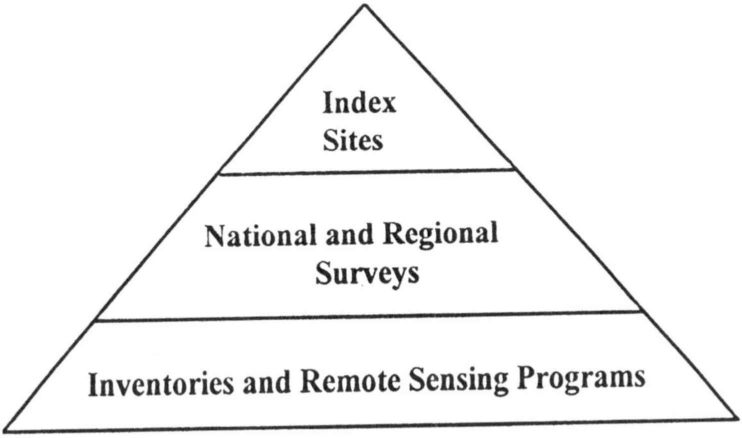

Fig. 1. Framework

An informal budget crosscut of these programs was conducted, providing a useful first approximation of the size of the federal effort -- about $640 million annually. By agency (Figure 2), the largest efforts are in the Department of Agriculture (37%), Department of the Interior (25%), the National Oceanographic and Atmospheric Administration (NOAA) (22%), and the Environmental Protection Agency (EPA) (11%). By tier (Figure 3), national and regional surveys receive most funding (52%), intensive monitoring and research sites nearly as much (45%), and inventories and remote sensing receive 3% (although most depend on satellite systems whose costs are not included). Some programs that should be included are missing from the list, such as those of the Departments of Energy and Defense, and there are some inconsistencies in funding estimates. Nevertheless, it is clear that the federal effort is large and dispersed. If it is to be as effective as possible, there must be collaboration that benefits all the agencies.

Most importantly, the nation's environmental monitoring and related research efforts must produce the information that we need to guide management and protection of the environment. Nationally, expenditures on pollution abatement and control totaled more than $120 billion in 1994 and have ranged between 1.7% and 1.8% of our Gross Domestic Product (GDP) since the mid-1970s (Figure 4)[2]. Monitoring and research must provide the needed information for decision-makers and objectively measure for citizens their return on this investment.

---

[2]Vogan, Christine R. (1996), "Pollution Abatement and Control Expenditures, 1972-94", Survey of Current Business, 74, 3, 48-67.

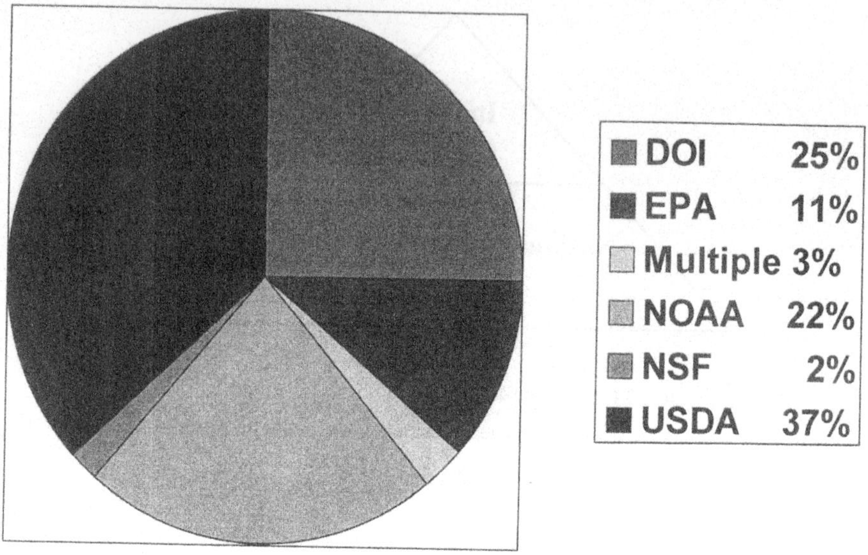

Fig. 2. Annual Federal Environmental Monitoring Budget (~$640 million) by Department/Major Agency.

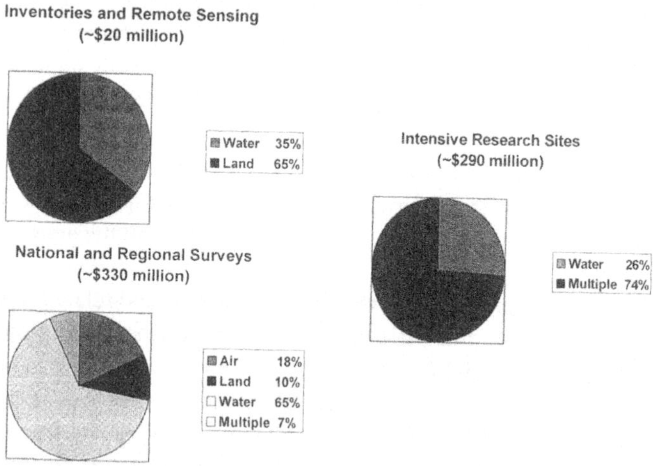

Fig. 3. Annual Federal Environmental Monitoring Budget (~$640 million) by Data Collection Method and Medium.

A national workshop was held in September of 1996, bringing together a wide range of interested parties to plan a course of action for the Environmental Monitoring and Research Initiative. The Vice President challenged the workshop participants to "work with the scientific community and other interested parties to produce a "report card" on the health of the Nation's ecosystems by 2001." By consensus, the national workshop recommended three actions. The workshop recommended:

- A first iteration toward the report card be completed in 24 months with a full report card to be completed by 2001.
- A regional pilot project or projects be undertaken—and that the mid-Atlantic be the place to begin.
- The concept of "index sites" be evaluated.

Efforts since the national workshop have focused on those three recommendations. The steering committee guiding the initiatives, which consists of senior-level officials of all the participating agencies, accepted the recommendations and set about to implement them.

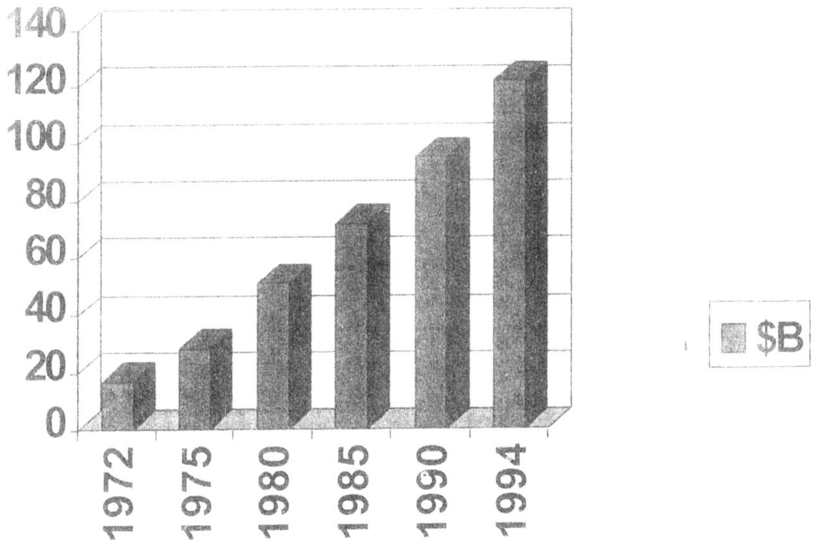

Fig. 4. Pollution Abatement and Control Expenditures.

3. Report Card

A small workshop in December of 1996 was devoted to developing an outline for the report card. The "columns," it was agreed, will be ecosystem types -- forests, croplands, rangelands, coastal/marine, fresh water, urban and possibly others. The focus of work on the first iteration should be on forests, croplands, and coastal/marine ecosystems since data may be more readily available and thinking more developed for those systems. The "rows" will reflect the status of the goods, services, and other valued attributes which ecosystems provide, including:

- Extent (such as total forest cover).
- Productivity (such as the ratio of total wood harvest to production).

- Ecosystem condition (measured by parameters such as fragmentation, extent of various age classes, and population of forest-dependent animal species).
- Recreation and aesthetics (measured by access and visitor use, extent of protected areas, etc.).
- Ecosystem services (such as cleansing air and water, providing flood and erosion protection and storing carbon).

Participants agreed that the initial effort should focus on the status and trends of ecosystem-provided goods, services, and valued attributes, and that subsequent work should include stresses with careful attention to the evidence of linkage to these goods and services. The form of the report card should be a small booklet with concise compilations of relevant data and facts.

To assemble the report card, the steering committee recognized that establishing a focal point outside the government would make it possible to engage the support and participation of the full range of interested parties including the federal government, state and local government, industry, academia and environmental organizations (Figure 5). We have had discussions with a number of interested organizations which meet the criteria of being expert in the environmental field, committed to a collaborative process, and viewed as impartial and objective. As a result of those discussions, the H. John Heinz III Center for Science, Economics, and the Environment has agreed to serve this function for the initial efforts.

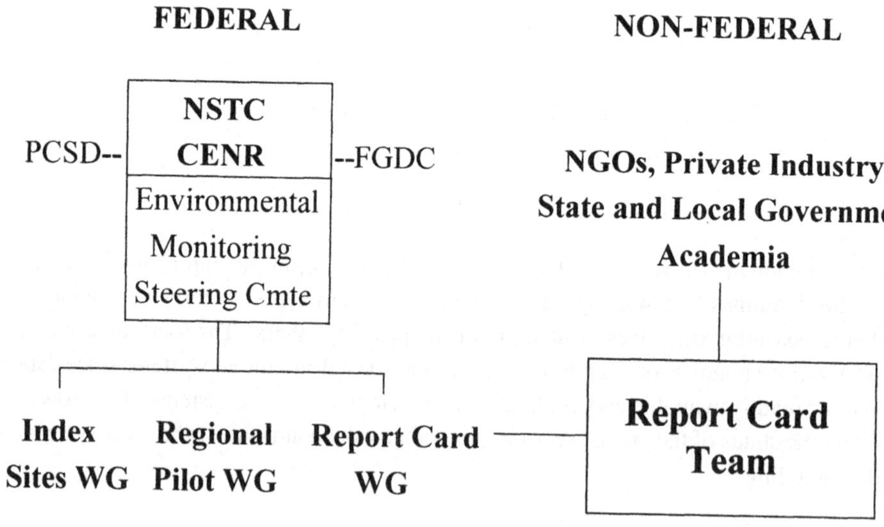

Fig. 5. Organization.

4. Regional Pilot

The second recommendation of the national workshop was to proceed with a pilot project or projects that would "take monitoring and synthesis to a more detailed level and address institutional issues." Recognizing the significant monitoring efforts underway in the mid-Atlantic region and that a regional workshop on the environmental monitoring initiative had been held in April of 1996, the national workshop recommended that the first regional pilot project be in the mid-Atlantic. A working group has been established to assemble a blueprint for work that will demonstrate the practicality and utility of integrated monitoring in the mid-Atlantic. An initial draft of the blueprint is nearing completion and, when completed, will be reviewed with as many of the regional stakeholders as possible.

5. Index Sites

The third recommendation of the national workshop was to evaluate the concept of "index" sites as a critical component of regional and national monitoring and research networks. There are many different viewpoints on the role of index sites and, at this point, no clear strategy for evaluation has yet emerged. A working group has been asked to pick a few, relatively broad science questions and examine the role of index sites in dealing with those questions. That examination may suggest ways in which the viewpoints can be brought together or seen to fit together and, as a result, provide a basis for evaluation of the concept.

6. Success Factors

With the foregoing as a status report, we would like to shift to some views of the important factors that will make this initiative a success. As background to that, recall that virtually every major review of environmental programs here in the United States has called for more comprehensive and effective environmental monitoring. For example, the National Research Council recently convened a national forum to forge a consensus on how to better link science and technology to society's environmental goals. Environmental monitoring was identified as one of six critical subject areas warranting high priority attention[3]. Simply put, environmental monitoring is recognized as something important to do better than we do now.

What is it that scientists, policy makers, resource managers and others call for environmental monitoring efforts to do that is not done today? There seem to be three pieces to the answer:

---

[3]National Research Council (1996), "Linking Science and Technology to Society's Environmental Goals" (National Academy Press, Washington, DC and World Wide Web at http://www.nas.edu).

- Support synthesis and scientifically-based prediction
- Convey degree of certainty
- Provide information to decision-makers

Following are examples of each of these factors.

First, environmental monitoring efforts should be designed and carried out to support synthesis of the information into an overall picture for a region. The New York metropolitan area regional plan[4] is an interesting example. It is the third plan from an independent planning organization that goes back to the 1930s. It covers a large enough area to contain most of the things that affect the quality of life of residents. The plan graphically depicts extrapolations of present practice into the future. The land-use pattern alone suggests that that future may not be appealing. The plan describes what might be a better alternative and outlines five campaigns to move in that direction: Greenswards, Centers, Mobility, Workforce, and Governance. It is an architectural drawing of that future, not an engineering design. It is comprehensive and plausible, but "the devil is in the details" and, certainly with respect to the air, water, forests, and other environmental aspects we care about, we don't have today either the information base or the ability to predict what that future could be.

A very similar discussion has been going on in Maryland recently around the governor's "Smart Growth" plan. The Washington Post ran a front page series last month[5] and the Washington Times ran one at the beginning of the year. The public has a great deal of interest in the way that the places in which we live can evolve—getting a picture of the environment, quality of life, development, economics, and other aspects in a comprehensive way and taking management actions based on that picture. Our environmental monitoring and research must develop the cause and effect understanding and provide the information to help in this process.

Second, environmental monitoring information must carry along with it clear expression of its degree of certainty. Take the climate change issue for example. The classic data set is Keeling's now almost 40-year record of CO_2 from Mauna Loa (Figure 6). It is rare that we can produce data that convey as clear a picture as this. However, the related issues that people are most concerned with today, the regional impacts of CO_2 accumulation in the atmosphere, are less clear. The data that we have about temperature, precipitation, extreme weather events, sea level rise, changes in wetlands extent, crop and forest productivity, and many other relevant parameters are noisier and less consistent over

---

[4]Yaro, Robert D. and Tony Hiss (1996), "A Region at Risk: The Third Regional Plan for the New York-New Jersey-Connecticut Metropolitan Area" (Island Press, Washington, DC).

[5]Frankel, Glenn and Stephen C. Fehr (1997), "Green, More or Less: Washington's Vanishing Open Space", Washington Post, March 23, 24, and 25.

time. In drawing those data together, we need to be very careful to convey what we know and how well we know it (and diligently work to improve the clarity of that picture).

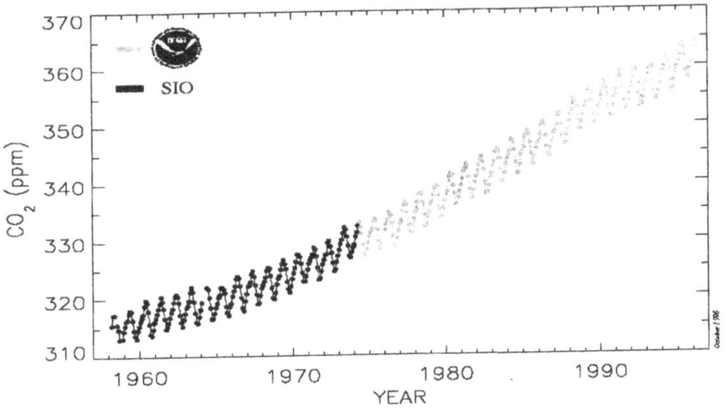

Atmospheric carbon dioxide mixing ratios. Data prior to May 1974 are from the Scripps Institution of Oceanography, data since May 1974 are from the National Oceanic and Atmospheric Administration. Principal investigators: Pieter Tans, NOAA/CMDL Carbon Cycle Group, Boulder, Colorado, (303) 497-6678, and Charles D. Keeling, SIO, La Jolla, California, (619) 534-6001.

Fig. 6. Mauna Loa Monthly Mean Carbon Dioxide

Water quality data is another example. It is very tempting to compare data from successive biennial reports of the National Water Quality Inventory[6] to examine the trends in water quality. Unfortunately, because of the differences in data from year to year, it is not valid to make that comparison. We do not have the data today to evaluate quantitatively whether much of our water is getting cleaner or not.

Thirdly, we need to ensure that the information we produce is useful to decision-makers. It must provide a solid basis for regulatory action; help with forest management plans, habitat conservation plans, environmental assessments and impact statements; be useful to community planning efforts and to individual citizens making their own decisions.

Nicolas Sanson d'Abbeville's map of North America of 1656 (Figure 7) illustrates this point as well as the previous two. His map shows California as an island. It was the focus of a great geographic controversy which raged throughout the 17th century and into the early 18th century. Not long after the map was produced, it was put to use by explorers. They disassembled their boats on the California coast and carried them as they traveled inland and over the mountains only to find there was no need—no Mar Vermeio separated California from the continent. Messages went back to Europe but were dismissed. The cartographers felt the explorers simply did not know where they were. It took 150 years to convince them.

---

[6]Environmental Protection Agency (1995), "The Quality of Our Nation's Water: 1994", Report No. EPA841-S-94-002 (EPA Office of Water, Washington, DC) is the latest of these reports.

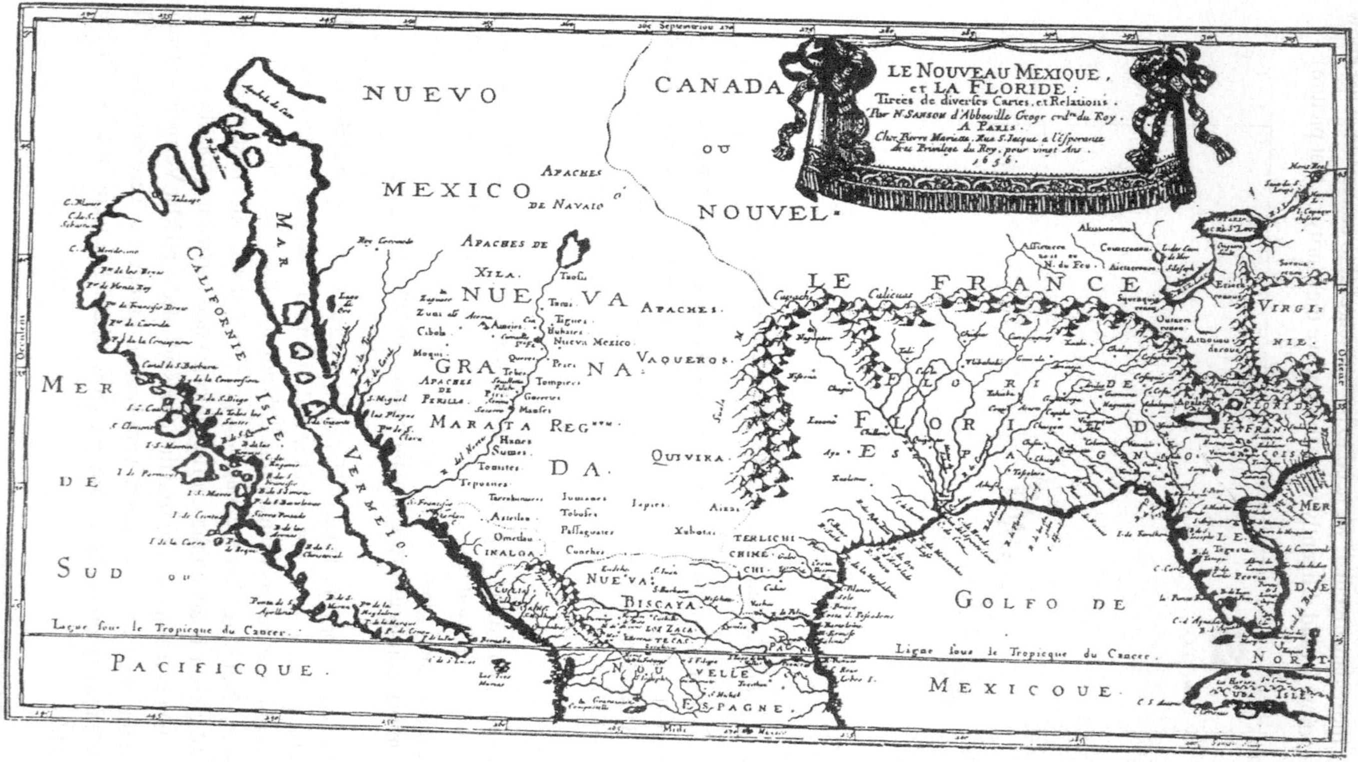

Fig. 7. D'Abbeville's Map of North America of 1656 Today's environmental monitoring and research results need to contribute to the synthesis of a picture that is more comprehensive than d'Abbeville's map even in its day. It needs to convey the degree of certainty we have that the picture is right. It needs to be useful to the decision-makers—and to incorporate new, solid, objective information as we learn more.

7. EMAP and the CENR Initiative

How then will EMAP relate to the coordinated, integrated Environmental Monitoring and Research Program that will result from the CENR initiative? A great many aspects will have to play out before there will be a clear answer to that question.

EPA's new EMAP strategy[7] recognizes the need for collaboration. The next steps for EMAP are to be built on an acknowledgment that "EMAP itself will not be the entire national monitoring network but will contribute components to it." The exact components that EMAP will implement are not yet determined within the agency or with potential CENR partners. A key recommendation common to the various reviews of the program to date is that it "should continue its efforts to develop close working relationships with the EPA Program Offices and other Federal Monitoring efforts."

An important element in the new strategy is regional-scale assessments. The mid-Atlantic region has been selected as the first regional-scale geographic study in EMAP. This effort should be an important contribution to the CENR initiative and that will become clear as the blueprint for the pilot emerges. The same three-tiered framework has been adopted by both CENR and EMAP. EMAP intends to conduct work at all three levels of the framework. A major function of the mid-Atlantic pilot is intended to be evaluation of the effectiveness and usefulness of this framework.

The new EMAP strategy calls for environmental report cards or "state of the region" assessments to be produced by the regional-scale geographic studies. These are described as "characterizing the ecological quality of the region and the important environmental stresses... of importance to the region." The goods and services perspective of the national report card effort may provide a useful guide to these studies and, in turn, these studies may provide useful information for the national effort.

EMAP's new strategy also calls for the establishment of a national network of index sites with the National Park Service to serve as "outdoor laboratories." This project has been dubbed "DISPro," the Demonstration of Intensive Sites Project. The strategy notes that, "consensus on how best to define an index site and how to locate sites in a network is still quite elusive. Because these questions are not likely to be resolved without research, EMAP will establish an effort to evaluate designs for index sites with respect to specific hypotheses and to evaluate the multiple options for linking survey networks with networks of intensive sites." Although the outline of this effort and the role of the demonstration project have not yet emerged, that work could be very helpful to the CENR activity on index sites.

---

[7]"Environmental Monitoring and Assessment Program (EMAP): Research Strategy 1997", Office of Research and Development, U.S. Environmental Protection Agency, January, 1997.

Finally, the EMAP strategy aims to "focus the next three years on the research and demonstration necessary to provide the scientific credibility for the monitoring network." During that time the CENR initiative envisions significant progress in reinventing our environmental monitoring and research programs. EMAP can be an important contributor to that progress.

ENVIRONMENTAL DATA IN DECISION MAKING IN EPA REGIONAL OFFICES

STANLEY L. LASKOWSKI[1] and FREDERICK W. KUTZ[2]

[1]*Deputy Regional Administrator, EPA Region 3, 841 Chestnut Building, Philadelphia, PA 19107,*
[2]*Office of Research and Development, Suite 200, 201 Defense Highway, Annapolis, MD 21401*

Abstract. The mid-Atlantic region of the United States has a wide diversity of natural resources. Human pressures on these natural resources are intense. These factors have resulted in the collection of substantial amounts of environmental information about the region by EPA (both Regional and Research Offices), other governmental agencies, industry, and environmental groups. EPA Regional Offices comprehend first hand the importance of environmental data and are extremely supportive of investments in these data. Environmental data are used prominently in a variety of strategic planning and resource management initiatives. In EPA Region 3, the use of scientifically-sound environmental data is, in fact, one of our strategic programmatic goals. Environmental information is captured and assessed continuously by Regional staff, sometimes working in partnership with other Federal and State agencies, to derive relevant resource management conclusions. The restoration goals for the Chesapeake Bay are based on environmental indicators and resulting data. Attainment of the water quality objectives for streams and coastal estuaries are predicated on monitoring data. Our initiative in the Mid-Atlantic Highlands area uses environmental indicators to measure the condition of forests and streams. Landscape-level indicators will provide unique opportunities for the use of data in planning and management activities in support of the principles of community-based activism and sustainable development. Significant value is added to these data during their use by Regional managers. Regional programs, such as the Chesapeake Bay Program and several National Estuary Programs, are founded in environmental data. Environmental information is used by the Regional program managers to ascertain whether programs are accomplishing their intended objectives. Finally, Regional programs provide a crucial means for disseminating this information to broad segments of the public, so that a better informed and educated client base for effective environmental protection will develop.

1. Introduction

EPA Regional offices are responsible for a wide variety of environmental programs within their respective jurisdictions. Many of these programs are cooperative efforts with federal, state, interstate, and local agencies, as well as with industry, academic institutions, and other private groups. These programs ensure that regional needs are addressed and that Federal environmental laws are upheld. Regional offices implement and enforce national environmental protection laws (in collaboration with States), provide guidance and oversight, technical assistance, and monitor trends and emerging problems.

EPA Regions, under the direction of a Regional Administrator, develop, propose, and implement programs for comprehensive and integrated environmental protection activities. Regional offices also manage effective regional enforcement and compliance programs. They support the Agency's overall mission by translating technical program direction and evaluation from various Headquarters Program Offices into effective operating programs at the regional level. In addition, Regional offices ensure that such programs are executed

Environmental Monitoring and Assessment **51:** 15–21, 1998.

efficiently, and that overall and specific evaluations of their programs are provided. EPA has ten Regions covering the United States with Regional offices located in metropolitan centers within their geographic area.

EPA Region 3 encompasses most of the mid-Atlantic area including five states (PA, MD, VA, WV, and DE) and the District of Columbia. Geographically, the Region contains about 3.4 percent of the land area with over ten percent of the population of the United States. A wide diversity of natural resources are present: forested mountains in the west; estuaries and open ocean along the east. Our Region includes most of the natural resources associated with the Chesapeake Bay watershed -- one of the largest estuarine watersheds in the world.

Our Region continues to experience intense human pressure. The State of Maryland estimates that its population will increase by one million people by the year 2025; this is an increase of about 20 percent making a population density of almost 500 people per square mile. This kind of growth and development means that a balance among natural resources protection and human uses is a constant challenge.

EPA Regions in collaboration with Headquarters Program Offices administer about 28 Federal environmental laws, many of which require collecting, reporting and synthesis of environmental data. Data collectors are other federal, state, and local agencies as well as industry and volunteer groups. Synthesis and interpretation of data are extremely important to EPA Regional offices, and represent an overwhelming requirement. In the past, much of the emphasis of EPA data collection had been on compliance and enforcement. The focus is now moving more toward using scientific data to answer other questions, such as "How are we doing?" and "Are environmental conditions getting better or worse?"

The objectives of this article are to describe the importance of environmental information in decision making in Regional offices, provide some examples of the kinds of information, and how we use this information. Additionally, some of the more promising future initiatives are described.

2. Sources of Environmental Information

A great deal of information already has been collected for the Region 3 area. In a recent inventory, 350 federal monitoring programs and 450 state, local, non-governmental environmental groups, and private or academic programs are involved in collecting data in our Region. Most of these databases are specialized —collected for particular purposes. The lack of standardization of data or monitoring methods make correlation across databases problematic.

Over the last few years and more recently with the help of the Office of Research and Development (ORD), the staff of our Region has analyzed much of the information available for the Region 3 area, and found significant data gaps. Little was known in terms of the ecological condition about some of our geographic area, for example, the Mid-Atlantic Highlands. Regional scientists and managers wanted to orient attention to critical assessment issues, such as the condition of plants and animals -- the living resources about which people really care. Our Region is committed to using scientific information in resource management and has made it one of our critical Regional strategic goals. Our Region has pioneered integrated approaches to studying natural resources; examples include the Chesapeake Bay Program, and now the Mid-Atlantic Integrated Assessment.

In 1995, our Region and the ORD jointly came together to establish the Mid-Atlantic Integrated Assessment (MAIA) initiative to enhance the information available for decision making. We formed the Community-Based Assessment Team in our Annapolis, MD, offices. This team was formed jointly by an agreement between the Assistant Administrator for Research and Development (then Dr. Robert Huggett) and the Regional Administrator (W. Michael McCabe). The creation of the team marked a new model for collaboration between the two offices for the identification of the greatest risks to our natural resources and the protection of critical ecosystems. This activity not only coordinates scientific data collection and assessment, but also outreaches to State and Local agencies to demonstrate how to use data in decision making processes. This team is supported by both Regional and ORD resources.

2.1. LITIGATION UNDER THE CLEAN WATER ACT

Currently, EPA is a party to litigation which offers a real world example of the importance of monitoring data. Under the Clean Water Act (§ 303), States in concert with the Regions are required to monitor stream segments to determine whether they are meeting their designated use category (fishable, swimmable, etc). Streams not meeting their designated use must be listed publicly and a plan prepared to restore them to their use category. The restoration process is known as Total Maximum Daily Loading (TMDL). As of April, 1997, about 20 lawsuits are active against EPA Regions and States for allegedly not accomplishing what this law requires. Several cases have already been settled out of court. In Region 3, lawsuits have been filed by environmental groups and private citizens against PA, DE, and WV.

3. Integrated Monitoring

When EPA was formed, most aquatic issues focused on water quality. Now key management issues involve living resources, such as fisheries, submerged aquatic vegetation, endangered species of plants and animals, habitat modification and change, etc. Biotic issues evoke a great deal of public attention. Traditional physical measurements of environmental stresses, i.e., pesticides, acid deposition, habitat loss, etc., still are important and critical, but now usually in the context of how they affect natural resources.

Some examples of the use of monitoring data in Regional programs include:

- The Chesapeake Bay Program has established a 40 percent reduction goal for nutrients entering the Chesapeake Bay. Monitoring data will be used directly to determine goal achievement.
- Habitat restoration programs have resulted in a noticeable increase in the stripped bass population in the Chesapeake Bay. Fishing restrictions also have enhanced this restoration goal —measured by monitoring fish populations and catches.
- In the Delaware Inland Bays, submerged aquatic vegetation has not been observed in the last 40 years. Monitoring data are being used to assist in the re-establishment of this important aquatic habitat component.

The ORD Environmental Monitoring and Assessment Program (EMAP) has proven of tremendous value to Regional decision makers because of its integrated approach to monitoring and assessment of natural resources. EMAP activities have met many of the needs of Regional programs for biological indicators. The merging of scientific and management questions into assessment issues in such diverse areas as land use and land cover data and interpretation, characterization of resource condition of estuaries, streams, forests, ground water and wetlands, is truly helpful in our approach to resource protection. As monitoring and assessment information becomes available, areas that are particularly vulnerable to degradation become evident and can be identified with known scientific precision. Regionally-sponsored community-based environmental protection programs are initiated in many of the areas that are so vulnerable. These programs have significant outreach to the general public. Convergence of environmental, social, economic and cultural considerations are important program components.

3.1. MID-ATLANTIC HIGHLANDS ASSESSMENT (MAHA)

An excellent example of one of the information gaps that existed in Regional data is the mid-Atlantic highlands area. The highlands occupies the broadest geographic area within Region 3 covering the mountainous parts of PA, MD, WV and VA (and NC which is in Region 4). In our Regional office, it is known as the "western jewel of Region 3." The area is of intense interest to Region 3 because it contains small streams that feed into the Chesapeake Bay. The health of these first and second order streams is crucial to the health of the Bay. In conjunction with ORD, other federal agencies, and the states, an intensive monitoring program was established looking at fish, bottom-dwelling organisms, habitat, water quality, and some other indicators of stress. Partnerships of this type are important to EPA Regional offices, since in most cases, resource management and decision making responsibilities are shared with others. Preliminary assessment of the MAHA database has estimated that over 50 percent of the streams in the area have impacted riparian zones which serve as buffer areas between the stream and adjacent fields. These are the kinds of findings that guide many of our management decisions aimed at resource restoration.

4. Future Directions

Future opportunities in this realm of environmental information are exciting indeed. As environmental managers and legislatures become more familiar with using environmental data, new requirements and uses for information will emerge. New scientific and engineering technologies, including using satellite imaging, are emerging for our use. For the first time the true mosaic of land use and cover will provide us with a better understanding of human influences on natural resources. Landscape pattern analysis of indicators such as forest fragmentation, patch size, and interior-to-edge ratios will foster predictions about suitability for wildlife. New statistical ways of doing comparative risk assessments will enhance our ability to manage resources and control undesirable impacts on them. As the Internet expands, better ways of disseminating our information can be created. EPA Regional programs already use the Internet for many useful purposes related to information dissemination.

EPA has taken several significant steps to facilitate the use of environmental information in our administrative processes. The National Environmental Goal Project (NEGP), initiated several years ago, is intended to establish goals based on environmental data for each Headquarters Program Office. Past efforts have been oriented toward regulatory measures, such as numbers of permits, numbers of regulatory actions, etc. Current progress of the NEGP includes using monitoring data to set actual goals for these offices to achieve, (e.g., reductions in the emissions of sulfur and nitrogen into the air).

Congress has also been active in focusing on measures of legislative efficacy. The Government Performance and Results Act requires EPA to develop performance measures to accompany its budget. These measures must relate the expenditure of resources to actual environmental results. The EPA submission to the Office of Management and Budget is due in August, 1997. This project, of course, is building upon the progress made by the NEGP.

Recently, Administrator Carol Browner has created a Center for Environmental Information and Statistics within the EPA organization. An agency-wide committee presently is developing an accurate mission statement for this entity. It is operational now, however, in a preliminary mode of collecting information and preparing reports on various environmental conditions. Internally, the Environmental Methods and Monitoring Council (EMMC) has been active in establishing common analytical methods to be used across all EPA programs and regions. EPA Regions play a major role in the EMMC, and serve as co-chairperson, with ORD officials.

The people of the United States are demanding national debt reduction and enhanced accountability for their tax investment. This movement has resulted in the election of officials at all levels of government who are promoting intense scrutiny of all government expenditures. Appropriations used for monitoring are not exempt from this oversight. Future data collection efforts must be efficient and made in partnership with federal, state, and local agencies, and many others. Our governmental programs simply can not collect information that is not used in decision making, unassociated with appropriate quality assurance, and not in a form that is easily integrated with other databases. Data collection and assessment are costly activities, and ways must be developed to make them as cost-effective as possible. Technological approaches, particularly remote sensing, appear to have promising prospects for the future in yielding a quantum leap in advances in data collection. In the words of our Deputy Administrator, Fred Hansen, our methods must move beyond "the bucket off the bridge" techniques.

As environmental managers, our vision must be oriented toward championing the use of information in decision making. Data are needed in real time. Past efforts too often have produced information after program decisions were made or court-ordered deadlines passed. Our vision should include means of integrating all available data which might impact a particular decision. Methods are also needed to display these data in an understandable format that can be shared with interested parties.

Governmental operations are moving to a new level of public openness and participation. As public participation grows, improved ways of describing the state of our environment are needed. Enhanced access to the Internet will no doubt provide a crucial mechanism for us to disseminate our information effectively. EPA encourages public participation in decisions. Many Regional programs, such as the National Estuary Programs, have specific mechanisms to foster public participation and outreach. One objective of our community-based environmental protection initiatives is to empower local municipalities to embark on public decision making. EPA Region 3 has provided some grants and technical assistance to localities as they approach this new way of doing business. Where Federal interests or more than one state is involved, EPA and Federal partners need to participate. However, when only local issues are at stake, the Federal role should be limited to providing technical assistance and other help as requested.

5. Summary

Environmental data are available, but the quality of some is unknown and large data gaps exist both in geographic area and scientific substance. EPA Regions continue to support and to pioneer the use of environmental information in decision making. Regional managers are strong advocates of getting "the right information in the right way." One way that EPA Regions add value to data is by using them as foundations for program initiatives—Mid-Atlantic Highlands, Chesapeake Bay Program, National Estuary Programs

—thereby providing multiple means of disseminating scientifically sound information to broad segments of the general public. This creates understanding and support for environmental protection.

Our Regional partnership with ORD has been particularly productive. Region 3 has made and continues to make tremendous progress in characterizing the condition of and trends in our ecological resources. A smooth continuation of our alliance with ORD is needed to insure appropriate completion of what has been started. Some of the future "fruits" of our partnership with ORD include:

- Complete State of the Resource Reports for the mid-Atlantic area.
- Demonstrate in the field to cooperators our unique monitoring and assessment approach to resource evaluation.
- Use landscape-level information in the assessment process.
- Develop a truly integrated monitoring and assessment approach for all resources in a Region.
- The future appears bright and exciting for further integration of environmental information into decision making. New approaches on the horizon for using satellite technologies, new ways of doing comparative risk assessments, promoting sustainable development through use of scientific data—all looking forward to building an environmentally-friendly bridge to the 21st century.

DEVELOPMENT AND VALIDATION OF ECOLOGICAL INDICATORS: AN ORD APPROACH

WILLIAM S. FISHER

ORD Ecological Indicators Working Group, U. S. Environmental Protection Agency, National Health and Environmental Effects Research Laboratory, Gulf Ecology Division, 1 Sabine Island Drive, Gulf Breeze FL 32561

Abstract. The U. S. Environmental Protection Agency's Office of Research and Development (ORD) is continuing research efforts initiated by the Environmental Monitoring and Assessment Program on ecological indicator development. An ORD Ecological Indicators Working Group has been formed with activities in three primary areas. (1) Guidelines and procedures are being developed to evaluate indicators for use in monitoring programs. Indicators will be evaluated on conceptual soundness, implementation, response variability, and interpretation/utility. The evaluation guidelines will be applied in peer review to endorse technically acceptable indicators and will provide research direction for improvements. (2) An ORD strategy for research in ecological indicators is being developed by the Working Group in collaboration with Division research scientists. The strategy will serve to prioritize research based on the greatest importance and uncertainty and identify goals for indicator development in both intramural and extramural programs. The research strategy includes application of the evaluation guidelines to identify relevant research questions. (3) Interactions with indicator client and user groups (states, program offices and regions) are actively being sought for successful development and implementation of indicators. Client indicator priorities are formally included in the research strategy and user feedback on indicators will help to identify relevant research questions. Consultations with users will serve to assist in evaluating, implementing, and interpreting indicators in monitoring programs.

1. Introduction

The theme of this session in the Symposium is the development and validation of ecological indicators. An indicator is a sign or signal that relays a complex message, potentially from numerous sources, in a simplified manner to convey useful information. However, the definition and application of indicators is dynamic, flexible, and influenced by project-specific factors, including purpose, scope and target audience. Environmental indicators have many definitions in the literature but, in general, environmental indicators are used to describe, analyze, summarize and present scientifically based information on environmental conditions, trends, and their significance (Florida State University 1995). Because of this focus on condition and trend, indicators have been inexorably linked with monitoring programs such as the Environmental Monitoring and Assessment Program (EMAP), coordinated by the U.S. Environmental Protection Agency's (EPA) Office of Research and Development (ORD).

Human impact on the environment has consistently increased with population growth, so it has become increasingly critical to manage and protect natural resources. As we have become more aware of the complexity and long-term consequences of environmental stress, management needs have expanded from stress-specific water quality standards to the much broader objectives of maintaining structural and functional integrity of entire ecosystems

Environmental Monitoring and Assessment **51**: 23–28, 1998.

(Cairns *et al.* 1993). The approach and application of environmental indicators has changed to reflect these new dictates of management. Indicators have progressed from chemical-specific, end-of-pipe effluent measures to measures that identify and characterize susceptible components of an ecosystem. Thus, *ecological indicators* is a term that has been applied to those environmental indicators that describe the condition of an ecosystem or one of its critical components or processes.

The changing emphasis of management toward ecosystem processes has greatly altered the perception of indicator development and generated unique scientific and social questions. Chemical-specific indicators or indicators that address specific and well-defined management goals (e.g., success of lake trout and walleye in the Great Lakes) require relatively little research. But the broadly based and complex goal of ecosystem integrity requires a comprehensive understanding of the components and dynamics of the ecosystem at risk. An ecosystem is extremely intricate with unique biotic and abiotic properties that fluctuate with both anthropogenic and natural events and cycles. Consequently, the measurable characteristics of an ecosystem are far too numerous to effectively quantify. Selected indicators are measured because they efficiently provide useful information about the condition of the ecosystem. One challenge for indicator research and development is to identify, from the multitude of possibilities, those characteristics that are critical to ecosystem integrity and responsive to stressors that threaten that integrity.

The ORD Ecological Indicators Working Group was formed in 1996 to continue the work in indicator development initiated by EMAP. The Working Group includes representatives from all four of ORD's Research Laboratories (National Exposure Research Laboratory, National Health and Environmental Effects Research Laboratory, National Center for Environmental Assessment, National Risk Management Research Laboratory). The Working Group has identified three major objectives: Establish procedures for indicator evaluation, outline a strategy for indicator research and development, and strengthen ties to client and user groups (EPA Program Offices and Regions, states). Each of these objectives is being pursued with the ultimate goal of providing the best indicators for the highest priority environmental needs.

2. Ecological Indicator Evaluation Guidelines

The Working Group recognized that a pressing problem in indicator research was the need to determine which measurement or group of measurements make good indicators. Since the measurable characteristics of an ecosystem are limitless, so are the number of possible indicators; many indicators have been proposed from diverse sources, including EMAP, for a variety of purposes. Yet, environmental measurements are sometimes used as indicators and environmental techniques are sometimes proposed as indicators with insufficient evidence to support their utility or potential when applied in a monitoring program. The Working Group has drafted the ***Ecological Indicator Evaluation Guidelines*** to assist researchers, program managers and resource managers in determining the capacity of environmental measurements to accurately reflect ecological condition and integrity.

Desirable attributes of environmental and ecological indicators have been discussed in the scientific and resource literature for many years (for example, Macek *et al.* 1978, Hammons 1981, Suter 1989, Kerr 1990, Kelly and Harwell 1990, and Cairns *et al.* 1993) including EMAP program development documents (Messer 1990, Hunsaker and Carpenter 1990, Knapp *et al.* 1991, Barber 1994) and recent reports from the Intergovernmental Task Force on Monitoring Water Quality (1995) and the International Joint Commission (1996). The attributes emphasized in the literature vary according to the type of indicator and the particular needs of the program, but many are common to all (Cairns *et al.* 1993). The Ecological Indicators Working Group has integrated most of the attributes described in the literature to develop the broadly-based *Evaluation Guidelines*. They are expected to be sufficiently flexible to meet the diverse needs of different programs.

The *Evaluation Guidelines* are grouped to address the four evaluation phases originally identified by EMAP (Barber 1994):

> Conceptual Soundness - does the indicator defendably link a critical ecological component and its stressor to the assessment question?
>
> Implementation - are the methods for sampling and measuring the environmental variables technically feasible, appropriate, and efficient for use in a monitoring program?
>
> Response Variability - are human errors of measurement and natural variability over time and space sufficiently understood and documented?
>
> Interpretation and Utility - when used in a monitoring program, will this indicator discriminate environmental differences that can be meaningfully applied to the assessment question?

It is anticipated that the *Evaluation Guidelines* will be used not only for technical and programmatic reviews, but as a tool by researchers in ORD and elsewhere to determine strengths and weaknesses of particular indicators.

3. ORD Ecological Indicator Research Strategy

During the last decade, EPA has progressed in the development of risk assessment guidelines for ecological effects that parallel risk assessment guidelines for human health effects. The framework for ecological risk assessment (Risk Assessment Forum 1992) consists of problem formulation, analysis, and risk characterization stages. The primary role of research and development is to address uncertainties in the problem formulation and analysis stages. Effective problem formulation must include a competent understanding of the progression from the source of a stressor, through exposure and biological effects, to potential adverse outcomes. Effective analysis requires a thorough characterization of exposure and ecological effects. ORD research in ecological indicators will directly support the needs and priorities of these two risk assessment stages.

One objective of the Ecological Indicators Working Group is to establish indicator research priorities. The Working Group recognizes that ORD has the mission and expertise to address some, but not all ecological indicator issues. Thus, the Working Group is preparing an ***ORD Ecological Indicator Research Strategy*** that will prioritize research based on ecological importance and greatest uncertainty. The *Research Strategy* will distinguish those priority areas to be performed by ORD in-house programs and those to be performed by academic colleagues through the grants program (Science to Achieve Results, or STAR, coordinated through the EPA National Center for Environmental Research and Quality Assurance). In the 1997 request for applications, the Working Group emphasized the need for indicators that span resource types. In-house expertise is organized and oriented toward research within a single resource type (i.e., estuary, forest), even though ecological condition is a consequence of multiple factors from multiple resource types. Thus, the need to bridge indicators from different resources was identified and entrusted to the extramural program.

The *Ecological Indicator Evaluation Guidelines* are an important part of the research strategy. The *Guidelines* will be informally applied by ORD scientists to guide their research toward improving indicators. They will also be applied formally in a peer-review process coordinated by the Working Group. The formal review will provide documentation of an indicator's strengths and weaknesses; weaknesses will be identified for further research and improvement. Ultimately, indicators found acceptable by peer review will be listed by the Working Group for potential application in a monitoring program.

4. User-Group Interactions

The third objective of the Ecological Indicators Working Group is to establish interactive ties with the user community, i.e., those involved in conducting ecological risk assessments and resource management. Many of these user groups wish to use ecologically-relevant indicators to move the focus of their environmental policies from program performance to environmental assessment. The Working Group views this interaction as critical to the successful development and implementation of indicators.

Interactions with user groups, or clients, are formally supported in the *Ecological Indicators Research Strategy*, where client perspectives on indicator priorities will be considered along with those priorities developed by ORD Divisional scientists. The Working Group will facilitate this formal interaction on an annual basis. The *Strategy* also provides user-feedback on individual indicators through the process of evaluation for programmatic objectives. Potential users will be included on ORD peer review evaluation panels and weaknesses can be identified for further research. Additionally, evaluations convened by clients can characterize indicator weaknesses to be addressed by investigators.

Other interactions with clients will consist of consultations on a variety of indicator issues. For example, it is anticipated that indicator users will want to convene independent panels to evaluate potential indicators for application to their specific objectives. The Working Group will consult with the clients to ensure that the intent of the *Evaluation Guidelines* is understood and meaningful results can be obtained. This consultation also benefits the Working Group by providing feedback on how to improve the *Guidelines* and the review process. Similarly, the Working Group will use consultations to assist clients with technical questions that arise during the evaluation or implementation of an indicator.

5. Summary

EPA's Office of Research and Development is continuing research in ecological indicator development initiated by the Environmental Monitoring and Assessment Program. An Ecological Indicators Working Group has been formed to develop guidelines and procedures to evaluate indicators for use in monitoring programs, provide an ORD strategy for research in ecological indicators, and foster interactions with indicator user groups. Indicator development is emphasized in this session of the Symposium, yet indicator research and application is evident throughout many of the different sessions. Due to the changing and complex requirements for managing ecological resources, the need for continued development and evaluation of indicators will remain an important priority for environmental protection.

Acknowledgements

The Ecological Indicators Working Group is ultimately responsible for developing and refining the goals and procedures described in this text. Special thanks are due to Kay Austin, Peter Beedlow, Steve Bradbury, Laura Jackson, Jan Kurtz, Jim Lazorchak, Dave Peck, Charles Strobel, and Ray Wilhour. This is Gulf Ecology Division Contribution No. 1014.

References

Barber, M. C. (ed.): 1994, *Environmental Monitoring and Assessment Program Indicator Development Strategy*. EPA/620/R-94/022, 74 pp.

Cairns, J. Jr., McCormick, P.V. and Niederlehner. B.R.: 1993, *Hydrobiologia* **263**:1-44.

Hammons, A. (ed.): 1981, *Methods for Ecological Toxicology*. Ann Arbor Science, Ann Arbor MI.

Hunsaker, C.T. and Carpenter, D.E. (eds.): 1990, *Environmental Monitoring and Assessment Program: Ecological Indicators*. EPA Office of Research and Development, Research Triangle Park, NC.

Intergovernmental Task Force on Monitoring Water Quality: 1995, *The Strategy for Improving Water-Quality Monitoring in the United States*. US Geological Survey, Office of Water Data Coordination. 117 p.

International Joint Commission: 1996, Indicators to Evaluate Progress under the Great Lakes Water Quality Agreement. Indicators for Evaluation Task Force of the International Joint Commission, ISBN 1-895085-85-3. 82 p.

Kelly, J.R. and Harwell, M.A.: 1990, *Environmental Management* **14**:527-545.

Kerr, A.: 1990, *Canada's National Environmental Indicators Project: Background Report.* Sustainable Development and State of the Environment Reporting Branch, Environment Canada, 12 pp.

Knapp, C.M., Marmorek, D.R., Baker, J.P., Thornton, K.W., Klopatek, J.M. and Charles, D.F.: 1991, *The Indicator Development Strategy for the Environmental Monitoring and Assessment Program.* EPA/600/3-91/023.

Macek, K., Birge, W., Mayer, F., Buikema, A. Jr. and Maki, A.: 1978, Discussion session synopsis, In: J. Cairns, Jr., K. Dickson and A. Maki, eds., *Estimating the Hazard of Chemical Substances to Aquatic Life, STP 657.* American Society for Testing and Materials, Philadelphia PA, pp. 27-32.

Messer, J.J.: 1990, EMAP indicator concepts. In: *Ecological Indicators for the Environmental Monitoring and Assessment Program,* C.T. Hunsaker and D. E. Carpenter, (eds.). EPA/600/3-90/060. pp. 2-1 through 2-26.

Risk Assessment Forum: 1992, Framework for Ecological Risk Assessment. U.S. Environmental Protection Agency, Washington, DC, EPA/630/R-92/001. 41 pp.

Florida State University: 1995, *Prospective Indicators for State Use in Performance Agreements.* State Environmental Goals and Indicators Project, Florida Center for Public Management, Florida State University.

Suter, G.: 1989, Ecological endpoints. In: W. Warren-Hicks, B. Parkhurst and S. Baker, Jr (eds.), *Ecological Assessment of Hazardous Waste Sites: A Field and Laboratory Reference.* EPA 600/3-89-013. Pp. 2.1 to 2.26.

A ZOOPLANKTON-N:P-RATIO INDICATOR FOR LAKES

RICHARD S. STEMBERGER[1] and ERIC K. MILLER[2]

Department of Biology[1] and Environmental Studies Program[2], Dartmouth College, Hanover, NH 03755, USA

Abstract. We develop the conceptual and empirical basis for a multi-level ecosystem indicator for lakes. The ratio of total N to total P in lake water is influenced or regulated by a variety of ecosystem processes operating at several organizational levels and spatial scales: atmospheric, terrestrial watershed, lake water, and aquatic community. The character of the pelagic zooplankton assemblage is shown to be well correlated with lake water N:P ratio, with species assemblages arrayed along the N:P gradient in accordance with resource supply theory. Features of specific zooplankton assemblages or deviations from expected assemblages can provide information useful for lake managers, such as the efficiency of pollutant transfer and biomagnification of toxins, loss of cool-water refuge areas, degree of zooplanktivory and food web simplification related to changes in fisheries, and assemblage changes due to anthropogenic acidification. Evaluation of the influence of watershed land use, forest cover and vegetation type, atmospheric deposition, and basin hydrology on the supply of N and P to lake ecosystems provides a means to couple changes in the terrestrial environment to potential changes in aquatic ecosystems. Deviations of lake water N:P values from expected values based on analysis of watershed and lake basin characteristics, including values inferred from appropriate diatom microfossil deposits, can provide an independent validation and baseline reference for assessing the extent and type of disturbance. Therefore, the N:P ratio of lake water can serve as a potentially useful and inexpensively obtained proxy measure for assessing changes or shifts in the biological and nutrient status of lakes.

1. Introduction

Biological communities and water quality of aquatic ecosystems in the northeastern USA are under increasing threat of change due to local and regional stressors, including, residential and recreational development, agricultural practices, exotic species introductions, over fishing, urbanization, and regional air pollution. Indeed, it is now generally recognized that most of earth's ecosystems are, at some level, influenced by human activities (Vitousek *et al.*, 1997a). Human alteration of the nitrogen (Vitousek *et al.*, 1997b) and phosphorus (Smil, 1990) cycles has generally increased the availability of these nutrients to aquatic ecosystems. Federal, state and local resource managers are in need of practical and economical methods for detecting and evaluating changes in the character and health of aquatic ecosystems that may occur in response to changes in aquatic and terrestrial environments. In this paper we develop the conceptual and empirical basis for a multi-tier ecosystem indicator for lakes. We discuss how factors regulating nitrogen (N) and phosphorus (P) export from the terrestrial environment can be linked to effects in the aquatic environment via resource supply theory (cf. Tilman, 1982) and how these relationships form a framework for evaluating the consequences of past, present, or potential land-use changes for aquatic ecosystems. We illustrate how consideration of factors which influence the relative availability of N and P (i.e., the N:P ratio) as well as the total loading of N and P to an aquatic system can be used to provide much information

Environmental Monitoring and Assessment **51:** 29–51, 1998.

about the character of an ecosystem (zooplankton community, suitability for specific fisheries, susceptibility to health risks).

Resource supply theory has helped explain community structure in phytoplankton (Tilman, 1982; Hecky and Kilham, 1988; Kilham and Kilham, 1990) and recently has been extended to include the freshwater zooplankton (Elser *et al.*, 1988; Elser *et al.*, 1996; Sterner and Hessen, 1994; Vanni *et al.*, 1997). Nitrogen and phosphorus are two elemental constituents of all cells that can impose significant limitations on an organism's potential for growth and reproduction (Tilman, 1982; White, 1993; Sterner and Hessen, 1994) and are important constituents of biomolecules like nucleic acids, proteins, lipids, ATP, and structural carbohydrates such as chitin (Elser *et al.*, 1996).

An organism's dietary needs for N and P are also associated with its life history. For example, opportunistic r-selected species such as *Daphnia*, small cladocerans, and probably rotifers (cellular N, P measurements have not been determined for this group) tend to have higher requirements for P relative to N (i.e., low N:P intracellular ratios). In contrast, slower growing k-selected competitive species, such as calanoid copepods and cyclopoids, tend to have higher intracellular requirements for N relative to P (high N:P intracellular ratios) and display complex life histories with multiple morphologically distinctive instars. Although nitrogen composition between cladocerans and calanoid copepods is similar (8-10% by weight), the difference in cellular P requirements is significant (2% in copepods vs. 10% in cladocerans). This difference is attributed to higher amounts of RNA in cladocerans—a molecule required for protein synthesis that is essential for rapidly growing r-selected organisms (Elser *et al.*, 1996). The dietary requirement among zooplankton for these elements also depends on size and trophic position in the food web. For example, omnivores and large herbivores require more N relative to P. Also, predatory cladocerans such as *Leptodora* and *Polyphemus* have higher intracellular N:P ratios than herbivorous relatives like *Daphnia* (Sterner and Hessen, 1994) (Figure 1).

In a survey of freshwater lakes, the growth rates of zooplankton with high elemental P demands like *Daphnia* may be constrained by P-limited phytoplankton (Elser and Hasslett, 1994). Hence, food quality (as determined by N and P supplied by phytoplankton and internal cytoplasmic elemental needs of the zooplankton) is a significant factor affecting not only growth and reproduction of zooplankton but also their ability to compete with other species. For instance, calanoid copepods should produce more biomass relative to *Daphnia* when feeding on P-limited food, thus potentially explaining their dominance in high N:P ratio ecosystems (Sterner and Hessen, 1994). Such important cellular-level constraints provide the mechanism by which watershed N:P supply to the lake can influence phytoplankton and zooplankton community composition. The in-lake mechanism underlying this linkage with the zooplankton is complex, being modified not only by the N:P supply in the phytoplankton food base but also by predatory and nutrient recycling interactions with fish and zooplankton (Vanni *et al.*, 1997; Sterner *et al.*, 1992).

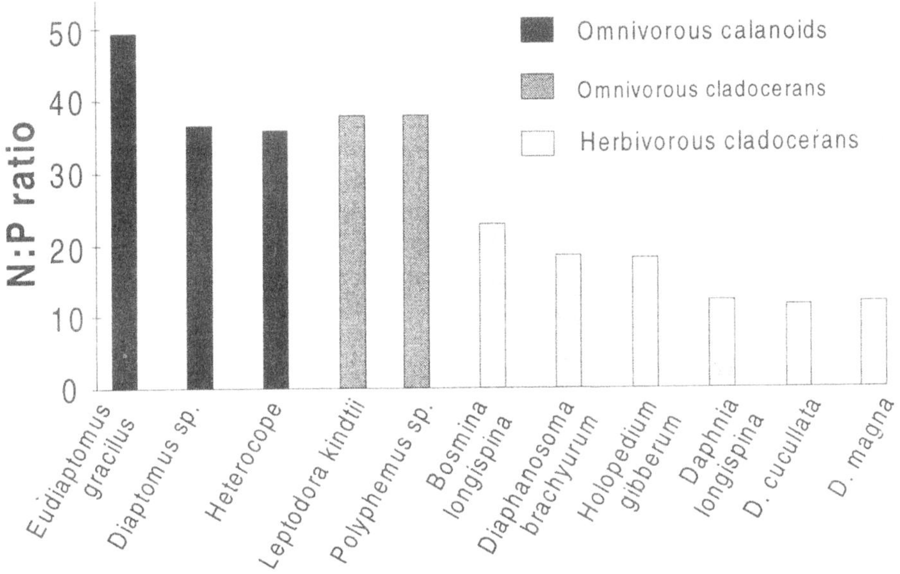

Fig 1. Intracellular N:P ratios of common freshwater zooplankton. Based on data from Sterner *et al.*,1992 and Sterner and Hessen 1994.

2. Methods

2.1 ZOOPLANKTON

Three hundred eighty-five lakes were sampled for biological and chemical variables during 4 consecutive summers (1991 to 1994) as part of the Environmental Monitoring and Assessment Program (EMAP) pilot survey (Larsen and Christie, 1993). We limited our analyses to a single lake visit, although many of the lakes had been sampled on 2 or more occasions (Stemberger *et al.*, 1996). Zooplankton were collected and identified using methods outlined in Stemberger and Lazorchak (1994). A single tow was hauled vertically from the deepest portion of the lake from 0.5 m off the bottom to the surface. Two nets were fitted to a single harness. A cone net (30 cm diameter, 202 µm mesh) was used for macrozooplankton (4:1 net length to diameter ratio) and a Wisconsin net (48 µm mesh) with a 15 cm diameter and 25 cm long reducing collar for the microzooplankton (7:1 net length to diameter ratio). The zooplankton were anesthetized for 1-3 minutes in CO_2-charged water and preserved in buffered formalin and sucrose (Baker *et al.*, in press; Stemberger and Lazorchak, 1994). Standardized species abundances (number/L) of each assemblage were aggregated into 17 new variables based on taxonomic association, feeding mode, and body size (Stemberger and Lazorchak 1994; Stemberger and Chen, in press).

Juvenile stages of copepods were categorized as nauplii or as large (>60 μm and <202 μm body length) and small (<60 μm body length) calanoid and cyclopoid copepodites (Stemberger and Lazorchak, 1994). Minor groups such as aquatic mites, tardigrades and gastrotrichs, and occasionally important groups like *Chaoborus* and ostracods were not effectively sampled and were excluded from gradient analysis.

We calculated structural variables (maximum chain length and total number of trophic links) of the predator-prey relationships among the 17 trophic variables which included presumed feeding relationships with fish zooplanktivores and piscivores (Stemberger and Chen, in press). Links represent the number of trophic interactions directed from prey to predators while chain length measures the number of consecutive links from producers, in this case the algae, to a consumer at the end of the chain (Sprules and Bowerman, 1988).

2.2 Water Chemistry, Watershed, and Fish Tissue Analysis

Water samples were collected at the deepest point in the lake (Baker *et al.*, 1995). Methods for selected water chemistry and geographic data are given in Baker *et al.*, (in press) and Larsen and Christie (1993). Methods for fish heavy metals and PCB's are given in US EPA (1991) and Yeardley (1994).

Nitrogen to phosphorus ratios from published data were calculated for different terrestrial ecosystems to predict expected values or ranges in total N and total P exports in the dissolved plus particulate fraction of stream and groundwater (cf. Likens and Bormann, 1995; Johnson and Lindberg, 1992; Waring and Schlesinger, 1985; Cole and Rapp, 1981; E.K. Miller, unpublished data).

2.3 Data Analysis

We partitioned the EMAP lakes into the three major geographic subregions--the Adirondack Mountains, New England Uplands, and the coastal, lowland, and plateaus areas (Larsen *et al.*, 1994). These divisions reflect major geological landscapes as well as different intensities and types of human disturbance. The number of lakes in the various analyses presented below differed according to the completeness of the data set and the inclusion of specific variables of interest. To reveal the integrating properties of the N:P ratio across geographic regions and spatial scales, we conducted statistical analyses of the full data set and of specific sub-regions.

The numerical abundance of specific zooplankton groups was expressed as a proportion of total crustaceans or as a proportion of taxonomic groups such as large cladocerans or macrozooplankton in relation to 3 ranges in lake N:P ratios. We then compared these zooplankton variables with the 3 geographic subregions. Expansion factors or sampling weights were used to give estimates for the population of lakes in each subregion (Larsen and Christie, 1993; Stemberger *et al.*, 1996).

We applied redundancy analysis (RA) (ter Braak, 1988) to zooplankton and limnological variables for 111 Adirondack lakes sampled during the 1991-1994 EMAP surveys. Zooplankton variables were $\log_e$ (x +1) transformed to stabilize variance and to adjust for 0 values. Physicochemical, fish metals, and land use (percentage cover) were not transformed. This constrained form of principle component analysis (PCA) detects the patterns of variation in the zooplankton variables that are best explained by the observed environmental variables. The results can be summarized easily in an ordination diagram that simultaneously expresses both the patterns of variation of the zooplankton variables and of the lakes in relation to the dominant environmental variables. The forward selection procedure of CANOCO was used to identify the environmental factors that are significantly correlated with the major principal component axes (ter Braak 1988). A final set of 26 variables was selected from an initial set of 71 variables based on the general criteria given in Stemberger and Lazorchak (1994).

Pearson product-moment correlations, linear regression, and ANOVA (SAS, JMP V3.1) were done on untransformed zooplankton abundances, structural variables, limnological, and land use variables. Specific analyses were limited by the inclusion in the database of the variables of interest. Stepwise regression using the N:P ratio as the dependent variable was applied to 261 lakes for which land use data was available. Pearson product moment correlations were done on 180 lakes sampled in the 1992-1994 survey which contained data on fish heavy metals and PCB concentrations.

Single and multiple linear regression analysis (JMP V3.1) was also conducted on 364 lakes from the 1991-1994 survey to explore associations between several landscape variables (elevation, latitude, longitude, atmospheric deposition of sulfur and nitrogen) and lake water N and P concentrations. This subset of lakes was selected to be free of geochemical outliers such as obvious road runoff contamination (Cl > 2000 μM/L) and TDS conditions (sum of cations > 4000 μM/L) associated with calcareous sediments in drainage basins.

3. Results

Our selected subset of lakes from the EMAP surface water survey in the northeastern US region represents 364 lakes ranging in size from 1 to 6560 ha with watershed areas from 6 to 362,000 ha. Seventy-five percent of these lakes were smaller than 100 ha and 50% were less than 26 ha with watershed areas less than 4130 ha and 706 ha, respectively. Total dissolved N concentrations ranged from 7.14 to 180 μM/L with 25-50-75 percentile concentrations of 18.8, 23.8, and 34.3 μM/L. Total dissolved P concentrations ranged from < 0.01 to 5.7 μM/L with 25-50-75 percentile concentrations of 0.19, 0.32 and 0.55 μM/L. Nitrogen to phosphorus ratios ranged from 13 to 1765. Twenty-five percent of lakes had an N:P ratio < 54, 50% < 74, 75% < 102, and 97.5% < 300. The relative abundance of zooplankton species varied widely across the sample lakes as described below.

3.1 Redundancy Analysis of Adirondack Lakes

Redundancy analysis of Adirondack lakes suggested important associations between zooplankton species assemblages and a suite of 26 lake environmental variables, including the N:P ratio of lake water. The first and second PCA axes explained 17.3 % (eigenvalue = 0.172) and 3.9 % (eigenvalue = 0.039), respectively, of the variance between zooplankton variables and environmental variables in 111 Adirondack lakes. Both canonical axes 1 and 2 were significant (p=0.01) using Monte Carlo permutation tests (99 unrestricted permutations) (ter Braak, 1988). Based on the redundancy analysis and the forward selection procedure, 7 of the 26 measured environmental variables made significant individual contributions to explain the gradients in the zooplankton trophic variables. Total phosphorus concentration in the water column contributed 8% of the explainable variance followed by the elevation (asl) of the highest point on the watershed (4%), total aluminum concentration in the water column (3%), maximum lake-water temperature (3%), minimum observed dissolved oxygen (4%), and NH_4 concentration in lake water (2%). Each variable was significant at p=0.01. Dissolved organic carbon (DOC) (2 %) was only marginally significant (p =0.08). These 7 variables explained 70.5 % and 15.8 % of the total variance in axes 1 and 2, respectively, for a total contribution of 86.3 % for both axes.

The first canonical axis contrasted lakes having the large-bodied omnivorous cladoceran, *Leptodora kindtii,* and large omnivorous calanoid copepods such as *Epischura lacustris* with small-bodied assemblages comprised of rotifers, nauplii, and other juvenile stages of cyclopoid copepods. PC axis 1 was highly correlated ($p<0.01$) with lake water N:P ratio (Figure 2; Table I). Low N:P ratios were associated with shallow warm water lakes having high chlorophyll *a* and high total-P and total-N concentrations. The second canonical axis contrasted lakes along a total aluminum gradient that is also highly correlated with pH (Figure 2; Table I). Low pH lakes had high total aluminum, high numerical abundances of calanoid copepods, low chlorophyll *a,* and low road density.

3.2 Comparisons of N:P Ratios and Zooplankton Assemblages

The proportion of calanoid copepods and omnivorous cladocerans increased with higher water column N:P ratios for Adirondack lakes. In contrast, cyclopoid copepods and rotifer abundances decreased with higher N:P ratios (Table II). Large cladoceran herbivores tended to be more prominent in assemblages at intermediate N:P ratios (Figure 2; Tables II, III). Zooplankton abundances also varied across geographic subregions in the northeast (Table III). Calanoids were more important contributors to the zooplankton assemblage in the Adirondack Province and New England Uplands, both subregions with higher N:P ratios. Lakes on the coastal plain and lowland plateaus exhibited greater contributions to the assemblage from cyclopoid copepods as well as lower N:P ratios (Table III).

The explanatory variables for lake N:P ratio in the stepwise regression model included zooplankton, site depth, pH, land use, and elevation (asl) of the highest point on the watershed (Table V). The elevation of the watershed high point may represent a surrogate for atmospheric loadings of NO_3 (cf. Miller *et al.*, 1993a). Calanoid assemblages were positively and significantly correlated with lake water N:P ratio while cladoceran assemblages were negatively correlated with the N:P ratio.

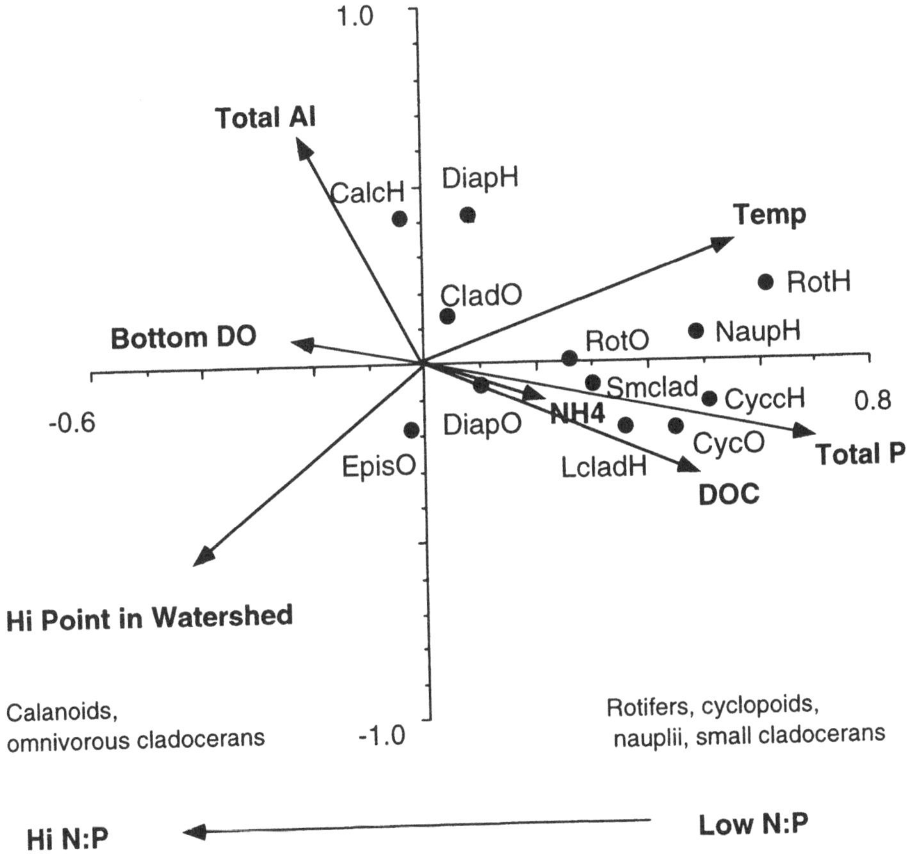

Fig 2. Redundancy analysis of zooplankton web variables and dominant environmental gradients that best explain variance patterns in the zooplankton assemblages of 107 Adirondack lakes for the 1991-1994 EMAP survey. Each arrow represents the direction of increasing magnitude of an environmental gradient relative to the 1st 2 axes. Length of arrow represents the amount of variance in the variable. Values at the origin represent the mean value for the variable. CalcH=calanoid grazers, EpisO=calanoid omnivores, DiapH=calanoid grazers, DiapO =calanoid omnivores, CladO= cladoceran omnivores, LcladH=large cladoceran grazers, ScladH=small cladoceran herbivore, CycO=cyclopoid omnivores, CycH=cyclopoid grazers, RotO=rotifer omnivores, RotH=rotifer grazers, naupH=nauplii grazers.

In the single and multiple linear regressions N and P were significantly correlated with several limnological and landscape parameters. Nitrogen to phosphorus ratios were negatively correlated with chlorophyll *a*, DOC, chloride, and road density, and were positively correlated with fish tissue Hg, chain length, and forest cover (Tables I, IV). The N:P ratio also was positively correlated with latitude ($p< 0.001$) and elevation ($p< 0.001$) and negatively correlated with longitude ($p < 0.01$, Table IV). Lake NO_3 was positively correlated with atmospheric nitrogen deposition ($p < 0.001$) estimated from the NADP network (Miller in preparation). Lake N:P was also positively correlated with estimated atmospheric nitrogen deposition ($p<0.001$). Lake P correlated positively with estimated atmospheric deposition of SO_4 ($p < 0.001$) and lake SO_4 ($p < 0.001$) and negatively with increasing latitude ($p < 0.001$).

Table I
Pearson product moment correlation coefficients 111 Adirondack lakes (based on data related to Figure 2). Bold type indicates p<0.01, * indicates p<0.05.

Variable	PC Axis 1	PC Axis 2	N:P ratio	NO_3	Total N	Total P
PC Axis 1	1					
PC Axis 2	0.0812	1				
N:P ratio	**-0.3920**	0.1262	1			
NO_3	**-0.3451**	0.1407	**0.7867**	1		
Total N	**0.2882**	-0.0048	-0.0053	-0.0187	1	
Total P	**0.4288**	**-0.2756**	**-0.3865**	**-0.4164**	**0.5207**	1
DOC	**0.3304**	-0.1882*	**-0.4024**	**-0.3241**	**0.7253**	**0.7465**
Chl a	**0.4534**	**-0.2712**	**-0.2982**	**-0.3585**	**0.4437**	**0.7655**
pH	0.0205	**-0.4597**	-0.1905*	-0.1718	-0.1770	-0.0297
Site Depth	**-0.3588**	-0.132	0.0319	0.1805	**-0.3735**	**-0.4552**
Hi-Point	**-0.4221**	-0.2388*	0.2388*	**0.4203**	-0.2061*	-0.2060*
% Calanoids	**-0.4118**	**0.2988**	0.1950*	0.1185	-0.0242	-0.1725
% Rotifers	**0.3843**	0.0850	-0.1967*	-0.1144	0.0370	0.1679
% Cal/ Lg Cladocera	-0.0701	-0.1360	**0.4744**	**0.5044**	0.0547	-0.1042

Table II

Means of major zooplankton assemblage and limnological variables associated with selected N:P slopes for 111 Adirondack lakes in Figure 2.

Variable	N:P < 50	N:P >50<100	N:P >100
No. lakes	11	53	43
% Calanoids excluding nauplii (based on total crustacean abundance)	<1	5	9
% Large cladocerans	3	4	2
% Cyclopoids excl. nauplii	5	4	3
% Small cladocerans	2	4	2
% rotifers	81	69	66
% Omnivorous cladocerans as a fraction of large cladocerans	0.3	0.6	5
Adult calanoid herb. (no. L^{-1})	4.7	4.2	3.1
Adult calanoid omnivores	<0.002	0.13	0.6
Cyclopoid omnivores	21.7	3.5	1.0
Large cladoceran herbivores	15.3	7.2	4.1
Small cladoceran herbivores	6.5	7.8	5.0
Rotifer herbivores	826	428	192
Rotifer omnivores	7.5	4.8	2.1
Chlorophyll *a* ($\mu g\ L^{-1}$)	8.39	5.5	2.32
Road density (km ha^{-1})	11.34	7.0	4.2
Fish tissue Hg (mg kg^{-1} wet weight)	0.13	0.20	0.42
Fish tissue Pb (mg kg^{-1} wet weight)	0.07	0.05	0.03
DOC (mg L^{-1})	7.45	5.78	4.22
Chloride ($\mu eq\ L^{-1}$)	80.0	56.5	39.3
Site depth (m)	4.96	8.03	14.3

Table III

Weighted median values of N:P ratios and zooplankton variables for the 1991-1994 EMAP lakes.

	Adirondack Province	New England Uplands	Coastal /Lowland Plateaus
Estimated number of lakes	1434	4131	4625
N:P ratio	76.1	70	42.4
% Large cladocerans /macrozooplankton	26.7	18	26.8
% Large Cyclopoids/macrozooplankton	10.8	8.6	13.7
% Calanoids/macrozooplankton	16.9	23.12	11.9

Table IV
Pearson product moment correlation coefficients of the N:P ratio of 180 northeastern US lakes sampled during the 1992-1994 EMAP survey with zooplankton food web variables and selected land use and physicochemical variables. Bold type indicates $p<0.01$ and * $p<0.05$.

Variable	N:P ratio	Web Links	Web Chain Length	DOC	Bottom DO	Temp	Chl *a*	Lat.	Lon.	Road Density
N:P ratio	1									
Web Links	0.1107	1								
Web Chain Length	**0.3240**	**0.6433**	1							
DOC	-0.1377	**-0.2079**	**-0.2526**	1						
Bottom DO	**0.240**	**-0.2524**	-0.1042	0.1521*	1					
Temp	**-0.3421**	**-0.4697**	**-0.4535**	**0.3412**	**0.3527**	1				
Chl *a*	**-0.2906**	-0.0828	-0.1772*	0.1708*	0.0422	**0.2808**	1			
Latitude	**0.3493**	0.0252	-0.0059	**0.2379**	**0.2382**	**-0.2313**	**-0.2856**	1		
Longitude	**-0.1859**	-0.0302	0.0322	**-0.2396**	**-0.2058**	0.0817	**0.222**	**-0.7129**	1	
Road Density	**-0.3278**	0.0219	-0.1023	-0.1164	**-0.2544**	0.0574	**0.2484**	**-0.5616**	**0.2447**	1
Forest Cover	**0.1904**	-0.0004	0.1071	0.0354	**0.2052**	-0.1195	**-0.4273**	**0.4299**	**-0.2786**	**-0.645**
Fish Hg	**0.2116**	0.0198	**0.2622**	-0.0903	0.1589*	-0.1267	-0.1282	0.0026	0.0369	**-0.1918**
Fish Pb	-0.1549 *	-0.1408	-0.0421	**0.3578**	0.0786	**0.2379**	**0.1977**	**-0.3479**	0.1379	**0.198**

Table V
Stepwise regression ANOVA model of the 1991-1994 EMAP northeastern lakes data that best explain the lake N:P ratios. N=261 lakes, R2= 0.26.

Variable	Slope	P
Total calanoids excluding nauplii	3.67	0.0003
Total Large cladoceran herbivores	-2.14	0.033
Site depth	5.10	<0.0001
pH	-3.01	0.003
% total forest on watershed	-1.94	0.053
% total urban on watershed	-2.31	0.022
% total agriculture on watershed	-1.66	0.098
High point on watershed	2.89	0.004

4. Discussion

There was a wide range of total-N, total-P, and N:P ratios observed in our lake sample. Of the 364 lakes included in our analysis, 50% can be considered oligotrophic ($P < 0.32$ μM/L), 41% mesotrophic ($P > 0.32$ and < 0.97 μM/L), and 9% eutrophic ($P > 0.97$ μM/L) based on the trophic state classification of Moore and Thornton (1988). The distribution of lake trophic state in our sample differs from the population distribution estimated for all northeastern lakes (38% oligotrophic, 40% mesotrophic, 22% eutrophic) by Larsen et al. (1994) because we eliminated the most severely impacted lakes and did not apply the probability sample weights. Our selection of lake data was intended to provide a range of environmental (e.g., N, P) and biological conditions, rather than to estimate the status of lakes in the region. Nonetheless, our sample included 364 out of 385 lakes in the original EMAP data set. In our sample only 2.5% of lakes exhibit N:P ratios less than 17 and are therefore likely to experience significant nitrogen inputs via N-fixation (cf. Howarth *et al.*, 1988). The remaining 97.5% of lakes receive the bulk of N inputs from atmospheric deposition and terrestrial sources. Although denitrification likely plays some role in regulating nitrogen supply in northeastern lake waters, both mass balance studies and the distribution of N:P ratios observed here and elsewhere indicate that other ecosystem processes such as atmospheric deposition, terrestrial ecosystem exports, and internal recycling are the dominant influences on nitrogen availability in the majority of northeastern lakes (cf. Molot and Dillon, 1993; Downing and McCauley 1992).

4.1 Evidence For Gradients in Lake-water N:P Ratios And Zooplankton Assemblages in Northeastern Us Lakes

The EMAP pilot survey of northeastern US lakes provides some of the first empirical support for the relationship of the N:P ratio of lake water with the major compositional and structural features of pelagic zooplankton food webs. These relationships are robust despite the great variation in geography, historical drainage influences on dispersal and colonization of aquatic organisms, and in the extent and type of anthropogenic impacts in the region.

The results of this analysis are consistent with the growing body of evidence that supports resource supply theory as a fundamental mechanism structuring phytoplankton and zooplankton assemblages in nature (Elser *et al.*, 1996; Mazumder *et al.*, 1990; Elser *et al.*, 1988; Shapiro and Wright, 1984; Smith, 1982). When redundancy analysis is applied to the 1991-1994 limnological EMAP data set, a recurrent pattern or contrast is produced irrespective of year or subregion (see Stemberger and Lazorchak, 1994; Stemberger and Chen, in press) (Figure 2). Axis 1 contrasts zooplankton assemblages dominated by calanoid copepods and omnivorous cladocerans (*Leptodora*) with cyclopoid copepods, nauplii, and rotifers.

The analysis of Adirondack lakes suggests that PC-axis 1 represents a gradient in lake N:P ratios. High ratios along PC-axis 1, associated with calanoid and omnivorous cladoceran assemblages, are consistent with theoretical expectation (Elser *et al.*, 1996; Sterner *et al.*, 1993; Sterner and Hessen, 1994) (Figure 1). Total nitrogen and total phosphorus (μg/L) were inversely correlated with the N:P ratio along PC-axis 1 (Table I; Figure 2). This correlation suggests that the ratio and not the ambient concentrations of N and P explains higher order compositional differences in zooplankton food webs. Furthermore, NO_3 is positively correlated with the N:P ratio (Table I), suggesting that free NO_3, probably from atmospheric sources, may exacerbate already high N:P ratios in some

Adirondack lakes. This is consistent with the loading or positive correlation of the watershed high point with calanoid and omnivorous cladoceran assemblages (Figure 2; Tables I and V).

The negative correlation of the N:P ratio of Adirondack lakes with pH (Tables I and V) suggests the possibility for an alternative interpretation of field studies that report a loss of small-bodied taxa such as rotifers and an increase in calanoids such as *Diaptomus minutus* with decreasing pH (Sprules, 1977; MacIsaac *et al.*, 1987; Siegfried *et al.*, 1987; Tessier and Horwitz, 1990). High N:P ratios of Adirondack lakes are consistent with observed high rates of NO_3 deposition (Stoddard, 1994; Miller *et al.*, 1993b; Aber *et al.*, 1989). Hence, NO_3 deposition may increase the N:P ratio and lower pH thus providing a competitive advantage to calanoids over herbivorous cladocerans (Sterner *et al.*, 1992). Overall, these relationships suggest a potential link between lake water chemistry and higher level taxonomic organization of the zooplankton assemblage.

4.2 Other Influences on Zooplankton Assemblages

Temperature, DOC, pH, and Al also clearly influence zooplankton community structure (Stemberger and Lazorchak, 1994; Williamson *et al.*, 1996; Havens, 1993). Many of these factors load with N, P, and the N:P supply ratio (Figures 1, 4; Tables I, IV), suggesting that the N:P ratio may be responsive to landscape factors.

In addition to watershed regulation of N and P supply, the intensity of zooplanktivory should also influence the N:P ratio. For example, high levels of predation leading to the removal of *Daphnia* and other large cladocerans should shift N:P ratios downward (Elser *et al.*, 1988). Lakes characterized by intensive zooplanktivory tend to be dominated by cyclopoid copepods and small-bodied zooplankton assemblages like rotifers and nauplii (Figure 2; Tables I, II) (Stemberger and Lazorchak, 1994). Cyclopoid copepods have high cellular N:P requirements but are prominent members of lakes with low N:P ratios (Figure 2; Table I). This apparent inconsistency with supply theory can be explained because these copepods have effective escape responses from most zooplanktivores and their diet is high in nitrogen due to omnivory (Stemberger, 1985; Stemberger and Lazorchak, 1994) (Table II). Hence, departures from expected zooplankton assemblages based on N:P ratios predicted from watershed influences can be accommodated and integrated within well-established top-down ecosystem processes (Carpenter *et al.*, 1985).

4.3 Linking Lakes with Terrestrial and Atmospheric Inputs

Terrestrial sources of N and P exhibit N:P ratios ranging from 9-800 for unmanaged forest lands to 17-44 for agricultural runoff and 6-20 for sewage (Table VI). Our preliminary evaluation of the EMAP data suggests that terrestrial influences on N and P supply may be best characterized in two separate landscape classes.

Table VI

Average N:P ratios by terrestrial ecosystem type and compartment for forest types and potential nutrient sources for freshwater lakes. Empty cells = no data. Groundwater/ streamwater N:P ratios were used to construct the slopes in Figure 3 (data from Basu and Pick, 1996; Likens and Borman; 1995; Johnson and Lindberg, 1992; Downing and McCauley, 1992; Waring and Schlesinger, 1985; Cole and Rapp, 1981; and E.K. Miller, unpublished data).

Ecosystem	Precipitation	Throughfall	Vegetation Debris/Litterfall	Soil Forest Floor	Forest Floor Soil Water	Groundwater and Streamwater
High Atmospheric N pollution impacted ecosystems						
Whiteface Mtn., NY Spruce/Fir Forest	799	94		45	364	**440**
Smokies Spruce Forest	488	362		48	187	**371**
Low or no Atmospheric N pollution						
Temperate Deciduous Forest			35	44		**181**
Temperate Coniferous	152	16	19	26	18	**76**
Boreal Deciduous			9	16		
Boreal Coniferous			10	12		
Other systems						
Rivers						**45-321**
Tundra						**54**
Agricultural Fields						**44**
Temperate Bog						**17**
Average Fertilizer						**18**
Domestic Sewage						**6 - 22**

Agricultural and urban influenced lakes were defined as those with < 80% forest cover on the watershed and comprised 42% of the 287 lakes in our target group for which land use data were available. The dominant landscape factors regulating N and P supply in these lakes are the proximity of intensive agriculture to the lake or streams feeding the lake and density of development with isolated septic systems with limited impact on N supply on regional gradients by nitrogen air pollution. In general, these lakes tend to have highest total-N, total-P and chlorophyll *a*, often with very low N:P ratios resulting from inputs of incompletely treated agricultural runoff and domestic waste water. Twenty-five percent of lakes in this class had total P concentrations > 1.3 μM/L, total-N > 46 μM/L, and N:P ratios < 72. Only 2.5% of agricultural and urbanized lakes had total-P concentrations < 0.14 μM/L and total-N concentrations <14 μM/L. Only 10% of these lakes had N:P ratios greater than 92. The mean N:P ratio in this class was 58.

Forested, rural lakes were defined as those with > 80% forest cover on the watershed and comprised 58% of the 287 lakes in our target group for which land use data were available. The extent of forest cover, forest species composition (which correlates with soil properties), forest age, hydrology, and regional and local gradients in atmospheric pollution all are likely to influence the relative supply of N and P in forested rural landscapes.

Forested rural lakes exhibited a wide range of total-N, total-P, chlorophyll *a* and N:P ratios. Fifty percent of these lakes were oligotrophic, 48% were mesotrophic, and 2% were eutrophic. Ninety percent of these lakes had total-N < 41 μM/L, with 50% measuring below 26 μM/L. Ninety percent of these lakes had an N:P ratio > 44, 50% > 80, 25% >104, and 10% > 159. The mean N:P ratio in this class was 115.

The results of the single and multiple regression analyses are consistent with expected landscape influences on lake water chemistry. For example, lake water N:P ratio is positively correlated with latitude and elevation. This correlation is thought to primarily reflect changes in the character of the landscape, shifting from greater to lesser agricultural and urban influence along these gradients. Percent forest cover was positively correlated, while percent agricultural and urban land were negatively correlated with latitude and elevation (all $p < 0.01$). The N:P ratio of stream water and ground water exported from terrestrial ecosystems tends to increase with increasing forest character, and varies with forest type (Table VI). Lake NO_3 and N:P are positively correlated with atmospheric nitrogen deposition. Lake P correlates positively with estimated atmospheric deposition of SO_4 and lake water SO_4. These correlations support the hypothesis that SO_4 pollution influences PO_4 availability. Phosphate may be released from anion retention sites in soils and lake sediments in exchange for SO_4 (Caraco *et al.*, 1989).

Lake P decreases with increasing latitude. Since P retention in soils is strongly correlated with organic matter content, it is likely that this correlation occurs in part from increased soil organic matter content that accompanies the shift in climate and forest type with increasing latitude in the survey region. The combination of cooler temperatures and the recalcitrant nature of coniferous forest litter result in slower decomposition rates and larger accumulations of soil organic matter at higher than at lower latitudes (cf. Van Cleave *et al.*, 1981). Lake water DOC and the DOC/N, DOC/P ratios all increase with latitude (all $p< 0.001$), reflecting the shift from temperate deciduous to boreal and coniferous forests with increasing latitude (Table VI). Other sources of organic matter, such as canopy throughfall, leaf litter, and forest floor, which may constitute a significant fraction of initial particulate C, N, and P inputs to streams (Meyer and Likens, 1979; Meyer *et al.*, 1981), also exhibit a range in N:P ratios (Table IV). The observed range of lake water N:P and total nutrient concentrations appears to be strongly influenced by variation in direct atmospheric deposition and the relative contributions of nutrients from landscape segments characterized by agricultural use, residential use, wetlands, and particular forest types between watersheds.

4.4 Conceptual Model for the Metric

The extent of anthropogenic impacts on lakes may be assessed using a combination of the measured water column N and P concentrations and the difference between the expected N:P ratio of the nutrient loading of the watershed and the N:P ratio measured in lake water. Using a graph of N and P (μM/L) space similar to that in Figure 3 we can estimate the expected N:P ratio for lake water based on terrestrial ecosystem characteristics (Table VI), assuming the system is in equilibrium with watershed loadings. We can evaluate the observed lake water N vs. P against both the expected N:P ratio and absolute N and P concentrations associated with the watershed's hydrology, land use, and forest types (Figure 3).

This analysis provides important information for interpreting alteration or change in fundamental lake ecosystem attributes. For example, if a lake's N:P ratio departs a determined amount from the expected value for its watershed type, we might interpret this deviation as indicating some level of disturbance. If the P concentration is in excess of 1 μM/L it very likely indicates anthropogenic disturbance in the watershed (Figure 3). In nonagricultural areas it would probably signify shoreline or riparian sources of P. Likewise, N loadings above 50 μM/L might be used to signify N pollution (Figure 3). The inferred zooplankton assemblage (Figure 3) would provide not only additional biological information but could also be used to evaluate ecosystem risks to correlative factors discussed above (Figure 4).

Practical application of the indicator could involve several or all of the following steps. An expected contemporary lake water N:P ratio can be estimated from watershed land use and forest ecosystem composition by appropriate measurements made in stream and groundwater inputs and consideration of lake hydrology such as water retention time (Table VI; Figures 3, 4). Furthermore, an expected "pre-settlement" N:P ratio also can be estimated from diatom assemblages observed in sediments from an appropriate depth core (Dixit *et al.*, 1992). These expected values can then be used to infer the character of the pelagic zooplankton community (i.e., the relative dominance of calanoids over cladocerans and microzooplankton) and associated ecosystem structural and functional features that would be in equilibrium with the watershed conditions. Both the inferred lake-water N:P ratio and zooplankton assemblage may be compared to the observed N:P ratio and zooplankton. An example of how the result of such comparisons might be evaluated is as follows.

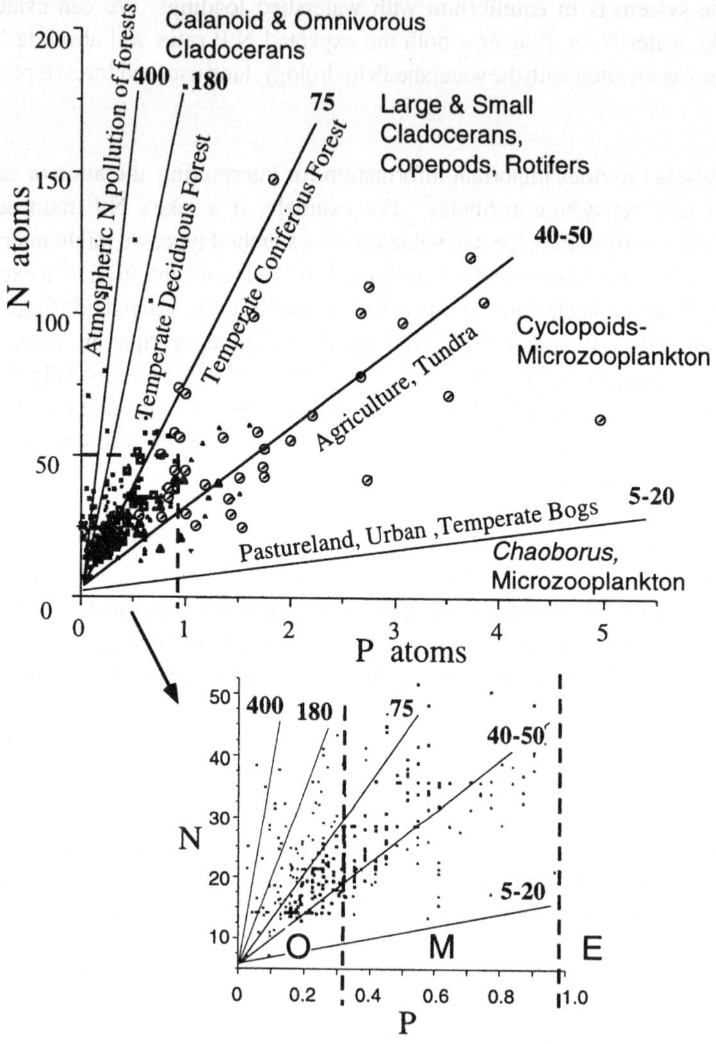

Fig 3. Plot of N and P space indicating how the zooplankton-N:P ratio indicator is interpreted within different classes of terrestrial watershed vegetation cover and land use. Open circles indicate lakes that have Chl *a* values > 12 µg L-1 . These systems are associated with P values > than 1 µmoles L-1 and reflect riparian, urban, or agricultural P pollution. N:P slopes were derived from data in Table VI. Dashed line in upper right panel delimits the N, P space (50 µmoles L-1 of N and 1 µmoles L-1 P) that characterizes the majority of northeastern lakes. Broken vertical lines of lower panel delimit lake trophic state based on total-P criteria given in Moore and Thornton (1988) but reexpressed as µmoles L-1 total P. O=oligotrophic lakes (<10µg L-1 total P), M=mesotrophic lakes (>10 < 30 µg L-1 total P), and E=eutrophic lakes (> 30 µg L-1 total P). Note that all zooplankton groups may potentially occur within each trophic state category.

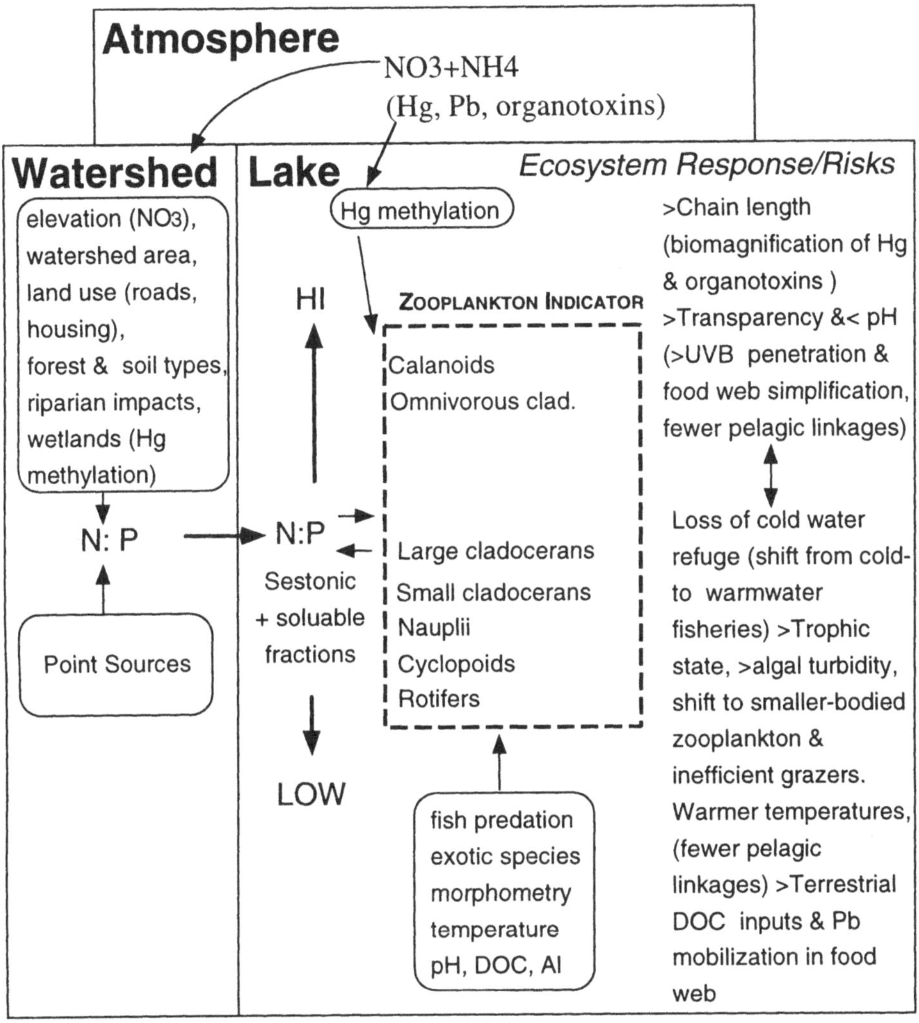

Fig 4. Conceptual diagram of linkages and risks between N:P ratios of watershed, lake water, and zooplankton indicators.

1) *When the observed N:P ratio in the lake is lower than the N:P loading from the watershed*

If the expected watershed contribution is significantly higher than the lake N:P ratio, then the indicator suggests potential dominance of P loadings from local sources. For example, lower N:P ratios could be caused by septic system failures, runoff from fertilized lawns or agricultural fields, and general disturbance of soils for roads and dwellings in the riparian zone.

Expected algal biomass would be high with dominance of small inefficient grazers. In dimictic lakes high P loadings could stimulate other effects from reproductive and recruitment failure of salmonid piscivores to the loss of cold water invertebrate omnivores (e.g., *Leptodiaptomus sicilis*, *Mysis relicta*, and *Limnocalanus macruus*), and other common temperature-sensitive invertebrates (cool-water copepods like *Epischura lacustris* and *Cyclops scutifer*, and a suite of cold-water rotifers) (Stemberger, 1995b).

Low N:P ratio systems are also likely to have shorter food chains with the loss or reduction of the cool water refuge (Figure 4; Table IV). Intensive zooplanktivory on larger omnivorous calanoid copepods may also promote dominance of smaller zooplankton groups. Soil disturbance may also increase mobilization of Pb as a function of higher organic components of soils especially for watersheds on the northeastern coastal plain (Miller and Friedland, 1994). High DOC loading associated with high road density poses greater risk of elevated Pb in fish tissues (Stemberger and Chen, in press) (Table IV).

2). *When the observed N:P ratio in the lake is higher than the N:P loading from the watershed*

This condition could indicate a watershed that is saturated with nitrogen from atmospheric or agricultural sources (Figures 3, 4). We expect omnivorous calanoid copepods and omnivorous cladocerans to be prominent in the assemblage along with smaller herbivorous calanoids that have high N:P ratio requirements (Figures 1, 3). These systems are characterized by long-chained omnivorous pelagic food webs that make them efficient in the biological transfer and biomagnification of Hg and organohalides like PCB's (Table IV). If anthropogenic acidification proceeds to low pH (i.e., below 6.0) we expect that calanoid omnivores will begin to disappear from lakes (Keller and Yan, 1991). Increasing acidification can lower DOC and increase UV-B and increase aluminum toxicity on zooplankton and other invertebrates (Williamson *et al.*, 1996; Yan *et al.*, 1996; Schindler *et al.*, 1996; Havens, 1993).

4.5 USE OF THE N:P INDICATOR IN CONJUNCTION WITH DIRECT ZOOPLANKTON MEASUREMENTS

In some instances, ancillary information may provide a reason to suspect that the zooplankton assemblage may differ from the one predicted by the observed lake water N:P ratio. Departures from the expected assemblage, such as the absence of calanoids in high N:P ratio systems or the absence of large cladocerans and dominance of rotifers and cyclopoids at intermediate N:P ratios, may indicate other disturbances related to human activities. For example, factors that decrease the abundance of large piscivores, such as excessive removal by sport fishermen or degradation of the cool water refuge, may increase abundance of forage fishes in response to lower predation. Microzooplankton-grazer related shifts in internal nutrient recycling may in turn lead to high algal turbidity. Factors reducing the predation efficiency of visual feeding piscivores may lower the lake N:P ratio (Elser *et al.*, 1996). Riparian disturbances that increase local P loading may produce changes that appear similar to the trophic cascade that can occur from removal of large piscivores, i.e., the zooplankton assemblage tends to be dominated by rotifers, small cladocerans, and cyclopoids (e.g., N:P ratios <50).

4.6 FUTURE RESEARCH

There are several lines of research that need to be refined for a more complete development of the N:P indicator. The EMAP data set probably contains adequate measurements of temporal variance of zooplankton populations and water chemistry, but more detailed time-series observations may be useful. An analysis of 76 lakes that were revisited twice within the sampling season and of 56 lakes that were sampled at least twice in different years suggests that variance in the N:P ratio within lakes is markedly stable (Stemberger and Miller, unpublished). On the other hand, measures of zooplankton species biomass may be more appropriate than simple species abundance data used in this analysis. While the EMAP data set has allowed us to present a case for the potential utility of the N:P ratio, limitations of the data set do not permit clear evaluations of the biogeochemical links between the terrestrial and aquatic systems. Such critical observations include separate measures of C, N, and P in the dissolved, fine, and coarse sestonic fractions of lake water; N, P, and chlorophyll analysis of the phytoplankton fraction; N and P analysis of the zooplankton size fraction; separate measures of C, N, and P in allochthonous particulate matter delivered at lake inlets; and classification of watershed segments by forest type with emphasis on riparian and littoral zones. This research is necessary to refine the indicator, by better quantifying and constraining the variance of its components.

5. Conclusions

We have described a conceptual model for linking terrestrial and aquatic ecosystem processes via resource supply theory. The proposed N:P ratio indicator provides a single measure proxy for several key aquatic ecosystem properties pertinent to lake management. These descriptors include food chain length, availability of oxic cool-water habitat, intensity of zooplanktivory, P-enrichment, algal turbidity, and correlative risks of food web structure to biologic transfer efficiency of various toxins.

At the first level of comparison an expected zooplankton assemblage and food web structure is associated with a range of N:P ratios. The species composition or food web structure of these assemblages can be directly related to a variety of potential risk factors. Second, an expected N:P ratio for lake water can be derived from analysis of forest cover type and land use and from basin hydrology and regional gradients in air pollution. This expected N:P ratio links conditions in the terrestrial environment to important measures of aquatic ecosystem state. A "pre-settlement" and contemporary N:P ratio also can be estimated from paleolimnological data of fossil diatom assemblages from appropriate sediment cores. Analysis of deviations of observed in-lake N:P from the expected N:P ratio provides a third level of function for the indicator. Deviations of in-lake N:P from the ratio expected provide information about the type and magnitude of other potential human disturbances, such as changes in fish assemblages and nutrient loading due to shoreline development. An expected diatom-inferred N:P ratio indicating "pre-development" conditions in a watershed can facilitate quantitative assessment of the effects of current land-use practices on the aquatic system. Similarly, an estimate of contemporary conditions from diatom microfossils could provide additional support for lake- and terrestrial-based N:P ratio measures, thus reinforcing confidence in the estimates.

The N:P ratio is a simple and robust ecological indicator that can provide a practical approach to evaluating the status, extent of change, and associated risks of aquatic ecosystems in the context of their associated terrestrial environments. Frequent and/or wide-spread samplings of lake water N and P can be conducted in order to quickly and inexpensively assess aquatic ecosystem changes in lakes. The N:P indicator can be used as a screening tool in extensive studies, permitting a cost-effective means of selecting a reasonable number of lakes for more intensive study based on several measures of ecosystem risk such as the relative likelihood of disturbance (observed vs. expected N:P based on watershed analysis), sensitivity to new disturbance (proximity of N:P ratio to the boundary of a species assemblage zone in N:P space, Figure 3), or the indicated score relative to management objectives for the lake (e.g., N:P ratio indicates zooplankton assemblage with no cool-water refuge when such a fishery is the desired management goal).

Acknowledgments

This study was supported by EPA Cooperative Agreement CR819689 and a USDA Forest Service Global Change Research Program Cooperative Agreement with Dartmouth College. We are grateful to the EMAP Surface Waters group making this data set available for our analysis. The comments of A. T. Herlihy, C. Y. Chen, M. M. Moore, and two anonymous reviewers improved the manuscript.

References

Aber, J.D., Nadelhoffer, J. K., Steudler, P., and Melillo, J.M.: 1989, 'Nitrogen saturation in northern forest ecosystems', *Bioscience* **39**, 378-386.

Baker, J.R., Peck, D.V., and Sutton, D.W. (ed.): in press, *Field Operations Manual for Lakes*, Environmental Monitoring and Assessment Program-Surface Waters: EPA 620/R-97/001, U.S. Environmental Protection Agency, Washington, DC.

Basu, B.K. and Pick, F.R..: 1996, 'Factors regulating phytoplankton and zoooplankton biomass in temperate rivers', *Limnol. Oceanongr.* **41,** 1572-1577.

Budd, L.F. and Meals, D.W.: 1994, 'Lake Champlain nonpoint source pollution assessment', *Technical Report No. 6B,* Lake Champlain Basin Program, South Hero, VT, USA.

Caraco, N.F., Cole, J.J., and Likens, G.E.: 1989, 'Evidence for sulphate-controlled phosphorus release from sediments of aquatic systems', *Nature* **341,** 316-318.

Carpenter, S.C., Kitchell, J.F., and Hodgson, J.R.: 1985, 'Cascading trophic interactions and lake productivity', *Bioscience* **35**, 634-639.

Cole, D.W. and Rapp, M.: 1981, 'Elemental cycling in forest ecosystems', in D.E. Rechle ed. *Dynamic Properties of Forest Ecosystems.* Cambridge University Press, New York, 683 p.

Dixit, S.S., Smol, J.P., Kingston, J.C.,and Charles, D.F.: 1992, 'Diatoms: Powerful Indicators of Environmental Change', *Environ. Sci. Tech.* **26**, 22-33.

Downing, J.A. and McCauley, E.: 1992, 'The nitrogen:phosphorus relationship in lakes', *Limnol. Oceanogr.* **37,** 936-945.

Elser, J.J., Elser, M.M. MacKay, N.A.,and Carpenter, S.R.: 1988, 'Zooplankton-mediated transitions between N and P limited algal growth', *Limnol. Oceanogr.* **33**, 1-14.

Elser, J.J., and Hassett, R.P.: 1994, 'A stoichiometric analysis of the zooplankton-phytoplankton interactions in marine and freshwater ecosystems', *Nature* **370,** 211-213.

Elser, J. J., Dobberfuhl, D.R., MacKay, N.E., and Schampel, J.H.: 1996, 'Organism size, life history, and N:P stoichiometry', *Bioscience* **46**, 674-684.

Havens, K.E.: 1993, 'Acid and aluminum effects on osmoregulation and survival of the freshwater copepod *Skistodiaptomus oregonensis*', *J. Plankton Res.* **15,** 683-691.

Hecky, R.A. and Kilham, P.: 1988, 'Nutrient limitation of phytoplankton in freshwater and marine environments: a review of the recent evidence on the effects of enrichment', *Limnol. Oceanogr.* **33,** 796-822.

Johnson, D.W. and Lindberg, S.E.: 1992, *Atmospheric Deposition and Forest Nutrient Cycling*, Springer-Verlag, New York, 707p.

Keller, W. and Yan, N.D.: 1991, 'Recovery of crustacean zooplankton species richness in Sudbury area lakes following water quality improvements', *Can. J. Fish. Aquat. Sci.* **48**,: 1635-1644.

Kilham, P. and Kilham, S.S.: 1990, 'Endless summer: internal loading processes dominate nutrient cycling in lakes', *Freshwat. Biol.* **23,** 379-389.

Larsen, D.P. and Christie, S.J. (ed.): 1993, EMAP-Surface Waters 1991 Pilot Report, EPA/620/R-93/003. U.S. Environmental Protection Agency, Office of Research and Development, Washington, DC.

Larsen, D.P., Thorton, K.W., Urquhart, N.S., and Paulsen, S.G.:. 1994, The role of sample surveys for monitoring the condition of the nation's lakes. Environ. Monit. Assess. **32**, 101-134.

Likens, G.E. and Borman, F.H.: 1974, 'Linkages between Terrestrial and Aquatic Ecosystems', *Bioscience* **24,** 447-456.

Likens, G.E. and Borman, F.H.: 1995, *Biogeochemsitry of a Forested Ecosystem* (2nd ed.). Springer-Verlag, New York, 159p.

MacIsaac, H.J., Hutchinson, T.C., and Keller, W.: 1987, 'Analysis of planktonic rotifer assemblages from Sudbury, Ontario, area lakes of varying chemical composition', *Can. J. Fish. Aquat. Sci.* **44,** 1692-1701.

Mazumder A., Taylor, W.D., McQueen, D.J., Lean, D.R.S., and Lafontain, N.R.: 1990, 'A comparison of lakes and lake enclosures with contrasting abundances of planktivorous fish', *J. Plank. Res.* **12**, 109-124.

Meyer, J.L. and Likens, G.E.: 1979, 'Transport and transformation of phosphorus in a forest stream ecosystem', *Ecology* **60,**1255-1269.

Meyer, J.L., Likens G.E., and Sloane, J.: 1981, 'Phosphorus, nitrogen and organic carbon flux in a headwater stream', *Arch. Hydrobiol.* **91,** 28-44.

Miller, E.K., Friedland, A.J., Arons, E.A., Mohnen,V.A., Battles, J.J., Panek, J.A., Kadlecek, J., and Johnson A.H.: 1993a, 'Atmospheric deposition to forests along an elevational gradient at Whiteface Mountain, NY, USA', *Atm. Environ.* **27,** 2121-2136.

Miller, E.K., Panek, J.A., Friedland, A.J., and Kadlecek, J.: 1993b, 'Atmospheric deposition to a high-elevation forest at Whiteface Mountain, NY, USA', *Tellus* **45B,** 209-227.

Miller, E.K. and Friedland, A.J.: 1994, 'Lead migration in forest soils: response to changing atmospheric inputs', *Env. Sci. Tech.* **28**, 662-669.

Molot, L.A. and Dillon, P.J.: 1993, 'Nitrogen mass balances and denitrification rates in central Ontario Lakes', *Biogeochemistry* **20**,195-212.

Moore, L., and Thornton, K. (eds): 1988, *The Lake and Reservoir Restoration Guidance Manual, 1st edition,* prepared by North American Lake Management Society, EPA 440/5-88-002, U.S. Environmental Protection Agency, Washington, D.C.

Shapiro, J. and Wright, D.I.: 1984, 'Lake restoration by biomanipulation: Round Lake, Minnesota- the first two years', *Freshwat. Biol.* **14**, 371-383.

Schindler, D.W., Curtis, J. P. Parker, B.R., and Stainton, M.P.: 1996, 'Consequences of climate warming and lake acidification for UV-B penetration in North American boreal lakes', *Nature* **379**, 705-708.

Siefried, C.A., Bloomfield, J.A., and Sutherland, J.W.: 1987, 'Acidification, vertebrate and invertebrate predators, and the structure of zooplankton communities in Adirondack lakes', *Lake Reserv. Manage.* **3**, 385-393.

Smil, V. : 1990, ' Nitrogen and Phosphorus', in B.L. Turner *et al.*, editors, *The Earth as Transformed by Human Action*, Cambridge University Press, Cambridge.

Smith, V.H.: 1982, 'The nitrogen and phosphorus dependence of algal biomass in lakes: an empirical and theoretical analysis', *Limnol. Oceanogr.* **27,** 1101-1112.

Sprules, W.G.: 1977, 'Crustacean zooplankton communities as indicators of limnological conditions: an approach using principal components analysis', *J. Fish. Res. Board. Can.* **34,** 962-975.

Sprules, W.G. and Bowerman, J. E.: 1988, 'Omnivory and food chain length in zooplankton food webs', *Ecology* **69,** 418-426.

Stemberger, R.S.:1985, 'Prey selection by the copepod *Diacyclops thomasi*', *Oecologia* **65**, 492-497.

Stemberger, R.S.: 1995a, 'Pleistocene refuge areas and postglacial dispersal of copepods of the northeastern United States', *Can. J. Fish. Aquat. Sci.* **52,** 2197-2210.

Stemberger, R.S.: 1995b, 'The influence of mixing on rotifer assemblages of Michigan lakes', *Hydrobiologia* **297,** 149-161.

Stemberger, R.S. and Lazorchak, J.M.: 1994, 'Zooplankton assemblage responses to disturbance gradients', *Can. J. Fish. Aquat. Sci.* **51,** 2435-2447.

Stemberger, R.S. and Chen, C.Y.: in press, 'Fish tissue metals and zooplankton assemblages of Northeastern US lakes', *Can. J. Fish. Aquat. Sci.*

Stemberger, R.S., Herlihy, A.T., Kugler, D.L., and Paulsen, S.G.: 1996, 'Climatic forcing on zooplankton richness of the northeastern United States', *Limnol. Oceanogr.* **41,** 1093-1101.

Sterner, R.W., Elser, J.J., Hessen, D.O.: 1992, 'Stoichiometric relationships among producers and consumers in food webs', *Biogeochemistry* **17,** 49-67.

Sterner, R.W., Hagemeier, D.D., Smith, W.L., and Smith, R.F.: 1993, 'Phytoplankton nutrient limitation and food quality for *Daphnia*', *Limnol. Oceanogr.* **38,** 857-871.

Sterner, R.W. and Hessen, D.O.: 1994, 'Algal nutrient limitation and the nutrition of aquatic herbivores', *Annu. Rev. Ecol. Syst.* **25,**1-29.

Stoddard, J.L.: 1994, 'Long-term changes in watershed retention of nitrogen: its causes and aquatic consequences', L.A. Baker, ed., *Environmental chemistry of lakes and reservoirs*, Advances in Chemistry Series No. 237, American Chemical Society, Washington, D.C. 223-284

Ter Braak, C.J.F.: 1988, 'CANOCO- A FORTRAN program for canonical community ordination by partial detrended canonical correspondence analysis, principal component analysis and redundancy analysis', TNO Institute of Applied Computer Science, Wageningen, The Netherlands. Technical Report LWA-88-02, Wageningen, 95 p.

Tessier, A. J. and Horwitz, R.J.: 1991, 'Influence of water chemistry on size structure of zooplankon assemblages', *Can. J. Fish, Aquat. Sci.* **47**, 1937-1943.

Tilman, D.: 1982, *Resource competition and community structure*, Princeton University Press, Princeton, NJ.

U.S. E.P.A: 1991, '*Methods for the determination of metals in environmental samples*', EPA/600/4-91/010, U.S. Environmenta; Protection Agency, Office of Research and Development, Washington, D.C.

Van Cleave, K., Barner, R., and Schlentner, R.: 1981, 'Evidence of temperature control of production and nutrient cycling in two interior Alaska black spruce ecosystems', *Can. J. For. Res.* **11,** 258-273.

Vanni, M.J., Layne, C.D., Arnott, S.E. 1997: ' "Top-down" trophic interactions in lakes: effects of fish on nutrient dynamics', *Ecology* **78,** 1-20.

Vitousek, P.M., Mooney, H.A., Lubchenco, J., and Melillo, J.M..: 1997a, ' Human Domination of Earth's Ecosystems', *Science* **277,** 494-499.

Vitousek, P.M., Aber, J.D., Howarth, R.W., Likens, G.E., Matson, P.A., Schindler, D.W., Schlesinger, W.H.., and Tilman, D.G..: 1997b, 'Human alteration of the global nitrogen cycle: sources and consequences', *Ecological Applications* **7**, 737-750.

Waring, R.H. and Schlesinger, WH.: 1985, *Forest Ecosystems Concepts and Management,* Academic Press, New York, 340 p.

Williamson, C.E., Stemberger, R.S., Morris, D.P., Frost, T.M., and Paulsen, S.G.: 1996, 'Ultraviolet radiation in North American lakes: attenuation estimates from DOC measurements and implications for plankton communities', *Limnol. Oceanogr.* **41,** 1024-1034.

White, T.C.R.: 1993, *The inadequate environment: nitrogen and the abundance of animals,* Springer-Verlag, New York.

Yan, N.D., Keller, W., Scully, N.M., Lean, D.R.S., and Dillon, P.J.: 1996, 'Increased UV-B penetration in a lake owing to drought-induced acidification', *Nature* **381,** 141-143.

Yeardley, R.B., Jr.: 1994, 'Fish Tissue Contaminants Indicator Laboratory Methods for Compositing Fish and Determining Target Analyte Concentrations', in D.J. Klemm and J.M. Lazorchak, ed., *Environmental Monitoring and Assessment Program Surface Waters and Region 3 Regional Environmental Monitoring and Assessment Program, 1994 Pilot Laboratory Methods Manual for Streams,* EPA/620/R-94/003, U.S. Environmental Protection Agency, Cincinnati, OH.

IMPLICATIONS OF SEASONAL AND REGIONAL ABUNDANCE PATTERNS OF *DAPHNIA* ON SURFACE WATER MONITORING AND ASSESSMENT

FRIEDA B. TAUB and CARMEN D. WISEMAN

School of Fisheries, University of Washington, Seattle, WA

Abstract. The seasonal dynamics of *Daphnia* populations vary regionally throughout the United States. Within the general pattern, *Daphnia* increase in abundance after the initiation of the spring algal bloom in all lakes, but their subsequent seasonal patterns differ in various climatic regions. Lakes in regions with cooler summers have large-bodied *Daphnia* populations that tend to persist throughout the summer, although the species dominance may shift. Regions with warmer summers tend to have large-bodied *Daphnia* populations that decline or are absent through much of the summer. Still warmer water bodies tend to have medium- to small-bodied species that are abundant during spring, but absent most of the summer. Many central Florida lakes lack *Daphnia*; if *Daphnia* species are present, they tend to be small-bodied. *Daphnia* abundance in these water bodies varies, but seems to be independent of temperature.

If surface water (lake, pond) sampling is done in all regions during July and August, the impression will be that *Daphnia* are absent from large segments of the United States. This would be erroneous, because *Daphnia* are important earlier during the spring and early summer but are likely to be absent during midsummer in some U.S. regions. Year-to-year variation will be superimposed on this regional pattern. Because there are differences in the dates when spring and summer occur, it would be useful to have an index period that would standardize the start of the growing season. The use of the terrestrial onset of greenness, based on remote sensing of the Normalized Difference Vegetation Index, is suggested as a possible index set point.

1. Introduction

There is an intersection between continuous reality and intermittent sampling. Only observations that are recorded become data for analysis. Frequent sampling of a single site allows observation of seasonal events (phenology), but limits the number of sites that can be sampled on limited budgets. When the intent of the sampling is to cover a wide area, as in the Environmental Monitoring and Assessment Program, it is anticipated that most sampling will take place once during the annual period or once every few years. For communities that are sampled infrequently, the timing of the sampling will influence our perception of reality.

On an annual scale, lakes and ponds do not exist in a steady state; they cycle through well-defined stages. Therefore, sampling at a single time provides only a single frame of a dynamic motion picture. The Lake Constance representation (Figure 1) gave the Plankton Ecology Group (PEG) a pattern for comparison with other water bodies (Lampert 1987, Sommer 1986). The PEG model demonstrates the brief, intense spring bloom of filtering zooplankton followed by a summer decline and then a second, more modest rise in autumn.

Environmental Monitoring and Assessment **51**: 53–60, 1998.

We suggest that regional variations could lead to misinterpretation of sampling data if these variations are ignored.

Schematic representation of the biotic interactions controlling the seasonal change of the biomass of phytoplankton and filter-feeding zooplankton in Lake Constance. The relative importance of different controlling mechanisms is indicated by the width of black horizontal bars. The phytoplankton biomass is divided into small ("edible") forms (*shaded*) and large forms (*hatched*) not edible for the zooplankton. *Dotted vertical line* indicates the predictable period of extremely transparent water in the beginning of June. *Arrows* symbolize causes and effects. Positive or negative interrelationships are indicated by the respective sign. *Bold arrows* represent a chain of interrelationships without a feedback loop (see text)

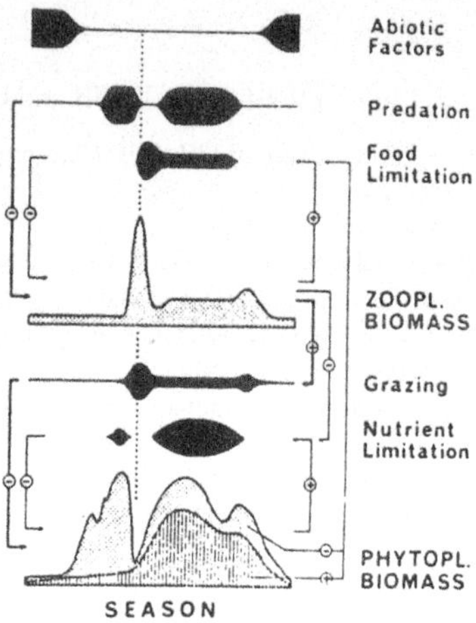

Fig. 1. Schematic representation of the seasonal abundances of phytoplankton and filter-feeding zooplankton in Lake Constance. (from Lampert 1987) [Reprinted with permission.]

Daphnia have been extensively used to measure the toxicity of chemicals. As grazers, they are acknowledged to be important ecological processors of primary productivity into larger packets more suitable for fish food. Examination of *Daphnia* distributions suggests that they are found in most of the conterminous United States (Brooks 1957, Hebert 1995). So, it might seem reasonable to expect to find *Daphnia* in lakes and ponds during the summer, and to potentially interpret their absence as response to a stressor. Surveys of the literature suggest that this interpretation would be an error. Some of the best studies of exposure and effects have been carried out in mesocosms. Examination of some of these data suggest that *Daphnia* declined in control mesocosms during the summer. As a result, negative impacts of toxicants on *Daphnia* populations could not be demonstrated during July and August in pesticide-treated mesocosms (Fairchild *et al.* 1992; Webber *et al.* 1992). Because these studies were conducted in mesocosms, however, it was not clear whether the *Daphnia* declines were artifacts of the small size of the water bodies and associated intense predation by fish.

Daphnia summer declines were described by Threlkeld (1979, 1985) for a few specific lakes, but it was not clear if this decline was a general property of water bodies, or if different patterns of decline existed. Previously collected data from original reports and the scientific literature were examined to determine the regional patterns of *Daphnia* summer declines (Wiseman 1996). The purpose of this paper is to consider these findings in the context of monitoring and assessment of ecological conditions in ponds and lakes.

2. Methods

Examination of more than 500 journal articles, dissertations, books, agency reports, and unpublished data sets on zooplankton yielded 16 sites that provided concurrent temperature and zooplankton abundance data; 18 additional sites provided summer maximum temperatures along with zooplankton abundance data. Zooplankton sizes, if not provided in the data set, were inferred from taxa, based on ranges given in taxonomic documents (e.g., Brooks 1957, Balcer *et al.* 1984). Adult sizes of the *Daphnia* reported in these studies were assumed to go from largest to smallest— *D. pulex/D. pulicaria, D. thorata, D. galeata mendotae, D. lumholtzi, D. retrocurva, D. laevis, D. rosea, D. parvula,* and *D. ambigua.* Only a few studies provided size distributions or distinguished between immature and mature animals. Therefore, we focused on adult sizes, although it was obvious that immature *Daphnia* often constitute the majority of the population.

Given these references and the general consideration of annual (once-yearly) sampling, the following implications for monitoring and assessment are inferred.

3. Results

In water bodies assigned to the Cool (North) regional group, all Daphnia species were relatively large and at least one daphnid species persisted through the summer (Figure 2). The Warm (North-Central) group comprised bodies with large daphnids that declined markedly or disappeared for most of the summer. In water bodies of the Hot (South) group, daphnid sizes ranged from medium to small and daphnids tended to be absent from the plankton during the summer. Daphnids were generally rare in Florida, and the small daphnids in two Florida ponds did not exhibit strong seasonality; therefore, Florida was characterized as a separate region. Statistical analyses of 34 water bodies showed a moderately significant ($r = -0.43$, $P < 0.05$) negative correlation between abundance of the largest daphnids (e.g., *D. pulex/pulicaria*) and highest surface water temperature. No significant correlation was obtained between abundance of small daphnids (e.g., *D. parvula)* and maximum water temperature, or between abundance of all daphnids at the maximum temperature and the depth or latitude of the water body (Wiseman 1996).

These findings suggest that daphnid declines have some predictability in U.S. lakes, ponds, and reservoirs, although multiple mechanisms of decline (e.g., predation, resource limitation, competition) may be involved. Moreover, daphnid seasonal patterns appear to be similar enough in some areas or regions of the conterminous United States to permit preliminary regional groupings. It would be useful to examine data from larger groups of lakes and ponds to confirm these patterns.

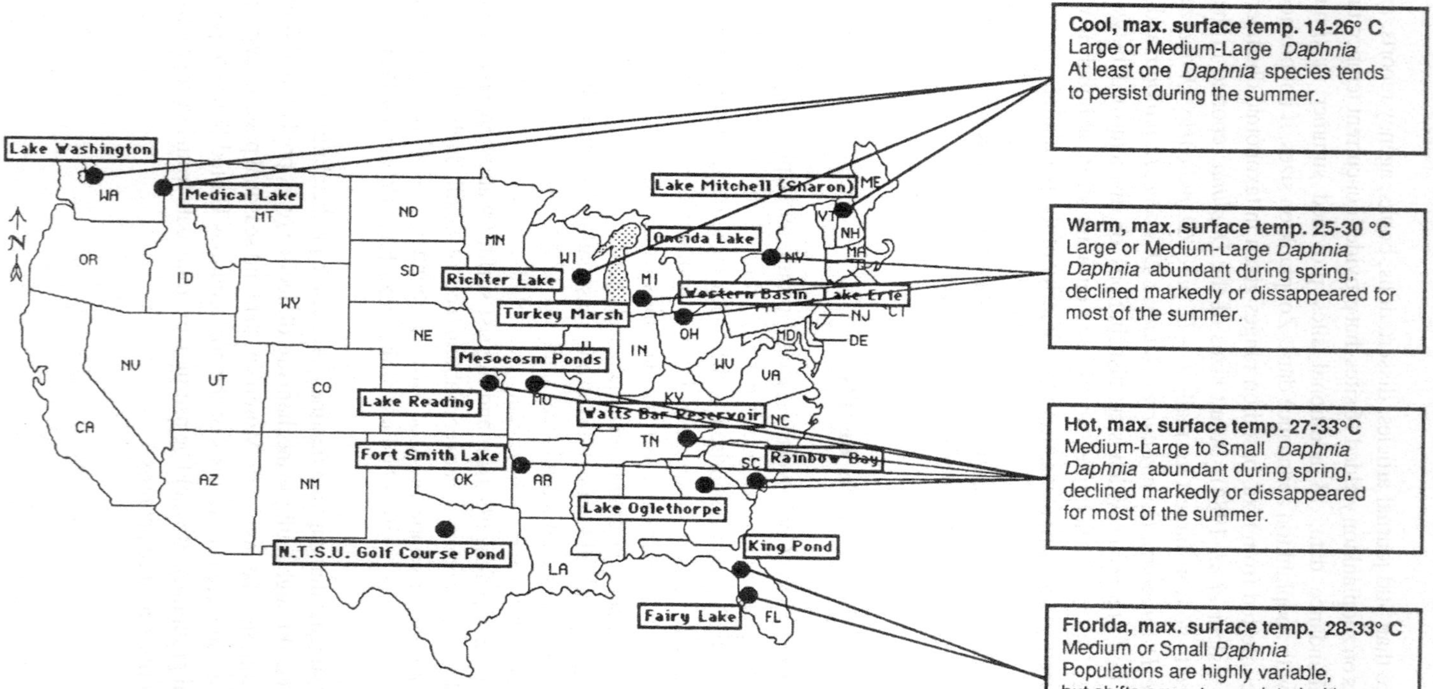

Note: N.T.S.U. Golf Course Pond was reported to have *D. pulex*, a larger *Daphnia* species than reported in other southern lakes and ponds; no size range was provided. Depending on the size, this could belong to the Warm or Hot Group.

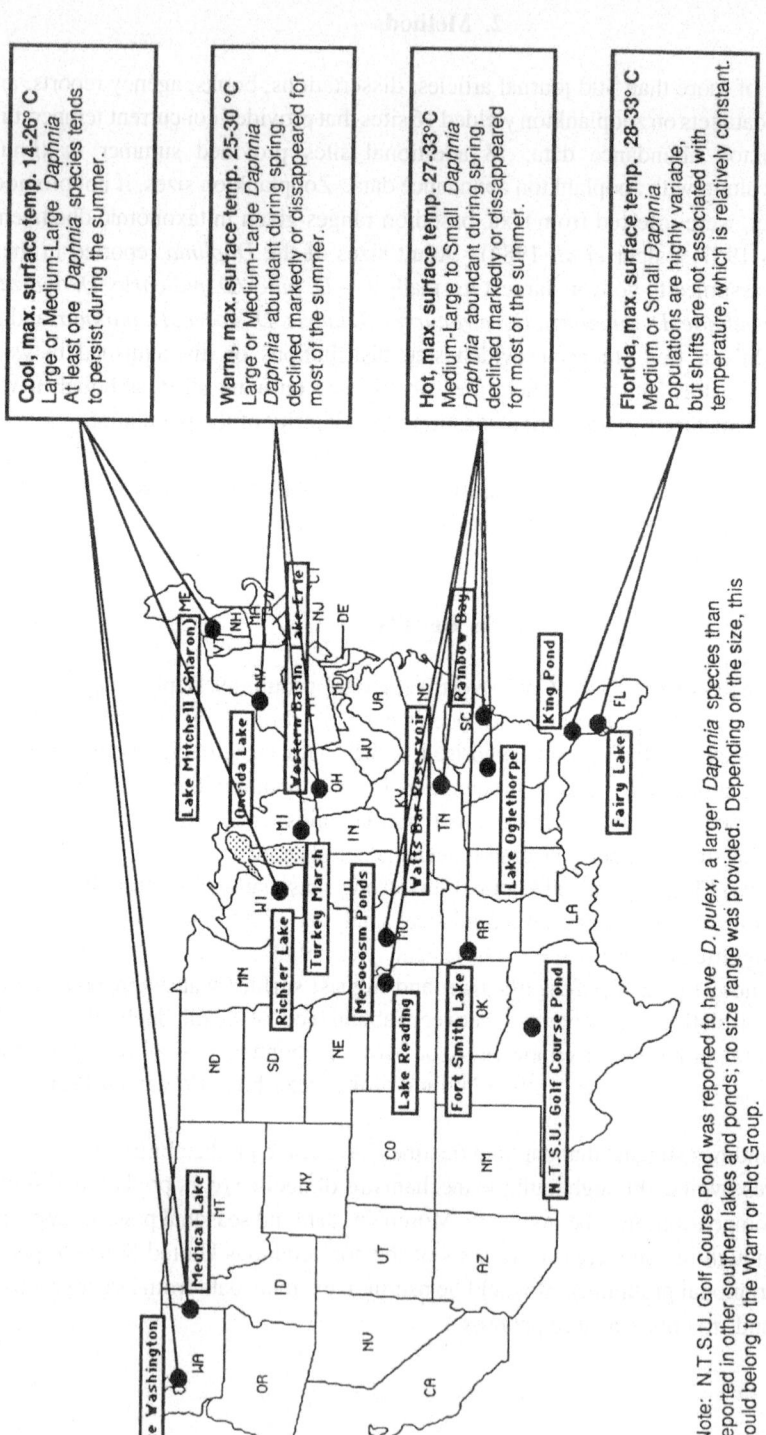

Fig. 2. Pattern of *Daphnia* body size and summer abundances as demonstrated by 16 lakes in the conterminous US.

4. Discussion

The midsummer declines or absences of *Daphnia* highlighted by Threlkeld (1979, 1985), but previously described by Hutchinson (1967) and first mentioned by Birge (1897), may exist over much of the conterminous United States, especially where water temperatures exceed 25° C and no oxygenated temperature refugia exist. An examination of mean monthly air temperatures during July and August shows that the entire center of the conterminous United States tends to have very warm periods during the summer. Northern areas have briefer periods of warm temperatures, but shallow water bodies may still become warm throughout their depth. In the South, not only does the warm period persist longer, but many of the water bodies are shallow ponds or artificial lakes; the few large, deep water bodies are reservoirs whose stratification and mixing are influenced by electrical production or irrigation needs. During the winter, air temperatures better reflect north (cold) to south (warm) differences, but latitude is a less-effective predictor of summer temperatures.

It may be important to recognize that *Daphnia* are not available in many warm environments throughout the entire summer irrespective of pollution inpacts. The *Daphnia* minimum or absence tends to occur during the time of warmest surface-water temperature. There are many mechanisms that could contribute to the absence of *Daphnia*, such as predation by fish or invertebrates, inappropriate food supply, increased metabolic demands, and competition from other taxa (Wiseman 1996). So, there may be many mechanisms for the summer declines of large-bodied *Daphnia*, and all tend to have similar effects.

If the presence of *Daphnia* were considered an attribute of a healthy aquatic environment, it would be necessary to monitor water bodies during the spring or very early summer before the *Daphnia* declined. Past limnological records could be used to estimate appropriate monitoring times for each region.

These cycles have implications for year-to-year variation and for lake-to-lake variability. The top part of Figure 3 shows a complex annual cycle; five such cycles with slight variations in timing are superimposed in the bottom part of the figure. If the water bodies depicted in the bottom of the figure were to be sampled at a given point along the *x*-axis (for example, if a lake were sampled on July 1 of every year, or if five lakes were sampled on July 1 of every year), the water bodies might be perceived as being very different from one another—although this simplistic example is the same pattern, just slightly displaced in time.

To minimize the apparent year-to-year differences that may result from year-to-year climatic variations or to better compare different lakes that have slightly different climatic zones, it would be useful to have a way to index the start point. A similar objective has been met by estimating the mean time of onset of greenness for Great Plains grasslands using Normalized Difference Vegetation Index data. Tieszen *et al.* (1997) were able to map the onset of greenness in an N-S strip from the northern border of North Dakota and

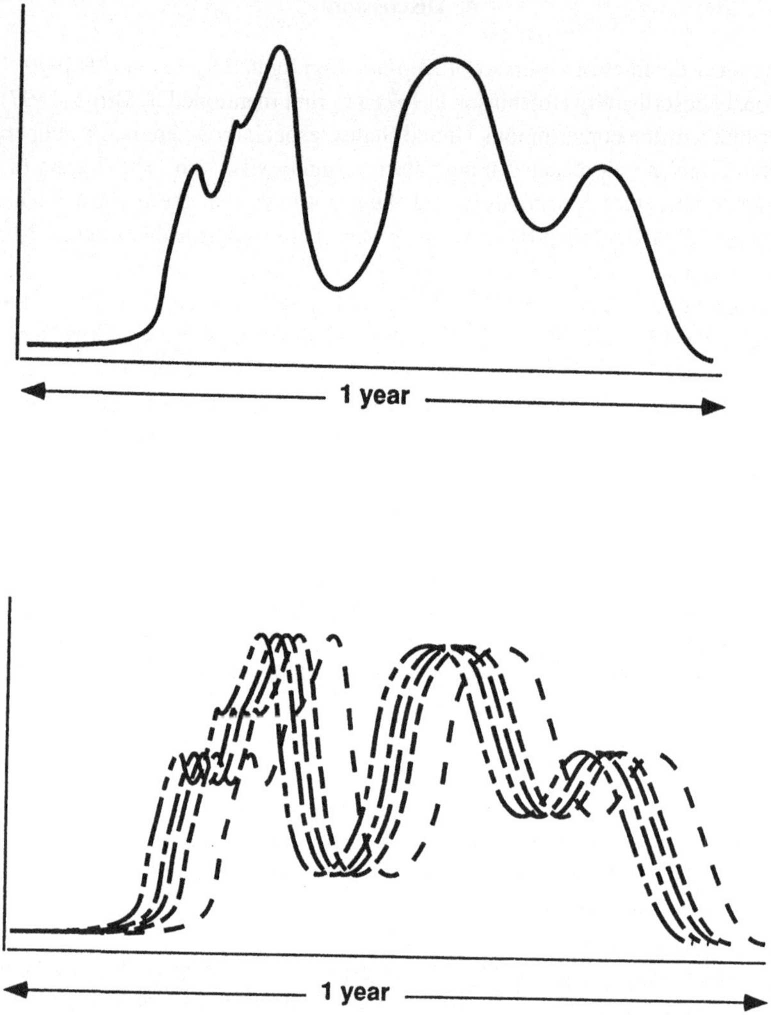

Fig. 3. Schematic representative of seasonal patterns: (top) seasonal pattern within a single lake; (bottom) five identical patterns, slightly offset in time. These could represent the year-to-year variations for a single lake, or the variations among similar lakes.

Montana to the southern border of Texas and part of New Mexico. Their analyses covered the years 1990, 1991, 1992, and 1993, and included the mean and coefficient of variation. The timing of the onset of greenness is presumably related to amount of sunlight as modified by cloud cover, and the temperature regime as modified by wind and other factors. These are the same factors that we would expect to initiate the spring bloom in ponds and lakes, perhaps with some lags for ice cover to melt and water to warm. Beaver *et al.* (1981) studied the thermal regimes of Florida lakes to relate the thermal boundaries of three lake thermal regimes (warm temperate, central transitional, and subtropical) to terrestrial floral and faunal transisitons within the state. This supports the hypothesis that land and aquatic phenological events are likely to be correlated in time.

In conclusion, if the surface waters of the entire United States were to be sampled, it would be important to note that the timing of sampling will influence whether *Daphnia* are observed. For mesocosm ponds such as those in Columbia, Mo., samples taken during early June would indicate *Daphnia* to be an important part of the community; samples taken in July and August would indicate their absence or rarity (Fairchild *et al.* 1992). A zooplankton sample containing no daphnids, or no large dapnids, might be misconstrued as evidence of toxicant exposure or other anthropogenic stress, since both pesticide exposure and acidification have been reported to reduce size or eliminate daphnids (Havens and Hanazato 1993; Odum 1985; Marmorek and Korman 1993). Depending on when zooplankton samples are obtained, abundant small taxa and reduced *Daphnia* numbers are likely to be the normal state of affairs in U.S. temperate lakes at midsummer. These findings should be taken into consideration when developing sampling designs for future environmental monitoring and assessment programs.

Acknowledgment

The zooplankton seasonal study was funded by award No. R821705-01-0 from the Environmental Protection Agency, U.S. Office of Exploratory Research in association with the Environmental Monitoring and Assessment Program.

References

Balcer, M. D., Korda, N. L., and Dodson, S. I.: 1984, "Zooplankton of the Great Lakes: a guide to the identification and ecology of the common crustacean species," University of Wisconsin.

Beaver, J. R., Crisman, T.L., and Bays, J.S.: 1981, "Thermal regimes of Florida Lakes: a comparison with biotic and climatic transitions,"*Hydrobiologia* **83**, 267-273.

Birge, E. A.: 1897, "Plankton studies on Lake Mendota. II. The Crustacea of the plankton from July, 1894, to December, 1896," *Trans. Wisc. Acad. Sci. Arts Lett.* **11**, 274 247-448.

Brooks, J. L.: 1957, "The systematics of North American Daphnia: Memoirs of the Connecticut Academy of Arts and Sciences, 13," Yale.

Fairchild, J. F., LaPoint, T. W., Zajicek, J. L., Nelson, M. K., Dwyer, F. J., and Lovely, P. A.: 1992, "Population-, community- and ecosystem-level responses of aquatic mesocosms to pulsed doses of a pyrethroid insecticide," *Environ. Toxicol. Chem.* **11**, 115-129.

Havens, K. E., and Hanazato, T.: 1993, "Zooplankton community responses to chemical stressors: a comparison of results from acidification and pesticide contamination research," *Environ. Pollut.* **82**, 277-288.

Lampert, W.: 1987, "Predictability in lake ecosystems: the role of biotic interactions," in E.-D. Schulze and H. Zwolfer (eds): Ecological Studies 61, Springer-Verlag, Berlin, pp. 333-346.

Marmorek, D. R., and Korman, J.: 1993, "The use of zooplankton in a biomonitoring program to detect lake acidification and recovery," *Water Soil Air Pollut.* **69**: 223-241.

Odum, E. P.: 1985, "Trends expected in stressed ecosystems," *BioScience* **35**, 419-422.

Threlkeld, S. T.: 1979, "The midsummer dynamics of two *Daphnia* species in Wintergreen Lake, Michigan," *Ecology* **60**, 165-179.

Threlkeld, S. T.: 1985, "Resource variation and the initiation of midsummer declines of cladoceran populations," *Arch. Hydrobiol.* **21**, 333-340.

Tieszen, L. L., Reed, B. C., Bliss, N. B., Wylie, B. K., and Dejong, D. D.: 1997, "NDVI, C3 and C4 production, and distributions in great plains grassland land cover classes," *Ecol. Appl.* **7**, 59-78.

Webber, E. C., Deutsch, W. G., Bayne, D. R., and Seesock W. C.: 1992, "Ecosystem-Level testing of a synthetic pyrethroid insecticide in aquatic mesocosms," *Environ. Toxicol. Chem.* **11**, 87-105.

Wiseman, C. D.: 1996, "Seasonal dynamics of cladoceran zooplankton in water bodies of the conterminous United States: do regional patterns exist?" M.S. thesis, University of Washington, 118 pp.

THE ROLE OF BIOLOGICAL INDICATORS IN A STATE WATER QUALITY MANAGEMENT PROCESS

CHRIS O. YODER and EDWARD T. RANKIN

Ohio Environmental Protection Agency, Division of Surface Water, 1685 Westbelt Drive, Columbus, Ohio 43228

Abstract: State water quality agencies are custodians of water quality management programs under the Clean Water Act of which the protection and restoration of biological integrity in surface waters is an integral goal. However, an inappropriate reliance on chemical/physical stressor and exposure data or administrative indicators in place of the direct measurement of ecological response has led to an incomplete foundation for water resource management. As point sources have declined in significance, the consequences of this flawed foundation for dealing with the major limitations to biological integrity (nonpoint sources, habitat degradation) have become more apparent. The use of biocriteria in Ohio, for example, resulted in the identification of 50% more impairment than a water chemistry approach alone and other inconsistencies of a flawed monitoring foundation are illustrated in the national 305(b) report statistics on waters monitored, aquatic life use attainment, and habitat degradation. Biological criteria (biocriteria) incorporates the broader concept of water resource integrity to supplement the roles of chemical and toxicological approaches and reduces the likelihood of making overly optimistic estimates of aquatic life condition. A carefully conceived ambient monitoring approach comprised of biological, chemical, and physical measures ensures all relevant stressors to water resource integrity are identified and that the efficacy of administrative actions can be directly measured with environmental results. New multimetric indices, such as the IBI, ICI, and BIBI represent a significant advancement in aquatic resource characterization that have allowed the inclusion of biological information into many States water quality management programs. Ohio adopted numerical biocriteria in the Ohio water quality standards regulations in May 1990 and, through multiple aquatic life uses that reflect a continuum of biological condition, represents a tiered approach to water resource management. Biocriteria provide the impetus and opportunity to recognize and account for natural, ecological variability in the environment, something which previously was been lacking in state water quality management programs. The upper Great Miami River in Ohio illustrates a case study where bioassessment data documented the efficacy of efforts to permit, fund, and construct municipal treatment systems in restoring aquatic life. In contrast, in the Mahoning River similar administrative actions were inadequate to restore aquatic life in an environment with severe sediment contamination and impacts from combined sewer overflows. A biocriteria-based goal of restoring 75% of aquatic life uses by the year 2000 in Ohio has led to the use of biological data to identify trends and forecast the status and the causes and sources of impairment to Ohio streams, an effort that should affect the strategic focus of our water resource management efforts. A biocriteria-based approach has profoundly influenced strategic planning and priority setting, water quality based permitting, water quality standards, basic monitoring and reporting, nonpoint source assessment, and problem discovery within Ohio EPA.

1. Introduction

State water pollution control agencies function as custodians of water quality management under the Federal Water Pollution Control Act (*i.e.*, Clean Water Act [CWA]). This role is delegated by U.S. EPA to qualifying States which then have the obligation to develop and maintain water quality standards, issue NPDES permits, lead in the development of basin-wide water quality management plans, and monitor the effectiveness of the overall water quality management program. It is in fulfillment of this latter function that the development of environmental indicators, which includes biological monitoring and assessment, has recently received renewed attention. While the implementation of a

Environmental Monitoring and Assessment **51**: 61–88, 1998.

consistent national framework is likely several years away, examples of nearly complete approaches exist within a few State programs.

The purpose of this paper is to describe how we developed and implemented a framework for using environmental indicators for surface waters, which is highlighted by the use of biological indicators and criteria, within Ohio's state water quality management process and to further suggest what roles these should have in risk management and policy applications. A principal objective of the CWA is to restore and maintain the physical, chemical, and biological integrity of the nation's surface waters (Clean Water Act Section 101[a][2]). Although this goal is fundamentally biological, the specific methods by which state and federal agencies have attempted to reach this goal have been predominated by such non-biological measures as chemical/physical water quality (Karr *et al.* 1986). The rationale for this process is well known. Chemical water quality criteria developed through laboratory toxicity tests on selected aquatic organisms serve as surrogates for determining attainment of the biologically-based goals of the Clean Water Act. These criteria are adopted in state water quality standards and hence serve as a basis for implementing Clean Water Act programs. However, the underlying presumption that improvements in chemical water quality will be followed by a restoration of biological integrity has increasingly come into question.

While this approach may give an impression of empirical validity and legal defensibility it does not incorporate direct measures of the ecological health and well-being of surface water resources. The almost exclusive emphasis on chemical water quality in general, and toxic substances in particular, has deterred the incorporation of broader ecological measures and indicators in state programs. This narrow focus leads to an incomplete foundation in water resource policy and legislation (*e.g.*, an emphasis on point sources and toxics). Some characteristics of incomplete approaches include:

- A reliance on prescriptive approaches to management and regulation.
- A reliance on anecdotal information.
- An emphasis on administrative activities at the expense of monitoring and assessment efforts.
- An emphasis on point sources to the near exclusion of other problems (*e.g.*, nonpoint sources, habitat, etc.).
- Inconsistent environmental statistics reported between States (*e.g.*, Clean Water Act sections 305[b] report statistics, 303[d] TMDL listings[1], and 304[l] toxic discharge lists, etc.).

[1] Recently several State 303[d] lists have been challenged by third parties. The remedy in some cases has included State agreements to increase the number of stream and river miles monitored and the inclusion of biological indicators.

In some cases the chemical criteria/point source emphasis has led to some well-intentioned, but flawed management strategies. In a few cases this has resulted in increased environmental degradation even though the original impetus was to protect the aquatic environment. One example of this was the accepted practice of placing sanitary sewers in stream beds to resolve regional water pollution problems in southwestern Ohio (Ohio Environmental Protection Agency [EPA] 1992). The ensuing habitat degradation caused by sewer construction and maintenance resulted in a net increase in the miles of degraded streams, some of which were permanently damaged. Thus water quality management efforts which principally rely on comparatively simple, surrogate frameworks (*e.g.*, "bean counts") carry a significant risk of incomplete success at best and outright failure at worst in attempting to achieve water quality goals. Therefore, ecological concepts, biological criteria, and the attendant biological and habitat monitoring and assessment tools must be further incorporated into and institutionalized within the routine management of surface water resources.

· Fortunately, the U.S. Environmental Protection Agency (EPA), an increasing number of States, and others are becoming cognizant of these shortcomings and are either initiating or are involved in efforts such as the Environmental Indicators initiative (U.S. EPA 1995a, 1995b), the State Environmental Goals and Indicators Project (Berquist *et al.* 1995), and the Intergovernmental Task Force on Monitoring Water Quality[2] (ITFM 1992, 1993, 1995). Each of these efforts have outlined partial frameworks for addressing the aforementioned deficiencies in water quality management. Taken together, these offer a more complete approach that should result in the better use of State resources and lead to solutions for the remaining and more complex water resource problems.

2. Biological Integrity: A Frequently Missing Concept

Multiple factors in addition to chemical water quality are responsible for the continuing degradation and decline of surface water resources in Ohio and elsewhere (Ohio EPA 1997; U.S. EPA 1994; Benke 1992; Judy *et al.* 1984). These include the modification and destruction of riparian and aquatic habitats, sedimentation of bottom substrates, and alteration of natural flow regimes, all of which can lead to the alteration of basic environmental processes on a watershed scale. Because biological integrity is influenced and determined by *multiple* chemical, physical, and biological factors, a singular strategy emphasizing the control of chemicals *alone* does not assure the restoration of biological integrity (Karr *et al.* 1986). A broader view of the water resource as a whole is needed if we are to restore existing damage and prevent the decline in the overall quality of surface water resources. Biological criteria automatically incorporate the broader concept of water resource integrity while preserving the roles of the chemical and toxicological approaches developed over the past three decades. The intent should be to supplement, not replace these latter tools and criteria.

---

[2]The ITFM is now the National Water Quality Monitoring Council.

MAJOR FACTORS THAT DETERMINE WATER RESOURCE INTEGRITY

The health and well-being of the aquatic biota is an important barometer of whether we are achieving the Clean Water Act goal of maintaining and restoring the biological integrity of the nations's surface waters. This concept underlies the basic intent of State water quality standards. Yet this tangible end-product of water quality management has either been overlooked or excluded altogether (Karr 1991). Thus opportunities to link measures of water quality program performance to environmental "end-products" has largely been missed. Simply stated biological integrity is the combined result of chemical, physical, and biological processes in the aquatic environment (Karr and Dudley 1981; Karr *et al.* 1986; Figure 1). It is the interaction of these processes which is readily apparent in the functioning of lotic ecosystems as exemplified by the quality and quantity of the biological resources that are produced and sustained.

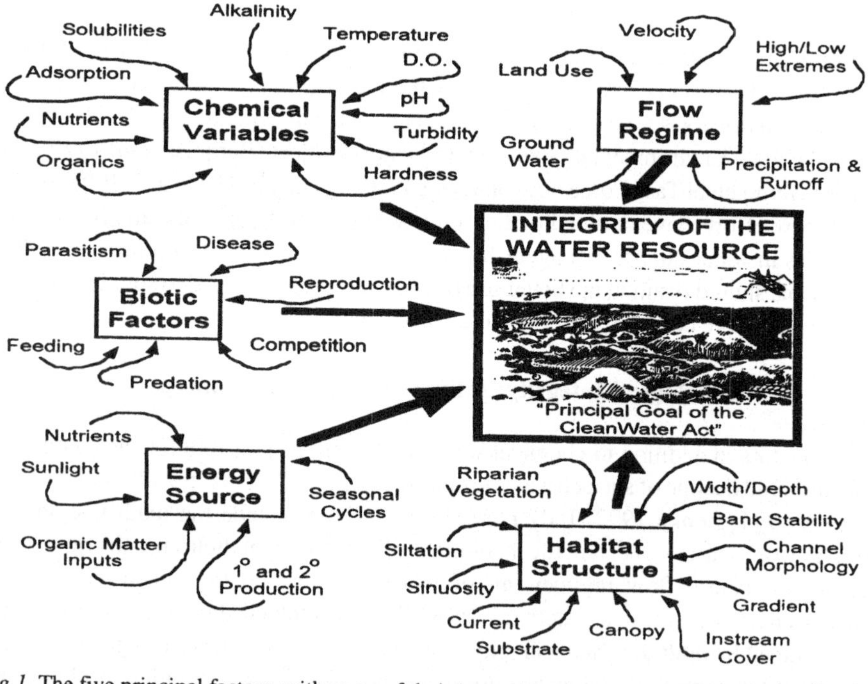

Fig 1. The five principal factors, with some of their important chemical, physical, and biological components that influence and determine the integrity of surface water resources (modified from Karr *et al.* 1986).

DISPARITIES IN THE USE OF INDICATORS: CASE EXAMPLES

The Clean Water Act requires U.S. EPA and the States to report on the condition of surface waters on a biennial basis (Section 305[b] report). The principal purpose is to report on the status and trends of the quality of the nation's waters. Perhaps the most frequently asked question that the 305[b] report endeavors to answer is: "Is water quality improving or getting worse?" The General Accounting Office (U.S. GAO 1986) criticized U.S. EPA in the past for not being able to adequately quantify improvements in water quality for the billions of dollars spent to improve WWTP effluents through the construction grants program. The historical failure to provide support for *adequate* State ambient monitoring programs has resulted in a lack of consistent and useable information to determine national trends. Nichols (1992) indicates that bioassessment information which documents changes over time could be found for only three water systems in the U.S. While this is likely a very conservative estimate of the national database, it nevertheless verifies the chronic problem of a lack of good ambient monitoring information.

A growing number of States and other organizations are relying primarily on biological indicators to assess the condition of water resources, but some continue to emphasize chemical and other indicators as surrogates of biological integrity. There are risks inherent to the latter in inaccurately portraying the condition of aquatic resources. Out of 645 stream and river segments analyzed in Ohio, impairment[3] revealed by biological indicators such as the Index of Biotic Integrity (IBI; Karr 1981) and the Invertebrate Community Index (ICI; DeShon 1995) was evident in nearly one-half (49.8%) of the stream and river segments where no impairments based on chemical indicators were observed (Rankin and Yoder 1990; Ohio EPA 1990). Instances of the biological indicators demonstrating full attainment when chemical water quality criteria exceedences were detected were much less frequent (2.8%). These latter cases were related to the inability of chemical water quality criteria to adequately account for different levels of ecological potential compared to the superior ability of the biological indicators to stratify background variability and hence define more appropriate levels of protection. While the discrepancy between the biological and chemical assessments may seem remarkable, the causes are related to the fundamental differences between what each indicator is actually capable of reflecting. As depicted in Figure 1, biological communities respond to and integrate a wide variety of chemical, physical, and biological factors in the environment whether of natural or anthropogenic origin. As such biological indicators more accurately reflect a wider range of environmental disturbances and quality gradients than chemical water quality alone, such as impacts to habitat. However, some of the biological impairment that was not detected by the chemical indicators was due to chemical associated causes (an inference made from source data and/or sediment chemistry data) that were beyond the capability of the grab sampling design widely employed by most State monitoring programs (Rankin and Yoder 1990).

Another example of deficiencies in State environmental indicators frameworks is in the proportion of waters reported by the States as impaired by habitat degradation (Figure 2) and summarized in the National Water Quality Inventory for 1994 (305[b] report; U.S. EPA 1994). Out of 58 states and territories which report such statistics on a biennial basis,

---

[3]Impairment is a term used to describe the failure to meet a use designated in State WQS. In this case it is the failure to meet either water quality or biological criteria for aquatic life uses as defined by the Ohio WQS.

nearly one-half (25) reported *zero* miles of rivers and streams impaired by habitat modification activities (which includes channelization, impoundment, riparian encroachment, hydrological modifications, or degradation of the substrate). Of the states that did report on this cause only 15 reported more than 100 miles of streams and rivers impaired due to degraded habitat and several others failed to report on common activities such as channelization. Such statistics are difficult to accept given the pervasiveness of habitat modifying activities for purposes such as flood control (channelization, riparian encroachment, impoundment), agricultural practices (channelization, riparian encroachment, sediment in runoff), resource extraction (mining, silviculture), and urbanization (watershed scale modifications) throughout the U.S., impacts which have been widely documented elsewhere (Judy *et al.* 1984; Benke 1992). The discrepancies between State 305[b] statistics for this major impairment category is likely related to the non-use of habitat sensitive indicators and residual State programmatic biases towards point source discharges and toxic chemicals.

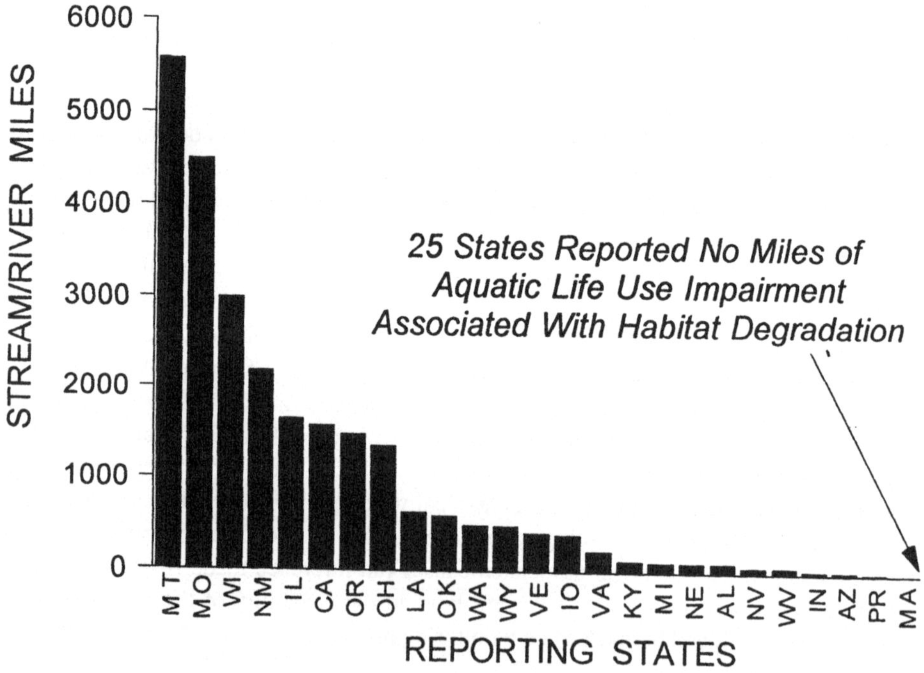

Fig 2. Miles of habitat impaired rivers and streams reported by the States to U.S. EPA as summarized in the *National Water Quality Inventory: 1992 Report to Congress* (U.S. EPA 1994).

The lack of a consistent approach to the use of indicators and monitoring networks by the States casts a great deal of uncertainty about what are the nation's most important water quality problems. Nowhere is this more evident than in the statistics reported to U.S. EPA by the States. These are used to produce the National Water Quality Inventory (305[b] report; U.S. EPA 1994). Full attainment of designated uses reported by individual States ranged from a high of 98% to a low of zero (0) while the proportion of river and stream miles considered to be fully assessed ranged from a high of 100% to a low of 5%. Adjoining states frequently have widely divergent estimates about their surface water quality. The variability and inconsistency among such statistics is attributable to fundamentally different frameworks for monitoring and assessment which includes the use of different classes (*i.e.*, stressor, exposure, response) of indicators for the same purposes. Most apparent is the inappropriate reliance on stressor and exposure indicators (*e.g.*, source information, pollutant loadings, chemical water quality criteria) as substitutes for the more direct response indicators (*e.g.*, biological criteria) in making the required statewide assessments of aquatic life use attainment and non-attainment. While this approach may have been adequate for characterizing the gross water pollution problems of previous decades it is simply insufficient for today's needs and frequently results in the gross under-reporting of problems (*e.g.*, the 25 states that reported no habitat impairments in Figure 2). An analysis of State 305[b] reports by U.S. EPA (1996) concluded that at least 30 states have some type of biological indicator data available for use in 305[b] reporting. However, only 12 states have developed the supporting assessment criteria needed to properly use this indicator. Even fewer states have developed biological criteria, but 22 states have the underlying research and development efforts well underway. Until more structured and consistent assessment frameworks are required, individual States will continue to approach monitoring and assessment quite differently, the end result being an "uneven playing field" and less than reliable national statistics about the condition of our waters. This also has significant ramifications well beyond the purview of EPA and State water quality agencies as 305[b] report statistics are used independently by a variety of other government and private organizations. Thus the potential to "export" the errors inherent to the national statistics is high and potentially costly both economically and environmentally. This is an issue in need of urgent attention at both the national and state levels.

The Role and Influence of Biological Criteria

The incorporation of numerical biological criteria into the Ohio EPA monitoring and assessment process has had major ramifications in our reporting on status and trends in water quality. This was first apparent in the marked change in the number of miles of Ohio streams and rivers that were reported as failing to attain CWA goals in the 1988 Ohio Water Resource Inventory (Ohio EPA 1988). The proportion of miles classified as exhibiting non-attainment increased from 9% in 1986 (based on a mix of chemical and *qualitative* biological indicators) to 44% in 1988 due primarily to the introduction and primacy of numerical biological criteria in the assessment and reporting process. The nearly five-fold "increase" in non-attainment illustrates the significant differences which can exist between states not only because of different indicators, but different frameworks for the biological assessment methodologies. These differences range from an exclusive reliance on chemical data to widely divergent bioassessment frameworks (*i.e.*, qualitative vs. quantitative bioassessments).

The preceding examples demonstrate that the risk in relying on chemical water quality indicators alone is towards making overly optimistic estimates of status which results from underestimating environmental degradation (*i.e.*, a type II error). Ironically much of the concern that was initially expressed about biological criteria has been over the risk of making type I errors, *i.e.*, becoming under-protective because of biological criteria. This concern clearly seems misplaced in light of the preceding examples and initially served as a deterrent to the wider acceptance of biological criteria. The implications of these findings to water resource management efforts are especially significant in that major environmental problems will either be underestimated, improperly characterized, or overlooked altogether. The cumulative influences of aquatic and riparian habitat, land use, and nutrient dynamics are particularly difficult to synthesize without using integrative indicators such as biological criteria.

3. A Framework of Environmental Indicators for Water

A carefully conceived ambient monitoring approach, using cost-effective indicators comprised of biological, chemical, and physical measures, can ensure that all relevant pollution sources are judged objectively and on the basis of environmental results. Ohio EPA relies on a tiered approach in attempting to link the results of administrative activities with true environmental measures. This integrated framework is outlined in Figure 3 and includes a hierarchical continuum from administrative to true environmental indicators. This framework was initially developed by U.S. EPA (1990a) as part of an agency initiative exemplified by the following vision statement:

> "EPA will use environmental indicators, together with measures of activity accomplishments, to evaluate the success of our programs. Working in partnership with others, we will be able to report status and trends of U.S. and global environmental quality to the public, Congress, states, the regulated community, and the international community. National program managers will use environmental indicators to determine where their programs are achieving the desired environmental results, and where inadequate results indicate strategies need to be changed. Over time, as more complete data are reported, environmental indicators will become the Agency's primary means of reporting and evaluating success."

This effort also responded to the later mandates of the Government Performance and Results Act which spurred the development of strategic goals by the Office of Water, national indicators for surface waters (U.S. EPA 1995a), and a conceptual framework for using environmental information in decision-making (U.S. EPA 1995b). While these are critical first steps in addressing some of the previously mentioned deficiencies, there remain large gaps between the vision statement and the support that is provided for adequate state monitoring programs, particularly for bioassessments and biological criteria. Such criticisms are not new (U.S. GAO 1986).

The framework includes six "levels" of indicators: level 1) actions taken by regulatory agencies (*e.g.*, permitting, enforcement, grants); level 2) responses by the regulated community (*e.g.*, construction of treatment works, pollution prevention); level 3) changes in discharged quantities (*e.g.*, pollutant loadings); level 4) changes in ambient conditions (*e.g.*, water quality, habitat); level 5) changes in uptake and/or assimilation (*e.g.*, tissue contamination, biomarkers, wasteload allocation); and, level 6) changes in health, ecology, or other effects (*e.g.*, ecological condition, pathogenicity). In this process the results of administrative activities (levels 1 and 2) can be linked to efforts to improve chemical water quality (levels 3, 4, and 5) which should translate into measurable environmental "results" (level 6). However, the process can be multi-directional with the level 6 indicators providing feedback about the completeness and accuracy of the preceding levels.

The information that is readily available to Ohio EPA is indicated for each level. Some of these are highly developed, some are in various stages of refinement, and others are in the initial stages of development. Further refinements of the supporting data sources to support each indicator level can be incorporated as each is developed through time. Thus the model serves both as an information tracking device and a feedback mechanism. For example we can now ascertain the aggregate effect of the billions of dollars spent on water pollution control since the early 1970s by comparing the implementation of level 1 administrative actions (*e.g.*, funding, permitting) with quantifiable measures of environmental condition (level 6). This hierarchy is also synonymous with the pressure-state-response paradigm frequently cited as part of the sustainable development framework (U.S. EPA 1995b).

The concept of stressor, exposure, and response indicators developed by the U.S. EPA Environmental Monitoring and Assessment Program (EMAP; U.S. EPA 1991a) is also woven into this framework (Figure 3). *Stressor* indicators generally include activities which have the potential to degrade the aquatic environment such as pollutant discharges, land use effects, and habitat modifications. *Exposure* indicators are those which measure the apparent effects of stressors and can include chemical water quality criteria, whole effluent toxicity tests, tissue residues, and biomarkers, each of which provides evidence of biological exposure to a stressor or bioaccumulative agent. *Response* indicators are generally composite measures of the cumulative effects of stress and exposure and include the more direct measures of biological community and population response that are represented here by the biological indices which comprise the Ohio EPA biological criteria. Other response indicators include target assemblages (*e.g.*, rare, threatened, endangered, special status, and declining species) or bacterial levels which serve as surrogates for the recreational uses. All of these indicators represent the essential technical elements for watershed-based management approaches. The key, however, is to use the different indicators *within* the roles which are most appropriate for each.

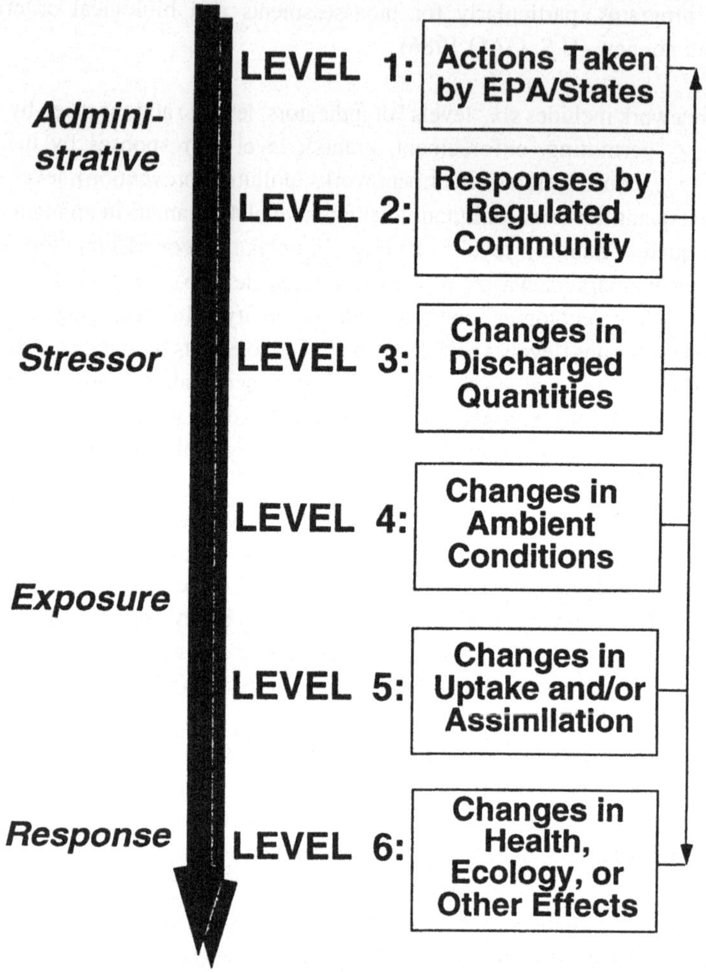

Fig 3. Hierarchy of administrative and environmental indicators used by Ohio EPA for monitoring, assessment, reporting, and evaluating program effectiveness. This is patterned after a model developed by the U.S. EPA, Office of Water.

The previous comparisons of chemical and biological indicator frameworks illustrate a national problem - the inappropriate use of stressor and exposure indicators as substitutes for response indicators. States which do not have well developed biological indicators must report on the status of their waters to U.S. EPA and are compelled to use whatever information is available. Usually the most readily available information is in the form of stressor or exposure indicators. While no single indicator measures everything, biological indicators are inherently better at evaluating biological integrity which is an over-arching goal of State water quality management programs. More accurately portraying the condition of the nation's surface waters depends on the wider use of biological indicators. However suites of chemical, physical, and biological indicators, each in their most appropriate role as stressor, exposure, or response, is needed to more effectively achieve the biological integrity goal of the CWA.

4. Biological Criteria: A Key Indicator End-Point

The Clean Water Act goal of maintaining and restoring the biological integrity of the nation's waters can be synonymized with the health and well-being of aquatic ecosystems which is the combined result of chemical, physical, and biological processes (Karr *et al.* 1986; Figure 1). To be truly successful in meeting these objectives, monitoring and assessment tools are needed that measure both the interacting processes and the integrated results of these processes (Karr 1991). With the recent emphasis on watersheds as the focal point of water quality management, the inherently watershed oriented approach of biological criteria and the attendant bioassessment methods present additional advantages. As such, biological criteria represent a key end-point for determining if water quality management actions have worked and, if not, where these actions might need to be modified.

THE MULTIMETRIC APPROACH

New methods and approaches have been developed over the past 20 years which have overcome many of the prior problems with using biological community data. The relatively recent development of multimetric indices represents a significant advancement in being able to utilize biological community information for aquatic resource characterization and as an arbiter of CWA goal attainment status. These include such innovations as the multimetric Index of Biotic Integrity (IBI), as originally developed by Karr (1981), and as subsequently modified by many others (Leonard and Orth 1986; Ohio EPA 1987; Steedman 1988; Miller *et al.* 1988; Lyons 1992; Oberdorff and Hughes 1992; Lyons and Wang 1996), the Index of Well-Being (Iwb) developed by Gammon (1976) and Gammon *et al.* (1981), the Invertebrate Community Index (ICI; Ohio EPA 1987; DeShon 1995), the U.S. EPA Rapid Bioassessment Protocols for macroinvertebrate assemblages (Plafkin *et al.* 1989), and the Benthic Index of Biotic Integrity (BIBI; Kerans and Karr 1992).

Multimetric indices not only make biological community data more understandable, but have fostered the inclusion of biological information as a more integral part of the State water quality management process. The theoretical underpinnings behind the IBI-type multimetric indices are more robust than previously available biological indices (Karr 1991) when used within a framework where regional faunal considerations, reference conditions, and background variability have been accounted for. Steedman (1988) described the IBI as being based on simple, definable ecological relationships which is quantitative as an ordinal, if not linear, measure and which responds in an intuitively correct manner to known environmental gradients. Further, when incorporated with mapping, monitoring, and modeling information, it has been shown to be valuable in determining management and restoration requirements for warmwater streams (Steedman 1988; Bennet *et al.* 1993).

As an aggregation of community information, the IBI-type indices provide a way to organize complex data and reduce it to a scale which is interpretable against communities of a known condition. Simply stated, multimetric indices can satisfy the demand for a straightforward evaluation that expresses a relative value of aquatic community health and well-being and allows program managers (who are frequently non-scientists) to, in effect, "visualize" relative levels of biological integrity. These measures also provide a means to establish numerical biological criteria. While this process is not without criticism (*e.g.*, Suter 1993), the development and merits of multimetric biological indices have been thoroughly detailed (Miller *et al.* 1988; Simon and Lyons 1995) and multimetric indices are being adopted by a growing number of States and monitoring organizations (U.S. EPA 1996).

BIOLOGICAL CRITERIA IN THE OHIO WATER QUALITY STANDARDS

Ohio EPA adopted numerical biological criteria in the Ohio water quality standards in 1990 following a seven-year developmental process and using a database spanning 10 years. Biological criteria in Ohio are based on measurable characteristics of fish and macroinvertebrate communities such as species richness, key taxonomic groupings, functional guilds, environmental tolerances, and organism condition which comprise the metrics of the IBI-type indices. Numerical biological criteria were derived using a regional reference site approach (Ohio EPA 1987; Yoder 1989; Yoder and Rankin 1995a) and are numerical benchmarks that reflect the health and well-being of aquatic communities (Figure 4). The biological criteria are further structured into the Ohio water quality standards within the existing system of aquatic life use designations (Yoder and Rankin 1995a).

Water quality standards (WQS) consist of designated uses and chemical, physical, and biological criteria designed to represent measurable properties of the environment that are consistent with the narrative goals specified by each use designation. Chemical, physical, and/or biological criteria are assigned to each use designation in accordance with the broad ecological goals defined by each. As such, the system of use designations employed in the Ohio WQS constitutes a stratified approach in that graduated levels of protection are provided by each and is especially apparent for the biological criteria (Figure 5). The narrative ratings of biological community performance (exceptional, good, fair, etc.) not only correspond to the continuum formed by the scale of numerical biological indices, but also to the designated uses (*i.e.*, EWH = exceptional , WWH = good, MWH = fair or poor, etc.). Additional stratification and refinement is provided by ecoregions (Omernik 1987) and stream and river size (Figure 4).

The process of adapting the numerical biological criteria (Figure 4) to the system of designated uses (Figure 5) also demonstrates the realities of reconciling the biological integrity ideal and the lasting effects of two centuries of intensive human use of the land and water resources of Ohio. In Figure 5 the attainment of the biological integrity ideal theoretically occurs only at the very upper end of the numerical index scale and likely includes less than 5% of the best performing sites in Ohio. Clearly what we are now considering to meet the goals of the CWA in terms of designated aquatic life uses falls well short of the biological integrity ideal. However, the system of numerical biological indices is able to track progress towards meeting these two goals simultaneously. Thus both goals are kept in perspective relative to the expectations over both the near and long term. U.S. EPA recognized this duality in adopting separate aquatic life use and biological integrity indicators as part of the national water indicators (U.S. EPA 1995a).

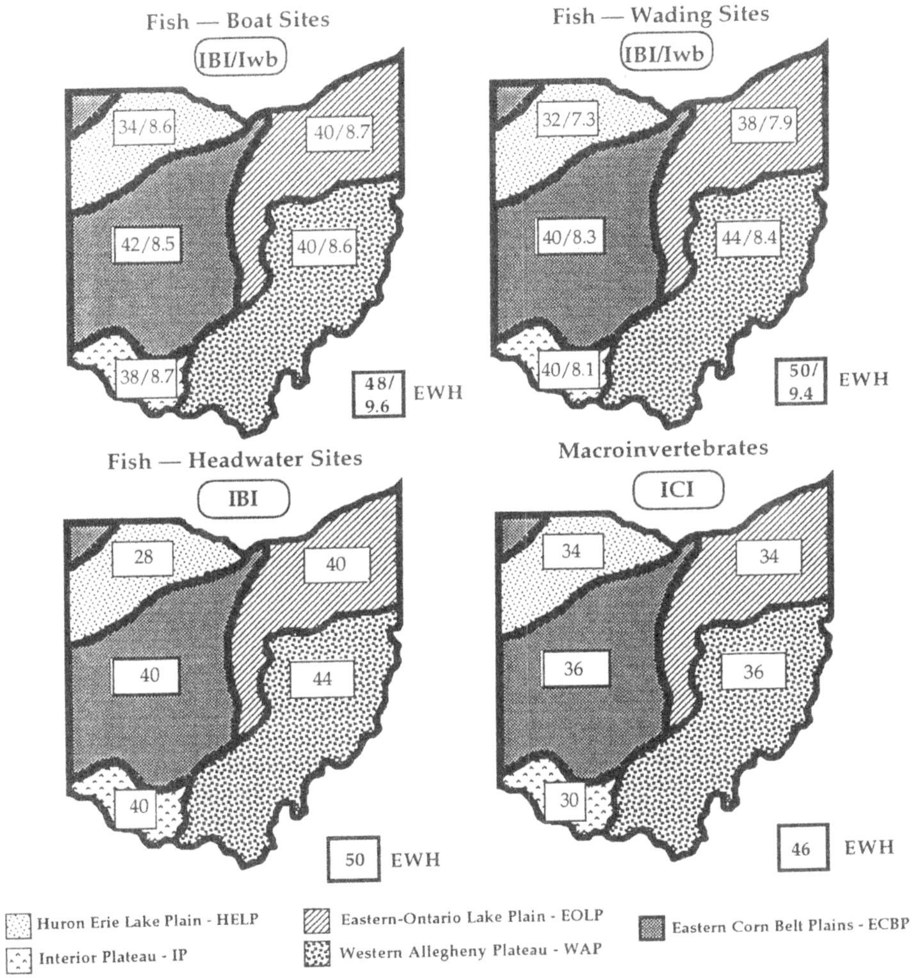

Fig 4. Biological criteria in the Ohio WQS for the Warmwater Habitat Use (WWH) and Exceptional Warmwater Habitat (EWH) use designations arranged by biological index, site type for fish, and ecoregion. The EWH criteria for each index and site type is located in the boxes located outside of each map.

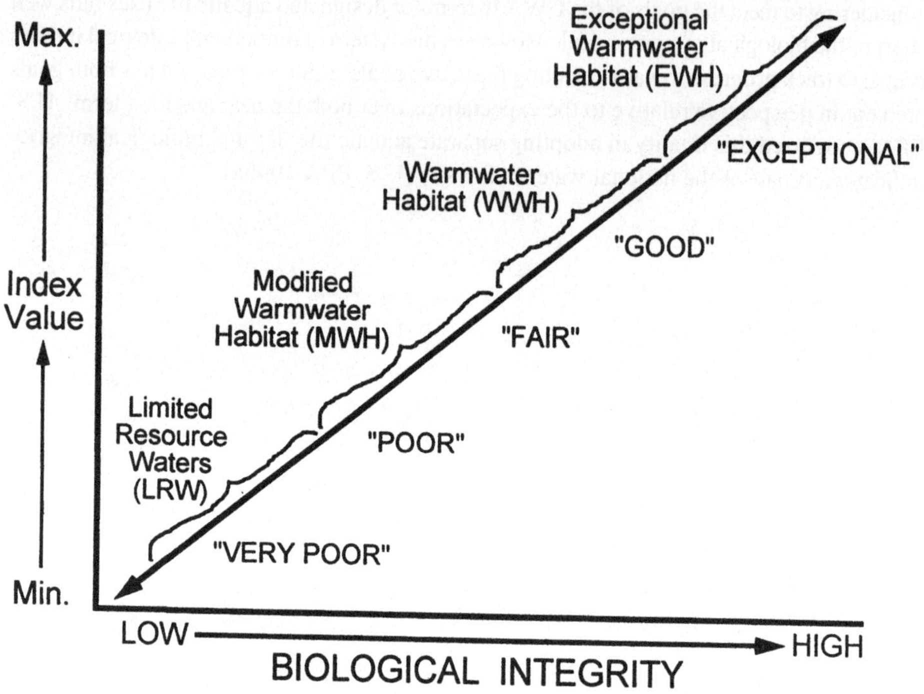

Fig 5. Relationship between the tiered aquatic life uses in the Ohio WQS and narrative evaluations of biological community performance and how this corresponds to a theoretical scale of biological integrity and measured biological index values (HELP = Huron/Erie Lake Plain ecoregion).

5. Applications of Biological Indicators

Biological criteria are used within the water quality management process as a level 6 indicator (Figure 3). They are employed principally as an ambient monitoring and assessment tool through biological surveys.

Ambient Monitoring: A Prerequisite for Using Biological Criteria

Monitoring is a tool, not a goal in itself. It provides information that allows managers to take accurate and justifiable actions and to evaluate the effectiveness of the overall water quality management program. In this process the collection of ambient data is linked to specific purposes relevant to decisions that are or will be made about the environment. A principal role of monitoring information is to drive the water quality management process. This includes the identification and characterization of problems, development of policies, regulation, and legislation, establishing strategies and requirements for pollution control, and reporting on the overall results. Following this approach ensures that: 1) limited State resources are devoted to resolving real problems; 2) criteria and regulations are founded on good science; and, 3) the results of water quality management are portrayed as environmental, not just administrative accomplishments (Yoder 1994). Despite the logic of such an approach, many States regard ambient monitoring as being less important than administrative actions such as issuing NPDES permits. A major impediment to the greater use of ambient monitoring in general, and biological assessment in particular, is the widely held notion that obtaining data is costly and time consuming, neither of which satisfies the regulatory urge for immediate results. However, our experience has shown biological data to be less costly than chemical/physical data and other allied information (Yoder and Rankin 1995a).

Biological surveys in Ohio are an interdisciplinary monitoring effort coordinated on a waterbody specific or watershed scale. These may involve a relatively simple setting focusing on one or two small streams, one or two principal stressors, and as few as 5-6 sampling sites or a much more complex effort including entire drainage basins, multiple stressors, and tens of sites. Each year Ohio EPA conducts biosurveys in 10-15 watershed areas with an aggregate total of 300-350 sampling sites. Biological, chemical, and physical monitoring and assessment techniques are employed in order to meet three major objectives:

1) Determine the extent to which use designations assigned in the Ohio WQS are either attained or not attained.
2) Determine if use designations assigned to a given water body are appropriate and attainable.
3) Determine if any changes in biological, chemical, or physical indicators have taken place over time, in response to point source pollution controls or best management practices for nonpoint sources.

The data gathered in a biosurvey is processed, evaluated, and synthesized in a biological and water quality report following the progression and hierarchy of indicators portrayed by Figure 3. Each biological and water quality report contains a summary of major findings and recommendations for revisions to WQS, future monitoring needs, or other actions which may be needed to resolve impairment of designated uses or facilitate the maintenance of high quality waters. While the principal focus of a biosurvey is on the status of aquatic life uses, the status of other uses such as recreation and water supply, as well as human health concerns, are also addressed. These reports and the underlying database serve to support all other water quality management activities including permitting, enforcement, planning, and reporting.

Describing the causes and sources associated with observed impairments revealed by the biological criteria and linking this with specific sources involves an interpretation of multiple lines of evidence including water and sediment chemistry data, habitat data, effluent data, whole effluent toxicity test results, land use data, and biological response signatures (Yoder and Rankin 1995b). The assignment of principal causes and sources of impairment represents the association of an impairment (as indicated by a failure to meet the biological criteria) with stressor and exposure indicators. While the principal reporting venue for this process is a biological and water quality report, the results provide the foundation for aggregated analyses such as the Ohio Water Resource Inventory (305[b] report), the Ohio Nonpoint Source Assessment, the many listing processes required by the CWA (*e.g.*, 303[d], 304[l] lists), and other technical bulletins. Such information later becomes important in formulating new policies and evaluating changes to existing ones.

LINKING INDICATORS ON A LONGITUDINAL REACH OR SUBBASIN BASIS

A longitudinal presentation of the biological sampling results is a commonly used method which provides an opportunity to visually describe and interpret the magnitude and severity of departures from the numerical biological criteria. This is accomplished by plotting the biological index results (Index of Biotic Integrity, Modified Index of Well-Being, or Invertebrate Community Index) by river mile for a study area. Major sources of potential impact and the applicable numerical biological criteria are indicated on each graph. The results of the fish community assessment in the upper Great Miami River during 1982 and 1994 is used here as an example (Figure 6). This type of analysis is used to demonstrate changes through both space (upstream/downstream trend) and time (differences between years). Unlike chemical parameters, biological indices integrate chemical, biological, and physical impacts to aquatic ecosystems and portray aquatic life use attainment/non-attainment in aggregate and direct terms. Frequency and duration considerations, which confound chemical/physical assessments, are integrated by the resident aquatic life in the receiving water body. Figure 6 represents the results of two bioassessments in a 75 mile segment of the Great Miami River, located in western Ohio, which was sampled before and after the imposition of water quality-based limitations for five major municipal wastewater treatment plants (WWTPs; Ohio EPA 1996a). The obvious improvements exhibited by the IBI in conjunction with the extensive reductions in the mass loadings of ammonia (a level 3 indicator) at the major WWTPs (represented in Figure 7 by the Piqua WWTP) illustrate the direct benefits of improved municipal wastewater treatment and instream water quality between 1982 and 1994. In this case example, 1982 precedes the improvements made to reduce water pollution at the major WWTPs and 1994 post-dates by 4-6 years the installation of treatment technology designed to comply with water quality based permit limits (a level 2 indicator). While the biological improvement that occurred between 1982 and 1994 is correlated with the overall reduction of pollutant loadings at the major WWTPs, the biological improvements were less dramatic in those areas in close proximity to past habitat modifications (a level 4 indicator) in the upper 20 miles of the study area. This example illustrates the results of Ohio EPA issuing NPDES permit limitations (a level 1 indicator) which were based on meeting instream chemical water quality criteria, awarding grants for the construction of new and upgraded treatment systems (a level 1 indicator), the response of the municipalities in constructing the new and upgraded systems (a level 2 indicator), a corresponding decrease in mass loadings of pollutants (a level 3 indicator), an improvement in ambient water quality (a level 4 indicator), and the corresponding response by the biological resources (a level 6 indicator). As such this

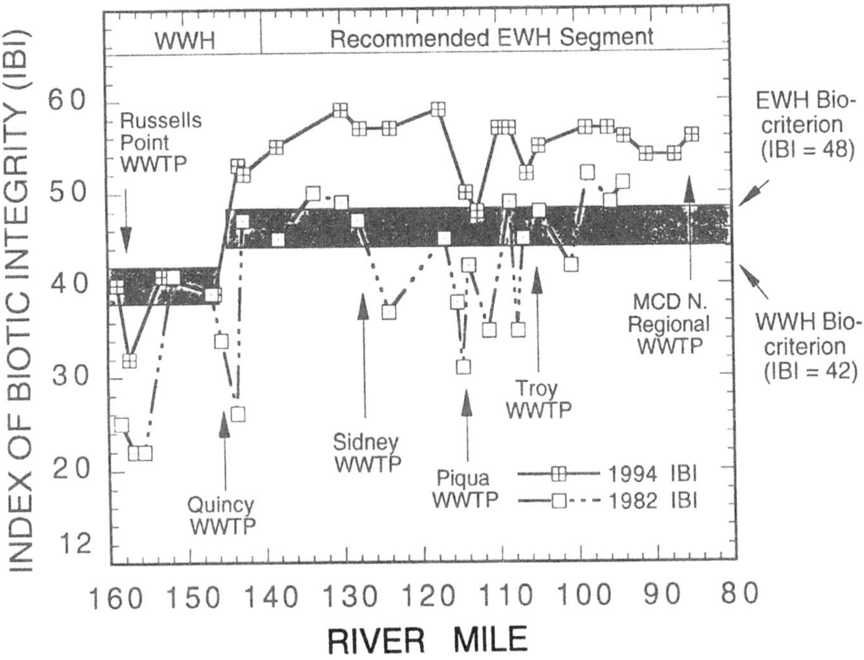

Fig 6. Index of Biotic Integrity results in a 75 mile segment of the upper Great Miami River during 1982 and 1994. Major point sources (vertical arrows) are indicated along with the applicable biological criteria (shaded bars).

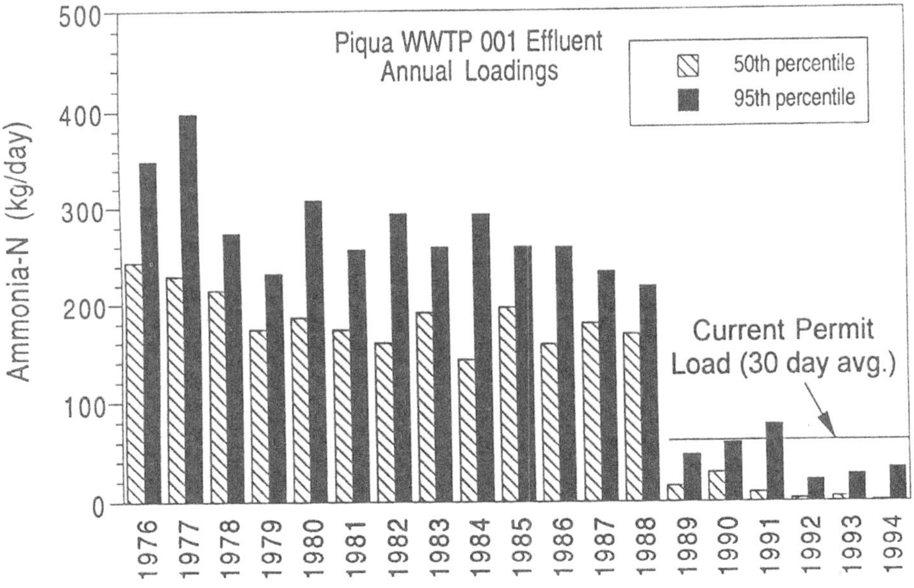

Fig 7. Mass loadings of ammonia-nitrogen discharged by the Piqua municipal wastewater treatment plant between 1976 and 1994. Solid bars represent 95th percentile values and shaded bars median values for each year based on daily measurements of the treated effluent.

represents a successful report on the water quality management efforts at the local, State, and federal levels. Approximately two dozen such case examples exist throughout Ohio. There are other areas where similar efforts to permit, fund, and construct new and improved municipal treatment systems have not been matched by similar improvements in biological quality. One such area is the lower 40 miles of the Mahoning River in northeastern Ohio. The same basic approach to water quality management and pollution control taken in the preceding example resulted in the construction of new and upgraded treatment systems and reduced loadings of pollutants like ammonia. However, the before and after biological results indicated only incremental changes in biological condition (Figure 8) which was rated as poor and very poor during both of the sampling years, 1980 and 1994 (Ohio EPA 1996b). The explanation lies partially in a very different setting with the Mahoning River having been the site of decades of intensive industrial uses, most of which was comprised of integrated steel production facilities. The residual effects of this past use are evident in highly contaminated bottom sediments (level 4 indicator) the effects of which have only just begun to lessen. In addition, some of the municipalities have serious problems with combined sewer overflows[4] which were largely absent in the preceding example from the Great Miami River. The pervasive influence of toxic effects were also apparent in the very high percentage of fish with external anomalies such as gross deformities, eroded fins, lesions, and tumors (DELT anomalies; Figure 8). Combined with the very poor IBI and Miwb scores this indicates a response to acutely toxic conditions (Yoder and Rankin 1995b). In this example the environmental exposure and response indicators suggest that water quality management strategies in addition to typical point source controls will be needed to successfully restore the designated aquatic life uses in the Mahoning River. If our water quality management program lacked the level 6 indicator information the seriousness of this shortfall would likely have been greatly underestimated.

STATEWIDE REPORTING AND ASSESSMENT

The Ohio Water Resource Inventory (CWA Section 305[b] report) represents the statewide aggregation of the data and information developed from the biological surveys and the rotating basin process. The previously mentioned problems with data and assessment inconsistencies are minimized by the standardized approach to ambient monitoring and long-term database development taken by Ohio EPA during the past 18 years. 305[b] report statistics are derived by the upwards aggregation of reach level information generated by the biosurveys. The importance of this information was further highlighted in 1994 by adoption of the Ohio 2000 goals for Ohio surface waters. This established an overall goal of reaching 75% full attainment of aquatic life uses in Ohio's surface waters by the year 2000. Thus the overall water quality management and strategic planning

---

[4]Combined sewer overflows (CSOs) are hydraulic relief points in a sanitary sewer system where a mixture of raw sewage and stormwater can discharge to rivers and streams during periods of rainfall and runoff. The purpose of CSOs is to prevent the hydraulic overloading of a sanitary sewer system.

processes within Ohio EPA are now driven by a goal which can only be evaluated and tracked with ambient monitoring results.

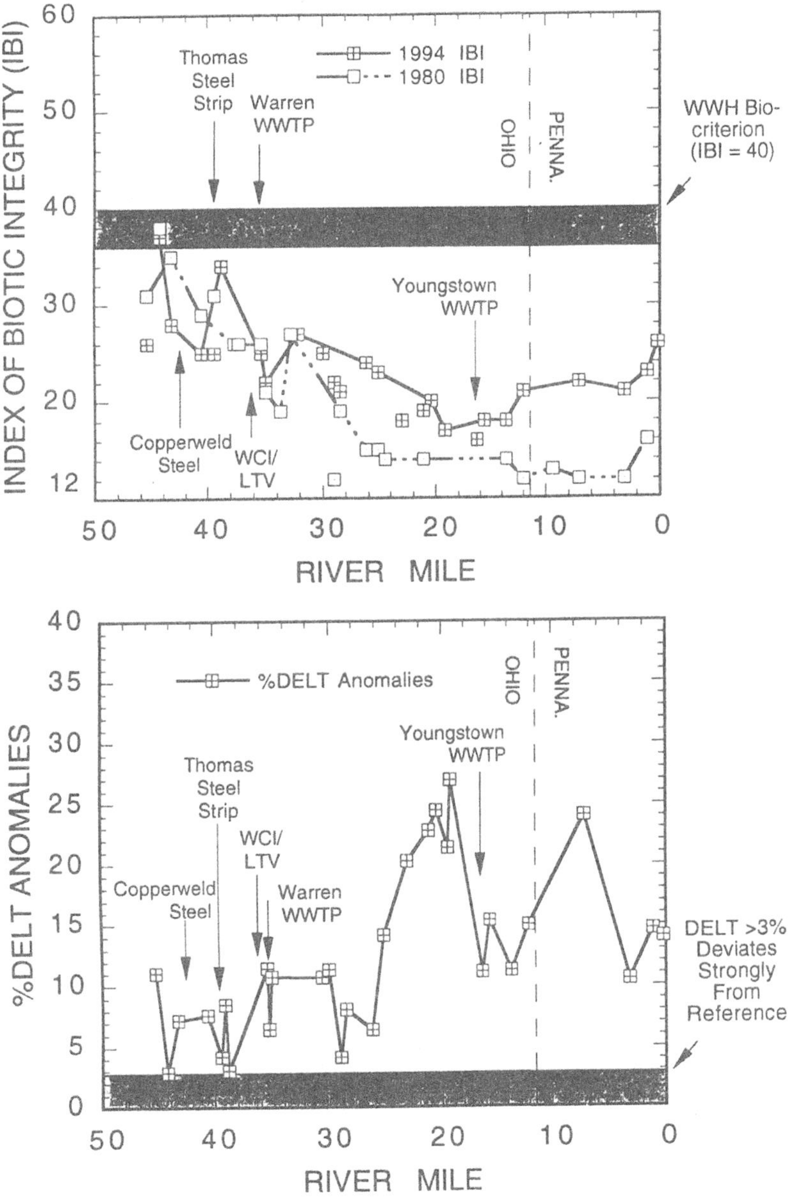

Fig 8. Index of Biotic Integrity results in a 50 mile segment of the lower Mahoning River during 1980 and 1994 (upper). Major point sources (vertical arrows) are indicated along with the applicable biological criteria (shaded bars). The percentage of fish with deformities, eroded fins, lesions, and tumors (%DELT) during 1994 appear in the lower graph.

In an attempt to answer the basic questions about the Ohio 2000 goals we performed different analyses of the statewide biological database as part of the 1996 305[b] reporting process (Ohio EPA 1997). One analysis consisted of comparing the direction and magnitude of change in biological index scores at sites with multiple years of data on a statewide basis. In this analysis trends represent the difference between the earliest and latest year results (most of which are approximately 10 years apart) before and after 1988. Sample sizes were 1160 sites, 845 sites, and 528 sites for the Index of Biotic Integrity, Modified Index of Well-Being, and Invertebrate Community Index, respectively. Table 1 summarizes pertinent percentile shifts in the biological indices between the earliest and latest time periods and the results of a paired t test (using a t statistic and Wilcoxon's Z test). The comparison of the results from the two time periods showed that the changes (increases) in biological index scores were highly significant ($p < 0.0001$). While the Ohio EPA database was not collected under a statistically random design and is spatially biased, the large number of sites sampled (>5000 locations[5]) and thorough coverage of the streams and rivers with drainage areas greater than 100 mi.^{2} (75% coverage statewide) makes valid statewide comparisons possible. For each index there has been significant improvement over time at most sites as demonstrated by a composite summary of changes between the earliest and latest sampling years before and after 1988 (Table 1) and as mapped information showing significant increases and declines in biological index scores by sampling location for the same periods of time (Ohio EPA 1997). The Invertebrate Community Index showed both the largest magnitude of increase and shift in the frequency of sites entering the good and exceptional performance ranges (*i.e.*, scores meeting the Warmwater Habitat and Exceptional Warmwater Habitat biological criteria), but the fewest sites exiting the poor and very poor ranges (scores <14). In contrast, the fish community IBI had the greatest number of sites exiting the poor and very poor ranges (scores >26-28), but the fewest sites entering the good and exceptional ranges (*i.e.*, scores meeting the Warmwater Habitat and Exceptional Warmwater Habitat biological criteria). The improvements which have been documented in several Ohio rivers and streams using biological criteria as the principal arbiter of CWA goal attainment are considerable and have recently included some examples of 100% of previously impaired rivers now meeting the biological criteria.

---

[5]The Ohio EPA database represents a sampling site density of more than 10 times that of the U.S. EPA EMAP and U.S. Geological Survey NAWQA national monitoring designs.

Table 1.

Summary of paired sites with at least two years of biological data from Ohio streams and rivers sampled between 1979 and 1994, at least once before and once after 1988. Data pairs represent earliest and latest index values for the ICI, MIwb, or IBI at each paired site (after Ohio EPA 1997).

	Index of Biotic Integrity		**Modified Index of Well-Being**		**Invertebrate Community Index**	
Category	Earliest	Latest	Earliest	Latest	Earliest	Latest
10th %ile	16	20	3.8	4.8	6	14
25th %ile	24	28	5.7	6.6	16	26
Median	32	36.5	7.4	8.1	32	38
75th %ile	41.5	45	8.6	9.2	42	46
90th %ile	48	50	9.3	9.9	46	52
Mean	32.2	36.2	6.91	7.72	28.9	35.5
Paired t-test	1159 *df*		844 *df*		527 *df*	
t value	-16.89		-14.34		-12.53	
Mean difference	3.95		0.80		6.65	
(L minus E)	p <0.0001		p <0.0001		p <0.0001	
Wilcoxon (Z)	-24.40		-13.50		-11.55	
	p <0.0001		p <0.0001		p <0.0001	

While substantial improvements have been observed statewide, it is clear that a significant proportion of Ohio's rivers and streams remain chemically too polluted and/or physically degraded to meet the applicable goals. Figure 9 illustrates the aggregate changes in aquatic life use non-attainment that took place between 1988 and 1996 and then projecting the trend through the year 2002. This, too, demonstrates the overall reduction in impairment noted statewide in Ohio since 1988 and tracks progress towards meeting the Ohio 2000 goals. However, the proportion of impairment that is associated primarily with point sources is declining at a much more rapid rate than the impairment associated with nonpoint sources. The latter includes the integrated effects of habitat modifications, nutrient enrichment, and sedimentation of lotic substrates via nonpoint source runoff and habitat alterations. The implications of this forecast analysis to the Ohio EPA surface water programs is significant in that more attention will need to be paid to nonpoint sources and watershed level effects (as opposed to a historical emphasis on site-specific, point source problems) if milestones such as the Ohio 2000 goal of 75% full attainment are to be realized. Accomplishing this will require a significant restructuring of Ohio's water quality management programs which remain heavily skewed towards point sources.

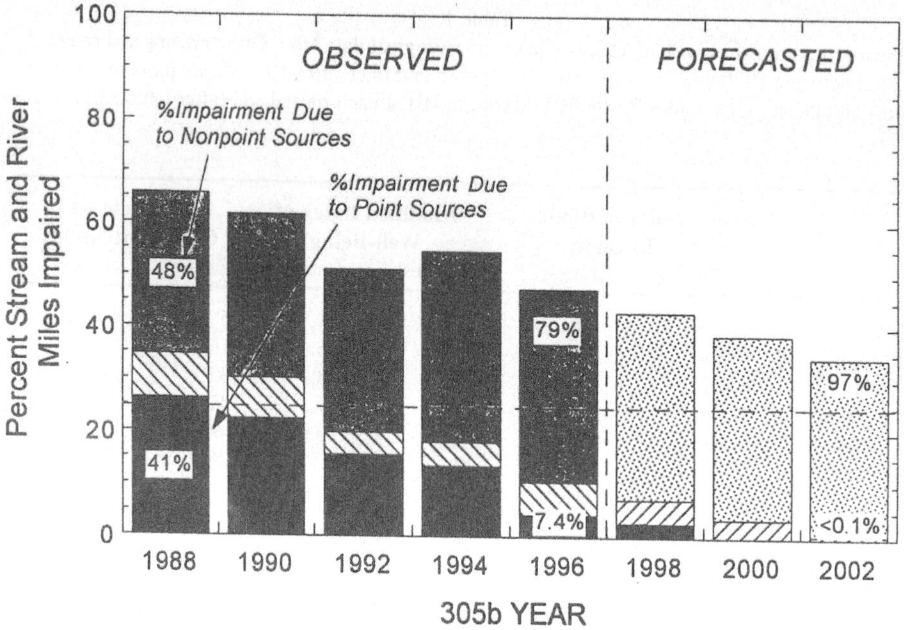

Fig 9. Observed reduction in the proportion of river and stream miles failing to attain criteria for designated aquatic life uses between 1988 and 1996 (left of dashed vertical line) and that forecasted through the year 2002 (right of dashed vertical line) based on the observed rate of restoration. The dashed horizontal line represents the Ohio 2000 goal (after Ohio EPA 1996a).

6. Overall Impact of Biological Indicators and Criteria

Having biological criteria in the Ohio WQS and as an ambient assessment tool has provided the Ohio EPA with some substantial advantages in surface water quality management. Some of these include being able to express the results of water quality management in terms of environmental results, rather than administrative accomplishments alone. Knowledge of the condition of the State's water resources coupled with an understanding of associated causes and sources of impairment can more effectively guide water quality management activities that might otherwise have relied on prescriptive approaches. Biological data and definition of regional reference conditions has led to a stratification of expectations for streams and rivers and a more refined classification framework. This in turn has lead to a more refined application of chemical water quality criteria and antidegradation procedures. Regulatory efforts which deal with habitat manipulation and degradation and nonpoint sources now have a firmer basis for decision making. Biological criteria offer States a vastly broader capability to define and discriminate impairment and attainment and the degree to which either exists. Implementation of biological indicators and criteria also fosters the inclusion of physical habitat and nontoxic chemical effects since each play important and interacting roles in determining the biological status and potential of surface water resources (Figure 1).

7. Remaining Challenges

While we have demonstrated some of the ways that biological indicators and criteria can be developed and used within a State water quality management framework, some important challenges remain. The cumulative costs associated with environmental mandates, many of which emanate from prescriptive regulations, have recently come into question. Both the regulated community and the public desire evidence of "real world" results in return for the expenditures made necessary by Federal and State mandated requirements. Biological criteria seem particularly well suited to meet some of these needs in that the underlying science and theory is robust (Karr 1991) and biocriteria certainly qualify as "real world."

There is an inherent resistance to the idea of incorporating additional tools in already burdened State programs. Even though biological criteria offer some substantial advantages to the States there are some important policy issues which remain as yet unresolved. Perhaps the most important of these is the policy of independent applicability developed by U.S. EPA (1990b, 1991b) to deal with the potential conflicts that might arise with the use of chemical water quality criteria, whole effluent toxicity test results, and biological criteria. The policy essentially states that the most stringent application of the three different tools will drive regulatory decisions and ambient assessments *independently*. For example if full attainment of WQS is suggested by the biological criteria and whole effluent toxicity test results, but an exceedence of a chemical criterion is either observed or predicted, the latter would automatically drive the management process. Most States and the regulated community advocate a different approach, termed weight of evidence, in which the circumstances of each tool are considered on a case by case basis with no one tool assumed to be either equal or superior *a priori*. Instead, an informed examination of the results and power of each tool are considered first. States and others view a strict adherence to the policy of independent application as a major disincentive to adopting biological criteria. Both U.S. EPA and the States would gain much by allowing prudent flexibility in this area. A more detailed treatment of this issue is available in U.S. EPA (1992) and Yoder (1995).

While no single environmental indicator can "do it all," particularly in the more complex situations (*i.e.*, multiple discharges, habitat alterations, presence of toxic compounds, etc.), it is obvious that biological criteria have much to contribute. A lack of information for or an over-reliance on any one indicator can result in environmental regulation that is less accurate and either under or overprotective of the water resource. Accounting for cost is not only a matter of dollars spent, but is also a question of environmental accuracy and technical validity. In short, a credible and genuinely cost-effective approach to water quality management should include an appropriate mix of chemical, physical, and biological indicators, each in their respective roles as stressor, exposure, and response indicators. Comprehensive monitoring designs that employ such cost-effective indicators must become a part of the "cost of doing business." This may need to happen at the expense of existing programs where environmental evidence suggests that the resources expended are disproportionate to the magnitude of the problem (*e.g.*, point sources vs. nonpoint sources).

Based on our experience over the past 20 years it is evident that including a biological indicators and criteria approach in the State water quality management process can foster a more complete integration of important ecological concepts, better focus water resource policy and management, maximize the use of limited resources, and enhance strategic planning. Some specific examples include:

1) *Watershed Approaches to Monitoring, Assessment, and Management*: The monitoring and assessment design inherent to biological criteria is fundamentally watershed oriented and will yield information on a watershed basis. The analysis and synthesis of both ambient and source data within the concept of the hierarchy of indicators (Figure 2) will lead to a more effective identification and prioritization of issues within specific watersheds.

2) *Integrated Point, Nonpoint, and Habitat Assessment and Management:* Biological criteria integrate the effects of all stressors over time and space, and the attendant use of chemical, toxicological, and physical tools enables the association of probable causes of observed impairments. This should provide a strong foundation for the collaborative use of the same information for the management and regulation of both point and nonpoint sources (including habitat), two disciplines which have thus far been operated as virtually independent programs.

3) *Cumulative Effects:* Biological communities inhabit the receiving waters all of the time and integrate the cumulative effects of multiple stressors. Such information provides a basis for management programs to prioritize problems and more effectively allocate scarce resources.

4) *Biodiversity Issues:* The biological survey data provides information about species, populations, and communities of concern and also affords the opportunity to incorporate this into water quality management issues.

5) *Interdisciplinary Focus:* Because of the inherently integrative character of the biological survey monitoring and assessment design, a biological criteria approach provides the opportunity to bring ecological, toxicological, engineering, and other sciences together in planning and conducting assessments, interpreting the results, and using the information in strategic planning and management actions.

Biological indicators and criteria are an emerging and increasingly important issue for U.S. EPA, the States, the regulated community, advocacy groups, and the public. The use of biocriteria and bioassessments is growing nationwide as State and local organizations shift their monitoring and assessment efforts in this direction. However, much remains to be done, particularly in the area of national and regional leadership in both policy and technical areas. Technical guidance and expertise is needed to ensure a nationally consistent and credible approach and to resolve outstanding technical concerns such as those listed by Yoder and Rankin (1995a). Resolving outstanding policy issues such as EPA's policy of independent applicability needs to be accomplished in such a manner as to encourage, not discourage, State participation. In an era of declining government resources ways to accomplish the "increases" needed in biological monitoring to support a biocriteria approach must be developed. Based on our experience in Ohio the staffing of State programs should include a minimum of one work year equivalent for every 1200 miles of perennial streams and rivers. This estimate may vary by region and should additionally incorporate lake acres and shoreline miles in States with a predominance of this water body type (Yoder and Rankin 1995a). The potential for biological indicators and criteria to modify the present capital and resource intensive system of tracking environmental compliance for point sources on a pollutant specific basis needs to be considered by U.S. EPA. This should prove to be a more cost and information effective approach to managing the nation's water quality programs.

There is an urgent need to shift the emphasis of water quality management from program activities to include a more resource focused approach. Program activities presently include the issuance of permits, taking of enforcement actions, awarding of grants, and similar types of activities where the overall goal is to improve the performance of the program. Indicators of program performance are referred to as "bean counts" and include items such as the proportion of backlogged permits, successful enforcement actions, grant dollars awarded, management plans developed and implemented, and pollution control facilities constructed, all of which are level 1 and 2 indicators in Figure 3. In this paradigm improving the performance of the programs is the overall goal. Shifting to a resource focused approach would also include the environmental measures that we have demonstrated here (levels 4-6) as indicators of overall program performance. In this paradigm the programs are tools to improve the environment. Accomplishing this shift does not mean abandoning program activity measures, but it does require that ambient monitoring and assessment become a more integral part of the overall water quality management process.

Acknowledgements

This paper would not have been possible without the many years of field work, laboratory analysis, and data assessment and interpretation performed by members (past and present) of the Ohio EPA, Ecological Assessment Unit. Several staff members contributed extensively to the development of the Ohio EPA biological assessment program and include the following: Dave Altfater, Randy Sanders, Marc Smith, and Roger Thoma (fish methods, MIwb, and IBI metrics development) and Mike Bolton, Jeff DeShon, Jack Freda, Marty Knapp, and Chuck McKnight (macroinvertebrate methods and ICI metrics development). None of this effort would have been possible without the excellent data management and processing skills contributed by Dennis Mishne. Other staff (past and present) who also made important contributions to the process include Paul Albeit, Ray Beaumier, Chuck Boucher, Kelly Capuzzi, Bernie Counts, Duane Davis, Beth Lenoble, and Paul Vandermeer. Dan Dudley and Jim Luey contributed extensively to the early development and review of the then emerging concepts of biological integrity, ecoregions, reference sites, and biological assessment in general. Charlie Staudt provided many hours of support in the development of the computer programs used for data analysis. Finally, Gary Martin and the late Pat Abrams are credited for their solid management support for the concept of biological criteria and biological assessment at the Ohio EPA.

References

Benke, A.C. 1990. "A perspective on America's vanishing streams." *J. N. Am. Benth. Soc.*, **9** (1): 77-88.

Bennet, M.R., J.W. Kleene, and V.O. Shanholtz. 1993. "Total maximum daily load nonpoint source allocation pilot project." File Report, Dept. of Agricultural Engineering, Blacksburg, VA. 49 pp.

Berquist, G.T., J. R. Bernard, and A.M. Paeble. 1995. "Prospective indicators for state use in performance partnership agreements." The State Environmental Goals and Indicators Project, Florida Center for Public Management, Florida State University, Tallahassee, FL. 46 pp. + appendices.

DeShon, J.D. 1995. "Development and application of the invertebrate community index (ICI)", pp. 217-243. in W.S. Davis and T. Simon (eds.). Biological Assessment and Criteria: Tools for Risk-based Planning and Decision Making. Lewis Publishers, Boca Raton, FL.

Gammon, J.R., Spacie, A., Hamelink, J.L., and R.L. Kaesler. 1981. "Role of electrofishing in assessing environmental quality of the Wabash River, in Ecological assessments of effluent impacts on communities of indigenous aquatic organisms," *in* Bates, J. M. and Weber, C. I., Eds., ASTM STP 730, 307 pp.

Gammon, J.R. 1976. "The fish populations of the middle 340 km of the Wabash River," *Purdue University, Water Resources Res. Cen. Tech. Rep.* **86**. 73 p.

Hughes, R. M., D. P. Larsen, and J.M. Omernik. 1986. "Regional reference sites: a method for assessing stream pollution," *Environmental Management*, **10**: 629.

ITFM (Intergovernmental Task Force on Monitoring Water Quality). 1995. "The strategy for improving water-quality monitoring in the United States." Final report of the Intergovernmental Task Force on Monitoring Water Quality. Interagency Advisory Committee on Water Data, Washington, D.C. + Appendices.

ITFM (Intergovernmental Task Force on Monitoring Water Quality). 1993. "Ambient water quality monitoring in the United States: second year review, evaluation, and recommendations." Interagency Advisory Committee on Water Data, Washington, D.C. + Appendices.

ITFM (Intergovernmental Task Force on Monitoring Water Quality). 1992. "Ambient water quality monitoring in the United States: first year review, evaluation, and recommendations." Interagency Advisory Committee on Water Data, Washington, D.C.

Judy, R. D., Jr., P. N. Seely, T. M. Murray, S.C. Svirsky, M. R. Whitworth, and L.S. Ischinger. 1984. 1982 national fisheries survey, Vol. 1. Technical Report Initial Findings. U.S. Fish and Wildlife Service, FWS/OBS-84/06.

Karr, J. R. 1991. "Biological integrity: A long-neglected aspect of water resource management." *Ecological Applications* **1**(1): 66-84.

Karr, J. R. 1981. "Assessment of biotic integrity using fish communities." *Fisheries* **6**(6): 21-27.

Karr, J. R., K. D. Fausch, P. L. Angermier, P. R. Yant, and I. J. Schlosser. 1986. "Assessing biological integrity in running waters: a method and its rationale." *Illinois Natural History Survey Special Publication* **5**: 28 pp.

Karr, J. R. and D. R. Dudley. 1981. "Ecological perspective on water quality goals." *Environmental Management*, **5**: 55.

Kerans, B. L., and Karr, J. R. 1992. "An evaluation of invertebrate attributes and a benthic index of biotic integrity for Tennessee Valley rivers," Proc. 1991 *Midwest Poll. Biol. Conf.*, EPA 905/R-92/003.

Leonard, P. M., and D.J. Orth 1986. "Application and testing of an index of biotic integrity in small, coolwater streams." *Trans. Am. Fish. Soc.* **115**: pp. 401.

Lyons, J. and L. Wang. 1996. "Development and validation of an index of biotic integrity for coldwater streams in Wisconsin." *N. Am. J. Fish. Mgmt.* **16**: 241-256.

Lyons, J. 1992. " Using the index of biotic integrity (IBI) to measure environmental quality in warmwater streams of Wisconsin." *Gen. Tech. Rep.* NC-149. St. Paul, MN: USDA, Forest Serv., N. Central Forest Exp. Sta. 51 pp.

Miller, D. L. and others. 1988. "Regional applications of an index of biotic integrity for use in water resource management." *Fisheries*: **13**: pp. 12.

Nichols, A.B. 1992. "It's clear, U.S. waters have improved." *Water Env. Technology.* Oct. 1992: 44-50.

Oberdorff, and R.M. Hughes. 1992. "Modification of an index of biotic integrity based on fish assemblages to characterize rivers of the Seine-Normandie Basin, France." *Hydrobiologia*, **228**: 116 - 132.

Ohio Environmental Protection Agency. 1997. "Ohio water resource inventory, volume I, summary, status, and trends." Rankin, E.T., Yoder, C.O., and Mishne, D.A. (eds.), *Ohio EPA Tech. Bull.* MAS/1996-10-3-I, Division of Surface Water, Columbus, Ohio. 190 pp.

Ohio Environmental Protection Agency. 1996a. "Biological and water quality study of the upper Great Miami River." *OEPA Tech. Rept.* MAS/1995-12-13.

Ohio Environmental Protection Agency. 1996b. "Biological and water quality study of the Mahoning River basin". *OEPA Tech. Rept.* MAS/1995-12-14.

Ohio Environmental Protection Agency. 1992. "Biological and habitat investigation of greater Cincinnati area streams: the impacts of interceptor sewer line construction and maintenance, Hamilton and Clermont Counties, Ohio." *OEPA Tech. Rept.* EAS/1992-5-1.

Ohio Environmental Protection Agency. 1990. "Ohio water resource inventory, volume I, summary, status, and trends." Rankin, E T., Yoder, C.O., and Mishne, D.A. (eds.), Division of Water Quality Planning and Assessment, Columbus, Ohio.

Ohio Environmental Protection Agency. 1988. "Ohio water quality inventory - 1988 305(b) report, volume I and executive summary." Rankin, E. T., Yoder, C. O., and Mishne, D. A. (eds.), Division of Water Quality Monitoring and Assessment, Columbus, Ohio.

Ohio Environmental Protection Agency. 1987. "Biological criteria for the protection of aquatic life: volume II. users manual for biological field assessment of Ohio surface waters," Division of Water Quality Monitoring and Assessment, Surface Water Section, Columbus, Ohio.

Omernik, J. M. 1987. "Ecoregions of the conterminous United States." *Ann. Assoc. Amer. Geogr.* **77**(1):118-125.

Plafkin, J. L. and others. 1989. "Rapid Bioassessment Protocols for use in rivers and streams: benthic macroinvertebrates and fish." EPA/444/4-89-001. U.S. EPA. Washington, D.C.

Rankin, E. T. 1995. "The use of habitat assessments in water resource management programs," pp. 181-208. in W. Davis and T. Simon (eds.). *Biological Assessment and Criteria: Tools for Water Resource Planning and Decision Making.* Lewis Publishers, Boca Raton, FL.

Rankin, E. T. 1989. "The qualitative habitat evaluation index (QHEI), rationale, methods, and application." Ohio EPA, Division of Water Quality Planning and Assessment, Ecological Assessment Section, Columbus, Ohio.

Rankin, E. T. and C. O. Yoder. 1990. "A comparison of aquatic life use impairment detection and its causes between an integrated, biosurvey-based environmental assessment and its water column chemistry subcomponent." Appendix I, Ohio Water Resource Inventory (Volume 1), Ohio EPA, Div. Water Qual. Plan. Assess., Columbus, Ohio. 29 pp.

Simon, T.P. And J. Lyons. 1995. "Application of the index of biotic integrity to evaluate water resource integrity in freshwater ecosystems," pp. 249-262. *in* W.S. Davis and T.P. Simon (eds.), *Biological assessment and criteria: tools for water resource decision-making.* Lewis Publishers, Boca Raton, FL.

Steedman, R.J. 1988. "Modification and assessment of an index of biotic integrity to quantify stream quality in southern Ontario." **Can. J. Fish. Aquatic Sci. 45**: 492-501.

Suter, G.W., II. 1993. "A critique of ecosystem health concepts and indexes." *Environmental Toxicology and Chemistry*, 12: 1533-1539.

U.S. Environmental Protection Agency. 1996. " Summary of state biological assessment programs for streams and rivers." EPA 230-R-96-007. U. S. EPA, Office of Policy, Planning, & Evaluation, Washington, DC 20460.

U.S. Environmental Protection Agency. 1995a. "Environmental indicators of water quality in the United States." EPA 841-R-96-002. Office of Water, Washington, DC 20460. 25 pp.

U.S. Environmental Protection Agency. 1995b. "A conceptual framework to support development and use of environmental information in decision-making." EPA 239-R-95-012. Office of Policy, Planning, and Evaluation, Washington, DC 20460. 43 pp.

U.S. Environmental Protection Agency. 1994. "National water quality inventory:" 1992 report to congress, EPA 841-R-94-001, U. S. EPA, Office of Water, Washington, D. C. 20460.

U.S. Environmental Protection Agency. 1992. "Water quality standards for the 21st century: proceedings of the third national conference." U. S. EPA, Office of Science and Technology, Washington, D. C. 20460. EPA 823-R-92-009.

U.S. Environmental Protection Agency. 1991a. "Environmental monitoring and assessment program." EMAP - surface waters monitoring and research strategy - fiscal year 1991, EPA/600/3-91/022. Office of Research and Development, Environmental Research Laboratory, Corvallis, OR. 184 pp.

U.S. Environmental Protection Agency. 1991b. "Policy on the use of bioassessments and criteria in the water quality program." Office of Science and Technology, Washington, D.C.

U.S. Environmental Protection Agency. 1990a. "Feasibility report on environmental indicators for surface water programs." U. S. EPA, Office of Water Regulations and Standards, Office of Policy, Planning, and Evaluation, Washington, D. C.

U.S. Environmental Protection Agency. 1990b. "Biological Criteria, national program guidance for surface waters." U. S. EPA, Office of Water Regulations and Standards, Washington, D.C. EPA-440/5-90-004.

U.S. General Accounting Office. 1986. "The nations water: key unanswered questions about the quality of rivers and streams." U.S. GAO, Prog. Eval. & Methods Div., Washington, D.C. GAO/PEMD-86-6.

Yoder, C.O. 1995. "Policy issues and management applications of biological criteria," pp. 327 - 344. *in* W. Davis and T. Simon (eds.). *Biological Assessment and Criteria: Tools for Water Resource Planning and Decision Making*. Lewis Publishers, Boca Raton, FL.

Yoder, C. O. 1994. "Toward improved collaboration among local, state, and federal agencies engaged in monitoring and assessment." *J. N. Am. Benth. Soc.* **13**(3): 391 - 398.

Yoder, C. O. 1989. "The development and use of biological criteria for Ohio rivers and streams." *in* Gretchin H. Flock, editor. *Water quality standards for the 21st century. Proceedings of a National Conference*, U. S. EPA, Office of Water, Washington, D.C.

Yoder, C.O. and E.T. Rankin. 1995a. "Biological criteria program development and implementation in Ohio", pp. 109-144. *in* W. Davis and T. Simon (eds.). *Biological Assessment and Criteria: Tools for Water Resource Planning and Decision Making*. Lewis Publishers, Boca Raton, FL.

Yoder, C.O. and E.T. Rankin. 1995b. "Biological response signatures and the area of degradation value: new tools for interpreting multimetric data", pp. 263-286. *in* W. Davis and T. Simon (eds.). *Biological Assessment and Criteria: Tools for Water Resource Planning and Decision Making*. Lewis Publishers, Boca Raton, FL.

MARYLAND BIOLOGICAL STREAM SURVEY: DEVELOPMENT OF A FISH INDEX OF BIOTIC INTEGRITY

N. ROTH[1], M. SOUTHERLAND[1], J. CHAILLOU[1], R. KLAUDA[2], P. KAZYAK[2], S. STRANKO[2], S. WEISBERG[3], L. HALL, JR.[4], and R. MORGAN II[5]

[1] *Versar, Inc., 9200 Rumsey Rd., Columbia, MD 21045, USA*
[2] *Maryland Department of Natural Resources, 580 Taylor Ave., Annapolis, MD 21401, USA*
[3] *Southern California Coastal Water Research Project, 7171 Fenwick Ln., Westminster, CA 92683, USA*
[4] *University of Maryland, Wye Research and Education Center, Box 169, Queenstown, MD 21658, USA*
[5] *University of Maryland, Appalachian Environmental Laboratory, Gunter Hall, Frostburg, MD 21532, USA*

Abstract. As a step towards determining the extent of degradation in non-tidal streams, a multi-metric Index of Biotic Integrity (IBI) based on fish assemblages was developed for the Maryland Biological Stream Survey (MBSS). The MBSS is a probability-based statewide sampling program designed to assess the status of biological resources and to evaluate the effects of anthropogenic activities. We used data from 419 MBSS sites sampled in 1994-95 to develop the IBI. Two distinct geographic strata, corresponding with ecoregional and physiographic boundaries, were identified via cluster analysis and multivariate analysis of variance (MANOVA) as supporting distinctly different species groups. Reference conditions were based on minimally degraded sites. We quantitatively evaluated the ability of various attributes of the fish assemblage (candidate metrics) to discriminate between these reference sites and sites known to be degraded, using statistical tests and classification efficiency. Provisional formulations of the IBI were selected for each region based on high classification efficiency and broad representation of fish assemblage attributes. Fish IBI scores for 1995 MBSS sites spanned a wide range of biological conditions, from good to very poor. Over all six basins sampled in 1995, half of the stream miles fell into the range of good to fair. Roughly 25% of stream miles showed some degradation. The IBI will be used in conjunction with physical and chemical data to answer critical questions about the health of Maryland streams and the relative impacts of human-induced stresses on the state's aquatic systems.

1. Introduction

Biological assessment techniques have been used successfully to measure the ecological health or condition of streams, lakes, and other aquatic systems. One widely accepted bioassessment method employs a multi-metric indicator known as the Index of Biotic Integrity (IBI, Karr *et al.* 1986), which has been adapted for a variety of regions (see Simon and Lyons 1995) and taxonomic groups (e.g., Fore *et al.* 1996, Barbour *et al.* 1996, Weisberg *et al.* 1996, 1997). Here we describe the development of an IBI for stream fishes, using data from the Maryland Biological Stream Survey (MBSS). The MBSS is a multi-year, probability-based sampling program designed to assess the status of biological resources in non-tidal streams of Maryland and to determine how biological resources are affected by acidic deposition and other human impacts (Volstad *et al.* 1996, Roth *et al.* 1997, Klauda *et al.* 1997). Indicators such as the fish IBI are being used to estimate the extent of stream degradation in the state and to examine relationships between various anthropogenic stresses and the condition of biological resources.

Building on previous efforts to apply and adapt IBIs in particular regions of the state (Hall *et al.* 1993, 1996a,b; Burkett and Morgan 1996; Van Ness *et al.* 1996, Kazyak *et al.* 1992; Jacobson *et al.* 1992), the MBSS represents the first development of a statewide fish

Environmental Monitoring and Assessment **51: 89–106, 1998.**

IBI for Maryland. Our IBI approach involved first establishing a reference set of minimally degraded streams based on physical habitat, land use, and water quality characteristics. Multivariate analyses were used to identify regional strata supporting different species groups, thus accounting for natural variability in biological assemblages. We then compared the ability of candidate metrics (describing attributes of the fish assemblage) to discriminate between these reference sites and sites known to be degraded. In addition to evaluating individual metrics, we tested the effectiveness of the overall index for discriminating reference and degraded sites and analyzed the performance of different metric combinations. Using this information to "build a better index," we developed an IBI formulation that yielded effective and ecologically meaningful results. This paper describes the development of the Maryland fish IBI, including results of metric and index evaluation, and addresses issues encountered in indicator development.

2. Methods

2.1. DEVELOPING THE DATA BASE

The data base for indicator development consisted of biological, chemical, physical habitat, and land use data from 419 stream sites sampled by the MBSS during 1994 and 1995. MBSS sites, located on first through third order (Strahler 1957), non-tidal streams, were selected using a probability-based sampling design (Volstad *et al.* 1996, Roth *et al.* 1997). The original sampling frame for the MBSS, based on a previous statewide stream chemistry survey (the Maryland Synoptic Stream Chemistry Survey, Knapp *et al.* 1988), was constructed by overlaying basin boundaries on a map of all blue line stream reaches in the study area as digitized on a U.S. Geological Survey 1:250,000-scale map. Seventy-five-meter stream segments served as the elementary sampling units for which biological, water chemistry, and physical habitat data were collected. At all sites, field data were collected using standard methods developed for the MBSS (Kazyak 1995, Kazyak 1994).

Fish were sampled during the summer index period (about June 1 to September 30) using quantitative, double-pass electrofishing of the 75-m stream segments. Block nets were placed at each end of the segment and one or more direct current, backpack electrofishing units were used to sample the entire segment. All fish captured (> 25 mm total length) were identified, counted, and weighed in aggregate; up to 100 individuals of each species were examined for external anomalies such as lesions and tumors.

During the spring index period (about March 1 to May 1), water samples were collected and pH, acid neutralizing capacity (ANC), conductivity, sulfate, nitrate, and dissolved organic carbon (DOC) were measured in the laboratory using standard methods (EPA 1987). During summer sampling, dissolved oxygen (DO), pH, temperature, and conductivity were measured *in situ*.

Physical habitat assessments were conducted at all stream segments using the procedures detailed in Kazyak (1995), which are largely patterned after other widely used assessment techniques (Plafkin *et al.* 1989, Barbour and Stribling 1991, Ohio EPA 1987, Rankin 1989, 1995). Instream habitat structure, bank stability, degree of channel alteration, and other physical habitat features were assessed qualitatively based on visual observations within each sample segment, and assigned scores (0-20 points) within four categories (optimal 16-20 points, sub-optimal 11-15, marginal 6-10, poor 0-5) following standard narrative guidelines for each category (Kazyak 1995). Observations of the surrounding area were used to assign similar ratings for aesthetic value and remoteness. Evidence of point sources, stream channelization, and other human impacts were recorded. Riparian vegetation width was estimated, up to 50 m from the stream.

Land-use information was extracted in digital format from Maryland Office of Planning 1990 land-use / land-cover maps (Fisher 1991). Catchments upstream from sample sites were digitized using contour lines from digital county topographic maps (1:62,500 scale). Following spatial definition of catchments, catchment areas for each site were calculated and the percentage of area as each major land use type (urban, agricultural, and forested) within each catchment determined.

2.2. IDENTIFYING REFERENCE AND DEGRADED SITES

Reference sites were defined as those with minimal anthropogenic disturbance, based on thresholds established for water chemistry, physical habitat, and land use (within the catchment upstream of the sample site). Reference site criteria eliminated sites impacted by extreme acidification, nutrient loading, or physical alteration. Water chemistry and riparian width thresholds were consistent with levels generally considered detrimental to streams (e.g., Baker *et al.* 1990, Osborne and Kovacic 1993). Other habitat criteria were based on interpretation of the MBSS protocols for habitat assessment (Kazyak 1995). Thresholds for catchment land use, which serve as overall estimates of human influence, eliminated the most extreme cases of land use alteration. Forty-five reference sites meeting all 12 of the following criteria were identified:

- pH $\geq$ 6 or blackwater stream (pH < 6 and DOC $\geq$ 8 mg/l)
- ANC $\geq$ 50 μeq/l
- DO $\geq$ 4 ppm
- nitrate $\leq$ 300 μeq/l
- urban land use $\leq$ 20% of catchment area
- forest land use $\geq$ 25% of catchment area
- remoteness rating: optimal or suboptimal
- aesthetics rating: optimal or suboptimal
- instream habitat rating: optimal or suboptimal
- riparian buffer width $\geq$ 15 m
- no channelization
- no point source discharges

We next identified a set of degraded sites exhibiting any of three types of anthropogenic stress: acidification, eutrophication, or physical habitat alteration. Seventy-five sites meeting any of the following criteria were designated as degraded:

- pH ≤ 5 and ANC ≤ 0 μeq/l (except for blackwater streams, DOC ≥ 8 mg/l)
- DO ≤ 2 ppm
- nitrate > 500 μeq/l and DO < 3 ppm
- instream habitat rating poor and urban land use > 50% of catchment area
- instream habitat rating poor and bank stability rating poor
- instream habitat rating poor and channel alteration rating poor

Streams impacted by physical habitat alteration were defined as those with poor instream habitat structure in combination with at least one other factor indicative of an anthropogenic source for the alteration: high degree of urban land use, poor bank stability, or indications of channel alteration. Poor instream habitat structure alone was not sufficient to designate a site as degraded, because some streams have little woody debris, boulder, or cobble substrate even under pristine conditions.

2.3. DETERMINING APPROPRIATE STRATA

To account for the natural variation in fish assemblage composition across the large area and diverse habitats covered by the MBSS, sites were stratified into groups based on naturally occurring biological assemblages and appropriate reference expectations were then set for each group. Cluster analysis using the Canberra metric and flexible sorting (β=-0.25) (Boesch 1977) on log-transformed percentages of species abundance was used to identify groups of sites based on assemblage similarity. Multivariate analysis of variance (MANOVA) was used to determine which groups had statistically different assemblages. Physical variables (stream order, catchment area, summer stream temperature, physiographic region) associated with significantly different biological assemblages were examined to ascertain which physical variables determined assemblage composition. To ensure sufficient sample size, a total of 183 sites were used for this analysis, including all 45 reference sites plus additional sites designated as not substantially degraded, meeting criteria slightly less stringent those used to define reference sites (e.g., pH ≥ 5.5 instead of ≥ 6, DO ≥ 3 instead of ≥ 4).

2.4. COMPILING CANDIDATE METRICS

A list of 39 candidate metrics was compiled, including each of the original 12 fish IBI metrics proposed by Karr *et al.* (1986) along with candidate metrics from earlier IBI investigations in Maryland and other states (Hall *et al.* 1996a,b; Burkett and Morgan 1996, Simon and Lyons 1995). Various formulations of some metrics were included as candidates. The full list of candidate metrics fell into five major groups: measures of species richness and composition, indicator species (based on tolerance), trophic function, fish abundance and condition, and reproductive function.

Fishes were classified into ecological categories, based on information in the literature, for the following characteristics: benthic species, tolerance, trophic status, native or introduced species, and lithophilic spawners. Benthic species are those fishes that reside primarily on the stream bottom. Benthic fishes can include all darters (*Etheostoma* spp., *Perca* spp.), sculpins (*Cottus* spp.), madtoms (*Noturus* spp.), and lampreys (*Petromyzon* spp., *Lampetra* spp.). Because many benthic fishes have relatively limited home ranges, they are potentially valuable indicators of local conditions. Tolerance of fishes to anthropogenic stress was determined using several sources (Plafkin *et al.* 1989, Hall *et al.* 1993, Hall *et al.* 1996a, Jacobson *et al.* 1992, Karr 1981, Karr *et al.* 1986, Lee *et al.* 1981, Ohio EPA 1987, EA 1993, Miller *et al.* 1988, Kazyak *et al.* 1988). Only those species that were considered tolerant or intolerant by a consensus of researchers were designated as tolerant or intolerant for this study. Future analysis of statewide fish and stream condition data may lead to later revision of tolerance designations. Nine trophic classifications for fishes were defined according to reported descriptions of fish diets (Jenkins and Burkhead 1993). For example, insectivores were defined as those that specialize primarily on insects, while invertivores eat insects and other invertebrates, including crustaceans, mollusks, and worms. Fishes were classified as native or introduced to the Chesapeake Bay drainage and to the Ohio River drainage (i.e., the Youghiogheny drainage in Maryland) (Lee *et al.* 1981, Jenkins and Burkhead 1993). Historically, streams in these two regions developed separate and distinct regional faunal pools of fish because of their separation by the Eastern Continental Divide. Fishes reported to use rock substrates for spawning (Jenkins and Burkhead 1993) were designated as lithophilic spawners. In addition, two metrics characterizing the health or condition of individual fish were developed. The percent occurrence of all anomaly types was calculated as the number of visible anomalies per fish examined; a second variation excluded blackspot and other visible external parasites.

Because some characteristics of fish assemblages, such as abundance or species richness, tend to vary with stream size, reference expectations and IBI metric scoring should account for this natural variability (Karr *et al.* 1986). To account for this variation, some IBIs use separate scoring criteria for each stream order; others adjust scores by catchment area (Ohio EPA 1987). To evaluate the potential effect of stream size on each of our metrics, we examined plots of each metric against log of catchment area for each stratum (Coastal Plain and non-Coastal Plain), using the 183 sites included in the cluster analysis. We then adjusted the metrics exhibiting a strong relationship with catchment area in both regions, as determined by significant Spearman correlations ($p < 0.05$ in both regions), strong linear relationship ($r^2 > 0.25$ in at least one region), and appearance of the plot. Adjusted values for metrics were calculated using the following equation, with values of m (slope) and b (intercept) derived from regression analyses:

adjusted value = observed value / expected value

where expected value = m * log (catchment area in acres) + b

The MBSS data set included a number of very small headwater streams, with correspondingly low species richness and fish abundance. If the expected total number of fish at a site (predicted for all streams of that size) fell below 100 or the expected number of fish species fell below 5, it was deemed to be practically impossible to characterize a reference condition accurately. For each stratum, log-linear relationships between total abundance, total number of species, and catchment area indicated that the expected values were below the minimum thresholds when the catchment area was less than 300 acres. Therefore, 21 sites on very small headwater streams (catchment area < 300 acres) were excluded from further analyses, reducing the number of available sites for IBI development to 40 reference and 59 degraded sites.

2.5. TESTING CANDIDATE METRICS

Two criteria were used to compare metric values at reference sites with those at degraded sites. The Mann-Whitney U test was used to test for differences in median, and distributions of reference and degraded site values were compared using the Kolmogorov-Smirnov test. Metrics were evaluated separately for each of the two strata. Metrics showing significant differences in these initial statistical tests were used in subsequent analyses.

The scoring of IBI metrics was based on the distribution of values observed at reference sites within each stratum. The IBI approach involves scoring each metric as 5, 3, or 1, depending on whether its value at a site approximates, deviates slightly from, or deviates greatly from conditions at the best reference sites (Karr *et al.* 1986). In other IBI applications (e.g., Fore *et al.* 1996, Lyons *et al.* 1996, Barbour *et al.* 1996), a number of different methods have been used to establish scoring thresholds, based on various subdivisions of observed values. In our case, threshold values for each selected metric were established as approximately the 10th and 50th (median) percentile values for reference sites (see example, Figure 1) and were established separately for each stratum. For each metric expected to decrease with degradation, values below the 10th percentile were scored as 1, as they showed the greatest deviation from most reference sites. Values between the 10th and 50th percentiles were scored as 3, as they fell short of median expected values for reference sites. Values above the 50th percentile were scored as 5. Scoring was reversed for metrics expected to increase with degradation (e.g., values below the 50th percentile were scored as 5, values above the 90th percentile were scored as 1). This method differs from other scoring systems in that both the upper and lower thresholds are independently derived from the distribution of reference site values. The 10th percentile threshold for designating scores of 1 represents our best attempt to identify values that are outside the natural expectation for reference sites.

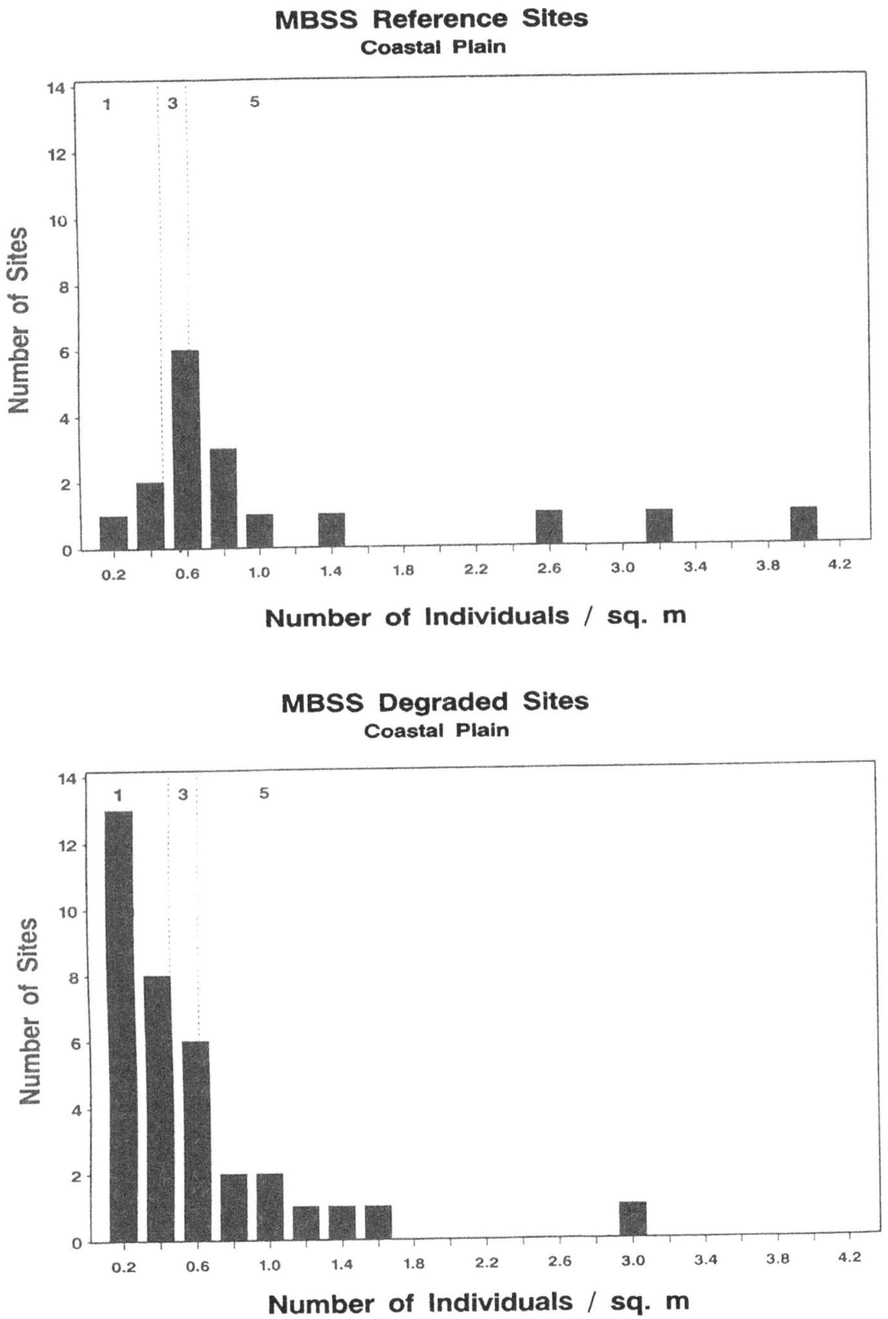

Fig. 1. Establishment of scoring thresholds based on the distribution of reference site values. This example shows the number of individuals per square meter at reference and degraded sites in the Coastal Plain. The upper threshold for IBI metric scoring was set at the 50th percentile of reference site values, and the lower threshold approximately the 10th percentile.

To test the discriminatory power of each candidate metric, we evaluated the degree of overlap between metric values at reference and degraded sites by examining the number of sites scoring above and below the lower threshold. A classification efficiency was calculated as the percent of reference sites with values scoring ≥ 3 plus the percent of degraded sites scoring < 3. Reference sites misclassified as degraded (score < 3) and degraded sites misclassified as reference (score ≥ 3) make up the remainder of the sites. A high classification efficiency indicates a small amount of overlap between values for reference and degraded sites.

2.6. COMBINING METRICS INTO AN INDEX

To develop an overall index, different combinations of metrics were constructed and the performance of each evaluated. For each combination, an index was calculated as the mean of the metrics selected. The resulting index was scaled from 1 to 5, as were individual metrics. Classification efficiencies of different metric combinations were calculated as above.

We first used a stepwise procedure to develop for each stratum the metric combination having the highest possible classification efficiency. This process began with the single best metric in each region; other metrics were then added one at a time (i.e., we added and removed each metric to determine which was the best 4-metric index, the best 5-metric index, and so on). After evaluating the performance and ecological relevance of each metric, we selected only one variation of a metric for inclusion in the index (for example, once the number of benthic species was added, the number of darter species could not be added). At the point that the overall classification efficiency of the index declined or plateaued, no other metrics were added.

Following this exploratory step, a core index of key metrics was identified as important to characterize biological integrity. From this core, we sequentially added additional metrics to maximize classification efficiency based on the core combination. We then modified these combinations for each region by adding sequentially other metrics representing additional aspects of assemblage structure and function not yet represented in the index. We evaluated whether adding each metric was warranted based on both ecological reasons and its effect on the overall classification efficiency of the index. Specifically, an attempt was made to include metrics representative of each major category. Finally, we evaluated redundancy among selected metrics by calculating the Spearman correlation coefficients between each metric pair.

3. Results

3.1. DETERMINATION OF STRATA

Two distinct geographic strata, corresponding with physiographic region boundaries, were identified via cluster analysis and MANOVA as having distinctly different naturally occurring species assemblages. None of the other physical variables showed any clear correspondence with groupings of sites characterizing streams by biological assemblage similarity. Based on these analyses, two strata were used throughout the remaining steps of the IBI development process: the Coastal Plain and non-Coastal Plain (including the Appalachian Plateau, Valley and Ridge, Blue Ridge, and Piedmont physiographic regions). These two geographic strata are coincident with aggregations of ecoregions (Omernik 1987) and the physiographic provinces (Maryland Geological Survey; Reger 1995) developed for Maryland. The cluster analysis did not support further subdivision of these site groupings, for example, into other ecoregions or subecoregions (White 1996). However, future MBSS indicator development efforts will consider whether further subdivisions are warranted. For example, increased geographic coverage with 1996 MBSS sampling, especially in the Piedmont region, may justify dividing the non-Coastal Plain stratum into Piedmont and western Maryland regions. A greater number of samples from western Maryland may also support the consideration of coldwater streams as a separate stratum.

3.2. METRIC EVALUATION

Among the 39 metrics tested, 28 exhibited a significant difference between reference and degraded sites, passing the Mann-Whitney U and Kolmogorov-Smirnov tests ($p < 0.05$) in the Coastal Plain and/or non-Coastal Plain (Table I). Metrics performing poorly in both regions in these initial tests were not used in subsequent analyses. Poor performers included four metrics based on single species (percent green sunfish, white sucker, or creek chub), percent white sucker and northern hogsucker, percent natives (species or individuals), occurrence of anomalies (with or without external parasites), percent hybrids, and percent invertivores.

Classification efficiencies for individual metrics ranged from 33 to 92% (Table I). Individual classification efficiencies for most metrics paralleled statistical performance. Fifteen candidate metrics exceeded 75% classification efficiency in at least one stratum, indicating a strong ability to discriminate between reference and degraded sites.

3.3. COMBINATION OF METRICS INTO AN INDEX

In combining metrics to ascertain the maximum possible classification efficiency, 87% was the highest classification efficiency reached for each of the two strata. Interestingly, a combination of just one to two metrics yielded classification efficiencies approaching this level. The addition of other metrics did not greatly affect the performance of the overall index.

Based on a combination of individual metric performance and best professional judgment, we decided that three metrics were critical to creating a useful index: the number of native species, number of benthic species, and percentage of tolerant individuals. The core indices comprising these metrics yielded a classification efficiency of 85% in the Coastal Plain and 74% in non-Coastal Plain. From this core, additional metrics were added to enhance classification efficiency. In the Coastal Plain, biomass per square meter and number of individuals per square meter were added, creating a 5-metric IBI with a classification efficiency of 87%. In the non-Coastal Plain, the maximum index classification efficiency was reached with four metrics, adding only the percentage of insectivores, for a classification efficiency of 83%.

As a final step, metrics were added to the index to fill out each of the major metric groups such as trophic composition, reproductive function, and fish abundance and condition. The effect of adding each new metric was evaluated, and additional metrics were retained only if they did not cause an appreciable decline in classification efficiency when added individually or in combination with other metrics. In the Coastal Plain, the percent generalists, omnivores, and invertivores; the percent abundance of dominant species; and the percent lithophilic spawners were added. In the non-Coastal Plain, these same three metrics were added along with the number of individuals per square meter.

Final formulations of the IBI (Table II) were selected for each region based on a combination of high classification efficiency and broad representation of fish assemblage attributes. In the Coastal Plain, this final eight-metric IBI classified sites correctly 81% of the time; misclassified sites (19%) were dominated by degraded (17%) rather than reference sites (2%). In the non-Coastal Plain, the resultant eight-metric IBI had a classification efficiency of 83%; misclassified sites (17%) were again dominated by degraded (13%) rather than reference sites (4%).

Table I.
Evaluating candidate metrics

Evaluations of individual candidate metrics used in IBI development. Results of Mann-Whitney U and Kolmogorov-Smirnov tests of differences between reference and degraded sites (+ indicates $p<0.05$). Individual classification efficiencies are the percent of reference and degraded sites correctly classified by each metric.

	Coastal Plain			Non-Coastal Plain		
	Mann-Whitney (p)	Kolmogorov-Smirnov (p)	Classification Efficiency (%)	Mann-Whitney (p)	Kolmogorov-Smirnov (p)	Classification Efficiency (%)
Species richness and composition						
Number of species*	<0.001 +	<0.001 +	75	<0.01 +	<0.01 +	74
Number of native species*	<0.001 +	<0.001 +	81	0.01 +	0.01 +	70
Number of darter species*	<0.01 +	<0.001 +	79	0.10	0.36	49
Number of darter and sculpin species*	<0.001 +	<0.001 +	79	<0.001 +	<0.001 +	79
Number of darter, sculpin, and madtom species*	<0.001 +	0.01 +	73	<0.001 +	<0.001 +	79
Number of benthic species*	<0.001 +	<0.001 +	83	<0.001 +	<0.001 +	79
Number of sunfish species*	<0.001 +	<0.01 +	65	0.38	0.97	49
Number of sucker species*	0.01 +	0.04 +	33	0.10	0.18	49
Percent round-bodied suckers	0.01 +	0.04 +	33	0.08	0.86	49
Indicator species						
Number of intolerant species	<0.001 +	<0.001 +	33	<0.01 +	0.02 +	49
Percent tolerants	<0.001 +	<0.001 +	71	<0.01 +	<0.01 +	74
Percent eastern mudminnows	0.01 +	<0.01 +	75	0.01 +	0.28	64
Percent pioneering species	<0.01 +	0.01 +	69	0.06	0.04 +	68
Percent abundance of dominant species	<0.001 +	<0.01 +	60	0.02 +	0.01 +	68
Trophic composition						
Percent omnivores	0.42	0.21	53	0.03 +	0.01 +	68
Percent generalists and omnivores	0.23	0.31	44	<0.01 +	<0.001 +	77
Percent generalists, omnivores, and invertivores	<0.01 +	0.01 +	63	<0.001 +	<0.001 +	85
Percent omnivores and invertivores	<0.001 +	<0.001 +	60	<0.001 +	0.01 +	72
Percent insectivorous cyprinids	0.28	0.96	33	0.04 +	0.02 +	49
Percent insectivores	0.45	1.00	33	<0.001 +	<0.001 +	87
Percent insectivores and invertivores	0.30	0.34	48	<0.001 +	<0.001 +	72
Percent top predators	0.02 +	0.02 +	65	0.11	0.38	49
Fish abundance and condition						
Number of individuals*	<0.001 +	<0.001 +	81	0.04 +	<0.01 +	72
Number of individuals per square meter	<0.01 +	0.01 +	69	0.32	0.01 +	72
Number of individuals, excluding introduced species*	<0.001 +	<0.001 +	75	0.04 +	<0.01 +	74
Number of individuals, excluding tolerants*	<0.001 +	<0.001 +	77	<0.001 +	<0.001 +	79
Biomass*	<0.001 +	<0.001 +	92	<0.001 +	<0.001 +	74
Biomass per square meter	<0.001 +	<0.001 +	83	0.04 +	0.01 +	68
Reproductive function						
Percent lithophilic spawners	0.01 +	0.07	33	0.22	0.04 +	70

* values were adjusted for watershed area

Table II. Metrics included in provisional fish IBI

Classification efficiencies reflect index performance. * indicates metrics adjusted for watershed area.

Coastal Plain	Non-Coastal Plain
Number of native species*	Number of native species*
Number of benthic species*	Number of benthic species*
Percent tolerant individuals	Percent tolerant individuals
Percent abundance of dominant species	Percent abundance of dominant species
Percent generalists, omnivores, and invertivores	Percent generalists, omnivores, and invertivores
Number of individuals per square meter	Percent insectivores
Biomass (g per square meter)	Number of individuals per square meter
Percent lithophilic spawners	Percent lithophilic spawners
Classification efficiency: 81%	Classification efficiency: 83%

4. Discussion

An effective indicator of environmental degradation is sensitive to different types of stress and yields results that are biologically meaningful (Fausch *et al.* 1990). Metrics in our IBI describe a variety of attributes of fish assemblages that respond to environmental stress. Each of our metrics on its own displays an acceptable level of effectiveness in detecting impairment, as was evident in statistical and other quantitative evaluations of metric performance. Each contributes unique ecological information to the overall assessment, although a few key metrics were sufficient to provide strong discriminatory power. Our use of quantitative methods to evaluate metric and index performance provided an objective basis for our decisions, although ecological knowledge and professional judgment also factored into the final index formulation.

Our approach builds on other recent indicator development efforts for stream fish (e.g., Paller *et al.* 1996), freshwater macroinvertebrates (e.g., Barbour *et al.* 1996, Fore *et al.* 1996), and estuarine benthic macroinvertebrates (Weisberg *et al.* 1996, 1997) that have used similar methods to evaluate candidate metrics based on their ability to discriminate between sites experiencing varying degrees of anthropogenic disturbance. A highlight of our approach is the use of statistical analyses and classification efficiencies to quantify this discriminatory power. These tools provided quantitative information on the performance of individual metrics and of various combinations of metrics in an index. We found classification efficiencies slightly lower than those reported for estuarine benthic macroinvertebrates (Weisberg *et al.* 1996); this may reflect greater patchiness inherent in stream fish distributions. Finally, the size of the MBSS data set afforded good coverage statewide, with a large enough sample size to support with good confidence the designation of scoring thresholds based on the distribution of scores at reference sites. In particular, there were sufficient numbers of sites to characterize the natural diversity of conditions in each region, an important consideration in establishing reference expectations (Yoder and Rankin 1995).

4.1. METRICS SELECTED

Our final suite of metrics corresponds well with those proving effective in other regional evaluations (Hall *et al.* 1996a, Paller *et al.* 1996, Burkett and Morgan 1996, Van Ness *et al.* 1996). For example, six of the ten metrics in an IBI developed for streams in Montgomery County, Maryland (Van Ness *et al.* 1996) were comparable to metrics selected for inclusion in our IBI. In addition, threshold values for most metrics were surprisingly similar across the two studies, despite the use of different methods to derive scoring criteria.

Biomass, a metric that demonstrated a strong ability to discriminate among our Coastal Plain sites, was not tested in previous evaluations examined (Hall *et al.* 1996a, Paller *et al.* 1996, Burkett and Morgan 1996). Although we included aggregate fish biomass in the Coastal Plain IBI, we recognize that total biomass could be influenced by the presence of a few tolerant individuals of large body size, such as common carp (*Cyprinus carpio*), in which case biomass would not accurately reflect high biotic integrity. This influence was not found in our dataset, but it could be important. A better metric might include only biomass for particular species or use log-transformed biomass values.

Although used in many IBIs, the number of anomalies was not effective in detecting differences between reference and degraded conditions, as was also reported by Paller *et al.* (1996). Ohio EPA (1987) includes in their IBI the proportion of individuals with deformities, eroded fins, lesions, and tumors ("DELT anomalies"), but reports that this metric is most sensitive at highly degraded sites subject to point source impacts. At MBSS sites, the number of anomalies was generally low, perhaps because samples were taken from relatively small streams. Potentially, a survey of fourth order and larger streams might involve more point source impacts and could profit from including anomalies as a metric. Inclusion of an anomaly metric, either within an IBI or as a separate indicator, may increase the general utility of biological assessment results in detecting many types of impacts, including future impacts not frequently encountered in the current dataset. Although we did not include the number of anomalies as a metric in the IBI, we recommend it be reported separately as an additional indicator of fish assemblage health.

Spearman correlations between metric pairs revealed little redundancy among the selected metrics. Only two metric pairs had Spearman correlation coefficients greater than 0.75: percent generalists, omnivores, and invertivores v. percent tolerants (r_S=0.76) and percent generalists, omnivores, and invertivores v. percent insectivores (r_S=-0.88), both in the non-Coastal Plain only. Although there may be some inherent redundancy in using these metrics, all three were retained because they represent different ecological aspects of the fish assemblage.

4.2. DEVELOPING AND APPLYING THE IBI

Interestingly, we found that once a few key metrics were included in the IBI, a high degree of discriminatory power was achieved. Our technique differs from many IBI development methods in that we started with a small number of metrics and then increased that number without sacrificing performance of the overall indicator. Many IBI applications have used a standard set of ten to twelve metrics, with minor regional modifications or substitutions. In our case, a four- to five-metric IBI was highly robust. Addition of further metrics did not have much effect on the classification efficiency of the index but might enhance discrimination in rare cases or address situations that may be encountered in future applications of the index. Similarly, Angermeier and Karr (1986) noted wide variation in individual metric contributions to total IBIs, with the relative importance of particular metrics differing among regions and degradation gradients. We suspect that IBIs developed for other regions also derive most of their discriminatory power from a relatively small subset of their metrics. While we acknowledge the value of retaining a sufficient number of metrics to address a variety of impacts, we also encourage careful selection of metrics based on evaluations of their individual and combined discriminatory power.

Many stream monitoring efforts have avoided sampling the smaller streams included in the probability-based MBSS design. Of the 419 sites used here, nearly 40% drained catchments smaller than 1,000 acres (405 hectares), while more than 90% drained areas less than 12,000 acres (4,856 hectares). In contrast, the majority of streams sampled by Ohio EPA's rotating basin monitoring program have catchments larger than 12,000 acres. Ohio also incorporates special provisions for small streams, recognizing both the fundamentally different fauna and utility of assessments in small streams. For example, Ohio routinely corrects metrics for sites with less than 200 fish per 300 m (Ohio EPA 1987). We recognized that developing IBIs for these small, less species-rich streams might be difficult, but we did not want to dismiss this widespread but often overlooked resource. Nonetheless, intuition argues for a lower limit on the number of fish and species that must be sampled at a site to produce a useful IBI. Indeed, variability in IBI scores has been shown to increase with low total abundance (<400 individuals per sample), raising concerns about applying the index in small streams (Fore *et al.* 1994). Based on our data, we designated a minimum catchment size requirement of 300 acres (121 hectares). Future MBSS efforts may generate alternative methods specifically for assessing these small headwaters.

Given the natural variability of fish assemblages, even the best index is unlikely to provide a 100% classification efficiency. Ideally, rates of misclassifying both reference and degraded sites should be minimized. However, even a highly reliable index will likely have some overlap between reference and degraded conditions based on the natural variability of fish assemblage composition. The best balance of these two kinds of "misclassifications" depends on the purpose of assessment. Programs to screen sites for potential problems would want a low misclassification of degraded sites, while efforts to identify specific sites for remedial actions would want to avoid mislabeling high quality sites. The Maryland IBI, which may be used to target future restoration efforts, favored the second strategy, rarely misclassifying reference sites.

Ultimately, determinations of degradation using the IBI rely on the ability of biota to respond to the full range of stressors, including those not initially used to define reference and degraded conditions. This is appropriate given that some factors used to develop the IBI (e.g., land use) are simply first-order approximations of human impacts. While the known array of physical and chemical factors was useful to define opposite ends of a scale of degradation and calibrate biological response, statements about the condition of streams should be based on the biology itself, as measured by the IBI.

The MBSS will continue to collect data in upcoming years and, as appropriate, use these new data to refine the IBI. A long-term bioassessment program is needed both for better characterizing reference condition and for monitoring and assessing environmental conditions statewide. In the near future, the addition of 200-300 sites sampled in 1996 will increase both sample size and geographic coverage (because the MBSS sampling is conducted in different river basins during each year of this multi-year survey). Addition of 1996 sites may support the further stratification of sites, given the increased coverage in western Maryland and the Piedmont area of the non-Coastal Plain. It may also be necessary to consider coldwater streams as a separate stratum. Generally, high-quality coldwater streams are dominated by salmonid species and have lower overall species richness than warmwater systems of the same area. In other parts of North America, fish IBIs for coldwater and coolwater streams have been tailored to account for their unique biological characteristics (e.g., Lyons *et al.* 1996, Leonard and Orth 1986).

Ultimately, this fish IBI will be used in conjunction with MBSS physical and chemical data to answer critical questions about the condition of Maryland streams and the relative impacts of human-induced stresses on the state's aquatic systems. Initially, the IBI was used in analyses of water quality, physical habitat, and land use influences on streams in 6 of Maryland's 18 basins sampled in 1995. The IBI was used to assess the range of stream biological conditions, the extent of degradation, and the geographic distribution of streams in varying biological condition. Fish IBI scores for 1995 MBSS sites spanned a wide range of biological conditions, from good to very poor. Over all six basins sampled in 1995, half of the stream miles fell into the range of good to fair. Roughly 25% of stream miles showed some level of degradation. Assessments will help to evaluate human impacts on biological resources, including acidification, habitat degradation, and watershed land use, and will be used as a screening tool to identify sites that may require further investigation to identify potential impacts.

Acknowledgements

The MBSS is a cooperative effort involving many individuals from the Maryland Department of Natural Resources, Versar, Inc., the University of Maryland's Appalachian Environmental Laboratory and Wye Research and Education Center, and Coastal Environmental Services. This project was supported in part by contract no. PR-96-055-001 from the Maryland Department of Natural Resources' Power Plant Research Program. We thank Chris Yoder and Paul Angermeier for their thoughtful comments on an earlier draft of this manuscript.

References

Angermeier, P.L. and Karr, J.R.: 1986, *N. Amer. J. Fish. Mgmt*, **6**, 418-429.

Baker, J.P., Bernard, D.P., Christensen, S.W., Sale, M.J., Freda, J., Heltcher, K.J., Marmorek, D.R., Rowe, L., Scanlon, P.F., Suter G.W. II, Warren-Hicks W.J., and Welbourn, P.M.: 1990, *Biological Effects of Changes in Surface Water Acid-Base Chemistry.* National Acid Precipitation Assessment Program. State of Science/Technology Report No. 13.

Barbour, M.T.,Gerritsen, J., Griffith, G.E., Frydenborg, R., McCarron, E., White, J.S., and Bastian, M.L.: 1996, *Journal of the North American Benthological Society,* **15**, 185-211.

Barbour, M.T. and Stribling, J.B. 1991, *Use of habitat assessment in evaluating the biological integrity of stream communities. In: Biological Criteria: Research and Regulation*, U.S. Environmental Protection Agency, EPA-440/5-91-005, 25-38.

Boesch, D.F.: 1977, *Application of numerical classification in ecological investigations of water pollution*, U.S. Environmental Protection Agency, EPA-600/3-77-033.

Burkett, M.J. and Morgan II, R.P.: 1996, *Metric selection for an index of biotic integrity: developing an IBI for the Upper Monocacy River Drainage,* Maryland Department of Natural Resources.

EA Engineering, Science, and Technology, Inc.: 1993, *Development and modification of an index of biotic integrity for New Jersey lotic waters,* EA project 60341.01, Hunt Valley, MD.

Fausch, K.D., Lyons, J., Karr, J.R., and Angermeier, P.L.: 1990, *American Fisheries Society Symposium,* **8**, 123-144.

Fisher, G.T.: 1991, *Preparation of 1990 land use/land cover maps and ARC/INFO digital data base,* Maryland Office of Planning.

Fore, L.S., Karr, J.R., and Conquest, L.L.: 1994, *Can. J. Fish. Aquat. Sci.*, **51**, 1077-1087.

Fore, L.S., Karr, J.R., and Wisseman, R.W.: 1996, *Journal of the North American Benthological Society,* **15**, 212-231.

Hall, L.W., Jr., Scott, M.C., and Killen, W.D. Jr.: 1996a, *Development of biological indicators based on fish assemblages in Maryland coastal plain streams,* Maryland Department of Natural Resources, CBWP-MANTA-EA-96-1.

Hall, L.W., Jr., Scott, M.C., Killen, W.D. Jr., and Anderson, R.D.: 1996b, *Environmental Toxicology and Chemistry,* **15**, 384-394.

Hall, L.W., Jr., Fischer, S.A., Killen, W.K. Jr., Scott, M.C., Ziegenfuss, M.C., and Anderson, R.D.: 1993, *A pilot study to evaluate biological, physical, chemical, and land-use characteristics in Maryland coastal plain streams,* Maryland Department of Natural Resources, CBRM-AD-94-1.

Jacobson, P., Kazyak, P., Janicki, A., Wade, D., Wilson, H., and Morgan, II R.P.: 1992, *Feasibility of using an Index of Biotic Integrity (IBI) approach for synthesizing data from a Maryland Biological Stream Survey,* Maryland Department of Natural Resources, CBRM-AD-93-11.

Jenkins, R.E. and Burkhead, N.M.: 1993, *Freshwater Fishes of Virginia, American Fisheries Society,* Bethesda, MD.

Karr, J.R.: 1981, *Fisheries* **6**(6), 21-27.

Karr, J.R., Fausch, K.D., Angermeier, P.L., Yant, P.R., and Schlosser, I.J.: 1986, *Illinois Natural History Survey Special Publication 5,* 28 pp.

Kazyak, P.F.: 1995, *Maryland Biological Stream Survey Sampling Manual,* Maryland Department of Natural Resources.

Kazyak, P.F.: 1994, *Maryland Biological Stream Survey Sampling Manual,* Maryland Department of Natural Resources.

Kazyak, P.F., Strebel, D.E., and Kou, J.:. 1992, *Assessment of the relationship between landuse patterns and biotic indices in the Gwynns Falls watershed,* Maryland Save Our Streams, Glen Burnie, MD.

Kazyak, P.F., Dwyer, R.L., and Weisberg, S.B.: 1988, Baseline characterization of the Gwynns Falls watershed prior to nonpoint source remediation, Maryland Department of Natural Resources.

Klauda, R., Kazyak, P., Stranko, S., Southerland, M., Roth, N., and Chaillou, J.: 1997, *Environ. Mon. and Assess.* This issue.

Knapp, C.M., Saunders, W.P., Heimbuch, D.G., Greening, H.S., and Filbin, G.J.: 1988, *Maryland Synoptic Stream Chemistry Survey: Estimating the Number and Distribution of Streams Affected by or at Risk from Acidification,* Maryland Department of Natural Resources, AD-88-2.

Lee, D.S., Platania, S., Gilbert, C.R., Franz, R., and Norden, A.: 1981, *Proceedings of the Southeastern Fishes Council,* **3**(3), 1-10.

Leonard, P.M. and Orth, D.J.: 1986, *Trans. Am. Fish. Soc,* **115**, 401-414.

Lyons, J., Wang, L., and Simonson, T.D.: 1996, *N. Amer. J. Fish. Mgmt* ,**16**, 241-256.

Miller, D.L., Leonard, P.M., Hughes, R.M., Karr, J.R., Moyle, P.B., Schrader, L.H., Thompson, B.A., Daniels, R.A., Fausch, K.S., Fitzhugh, G.A., Gammon, J.R., Halliwell, D.B., Angermeier, P.L., and Orth, D.J.: 1988, *Fisheries* **13**(5), 12-20.

Osborne, L.L. and Kovacic, D.A.: 1993, *Freshwater Biology,* **29**, 243-258.

Ohio Environmental Protection Agency (Ohio EPA): 1987, *Biological criteria for the protection of aquatic life. Volumes I-III.*

Omernik, J.M.: 1987, *Annals of the Association of American Geographers,* **77**,118-125.

Paller, M.H., Reichert, M.J.M., and Dean, J.M.: 1996, *Trans. Am. Fish. Soc.*, **125**, 633-644.

Plafkin, J.L., Barbour, M.T., Porter, K.D., Gross, S.K., and Hughes, R.M.: 1989, *Rapid bioassessment protocols for use in streams and rivers: benthic macroinvertebrates and fish,* U.S. Environmental Protection Agency, EPA/440/4-89/001. 162 pp.

Rankin, E.T.: 1995, *Habitat indices in water resource quality assessments. In:* Davis, W.S. and Simon, T.P., eds. *Biological assessment and criteria: tools for water resource planning and decision making,* Lewis Publishers, Boca Raton, FL.

Rankin, E.T.: 1989, *The Qualitative Habitat Evaluation Index (QHEI): Rationale, methods, and application,* Ohio EPA.

Reger, J.: 1995, *Memorandum from J. Reger, Maryland Geological Survey, Environmental Geology and Mineral Resources, to J. Perdue, A. Raspberry, and E. Bradley, Digital files and metadata for draft physiographic map of Maryland.* 14 July 1995.

Rosgen, D.L.: 1994, *Catena,* **22**, 169-199.

Roth, N.E., Southerland, M.T., Chaillou, J.C., Volstad, J.H., Weisberg, S.B., Wilson, H.T., Heimbuch, D.G., and Seibel, J.C.: 1997, *Maryland Biological Stream Survey: Ecological Status of Non-Tidal Streams in Six Basins Sampled in 1995,* Maryland Department of Natural Resources.

Simon, T.P. and Lyons, J.: 1995, *Application of the Index of Biotic Integrity to evaluate water resource integrity in freshwater ecosystems. In:* Davis, W.S. and Simon, T.P., eds. *Biological assessment and criteria: tools for water resource planning and decision making*, Lewis Publishers, Boca Raton, FL.

Strahler, A. N.: 1957, *Trans. of the American Geophysical Union*, **38**, 913-920.

U.S. Environmental Protection Agency (EPA).: 1987, *Handbook of methods for acid deposition studies: Laboratory analysis for surface water chemistry*, EPA-600//4-87/026.

Van Ness, K., Brown, K., Haddaway, M.S., Marshall, D., Jordahl, D.: 1996, *Montgomery County water quality monitoring program stream monitoring protocols*, Montgomery County Department of Environmental Protection.

Volstad, J.H., Southerland, M.T., Weisberg, S.B., Wilson, H.T., Heimbuch, D.G., and Seibel, J.C.: 1996, *Maryland Biological Stream Survey: the 1994 Demonstration Project.* Maryland Department of Natural Resources

Weisberg, S.B., Ranasinghe, J.A., O'Connor, J.S., and Adams, D.A.: 1996 (in review), *A benthic index of biotic integrity (B-IBI) for the New York/New Jersey Harbor*, Ecological Applications.

Weisberg, S.B., Ranasinghe, J.A., Dauer, D.M., Schaffner, L.C., Diaz, R.J. , and Frithsen, J.B.: 1997, *Estuaries*, **20**, 149-158.

White, J.S.: 1996 (draft), *Subregionalization of the Northern Piedmont and Mid-Atlantic Coastal Plain ecoregions within Maryland*, University of Maryland, and Tetra Tech, Inc.

Yoder, C.O. and Rankin, E.T.: 1995 Biological criteria program development and implementation in Ohio. *In:* Davis, W.S. and Simon, T.P., eds. *Biological assessment and criteria: tools for water resource planning and decision making*, Lewis Publishers, Boca Raton, FL.

DIATOM INDICATORS OF STREAM AND WETLAND STRESSORS IN A RISK MANAGEMENT FRAMEWORK

R. JAN STEVENSON

Department of Biology, University of Louisville, Louisville, KY 40292

Abstract. Ecological risk assessment and risk management call for "state-of-the-science" methods and sound scientific assessments of ecosystem health and stressor effects. In this paper recent developments of periphyton indicators of biotic integrity and ecosystem stressors of streams and wetlands are related in a framework of ecological metrics that can be used to quantify risk assessment and risk management options. Many periphyton metrics have been employed in past assessments of water quality and a periphyton indices of biotic integrity has been applied by the state of Kentucky. In addition, the sensitivity of species composition of periphytic diatom assemblages has been shown to respond predictably to ecological stressors so that specific pH, conductivity, and total phosphorus in wetlands and streams can be inferred with weighted average indices. Inference of nutrient conditions by diatom indicators of total phosphorus is shown to have sufficient precision to be a valuable complement to one-time measurement of highly variable total phosphorus in streams. Quantitative indices of sustainability and restorability of ecosystem integrity are proposed, respectively, as the changes in ecological conditions that can occur without significant change in ecological integrity or changes that are necessary to restore ecological integrity.

1. Introduction

Ecological risk assessment is a multistage process in which characteristics of ecosystems are measured, ecological integrity is assessed, and stressors are identified (Figure 1). Many physical, chemical, and biological characteristics of ecosystems can be measured in ecosystem surveys, monitoring programs, or specific ecosystem assessments. In aquatic ecosystems, water and sediment chemistry, temperature, substratum condition, and bank stability, for example, can be used to evaluate abiotic habitat integrity (e.g., Plafkin *et al.*, 1989; Meador *et al.*, 1993b; Klemm and Lazorchak, 1994). In addition, the biomass, species composition, and physiology (e.g., production, respiration, stress responses) of fish, macrobenthic invertebrates, algae, plants, amphibians, and reptiles can be assessed to characterize biotic integrity of aquatic ecosystems (e.g., Hughes, 1993; Cuffney *et al.*, 1993; Meador *et al.*, 1993b; Porter *et al.*, 1993). Measured ecosystem characteristics are then used to determine ecosystem integrity and the stressors affecting integrity.

Many goals of EPA's Office of Research and Development call for relating risk management decisions to risk assessments that are based on scientific evidence of stressor effects and "state-of-the-science" methods and models (USEPA, 1996). Ecological risk assessment is linked to the development of management decisions in a framework that emphasizes the importance of public health risks, statutory and legal considerations, social, political, and economic factors, and risk management options (USEPA, 1996). Identifying the options for risk management calls for a comparative risk approach (USEPA, 1993) in which: exposure and response of ecosystems to different stressors are quantified and

Environmental Monitoring and Assessment **51: 107–118, 1998.**

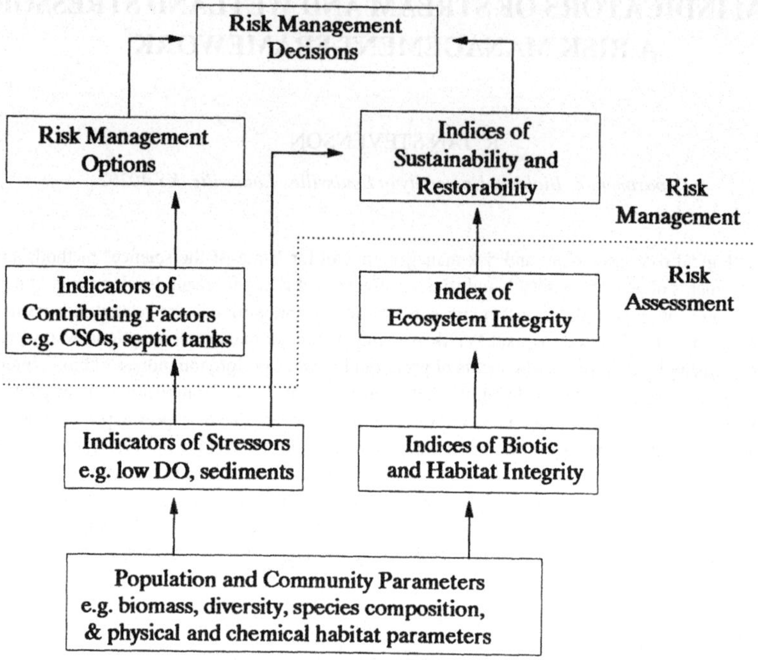

Fig. 1. Organizational framework for ecosystem assessment, a suite of ecological indicators, and risk management.

compared (USEPA, 1992); the human and natural factors that cause the stressors must be identified; the effect of regulating stressors on protecting or restoring ecosystem integrity is determined; and the costs and benefits of management options are evaluated. Thus, better risk management decisions could be made if we had more quantitative metrics for assessing the benefits of regulating stressors (such as indices of sustainability or restorability, Figure 1) and for identifying the importance of different human activities that produce stressors (such as indices of contributing factors, Figure 1).

The objectives of this paper are to review recent developments in diatom assessment of ecological conditions in streams and wetlands, and to propose a framework (Figure 1) for a suite of assessment metrics (i.e., ecological attributes that respond to human influence, *sensu* J. R. Karr, U. Washington, personal communication) and multi-metric indices that will facilitate risk management decisions. In particular, recent developments of quantitative indices of ecological integrity and stressors with diatoms and other algae will be reported. The proposed framework for assessment indices complements the risk assessment and management framework and should facilitate management decisions. The proposed framework calls for use of metrics of ecosystem stressors to infer the human activities that stress ecosystems and metrics of ecosystem sustainability and restorability that relate stressor levels, human activities, and ecosystem integrity. This framework provides objectives for indicator/metric development with which management decisions can be expected to better protect and restore ecosystem services.

2. Assessment of Ecosystem Integrity

Biological integrity can be defined conceptually as "the capability of supporting and maintaining a balanced, integrated, adaptive community of organisms having a species composition, diversity, and functional organization comparable to that of natural habitat of the region" (Karr and Dudley, 1981). In a broader definition, ecosystem integrity could incorporate measurements of water quality (chemical integrity of the ecosystem) and physical habitat integrity. However, biological assessments of aquatic ecosystems and a multi-metric characterization of biological integrity are repeatedly emphasized as the foundation of ecological risk assessment (Angermeier and Karr, 1994; Barbour, 1995; Karr and Chu, in press).

Many indices of biotic integrity (IBI) of stream ecosystems have been developed for fish and invertebrates (Karr 1981; Hilsenhoff, 1988; Plafkin *et al.*, 1989; Lenat, 1993; Kerans and Karr, 1994; DeShon, 1995). These indices are based on readily measurable characteristics of fish and invertebrate assemblages in streams (Figure 1), and they can be modified for use in different regions (Steedman, 1988; Oberdorff and Hughes, 1992; Minns *et al.*, 1994; Barbour *et al.*, 1995; Paller *et al.*, 1996). To address the demand for an integrated assessment of stream ecosystem integrity, the scientists with the Kentucky Division of Water (KYDOW, 1994) use the average of the four indices of habitat integrity and biotic integrity of fish, invertebrate, and algal assemblages. Indices of biotic integrity of wetlands with wetland fauna are currently being developed and tested (Helgen, personal communication).

Periphyton bioassessment is starting to be used or in place in at least seven states (Davis *et al.*, 1996). Periphyton indices of biotic integrity were developed for use in Kentucky by Metzmeier (KYDOW 1994) and similar multi-metric approaches have been used in Montana (Bahls 1993). Metzmeier's index of biotic integrity (KYDOW, 1994) includes taxa richness, Shannon diversity (Shannon, 1948), the pollution tolerance index (Lange-Bertalot, 1979), and proportion of species sensitive to pollution (Table I). These metrics are then translated to a score ranging from 1 to 5, i.e., poor to excellent. The average score for the metrics is Metzmeier's Diatom Bioassessment Index.

Many periphyton metrics could be used in modifications of Metzmeier's biotic index. The proportions of communities composed of filamentous green algae or cyanobacteria could be used as indicators of low biotic integrity because these often form nuisance growths and may not be as readily available to herbivores as diatoms. In contrast, even though red algae are not readily available to herbivores, they indicate high biotic integrity because many are highly sensitive to pollution. Other metrics that could be used in periphyton IBI could relate to species abundances and their autecologies, such as the proportion of communities made up of taxa tolerant to specific pollutants or environmental extremes (high sedimentation, acidity, alkalinity, salinity, heavy metal contamination, BOD, or nutrient concentrations). Algal biomass and mass and chemical ratios (such as chl a:ash-free dry mass(AFDM), phaeophytin:phaeophytin+chl a, TP:AFDM, total nitrogen:AFDM,

Table I.
Metrics and scoring ranges for Metzmeier's diatom bioassessment index (DBI) (Kentucky Division of Water, 1994; Metzmeier, personal communication). Taxa richness is the number of taxa observed in a 500-1000 valve (half of a cell wall) count. Diversity is Shannon (1948) diversity. Diatom tolerance index (DTI) is calculated with the equation å$p_i v_i$/åp_i in which proportional abundances (p_i) of all taxa (for i=1,2,... S, where S is the Taxa Richness) are known and all taxa are assigned a pollution tolerance rank (v_i) from lowest tolerance (4) to highest (1). RA(s) is the sum of the % relative abundance of all sensitive species (for which v_i=4). PSc is the percent similarity of the assessed and reference site assemblages (100-0.5 S|r_{ia}-r_{ir}| where r_{ia} and r_{ir} are the % relative abundances (0-100%) for i=1,2,... S species in the assessed and reference sites, respectively). Scores are averaged for all metrics to calculate the DBI. Scores from 1-2 indicate severe impairment, from 2-3 indicate moderate impairment, from 3-4 indicate good ecological integrity, and from 4-5 indicate excellent ecological integrity.

Score	Taxa Richness	Diversity	DTI	RA(s)	PSc
1	<20	<1.5	1.0-1.5	<0.1	<10
2	20-30	1.5-2.5	1.5-2.0	0.1-1	10-30
3	30-50	2.5-3.5	2.0-2.5	1.0-5.0	30-50
4	50-70	3.5-4.5	2.5-3.0	5.0-20	50-75
5	>70	>4.5	>3.0	20-100	75-100

ash mass:dry mass) could be used as indicators of ecosystem integrity or impairment. Functional measurements of photosynthesis, respiration, and phosphatase activity could also be incorporated into a periphyton IBI. Future research is planned (B. Hill, pers. comm.) to test many of these parameters for sensitivity, accuracy, and precision and for use in a periphyton IBI as was recently done for invertebrate IBI (e.g., Barbour *et al.*, 1992; Kerans and Karr, 1994; Fore *et al.*, 1994).

3. Assessment of Ecosystem Stressors

Most risk assessments call for direct measurement of ecosystem stressors, which are physical, chemical, or biological entities that negatively affect ecosystem integrity (USEPA, 1992). Then correlations between stressors and ecosystem integrity, plus evidence gathered in other research, help us determine which environmental stressor is impairing an ecosystem or making it most susceptible to pollution.

Another approach for assessing ecosystem stressors is to use metrics based on the autecologies of the organisms in the habitat. Changes in species composition of algal assemblages have been used to infer habitat characteristics for a long time (see Lowe 1974 for an older review). Relatively recently, weighted average metrics have been developed that use diatom species autecologies to infer specific environmental conditions, such as pH (ter Braak and van Dam, 1989). These autecological indices of environmental stressors (AIES) are based on the assumption that species have specific optima and tolerances for environmental conditions. These optima and tolerances are evident in a pattern of species relative abundances in different habitats with a broad range of environmental condition. A species will have its maximum relative abundance in habitats with optimum environmental conditions, and relative abundances of the species will decrease in habitats in which the tolerance limits of the species are approached. The power of AIES and this statistical approach for inferring environmental conditions is that autecologies of many species can be used to provide an integrated assessment of specific environmental conditions.

Diatom AIES have been particularly useful in paleoecology for inferring historical changes in lake pH, but they have also been used successfully in studies of salinity, eutrophication, and many other factors (e.g., Kingston and Birks, 1990; Fritz, 1990; Fritz *et al.*, 1991; Sweets, 1992; Kingston *et al.*, 1992; Hall and Smol, 1992; Cumming *et al.*, 1992; Cumming and Smol, 1993; Reavie *et al.*, 1995). These inference models are constructed by sampling diatoms in the surface sediments of a large number of lakes to determine the autecologies of species. Then specific statistical programs are used (WACALIB, Line *et al.*, 1994; CALIBRATE, Juggins and ter Braak, 1992) to develop and test the AIES using bootstrap or jacknifing techniques to eliminate circularity in evaluation of the index (by inferring conditions in a habitat when that sample was left out of the AIES development).

Some doubt existed about whether AIES could be developed in habitats with high spatial and temporal variability, such as streams and wetlands. Using the diatom species composition in streams of the Mid-Atlantic highlands, Pan *et al.* (1996) developed indices of pH and TP. Using the diatom species composition of wetlands in Kentucky and Michigan, AIES for conductivity, pH, and TP were constructed (Pan and Stevenson, 1996; Stevenson *et al.*, submitted). The precision of these indices is usually measured with the correlation coefficient (r^2) and root mean square error (RMSE) of the relationship between the stressor condition (e.g. pH, conductivity, or TP concentration) inferred by the diatom AIES and the observed AIES. When compared to the diatom AIES developed in lakes, the r^2 and RMSE of the AIES in streams and wetlands after jacknifing were similar to those developed in lakes (r^2>0.20).

One reason for using AIES based on diatoms, invertebrates, or other organisms is that they provide corroborative evidence of stressors in ecosystems that complement direct measurement of physical, chemical, and biological conditions. In addition, changes in composition of assemblages occur over extended periods and respond to environmental changes at various spatial scales. Thus AIES should provide a more spatially and temporally integrated assessment of environmental stressors than simple one-time direct measurement of physical or chemical conditions in a habitat. Spatial and temporal integration may be very important in assessment of biologically active stressors (such as nutrients, DO, pH), particularly in shallow water habitats like streams and wetlands, where high concentrations of microorganisms occur.

Evidence to support the hypothesis that AIES provide a more reliable assessment of environmental conditions than simple one-time sampling of physical or chemical conditions has been found in lakes, streams, and wetlands. The RMSE for a diatom AIES of log(TP mg/L) was 0.32 for the Mid-Atlantic Highlands (Pan *et al.*, 1996). Data for TP variation in streams around Louisville, KY, show that the ranges of TP varies between 1.0 and 4.0

log (TP mg/L) during a 9 week period in specific streams. Thus an AIES with an RMSE of 0.32 could provide a valuable assessment of P conditions in streams that would complement direct TP measurement.

A pattern of increased precision in TP indicators from streams to wetlands also indicates that diatom AIES of TP may be better indicators of P conditions in streams and wetlands that affect algae than simple one-time measurement of TP. There are two basic sources of error in an AIES: diatom autecologies vary among habitats or habitat conditions vary so much that they were not measured precisely. Thus, one way to improve precision of an AIES is to increase the precision in estimates of habitat conditions. If we assume that P conditions become more predictable as we go from streams with varying discharge, to wetlands with primarily diurnal variability in P availability, to experiments in which P is manipulated, then we would expect successively more precise diatom AIES of TP from streams, to wetlands, to experiments. Indeed, indices of precision (RMSE and r^2 after jackknifing indicate this pattern of increasing precision with r^2=0.27 and RMSE=0.32 in streams (Pan *et al.*, 1996), r^2=0.28-44 and RMSE=0.12-0.40 in wetlands (Pan and Stevenson, 1996; Stevenson *et al.*, submitted; Slate and Stevenson, unpublished data), to r^2=0.81 and RMSE=0.13 in experiments (Slate and Stevenson, unpublished data).

4. A Metric Framework for Risk Assessment and Management

A framework of ecosystem metrics is proposed that should facilitate ecological risk assessment and risk management (Figure 1). This framework builds on the basic risk assessment framework (USEPA, 1992) and incorporates the risk management framework (USEPA, 1996). The proposed framework synthesizes the concepts that have been developed in other indicator and comparative risk assessment frameworks (Paulsen *et al.*, 1991; USEPA, 1992, 1993, 1996). This framework (Figure 1) emphasizes a distinction between stressors that affect ecological integrity and the human activities that indirectly affect ecosystem integrity by causing the stressors (contributing factors, KYNREPC, 1996). The proposed framework also includes quantitative linkage between effects of stressors and ecological integrity in indices of sustainability and recoverability.

The terminology of the EMAP Surface Waters indicator model with response, exposure/habitat, and stressor indicators (Paulsen *et al.*, 1991) has been modified in the proposed framework to distinguish factors that have direct and indirect effects and to follow the terminology that is consistent with risk assessment (USEPA, 1992) and comparative risk assessment (KYNREPC, 1996) definitions. This is a framework for making information as valuable as possible; the choice of terms used is not nearly as important as what we do with the data.

The proposed framework has two categories of indicators that use the physical, chemical, and biological information attained in ecosystem sampling. One category uses this basic information to determine habitat, biotic, and ecosystem integrity and the other category uses the same information to infer the stressors that are affecting the organisms in a habitat. Development of some multimetric indices of ecological integrity and AIES were discussed above.

The framework also proposes that the information obtained during assessment can be used to determine the human activities that cause stressors. Activities that cause the stressors that directly affect ecosystems are referred to as contributing factors (KYNREPC, 1996). Knowledge of the contributing factors causing stressors provides a set of risk management options, for example, eliminating combined sewer overflows to protect ecosystems that are susceptible to degradation or to restore ecosystems that are impaired.

Distinguishing contributing factors from stressors is important because contributing factors are human activities that can be regulated and they indirectly affect ecosystem structure and function. Distinguishing contributing factors from stressors is important to reduce ambiguity of management options when stressors have multiple contributing factors. A given stressor may have many contributing factors and a given contributing factor may produce many stressors. For example, agricultural development of watersheds may affect sedimentation in streams and pesticide and nutrient loading. Nutrient loading, however, may arise from many alterations of a watershed, such as fertilization of crops or lawns or forestry.

To further reduce ambiguity of management options when stressors have multiple contributing factors, biological indicators of contributing factors could be developed to complement direct measurement of their presence. Changes in invertebrate communities have been used recently to infer the human activities that impaired ecosystem integrity. For example, Yoder and Ranking (1995) developed "biological response signatures" with composition of invertebrate communities to determine whether toxics, combined sewer overflows, or channelization was affecting invertebrate communities. Richards *et al.* (1996) related changes in invertebrate functional groups to changes in physical habitat variables and geologic and anthropogenic characteristics of watersheds. More complex multimetric indices of contributing factors could be developed with suites of physical, chemical, periphyton, invertebrate, and fish metrics.

Sustainability and restorability could be quantified and developed into a metric that would quantitatively link assessment of ecosystem integrity and stressors. These metrics could also help in comparative risk assessment and in risk management decisions. If we knew the relationship between ecological integrity and environmental stressors, we could define sustainability, quantitatively, as the difference between present environmental conditions in a habitat and the environmental criterion (Stevenson 1997, Figure 2), which was presumably set above or at least at the environmental conditions that are known to lower ecosystem integrity to unacceptable levels. Unimpaired ecosystems in which environmental conditions were close to criteria would have lower sustainability than ecosystems in which conditions were farther from criteria.

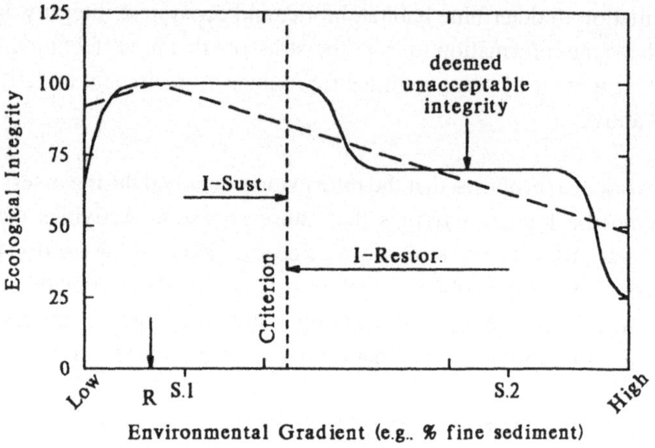

Fig. 2. Hypothetical changes in ecological integrity (where the maximum of 100 is based on condition in reference sites, indicated by R) along environmental gradients are plotted as solid and long-dash lines. S.1 and S.2 indicate the environmental conditions of two assessed sites. R marks environmental conditions at a reference site. Length of the horizontal solid arrows represent the magnitude of differences in environmental conditions that would be used to quantify indices of sustainability and restorability.

Restorability could, in a similar manner, be related to the difference between conditions in an impaired ecosystem and an environmental criterion (Stevenson, 1997). However restorability would be inversely related to the difference in environmental conditions and an environmental criterion (versus the direct relationship between that difference and sustainability). Thus impaired ecosystems with environmental conditions that were close to environmental criteria would have higher restorability than impaired ecosystems with conditions farther from the criterion. These indices of sustainability and restorability are based on responses of single or multiple elements of ecosystem integrity to specific environmental stressors.

The stressors of an ecosystems are usually complex and are usually inferred by correspondence between changes in abiotic and biotic characteristics that have been related to previous research. Comparison of sustainability and restorability metrics for specific stressors could quantify the linkage between assessments of stressors and ecological integrity and may help determine which stressors most regulate ecosystem integrity (Figure 1). If we knew, for example, that nuisance growths of algae were caused by high light and high nutrients, we could look at the cost of regulating each of these stressors to restore periphyton biotic integrity to acceptable levels. We would probably find that planting trees along the stream that reduced light to intensities that limit nuisance algal growths would be much less expensive than reestablishment of forests throughout a watershed to reduce nutrients to concentrations that limit nuisance growths.

Often, however, many stressors have both individual and interactive effects on ecosystem integrity and solving environmental problems is not a matter of remedying only one stressor. Multivariate indices, such as scores along ordination axes, could be used to characterize and quantify the levels of more complex ecological stresses. In addition, direct measurement of ecological stressors as well as algal or invertebrate-based AIES could be used to quantify stressor levels and assess stressor effects on all forms of biotic integrity.

Sustainability and restorability metrics could provide a quantitative, scientific foundation to comparative risk assessment (USEPA, 1993). Although these metrics have not been used in the past, the information to calculate these metrics is often collected and application of the metrics in risk management should be investigated.

5. Conclusions

Metrics of periphyton integrity and algae-based indices of ecosystem stressors could be valuable elements in a set of many environmental metrics that facilitate risk assessment and risk management. Future research should be directed to testing the accuracy and precision of biological indicators of ecological integrity and of stressors and should better assess the specific effects of stressors along environmental gradients. Better assessments of specific effects of stressors along environmental gradients will help to establish environmental criteria that optimize human use of watersheds and protection of stream, lake, and wetland integrity. Future research should develop metrics with regard to needs of risk assessment and risk management so that these metrics are as useful and effective as possible.

Acknowledgments

I would like to thank Yangdong Pan, Roger Sweets, Jennifer Slate, and many of my graduate students for the assistance with the streams and wetlands projects. Don Charles, Bill Fisher, Brian Hill, and Jim Karr provided thoughtful reviews of the paper. This research was funded with a grant to Curtis J. Richardson (Duke University) from the Everglades Agricultural Area Environmental Protection District, a contract through Gary Collins and Brian Hill and the USEPA in Cincinnati, Ohio, a USEPA Exploratory Research Grant (R 821627), and an NSF/Water and Watersheds Grant (R 824783).

References

Angermeier, P. L. and J. R. Karr. 1994. "Biological integrity versus biological diversity as policy directives", *BioScience* **44**, 690-697.

Bahls, L. L. 1993. Periphyton Bioassessment Methods for Montana Streams. Water Quality Bureau, Department of Health and Environmental Sciences, Helena, Montana, USA.

Barbour, M. T., J. B. Stribling, and J. R. Karr. 1995. "Multimetric approach for establishing biocriteria and measuring biological condition", in W. S. Davis and T. P. Simon, ed., *Biological Assessment and Criteria: Tools for Water Resource Planning and Decision Making*, Lewis Publishers, Boca Raton, Florida, USA. pp. 63-77.

Barbour, M. T., J. L. Plafkin, B. P. Bradley, C. G. Graves, and R. W. Wisseman. 1992. "Evaluation of EPA's rapid bioassessment benthic metrics: metric redundancy and variability among reference stream sites", *Environmental Toxicology and Chemistry* **11**, 437-449.

Cuffney, T. F., M. E. Gurtz, and M. R. Meador. 1993. "Methods for collecting benthic invertebrate samples as part of the National Water-Quality Assessment Program" , U. S. Geological Survey, Report 93-406. Raleigh, North Carolina, USA.

Cumming, B. F. and J. P. Smol 1993. "Development of diatom-based salinity models for paleoclimatic research from lakes in British Columbia (Canada)", *Hydrobiologia* **269/270**, 575-586.

Cumming, B. F., J. P. Smol, J. C. Kingston, D. F. Charles, H. J. B. Birks, K. E. Camburn, S. S. Dixit, A. J. Uutala, and A. R. Selle. 1992. "How much acidification has occurred in Adirondack region lakes (New York, USA) since pre-industrial times?", *Canadian Journal of Fisheries and Aquatic Sciences* **49**, 128-141.

Davis, W. S., B. D. Snyder, J. B. Stribling, and C. Stroughton. 1996. "Summary of State Biological Assessment Programs for Streams and Wadeable Rivers", EPA 230-R-96-007. U. S. Environmental Protection Agency; Office of Policy, Planning, and Evaluation; Washington, D.C.

DeShon, J. E. 1995. "Development and application of the invertebrate community index (ICI)", in W. S. Davis and T. P. Simon, ed., *Biological Assessment and Criteria: Tools for Water Resource Planning and Decision Making*, Lewis Publishers, Boca Raton, Florida, USA. pp. 217-244.

Fore, L. S., and J. R. Karr, and L. L. Conquest. 1994. "Statistical properties of an index of biological integrity used to evaluate water resources", *Canadian Journal of Fisheries and Aquatic Sciences* **51**, 1077-1087.

Fritz, S. C. 1990. "Twentieth-century salinity and water-level fluctuations in Devils Lake, North Dakota, test of a diatom-based transfer function", *Limnology and Oceanography* **35**, 1771-1781.

Fritz, S. C., S. Juggins, R. W. Battarbee, and D. R. Engstrum. 1991. "Reconstruction of past changes in salinity and climate using a diatom-based transfer function", *Nature* **352**, 706-708.

Hall, R. I. And J. P. Smol. 1992. "A weighted-averaging regression and calibration model for inferring total phosphorus from diatoms from British Columbia (Canada) lakes", *Freshwater Biology* **27**, 417-437.

Hilsenhoff, W. L. 1988. "Rapid field assessment of organic pollution with a family level biotic index", *Journal of the North American Benthological Society* **7**, 65-68.

Hughes, R. M. (ed). 1993. Stream indicators and design workshop. EPA/600/R-93/138. U. S. Environmental Protection Agency, Corvallis, Oregon. 84pp.

Juggins, S. and C. J. F. ter Braak. 1992. CALIBRATE - a program for species-environment calibration by [weighted averaging] partial least squares regression. Environmental Change Research Center, University College, London.

Karr, J. R. 1981. "Assessment of biotic integrity using fish communities", *Fisheries* **6**, 21-27.

Karr, J. R. and E. W. Chu. In press. "Biological monitoring: essential foundation for ecological risk assessment", *Human and Ecological Risk Assessment* **3**:

Karr, J. R. and D. R. Dudley. 1981. "Ecological perspective on water quality goals", *Environmental Management* **5**, 55-68.

Kerans, B. L. and J. R. Karr. 1994. "A benthic index of biotic integrity (B-IBI) for rivers of the Tennessee Valley", *Ecological Applications* **4**, 768-785.

Kentucky Division of Water (KYDOW). 1994. "Pond Creek Drainage (Ohio River - Oldham County) Biological and Water Quality Investigation" Technical Report No: 51, Frankfort, Kentucky.

Kentucky Natural Resources and Environmental Protection Cabinet. (KYNREPC). 1997. "Kentucky Outlook 2000: A Strategy for Kentucky's Third Century", Executive Summary and Guide to the Technical Committee Reports. Frankfort, Kentucky.

Kingston, J. C. and H. J. B. Birks. 1990. "Dissolved organic carbon reconstructions from diatom assemblages in PIRLA project lakes, North America", Philosophical Transactions of the Royal Society of London, *B* 327:279-288.

Kingston, J. C., H. J. B. Birks, A. J. Uutala, B. F. Cumming, and J. P. Smol. 1992. "Assessing trends in fishery resources and lake water aluminum from paleolimnological analyses of siliceous algae", *Canadian Journal of Fisheries and Aquatic Sciences* **49**, 127-138.

Klemm, D. J. and J. M. Lazorchak, eds. 1994. "Pilot field operation and methods manual for streams", EPA/620/R-94/004. Environmental Monitoring Systems Lab. Office of Research and Development. U. S. Environmental Protection Agency. Cincinnati, Ohio.

Lange-Bertalot, H. 1979. "Pollution tolerance of diatoms as a criterion for water quality estimation", *Nova Hedwigia* **64**, 285-304.

Lenat, D. R. 1993. "A biotic index for the southeastern United States: derivation and list of tolerance values, with criteria for assigning water quality ratings", *Journal of the North American Benthological Society* **12**, 279-290.

Line, J. M., C. J. F. ter Braak, and H. J. B. Birks. 1994. "WACALIB version 3.3 - a computer program to reconstruct envrionmental variables from fossil assemblages by weighted averaging and to derive sample-specific errors of prediction", *Journal of Paleolimnology* **10**, 147-152.

Lowe, R. L. 1974. "Environmental Requirements and Pollution Tolerance of Freshwater Diatoms", EPA-670/4-74-005. U.S. Environmental Protection Agency, Cincinnati, Ohio, USA.

Meader, M. R., C. R. Hupp, T. F. Cuffney, and M. E. Gurtz. 1993a. "Methods for characterizing stream habitat as part of the National Water-Quality Assessment Program", U. S. Geological Survey, Report 93-408. Raleigh, North Carolina, USA.

Meador, M. R., T. F. Cuffney, and M. E. Gurtz. 1993b. "Methods for sampling fish communities as part of the National Water-Quality Assessment Program", U. S. Geological Survey, Report 93-104. Raleigh, North Carolina, USA.

Minns, C. K., v. w. Cairns, R. G. Randall, and J. E. Moore. 1994. "An Index of Biotic Integrity (IBI) for fish assemblages in the littoral zone of Great Lakes' Areas of Concern", *Canadian Journal of Fisheries and Aquatic Sciences* **41**, 1844-1822.

Oberdorff, T. and R. M. Hughes. 1992. "Modifications of an index of biotic integrity based on fish assembalges to characterize rivers of the Seine Normandie Basin, France", *Hydrobiologia* **228**, 117-130.

Paller, M. H., M. J. M. Reichert, and J. M. Dean. 1996. "Use of fish communities to assess environmental impacts in South Carolina Coastal Plain streams", *Transactions of the American Fisheries Society* **125**, 633-644.

Pan, Y. and R. J. Stevenson. 1996." Gradient analysis of diatom assemblages in western Kentucky wetlands", *Journal of Phycology* **32**, 222-232.

Pan, Y., R. J. Stevenson, B. H. Hill, A. T. Herlihy, and C. B. Collins. 1996. "Using diatoms as indicators of ecological conditions in lotic systems: a regional assessment", *Journal of the North American Benthological Society* **15**, 481-494.

Paulsen, S. G., D. P. Larsen, P. R. Kaufmann, T. R. Whittier, J. R. Baker, D. V. Peck, J. McGue, R. M. Hughes, D. McMullen, D. Stevens, J. L. Stoddard, J. Lazorchak, W. Kinney, A. R. Selle, and R. Hjort. 1991. "Environmental Monitoring and Assessment Program: EMAP-Surface waters monitoring and research strategy - fiscal year 1991", EPA/600/3-91/022. U. S. Environmental Protection Agency, Office of Research and Development, Washington, D.C.

Porter, S. D., T. F. Cuffney, M. E. Gurtz, and M. R. Meador. 1993. "Methods for Collecting Algal Samples as Part of the National Water-Quality Assessment Program", U. S. Geological Survey, Report 93-409. Raleigh, North Carolina, USA.

Reavie, E. D., R. I. Hall, and J. P. Smol. 1995. "An expanded weighted-averaging model for inferring past total phosphorus concentrations from diatom assemblages in eutrophic British Columbia (Canada) lakes", *Journal of Paleolimnology* **14**, 49-67.

Richards, C., L. B. Johnson, and G. E. Host. 1996." Landscape-scale influences on stream habitats and biota", *Canadian Journal of Fisheries and Aquatic Sciences* **53** (Suppl 1), 295-311.

Steinberg, C., H. Hartmann, K. Arzet, and D. Krause-Dellin. 1988. "Paleoindication of acidification in Kleiner Arbersee (Federal Republic of Germany, Bavarian Forest) by chydorids, chrysophytes, and diatoms", *Journal of Paleolimnology* **1**, 149-57.

Steedman, R. J. 1988. "Modification and assessment of an index of biotic integrity to quantify stream quality in Southern Ontario", *Canadian Journal of Fisheries and Aquatic Sciences* **45**, 492-501.

Stevenson, R. J. 1997 (in press). "Resource thresholds and stream ecosystem sustainability", *J. N. Am. Benthol. Soc.*

Stevenson, R. J., P. R. Sweets, and Y. Pan. submitted. "Algal community patterns in wetlands and their use as indicators of ecological conditions", Proceedings of INTECOL's Vth International Wetland Conference.

Sweets, P. R. 1992. "Diatom paleolimnological evidence for lake acidification in the Trial Ridge region of Florida", *Water, Air and Soil Pollution* **65**, 43-57.

ter Braak, C. J. F. and H. van Dam. 1989. "Inferring pH from diatoms: a comparison of old and new calibration methods", *Hydrobiologia* **178**, 209-223.

United States Environmental Protection Agency. (USEPA). 1992. "Framework for Ecological Risk Assessment", EPA 630/R-92/001, Washington, D. C.

United States Environmental Protection Agency. (USEPA). 1993. "A Guidebook to Comparing Risks and Setting Environmental Priorities", EPA 230-B-93-003, Washington, D. C.

United States Environmental Protection Agency (EPA). (USEPA). 1996. "Strategic plan for the Office of Research and Development", EPA/600/R-96/059. Washington, D. C.

Yoder, C. O. and E. T. Rankin. 1995. "Biological response signatures and the area of degradation value: new tools for interpreting multimetric data", in W. S. Davis and T. P. Simon, ed., *Biological Assessment and Criteria: Tools for Water Resource Planning and Decision Making*, Lewis Publishers, Boca Raton, Florida, USA. pp. 263-286.

THE OCCURRENCE AND IMPACT OF SEDIMENTATION IN CENTRAL PENNSYLVANIA WETLANDS

DENICE H. WARDROP and ROBERT P. BROOKS

Penn State Cooperative Wetlands Center, The Pennsylvania State University, University Park, PA 16802

Abstract: Sedimentation rates and deposited sediment characteristics in twenty-five wetlands in central Pennsylvania were measured during the period Fall 1994 to Fall 1995. Wetlands were located primarily in five watersheds, and represented a variety of hydrogeomorphic (HGM) subclasses and surrounding land use. Sedimentation rates were measured via the placement of 135 Plexiglas disks. Annual organic and inorganic loadings were determined. Sedimentation rates ranged from 0 to 8 cm/year, with sedimentation rates significantly correlated with surrounding land use and HGM subclass. Overall mean mineral and organic accretion rates were 778 g m^2 yr^{-1} (+/- 1417) and 550 g m^2 yr^{-1} (+/- 589), respectively. Mean mineral and organic accretion rates were significantly different by HGM subclass. The highest mineral accretion rates were for headwater floodplains, followed by impoundments, riparian depressions, mainstem floodplains, and slopes. The highest organic accretion rates were for riparian depressions, followed by impoundments, slopes, headwater floodplains, and mainstem floodplains. The potential effects of landscape disturbance on these sedimentation rates was also investigated, in order to develop a conceptual model to predict sedimentation rates for a given wetland in a variety of landscape settings. Different HGM subclasses exhibited significantly different mineral and organic accumulation rates, and varied in their responses to landscape disturbance and spatial variability in sedimentation patterns. Characterization of wetland plant communities in these same wetlands showed clear associations between individual plant species and ability to tolerate sediment. Species were categorized as very tolerant, moderately tolerant, slightly tolerant, and intolerant based on their association with environments of varying sedimentation magnitude. In general, species that were categorized as very tolerant or moderately tolerant increased their percent cover (dominance) over the sedimentation gradient. These observations were supported by greenhouse germination trials of eight species of wetland plants under a variety of sediment depths, ranging from 0 to 2 cm.

1. Introduction

The most obvious change in landscapes surrounding wetlands in central Pennsylvania has been land use conversion with accompanying landscape fragmentation (Brooks *et al.*, 1996). While many studies have investigated the biological impacts of landscape change and fragmentation (Robinson *et al.*, 1992), few have documented the change in physical flows or fluxes between landscape elements that may be the initial forcing function of a biological response. Even less frequent is any study of the impact of landscape change on wetland ecosystems, in particular. One potential flux between landscape elements (including wetlands) is sediment transported by water. The ability of flowing water to transport sediment is dependent upon both its velocity and the size of the particles being transported. Changes in land use and fragmentation of the landscape can alter the velocity of water, whether in the form of streams or overland flow, by changing the slope or gradient and the roughness encountered by the flow. Such landscape changes may also alter the size and/or amount of available sediment. Elevated sedimentation can be associated with such activities as agriculture, hydrologic modification, urban runoff, unsatisfactory waste water treatment, deposition of fill material, and erosion from mining

Environmental Monitoring and Assessment **51:** 119–130, 1998.

and construction sites (Adamus and Brandt, 1990). Sedimentation can, thus, be viewed as a potential marker of a wide range of physical disturbances.

This study was an attempt to trace the evolution of the impact of one selected stressor (sedimentation) from its origin, through the mechanisms that ultimately create an impact on the herbaceous vegetation in a wetland, taking into consideration the type and position of the wetland itself. The objectives of the study were as follows:

- Characterize sedimentation, as it occurs in a variety of wetland settings, and the associated plant communities of the individual wetlands.
- Use the relationship between sedimentation and associated plant communities to build stressor-impact curves.
- Develop a predictive model of amount of sedimentation to an individual wetland, utilizing landscape-level indicators of potential disturbance (i.e., land use, agricultural practices).

These objectives were articulated as the following specific hypotheses:

1. Discrete plant communities or individual species differ in their ability to occupy space in environments with increasing levels of sediment accumulation.

2. The ability to occupy space in environments with increasing levels of sediment accumulation varies according to the hydrologic regime and geomorphic setting of a wetland, as described by hydrogeomorphic (HGM) subclass.

3. The ability to occupy space in high sediment accumulation environments is dependent on differential abilities to germinate under high loads of sediment.

4. Sedimentation loads experienced by an individual wetland can be predicted in a relative sense using landscape-level indicators.

2. Methods

2.1 Site Selection

A total of 28 sites in central Pennsylvania were selected for investigation, and chosen to represent a range of wetland types and surrounding land uses. These sites were a subset of 51 reference wetlands studied by Brooks (1996). All but two of the sites were located in five watersheds. General land use characteristics in the five watersheds are as follows:

Watershed	Area (km^2)	Level and Type of Disturbance
Spring Creek	370	High; Urban
Shavers Creek	163	Intermediate; Agriculture
Little Fishing Creek	109	High; Agriculture (in lower reaches)
White Deer Creek	117	Low; Recreational
Bald Eagle Creek	2137	Intermediate; Agriculture

The first four watersheds are located in the Ridge and Valley physiographic province, and Bald Eagle watershed is located between the eastern edge of the Allegheny Plateau (Allegheny Front) and the Ridge and Valley province. Wetlands were selected within the watershed to represent four classes of landscape position and four wetland types. Landscape positions were designated as headwater (contributing to, or located on, a first or second order stream), mainstem (contributing to, or located on, a stream of order greater than two), riparian (within the riparian corridor), and isolated. Primary wetland types consisted of those described by Brinson (1993): depressional, riverine, slope, flats, and fringing.

A new hydrogeomorphic (HGM) classification key for Pennsylvania was used to classify potential sites, and led to the recognition of six primary HGM subclasses in central Pennsylvania; riparian depression, headwater and mainstem floodplains (riverine), slope, and human and beaver impoundments (fringing). Site maps and descriptions are located in Brooks (1996). Initially, 13 sites were selected at random from National Wetlands Inventory (NWI) maps. The remaining 15 sites were chosen specifically to complete experimental categories, such as hydrogeomorphic type, disturbance category, and watershed. The resulting study sites represented the six HGM subclasses fairly evenly, with the exception of impoundments. Eight of the sites are classified as riparian depressions, representing pristine and moderate land use disturbance and located in four watersheds. Seven slope wetlands are included in the study, representing pristine, moderate, and severe surrounding land use disturbances, and covering four watersheds. Ten of the study sites are classified as riverine (four headwater and six mainstem sites), representing the full range of landscape disturbances and located in five watersheds. The least represented HGM category, impoundments, is represented by four sites (three human impoundments, one beaver impoundment) with pristine and moderate landscape disturbance, and located in three watersheds.

2.2 Sedimentation Measures

Annual sedimentation rates were measured via the installation and subsequent recovery of sediment disks (Kleiss, 1993). Sediment disks were constructed of a 0.5 cm thick plexiglas disk, 20 cm in diameter, sanded on one side to create a rough surface that was able to retain sediment particles. To install a disk, a 30-cm long steel rod, 0.6 cm in diameter, was pushed into the surface with approximately 2-5 cm remaining above the surface. The disk was placed on the rod, seated into the sediment until the top of the disk was flush with the wetland surface, and secured on both sides with wing nuts. The disk was then stable on the wetland surface and was resistant to movement.

Sediment disks were installed at all sites primarily during the period May-August 1994, but disks at three sites were installed in October 1994. The number of sediment disks installed per wetland was a function of wetland size and expected local variability in sedimentation patterns. Disks were placed on sampling grid points so that data obtained from a given sediment disk could be directly related to the plant community at that location. A total of 139 disks were installed between May and October 1994. It should be noted that the period of May 1994-August 1995 represented an historically dry period. Precipitation for the period was 29 inches, compared to an overall annual average of 38 inches for the State College area (personal communication, Accu-Weather, Inc., State College, PA).

Disks and associated sediment were collected during summer and fall of 1995. Disks were collected roughly in the order that they were installed, to ensure that all disks were in place for a period of one year. All fractions were placed into labeled, resealable plastic bags and transported to the lab. All coarse vegetative material was dried and weighed, and contributing types were recorded (e.g., leaves, twigs, senescent herbaceous, etc.). The remaining material was air-dried and weighed. Organic fraction was determined by ashing oven-dried samples to a constant weight at 450 degrees Celsius in a muffle furnace (Storer, 1984).

2.3 Plant Community Measures

Plant communities were characterized according to a standard protocol (Brooks 1996). Each study site was first delineated, a baseline was established, and transects were established at uniform intervals (with a random starting point) along the baseline (typically 20 m apart, or 40 m on larger sites) to create a sampling grid. Sampling plots were located at grid points along each transect. For estimation of percent cover, a 2 m x 0.5 m plot was placed immediately adjacent to the grid point. Percent cover was visually estimated to the nearest 5% for all species with greater than 5% cover in the plot. Species lists were constructed by recording the presence of all species located in a circular plot 6 m in diameter, centered on the grid point. All plant species were identified to the species level when possible (e.g., Fassett, 1957, Newcomb, 1977, Brown, 1979), but only herbaceous information was used in this study. Because plant sampling occurred during the period June through August 1994, with some sites being characterized early in the growing season, some grass and sedge species could only be identified to the genus level due to lack of flowering bodies or other identification characteristics. Plots with sediment disks were re-sampled when the sediment disks were retrieved during fall of 1995, and any species identifications made in the previous sampling were verified.

For analyses of functional group coverage differences, plants were initially assigned to the seven major functional groups of Boutin and Keddy (1994); ruderal (subclasses of obligate annuals and facultative annuals), interstitial perennials (subclasses of reed, clonal, and tussock), and matrix (subclasses of clonal dominants and clonal stress-tolerators). Of the 112 dominant species found on the 25 sites, 17 (15%) were classified as ruderals. Interstitial species accounted for 70 (63%) of the dominant species, and matrix species accounted for 25 (22%) of the dominant species. An eighth functional group of vines was added.

2.4 Germination Trial

An extensive screening procedure was used to initially select species for germination trials, based upon the occurrence of species in various sedimentation environments and life history strategies. The following species were selected: *Carex vulpinoidea, Impatiens capensis, Hypericum pyramidatum, Typha latifolia, Carex stricta, Leersia oryzoides, Polygonum persicaria,* and *Eupatorium perfoliatum.* Seeds were purchased from fall stock from commercial vendors.

In order to incorporate co-occurring stressors in the experimental design, two moisture regimes and three chill treatments were included. The chill treatments were selected to represent a range of stratification conditions in wetlands of varying HGM subclass in Pennsylvania: wet chill, moist chill, dry chill, and dry no chill (control). Stratification is necessary for breaking of dormancy in a wide variety of wetland species (Garbisch and McIninch, 1992). Fifty seeds of each species were counted out as a sample, and mixed with approximately 60 cm^3 of potting soil. For wet chill, the seed/soil mixture was completely submerged. For moist chill, the seed/soil mixture was wetted until moist. All three chill treatments were stored separately in closed containers at 4 degrees Celsius on 8 March 1996. All chill samples remained in refrigeration for a period of 6.5 weeks.

Four sediment depths, or loadings, were selected for testing. These were based on mid-year measurements, and represented the full range of conditions encountered in the field. The sediment depths selected were 0 cm, 0.5 cm, 1.0 cm, and 1.5 cm, which represented mineral loadings of 0 g m^{-2}, 5,833 g m^{-2}, 11,806 g m^{-2}, and 17,847 g m^{-2}, respectively. Play sand was selected as the sediment because it was consistent in size, was of intermediate size as compared to the range of size classes present in wetland soils (Bishel-Machung, 1996), and no additional seeds would be introduced into the experiment. Two watering regimes were selected to represent variations in wetness during germination: saturated and moist. Germination trials were structured as a 4 x 4 x 2 design for each species, consisting of four sediment depths (0, 0.5, 10.0, and 1.5 cm), four storage regimes (dry cold, moist cold, wet cold, and dry no chill), and two watering regimes (moist and wet).

Each seed/soil mixture was spread out evenly on the soil surface of a 12 cm x 12 cm flat. Each flat had been prepared by filling it with approximately 720 cm^3 of commercial potting soil to a depth of 5 cm. Flats were placed in a randomized order on a single bench in the School of Forest Resources greenhouse, Pennsylvania State University, University Park; this experimental design was repeated on two more benches, resulting in triplicates of all treatments, with each treatment containing 50 seeds. A total of 3,600 seeds of each species were tested in a single trial. Flats were watered every 1, 2, or 3 days to achieve the desired moisture level. Flats were subjected to ambient light levels. Air temperatures ranged from 20 degrees to 25 degrees Celsius. Seedling emergence was monitored at 1-2 day intervals, and seedlings were formally counted at 3, 6, and 9 weeks after planting.

2.5 LANDSCAPE MEASURES

A 1-km radius circle around each site was selected to provide the necessary landscape information. This size area was selected because it readily encompassed contributing watershed area for most wetlands, data were relatively easy to obtain, and a 1-km circle matches many biological sampling regimes. The entire area of the circle was assumed to be the contributing watershed. While this assumption held true for the sites in this study, it should be recognized that this assumption may be inappropriate in other study areas. The landscape within a 1-km radius circle of each wetland was characterized based on interpretation of 1:40,000, color infrared, aerial photographs. Areas were mapped using seven cover type categories: developed, agricultural, barren, shrub, forest, wetland, and open water. Data were digitized by personnel at the Penn State Office of Remote Sensing of Earth Resources (ORSER). A modified version of the SPAN (Spatial Analysis) computer program was used to process and quantify a number of landscape characteristics: diversity, dominance, contagion, length of forest/nonforest edge, and percent cover, average patch size, and number of patches for each cover type category. Cluster analysis was used to group landscapes into four classes of land use/fragmentation.

3. Results and Discussion

The first hypothesis involved testing a number of plant community descriptors over a range of scales, from general community measures to individual species. General measures of community response, such as richness, diversity, and evenness, when taken over all wetland types, did little to establish general patterns of response to sedimentation. The functional groups used in this study did not provide the necessary resolution for establishment between general life history strategies and sedimentation environments. The basis of the functional groups was not appropriate given the stressor of interest. The use of functional groups based on traits that exemplify a species' germination capabilities as well as its ability for clonal growth is recommended in the future. Responses did occur, however, at the level of individual species, and species were categorized as very tolerant, moderately tolerant, slightly tolerant, and intolerant based on their association with environments of varying sedimentation magnitude. Inspection of mean percent covers of individual species across the sedimentation gradient clearly indicated that, for some species, hypothesis 1 was not rejected, and the differential in the ability to occupy space over the sedimentation gradient is demonstrated (Table 1). In general, species that were categorized as sediment tolerant or moderately intolerant increased their percent cover (dominance) over the sedimentation gradient. Mean percent cover, when plotted versus sediment accumulation, provides a stressor-impact curve for an individual species (Figure 1).

Table I
Mean percent cover of selected wetland plant species in central Pennsylvania wetlands over a sedimentation gradient

	Relative Annual Sedimentation Rate			
Species	**Negligible**	**Low**	**Medium**	**High**
	n=36	n=44	n=18	n=4
Very Tolerant				
Leersia oryzoides	90	39	0	65
Dipsacus sylvestris	19	0	0	60
Impatiens capensis	12	24	55	60
Polygonum sagittatum	5	0	0	30
Dulichium arundinaceum	5	10	0	30
Carex lurida	0	20	0	20
Aster novae-angliae	5	0	0	10
Solidago patula	23	0	20	10
Solidago uliginosa	22	10	20	10
Moderately Tolerant				
Phalaris arundinacea	46	54	67	0
Thelypteris noveboracensis	15	10	60	0
Carex emoryi	0	0	50	0
Carex retroflexa	0	0	50	0
Carex stricta	0	60	40	0
Symplocarpus foetidus	25	40	40	0
Brachyelytrum erectum	0	43	35	0
Triadenum virginicum	10	0	30	0
Carex prasina	10	20	30	0
Solidago sp.	21	10	30	0
Carex vulpinoidea	17	20	25	0
Carex folliculata	20	10	20	0
Slightly Tolerant				
Juncus canadensis	30	48	0	0
Euthamia graminifolia	65	23	0	0
Sagittaria latifolia	0	18	0	0
Eleocharis sp.	0	15	0	0
Verbena hastata	0	12	0	0
Equisetum arvense	23	12	0	0
Carex intumescens	0	10	0	0
Solidago canadensis	23	10	0	0
Urtica dioica	77	10	0	0
Intolerant				
Poa pratensis	90	0	0	0
Lysimachia nummularia	13	0	0	0
Mentha arvensis	10	0	0	0
Aster vimineus	7	0	0	0
Cirsium arvense	5	0	0	0
Asclepias syriaca	5	0	0	0

The tolerance for sedimentation in most species, however, is dependent upon the overlay of sedimentation onto other possible co-occurring stressors, such as wetting and drying cycles. HGM subclass was used as an indicator of some of these co-occurring stressors (i.e., parent material and wetness). It was expected that the ability of a plant community or individual species to occupy space along a gradient of increasing sediment accumulation would be different between HGM subclasses. This was borne out in shifts of some species between sediment tolerance groups between wetlands of different HGM types, or drastic reductions in the mean percent cover they demonstrated between HGM subclasses. An individual species' stressor-impact curve thus varied with HGM subclass. In this context, HGM classification did much to establish the range of conditions of other co-occurring stressors, and thus, provided a constrained condition for examining the effects of sedimentation. The strong presence of an HGM gradient in these results indicated that the underlying mechanism of differential plant response to sedimentation is based on some physical characteristic embodied in the HGM classification.

This study required the performance of the germination trial, since seed germination requirements of only a few of the species studied here have been published (e.g., Kadlec and Wentz, 1974, Grime 1981). Germination trials of *Carex vulpinoidea*, *Eupatorium maculatum*, *Hypericum pyramidatum*, *Polygonum persicaria*, and *Typha latifolia* showed a significant decrease in germination with as little as 0.5 cm of sediment. Only *Leersia oryzoides* showed no significant response to as much as 1.5 cm of sediment. Germination of *Carex stricta* and *Impatiens capensis* were insufficient for interpretation. The germination responses are consistent with the sediment tolerance categorizations of the individual species (Figure 2). Species categorized as very tolerant (based on field observations) showed no significant response to high levels of sediment, while species categorized as moderately tolerant or slightly tolerant showed a significant decrease in germination with increasing sediment load.

It was also necessary to investigate the ability of an individual species to successfully germinate along both a gradient of increasing sediment accumulation and a hydrogeomorphic one, as the field observations had suggested. As an experimental representation of the HGM gradient, this study examined the impact of stratification and water regimes on the germination of wetland species with associations to varying sedimentation environments. Significant interactions of chill treatment and sediment depth, and watering regime and sediment depth, were demonstrated, which may be an approximation of annual conditions in wetlands of various HGM types. This finding supports differences in individual species' response to sedimentation in wetlands of various HGM subclasses. *Carex vulpinoidea*, *Hypericum pyramidatum*, *Typha latifolia*, and *Leersia oryzoides* exhibited germination responses affected by wetness regimes, and these responses were consistent with field observations.

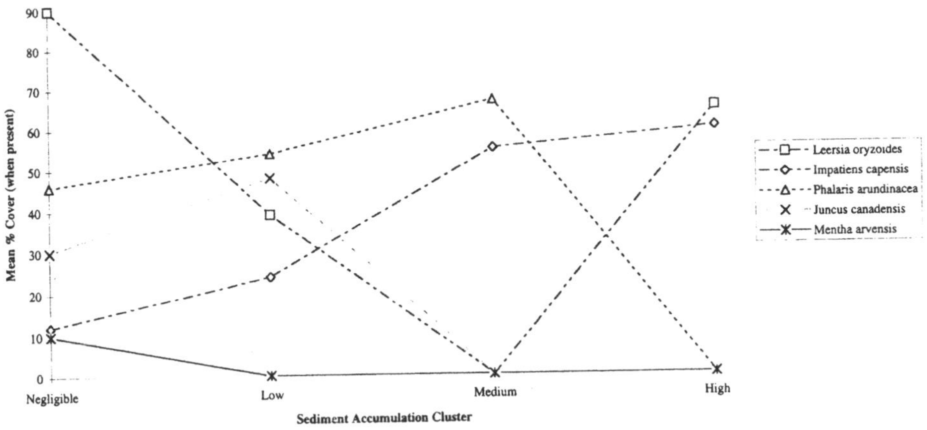

Fig. 1. Mean percent cover of representative wetland plant species from each sediment tolerances group over the sedimentation gradient.

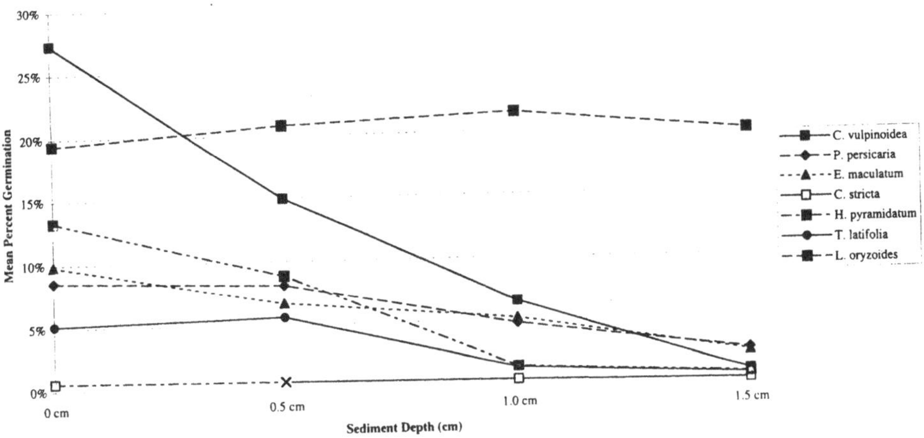

Fig. 2. Mean percent germination of seven wetland plant species of various sediment tolerance groups by sediment burial depth.

If the stressor-impact curves were to be useful, some measure of sedimentation rates in central Pennsylvania wetlands was necessary, as well as a predictive model of sedimentation. In this manner, a prediction of stressor level (sedimentation) could allow a prediction of impact (plant community change). This study established a robust set of sedimentation measurements in central Pennsylvania wetlands, covering a variety of HGM subclasses and landscape disturbance levels. Sediment depths ranged from 0-8 cm of accumulated material in a 12-month period. Mean measured mineral and organic accretion rates were 778 g m^2 yr^{-1} (+/- 1417) and 550 g m^2 yr^{-1} (+/- 589), respectively. To additionally characterize sedimentation rates, this study sought to test the hypothesis that wetlands of different HGM types would exhibit different characteristic sedimentation rates. This hypothesis was accepted; mean mineral and organic accretion rates for both sites (all plots within a site were averaged) and plots were significantly different by HGM class. The highest mineral accretion rates were for headwater floodplains, followed by impoundments, riparian depressions, mainstem floodplains, and slopes. The highest organic accretion rates were for riparian depressions, followed by impoundments, slopes, headwater floodplains, and mainstem floodplains. These patterns were identical when considering either plot results or site averages. The mineral and organic accumulation rates found in this study are not entirely consistent with published results. Differences in wetland type account for the difference, and indicate that relevant comparisons to literature values are limited due to insufficient classification of wetland type in current literature.

The potential effects of landscape disturbance on these sedimentation rates were also investigated, in order to develop the conceptual model to predict sedimentation rates for a given wetland in a variety of landscape settings. Different HGM subclasses exhibited significantly different mineral and organic accumulation rates, and varied in their responses to landscape disturbance and spatial variability in sedimentation patterns. Therefore, in constructing the applied conceptual model of sedimentation, it was most appropriate to organize the model initially on the basis of HGM. Four primary HGM classes were represented in this study; riparian depressions, headwater and mainstem riverine, slope, and human impoundment. After segregation by HGM type, other model factors of landscape disturbance and within-wetland location were investigated. The model serves as a guide to expectations concerning mineral and organic accretion rates in a variety of wetland types, in a variety of landscape settings, and at various locations within the wetland itself. It does not cover all possible conditions, nor does it provide an analytical model for the prediction of accretion rates. It does demonstrate that wetlands in different hydrogeomorphic settings, as described by the HGM class, both behave differently in terms of physical processes and react differently to landscape disturbance. The conceptual model indicates that riparian depressions do not respond to landscape disturbance, while riverine, slope, and impoundments do. Conversely, the within-wetland location (edge, interior, or channel, if appropriate) has significant implications for predicting sedimentation only in riparian depressions and slope wetlands. There are obvious transport vectors in these environments, and sedimentation is significantly elevated in them.

4. Conclusions

The largest data gap preventing wide application of any predictive model of spatial and temporal vegetation patterns is germination requirements of wetland plant species under various sedimentation conditions, and with various co-occurring stressors such as those embodied in the HGM classification. In addition, this same germination information must be incorporated into construction of functional groups, if such groups are to provide a reliable descriptor of plant community response under sedimentation stress.

This study demonstrates the need for buffer requirements for riparian depression, headwater floodplain, slope, and impoundment wetland types, although the appropriate size of this buffer is not determined herein. Land use management of contributing drainage areas to mainstem floodplains is required for their protection. Future work should: 1) implement the HGM classification in experimental designs; 2) investigate the relationship of sedimentation effect with distance from the wetland itself to establish necessary buffer recommendations; and 3) develop an analytical model for prediction of sedimentation in freshwater wetlands.

Acknowledgments

The authors wish to thank Dr. C. Andrew Cole, Laura E. Jackson, and James Kooser for their thoughtful comments on the manuscript. We also gratefully acknowledge the funding resources of the U.S. Environmental Protection Agency Region 3 and the Pennsylvania Department of Environmental Protection.

References

Adamus, P. R., and Brandt, K.: 1990, *Impacts on quality of inland wetlands of the United States: A survey of indicators, techniques, and applications of community-level biomonitoring data*, EPA/600/3-90/073. U. S. Environ. Prot. Agency, Environ. Res. Lab., Corvallis, OR. 406pp.

Bishel-Machung, L., Brooks, R. P., Yates, S. S., and Hoover, K. L.: 1996, "Soil properties of reference wetlands and wetland creation projects in Pennsylvania", *Wetlands* **16**(4), 532-541.

Boutin, C., and Keddy, P. A.: 1994, "A functional classification of wetland plants", *J. Veg. Sci.* **4**, 591-600.

Brinson, M. M.: 1993, *A hydrogeomorphic classification for wetlands*, Tech. Rep. WRP-DE-4, U. S. Army Corps of Engineers, Washington, DC. 59pp.

Brooks, R.P., Cole, C.A., Wardrop, D.H., Bishel-Machung, L., Prosser, D.J., Campbell, D.A., and Gaudette, M.T.: 1996, *Wetlands, Wildlife, and Watershed Assessment Techniques for Evaluation and Restoration: Final Report for the Project: Evaluating and Implementing Watershed Approaches for Protecting PennsylvaniaÕs Wetlands*, Report No. 96-2, Penn State Cooperative Wetlands Center, University Park, PA. Volumes I and II.

Brown, L.: 1979, *Grasses, an identification guide*, Houghton Mifflin Co., New York, NY. 240 pp.

Fassett, N.C.: 1957, *A Manual of Aquatic Plants*, University of Wisconsin Press, Madison, WI. 405pp.

Garbisch, E. W., and McIninch, S.: 1992, "Seed information for wetland plant species of the northeast United States", *Restoration and Management Notes* **10**(1):85-86.

Grime, J.P., Mason, G., Curtis, A.V., Rodman, J., Band, S.R., Mowforth, M.A.G., Neal, A.M., and Shaw, S.: 1981, "A comparative study of germination characteristics in a local flora", *Journal of Ecology* **69**: 1017-1059.

Kadlec, J. A., and Wentz, W. Alan: 1974, *State of the Art Survey and Evaluation of Marsh Plant Establishment Techniques: Induced and Natural, Volume 1, Report of Research, Michigan University,* prepared for the Army Corp of Engineers. AD-A012 834, pp. 231.

Kleiss, B.: 1993, *Methods for measuring sedimentation rates in bottomland hardwood (BLH) wetlands,* U.S. Army Engineer Waterways Experiment Station, Vicksburg, MS. WRP Technical note SD-CP-4.1.

Newcomb, L.: 1977, *Newcomb's Wildflower Guide*, Little, Brown, and Company (Canada) Limited. 490 pp.

Robinson, G.R., Holt, R.D., Gaines, M.S., Hamburg, S.P., Johnson, M.L., Fitch, H.S., and Martinko, E.A.: 1992, "Diverse and contrasting effects of habitat fragmentation", *Science* **257**: 524-526.

Storer, D.A.: 1984, "A simple high sample volume ashing procedure for determination of soil organic matter content", *Commun. in Soil Sci. Plant Anal.* **15**(7): 759-772.

TOWARDS A REGIONAL INDEX OF BIOLOGICAL INTEGRITY: THE EXAMPLE OF FORESTED RIPARIAN ECOSYSTEMS

ROBERT P. BROOKS[1], TIMOTHY J. O'CONNELL[1], DENICE H. WARDROP[1], LAURA E. JACKSON[2]

[1]*Penn State Cooperative Wetlands Center, Forest Resources Laboratory, Pennsylvania State University, University Park, PA 16802 USA*, [2]*U.S. Environmental Protection Agency, Office of Research and Development, Research Triangle Park, NC 27711 USA*

Abstract: Our premise is that measures of ecological indicators and habitat conditions will vary between reference standard sites and reference sites that are impacted, and that these measures can be applied consistently across a regional gradient in the form of a Regional Index of Biological Integrity (RIBI). Six principles are proposed to guide development of any RIBI: 1) biological communities with high integrity are the desired endpoints; 2) indicators can have a biological, physical, or chemical basis; 3) indicators should be tied to specific stressors that can be realistically managed; 4) linkages across geographic scales and ecosystems should be provided; 5) reference standards should be used to define target conditions; and 6) assessment protocols should be efficiently and rapidly applied. To illustrate how a RIBI might be developed, we show how four integrative bioindicators can be combined to develop a RIBI for forest riparian ecosystems in the Mid-Atlantic states: 1) macroinvertebrate communities, 2) amphibian communities, 3) avian communities, and 4) avian productivity, primarily for the Louisiana waterthrush (*Seirius motacilla*). By providing a reliable expression of environmental stress or change, a RIBI can help managers reach scientifically defensible decisions.

1. Introduction

How can one provide scientifically reliable information to the environmental decision maker who must operate at a regional or national level, when data are collected at the site level and when most protective and restorative actions occur at a local level? Finding ecological measures that behave predictably across scales, and that can be aggregated to assess regional ecological trends, poses a serious dilemma. Given limited resources to assess and protect ecosystem health, a Regional Index of Biological Integrity (RIBI), if properly constructed and evaluated, can help scientists, managers, and policy makers document trends in ecosystem degradation, prioritize management issues, and target restoration activities for appropriate locations. By providing a reliable expression of environmental stress or change, a RIBI can integrate impacts that are spatially and temporally disparate.

The concept of using indicators to assess ecological integrity was summarized by McKenzie *et al.* (1992). Messer (1992) discussed concerns regarding the development of regional indicators. The process of developing and evaluating regional ecological indicators is a scoping one, varying by region, and requiring at least conceptual links, and preferably causal links, between a stressor and the resultant ecological change. When constraining factors are strong and causal relationships are well established, the risk of using indicators to draw conclusions will be relatively low. As predictive capability wanes, and as correlation replaces cause-and-effect understanding, the risk increases. In either

Environmental Monitoring and Assessment **51: 131–143, 1998.**

case, these relationships should be documented to improve the knowledge base upon which decisions are made. For a RIBI to be useful, it must relate closely to societal concerns and be defensible for the decision-makers (Brooks *et al.*, 1991, Messer 1992, Angermeier and Karr, 1994). Noss (1990) described four levels of organization that should be addressed by ecological indicators: regional landscape, community/ecosystem, population/species, and genetic. In this paper, our objective is to demonstrate how the proposed RIBI could encompass all four levels of organization, and therefore, be used for making decisions or assessing risk on a regional basis.

To illustrate how a RIBI might be developed, we present an example for forested riparian ecosystems of the Mid-Atlantic states. Forested headwater streams, and their associated riparian wetlands, represent the reference condition for ecological integrity for this ecosystem type throughout this region. Headwater streams (first and second order) contribute 60-75% of the total stream length and total drainage area of watersheds in the Mid-Atlantic states. The ecological integrity of headwaters is important to the region (e.g., Sweeney, 1992), but they are significantly impacted by a variety of environmental stressors.

Maintaining and restoring the ecological integrity of forested riparian buffers have been identified as important strategies to protect the water quality and living resources of the Chesapeake Bay/Susquehanna Basin. A recent study used satellite imagery (Day *et al.*, 1997) to evaluate the extent of forested riparian buffers in a portion of the proposed study area, the Chesapeake Bay/Susquehanna Basin. That study found about one-third of the streams had buffers >100 m and about half had buffers of 30 to 100 m wide. The effects of forest buffer width on biotic communities had been studied (e.g., Brooks *et al.*, 1991, Croonquist and Brooks, 1991, 1993), but the implications for maintaining biological integrity are uncertain.

By definition, a RIBI must integrate a variety of stressors, and this one focuses on several stressors of high priority to the U.S. Environmental Protection Agency, both nationally and in the target area, Region 3: habitat loss and fragmentation, acidification, and sedimentation (Adamus and Brandt, 1990, USEPA, 1990, Kepner *et al.*, 1995, Magnien *et al.*, 1995). Given the cumulative importance of headwaters to the ecological integrity, recreational quality, and human food production of riparian and estuarine ecosystems throughout the region, it is prudent to develop an index that managers and decision-makers can use easily and confidently to target protection and restoration efforts.

2. Principles for RIBI Development

Before describing the RIBI, we will outline six general principles of RIBI development that have universal application, regardless of the targeted ecosystem or region.

1. Biological communities with high integrity are desired endpoints. We define high biological integrity as a condition where the biota represent the full spectrum of community, population, and genetic biodiversity for a targeted ecosystem (Salwasser, 1991). If an expected biological community is comparable to reference standard conditions in natural habitats (e.g., Smith *et al.*, 1995), then the physical and chemical attributes of the environment that supports that community should be intact. As that spectrum of biodiversity declines, biological integrity should change measurably. Detection of this change would signal that some stressors may be impacting the community, thereby causing a shift in its structure, composition, and functional organization (e.g., Angermeier and Karr, 1994). There are many ways to characterize biological communities, as portrayed below. As one moves down this list, the data become increasingly more difficult and expensive to collect, particularly when the goal is a regional assessment.

SPECIES RICHNESS
SPECIES DIVERSITY

RESPONSE GUILDS
FUNCTIONAL GROUPS

POPULATION DEMOGRAPHICS
GENETIC PROFILES
INDIVIDUAL HEALTH PROFILES

We have found that traditional measures of species richness and diversity are not sensitive to the stressors of interest, yet we can not afford to collect detailed population, genetic, or health data across an entire region. So, we have found considerable utility in constructing response guilds explicitly designed to address specific stressors (e.g., Brooks *et al.*, 1991, Croonquist and Brooks, 1991, Miller *et al.*, 1997, O'Connell *et al.*, 1997). Response guilds and functional groups can be generated from species lists, thus, describing what species are present, not just how many are there (Brooks and Croonquist, 1990).

An example of how response guilds can be useful is provided from bird data collected from two emergent wetlands (Table 1, data from Brooks *et al.*, 1996). Mothersbaugh Swamp is relatively undisturbed, whereas Millbrook Marsh is located in an urbanizing area. Both wetlands are about 20 ha in area. Bird data were collected from several plots during 10-min point counts for one day in June. Measures of species richness, Shannon diversity, and the number of individuals suggest there is no difference between the bird communities, yet percent similarity between the species lists is only 35%. A habitat response guild shows that interior forest species and species with large ranges are poorly represented in the disturbed wetland, where edge species predominate.

Table I
A comparison of bird data from two emergent wetlands in central Pennsylvania

	Mothersbaugh Swamp		Millbrook Marsh
SPECIES RICHNESS	21		18
NUMBER OF INDIVIDUALS	62		67
SHANNON DIVERSITY (H')	-2.59		-2.64
PERCENT SIMILARITY		35%	
RESPONSE GUILD -HABITAT USE			
Interior forest species (%)	19		6
Large range species (%)	19		0
Edge species (%)	62		94

2. Indicators can have a biological, physical, or chemical basis, hence, ecological indicator is commonly used as a collective term for all types of indicators (McKenzie *et al.*, 1992). We define ecological indicators as measures, variables, or indices that represent or mimic either the structure or functions of ecological processes and systems across a disturbance gradient.

3. Indicators should be tied to specific stressors that can be realistically managed. Although it may be intellectually satisfying to identify clear linkages between indicators and a variety of stressors, if the stressor of concern cannot be realistically managed, there will be minimal opportunities to improve the health of the targeted ecosystem. Thus, the search for indicators should be conducted according to a prioritized list of stressors.

4. Linkages across geographic scales and ecosystems should be provided. No single indicator works across all scales, so one must be aware of the appropriate ranges of scale for each indicator and how there are linkages among indicators and across scales. In the example provided, we postulate linkage across scales from plot level to landscape for a single ecosystem type, forest riparian headwaters. The proposed RIBI could be applied to all linear segments of this ecosystem type, and although they are found throughout the region of concern, the assessment would not encompass all of the land area in that region. Another paper in this volume addresses how one indicator, bird communities, can be used to make an integrated assessment across varied ecosystems for an entire region (e.g., O'Connell *et al.*, 1997).

5. Reference standards should be used to define target conditions. The use of reference sites has become increasingly more common as ecologists and regulators search for reasonable and scientifically based methods to measure and describe the inherent variability in natural systems (e.g., Kentula *et al.*, 1992). The primary reason to include sites

designated as reference standards (Smith *et al.*, 1995) in a RIBI is the need to compare impacted or degraded sites to a standard set of conditions or benchmarks, defined by the best possible sites in the region (Hughes *et al.*, 1986). These benchmarks can represent a starting point in time for trend analyses performed on the same sites repetitively. Reference sites can also serve as alternatives to standard experimental controls which are seldom available in large-scale field studies.

The criteria for selecting reference sites can range from choosing ideal, pristine conditions represented by the least disturbed sites available, to simply best attainable conditions for a particular region. Although reference sites often represent areas of minimal human disturbance (i.e., reference standards for wetlands; Smith *et al.*, 1995), in many instances it is more useful to select a set of reference sites that represents a range of environmental conditions across a landscape. Sites within the reference set used to develop a RIBI should span several gradients. They should include, at a minimum, the common types of sites found within a region for the targeted ecosystem type, and the range of conditions from relatively pristine (ecologically intact or reference standards) to severely disturbed sites (degraded ecological integrity and functions). For example, using reference wetlands from a wide variety of vegetation types, disturbance regimes, and landscape positions allows for characterization of the inherent variability found in wetlands (Brooks *et al.*, 1996). Throughout this paper, we use the term reference sites to connote naturally occurring sites composed of wetland, stream, and riparian components of headwaters. Reference standards represent the most ecologically intact sites in the reference set.

6. Assessment protocols should be efficiently and rapidly applied. It is entirely appropriate to conduct intensive studies to develop and test indicators. This level of effort is not appropriate, however, when collecting data across a region. Whenever possible, there should be an attempt to develop a rapid assessment protocol from more intensive studies.

3. A RIBI for Forested Riparian Ecosystems

Although regulatory scrutiny and restoration efforts in the eastern states have focused on wetlands, contrasting with a riparian restoration focus in many western states, it is timely to begin integrating wetland and riparian restoration in the context of watershed protection. The need for strategic restoration of aquatic ecosystems is strongly supported by the National Research Council (1995), which has called for integrated approaches. This can be accomplished only if we consider wetlands, streams, floodplains, and riparian areas to be gradations along an ecological continuum, rather than individual entities.

During a previous series of studies, Brooks *et al.* (1996) and Miller at al. (1997) developed and evaluated a series of tools for assessing cumulative impacts on wetlands and associated streams and riparian areas by characterizing their current structure, potential functions, and restoration potential in a watershed context. This research focused primarily on wetlands and riparian areas associated with streams equal to or lower than third order, or headwaters. Individually, headwater streams and wetlands are smaller in scope than the more expansive areas of forested floodplains found downstream. Their position within the watershed, however, suggests these areas assume a relatively more important role in maintaining in-stream water quality, because proportional to size, more overland flow passes through these low order riparian wetlands than through bottomland forests (Brinson, 1993). In most watersheds, there are more headwater streams, with a larger cumulative length, than mainstem rivers (Leopold, 1974).

We propose to use four integrative bioindicators to develop a RIBI for forest riparian ecosystems in the Mid-Atlantic Integrated Assessment area (MAIA) of the U.S. Environmental Protection Agency (USEPA): 1) macroinvertebrate communities, 2) amphibian communities, 3) avian communities, and 4) avian productivity, primarily for the Louisiana waterthrush (*Seirius motacilla*). Previous studies suggest that these bioindicators are directly related to the ecological condition of associated habitat components at one or more scales [e.g., in-stream conditions (Plafkin *et al.*, 1989, Brooks *et al.*, 1991), riparian habitat (Croonquist and Brooks, 1993, Fearer *et al.*, in prep.), and landscape patterns (O'Connell *et al.*, 1997)]. The fourth bioindicator, Louisiana waterthrush productivity, density, and abundance, provides a means to link the other three. Population measures of the Louisiana waterthrush, span the widest range of scale. They provide a means to calibrate the other indicators, and link them across scales. Thus, a calibrated RIBI could be developed to identify and refine thresholds of environmental disturbance across multiple stressors that are of concern to the USEPA in the MAIA.

The Louisiana waterthrush represents both the biological resource of concern and the means to calibrate this index. Previous work suggests that the Louisiana waterthrush, is an excellent indicator of healthy forested riparian ecosystems in the eastern U.S. (D. Prosser, R. Brooks, R. Mulvihill, T. Master, unpublished data). This species depends on stream macroinvertebrates for food and forest riparian habitats for nesting. As a common top predator and the only obligate avian species of this ecosystem in the eastern U.S., it is an ideal calibrator for an index of headwater ecosystems. Measuring the population parameters of the Louisiana waterthrush requires a substantial investment, but once completed, provides a means to calibrate the other indicators, thus linking them across scales. Once tested between reference and impacted sites, and then calibrated across scales, a set of indicators could be combined into a RIBI for the MAIA.

To construct the RIBI, field sampling in headwater streams and riparian habitats would occur over three ecoregions in the MAIA at different scales (e.g., 2.5 ha/territory, 25 ha/reach, 250 ha/watershed, and 2,500 ha/landscape).

Each bioindicator is most strongly associated with measures of habitat at a particular scale. Measuring productivity for the Louisiana waterthrush relates primarily to quality of riparian habitat, but it is also dependent on the availability of macroinvertebrates as food. Biomass and composition of macroinvertebrate communities relate to instream and wetland habitat and measures of water chemistry and sedimentation. Avian communities relate primarily to landscape metrics. However, by combining measures of nest productivity, territory density, and survey abundance, attributes of the Louisiana waterthrush spans the widest range of scale, as seen in Figure 1.

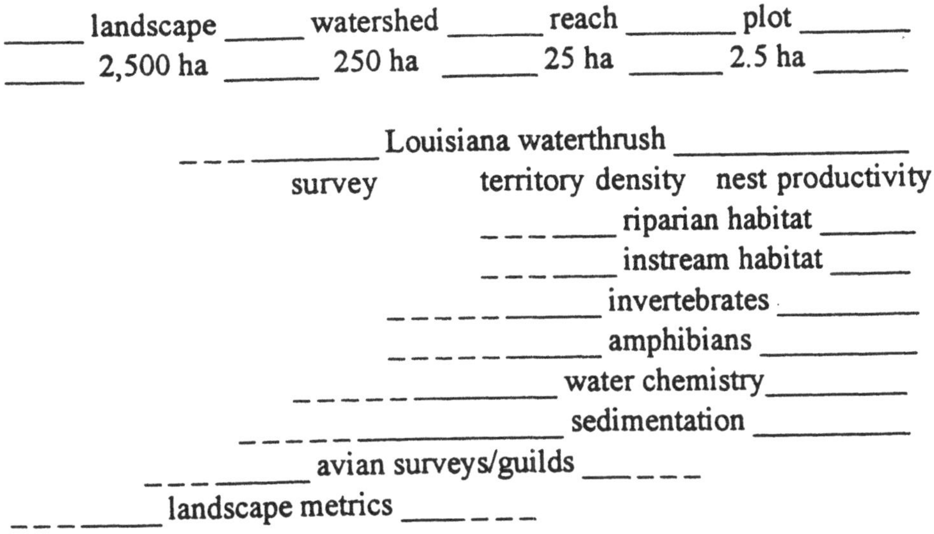

Fig. 1. Relative overlap of bioindicators and habitat measures for determining ecological condition as a function of scale (____ solid lines represent confirmed relationships, _ _ _ dashed lines represent possible relationships) for a RIBI of forest riparian ecosystems. The waterthrush spans the widest range of scales, plot to watershed, and thus, would be used to link plot-level measures to landscape metrics.

4. Rationale for Selection of Indicators

Each indicator operates at a different scale, but these can be nested within each watershed. Breeding territories of the Louisiana waterthrush are about 250 m in length, and riparian habitat can be assessed within territories and at nest sites to a width of about 100 m, or about 2.5 ha/plot. The integrity of headwater streams, including measures of macroinvertebrates, water quality, channel morphometry, bottom substrate, and sedimentation, should be assessed along a reach of about 2.5 km in length to a width of 100 m, or about 25 ha ha/reach. The landscape assessment will be made within circular subsets of 2-km and 4-km radius circles within a Breeding Bird Atlas (Brauning, 1992) block (25 km^2), or about 1,250-5,000 ha per study site . Thus, sampling and data evaluation will occur over several orders of magnitude, but within a logical, nested design that addresses scale issues.

4.1. Macroinvertebrate and Amphibian Communities and Stream Assessments

Aquatic macroinvertebrates and breeding amphibians provide the best alternative to fish for assessing headwater streams and vegetated wetlands where fish are seldom present. Aquatic invertebrate communities are known to change in response to a variety of stressors (Adamus and Brandt, 1990, Brooks *et al.*, 1991, Garano and Kooser, 1994). A significant effort has been made to integrate chemical, biological, and physical parameters for assessing the ecological integrity of streams (e.g., Plafkin *et al.*, 1989, USEPA, 1991, Sweeney, 1992), resulting in satisfactory predictions of the health and condition. Research on amphibians as indicators has found there are distinct species assemblages associated with different levels of watershed disturbance (Fearer *et al.*, in prep.). Linkage between aquatic and upland habitats is strong due to their typical aquatic breeding and upland dispersal behaviors.

The assessment of sedimentation is a major part of the characterization of in-stream habitat. The negative impacts of excessive sedimentation in aquatic systems are well documented. Measurement of sedimentation rates provide a surrogate measure for contaminants that are attached to or carried with water-borne sediments. Although the measurement of sedimentation rates varies with scale, we have developed ways to link these measures to ascertain the direct impacts on stream macroinvertebrates (Brooks *et al.*, 1996, Wardrop, 1997).

4.2. Avian Community and Landscape Pattern Assessment

Use of bird communities as a regional indicator is easily justified. Due to their mobility, birds may respond to a wide range of stressors affecting both terrestrial and aquatic habitats (although local responses must be segregated from external factors affecting migratory species). Predictions regarding bird community responses to changes in land cover and connectivity are based on this type of information, and have proven to be reliable (e.g., Robbins *et al.*, 1989, Croonquist and Brooks, 1991). Although census data are usually site-specific, they can be aggregated at least to a landscape scale, and perhaps to a region. Trends in songbird populations are reported both regionally and nationally, and their suitability as a regional indicator is currently being tested (O'Connell *et al.*, 1997). Measurement of bird communities is relatively simple, a volunteer data collection network is in place, and historic databases exist. This information can be used, in conjunction with on-site avian censuses, as a coarse indicator of landscape condition within a watershed. Watersheds can be aggregated across a region.

Analysis of landscape patterns is appropriate for a regional indicator because: 1) measures of landscape pattern occur at a large scale, so they can be easily aggregated to a regional scale; 2) databases compiled from recent remotely sensed data are available, and updates in the future are likely; 3) local planners and managers are familiar with this tool and its applications; and 4) a landscape context for site-specific locations can be developed, although processing of digital landscape data at small scales is limited by computer processing time.

4.3. AVIAN PRODUCTIVITY AND RIPARIAN HABITAT ASSESSMENTS

During prior studies, the Louisiana waterthrush (*Seiurus motacilla* was identified as the best avian indicator of healthy forested headwater riparian ecosystems (Prosser and Brooks, 1998)). The Louisiana waterthrush is a member of the avian guild of area-sensitive, neotropical wood warblers, and hence, the population status of the Louisiana waterthrush can provide a surrogate measure of the status of many other similar species of concern. The species is an insectivorous, ground-nesting, single-brooded, neotropical migrant species that is area sensitive for large, intact forests (Brauning, 1992, Robinson, 1995). Because they rely on forested headwater streams and wetlands for foraging and nest sites, they are sensitive to stressors that impact both terrestrial habitat condition and water quality.

The ability of the Louisiana waterthrush to occupy, survive, and reproduce in forested riparian ecosystems appears to relate directly to the ecological integrity of these systems. Standard bird censuses can detect the presence or absence of Louisiana waterthrush from a given habitat. Although the absence of Louisiana waterthrush infers major environmental degradation, presence without reproduction or reduced productivity are likely to signal more subtle changes in ecological integrity in forest riparian ecosystems.

If the genetic variability of the waterthrush among sites of differential ecological conditions if of interest, feathers can be collected and analyzed from all relevant ecoregions. If differences do or do not occur among ecoregions or within populations, there may be implications regarding avian productivity and isolation of regional populations that suggest further investigation.

5. Development of the RIBI

Once data are compiled, analysis will consist of a series of comparisons outlined below. Analyses will begin with the smallest scale with sites within a study area and expand, as appropriate, to larger scales and across ecoregions as follows:

Within reference sites, within ecoregions, across ecoregions

Within impacted sites, within ecoregions, across ecoregions

Between reference and impacted sites, within ecoregions, across ecoregions

Comparisons could include the following:

1) Louisiana waterthrush productivity and riparian habitats

Density of territorial pairs (e.g., number of singing males, number and length of territories)

Nesting success (e.g., number of eggs laid, nestling survivorship, number fledged)

Successful vs. unsuccessful nest sites (vegetation and bank characteristics)

Riparian habitat (e.g., relative scores of habitat suitability index models)

Site fidelity (e.g., number or proportion of returning adults in subsequent years)

2) Macroinvertebrate and amphibian communities and instream habitats

Biomass and community composition of macroinvertebrates (e.g., g/m^2, functional groups; by sampling method, by foraging site)

Presence/absence and abundance data for amphibians

Sedimentation (e.g., % embeddedness in bottom substrate, accumulation rates)

Water chemistry (e.g., pH, conductivity, total alkalinity, total nitrogen, total phosphate)

Stream habitat assessment (e.g., relative scores, wetted perimeter/mean depth ratio, proportion of riparian wetlands in the stream corridor)

3) Avian communities and landscape patterns

Avian community (e.g., richness, diversity, response guilds)

Habitat metrics from survey points (e.g., vertical stratification of vegetation, % canopy cover, % ground cover)

Landscape metrics (e.g., contagion, % cover by type, patch areas by cover type)

To develop the index of regional ecological integrity, we will perform a series of single and multivariate correlations among measures of bioindicators, among habitat variables, and between bioindicators and habitat variables. These comparisons will follow a nested design across scales. The objective will be to assemble a set of measurements into an index that is calibrated against the population parameters of the Louisiana waterthrush or other bioindicators. For example:

Scale	Calibration measures	Index measures
plot	waterthrush fledging rate	macroinvertebrate biomass by functional group, sedimentation rate
reach	waterthrush density	width of forested riparian corridor, stream habitat assessment or HSI scores
watershed	waterthrush abundance at survey points	proportion of forest cover, proportion of exotic bird species at survey points
landscape	occurrence of similar species in response guilds	mean forest patch area, contagion, dominance

Once the index is developed, an independent test should be conducted on a separate set of data. After the RIBI is validated, then it could be used in general practice. During a regional analysis, a genetic component could be added if it were hypothesized that there was a decrease in fitness or survivorship due to genetic impairment (contaminants) or genetic isolation (habitat fragmentation). To add the genetic component to the RIBI, it would be necessary to collect tissue samples from individuals and use a suitable analytical procedure to detect differences. This component of the work is expensive, and should be restricted to suspected problem areas.

6. Summary

In the example presented, we show how a RIBI for forested riparian headwater ecosystems could be applied consistently across a regional gradient as large as the Mid-Atlantic states. We illustrated the development of this RIBI based on how four integrative bioindicators could be combined and calibrated: 1) macroinvertebrate communities, 2) amphibian communities, 3) avian communities, and 4) avian productivity, primarily for the Louisiana waterthrush (*Seirius motacilla*). Based on six principles, we provide guidance for the development of any RIBI: 1) biological communities with high integrity are the desired endpoints; 2) indicators can have a biological, physical, or chemical basis; 3) indicators should be tied to specific stressors that can be realistically managed; 4) linkages across geographic scales and ecosystems should be provided; 5) reference standards should be used to define target conditions; and 6) assessment protocols should be efficiently and rapidly applied. By providing a reliable expression of environmental stress or change, RIBIs can help managers reach scientifically defensible decisions.

Acknowledgments

This work was made possible through support by the Penn State Cooperative Wetlands Center of the Environmental Resources Research Institute and School of Forest Resources of The Pennsylvania State University. The manuscript was improved by comments on an earlier draft by P. L. Angermeier, M. J. Casalena, and an anonymous reviewer.

References

Adamus, P. R., and Brandt, K.: 1990, "Impacts on quality of inland wetlands of the United States: A survey of indicators, techniques, and application of community-level biomonitoring data", U. S. Environ. Prot. Agency, Environ. Res. Lab., Corvallis, OR, EPA/600/3-90/073.

Angermeier, P. L., and Karr, J. R.: 1994, "Biological integrity versus biological diversity as policy directives", *Bioscience* **44**, 690-697.

Brauning, D. W., ed: 1992, Atlas of breeding birds in Pennsylvania, Univ. Pittsburgh Press, Pittsburgh, PA, 484pp.

Brinson, M. M.: 1993, "Changes in the functioning of wetland along environmental gradients",

Wetlands **13**, 65-74.

Brooks, R. P., and Croonquist, M. J.: 1990, "Wetland, habitat, and trophic response guilds for wildlife species in Pennsylvania", *J. PA Acad. Sci.* **64**, 93-102.

Brooks, R. P., Croonquist, M. J., D'Silva, E. T., Gallagher, J. E., and Arnold, D. E.: 1991, Selection of biological indicators for integrating assessments of wetland, stream, and riparian habitats, Pages 81-89 *in* Biological Criteria: Research and Regulation, U.S. Environ. Prot. Agency, Office of Water, EPA-440/5-91-005, Washington, DC, 171pp.

Brooks, R. P., Cole, C. A., Wardrop, D. H., Bishel-Machung, L., Prosser, D. J., Campbell, D. A., and Gaudette, M. T.: 1996, Wetlands, Wildlife, and Watershed Assessment Techniques for Evaluation and Restoration (W^3ATER), Volumes 1, 2A, 2B, Final Report to PA Dep. Environ. Prot. and USEPA Region 3, ERRI Rep. No. 9609.

Croonquist, M. J., and Brooks, R. P., 1991, "Use of avian and mammalian guilds as indicators of cumulative impacts in riparian-wetland areas", *Environ. Manage.* 15, 701-714.

Croonquist, M. J., and Brooks, R. P.: 1993, "Effects of habitat disturbance on bird communities in riparian corridors", *J. Soil Water Conserv.* **48**, 65-70.

Day, R. L., Richards, P. L., and Brooks, R. P.: 1997, Chesapeake riparian forest buffer inventory, Final Rep. to Chesapeake Bay Program Office, Annapolis, MD.

Fearer, T. M., Brooks, R. P., and Bradford, D. F.: (in prep.), "Amphibians as ecological indicators for freshwater wetlands and streams in the Ridge and Valley region of Pennsylvania".

Garano, R. J., and Kooser, J. G.: 1994, "Ordination of wetland insect populations: evaluation of a potential mitigation monitoring tool", Pages 509-516 *in* W. J. Mitsch (ed.), Global wetlands, Old world and new, Elsevier, Amsterdam, 967pp.

Hughes, R. M., and Larsen, D. P., and Omernik, J. M.: 1986, "Regional reference sites: a method for assessing stream potentials", *Environ. Manage.* **10**, 629-635.

Kentula, M. E., Brooks, R. P., Gwin, S. E., Holland, C. C., Sherman, A. D., and Sifneos, J. C.: 1992, An approach to improving decision-making in wetland restoration and creation, Island Press, Washington, DC, 151pp.

Kepner, W. G., Jones, K. B., Chaloud, D. J., Wickham, J. D., Ritters, K. H., O'Neill, R. V.: 1995, Mid-Atlantic landscape indicators project plan, U. S. Environ. Prot. Agency, Environmental Monitoring and Assessment Program, EPA/620/R-95/003, 37pp.

Leopold, L. B.: 1974, Water, A primer, W. H. Freeman and Co., San Francisco, 172pp.

Magnien, R., Boward, D., and Bieber, S.: 1995, The state of the Chesapeake Bay 1995, U.S. Enviorn. Prot. Agency, Cheaspeake Bay Program, Annapolis, MD, 45pp.

McKenzie, D. H., Hyatt, D. E., and McDonald, V. J. (eds.), 1992, Ecological indicators, Proc. Int. Symp. Ecological Indicators, 16-19 October 1990, Fort Lauderdale, FL, Elsevier Science Publishers, Ltd., Esses, England, Vols. 1 & 2.

Messer, J. J.: 1992, "Indicators in regional ecological monitoring and risk assessment", Pages 135-146 *in* McKenzie, D. H., Hyatt, D. E., and McDonald, V. J. (eds.): Ecological indicators, Proc. Int. Symp. Ecological Indicators, 16-19 October 1990, Fort Lauderdale, FL, Elsevier Science Publishers, Ltd., Esses, England, Vols. 1 & 2.

Miller, J. N., Brooks, R. P., and Croonquist, M. J.: 1997, "Effects of landscape patterns on biotic communities", *Landsc. Ecol.*, **12**, 137-153.

National Research Council: 1995, Wetlands, Characteristics and boundaries, National Academy Press, Washington, DC, 307pp.

Noss, R. F.: 1990, "Indicators for monitoring biodiversity: a hierarchical approach", *Conserv. Biol.* **4**, 355-364.

OÕConnell, T. J., Jackson, L. E., and Brooks, R. P.: 1997, "A bird community index of biotic integrity for the Mid-Atlantic Highlands", Environ. Monitoring and Assessment, (this issue).

Plafkin, J. L., Barbour, M. T., Porter, K. D., Gross, S. K., and Hughes, R. M.: 1989, Rapid bioassessmet protocols for use in streams and rivers: benthic macroinvertebrates and fish, EPA/444/4-89-001,U. S. Environ. Prot. Agency, Washington, DC.

Prosser, D. J. and Brooks, R. P. 1998, "A verified habitat suitability index for Louisiana waterthrush", *J. Field Ornith.* **69**, in press.

Robbins, C. S., Dawson, D. K., and Dowell, B. A.: 1989, "Habitat area requirements of breeding forest birds of the Middle Atlantic State", *Wildl. Monog*r. **103**.

Robinson, W. D.: 1995, Louisiana waterthrush, No. 151, A. Poole and F. Gill, eds., The birds of North America, Acad. Nat. Sci. Philadelphia.

Salwasser, H.: 1991, "Roles for land and resource managers in conserving biological diversity", Pages 11-31 *in* D. L. Decker *et al.* (eds.), Challenges in the Conservation of Biological Resources: A Practioner's Guide, Westview Press, Boulder, CO, 402pp.

Smith, R. D., Ammann, A., Bartoldus, C., and Brinson, M. M: 1995., An approach for assessing wetland functions using hydrogeomorphic classification, reference wetlands, and functional indices, U. S.Army Corps Engin., Waterways Exp. Stn., Wetlands Res. Prog. Tech. Rep. WRP-DE-9, Washington, DC, 79pp.

Sweeney, B. W.: 1992, Streamside forests and the physical, chemical, and trophic characteristics of Piedmont streams in eastern North America, *Wat. Sci. Tech.* **26**, 2653-2673.

U.S. Environmental Protection Agency: 1990, Reducing risk: Setting priorities and strategies for environmental protection, Report of the Science Advisory Board: Relative Risk Reduction Strategies Committee, Washington, DC.

U.S. Environmental Protection Agency: 1991, Biological criteria: research and regulation, Proc. Symp., 12-13 December 1991, Arlington, VA. Washington, DC. 171pp.

Wardrop, D. H.: 1997, The occurrence and impact of sedimentation in central Pennsylvania wetlands, Ph.D. Thesis, Pennsylvania State University, University Park, PA, 189pp.

A BIRD COMMUNITY INDEX OF BIOTIC INTEGRITY FOR THE MID-ATLANTIC HIGHLANDS

T. J. O'CONNELL[1], L. E. JACKSON[2], and R. P. BROOKS[1]

[1]*Forest Resources Laboratory, Penn State Cooperative Wetlands Center, Pennsylvania State University, University Park, PA 16802,* [2]*Office of Research and Development, U. S. Environmental Protection Agency, Research Triangle Park, NC 27711*

Abstract. We report on the development and preliminary application of a songbird community-based index of biotic integrity. The bird community index (BCI) sorts bird species found at sample sites into a series of values representing the proportional species richness of 20 behavioral and physiological response guilds. Relative proportions of specialist and generalist guilds are used to assign a composite score to each site. Scores from multiple sites indicate the overall biotic integrity of the study area. The BCI is intended to function as a landscape-scale indicator of biotic integrity, integrating conditions across large sample sites containing diverse ecological resources and intensities of human use. We developed the BCI with data from a 1994 pilot study in central Pennsylvania, then applied our preliminary index in 1995 and 1996 to independent samples of sites across the Mid-Atlantic Highlands Assessment area (MAHA). The 1995 and 1996 sample sites were selected using the probability-based sampling design of the Environmental Monitoring and Assessment Program (EMAP), and therefore represent the total land area in MAHA. Our preliminary assessment indicates that MAHA exhibits six categories of biotic integrity, and that more than 40% of the land area supports the two highest biotic integrity categories. Pending BCI refinement and incorporation of landscape and vegetation explanatory variables, the BCI will be included in a suite of indicators designed to provide an assessment of overall ecological condition in MAHA.

1. Introduction

The U.S. Environmental Protection Agency's (EPA's) Environmental Monitoring and Assessment Program (EMAP) is developing tools to assess ecological condition at regional and national scales. Among other research, EMAP is evaluating indicators which reflect key elements and processes of natural systems. Desirable indicators are differentially sensitive to a variety of stressors so that numerous impacts to the resource of concern may be detected. Further discussion of the indicator concept and development strategy within EMAP can be found in Hunsaker and Carpenter (1990) and Barber (1994).

Important components of EMAP indicator research are indices of biotic integrity (IBIs), originally developed to assess the condition of riverine systems (Karr, 1991, 1993; Fore *et al.*, 1996) and now being adapted for use in terrestrial environments (e.g., Bradford *et al.*, in press). We describe here the development of a bird community index (BCI) for the Mid-Atlantic Highlands Assessment area (MAHA). The BCI is modeled after previously-developed IBIs, but it is intended to function as a multiple-resource indicator of biotic integrity. We selected songbird communities for index development because songbirds are conspicuous components of all terrestrial environments in MAHA, and can integrate conditions across major habitat types that are typically assessed individually. In addition, many songbirds require specific vegetation structure, habitat patch size, and prey base

Environmental Monitoring and Assessment **51:** 145–156, 1998.

conditions, so the occurrence of these species can indicate other specific attributes of the ecosystem.

The biotic integrity concept provides an ecologically-based framework in which species assemblage data can be ranked on a qualitative scale. Biotic integrity can be defined as the capability of supporting and maintaining "a balanced, integrated, adaptive community of organisms having a species composition, diversity, and functional organization comparable to that of natural habitat of the region" (Karr and Dudley, 1981). Use of the integrity concept permits a ranking of site quality that is more descriptive than traditional measures such as richness and diversity (Brooks *et al.*, 1997).

The BCI we have developed is based on an analysis of songbird response guilds. Response guilds can be defined as groups of species which require similar habitat, prey, or other conditions for survival (Verner, 1984; Szaro, 1986; Brooks and Croonquist, 1990). For example, the loss of snags in a forest stand can result in a decrease in the guild of bark-probing insectivores. A bird community can be described in terms of response guilds that are based on life history traits of the component species. Croonquist and Brooks (1991) demonstrated that bird response guilds can be used to produce an effective indicator of habitat disturbance.

We envision bird communities as supporting varying proportions of common and ubiquitous species (generalists) and less common species which may be limited by specific habitat requirements and/or relatively low intrinsic rates of population increase (specialists). A BCI score indicating high integrity describes a community in which specialists are well represented relative to generalists. Our assessment defines specialists and generalists for 20 response guilds in eight guild categories: trophic, foraging substrate, nest placement, primary habitat, patch size, number of broods, migratory status, and nest predators/nest parasites. These response guild categories are based on habitat attributes, system productivity, and population and interspecific dynamics. The BCI, therefore, reflects structural, functional, and compositional elements of the system, while the relative proportions of specialists and generalists serve as indicators of system condition.

It is important to recognize that the BCI in its current formulation is intended solely for use in MAHA; both the reference high-integrity condition and the specific guilds used to formulate the index may differ in another region (Bradford *et al.*, in press). Furthermore, the BCI reflects biotic integrity at a fairly coarse level of resolution. In MAHA, the land cover and use types that are most likely to be encountered in a random sample are mature and regenerating forest, pasture and row crops, urban and suburban areas, and mined lands (Bailey, 1980). In the absence of irreversible anthropogenic disturbance, most of the region would succeed to forest. However, a minority of natural habitats in MAHA, such as shale barrens and cedar groves, are perpetually maintained by edaphic factors in states that resemble early-successional old-field or shrublands. These rare habitats, while entirely natural, may support native songbird communities that do not conform to our general definition of high biotic integrity for the region. Therefore, the BCI we ascribe to our sample sites is not intended for use in rare or isolated habitats.

Our specific research objectives are to: 1) develop a regional index of biotic integrity based on songbird community composition, and 2) apply the index to a probability-based sample of field sites to determine the proportion of MAHA exhibiting various categories of biotic integrity. This research represents the first attempt to develop and apply a field indicator of ecological condition across the full extent of an EMAP reporting region, without stratifying by resource type. When further refined, the BCI will be applied in conjunction with indicators at other scales to produce an overall assessment of ecological condition for the region.

2. Methods

Sampling. MAHA encompasses the mountainous portions of EPA Region 3, including most of Pennsylvania and the entire state of West Virginia. Within MAHA, we collected data from 34 sites in 1994, 58 sites in 1995, and 68 sites in 1996 (Figure 1). The 1994 sites constitute a reference gradient of ecological condition from nearly pristine to severely degraded. Site classifications were based on previous studies of vegetation, hydrology, soils, amphibians, and professional judgment (Brooks *et al.*, 1996). We developed the BCI from bird data collected at the 1994 sites, and then applied the BCI to two independent samples (1995 and 1996) from the EMAP probability-based sampling grid (Overton *et al.*, 1990). Use of the probability-based design permits an estimate of condition across the entire study region, with known statistical confidence. Confidence intervals for estimates of condition were calculated based on the Yates-Grundy variance formula, with an edge correction as described in Stevens and Kincaid (1997).

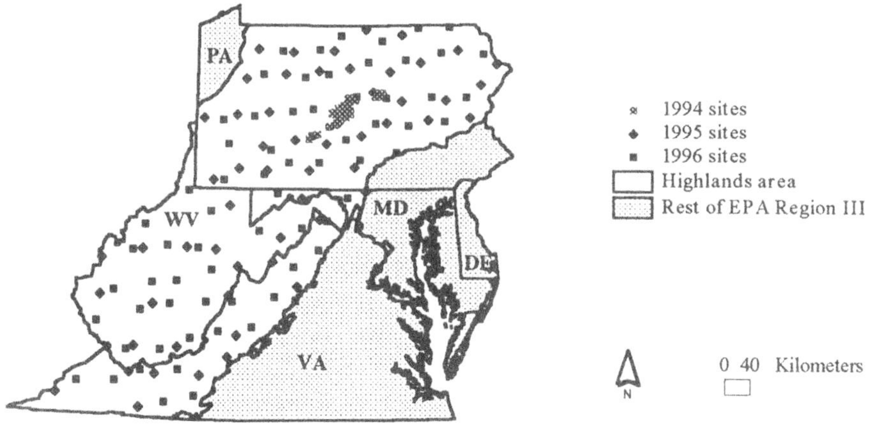

Fig. 1. Location of the mid-Atlantic highlands study area within EPA Region 3, and songbird sampling sites.

Sample sites in 1994 consisted of a variable number of plots (3-11) placed every 50-200m along a transect of up to 2km. In 1995 and 1996, each site consisted of five plots spaced every 200m along a randomly-oriented 1km transect. At each plot along a transect, we sampled songbirds with a 10-minute, 30m-radius point count between sunrise and 10:00hrs EDT (Hutto *et al.*, 1986; Manuwal and Carey, 1991; Ralph *et al.*, 1993). Sampling took place within the "safe dates" for breeding birds to ensure that any birds detected would be resident at each site throughout the breeding season (Brauning, 1992).

Response Guilds. We define the term "songbirds" to include the Passeriformes (perching birds), Piciformes (woodpeckers), Cuculiformes (cuckoos), and Columbiformes (doves) we documented in MAHA from 1994-6 (112 total species). Songbirds were assigned to behavioral and physiological response guilds based on a literature review (Harrison, 1975; Blake, 1983; DeGraaf *et al.*, 1985; Freemark and Collins, 1992; Brooks and Croonquist, 1990; Santner *et al.*, 1992). Table I lists the 20 guilds in eight guild categories used in our analysis. Because we selected these guilds specifically to reflect different aspects of each species' life history traits, each species belongs simultaneously to several guilds. However, guild assignments within each of the eight guild categories are mutually exclusive, so species belong to no more than eight guilds. For example, in the Trophic Level category, species are classified as either omnivores or insectivores (Table I). Also, these guild assignments apply only to breeding season life history traits: For example, we consider the Eastern Kingbird (*Tyrannus tyrannus*) to be an insectivore in MAHA, even though this species subsists largely on fruit in its wintering range (Terborgh, 1989).

We categorized individual guilds as "specialist" or "generalist" based on each guild's relationship to specific elements of ecosystem structure, function, and composition. For example, the Nest Placement guilds relate directly to the availability of appropriate nesting substrate (a structural element). We consider shrub nesters to be generalists because shrubs are widely distributed throughout the MAHA in mature and regenerating forests, in agricultural hedgerows, and in suburban areas. In contrast, we consider cavity nesters to be specialists because snags tend to be limited in distribution to mature forests in MAHA. The Trophic Guilds reflect aspects of ecosystem function. We label insectivores as specialists relative to omnivores because insectivores are obligate secondary or higher consumers, whereas omnivores can function as primary or higher consumers and therefore exploit a wider range of energy sources. The nest predator/nest parasite guild is an example of a guild with influence over the compositional elements of an ecosystem. Species in the nest predator/nest parasite guild can affect the abundance and distribution of other species. We consider nest predators and nest parasites to be generalists due to their relatively indiscriminant exploitation of other species as sources of food or surrogate parents.

Table I
Response guilds included in BCI development

Guild Category	Guild	Specialist	Generalist
Trophic Level	omnivore		X
	insectivore	X	
Insectivore Feeding Behavior	bark prober	X	
	ground gleaner	X	
	upper canopy gleaner	X	
	lower canopy gleaner	X	
	aerial sallier	X	
Nest Placement	native cavity nester	X	
	canopy nester	X	
	forest ground nester	X	
	shrub nester		X
Primary Habitat	edge species		X
	forest generalist		X
Patch Size Dependent	forest area sensitive	X	
Number of Broods	single brooded	X	
	double brooded		X
Migratory Status	resident		X
	temperate migrant		X
	tropical migrant	X	
Predator/Brood Parasite	predator/brood parasite		X

It is important to recognize not only that species are assigned to several guilds simultaneously, but that the individual guilds to which a species is assigned are not necessarily all specialist or all generalist. For example, the Downy Woodpecker (*Picoides pubescens*) is a generalist according to primary habitat and migratory status, but a specialist according to its membership in four other guild categories.

With species assigned to guilds, we quantified the bird community at each site in terms of its component guilds. We used presence/absence data to determine the proportion of species at each site belonging to each guild. The result is a matrix of sites and guilds, expressing the percentage of the total number of species at a site in each of 20 guilds. Because species are assigned to multiple guilds, the percentages for several guilds are correlated and the individual percentages for each site will sum to greater than 100.

BCI Development. Construction of the BCI is a multi-step process. The first step is to identify statistically separable levels of proportional occurrence of each guild from the 1994 sample data. We use the bark prober guild to illustrate the process here. In this example, the proportion of bark probers in our sample ranges from near zero to approximately 50% of the species at a site. We applied a combination of multivariate and univariate procedures to determine the number of categories of bark prober occurrence that could be separated statistically. We first performed cluster analyses on the 34 sites from

1994, clustering sites according to their similarity in bark prober proportion (Minitab, 1995). We used the cluster analysis iteratively, to obtain several cluster dendrograms with different numbers of clusters identified. We then applied one-way analysis of variance (ANOVA) with Tukey's multiple comparisons to the bark prober data using the various cluster memberships as factors in the ANOVA model (Neter *et al.*, 1990). This procedure allowed us to establish the number of statistically separable levels of bark prober occurrence at alpha = 0.05. Differences in bark prober proportion were statistically significant between two, three, and four clusters, but the significant difference between categories was lost as we increased the number of clusters to five. Therefore, we were able to identify four statistically separable categories of bark prober occurrence. We repeated this process of iterative clustering and ANOVA to determine the number of statistically separable categories of occurrence for all 20 guilds.

The next step in BCI development is to identify which levels of proportional occurrence for each guild indicate a low integrity condition, and which are indicative of high integrity. We ranked each category of occurrence for each guild on a scale of high integrity to low integrity. For specialist guilds, we ranked the highest occurrence category a "1," the next highest a "2," etc. For generalist guilds, we reversed the ranking, assigning "1s" to the lowest occurrence category. Therefore, a site can receive a rank of "1" for a guild if the site supports the highest category of occurrence for a specialist guild or the lowest category of occurrence for a generalist guild; a theoretical maximum-integrity site would have a "1" rank entered in every guild column. Returning to our example, bark probers exhibit four categories of occurrence in our sample: 0-10%, 10.5-30%, 30.5-50%, and >50% of the species at a site. Because "bark prober" is a specialist guild (Table I), we ranked the >50% category = 1, 30.5-50% = 2, 10.5-30% = 3, and the 0-10% category = 4. After applying ranks to occurrence categories for all 20 guilds, we obtained a new matrix of the 34 sites and 20 guilds with a numerical rank for each guild at each site corresponding to the proportion of species in each guild at each site. The actual BCI score assigned to a site is the series of ranks for that site's component guilds. In practice, it is easier to compare BCI scores between sites using a composite of the individual guild ranks for that site.

BCI Application. Once we had derived BCI scores for the 34 sites sampled in 1994, our next step was to group sites with similar composite BCI scores into statistically separable categories. Because all the individual response guild ranks were assigned so that the highest-integrity condition received the lowest rank, the clusters of sites containing the lowest BCI scores are the clusters indicating the highest biotic integrity. To identify categories of BCI scores, we again used cluster analysis, grouping sites according to the similarity in a composite of their individual guild ranks and by the composite score's absolute rank within the sample (1-34). We found that the sum of the squares of individual guild ranks functioned well as a composite score in our sample by ensuring that no two sites received the same composite score. We used a 95% similarity among sites as our cut-off point between clusters. The total number of clusters determined to be statistically separable through ANOVA with Tukey's procedure is the number of categories of biotic integrity in the sample. Because sites exhibit a mix of high and low integrity ranks for their 20 guilds, there is no *a priori* reason to expect that the categories of integrity obtained through the BCI application will mirror the categories of proportional guild occurrence incorporated into the BCI development.

To apply the BCI to independent sets of data collected in 1995 and 1996, we first constructed matrices of sites and guilds with the percentage of the total number of species at a site in each of 20 guilds. We then substituted the corresponding rank determined from the 1994 data for the appropriate level of proportional guild occurrence in the 1995 and 1996 samples. For example, a site sampled in 1995 at which 35% percent of the species are bark probers would receive a rank of “2" for the bark prober guild. Once ranks had been assigned for every guild at every site, we applied a combination of cluster analysis and ANOVA to determine the number of statistically separable categories of biotic integrity in the new samples. To provide a bird community-based assessment of biotic integrity for the MAHA, we pooled data from the 126 randomly-selected sites we sampled in 1995 and 1996 and applied the BCI.

3. Results

When we applied the 1994 guild ranks to the raw data from 1995 and 1996, and clustered with a 95% similarity cut-off, we obtained seven ($F=315.55$, $p=0.000$) and six ($F=354.53$, $p=0.000$) categories, respectively, of biotic integrity. Pooled 1995 and 1996 data (126 total sample sites) break into six statistically separable categories ($F=684.30$, $p=0.000$). Our results from the pooled sample indicate, with 95% confidence, that 17 (±6) % of the MAHA area supports the highest integrity bird community (category 1); 26 (±7) % is high integrity (category 2); 13 (±5) % is medium-high (category 3); 17 (±6) % is medium-low (category 4); 20 (±6) % is low (category 5); and 7 (±4) % supports the lowest integrity bird community (category 6) (Figure 2).

Table II lists the mean proportional guild occurrence of each guild in each category of biotic integrity. With the exception of native cavity nesters ($F=1.19$, $p=.320$), all 20 guilds exhibit a statistically significant difference in proportional guild occurrence between at least two categories of biotic integrity (alpha=0.05). Mean proportional guild occurrences for four of the response guilds (omnivores, insectivores, single-brooded species, and forest area-sensitive species) are statistically separable between all six integrity categories (ANOVA with Tukey multiple comparisons).

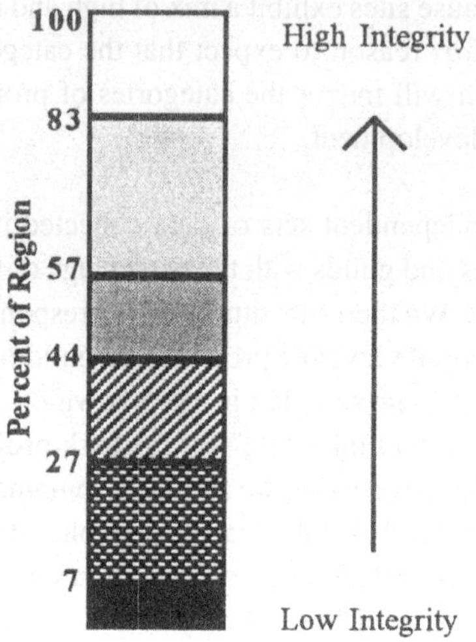

Fig. 2. Percent of study region in six categories of biotic integrity.

Table II
Mean percentage of the total number of species in each guild by integrity category. Guilds marked with an asterisk exhibit statistically separable values across all six categories at alpha = 0.05.

Response Guild	highest integrity (1)	high integrity (2)	medium-high integrity (3)	medium-low integrity (4)	low integrity (5)	lowest integrity (6)
bark prober	15%	11%	9%	9%	4%	3%
ground gleaner	12%	9%	9%	5%	3%	4%
aerial sallier	10%	8%	8%	8%	7%	4%
upper canopy gleaner	17%	16%	10%	10%	5%	2%
lower canopy gleaner	20%	17%	18%	17%	14%	10%
*insectivore	75%	64%	57%	52%	40%	27%
*omnivore	26%	36%	41%	46%	55%	65%
predator/brood parasite	7%	10%	11%	11%	15%	20%
resident	29%	32%	36%	36%	44%	54%
temperate migrant	16%	19%	22%	26%	30%	31%
tropical migrant	55%	49%	42%	37%	25%	15%
native cavity nester	22%	19%	18%	18%	19%	14%
canopy nester	37%	37%	35%	28%	26%	26%
shrub nester	19%	22%	25%	29%	27%	26%
forest ground nester	20%	16%	15%	8%	5%	2%
*single brooded	73%	66%	60%	53%	40%	32%
double brooded	24%	28%	33%	38%	46%	52%
edge species	18%	26%	35%	47%	57%	63%
forest generalist	36%	39%	39%	34%	30%	27%
*forest area sensitive	61%	48%	38%	27%	16%	8%

4. Discussion

We use the cumulative BCI score to assign categories of biotic integrity. However, the additional value of the BCI approach is in the descriptive information conveyed by the levels of representation of specific guilds in each category. While Figure 3 illustrates the stepwise reduction in the proportion of all insectivorous species from a high-integrity to a low-integrity condition, Figure 4 reveals the proportional reduction of specific insectivorous guilds at each step. Note in Figure 3 that insectivores outnumber omnivores in categories 1 through 4. Although proportionally less common in category 4 than in categories 1-3, insectivores still comprise more than 50% of the species at category 4 sites. However, examination of Figure 4 reveals that category 4 sites support significantly fewer bark probers, ground gleaners, and upper-canopy foragers than the highest-integrity category 1 sites. Thus, the BCI indicates not only the wholesale loss of insectivores between categories 1 and 4, but also reveals a compositional change in the insectivore community. This in turn helps to narrow the range of possible explanatory factors.

For example, the continued strong presence of lower-canopy foragers in category 4 sites would weaken a theory that total insect biomass is reduced from widespread pesticide use at these sites. Rather, the plot-level vegetation structure in category 4 sites may not provide appropriate foraging substrate for bark probers, ground gleaners, and upper-canopy foragers. In the next phase of our research, we will explore the relationships between landscape pattern, vegetation structure, and bird community composition to pursue associations with potential sources of disturbance in each biotic integrity category.

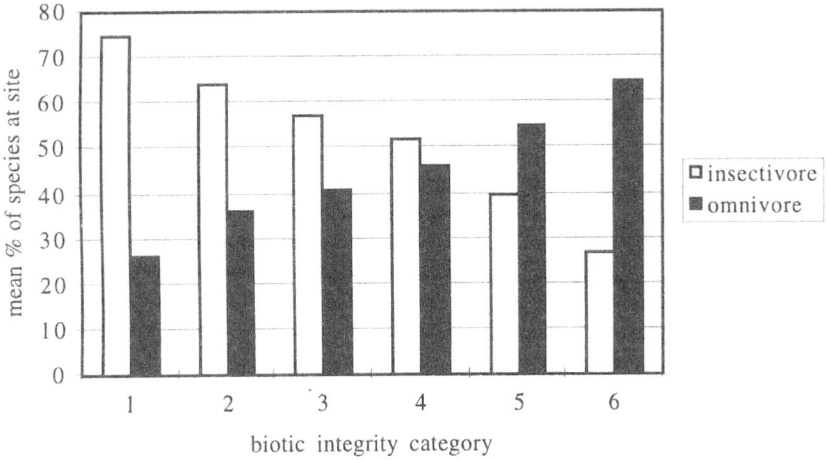

Fig. 3. Trophic guilds by integrity category.

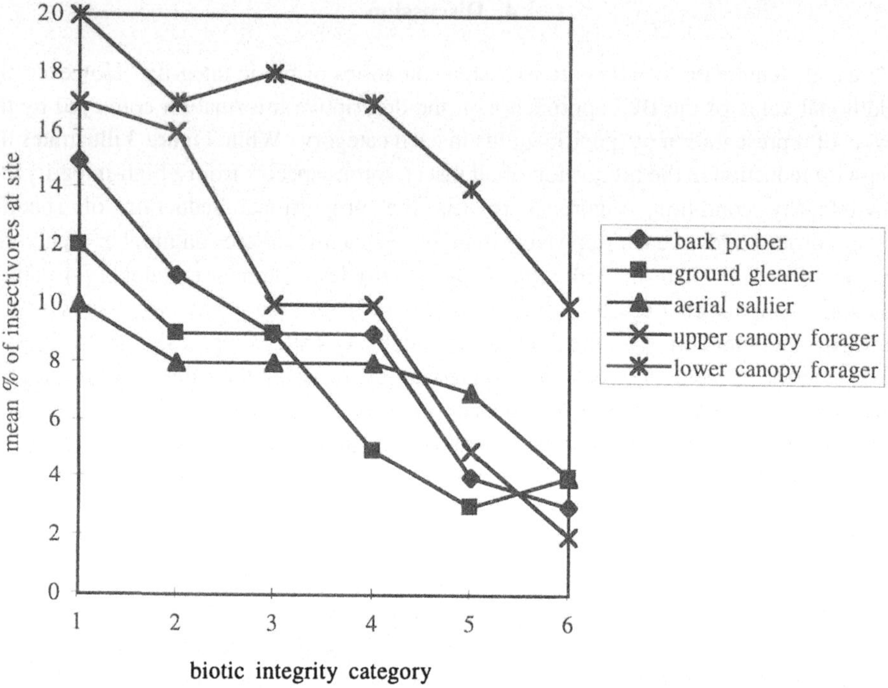

Fig. 4. Insectivore feeding behavior by integrity category.

Figure 4 also illustrates that response guilds are complementary in their ability to discriminate among biotic integrity categories. The bark probers, for example, show a significant reduction between category 1 and categories 2-4, and another between categories 2-4 and 5-6. However, we need data from the upper-canopy foragers to distinguish between categories 2 and 3, and from the ground gleaners to distinguish between categories 3 and 4. This differential sensitivity provides information beyond the aggregate BCI score that helps us to interpret the conditions that define our distinct categories of biotic integrity.

The BCI is one component towards an integrated, region-wide assessment for MAHA. The response guilds reflect a variety of physiological and behavioral aspects of the species which breed in the region, and thus are valuable as indicators of ecological condition. Several of the guilds, however, are correlated, and our ultimate BCI may contain fewer guilds. Furthermore, our preliminary results may differ from those of other indicators such as the fish IBI for MAHA streams. The ultimate assessment of ecological condition in MAHA may therefore include the results of multiple indicators.

To date, our research has focused on developing and applying the BCI to determine if bird species assemblage data can discriminate categories of biotic integrity. Pending refinement of the BCI, we will establish the association between integrity defined by the bird community and the suite of landscape and vegetation variables which support that community. Our objective will be to identify thresholds of land cover and vegetation modification at which significant changes in bird community composition occur.

Acknowledgments

We wish to thank Don Stevens of Dynamac Corporation for his assistance with the survey design-based statistical analysis of the data. We also thank Raymond O'Connor of the University of Maine, Mary Jo Casalena (nee Croonquist) of the Pennsylvania Game Commission, and David Bradford and Kevin Summers of the U.S. Environmental Protection Agency for their insightful comments during manuscript review.

References

Bailey, R. G.: 1980, 'Description of the ecoregions of the United States,' USDA Forest Service Misc. Pub. 1391, 77 pp.

Barber, M. C. (ed.).: 1994, 'The Environmental Monitoring and Assessment Program Indicator Development Strategy,' EPA/620/R-94/022, U.S. Environmental Protection Agency, Office of Research and Development, EMAP-Center, Research Triangle Park, NC.

Blake, J. G.: 1983, 'Trophic structure of bird communities in forest patches in east-central Illinois,' *Wilson Bulletin* **95**, 416-430.

Bradford, D. F., Franson, S. E., Neale, A. C., Heggem, D. T., Miller, G. R., and Cantebury, G.E.: (In press), 'Bird species assemblages as indicators of biological integrity in Great Basin rangeland,' *Environmental Monitoring and Assessment.*

Brauning, D.W. (ed.): 1992, *Atlas of Breeding Birds in Pennsylvania*, University of Pittsburgh Press, Pittsburgh, PA.

Brooks, R. P. and Croonquist, M. J.: 1990, 'Wetland, habitat, and trophic response guilds for wildlife species in Pennsylvania,' *Journal of the Pennsylvania Academy of Science* **64**, 93-102.

Brooks, R. P., Cole, C. A., Wardrop, D. H., Bishel-Machung, L., Prosser, D. J., Campbell, D. A., and Gaudette, M. T.: 1996, 'Wetlands, Wildlife, and Watershed Assessment Techniques for Evaluation and Restoration: Final Report for the Project: Evaluating and Implementing Watershed Approaches for Protecting Pennsylvania's Wetlands,' Report No. 96-2, Penn State Cooperative Wetlands Center, University Park, PA. Volumes I and II.

Brooks, R. P., O'Connell, T. J., Wardrop, D. H., and Jackson, L. E.: 1997, 'Towards a Regional Index of Biological Integrity: The Example for Forested Riparian Systems,' *Environmental Monitoring and Assessment*, this volume.

Croonquist, M. J. and Brooks, R. P.: 1991, 'Use of avian and mammalian guilds as indicators of cumulative impacts in riparian-wetland areas,' *Environmental Management* **15**, 701-714.

DeGraaf, R. M., Tilghman, N. G., and Anderson, S. H.: 1985, 'Foraging guilds of North American birds,' *Environmental Management* **9**, 493-536.

Fore, L. S., Karr, J. R., and Wisseman, R.W.: 1996, 'Assessing invertebrate responses to human activities: evaluating alternative approaches,' *Journal of the North American Benthological Society* **15**, 212-231.

Freemark, K. and Collins, B.: 1992, 'Landscape ecology of birds breeding in temperate forest fragments,' pages 443-454 in *Ecology and Conservation of Neotropical Migrant Landbirds* (J.M. Hagan III and D.W. Johnston, Eds.), Smithsonian Institution Press, Washington, D.C.

Harrison, H. H.: 1975, *A Field Guide to the Birds' Nests: United States east of the Mississippi River*, Houghton Mifflin Co., Boston, MA.

Hunsaker, C. T. and Carpenter, D. E. (eds.): 1990, 'Ecological Indicators for the Environmental Monitoring and Assessment Program,' EPA/600/3-90/060. U.S. Environmental Protection Agency, Office of Research and Development, Research Triangle Park, NC.

Hutto, R. L., Pletschet, S. M., and Hendricks, P.: 1986, 'A fixed-radius point count method for nonbreeding and breeding season use,' *Auk* **103**, 593-602.

Karr, J. R. : 1991, 'Biological integrity: a long-neglected aspect of water resource management,' *Ecological Applications* **1**, 66-84.

Karr, J. R.: 1993, 'Defining and assessing ecological integrity: beyond water quality,' *Environmental Toxicology and Chemistry* **12**, 1521-1531.

Karr, J. R. and Dudley, D. R.: 1981, 'Ecological perspective on water quality goals,' *Environmental Management* **5**, 55-68.

Manuwal, D. A. and Carey, A. B.: 1991, 'Methods for measuring populations of small, diurnal forest birds.' Gen. Tech. Rep. PNW-GTR-278, Portland, OR: U. S. Department of Agriculture, Forest Service, Pacific Northwest Research Station, 23 p.

Minitab, Inc.: 1995, *Minitab Reference Manual: Release 10Xtra*, Minitab, Inc., State College, PA.

Neter, J., Wasserman, W., and Kutner, M. H.: 1990, *Applied Linear Statistical Models*, Richard D. Irwin, Inc., Homewood, IL.

Overton, W. S., White, D., and Stevens, D. L. Jr.: 1990, 'Design Report for EMAP,' EPA/600/3- 91/053, U.S. Environmental Protection Agency, Office of Research and Development, Washington, DC.

Ralph, C. J., Geupel, G. R., Pyle, P., Martin, T. E., and DeSante, D. F.: 1993, 'Handbook of field methods for monitoring landbirds,' Gen. Tech. Rep. PSW-GTR-144, Albany, CA: Pacific Southwest Research Station, Forest Service, U. S. Department of Agriculture, 41 p.

Santner, S. J., Brauning, D. W., Schwalbe, G., and Schwalbe, P. W.: 1992, 'Annotated List of the Birds of Pennsylvania,' Pennsylvania Biological Survey No.4, Ornithological Technical Committee.

Stevens, Jr., D. L. and Kincaid, T.: (In prep., 1997.) 'Variance estimation for subpopulation parameters from samples of spatial environmental populations,' Invited paper to be presented at the annual meeting of the American Statistical Association, Aug., 1997, and to appear in the Proceedings of the Environmental Statistics section.

Szaro, R.: 1986, 'Guild management: an evaluation of avian guilds as a predictive tool,'*Environmental Management* **10**, 681-688.

Terborgh, J.: 1989, *Where have all the birds gone?*, Princeton University Press, Princeton, N. J.

Verner, J.: 1984, 'The guild concept applied to management of bird populations,' *Environmental Management* **8**, 1-14.

ACID RUNOFF CAUSED FISH LOSS AS AN EARLY WARNING OF FOREST DECLINE

WILLIAM E. SHARPE and MICHAEL C. DEMCHIK

The Pennsylvania State University; Land and Water Res. Bldg.; University Park, PA 16802

Abstract: Sulfate, nitrogen, and hydrogen ion deposition in the Laurel Hill region of the Appalachian Plateau province in Pennsylvania has been very high. Records indicate that losses of rainbow trout (*Oncorhynchus mykiss*) first occurred about 1960, although unrecorded losses probably preceded that date. Research has also attributed loss of brook trout (*Salvelinus fontinalis*) in this region to chronic and episodic stream acidification. Relatively recently, mortality of northern red oak has become a problem in parts of the region with mortalities as high as 60 percent of standing trees evident in some areas. Preliminary analysis indicates that soil acidification may play a significant role in the observed mortality. If this is the case, it would appear that fish losses due to watershed acidification in the region were evident about 30 years prior to the current mortality of northern red oak. Therefore, fish loss caused by acidification may be a prelude to more widespread ecosystem damage as a consequence of chronic deposition of acidifying elements.

1. Introduction

The Laurel Hill region of southwestern Pennsylvania (Figure 1) has received large amounts of acidic deposition since the beginning of deposition monitoring activities (Sharpe *et al.*, 1984; DeWalle and Sharpe, 1985; Lynch *et al.*, 1996). The water quality of several streams draining the Laurel Hill anticline has been intensively monitored and the existence of episodic acidification of these streams has been documented (Sharpe *et al.*, 1984; Wigington *et al.*, 1996). Episodic acidification has been shown to cause fish (primarily salmonids) mortality in numerous streams (Sharpe *et al.*, 1983; Sharpe et al., 1984; Gagen and Sharpe, 1987; Van Sickle et al., 1996, and Baker *et al.*, 1996). An extensive survey of streams draining the Laurel Hill (Sharpe *et al.*, 1987) revealed that 26 percent were fishless or had remnant fish populations. Analysis of northern red oak (*Quercus rubra*) tree wood chemistry for trees growing in this area was reported by DeWalle *et al.* (1991). Their data revealed calcium declined slightly in the tree wood of northern red oak since the beginning of this century. In addition, unpublished data for these same trees indicated increasing trends in manganese concentration. Both of these trends are consistent with increasing soil acidification in this region.

More recently, acidification research in the Laurel Hill region has focused upon relationships between acidification and forest decline. Decline of northern red oak has occurred in the region (Demchik and Sharpe, 1996; McClenahen *et al.*, 1997). Soil acidity, more specifically, unfavorable ratios of calcium to aluminum (Ca:Al), has been shown to be characteristic of mineral soils in the declining northern red oak stands that have been studied (Demchik and Sharpe, 1996). Regeneration of northern red oak following timber harvest in this region has also been problematic (Lyon and Sharpe, 1995).

Environmental Monitoring and Assessment **51:** 157–162, 1998.

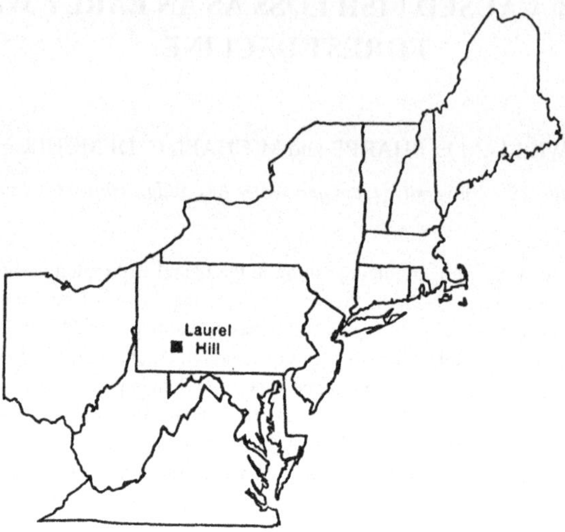

Fig. 1. Location of the Laurel Hill study site in the northeast United States.

Environmental monitoring is conducted for a number of reasons, one of which is to establish trends that may foretell environmental problems. In this way, proper corrective measures (e.g., pollution control) may be instituted to control or limit damage. In the case of environmental damage as a consequence of acidic deposition, degradation is a slow process that occurs over many decades (Ulrich, 1984). Had appropriate monitoring been in place at the beginning of this century, some of the damage from acidic deposition may have been avoided. A review of the literature and data concerning acidic deposition in the Laurel Hill region was conducted to ascertain the sequence of important environmental changes there. The intent was to gain insight into forest ecosystem response to continued, long-term inputs of acidic deposition.

2. Methods

A literature review was conducted to determine the chronology of events related to acidification of soils and water and associated biotic impacts. A combination of agency reports and anecdotal information were utilized to establish the dates of fish kills on the Linn Run Watershed in Westmoreland County, Pennsylvania. These dates were compared to radial growth trends in northern red oak trees in declining and non-declining stands of trees to determine their possible coincidence with the approximate time of radial growth decline. Onset of tree mortality was not accurately recorded; consequently, we used the approximate year of death of standing dead trees on our study plots as determined by tree ring analysis to determine the time of red oak mortality.

Twenty living and ten dead, standing northern red oak trees were cored at breast height with a 4 mm Teflon™ - coated increment bore in the summer of 1995. Ten trees were from a healthy stand on the Linn Run watershed and ten living and ten dead trees were from a declining stand (40 - 60% mortality of northern red oak). The cores were mounted in grooved hardwood blocks and sanded to improve individual ring clarity. Cores were scanned and ring widths measured to the nearest 0.01 mm with a Hewlett-Packard Scan Jet IIC and processed with Mac/Win DENDRO v3.3.2 software. Stands with no known prior defoliation by gypsy moth (*Lymantria dispar L*) were selected.

Soil was collected from a hand-excavated shallow soil pit in the rooting zone of each tree and extracted with 0.01 M $SrCl_2$ following the procedure of Joslin and Wolfe (1989). Extracted elements were determined by atomic adsorption spectroscopy using a Thermo-Jarrel-Ash (IL) Video 22 spectrometer.

3. Results and Discussion

A review of pertinent literature and state agency reports on fish kills in the Laurel Hill region was previously completed by Sharpe (1989). From this, a chronology of fish kills in Linn Run was prepared (Figure 2). The basal area increment (area of each concurrent tree ring at the point of measurement) of the northern red oak trees was determined over their life span at the time of measurement. These data are displayed as mean annual basal area increment over time in Figure 2.

It would appear from the data presented that there are no declines per se in the basal area increment of the northern red oak trees sampled. However, there does appear to be a large downward step in basal area increment in the live tree population on the declining site starting about 1952. This change is not evident in the dead tree population from the same site. The dead tree population shows a decline in the last few years prior to death. Other than this decline, which occurred in about 1988, there is no indication of a declining trend in basal area growth in either of the other two northern red oak populations sampled. Dead trees died in the period from 1988-94; 28-34 years after the first fish kills were reported in the Linn Run watershed.

Soil chemical analysis indicated that molar ratios of calcium to aluminum were quite low (0.7) in the mineral soil at the site with declining red oak. Cronan and Grigal (1995), Tomlinson and Tomlinson (1990) and Ulrich (1989) have reported that Ca:Al ratios less than one are indicative of aluminum stress to trees. Swistock *et al.*, (1989) showed that soil water with high Al concentrations entering surface water on the Linn Run watershed was responsible for the high Al concentrations during acidic runoff episodes. We assume from this and the previously referenced studies on fish loss in Linn Run that acidification and, more specifically, Al stress has played a significant role in the demise of both fish and northern red oak on the Linn Run watershed. Obviously, where shallow water flow paths do not deliver significant amounts of Al to surface streams, forest decline may be present

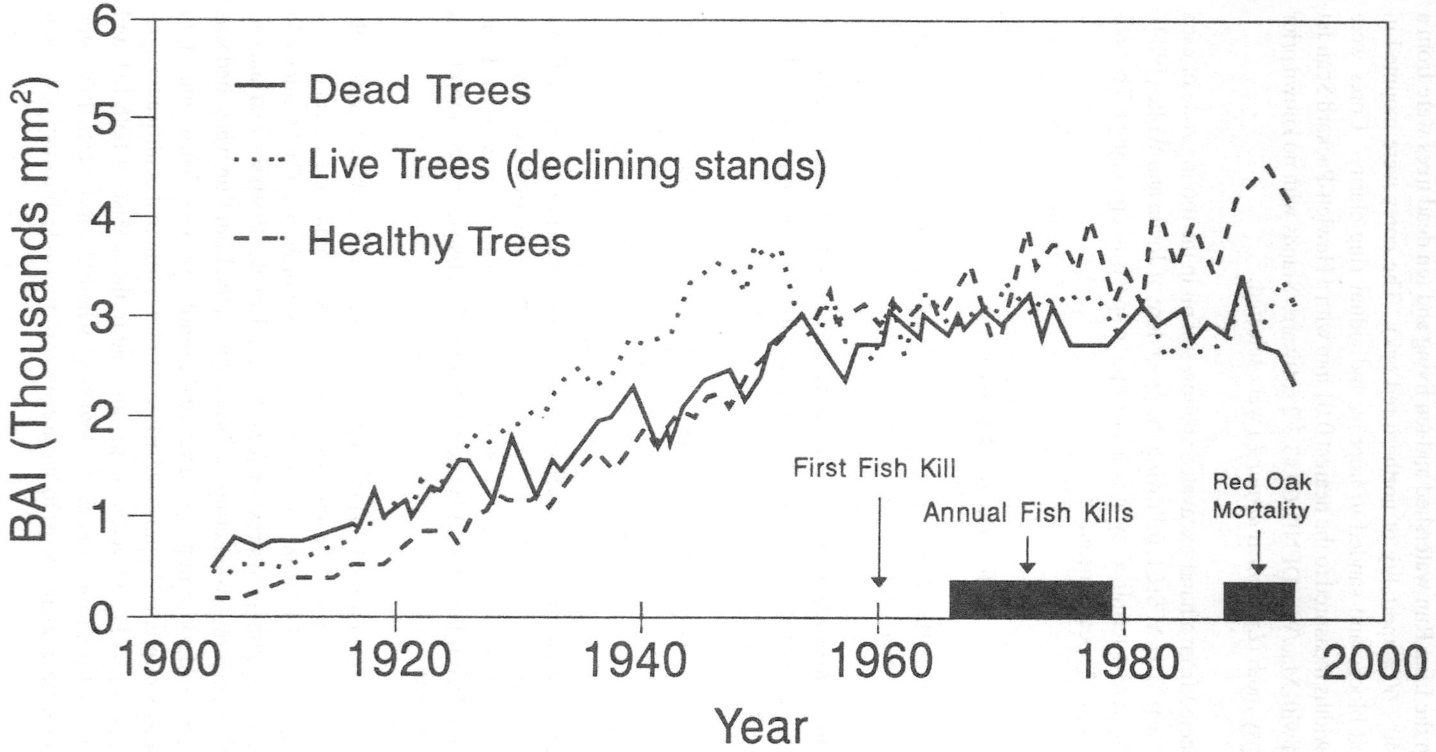

Fig. 2. Annual mean basal area increment for living and dead northern red oak from the decline site and relatively healthy northern red oak from the healthy site on the Linn Run watershed and the reported dates of fish kills in Linn Run.

in the absence of effects to aquatic biota. This is probably the case for many watersheds in the Laurel Hill region.

Although some northern red oak stands on the Linn Run watershed are obviously growing on soils that have more favorable Ca:Al ratios (the Ca:Al ratio for the healthy stand in Figure 2 was 2.1), others, are not. The ten recently deceased northern red oaks growing on soils with Ca:Al ratios less than one died from 1988-94, about 30 years after the first fish were reported to have died during an acid runoff episode on Linn Run. Thus, it would appear that transport of Al to stream water during acid runoff episodes preceded tree mortality by about three decades.

Interestingly, northern red oak growing on the low Ca:Al sites showed a step change in basal area growth during the 1950's. For these trees, basal area growth fluctuated around 3000 mm^2 until the present, while trees at the healthy site continued to increase in growth to 4000 mm^2 over this same time period. This change in growth was roughly coincident with the onset of fish loss in the Linn Run watershed and would be consistent with a shift of the more poorly Ca buffered soils on the watershed to predominantly Al buffering and the subsequent dominance of Al in soil water on the Laurel Hill (DeWalle and Sharpe, 1985). Thus, step changes in basal area increment growth could signal major environmental change, particularly where soil acidification related perturbations to tree nutrient supply are involved. In this case such perturbations would include the well known interference by Al with the uptake of Ca and magnesium (Tomlinson and Tomlinson, 1990). We suggest that these relationships be examined in other watersheds (regions) to determine if, indeed, fish kills owing to acidification are indicative of future forest decline.

Acknowledgements

Funding for this work was provided, in part, by the School of Forest Resources and the Environmental Resources Research Institute at The Pennsylvania State University and in part, by the Allegheny Foundation. We are indebted to Bryan Swistock for designing the figures and to Bob Carline for reviewing a preliminary draft of the manuscript.

References

Baker, J. P., Van Sickle, J., Gagen, C. J., DeWalle, D. R., Sharpe, W. E., Carline, R. F., Baldigo, B. P., Murdoch, P. S., Bath, D. W., Kretser, W. A., Simonin, H. A., and Wigington Jr., P. J. 1996. "Episodic Acidification of Small Streams in the Northeastern United States: Effects on Fish Populations". *Ecol. Appl.* **6**, 422-437.

Cronan, C. S., and Grigal, D. F. 1995. "Use of Calcium/Aluminum Ratios as Indicators of Stress in Forest Ecosystems". *J. Environ. Qual.* **24**, 209-226.

Demchik, M. C. and Sharpe, W. E. 1996. "*Quercus rubra* Decline in Southwestern Pennsylvania." *Supplement to Bull. of the Ecol. Soc. of Amer. Program and Abstracts, Part* 2, **77**, 109.

DeWalle, D. R. and Sharpe, W. E. 1985. "Biogeochemistry of Three Appalachian Forest Sites in Relation to Stream Acidification." *The Pennsylvania State University, Environmental Resources Research Institute, Final Rept. to U.S.D.I. Geological Survey* (LW8510), 37p.

DeWalle, D. R., Swistock, B. R., and Sharpe, W. E. 1991. "Radial Patterns of Tree Ring Chemical Element Concentration in Two Appalachian Hardwood Stands." *In Proc. 8th Central Hardwood Forest Conference.* (Eds.) L. H. McCormick and K. W. Gottschalk, *USDA For Serv. Gen. Serv. Gen. Tech. Rep. NE* **148**, 459-474.

Gagen, C. J. and Sharpe, W. E. 1987. "Net Sodium Loss and Mortality of Three Salmonid Species Exposed to a Stream Acidified by Atmospheric Deposition." *Bull. of Env. Cont. and Tox.*, **39**, 7-14.

Joslin, J. D. and Wolfe, M. H. 1989. "Aluminum Effects on Northern Red Oak Seedling Growth in Six Forest Soil Horizons." *Soil Sci. Soc. Am. J.*, **53**,274-281.

Lynch, J. A., Horner, K. S. and Grimm, J. W. 1996. *Atmospheric Deposition: Spatial and Temporal Variations in Pennsylvania* - 1995, (ER9607) 275p.

Lyon, J. and Sharpe, W. E. 1995. "Impacts of Electric Deer Exclusion Fencing and Soils on Plant Species Abundance, Richness, and Diversity following Clearcutting in Pennsylvania." *In Proc. 10th Central Hardwood Forest Conf.* (Eds.) K. W. Gottschalk and S. L. C. Fosbroke, *USDA for. Serv. Gen. Tech. Rep. NE-***197**, 47-59.

McClenahen, J. R., Hutnik, R. J., Davis, D. D. 1997. "Patterns of Northern Red Oak Growth and Mortality in Western Pennsylvania." *In Proc. 11th Central Hardwood Forest Conf.*, (Eds.) S. G. Pallardy, R. A. Cecich, H. G. Garrett and P. S. Johnson, *USDA For. Serv. Gen. Tech. Rep. NC*-188, 387.

Sharpe, W. E., Young, E. S., Kimmel, W. G. and DeWalle, D. R. 1983. "In-situ Bioassays of Fish Mortality in Two Pennsylvania Streams Acidified by Atmospheric Deposition, Northeastern." *Env. Sci.* **2**, 171-178.

Sharpe, W. E., DeWalle, D. R., Leibfried, R. T., Dinicola, R. S., Kimmel, W. G. and Sherwin, L. S. 1984. "Causes of Acidification of Four Streams on Laurel Hill in Southwestern Pennsylvania." *Journ. Env. Qual.* **13**, 619-631.

Sharpe, W. E., Leibfried, V. G., Kimmel, W. G. and DeWalle, D. R. 1987. "The Relationship of Water Quality and Fish Occurrence to Soils and Geology in an Area of High Hydrogen and Sulfate Ion Deposition." *Water Res. Bull.* **23**, 37-46.

Sharpe, W. E. 1989. "Impact of Acid Precipitation on Pennsylvania's Aquatic Biota: An Overview." *In Proc. Of the Conference on Atmospheric Deposition in Pennsylvania: A Critical Assessment.* (Eds.) J. A. Lynch, E. S. Corbett, and J. W. Grimm, Environmental Resources Research Inst., The Pennsylvania State Univ., University Park, PA 16802, **98**-107.

Tomlinson, G. H. and Tomlinson, M. F. 1990. *Effects of Acid Deposition on the Forests of Europe and North America.* CRC Press. Boca Raton, FL. 281 p.

Swistock, B. R., DeWalle, D. R., and Sharpe, W. E. 1989. "Sources of Acidic Stormflow in an Appalachian Headwater Stream." *Water Res. Resch.* **25**, 2139-2147.

Ulrich, B. 1984. "Ion Cycle and Forest Ecosystem Stability. In State and Change of Forest Ecosystems - Indicators in Current Research." (Ed.) *G. I. Agren, Swed. Univ. Agric. Sci. Rep. No. 13*, 207-233.

Ulrich, B. 1989. "Effects of Acidic Precipitation on Forest Ecosystems of Europe." p. 189-272. *In Advances in Environmental Science - Acid Precipitation Series.* (Eds.) D. C. Adriano and M. Havas. Vol. 1, Springer-Verlag, NY.

Van Sickle, J., Baker, J. P., Simonin, H. A., Baldigo, B. P., Kretser, W. A., and Sharpe, W. E. 1996. "Episodic Acidification of Small Streams in the Northeastern United States: Fish Mortality in Field Bioassays." *Ecol. Appl.* **6**, 408-421.

Wigington, Jr., P. J., DeWalle, D. R., Murdoch, P. S., Kretser, W. A., Simonin, H. A., Van Sickle, J., and Baker, J. P. 1996. "Episodic Acidification of Small Streams in the Northeastern United States: Ionic Controls of Episodes." *Ecol. Appl.* **6**, 389-407.

FOREST INTEGRITY AT ANTHROPOGENIC EDGES: AIR POLLUTION DISRUPTS BIOINDICATORS

MARIAN G. GLENN[1], SARA L. WEBB[2], and MARIETTE S. COLE[3]

[1] *Biology Department, Seton Hall University, South Orange NJ 07079,* [2]*Biology Department, Drew University, Madison NJ 07940,* [3]*Biology Department, Concordia College, St. Paul MN 55104*

Abstract. The response of corticolous lichens, bryophytes, and vascular plants to anthropogenic edges in northern hardwood forest preserves is compared in east-coast and mid-west (NW Minnesota) sites, using micro-epiphytes on red oak (*Quercus rubra*) and sugar maple (*Acer saccharum*). The drastically attenuated lichen flora in the east, apparently due to regional air pollution, restricts the usefulness of these bioindicators, even 120 km from New York City. The forest edge is not necessarily equated with increased light. Established edges may have pronounced shoot growth that shades epiphytes. In the absence of air pollution, lichen and bryophyte species exhibit individual responses to light, humidity, and substrate chemistry. Thus summary variables such as total cover or species richness have limited value as bioindicators of forest integrity.

1. Introduction

This project explored the utility of corticolous (bark-dwelling) lichens, bryophytes, and vascular plant assemblages, separately and together, as potential bioindicators of forest integrity. Lichens seem promising because of their extreme sensitivity to environmental conditions, such as light, humidity, and substrate chemistry (Barkman, 1958; Culberson, 1955; Hale, 1955; Jesberger and Sheard, 1973). Lichens are well known as pollution indicators (e. g. The *Lichenologist*'s serialized bibliography Literature on air pollution and lichens). However, our project probes lichens as more general indicators of forest integrity. Bioindicators are most practical if simple measurements taken by nonexperts are ecologically meaningful. Estimated total cover is the simplest measurement for non-experts; species richness is more meaningful, yet more difficult, since species identifications of lichens and bryophytes require a laboratory and expert knowledge.

Moreover, the value of these general, summative measures is questioned because rarely are the various species equivalent ecologically. Disturbed, low-integrity habitats can exhibit high species richness but with different species predominating over those of intact forests; likewise, a large cover value may be due to a few disturbance-adapted species. Also, reciprocal changes within summed groups can mask important differences. For example, one foliose lichen *(Phaeophyscia rubropulchra)*, is abundant only at the forest edge, while another, (*Flavoparmelia caperata*) rare at the edge, increases toward the humid interior.

Forest integrity is a general concept, subsuming multiple concepts such as forest age, forest continuity, and disturbance regime. The work described here characterizes two types of forest integrity gradients: (1) from forest edges, where fragmentation generates large

Environmental Monitoring and Assessment **51:** 163–169, 1998.

zones of edge-influenced habitat, to forest interiors, and with yet further comparison with similar forest stands without anthropogenic edges; and (2) from pollution-impacted sites approximately 100 km from New York City to remote sites in northern Minnesota, with shared tree species for control of the epiphytes' bark substrate.

2. Study Areas

The portion of our project reported here is based on intensive sampling in three northern hardwood forests with anthropogenic edges, one each in New Jersey (Stokes State Forest, Sussex County, 41° 13' N. lat., 74° 44' W. long.), New York (Mohonk Preserve, Ulster County, 41° 47' N. lat., 74° 24' W. long.), and Minnesota (Itasca State Park, Clearwater County, 47° 11' N.lat., 95° 11' W. long.). We also report on microepiphytes in two other Minnesota hardwood forests without edges (locations very close to the latter). This exploratory project surveyed a variety of edge types. The New Jersey site (NJ) faced a 4-year old clearcut; the New York and Minnesota sites (NY and MN) faced mowed fields maintained for decades. Thus comparable old edges were located in both regions. NJ and MN face south, NY faces north. Other edges studied, in Minnesota pine forests, are discussed in Glenn and Webb (1997).

3. Methods

At each study area, we established a set of plots measuring 5x20 m, within which trees were sampled for micro-epiphytes and light measurements as well as stand attributes. Also within these plots, vascular plants were sampled quantitatively in two strips of five 1x1m subplots, laid out along the width of the plot. At sites with edges, plots were lined up in three belts parallel to the edge: edge belts (E) were 0-5 m from the opening, middle belts (M) at 20-25 m from the opening, and interior belts (I) at 40-45 m from opening. Each belt was a line of three or four plots, depending upon the size of the edge. For intact sites (no anthropogenic edge within 100 m) identical analysis was carried out in three plots (each 100 m^2) per stand.

Lichens and bryophytes were sampled in a rectangular template (10x25 cm) placed on eight points for each tree in each plot: at the base and at 1.5 m above ground, for each of the four cardinal compass directions. In this analysis, the 4 plots at each height were combined, and only red oak and sugar maple, trees well-represented in both regions, are discussed. All microepiphytes were given cover estimates and species-level identifications, to the extent possible; collections were made for laboratory confirmation. Inclusion of the sterile crustose lichens and *Cladonia* squamules was made possible by use of a HPTLC analytical system (CAMAG, North Carolina), to characterize the lichen secondary chemicals necessary for species identification. Light measurements were taken at each upper lichen plot between 10:00 and 14:00 hr, using a multi-sensor ceptometer ("Accupar,"

Decagon Devices, Pullman, WA). Data analysis utilized SPSS for Windows (Norusis, 1993).

4. Results and Discussion

Figures 1-3 present a comparison between the micro–epiphytes and vascular plants found in the two eastern forest stands and those found in the three midwestern stands, using as variables the cover and species richness of all bryophytes, of all lichens and of vascular plants divided by growth habit. Differences discussed in the text are significant at the level of $P < 0.05$, unless noted otherwise. A number of general observations emerge.

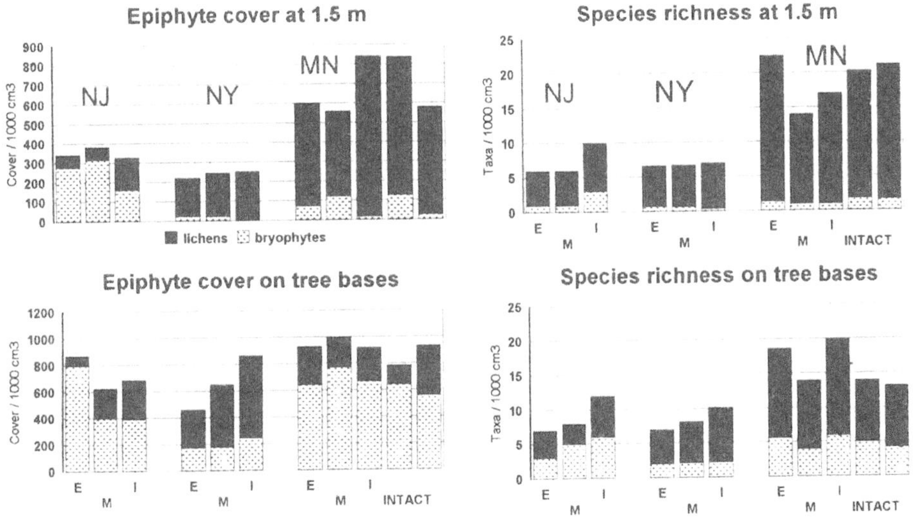

Fig. 1. Micro–epiphytes on red oak. E=edge, M=middle, I=interior. Intact stands had no edge

1. The east-coast sites were surprisingly depauperate in both cover and richness of lichens (Figures 1 and 2), a result not explained by light levels (Figure 3) but almost certainly due to regional air pollution (Garner et al., 1989). The warmer, more humid eastern region would otherwise lead to a richer lichen and bryophyte flora. Historical records show 64% of lichen species reported for the area 100 years ago have disappeared (Wetmore, 1987). At the NY site paired photographs document a similar decline in rock-dwelling lichens (Smiley and George, 1974). Our measurements of airborne particles and heavy metals regionally show that this area falls intermediate between the urban readings c. 20 km from New York City, where lichen flora is restricted to pollution tolerant species in protected microhabitats, and measurements from the pristine Adirondacks that have a lush and diverse micro-epiphyte flora (Glenn et al., 1991; Orsi and Glenn, 1991).

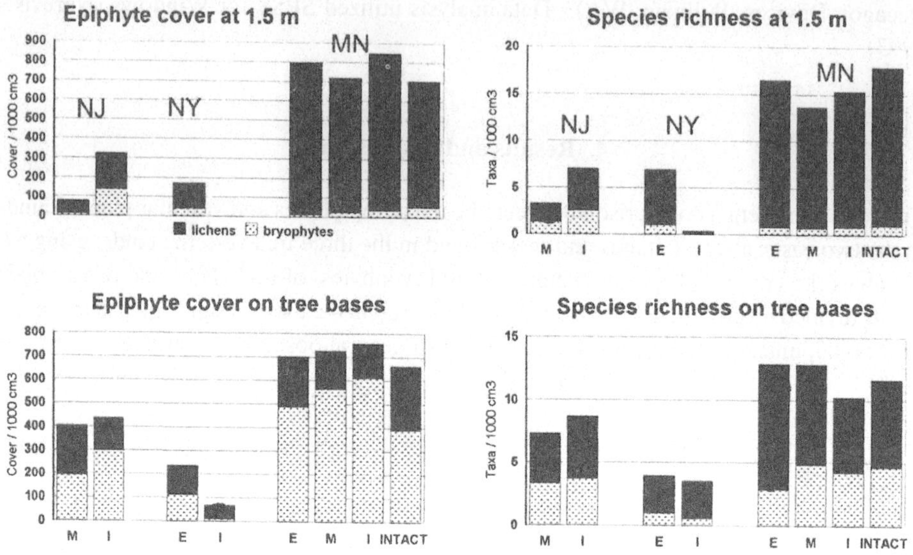

Fig. 2. Micro-ephiphytes on sugar maple. E=edge, M=middle, I=-interior. Intact stand had no edge. NJ edge and NY middle had no sugar maples present.

Our results also suggest that pollution interacts with fragmentation. Pollution deposition is intercepted first and in greatest amounts at forest edges and openings. Lichens that account for the higher species richness at the MN edge are absent from both the bright NJ and the shaded NY edges (e. g. Physciaceae spp.) The large foliose lichens (Parmeliaceae spp.) that contribute to the high cover values in MN are present but small and sparse, in the eastern sites. This disturbing loss of diversity and cover by micro-epiphytes in the East has implications for biomonitoring efforts. With the effects of pollution swamping lichen distribution patterns there remains little potential for detecting other aspects of forest integrity from lichen assemblages in these regions, even rural areas 120 km from New York City.

At the base of trees, bryophytes, the predominate epiphyte, showed highest cover in MN and lowest cover in NY (Figures 1 and 2). Also, mosses in both NJ and NY were almost always sterile and many in NY were small and malformed. Higher on the trunk, NJ trees had abundant *Frullania eboracensis,* a corticolous liverwort frequent at the other sites as well, but not so abundant. Edge effects, such as the increase in bryophyte cover seen on tree bases in Figure 2, are not significant (P= .16), nor is the decrease in lichen cover (P=.11 for NJ and MN).

Significant differences were found, however, in the cover of *Cladonia* spp. squamules at the base of sugar maples in NJ and red oaks at NY. Interestingly, *Cladonia* squamules were more abundant in NY in the interior, but in NJ the *Cladonia* squamules were more

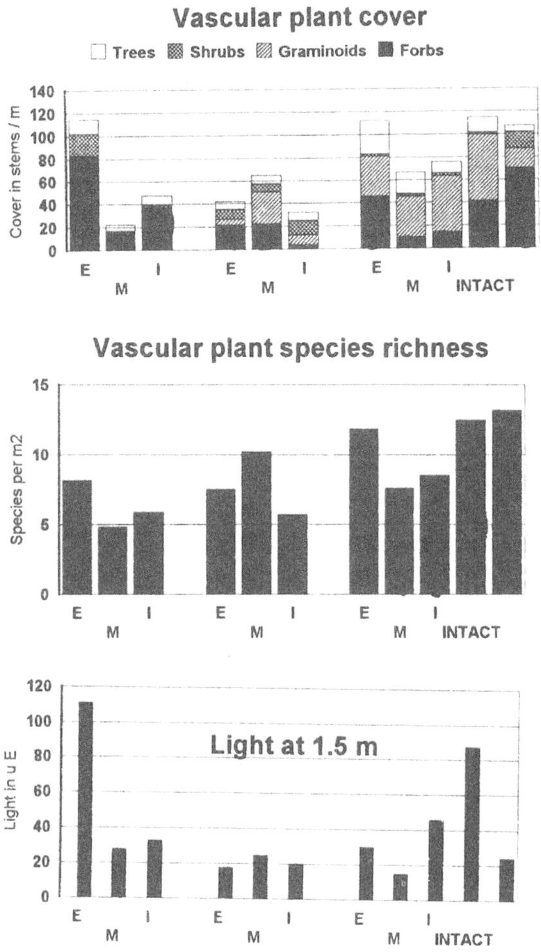

Fig. 3. Edge effects on plants and light. E=edge, M=middle, I=interior. Intact stands had no edge.

abundant at the edge. These and other individual responses of micro-epiphyte species make clear the limitations in using summary variables such as total cover as a monitoring variable.

The richness and cover of vascular plants were generally higher in MN than the eastern sites; but this regional difference was not as dramatic as for the epiphytes (Figure 3). The edge effect for vascular plants varied among sites. When edges had higher plant cover and richness, this was explained by higher light levels.

2. The summary variables of total cover and species richness, useful for comparing pollution effects, are unreliable predictors of forest integrity, and thus biomonitoring requires taxonomic and ecological expertise.

When fragmentation and forest management change the distribution of light and humidity, the elevated heterogeneity should theoretically increase the numbers of species present. Our findings with both micro-epiphytes and vascular plants show that species that are added, or are predominate, in openings are not those of the forest interior which may, in fact, decline or drop out of the assemblage. Thus valid assessments of forest integrity require an understanding of the ecolgical requirements of individual species, and the ability to make precise taxonomic identifications. Counting numbers of species will not alone elucidate the condition of the forest as a community, although these general measures are useful in regional comparisons.

3. For corticolous micro-epiphytes, the effect of forest fragmentation does not fit the classical scenario of increased light at the edge. In MN and NY, which had long-established edges, trees had produced a dense growth of lower branches. This resulted in lower light reaching lichens at the upper trunk plots on edge trees than on some interior trees exposed to canopy openings, as seen in MN. Thus the age of the edge is an important factor in the response of the micro-epiphyte community.

Much recent work demonstrates the ecological problems that edges pose for forest communities (Wilcove, 1987; Alverson et al., 1988, 1994). Although edges can increase species richness, particularly for deer and other game species, the intact forest assemblage (of the "high-integrity" forest) is damaged. These negative edge effects have important implications for natural lands management, which often reflects old favorable views of edges, and also for the wisdom of using total cover or species richness as a predictor for ecological integrity. The species promoted by edges may be invasive weeds and opportunisitic animals such as cowbirds, at the expense of perhaps fewer but native species that indicate forest integrity.

There are several mechanisms by which edges alter epiphyte communities: increased light, lowered humidity (Gustafsson, et al. 1992; Sillett, 1994, 1995), and enhanced deposition of atmospheric pollution (Farmer, 1993). Epiphyte species respond in an individual manner to these factors, according to other studies not explicitly focused on edges (Hyvarinen et al., 1992; Jesverger and Sheard, 1973; McCune and Antos, 1982; Rose, 1992; Slack, 1977; Schmitt and Slack, 1990). The use of micro-epiphytes as indicators of edge conditions is complicated by the variety of these sensitivities, along with the difficulty of making species-level identifications.

Acknowledgments

This work was supported by the U.S. Environmental Protection Agency Office of Exploratory Research, through grant number R82-1641-010. The authors also thank Angela Kociolek, Kathy Helbling, Arek Nowak, Susan Burnham, and Johanna Szillery for tireless assistance in field and lab, Jack Shuart of Stokes State Forest for help in locating field sites, Paul Huth, and the Mohonk Preserve for field support, Richard Harris, Georgeann Maxson and Benito Tan for taxonomic consultation, and Laura Jackson and the anonymous reviewers for helpful suggestions in presenting the work.

References

Alverson, W.S., Waller D. M. and Solheim, S.L.: 1988, *Conservation Biology*. **2**, 348-358.

Alverson, W.S., Kuhlman, W. and Waller, D. M.: 1994, *Wild forests, conservation biology and public policy*. Island Press, Washington, D.C.

Barkman, J. J.: 1958. *Phytosociology and ecology of cryptogamic epiphytes*. Assen, (Van Gorcum) 628 pp.

Culberson, W. L.: 1955, *Ecological Monographs*. **25**, 215-231.

Farmer, A. M.: 1993, *Environmental Pollution*. **79**, 63-65.

Garner, J.H.B., Pagano, T. and Cowling, E. B. : 1989, *An Evaluation of the role of ozone, acid deposition, and other airborne pollutants in the forests of Eastern North America*, USDA General Technical Report SE-59, Asheville, NC.172 pp.

Glenn, M.G., Orsi, E. V. and Hemsley, M. E.: 1991, *Grana* **30**, 47-47.

Glenn, M.G. and Webb, S. L.: 1997, *Bibliothecha lichenologia* In press.

Gustafsson, L., Fiskesjo, A., Ingelog, T. *et al.*: 1992, *Lichenologist* **24**, 255-266.

Gustafsson, L., Fiskesjo, A., Hallingback, T. *et al.*: 1992, *Biol. Conservation* **59**, 175-181. Hale, M.E.: 1955, *Ecology* 36, 45-63.

Hyvarinen, M. Halonen, P. and Kauppi, M.: 1992, *Lichenologist* **24**, 165-180.

Jesberger, J.A. and Sheard, J. W.: 1973, *Canadian Journal of Botany* **51**, 185-201.

McCune, B. and Antos, J. A.: 1982, *Bryologist* **85**, 1-12.

Orsi, E. V. and Glenn, M. G.: 1991. *Grana* **30**, 51-58.

Norusis, M.J.: 1993, *SPSS for Windows*, Release 6.0. SPSS, Inc., Chicago IL.

Rose, F.: 1992, In: Bates, J. W. and Farmer, A. M. (eds) *Bryophytes and lichens in a changing environment*, pp. 211-233. Oxford (Clarendon Press).

Schmitt, C. K. and Slack, N. G.: 1990, *Bryologist* **93**, 257-274.

Sillett, S. C.: 1994, *Bryologist* **97**, 321-324.

Sillett, S. C.: 1995, *Bryologist* **98**, 301-312.

Slack, N. G.: 1977, *Species diversity and community structure in bryophytes:New York State Studies*. N. Y. State Museum, Bull. No. 428, Albany, NY.

Smiley, D. and George, C. J.: 1974, *Bryologist* **77**, 179-187.

Wetmore, C.M.: 1987, Lichens and air quality in Delaware Water Gap Recreation Area: Final Report to Natl Park Service. Botany Depart., Univ. Minnesota, St. Paul, MN.

Wilcove, D.S.: 1987, *Natural Areas Journal* **7**, 23-29.

Woodley, S., Francis, G. and Kay, J.: 1993, *Ecological integrity and the management of ecosystems* St. Lucie Press, Delray Beach FL.

COMMON PATTERNS OF ECOSYSTEM BREAKDOWN UNDER STRESS

DAVID J. RAPPORT[1], WALTER G. WHITFORD[2], and MIKAEL HILDÉN[3]

[1] *Faculty of Environmental Sciences, University of Guelph,* [2]*United States Environmental Protection Agency, P.O. Box 93478, Las Vegas, NV,* [3] *Finnish Environment Institute, P.O. Box 140, Helsinki, Finland*

Abstract. Ecosystems, despite their diversity, respond to stress in similar ways. The major pressures which cause the transformation of systems from healthy states to pathological states are classified into four groups: physical restructuring, overharvesting, waste residuals, and the introduction of non-native species. Signs of Ecosystem Distress Syndrome (EDS) are briefly examined in three contrasting ecosystems: desert grasslands, the Great Lakes, and the Baltic Sea. The issue is raised as to the difficulty in discerning between healthy ecosystems, recovering from a natural disturbance, and those ecosystems that have lost their original resilience due to anthropogenic stress. Knowledge of site history and a rigourous monitoring program are important in the evaluation of EDS. An assessment of how ecosystem services are affected is indicative of the consequences to the human component of ecosystems. Management strategies which are employed to mitigate the signs of EDS are usually initiated after resilience is lost or the ecosystem has transformed to an alternate, stress-induced, stable state. It is proposed that preventive strategies measure signs of EDS that serve as early warning signals, combined with "fitness tests" that measure ecosystem response to natural perturbations. The fitness test for ecosystems is based on the premise that unstressed systems are more resilient to natural disturbances than stressed systems.

1. Introduction

It is one of those refreshing simplifications that natural systems, despite their diversity, respond to stress in very similar ways. Partly, we suspect, this is owing to similar stress pressures that impact nearly all of the Earth's ecosystems. But partly, it also reflects the fact that ecosystem breakdown naturally has manifestations in all of the ecosystem properties. Thus, those features that are commonly measured and reflect ecosystem condition, also reflect stress pressures that impinge upon ecosystems.

In this paper, we briefly review the major forces which serve to transform ecosystems from healthy states to those exhibiting pathological conditions. We outline the major signs of ecosystem distress; we discuss the impacts of ecosystem disfunction on the supply of ecosystem services; and we discuss the implications of this for management options.

In developing this overview on the transformation of ecosystems under stress pressures, we draw upon our investigations of three specific ecosystems: the desert grasslands of the southwest U.S.A. (Whitford, 1995), the Great Lakes (Rapport, 1992) and the Baltic Sea (Rapport, 1989; Hildén and Rapport, 1993).

Environmental Monitoring and Assessment **51:** 171–178, 1998.

2. Stress Pressures

Four major pressures have affected, to some degree, all three of the ecosystems investigated. One of the most pervasive stress pressures is physical restructuring. Here we speak of wetland drainage, restructuring shorelines, damming rivers, clearing of rapids, and the like. These practices are commonplace in both the Great Lakes and the Baltic Sea. In the desert grasslands, physical restructuring takes place with the loss of vegetation due to overgrazing. This causes the redistribution of topsoil, losses of soil silt fraction, dune formation, and channel cutting as a result of flash floods.

Another pressure that is widely present is overharvesting. Overharvesting has played a significant role in the Great Lakes, contributing to the extirpation of some fish species, particularly the large nearshore benthic species such as sturgeon (*Acipenser fulvescens*) and whitefish as well as valued offshore deep water species, such as the ciscoes (*Coregonus* spp.) (Regier and Baskerville, 1986). In the desert grasslands, it is the native grasses themselves that have been overharvested by the overstocking of cattle.

Pollution, or what may be more appropriately referred to as the generation of waste residuals, also takes its toll in all three ecosystems. The pervasive influence of toxic substances from industrial discharge as well as sewage discharge and runoff from agricultural lands has impacted waterbodies worldwide and is particularly prevalent in the Baltic Sea and the Great Lakes (Harris *et al.*, 1988). Waste residuals in desert grasslands are confined, by and large, to livestock wastes which impact ephemeral lakes.

The introduction of exotic species has played a major role in the Great Lakes basin in which the sea lamprey (*Petromyzon marinus*), alewife (*Alosa* sp.), and rainbow smelt (*Osmerus* sp.) have displaced native species. Pacific salmon (*Oncorhynchus* sp.) was purposefully introduced into these waters in an effort to control the smelt and the alewife. The introduction of zebra mussels (*Dreissena* sp.) has also served to transform the system. In contrast, the Baltic has few introduced species and none of them have had a serious impact on ecosystem structure or function. In the desert grasslands, an African grass, Lehmann lovegrass (*Eragrostis lehmanniana*), was introduced in order to control erosion, but it has displaced many of the native grasses. Likewise, gemsbok and cattle have displaced the indigenous prong-horned antelope.

These pressures act synergistically with natural stresses such as extreme storm events, periods of drought, *etc.*, resulting in the transformation of ecosystems from one dynamic state of equilibrium to another. However, these transformations result in alterations of ecosystem function which impact ecosystem services and ultimately the quality of life for the human component of the system.

3. Profile of Response

Elsewhere (Rapport *et al.*, 1985; Odum, 1985; Whitford, 1995; Whitford, Rapport and Groothousen, 1996), we have documented what has become known as the "Ecosystem Distress Syndrome" (EDS), a group of signs of ecosystem dysfunction in response to stress. These signs have been confirmed in a number of ecosystems in addition to the three that are our present focus. However, within these three, it is clear that the signs shown in Table 1 consistently appear.

Thus, we have a group of common signs of ecosystem dysfunction, basically pointing to the fact that ecosystems under stress differ in a number of features from the unstressed state. However, one complication here is that many of the differences are also characteristics of ecosystems at early stages of secondary succession. How, then, can one distinguish between a healthy system that is in a stage of regeneration or recovery from disturbance and an unhealthy system that is in a state of stress?

To do this requires knowledge of the history of the system, knowledge of the existing or potential stress forces that are acting on the system, and careful monitoring of how the system is evolving. Stressed ecosystems not only reflect the signs of ecosystem distress, but they continue to degrade. That is, they run downhill. Healthy, regenerating systems, while exhibiting signs of disorganization at early stages of succession, evolve to more integrated, functional systems. That is, they run uphill. Rates of change, however, may be so slow that it may take decades before there is evidence of which "direction" the system is going. In practice, diagnostics draw upon an evaluation of risks (risk assessment) and ecosystem response. The EMAP program has tended not to use risk indicators, however, but their utility in an early warning system is acknowledged here (Rapport 1989).

4. Impacts on Ecosystem Services

Ecosystem services may be defined as all functions of the ecosystem which have some value to human society. The services that are most often discussed are the obvious functions of ecosystems, such as, the provision of renewable resources (fish, fiber, wildlife), clean air and drinkable water. However, there are many other ecosystem functions that are of direct and indirect benefit to humans, as these are involved in the control of pest species, maintenance of soil fertility, provision of a reservoir of biodiversity, *etc.*

TABLE I
Changes in Ecosystem Properties in Response to Stress Shown by Desert Grasslands of US, Great Lakes and Baltic Sea.

System Property	Increase/decrease
Primary productivity	+/-[1]
Horizontal nutrient transport	+
Species diversity	-
Disease prevalence	+
Population regulation	-
Reversal of succession	+
Metastability	-
Community structure:	
r-selected species	+
Short lived species	+
Smaller biota	+
Exotic species	+
Mutalistic interactions between species	-
Boundary linearity	+
Extinction of habitat specialists	+

It is the loss of services in many ecosystems (Daily, 1997) that has galvanized renewed interest by the public in the whole issue of ecosystem health and integrity. For these losses have more immediate consequences than merely aesthetic or sentimental value. They directly effect economic opportunity (Costanza *et al.*, Nature, 1997) and human health (Epstein & Rapport, 1996). Stressed ecosystems directly result in loss of these valued services. In the Baltic Sea basin, there has been deterioration of water quality resulting in a water supply unsuitable for human consumption, losses of valued fisheries (e.g., salmonides), deterioration in air quality, and soil erosion. In the Great Lakes, loss of ecosystem services involve deterioration in water quality, loss of commercial fisheries in preferred species, reduction of air quality resulting in risks to human health, *etc*. In the desert grasslands, there has been a decrease in secondary productivity which has affected

[1] + implies an increase in aquatic systems when nutrient stressed; - implies a decrease in terrestrial systems

the economics of the region in that it no longer supports cattle grazing on the scale that was possible a century ago.

It thus appears that not only do ecosystems under stress often exhibit similar signs of distress, but they also have broadly similar consequences (while recognizing the inherent variability of response to stress in particular systems) for the human component of the system in terms of ecosystem services. It is these consequences that suggest a variety of management strategies that might be considered for more effective management of human activities to promote healthy ecosystems.

5. Management Strategies

One strategy that has proven time and time again to be largely ineffective and, unfortunately, it is often the dominant strategy, is "after the fact" (after degradation is obvious) attempts to restore ecosystem functions one by one. This often leads to further disruption of ecosystem functions. For example, in the Great Lakes, exotic species such as the Pacific salmon were introduced to control yet other exotic species (smelt and alewife), since the sea lamprey and overfishing pressures had led to the demise of lake trout, the natural predator of pelagic species. However, the Pacific salmon itself precludes the re-establishment of a natural, self-reproducing community to perform the function of predator control and therefore perpetuates a state of degradation in which a dominant predator has to be maintained through hatchery operations. Another example, is the introduction of Lehmann lovegrass in the southwest U.S.A. in order to control erosion resulting from desertification. The lovegrass does indeed stem erosion, but at the same time this South African exotic outcompetes the native grasses, thus lowering the quality of forage for cattle and other grazers.

Another strategy that has proven time and time again to be ineffective is the attempt to restore ecosystems once they have become highly pathological. Inevitably, such efforts are too little, too late. Experience in the Great Lakes might give the appearance of giving a counter-example, where eutrophication has been arrested and in some cases reversed, and where the recovery of some stocks (walleye in the Lower Great Lakes) gives grounds for optimism. But these successes are only with respect to a narrow range of ecosystem attributes. The overall health of the Great Lakes is far from showing improvement -- habitat loss continues unabated, contaminant problems are severe, many native species have been extirpated, air quality and threats to human health are increasing. All of this has occurred despite the billions of dollars that have been invested in efforts to restore ecosystem health and integrity.

The failures here reflect the fact that degraded ecosystems can reach new domains of relative stability which defy efforts to alter these states and restore the system to resemble previous conditions which offered a more full range of ecosystem services. A vivid example of this is the multiple efforts in the southwest U.S.A. to rid the area of the invasive shrubs and restore it to a grassland. It turns out that the shrubs are very resilient and even bulldozing them out has proved ineffective.

What management strategies of a more positive nature might be suggested? Preventative measures come quickly to mind. Preventative measures, of course, require the selection of key indicators which require constant monitoring and at the earliest signs of deviations from established norms, monitoring would need to call into play positive actions. The same set of variables (or a subset of these) that comprise EDS could, in theory, also be those attributes that would serve as early warning indicators. However, many of the signs of EDS occur rather late in the degradation process and so it would appear that an entirely different set of indicators or tests for the maintenance of health -- as opposed to the detection of pathology -- is required.

One promising avenue here is to devise the equivalent of a "fitness test" for ecosystems. In public health, there has been a revolution in health maintenance by focusing on personal fitness. Here there are well established indicators by which individuals can gauge their state of health, at least with respect to cardiovascular functions, muscular functions, *etc.* It may be speculation at this stage, but might not a similar approach be appropriate for monitoring the health of ecosystems?

Ecosystems are constantly being challenged by natural perturbations, such as storm events, fire, drought, pests, *etc.* Healthy systems might be hypothesized to be more resistant to the consequences of these perturbations and/or recover more quickly from the perturbations. If this hypothesis bears out, then we have in fact a basis for measuring the fitness of ecosystems. Ecosystems that show unusually slow recovery from perturbations, even if they have the appearance being in healthy condition, would be screened out as warranting special attention. Additional tests could then be applied to determine if some of the functions of the system have been lost or compromised. One drawback to this approach is that, in many cases, recovery times are very long. A more useful approach may be based on "cumulative stress loads" or "critical loads", from which models for key functions and their "breaking points" can be developed. For some cases, this can be done on a fully quantitative basis (e.g. load of acidifying substances), but in other cases, we are only approaching ways of measuring pressures (e.g. habitat fragmentation).

This approach requires devising limits to cumulative stress loads for particular ecosystems based on experience with similar systems elsewhere. This may be a crude management strategy, but nevertheless it might result in "rules of thumb" that provide a margin of safety so that ecosystems do not become overstressed and thus their health status maintained. With respect to stresses such as nutrient loading, overharvesting, and physical restructuring, there are sufficient case histories to indicate, with respect to specific ecosystem types, some of the boundaries for acceptable stress loads.

6. Discussion

Ecosystem health is a wide domain and we have approached the topic here from a subset of considerations that might go into a holistic assessment. We have not, for example, dealt with questions of human health implications of ecosystem change, nor have we dealt with questions of altered social structures and security issues, which are intimately connected with stress on ecosystems (Homer-Dixon,1993) and we have given short consideration to the economic aspects of ecosystem degradation. Rather we have focused our attention on the main elements of the biophysical transformations, which are the central elements of the EMAP program. We simply raise this issue here to suggest that a more comprehensive assessment of ecosystem health and its consequences for humankind needs to go beyond the biophysical aspects.

With respect to the biophysical consequences, we have shown that common stress pressures acting on a variety of different ecosystems invariably lead to common signs of ecosystem distress and similar losses in ecosystem services. We have also shown that some of the main strategies which are called into play to address degraded ecosystems are ineffective. Methodologies need to be explored which address the preventative rather than the post mortem side of the equation.

Acknowledgements

This work was sponsored by the Eco-Research Chair in Ecosystem Health, which is supported by three research councils of Canada (Medical, Social, and Natural Sciences) and their partner Federal and Provincial agencies. Appreciation is also extended to three reviewers, Drs. Barbara Brown, Anne Hellcamp, and Kevin Summers.

References

Costanza, R., d'Arge, R., de Groot, R., Farber, S., Grasso, M., Hannon, B., Limburg, K., Naeem, S., O'Neill, R.V., Paruelo, J., Raskin, R.G., Sutton, P. and van den Belt, M.: 1997, *Nature* **387**(6630): 253-260.

Daily, G.: 1997, *Nature's Services*. Island Press, Washington, D.C.

Epstein, P.R. and Rapport, D.J.: 1996, *Ecosystem Health* **2**(3): 166-176.

Harris, H.J., Harris, V.A.. Regier, H.A. and Rapport, D.J.: 1988, *Ambio* **17**(2): 112-120.

Hildén, M. and Rapport, D.J.: 1993, *J. Aquatic Ecosystem Health* **2**, 261-275.

Homer-Dixon, T.F., Boutwell, J.H. and Rathjens, G.W.: 1993, *Sci. Am.* **268**(2): 38-45.

Odum, E.P. 1985. Trends expected in stressed ecosystems. *Bioscience* **35**: 419-422.

Rapport, D.J.: 1989, *Perspectives in Biology and Medicine* **33**, 120-132.

Rapport, D.J.: 1992, *J. Aquatic Ecosystem Health* **1**: 15-24.

Rapport, D.J.:1985, *The American Naturalist* **125**: 617-640.

Regier, H.A. and Baskerville, G.L.: 1986, *Sustainable Development of the Biosphere*. Cambridge University Press, London.

Whitford, W.G., Rapport, D.J. Groothousen, R.M.: 1996, *GIS World*. **9**(12): 60-62.

Whitford, W.G.: 1995, *Evaluating and Monitoring the Health of Large-scale Ecosystems*. Springer-Verlag, Heidelberg.

VEGETATION, SOIL, AND ANIMAL INDICATORS OF RANGELAND HEALTH

WALTER G. WHITFORD[1], AMRITA G. DE SOYZA[2], JUSTIN W. VAN ZEE[2], JEFFERY E. HERRICK[2], and KRIS M. HAVSTAD[2]

[1]*Senior Research Ecologist, U.S. EPA, ORD, NERL, Environmental Sciences Division-Las Vegas, PO Box 93478, Las Vegas, NV 81913,* [2]*Research Scientists, USDA-ARS, Jornada Experimental Range, Dept. 3JER, New Mexico State University, Las Cruces, NM 88003*

Abstract. We studied indicators of rangeland health on benchmark sites with long, well documented records of protection from stress by domestic livestock or histories of environmental stress and vegetation change. We measured ecosystem properties (metrics) that were clearly linked to ecosystem processes. We focused on conservation of soil and water as key processes in healthy ecosystems, and on maintenance of biodiversity and productivity as important functions of healthy ecosystems. Measurements from which indicators of rangeland health were derived included: sizes of unvegetated patches, cover and species composition of perennial grasses, cover and species composition of shrubs and herbaceous perennials, soil slaking, and abundance and species composition of the bird fauna. Indicators that provided an interpretable range of values over the gradient from irreversibly degraded sites to healthy sites included: bare patch index, cover of long-lived grasses, palatability index, and weighted soil surface stability index. Indicators for which values above a threshold may serve as an indicator of rangeland health include: cover of plant species toxic to livestock, cover of exotic species, and cover of increaser species. Several other indicator metrics were judged not sensitive nor interpretable. Examples of application of rangeland health indicators to evaluate the success of various restoration efforts supported the contention that a suite of indicators are required to assess rangeland health. Bird species diversity and ant species diversity were not related to the status of the sample site and were judged inadequate as indicators of maintenance of biodiversity.

1. Introduction

There is a clear need to develop assessment and monitoring systems for the rangelands of North America (National Research Council, 1994). Rangeland health assessments are needed to evaluate the status of the component ecosystems of the rangelands. Monitoring systems are needed to detect changes that are related to the probability of degradation (i.e. risk of crossing a threshold to an alternate but less desirable state (Westoby *et al.* 1989). Monitoring is also needed to track the success or failure of restoration efforts. For both assessment and monitoring, it is necessary to have a suite of indicators that are sensitive to environmental stress, that focus on risk of degradation, and that are related to ecosystem function (Herrick *et al.* 1995).

"Rangeland health may be defined as the degree to which the integrity of the soil and the ecological processes of rangeland ecosystems are sustained" (National Research Council 1994). While this definition is useful, the concept of integrity is difficult to quantify. Examination of other descriptions of ecosystem health (where ecosystems are broadly defined to include humans and all of their activities) reveals a common set of characteristics of healthy ecosystems: (1) they are free from ecosystem distress syndrome

Environmental Monitoring and Assessment **51:** 179–200, 1998.

(a common set of signs that characterize the most heavily damaged ecosystems (Rapport *et al.* 1985), (2) they are resilient (rebound quickly to pre-disturbance state) following normal perturbations (i.e. disturbances normally encountered in their evolutionary history such as fire, floods, and drought), (3) they are self-sustaining (i.e. they can be perpetuated without subsidies or drawing down natural capital), (4) management practices and ecosystem processes do not impair adjacent systems or systems at some distance, (5) they are economically viable and (6) they sustain healthy human communities. For rangeland health the indicators must provide information which can be used to evaluate the capacity of the system to conserve and efficiently use water and nutrients, and to support biodiversity, economic production, and recreational uses.

The indicators of rangeland ecosystem health must meet criteria similar to those applied to judgments of which tests should be done in an assessment of human health. Indicators of rangeland health should (1) reflect the status of a critical ecosystem process, important ecosystem property, or an economic-social value, (2) be unambiguous (i.e., the trajectory of the measure is unidirectional in response to ecosystem stressors of increasing intensity), (3) be applicable in the range of ecosystems encountered in rangeland landscapes, and (4) be readily and inexpensively measured. There are ecosystem properties that provide information on the characteristics of most ecosystem processes. Ecosystem properties such as structural characteristics of the vegetation, spatial distribution patterns of plant species, morphological characteristics of plants, and physical and biological characteristics of the soil surface do not change rapidly over time and the quantitative values of measurements of these properties are frequently directly related to one or more ecosystem processes. The properties chosen as indicators in our studies met the criteria listed above and there was sufficient information in the literature to clearly link indicators and ecosystem processes.

Potential indicators need to be evaluated for their sensitivity to disturbance and stress. Disturbance is used here as a variable to which the ecosystem has been exposed over evolutionary time while a stress is a variable to which the ecosystem has had no evolutionary experience. In rangelands in the western U.S., drought is a disturbance and poorly managed grazing by domestic livestock is a stress. A sensitive indicator is one that yields very different quantitative values when measured at locations that are known to be degraded and at locations that are known to be in healthy condition (deSoyza *et al.* 1997). A sensitive indicator also yields very different quantitative values when measured at locations known to be exposed to very different levels of environmental stress. If the indicator is sufficiently sensitive, it should be possible to use that same indicator to evaluate the relative success of restoration efforts. Our studies were designed to test the sensitivity of a series of indicators derived from measurements of ecosystem properties for use in assessing rangeland health and in evaluating the health status of restoration efforts. For each indicator we hypothesized the response to stress or change in state (Table I).

Table I

Hypothesized responses of indicator metrics to exposure to environmental stress or ecosystem change resulting from exposure to stress and as measures of recovery resulting from restoration efforts.

Indicator	Exposure to Stress/Change	Restoration Efforts
Water and Nutrient Conservation		
percent cover long-lived grasses	<	>
bare patch size (bare patch index)	>	<
percent cover vegetation	<	>
soil surface stability - slake tests	<	>
percent cover "increasers"	>	<
percent cover shrubs	>	<
Productivity (commodity yield)		
percent cover species preferred by livestock	<	>
percent cover toxic species	<	>
percent cover shrubs	>	<
percent cover exotics	?	?
Biodiversity		
perennial plant species richness	<	>
perennial plant species diversity (H')	<	>
breeding bird species richness	<	>
breeding bird species diversity (H')	<	>

2. Methods

The study sites were located in the Jornada Basin including the slopes of the San Andres mountains and Dona Ana mountains and the Jornada plain on the USDA-ARS Jornada Experimental Range approximately 40 km north of Las Cruces, N.M. Measurements were made on 44 sites. Seven sites were selected on the basis of historical records of vegetation change and land use (Buffington and Herbel 1965, Gibbens and Beck 1988, Gibbens *et al.* 1992). Four sites that had been subjected to different restoration practices in the mid to late 1970's (Herbel *et al.* 1958, 1973, 1983, 1985) were paired with sites that were grazed and also with plots within grazing exclosures. We included plots established at varying distances from stock watering points. These provide a distinct disturbance gradient from very intense at the water point to no perceptible disturbance at distances greater than 1 km

(Andrew and Lange 1986, Fusco *et al.* 1995). We made measurements outside and within three 50 year old grazing exclosures. Measurements were also made on four sites in southeastern Arizona.

At each site, we established a 1 ha plot with the baseline centered on a randomly selected point. Vegetation and soils were measured along ten, 100 m lines spaced at 10 m intervals along the baseline. At sites with a perceptible slope, the sample lines were oriented with the slope. On the flat areas, the transect lines were parallel to the disturbance gradient (if present). We recorded the intercept length of canopy cover by species and the length of bare (unvegetated) patches (Canfield 1941).

A slake test (Tongway 1994) modified to account for differences in wet aggregate strength, was used to measure soil crust stability. Soil stability was measured from three different strata (bare soil, grass clump, and under shrub canopy). At each site we selected three transects at random and for each transect generated 25 random numbers between 0 and 100. We sampled the soil at each point corresponding to a random number until we had obtained three samples for each stratum. If the soil at a point did not yield a soil fragment (i.e. the soil disintegrated), it was assigned a value of 1.0. If fewer than 3 samples were obtained for a stratum, an additional 25 random numbers were generated and the transect was repeated for that stratum only. Soil was tested from surface (0 - 3 mm depth). The slake values provide a measure of current stability. The soil stability test was done only on air dried soils because moist soil tends to yield an overestimate of stability. Stability was measured on soil fragments (6mm - 8mm in diameter) that were 2 - 3 mm thick. Soil fragments were carefully placed in a small, 25 mm diameter PVC basket with a wire mesh bottom (1.5 mm mesh size). The baskets were slowly lowered into a reservoir of distilled water. The disintegration of each fragment was observed for 5 minutes. If the soil fragment remained intact at the end of this time, the basket was raised and lowered three times. Soil stability was ranked according to the time required for the fragment to disintegrate during the five minute immersion or on the proportion of the fragment remaining intact after three immersion cycles.

Breeding birds were censused in areas centered on sites where detailed vegetation and soil indicator measurements were made. Breeding birds were censused in early June, the peak of the breeding season in the Chihuahuan Desert. Nine circular plots of 50 m radius were established in good condition grassland, a mixed grass - mesquite shrub mosiac, creosotebush shrubland, tarbush shrubland, and mesquite coppice dune area. Based on historical records, these areas provided a gradient from minimally changed (healthy rangeland) to maximally degraded (changed) rangeland. Breeding birds were also censused in southeastern Arizona on landscape units dominated by Lehmann lovegrass and compared with censuses conducted on native grasslands. The center of each circular plot was 300 m from the center of all other plots. Plots were arranged in a 3 x 3 grid. Birds that were seen or heard singing within the 50 m circle were recorded by two observers standing at the center point. Observations were made for 10 minutes at each center point.

Indicator metrics applicable to assessment of rangeland health were plotted against an axis of percent grass cover. All of the sites which were selected for these measurements were in areas on the Jornada Basin that were classified as grassland in the original surveys and which have been documented to have undergone some change during the past 150 years (Buffington and Herbel 1965). By selecting grass cover as the independent variable, we provide a measure of departure from the least disturbed (least changed) landscape units. If the indicator metrics that we have selected are sensitive measures of the health of these rangeland ecosystems, they should vary in a systematic way on a gradient of change measured as percent grass cover.

3. Results

Several of the hypothesized indicators failed to change in a systematic fashion on a gradient from most changed to least changed ecosystem. Although there was a cluster of sites with low percent of vegetative cover on the maximally changed end of the gradient, there were 5 sites that had vegetative cover equivalent to sites in the intermediate change category (Figure 1). Total vegetative cover is therefore not a sensitive indicator of exposure to stress. Two other indicators that produced patterns that were not consistent with the hypothesized responses and that were not easily interpreted were perennial plant species richness and Shannon's diversity index for perennial plant species (Figures 2 and 3). Percent shrub cover was hypothesized to increase as sites were exposed to increased stress since the historical trend in the degradation (desertification) of Chihuahuan Desert rangelands has been "invasion" by woody shrubs (Buffington and Herbel 1965, Grover and Musick 1990). Shrub cover varied from zero to twenty percent on degraded sites and one of the "healthy" sites had shrub cover of fifteen percent (Figure 4). Thus shrub cover is not a sensitive indicator for incorporation into a rangeland health assessment system for Chihuahuan Desert rangelands.

An indicator based on the size of unvegetated patches that is identified as a bare patch index was calculated as: bare patch index = mean bare patch size x proportion of line that is bare. The bare patch index provided one of the most sensitive metrics for assessing change (Figure 5). The inflection point in the bare patch index at a value of 80 may represent the threshold value for this metric (Figure 5). A vegetation metric that was very sensitive for assessing change was percent cover of long-lived grasses. These grasses are species that are very drought resistant and that live for several decades. The percent cover of these grasses decreased to zero or virtually zero on the maximally changed sites (Figure 6). The surface soil stability index was calculated from the slake test data by multiplying the mean slake value for grass, shrub, and bare soil by the cover of each of those types to obtain a weighted mean for each of the three sub-locations sampled. The three weighted values were summed to obtain the surface soil stability index for the site. This index provided a clearly interpretable pattern that was consistent with the hypothesized response (Figure 7). The surface soil stability index combined with the bare patch index provide

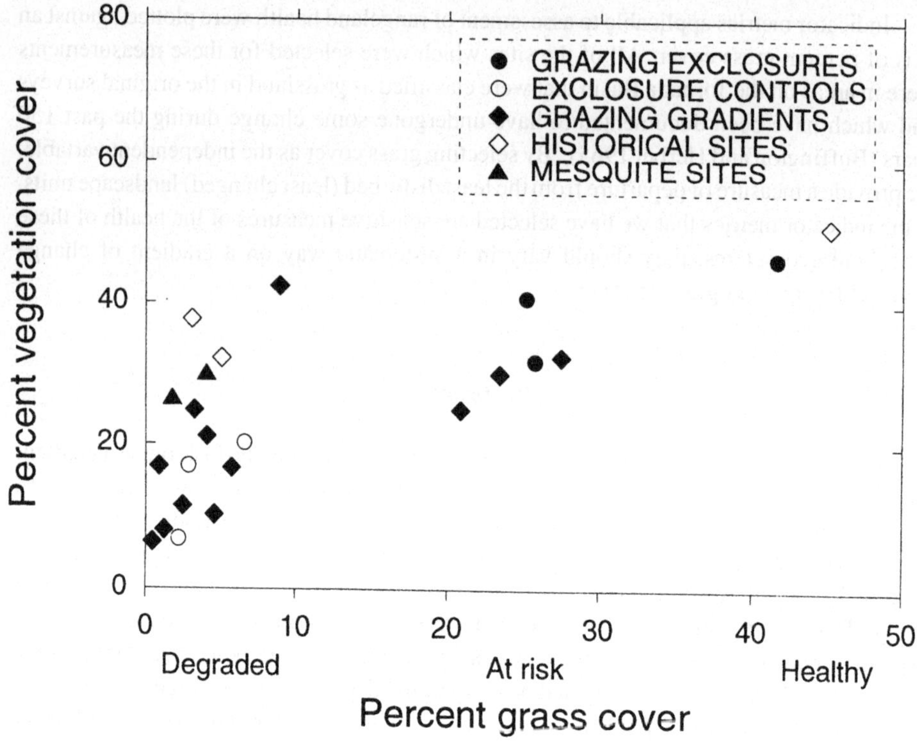

Fig. 1. Average total vegetative cover at sites ranging from irreversibly degraded to healthy.

metrics that are related to erosion potential of a site.

The indicator that provides information on the productive capacity of the system, the index of relative preference of vegetation by livestock, incorporates data on cover by species and the relative palatability of that species to livestock. The relative preference index is calculated as: percent cover (by species) x forage quality (good = 1.0, fair = 0.75, poor = 0.25, toxic or not eaten = 0.0) x fraction of the year the species is eaten by cattle (primary data source, Stubbendieck *et al.* 1993). The relative preference index is a very sensitive indicator of one function related to the health of rangeland ecosystems (Figure 8). Other metrics that can be derived from the basic data set include: cover of toxic species, cover of exotic (non-native) species and cover of invader species, i.e. those species that rapidly occupy sites outside their pre-stress habitats (examples are mesquite, Prosopis glandulosa and creosotebush, Larrea tridentata)(Gardiner 1951, Buffington and Herbel 1965, Grover and Musick 1990, Gibbens *et al.* 1992, Dick-Peddie 1993). All of these metrics are related to the productivity of rangelands. The percent cover of perennial plant species that are toxic to livestock was similar on many of the degraded, at risk, and healthy sites (Figure 9). The only sites where cover of toxics exceeded 5% were sites currently exposed to activity of large numbers of livestock for a large part of each year (Figure 9).

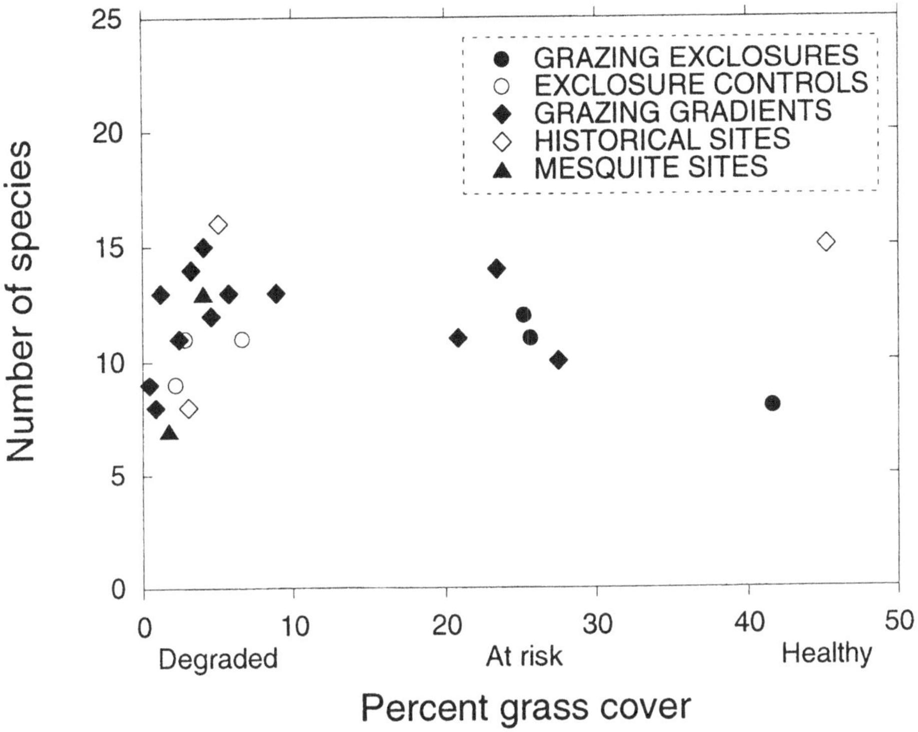

Fig. 2. Perennial plant species richness at sites ranging from irreversibly degraded to healthy.

A measurement with these characteristics may be used as an indicator of exposure to grazing stress but not as an indicator of desertification. The distribution of values of cover for toxic species suggests that cover of toxics exceeding a 5% cover threshold may be considered as an indicator of degradation with the caution that values below 5% should not be considered as an indicator of a healthy site. Here we identify the threshold as values above those recorded on sites at the healthy end of the spectrum.

The percent cover of increasers (species that establish in areas not occupied before disturbance and increase in density and cover following establishment) was higher at most of the degraded sites but there were degraded sites where cover of increasers was less than 5% (Figure 10). This is another indicator where values above a threshold, i.e., cover of 10%, may be a useful indicator of degradation in a multi-metric assessment. However cover of increasers less than 10% must not be judged as an indicator of a healthy rangeland site. None of the sites in the Jornada Basin had significant cover of exotics with the exception of one of the restoration site controls. We have data on exotics from only two sites in southeastern Arizona where there was documentation of recent expansion of the

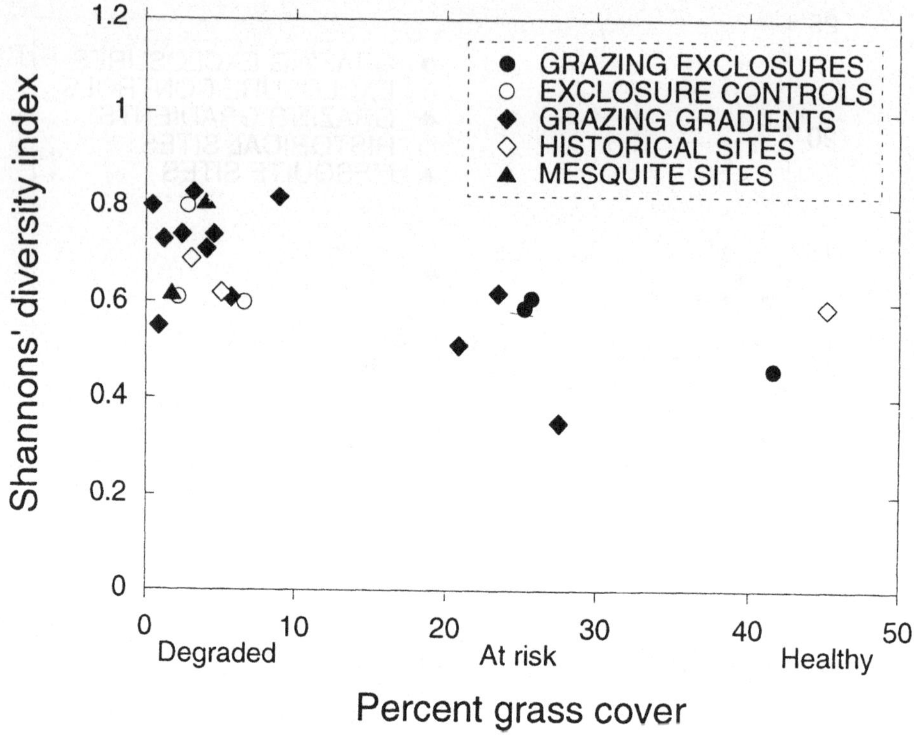

Fig. 3. Shannon's species diversity index (H') at sites ranging from irreversibly degraded to health.

South African Lehmann's lovegrass, *Eragrostis lehmanniana* (Anable *et al.* 1992). At these sites E. lehmanniana accounted for 88-97 % of the vegetative cover. On native grassland sites in the same areas, grass cover was 10.4-19.1% and Lehmann lovegrass accounted for less than 5% of that cover. Percent cover of exotic species is an indicator like the cover of increasers and cover of toxics where a cover value above some threshold may be considered an indicator of degradation.

INDICATORS OF RESTORATION SUCCESS

The applicability of rangeland health indicators as measures of restoration success was examined on a series of paired sites in mesquite coppice dunes and creosotebush-tarbush shrub sites. Restoration efforts in the Chihuahuan Desert have focused on reducing or eliminating shrubs. These efforts have included treatment with chemical herbicides and mechanical treatments such as root plowing and bulldozing (Herbel *et al.* 1973, Herbel *et al.* 1983, Herbel *et al.* 1985). All of the restoration treatments had been in place for more than 25 years when the indicator measurements were made. In the mesquite coppice dunes, the bare patch index was greatly reduced by bulldozing that flattened the dunes and

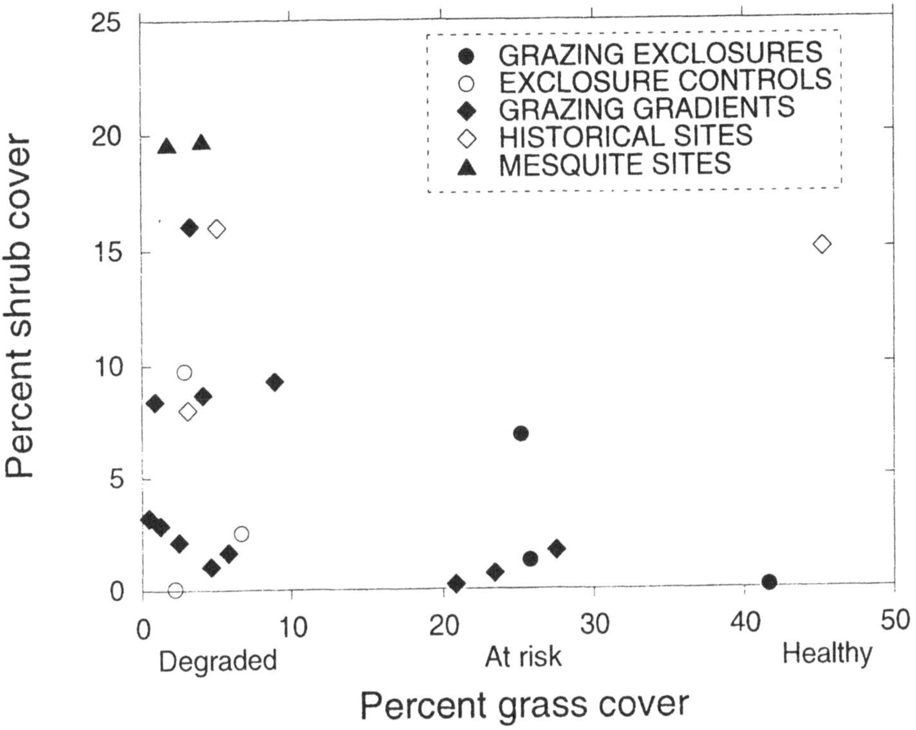

Fig. 4. Average shrub cover at sites ranging from irreversibly degraded to healthy.

removed the coppiced shrubs and the percent grass cover was increased (Figure 11). Bulldozing the dunes did not result in increased productivity as measured by the relative preference index despite the reduction in shrub cover and increase in grass cover (Figure 11).

Herbicide treatment of mesquite coppice dunes did not change the bare patch index nor the percent grass cover and had only a small effect on the relative preference index (in the direction opposite that hypothesized). Herbicide application in tarbush dominated sites produced mixed results (Figure 12). Bare patch index was higher on the grazed sites and lowest on the ungrazed, herbicide treated plot. Grass cover was equivalent to the best condition sites on the ungrazed, herbicide treated plots. Shrub cover was the same on the grazed sites and remained significantly depressed only on the ungrazed, herbicide treated plots (Figure 12). Despite changes in some of the key indicators, the productivity indicator (relative preference index) was not dramatically increased by the herbicide treatment. Root-plowing and seeding of creosotebush slope was relatively more effective than herbicide treatment in shifting indicators toward the healthy end of the spectrum. With the exception of percent long-lived grasses, the indicator values for the root-plowed site were healthy especially when compared to the untreated sites (Figure 13). The high cover of

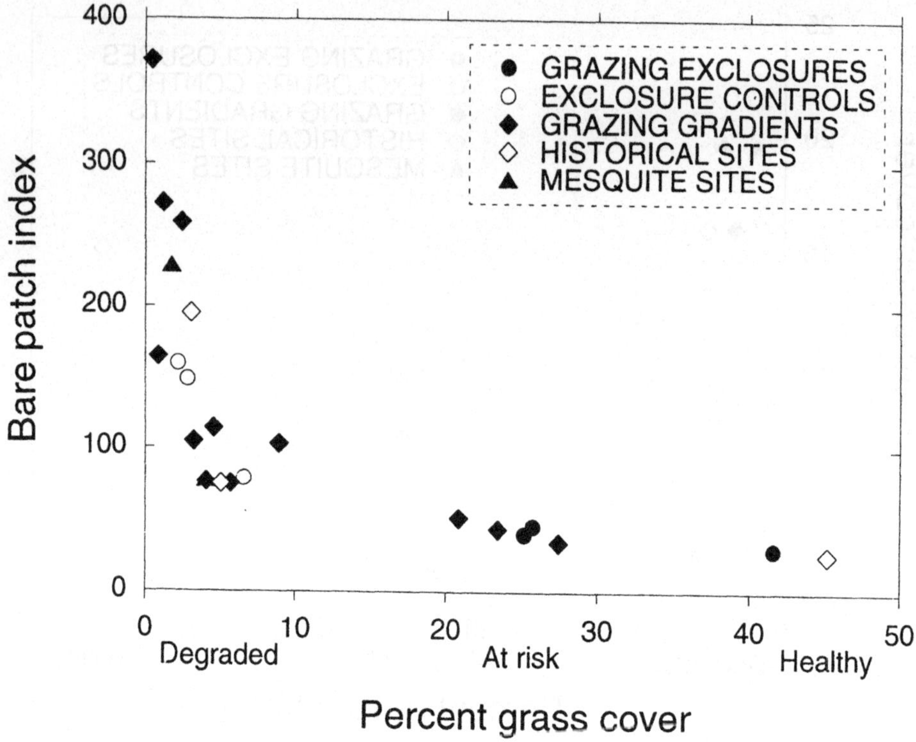

Fig. 5. Bare patch index at sites ranging from irreversibly degraded to healthy.

long lived grasses and low shrub cover on the ungrazed site on the creosotebush slope was due to bush muhly grass, *Muhlenbergia porteri*, which grew around the base of virtually every shrub in the exclosure.

BIODIVERSITY INDICATORS

The diversity indices based on breeding birds reflected the vegetation structural diversity and not the relatively recent change from the pre- grazing stress vegetation. The lowest species richness was recorded in the black grama (Bouteloua eriopoda) grassland and the lowest diversity index was in the tarbush (Flourensia cernua) (Figure 14). These landscape units are dominated by shrubs or grasses that are less than 1 m in height. In each unit, there was one plot that had taller shrubs or yucca (Yucca elata). The highest diversity of breeding birds was in the mesquite - grass mosaic landscape unit that was centered on a small grass covered dry lake with a fringe of mesquite shrubs that were between 5m and 7 m tall (Figure 14) The breeding bird species richness and diversity was very similar in the other degraded landscape units. The pattern of abundance of breeding birds (no.. ha^{-1}) was the same as species richness. Species richness and diversity were higher in the area

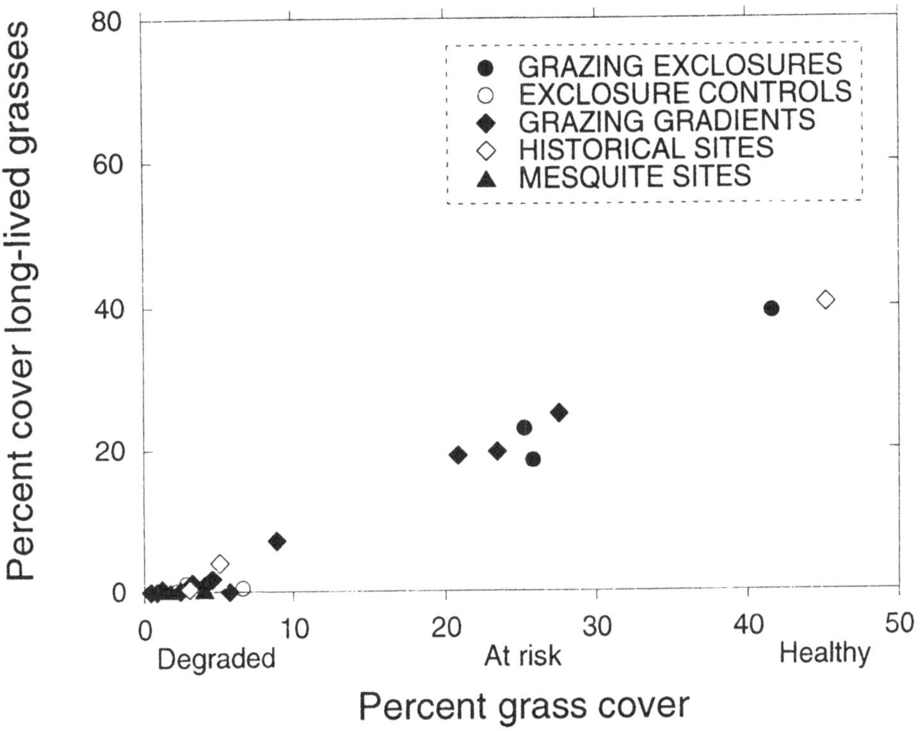

Fig. 6. Average percent cover of long-lived grasses at sites ranging from irreversibly degraded to healthy.

dominated by Lehmann lovegrass at the Empire Cienega Ranch but were higher in the native grassland at the Santa Rita Experimental Range (Figure 14).

4. Discussion

The National Research Council (1994) report on rangeland health emphasized the need for multiple indicators and for the use of benchmark sites for evaluating the efficacy of indicators. Our studies approached these needs by screening a suite of indicators on sites with long term vegetation cover records and on sites that had been subjected to one of several restoration efforts. These studies demonstrated that some potential indicators did not provide interpretable responses across a series of sites representing a gradient from healthy rangeland to degraded rangeland.

The National Research Council (1994) described the problems encountered in comparisons of plant composition to desired or site potential vegetation. They also cautioned about extrapolating from benchmark sites to all rangeland sites in a region. By focusing on indicators of ecosystem function rather than on indicators of desired or

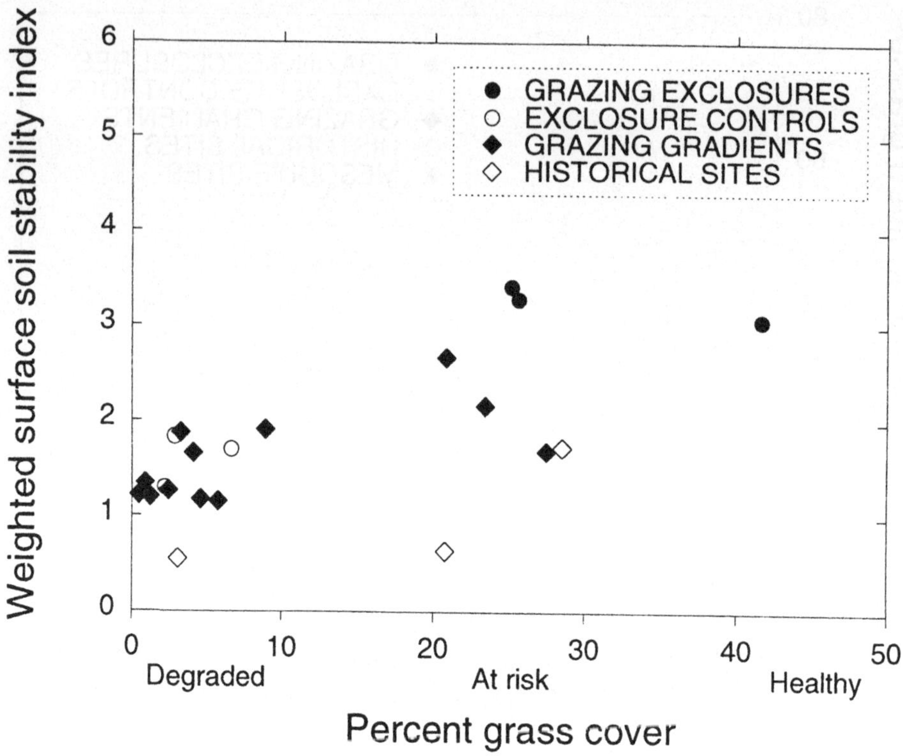

Fig. 7. Soil surface stability index at sites ranging from irreversibly degraded to healthy.

potential state, we avoid the problem of pre-judging the potential of a site. Indicators of ecosystem function allow a manager to evaluate the health of an ecosystem by assigning variable weights to the indicators that reflect management goals or community values (Herrick *et al.* 1995). Indicators of ecosystem function are general indicators because they are directly linked to ecosystem processes and are applicable to most if not all rangeland ecosystems. Since the most important feature of the indicators presented here is the linkage to ecosystem processes, we provide a discussion of those linkages.

LINKAGE TO ECOSYSTEM PROCESSES: The indicators for retention and use of soil and water resources *in situ* and the indicators for economic productivity are interdependent. Proportional vegetative cover of shrubs and grasses is directly related to the degree of linkage between precipitation and nitrogen mineralization and nitrogen immobilization. In grasses there is a direct coupling of water availability and the availability of nitrogen and other nutrients resulting in a proportional increase in the production response following rainfall (Stephens and Whitford 1993, Whitford and Herrick 1996). In shrub systems, there is frequent decoupling of rainfall and plant growth resulting from immobilization of nutrients in decomposing roots of ephemeral plants that grow in high densities under the shrub canopies (Whitford and Herrick 1996). Redistribution of rainfall also varies with the

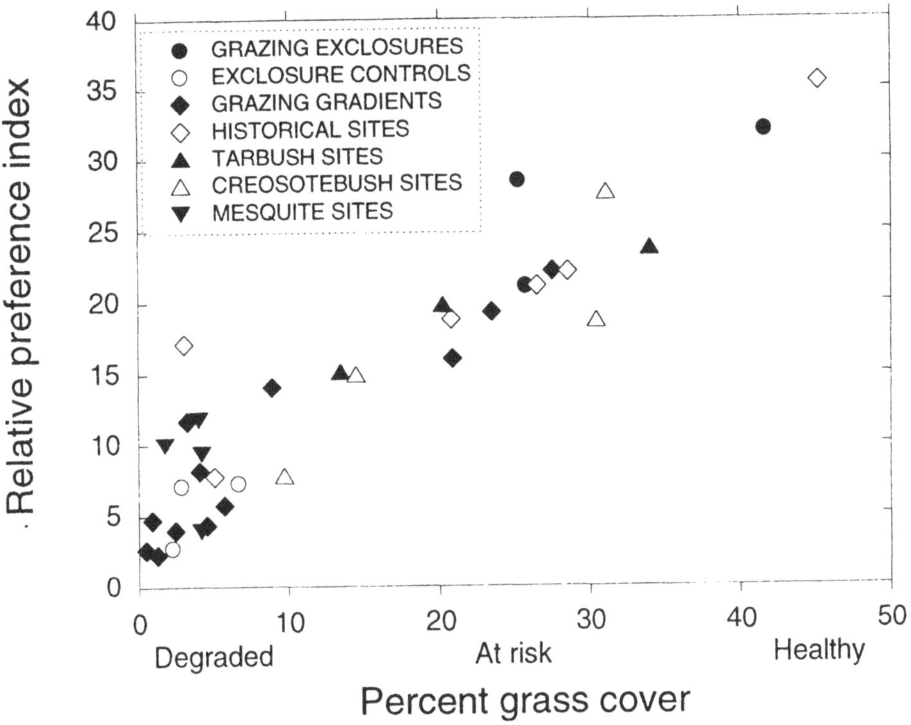

Fig. 8. Relative preference index (by livestock) at sites ranging from irreversibly degraded to healthy.

proportional cover of shrubs and grasses. Water is translocated deeper into the soil profile by shrubs than by grasses (Martinez-Meza and Whitford 1996). Preference of vegetation by livestock is in general related to growth form with grasses being generally more preferred than shrubs (Stubbendieck *et al.* 1993). However because of the differences in seasonal preference for different species of grasses, grass cover alone is not a good measure of the potential for livestock production. The relative preference index uses the information on seasonal use of each plant species at a site, thereby incorporating this variability into the calculation of the indicator.

An important component of retention of water and soil resources *in situ* is the efficacy of the vegetation in disrupting overland flow during intense rainfall events and of the vegetation to disrupt airflow during high winds. Grass clumps are generally far more effective in disrupting overland water movement than are shrubs or trees (Tongway and Ludwig 1997) and the efficacy of the grasses in slowing water movement is directly proportional to the grass cover. In desert rangelands where periodic drought is a characteristic of the climate, long-lived grasses afford greater protection to the soil than do the short lived grasses (i.e. perennial grasses that live less than a decade). Short lived grasses exhibit large swings in cover values in wet years following drought (Herbel and

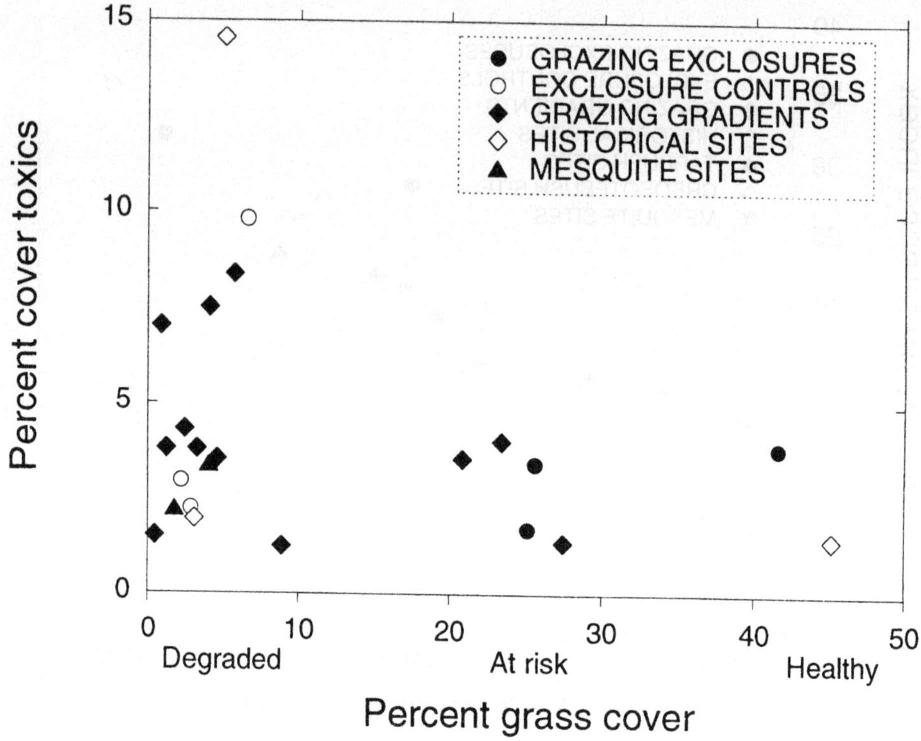

Fig. 9. Average percent cover of perennial plant species that are toxic to livestock at sites ranging from irreversibly degraded to healthy.

Gibbens 1996). Cover values of long-lived grasses are also related to the long-term protection of the soil surface from wind erosion.

The resistance of rangeland ecosystems to erosion is a function of a number of ecosystem properties. Both vegetation cover and bare patch size are important variables in the WEPS (Wind Erosion Prediction System) model. The functional model for wind erosion is: $E = F(I, K, C, L, V)$ where E is potential annual soil loss per unit area, I is a soil erodibility index, K is a soil roughness factor, C is a climatic factor, L is the unsheltered median travel distance of wind across an unvegetated space and V is the quantity of vegetative cover (Skidmore 1986). Included in the soil erodibility index (I) are a number of other soil properties that affect wind erosion: crust properties, crust cover fraction, loose erodible material and soil bulk density (Zobeck 1991). This set of properties is related to the rangeland health soil stability indicator and by measures of cover of cryptogams, stones, and litter (variables measured on detailed lines not reported here). The most important variable in the WEPS model is the wind fetch, or unvegetated patch size. Unvegetated patch size is also an important variable with respect to water infiltration and storage and soil nutrients. Patches devoid of vegetation have very low soil organic matter,

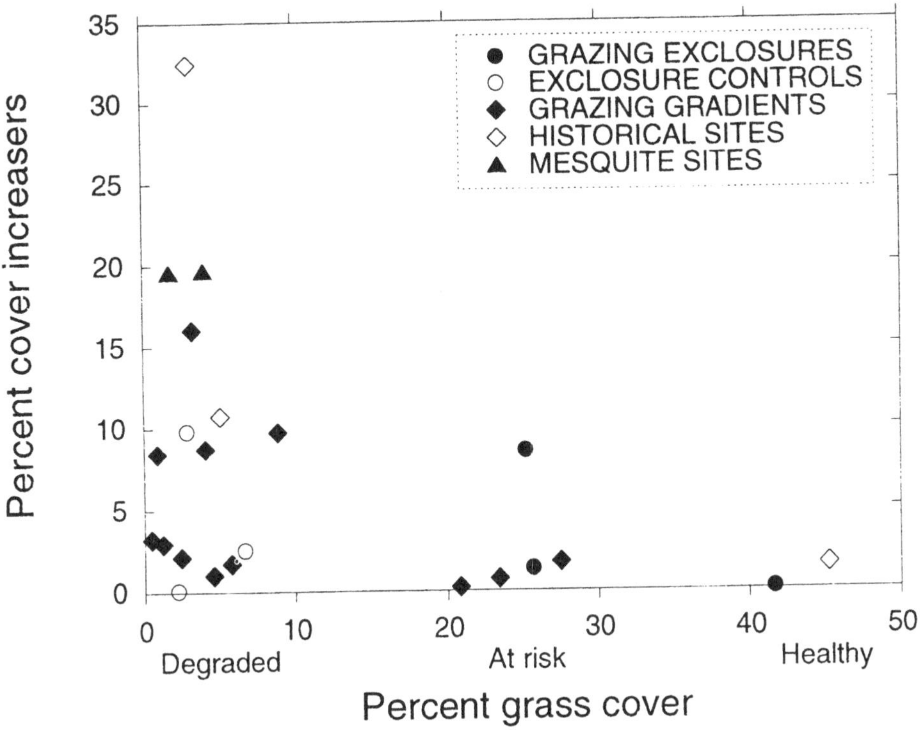

Fig. 10. Average percent cover of shrubs that are increasers at sites ranging from irreversibly degraded to healthy.

low total nitrogen and low rates of mineralization (Fisher *et al.* 1990). Unvegetated patches are also unsuitable habitat for most soil organisms (Santos *et al.* 1978, Elkins *et al.* 1986, and unpublished observations). For example subterranean termites and ants were absent from large unvegetated patches in Australian mulga woodlands (Whitford *et al.* 1992). Since soil organisms are responsible for the production of soil macropores which enhance water infiltration, the size and frequency of unvegetated patches also have a direct effect on infiltration and water storage (Bevan and Germann 1982).

The preceding discussion, while not an exhaustive review of linkages between the indicators based on ecosystem properties and ecosystem processes, does serve to demonstrate that properties that are rapidly and easily quantified in the field can be used as indicators of ecosystem processes.

ECONOMIC HEALTH AND PRODUCTIVITY: In arid lands around the world, productivity that is usable by humans is in the form of meat, fiber, or milk products from domestic livestock. Low quantity and temporally unpredictable rainfall precludes rain-fed crop production. Thus most of the arid and semi-arid lands of the world are referred to as rangelands where livestock production is virtually the only industry. The productivity

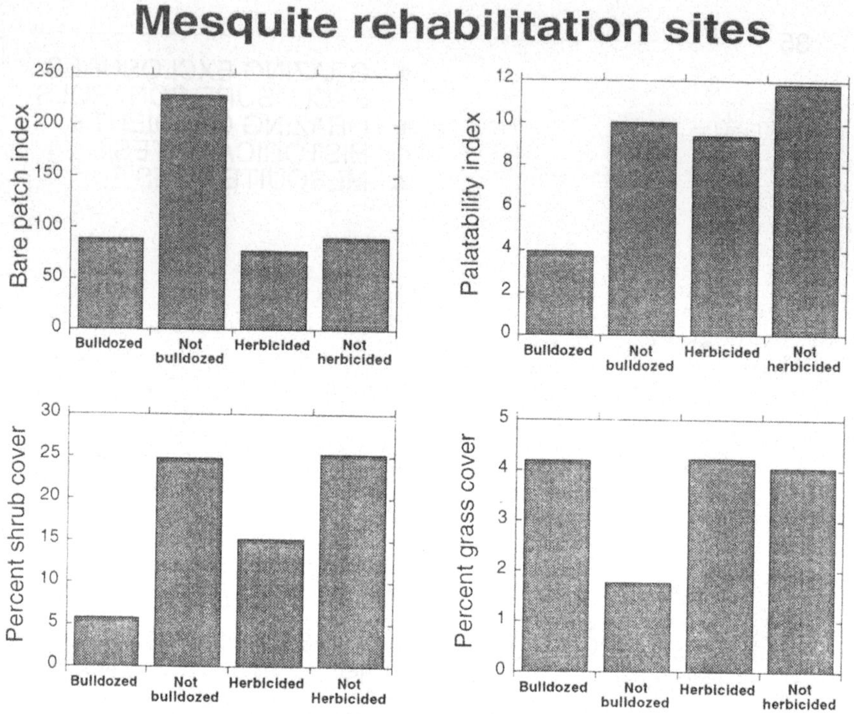

Fig. 11. Comparisons of values of selected indicators at mesquite, Prosopis glandulosa, coppice dune sites subjected to herbicide application restoration efforts and untreated sites.

measure that we chose to investigate is one that relates directly to livestock production: cover of preferred vegetation. This indicator was very sensitive to vegetation change due to desertification and to acute exposure to grazing by domestic livestock.

BIODIVERSITY: The lack of a pattern of reduction in biodiversity of perennial plants related to the change in vegetation resulting from exposure to grazing or as the result of long-term desertification processes may be in part attributable to the sampling technique used. The frequency of plants encountered along a line is not as good a measure of species richness as is obtained by careful census of nested quadrats (T. J. Stohlgren, personal communication).

The lack of effects of vegetation change and stress on ecosystems from the grazing of domestic livestock was clearly documented in a study of ant communities on all of the sites sampled in testing indicators of rangeland health. There were no differences in species richness nor H' among protected, healthy sites and sites that were exposed to grazing stress or that had experienced dramatic change in composition and cover of the vegetative community within the past century (Whitford *et al.* in press). In addition to the absence of effects on diversity, there were no interpretable patterns of difference in a variety of metrics based on ant species feeding and life history characteristics.

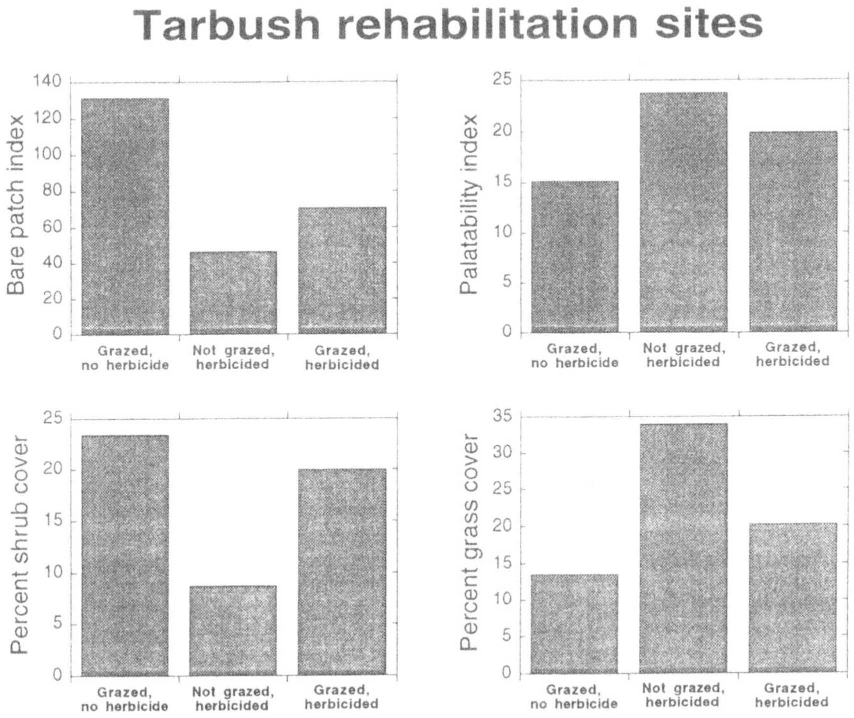

Fig. 12. Comparisons of values of selected indicators at tarbush, Flourensia cernua, dominated sites subjected to herbicide application restoration efforts and untreated sites.

Our data on rangeland health and breeding birds are similar to data from other studies in desert shrubland-grassland in the western U.S.. In a study of bird species assemblages as indicators in Great Basin rangelands, Bradford *et al.* (1997) found that two metrics (species richness and dominance) exhibited little overlap between values for heavily impacted sites and relatively unimpacted sites. They concluded that these measures were potentially good indicators of biological integrity. Bradford *et al.* (1997) added the caution that the sensitivity of these metrics suggested that they may be of limited usefulness in distinguishing between sites with light to moderate impacts. Our data on breeding birds demonstrated that species richness did not distinguish between sites except for the extreme example of a site dominated by an exotic grass species.

There are few species of breeding birds in desert grasslands and several of these species utilize the inflorescence stalks or leaf crowns of soaptree yucca, *Yucca elata*, as nesting sites (Wiens 1973, Naranjo and Raitt 1993). In 1970, Wiens (1973) reported 6 species of breeding birds from the desert grassland on the Jornada which we sampled in 1994 in this study. Of the 6 species that he reported, only four of those species were recorded in our census and we recorded 2 species that Wiens did not report (Scott's Oriole and Black-throated Sparrow). The higher diversity of breeding birds in the shrub

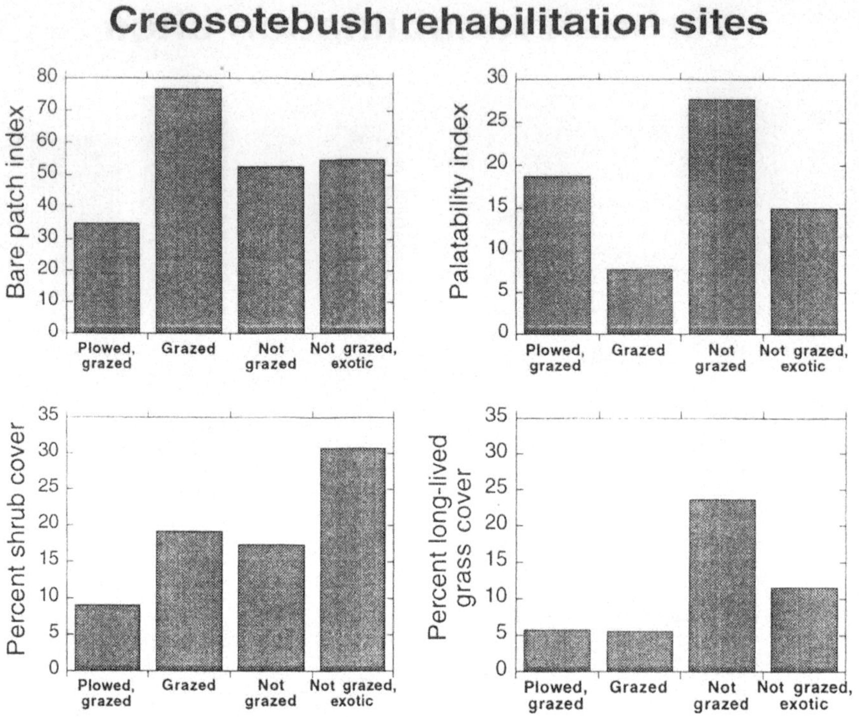

Fig. 13. Comparisons of values of selected indicators at creosotebush, Larrea tridentata, dominated sites subjected to root-plowing restoration efforts and untreated sites.

dominated habitats appears to be related to the architectural complexity of the vegetation. The highest species richness and diversity were recorded at a site with large mesquite trees, clumps of mesquite shrubs, scattered creosotebushes and dense patches of grass. The creosotebush, and tarbush shrub sites included patches of taller shrubs and patches of grasses within the relatively uniform height and density stands of the dominant shrubs. All of the sites were grazed at approximately the same stocking rates; hence, we cannot separate the responses of the avifauna to exposure to grazing from their responses to vegetation change resulting from desertification. The clear differences in species richness and diversity on the Lehmann lovegrass dominated site compared with the native grass site at the Santa Rita Range in southeastern Arizona is instructive. The reduction in diversity and abundance in the Lehmann lovegrass area at the Santa Rita is probably the result of changes in cover characteristics and food availability. At the Empire Cienega site, Lehmann lovegrass cover was not as high nor was the lovegrass distributed widely over all of the 50 m radius census plots. At Empire Cienega, four of the census circles in the native grassland had dense stands of large mesquite which were absent on the Lehmann lovegrass plots. At the Santa Rita Experimental Range, the Lehmann lovegrass had occupied the area for years while at the Empire Cienega, the Lehmann lovegrass was mixed with native grasses and had established less than 10 years prior to our studies.

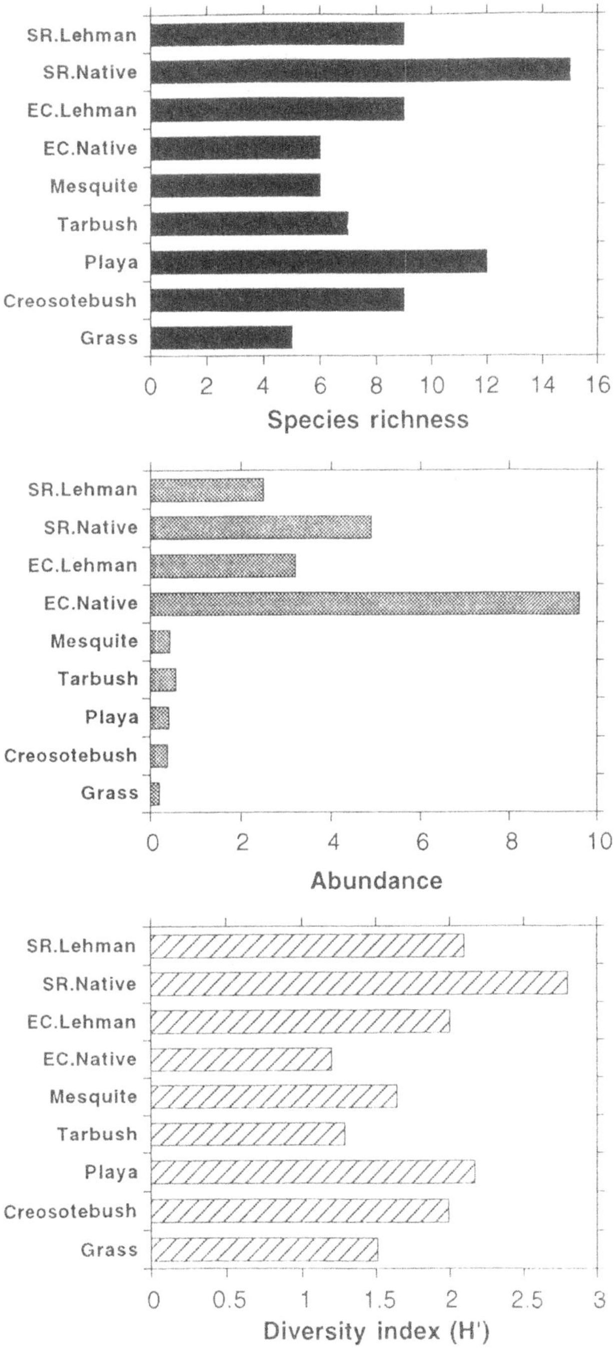

Fig. 14. Mean species richness, abundance, and Shannon's diversity index for breeding birds at a variety of sites exposed to environmental stressors and sites that exhibit few signs of environmental stress.

RESTORATION: The use of rangeland health indicators to examine the recovery of degraded sites subjected to restoration efforts supported the contention that the mesquite coppice dune sites were changed sufficiently to be classified as irreversibly degraded (Whitford 1995). Only a small fraction of the indicators showed change toward the healthy values in the mesquite coppice dune restoration sites when compared with the untreated controls. Those few indicators were primarily associated with reduction in shrub cover but not with soil surface stability, resistance to erosion, or restoration of productivity. The rangeland health indicators also demonstrated that the beneficial effects of herbicide application to mesquite coppice dunes were generally short-lived and that the herbicided dunes exhibited few indicators of good rangeland health. In the creosotebush and tarbush dominated degraded sites, the indicators suggested that herbicide application and root-plowing restored some of the ecosystem functions of healthy rangelands but did not completely restore the areas to a healthy state. The suite of indicators examined in this study do provide a means of quantifying the relative success of restoration efforts and demonstrating what functions of the ecosystem were improved by restoration.

GENERAL CONCLUSIONS: Our studies demonstrate that there are a number of indicators of rangeland health that can be derived from a simple, rapid, quantitative set of field measurements. Not all of the hypothesized indicators are sufficiently sensitive to environmental stress or to recent degradation of the system to be useful in an assessment. Some of the indicators that we examined were not easily interpreted and therefore not recommended for inclusion in an assessment system. Most of the sensitive indicators provided measures of the capacity of the patch of rangeland sampled to conserve soil and water resources. There was at least one sensitive indicator of productivity and of potential economic sustainability (relative preference index). However, none of the indicators of biodiversity examined to date were sensitive to ecosystem stress resulting from livestock grazing nor to recent large changes in vegetation and soils. Measures other than species richness or species diversity indices need to be examined as potential indicators of biodiversity for rangeland health assessments.

Herrick *et al.* (1995) listed a number of potential indicators for use in assessing and/or monitoring rangeland health. Some of the indicators that they listed are more appropriate for a monitoring system than for assessment because of the time requirements to make the measurements (i.e. infiltration capacity, soil aggregate stability, and soil texture). Other indicators are derived from a set of detailed measurements made on randomly selected small segments of the lines used to collect the data reported here. The sensitivities of indicators based on these detailed measurements have yet to be analyzed. When the complete suite of indicators has been tested, it will be possible to examine how these indicators can be combined and incorporated into a scoring system for evaluating rangeland health (Herrick *et al.* 1995).

Acknowledgments

Philip Alkon and Walter Smith assisted with field sampling. The U. S. Environmental Protection Agency (EPA) through its Office of Research and Development funded and collaborated in the research reported here. It has been subjected to the Agency's peer review and approved for publication.

References

Anable, M. E., M. P. McClaran, Ruyle, G. B. 1992. "Spread of introduced Lehmann lovegrass (Eragrostis lehmanniana Nees.) in southern Arizona, USA." *Biol. Cons.* **61**:181-188.

Andrew, M. H., Lange, R. T. 1986. "Development of a new piosphere in arid chenopod shrubland grazed by sheep. 2. Changes to the vegetation". *Australian J. Ecology* **11**: 411-424.

Bevan, K., Germann, P. 1982. "Macropores and water flow in soils." *Water Resources Research* **18**:1311-1325.

Bradford, D. F., Franson, S. E., Neale, A. C., Heggem, D. T., Miller, G. R., Canterbury, G. E. 1997. "Bird species assemblages as indicators of biological integrity in Great Basin rangeland." *Environ. Monitor.Assess.* (In Press)

Buffington, L. C., Herbel, C. H. 1965. "Vegetation changes on a semi-desert grassland range from 1858 to 1963." *Ecol. Monogr.* **35**:139-164.

Canfield, R. H. 1941. "Application of the line interception method in sampling range vegetation." *J. Forestry* **39**: 388-394.

De Soyza, A. G., Whitford, W. G., Herrick, J. E. 1996. "Sensitivity testing of indicators of ecosystem health." *Ecosystem Health* (In press).

Dick-Peddie, W. A. 1993. "New Mexico Vegetation, Past, Present and Future." University of New Mexico Press. Albuquerque, N.M.

Elkins, N. Z., Sobal, G. V., Ward, T. J., Whitford, W. G. 1986. "The influence of subterranean termites on the hydrological characteristics of a Chihuahuan Desert ecosystem." *Oecologia* **68**:521-528.

Fisher, F. M., Freckman, D. W., Whitford, W. G. 1990. "Decomposition and soil nitrogen availability in Chihuahuan Desert field microcosms." *Soil Biol. Biochem.* **22**:241-249.

Fusco M., Holachek, J., Tebo, A., Daniel, A., Cardenas, M. 1995. "Grazing influence of watering point vegetation in the Chihuahuan Desert." *J. Range Mgmt.* **48**: 186-192.

Gardiner, J. L. 1951. "Vegetation of the creosotebush area of the Rio Grande valley in New Mexico." *Ecological Monographs* **21**:397-403.

Gibbens, R. P., Beck, R. F. 1988. "Changes in grass basal area and forb densities over a 64-year period on grassland types of the Jornada Experimental Range." *J. Range Mgmt.* **41**:186-192.

Gibbens, R. P., Beck, R. F., McNeely, R. P., Herbel, C. H. 1992. "Recent mesquite establishment in the northern Chihuahuan Desert." *J. Range Mgmt.* **45**:585-588.

Grover, H. D., Musick, H. B. 1990. "Shrubland encroachment in southern New Mexico, U. S. A.: An analysis of desertification processes in the American southwest." *Climatic Change* **17**:305-330.

Herbel, C. H., Ares, F., Bridges, J. 1958. "Hand-grubbing mesquite in the semidesert grassland." J. Range Mgmt. **11**: 267-270.

Herbel, C. H., Abernathy, G. H., Yarbrough, C. C., Gardner, D. K. 1973. "Rootplowing and seeding arid rangelands in the southwest." *J. Range Mgmt.* **26**: 193-197.

Herbel, C. H., Gould, W. L., Leifeste, W. F., Gibbens, R. P. 1983. "Herbicide treatment and vegetation response to treatment of mesquites in southern New Mexico." *J. Range Mgmt.* **36**: 149-151.

Herbel, C. H., Morton H. L., Gibbens, R. P. 1985. "Controlling shrubs in the arid southwest with tebuthiuron." *J.Range Mgmt.* **38**: 391-394.

Herbel, C. H., Gibbens, R. P. 1996. "Post-drought vegetation dynamics on arid rangelands in Southern New Mexico. New Mexico State Univ. Agric. Exper. Sta. Bull. 776. Las Cruces, NM.

Herrick, J. E., Whitford, W. G., deSoyza, A. G., Van Zee, J. 1995. "Soil and Vegetation Indicators for Assessment of Rangeland Ecological Condition." (pp.157-166) In: Celedonio, A. G., ed. *North American Workshop on Monitoring for Ecological Assessment of Terrestrial and Aquatic Ecosystems. USDA Forest Service, Rocky Mountain Forest and Range Experiment Station.* 305 p.

Martinez-Meza, E., Whitford, W. G. 1996. "Stemflow, throughfall and channelization of stemflow by roots in three Chihuahuan Desert shrubs." *J. Arid Environ.* **32**:271-287.

Naranjo, L. G., Raitt, R. J. 1993. "Breeding bird distribution in Chihuahuan Desert habitats." *Southwest. Nat.* **38**:43-51.

National Research Council. 1994. "Rangeland Health: New Methods to Classify, Inventory, and Monitor Rangelands." National Academy Press. 180 p.

Rapport, D. J., Regier, H. A., Hutchinson, T. C. 1985. "Ecosystem behaviour under stress." *Am. Nat.* **125**:617-640.

Santos, P. F., Depree, E., Whitford, W. G. 1978. "Spatial distribution of litter and microarthropods in a Chihuahuan Desert ecosystem." *J. Arid Environ.* **1**:41-48.

Skidmore, E. L. 1986. "Wind erosion control." *Climatic Change* **9**:209-218.

Stephens,G., Whitford, W. G. 1993. "Responses of Bouteloua eriopoda to irrigation and nitrogen fertilization in a Chihuahuan Desert rassland." *J. Arid Environ.* **24**:415-421.

Stubbendieck, J., Hatch, S. L., Butterfield, C. H. 1993. North American Range Plants (4th ed.) University of Nebraska Press, Lincoln

Tongway, D. J., Ludwig, J. A. 1997. "The conservation of water and nutrients within landscapes." In. Ludwig, J., Tongway D., reudenberger D., Noble, J., Hodgkinson, K. (eds.) *Landscape Ecology: Function and Management.* CSIRO Publishing, Collingwood, Victoria

Tongway, D. 1994. "Rangeland Soil Condition Assessment Manual." CSIRO, Division of Wildlife and Ecology.Australia.

Westoby, M., Walker, B., Noy-Meir, I. 1989. "Opportunistic management for rangelands not at equilibrium." *J. Range Manage.* **42**:266-274.

Whitford, W. G., Ludwig, J. A., Noble, J. C. 1992. "The importance of subterranean termites in semi-arid ecosystems in south-eastern ustralia." *J. Arid Environ.* **22**:87-91.

Whitford, W. G. 1995. Desertification: Implications and limitations of the ecosystem health metaphor. In: Rapport, D. J., Gaudet, C. L., alow, P. (eds) Evaluating and Monitoring the Health of Large-Scale Ecosystems. NATO ASI Series, Vol. 128. Springer-Verlag, Berlin pp.271-296.

Whitford, W. G., Herrick, J. E. 1996. "Maintaining soil processes for plant productivity and community dynamics." In.West, N. E. (ed) Rangelands in a sustainable biosphere (vol. 2) Proceedings of the Fifth International Rangeland Congress. Society for Range Management. Denver, CO pp. 33-37.

Whitford, W. G., Van Zee, J., Nash, M. S., Smith, W. E., Herrick, J. E. 1997. "Ants as indicators of exposure to environmental stressors in North American desert grasslands." *Environ. Monitor. Assess.* In Press.

Wiens, J. A. 1973. "Pattern and process in grassland bird communities." *Ecol. Monogr.* **43**:237-270.

Zobeck, T. M. 1991. "Soil properties affecting wind erosion." *J. Soil Water Conserv.* **46**:112-118.

MONITORING CHANGES IN STRESSED ECOSYSTEMS USING SPATIAL PATTERNS OF ANT COMMUNITIES

MALIHA S. NASH[1*], WALTER G. WHITFORD[1], JUSTIN VAN ZEE[2], and KRIS HAVSTAD[2]

[1]*US EPA, NERL, ESD, PO Box 93478, Las Vegas NV 89193-3478 (*current mailing address);* [2]*USDA-ARS, Jornada Experimental Range, 3JER, New Mexico State University, Las Cruces NM 88003.*

Abstract. We examined the feasibility of using changes in spatial patterns of ants-distribution on experimental plots as an indicator of response to environmental stress. We produced contour maps based on relative abundances of the three most common genera of ants based on pit-fall trap captures. Relative abundance of *Conomyrma* spp. decreased, relative abundance of *Solenopsis* spp. increased, and relative abundance of *Pogonomyrmex* spp. remained relatively unchanged. The contour maps showed long-term changes in foraging activity and/or distribution of colonies of ants in response to grazing by domestic livestock. This study demonstrated that analysis of spatial patterns of ant activity derived from relative abundances of ants in pit-fall traps provided interpretable data for developing an indicator of exposure to ecosystem stress.

1. Introduction

Assessing and/or monitoring the health of arid ecosystems requires a suite of indicators. That suite of indicators should include measurements that relate to important ecosystem functions such as the redistribution of water and nutrients, productivity, and the maintenance of biodiversity (Herrick *et al.*, 1995). The measurements that are incorporated into the calculation of indicators may provide information on all or a few of these ecosystem functions. In arid ecosystems, ants affect ecosystem functions such as water infiltration, soil nutrient distributions, and composition of the soil seed bank (Whitford and DiMarco, 1995; Carlson and Whitford, 1991; J. Herrick, unpublished data). Ants have also been used to examine the effects of changes in ecosystems on biodiversity (Roth *et al.* 1994, Perfecto and Snelling, 1995). However, in some arid ecosystems, ants were poor indicators of exposure of the ecosystems to stress (Whitford *et al.*, in press). The results of those studies suggested that changes in spatial patterns resulting from altered behavior of selected genera may be better indicators of exposure to stress than quantitative estimates of relative abundances of species and/or species richness.

Ant species differ with respect to the physical characteristics of the sites chosen for construction of nests. For some species, the distribution and characteristics of the vegetation may be less important than soil characteristics such as depth and texture. For other species, the presence of tall plants that provide shade may be an important environmental variable determining where nests are constructed (Burbidge *et al..*, 1992, Roth *et al.*, 1994, Perfecto and Snelling, 1995). We hypothesized that some ant species would respond to changes in vegetation characteristics by relocating their colonies, or by modifying their foraging behavior (Whitford and Ettershank, 1975).

Environmental Monitoring and Assessment **51:** 201–210, 1998.

In this paper we present the results of a multi-year spatial analysis of ant communities exposed to intensive, short-term grazing by cattle and to vegetation restructuring resulting from removal of woody shrubs in a shrub grassland mosaic. We hypothesized that analysis of spatial patterns would provide information that could be used in interpreting the effects of stressors on ecosystem function. This study provided a test of the feasibility of using spatial patterns of ant abundances as an indicator.

2. Study Site and Methods

The study site was located on the Jornada Experimental Range, 40 km northeast of Las Cruces, NM. Eighteen 0.5 ha plots were established in a grassland pasture that had been grazed at light stocking rates, only during the winter, for more than 30 years. The study plots were arranged in two rows of nine plots that were blocked along the long axis (Figure 1). The treatments were a combination of two factors: (1) removal and non removal of the mesquite shrubs (*Prosopis glandulosa*) and (2) winter grazing, summer grazing, and ungrazed control. These treatments were used in a randomized complete block design (Figure 1). The study area was fenced in 1993. Mesquite shrubs were removed from nine plots in January and February 1994, and there was no grazing in 1994. Winter-grazed plots were grazed on 15-16 February 1995 and 9-12 January 1996. The plots were grazed at a stocking rate that removed 65-80% of the vegetation in a 24-hour period. Ant communities were sampled in September of each year beginning 1993. In 1996, we sampled ant communities one week prior to summer grazing (August 16) and again on September 10, three weeks post summer grazing.

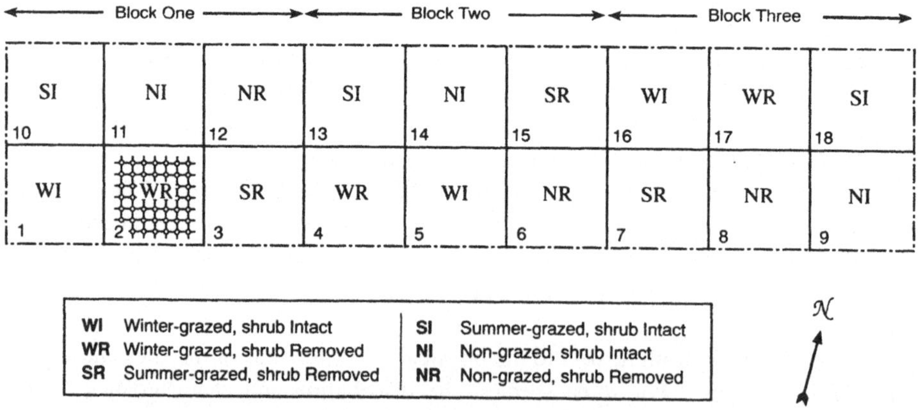

Fig. 1. Multiple Stressor Exclosure plots and grid layout for pitfall sampling of ants. Numbers in the left lower corners denote the plot number. Pitfall ants sampling grids is given in plot 2.

Ants were sampled by pitfall traps arranged in 7 x 7 trap arrays with 9.14 m spacing between traps on each of the 0.5 ha plots (see plot 2, Figure 1). Pitfall traps (38 x 70 mm tall plastic vials) were filled to a depth of 30 mm with a mixture of 70 percent ethanol and 30 percent glycerol (Greenslade and Greenslade, 1971). Traps were buried in the soil flush with the surface and left in place for 24 hours. Traps were retrieved, labeled with grid location information and stored prior to processing.

All ants in a vial were identified by species or by operational taxonomic unit (e.g. *Pheidole* spp. with no major workers) and counted. The data used in the analysis were a) percentage of traps in which selected genera were found and b) number of individuals of a species in a vial (relative abundance). The feasibility of using spatial analysis was examined by using data from only the winter-grazed plots.

3. Data Analysis

We analyzed the spatial distribution patterns of only the three most common genera (*Solenopsis* spp., *Conomyrma* spp., and *Pogonomyrmex* spp.). These genera differ sufficiently in foraging behavior and nest placement (Whitford 1978, Holldobler and Wilson 1990) to provide a reasonable test of the feasibility of this analytical approach. We used contour mapping to look for the spatial distribution patterns with the time. We used Surfer software (Golden Software, Inc. 809 14th st., Golden CO 80401-1866) for contour mapping. Mapping was done using kriging when the number of traps were ≥ 20. Mapping was done using an inverse distance technique when the number was < 20.

4. Results and Discussion

In 1993, before any of the treatments was imposed, *Conomyrma* spp. accounted for 96.6 percent of all the ants trapped on the plots and were found in 96.3 percent of the traps that were recovered. In each of the winter-grazed shrub-removed plots, there were dramatic reductions in relative abundance of *Conomyrma* (Figure 2). Fire ants, *Solenopsis,* which occurred at low abundances prior to the imposition of treatments, showed a dramatic increase in relative abundance (Figure 2). There was little change in the relative abundance of seed-harvester ants, *Pogonomyrmex*, following the shrub-removal and grazing treatments (Figure 2).

The response of *Conomyrma* is what is expected for a liquid-feeding ant that tends homopteran insects on vegetation. Shrub removal during the winter 1994 eliminated mesquite (*Prosopis glandulosa*), a host plant for homopterans that is phenologically predictable and that provides a food source of cell contents and phloem sap for sucking herbivores such as homopteran throughout the growing season. The growing seasons of 1994 and 1995 was characterized by drought with virtually no leaf production by the grasses. Thus, during 1994-1995 there was virtually no foliage for the sucking insects that *Conomyrma* rely upon for food. An examination of the spatial distribution of *Conomyrma*

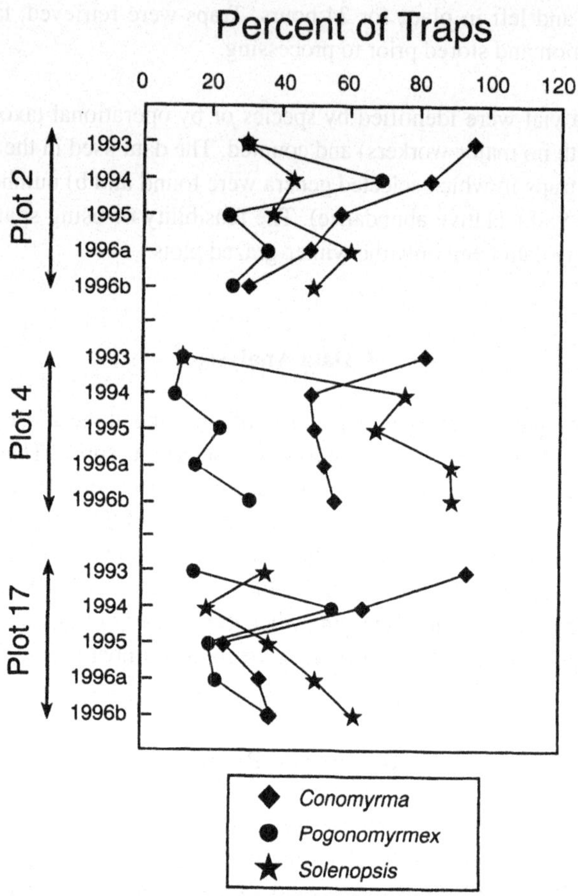

Fig. 2. Percent of traps of *Conomyrma, Solenopsis,* and *Pogonomyrmex* in winter-grazed shrub-removed plots with time.

in plot 2 over time showed that their activity was concentrated at the edges of the plots. This pattern is clearly the result of *Conomyrma* foraging out from the nests located within plot 2 to plants located in adjacent plots where shrubs were not removed (compare Figure 3a vs. 3b, 3c, and 3d). This spatial pattern became more pronounced following removal of grass and forb foliage by livestock grazing. *Conomyrma* are known to recruit rapidly to food sources and to exclude other species from food sources (Holldobler and Wilson, 1990). Thus, the change in spatial patterns of activity is consistent with the biology of the species and should not be interpreted as resulting from the death of colonies at the center of the plots.

The increase in abundance of *Solenopsis* occurred spatially in a pattern that was distinctly different from the pattern seen in *Conomyrma*. The *Solenopsis* tended to clump with higher abundance in the center of the plots. *Solenopsis* are generalists that feed on seeds as well as on liquids (field observation). This generalist feeding behavior should

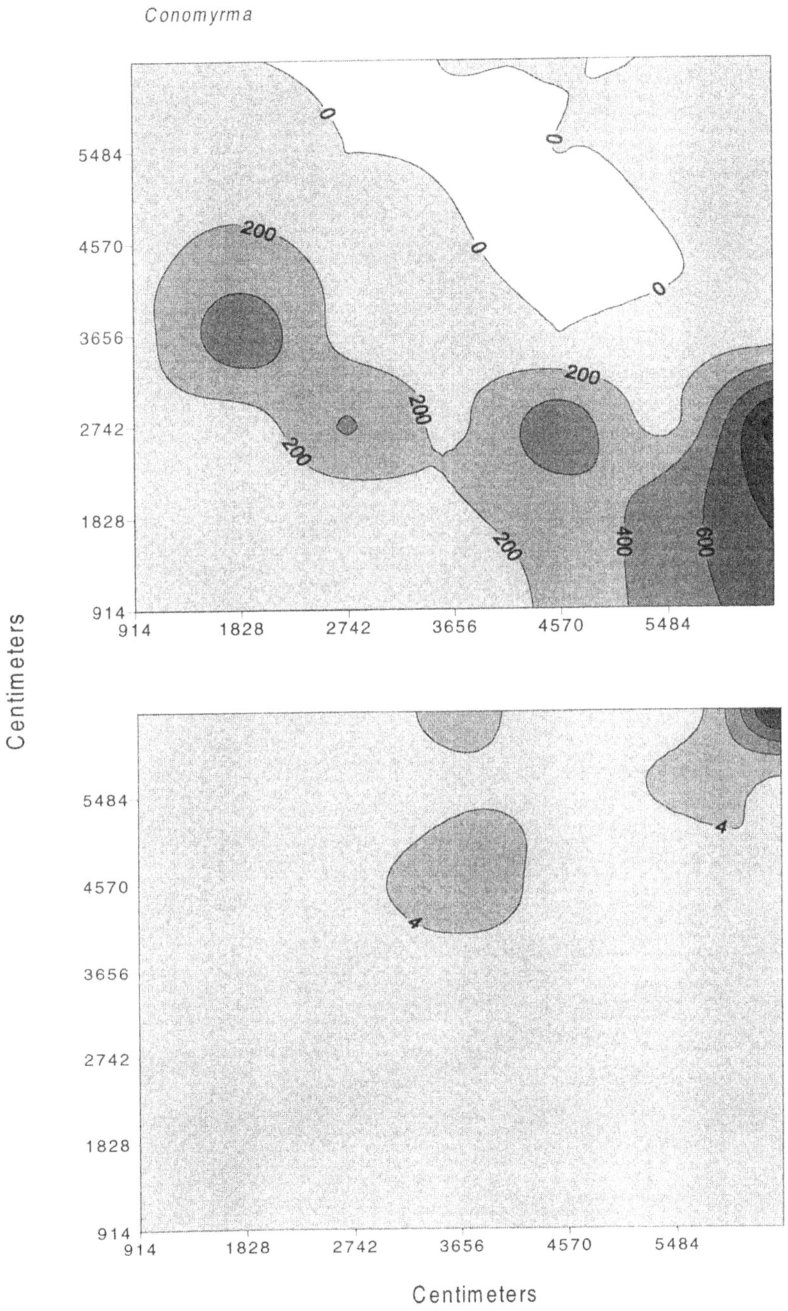

Fig. 3. Distribution of relative abundance (counts) of *Conomyrma* and *Solenopsis* for plot 2 in a) 1994, 20 weeks before grazing, b) 1995, 26 weeks after grazing, c) 1996a, 28 weeks after grazing, and d) 1996b, 31 weeks after grazing. (Increasing relative abundance values from clear to dark).

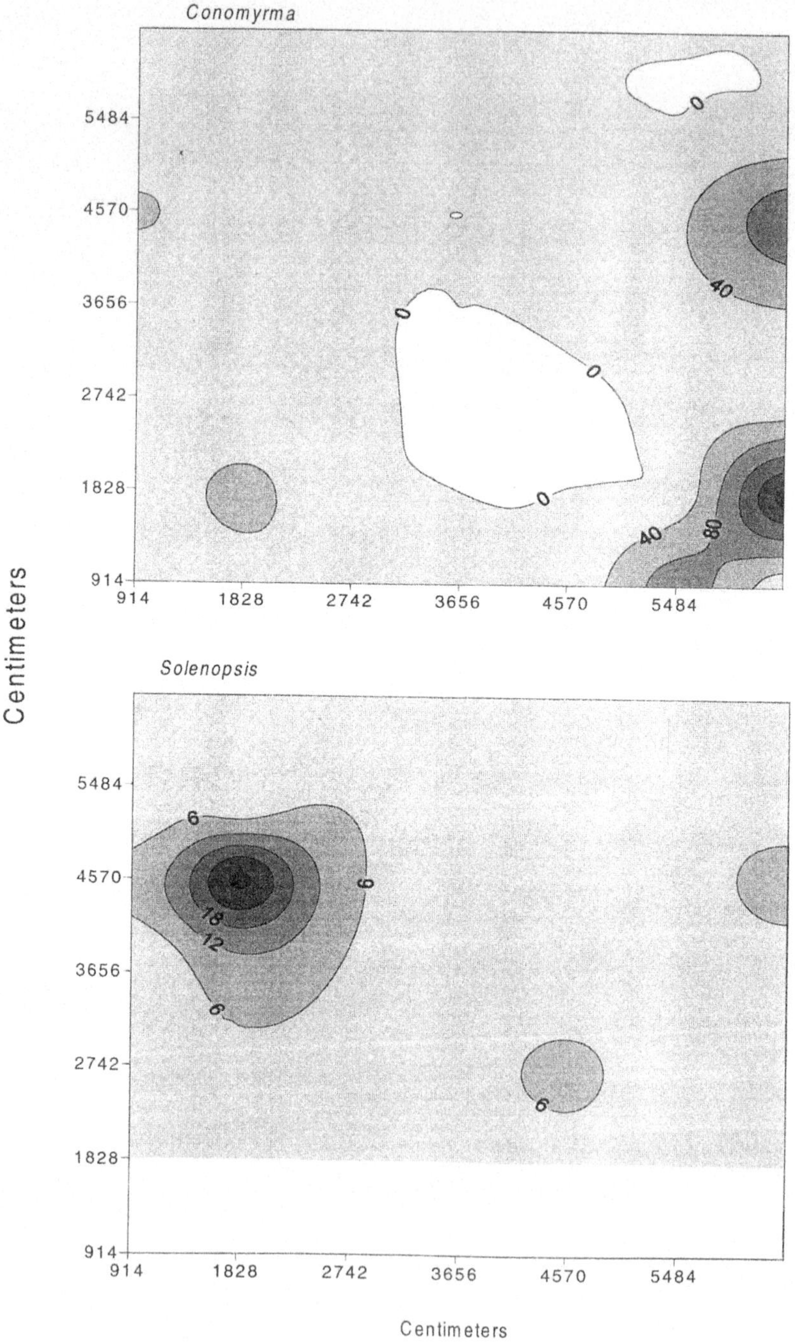
Conomyrma
5484
4570
3656
2742
1828
914
914
1828
2742
3656
4570
5484
0
0
0
0
40
40
80
Centimeters
Solenopsis
5484
4570
3656
2742
1828
914
914
1828
2742
3656
4570
5484
6
6
6
6
18
12
Centimeters

Fig. 3b

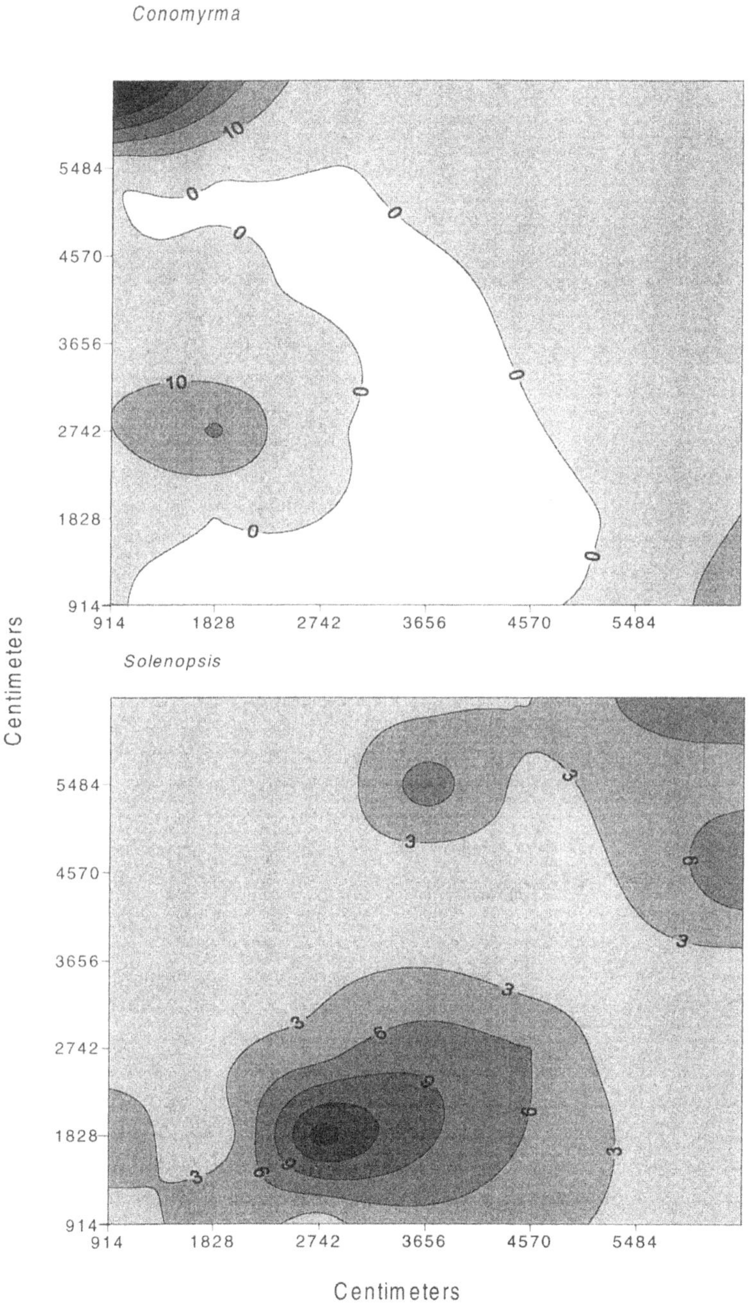

Conomyrma
10
0
0
0
0
10
0
0
0
5484
4570
3656
2742
1828
914
914
1828
2742
3656
4570
5484
Centimeters
Solenopsis
3
3
6
3
3
3
6
9
9
6
6
3
3
5484
4570
3656
2742
1828
914
914
1828
2742
3656
4570
5484
Centimeters

Fig. 3c

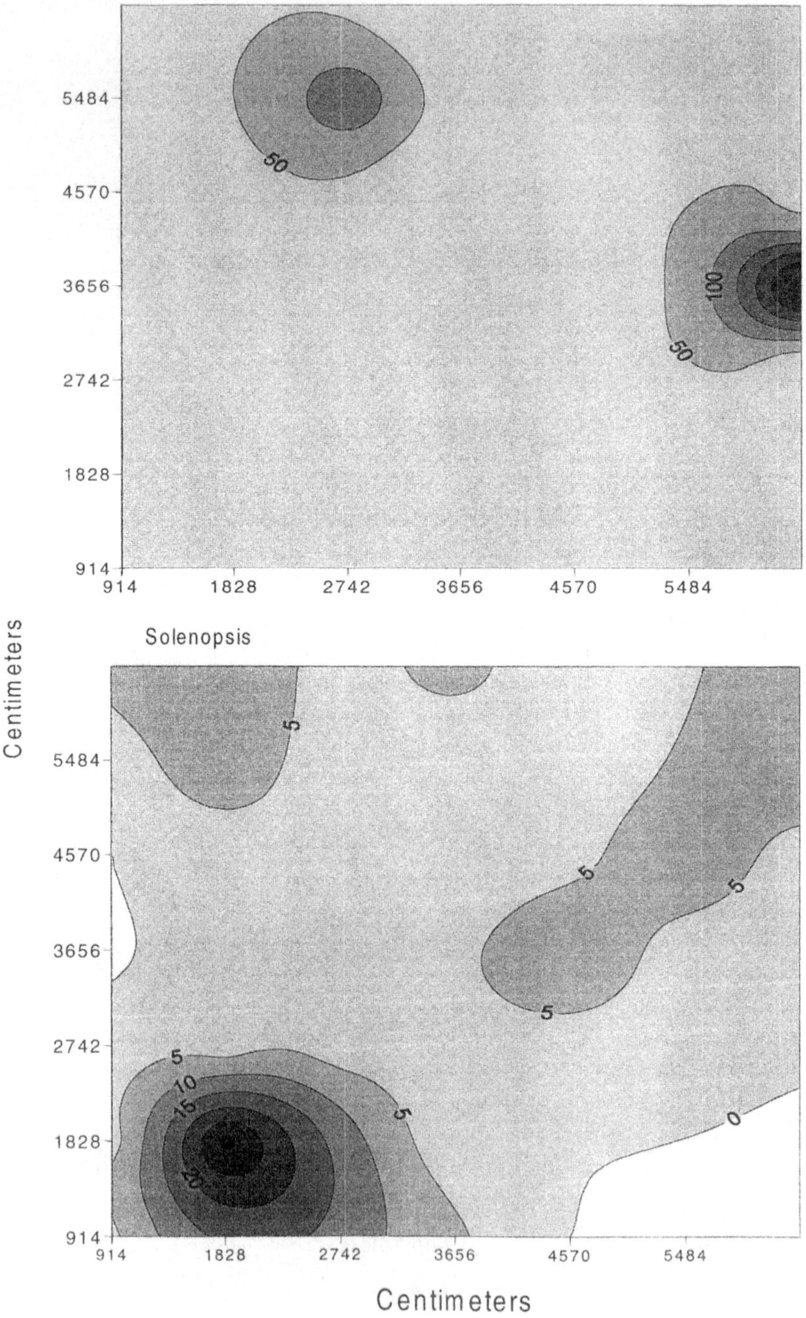
5484
4570
3656
2742
1828
914
914
1828
2742
3656
4570
5484
50
100
50
Centimeters
Solenopsis
5
5
5
5
5
10
15
20
5
0
5484
4570
3656
2742
1828
914
914
1828
2742
3656
4570
5484
Centimeters

Fig. 3d

result in a more random foraging pattern and relatively equal probability of capture in traps anywhere on the plot. However, *Solenopsis* have colonized soil beneath the dung pats which explains their rapid increase in abundance following grazing (Figure 2). The dung pats provide a sheltered environment for shallow nesting species such as *Solenopsis*.

The lack of change in the abundance of *Pogonomyrmex* reflects the resilience of these ants to disturbance and stress. Despite the drought conditions during the summer growing season, there were sufficient winter rains for germination and growth of spring ephemerals. Seeds of spring ephemerals are harvested by these ants, thus providing food resources that did not change with habitat modification or livestock grazing (Whitford, 1978). It is also likely that habitat modification and livestock grazing did not cause the death of colonies and/or induce *Pogonomyrmex* to move their nests.

The spatial pattern analysis showed that there were consistent and interpretable changes in the behaviors of the ant genera selected for study. The changes in spatial pattern exhibited by the ant genera examined in this study demonstrate the feasibility of using spatial pattern analysis of selected genera of ants for monitoring and assessing the effects of chronic and acute environmental stress. The short-term changes recorded in the pre-grazing and post-grazing comparison in 1996 are consistent with changes in foraging behavior and not the result of nest re-location. The behaviors of ants that may be responsible for differences in spatial patterns derived from pitfall trap data are: (1) cessation of surface activity by some or all of the colonies, (2) moving the nest entrance (3) changing direction, duration, and intensity of foraging. Some ant species move nest entrances frequently (*Conomyrma* and *Pogonomyrmex desertorum;* unpublished data). Since nest entrances are generally moved less than 2 m, that behavior is the least likely to be detected by the pattern analysis used in this study. Large changes in spatial pattern are probably the result of cessation of activity by colonies or marked changes in foraging intensity and direction. Nest relocation may contribute to the spatial patterns recorded by analysis of pitfall trap data but actual mapping of nests is required to confirm this.

Analysis of spatial patterns of selected ant genera demonstrated that some genera respond to soil surface changes resulting from livestock grazing (*Solenopsis*), some respond to changes in composition and structure of the vegetation (*Conomyrma*), and some are resistant to the short-term changes resulting from grazing and vegetation manipulation (*Pogonomyrmex*). The analysis of spatial patterns of selected genera of ants also provides data on changes in spatial patterns of ecosystem functions, e.g. water infiltration, wind transport of soil, and patterns of soil nutrients (Whitford 1994).

This study shows that spatial analysis can be used to detect changes in behavior of species that respond to environmental stressors. These patterns can be used in developing indicators of exposure to environmental stress. However, we will have to develop statistical tools for analyzing the patterns, and that is beyond the scope of this paper.

Notice: this study was funded and performed in part by the U.S. Environmental Protection Agency (EPA), through its Office of Research and Development (ORD). This manuscript has been reviewed and approved for publication. Mention of names or commercial products does not constitute endorsement or recommendation for use.

Acknowledgment

The authors would like to thank Dr. Llewellyn Williams and the two anonymous reviewers for their valuable comments and input for this paper.

References

Burbidge, A. H., K. Leicester, S. McDavitt, and J. D. Majer.: 1992, Ants as indicators of disturbance of Yanchep National Park, Western Australia. Journal of the Royal Society, West Australia. **75,** 89-95.

Carlson S. R., W. G. Whitford (1991) Ants mound influence on vegetation and soils in a semiarid mountain ecosystem. American Midland Naturalist 126:125-139.

Greenslade, P. J. M. and P. Greenslade.: 1971, The use of baits and preservatives in pitfall traps. Journal of the Australian Entomological Society **10,** 253-260.

Herrick, J. E., W. G. Whitford, A. G. deSoyza, J. Van Zee: 1995. Soil and vegetation indicators for assessment of rangeland condition (pp. 157-166) In: Celedonio AB, ed. North American Workshop on Monitoring of Ecological Assessment of Terrestrial and Aquatic Ecosystem. US Forest Service, Rocky Mountain Forest and Range Experiment Station, Ft, Collins, Co.

Holldobler, B. and E. O. Wilson.: 1990, The Ants, Bellnap Press. Cambridge, MA.

Perfecto, I. and R. Snelling.: 1995, Biodiversity and the transformation of a tropical agroecosystem: ants in coffee plantation. Ecological Application **5,** 1084-1097.

Roth, D. S., I. Perfecto, and B. Rathcke.: 1994, The effects of management systems on ground-foraging ant diversity in Costa Rica. Ecological Application **4,** 423-426.

Whitford, W. G.: 1978, Foraging in seed-harvester ants, *Pogonomyrmex spp.* Ecology **59**:185-189.

Whitford W. G. and R. DiMacro(1995). Variability in soils and vegetation associated with harvester ant (*Pogonomyrmex rugosus*) nests on a Chihuahuan Desert watershed. Biological and Fertility of Soils 20:169-173.

Whitford W. G. and G. S. Forbes, G. I. Kerley (1994). Diversity, spatial variability and functional roles of invertebrates in desert grassland ecosystems. Pp. 152-195. In McClaran MP, Van Devender TR (eds.) The Desert Grassland. University of Arizona Press, Tucson.

Whitford, W. G., J. Van Zee, M. S. Nash, W. E. Smith, and J. E. Herrick.: In press, Ants as indicators of exposure to environmental stressors. Environmental Monitoring Assessment.

Whitford, W. G. and G. Ettershank.: 1975, Factors affecting foraging activity in Chihuahuan Desert harvester ants. Environmental Entomology **4**: 689-696.

PARASITES OF FISH AS INDICATORS OF ENVIRONMENTAL STRESS

J. H. LANDSBERG, B. A. BLAKESLEY, R. O. REESE, G. MCRAE, and P. R. FORSTCHEN

Florida Department of Environmental Protection, Florida Marine Research Institute, 100 Eighth Ave S.E., St. Petersburg, FL 33701, USA

Abstract. During the 1994/1995 EMAP-Estuaries program in the Carolinian Province we investigated the feasibility of using parasites of fish as response indicators. Parasites of fish are an indigenous component of healthy ecosystems. Within the EMAP-E design, the suite of environmental parameters which may affect parasite abundance, richness, prevalence, and diversity can be divided into three categories: 1) the physical and chemical characteristics of the water and sediment (including contaminants) external to the fish; 2) the internal environment defined by the physical condition (physiological) of individual fish; and 3) the presence and relative abundance of benthic macroinvertebrates, many of which serve as intermediate hosts. The biotic response of parasites to environmental stressors is also reflected in the health of fish. Parasite assemblages of silver perch *Bairdiella chrysura* respond to both natural and anthropogenic stressors. Our results showed that particular environmental stressors and specific parasites that respond include: temperature and monogeneans; contaminants and nematodes; low dissolved oxygen and protists; and salinity, together with a mixture of metal and organic contaminants and crustacea. Parasites of fish are useful biomarkers and appear to be more sensitive to environmental stressors than are the fish themselves. Parasite responses to selected environmental stressors may be used to discriminate polluted and unpolluted sites. The use of parasites of fish as biomarkers has relevant application to fisheries management and coastal monitoring programs.

1. Introduction

Ecosystems with a high degree of biotic integrity (healthy ecosystems) are comprised of balanced populations of indigenous organisms with diverse structural and functional organizations (Summers *et al.*, 1992). Healthy ecosystems have a complex trophic structure with many species forming the food web. In the estuarine environment, fish are at or near the top of that food web, and their parasites are also an indigenous component of the web. The parasites of fish reflect the life habits of the fish, including their interactions with the benthic, planktonic, and fish communities. Evaluation of the parasitic fauna of selected fish species within a particular estuary may provide both qualitative and quantitative bioindicators indicative of, and sensitive to changes in, the overall health of that estuary. The parasite assemblages of fish can have a potential role as response indicators to environmental stress in relation to other identified response indicators used in the EMAP-E program. These parasites may have life cycles involving several host organisms that are found at different trophic levels. Such parasite assemblages can be evaluated not only for their influence on the health of the fish but also for their own changes in diversity, prevalence, richness, and abundance in response to environmental stressors. Parasites of fish could therefore be a more sensitive indicator of environmental stressors than the fish themselves. An index based on the parasite assemblages of fish could also be integrated with other indices of biological integrity.

Environmental Monitoring and Assessment **51:** 211–232, 1998.

A variety of anthropogenic pollutants are known to acutely or chronically influence parasites of fish. These pollutants include heavy metals, petroleum hydrocarbons, pesticides, pulp mill and thermal effluents, domestic sewage and waste, sedimentation, and agricultural and industrial toxins (Esch *et al.*, 1976; Boxrucker, 1979; Möller, 1986; Khan, 1987, 1990; McVicar *et al.* 1988; Khan and Thulin, 1991; Overstreet, 1988; 1993; Overstreet *et al.* 1996). The levels of some of these contaminants in tissues of fish are often measured and correlations with gross pathological changes tested. However, a pathological response to contaminant stress may only be manifested after chronic exposure. This integrated and cumulative pathological response in fish is a specific measure of fish health that is often used during environmental monitoring surveys such as EMAP (Hyland *et al.* 1996). Such pathological responses are usually only detectable in a small percentage of the fish population (Ziskowski *et al.* 1987). However, in healthy environments, parasites are usually present in individual fish and their response to environmental stress is measurable.

In fish populations parasite species have both a temporal and a spatial sensitivity to pollutants and other water quality parameters that can be measured instantaneously, both in the short-term and in the long-term. The sensitivity of the parasite fauna of fish to environmental stressors may vary at different stages in their life cycle prior to, or after entry into, the fish population. The timing of parasite immigration into or emigration from the fish population will influence the diversity and abundance of the parasite fauna of fish at any one sampling period. Because fish are mobile, there may not be any apparent correlation with measured environmental variables, tissue contaminant levels, or gross pathology at the time of sampling. However, the parasite fauna of fish can be a reflection of prior exposure to pollutants and other stressors. Despite many recent studies to establish and evaluate different indices of biotic integrity, a parasite index has not been developed for use in environmental monitoring, nor has its potential been fully determined. Like other organisms, parasites can be used as biomarkers and their biological responses related to changes in overall health from the ecosystem level down to the cellular level. Information concerning parasites of fish and environmental degradation is usually applied at the level of the individual fish and related to pathological effects associated with parasitism. Parasite studies of fish at the community level have identified parasite species assemblages *per se* (Leong and Holmes, 1981; Thoney, 1991,1993), but they have only recently been applied to investigate the parasite response to pollutants (Overstreet *et al.* 1996). A parasite index can be developed to enhance the relevance of measures of other biotic components, such as the benthic infauna and demersal fish, and can also be directly linked with the fish pathology component and other measures of fish health.

The goal of this study was to investigate the feasibility of developing a cost-effective index using parasites of fish as response indicators and determining if this index could be used to monitor and assess estuarine health. The objectives were to 1) evaluate the use of parasites of fish as response indicators of environmental quality, 2) establish measurable criteria for a parasite index, 3) use the parasite index to discriminate between polluted and unpolluted sites, and 4) monitor temporal changes in estuarine health. The present communication describes the parasite assemblages of the estuarine fish, silver perch,

(*Bairdiella chrysura*), and examines how these assemblages appear to respond to the particular environmental stressors that were measured during the 1994 EMAP-E project in the Carolinian Province.

2. Materials and Methods

2.1. FISH HEALTH AND PARASITOLOGY

A total of 19 sites were sampled by the EMAP-E core program (Hyland *et al.* 1996) in the Florida portion of the Carolinian Province during August, 1994. Sampling for the fish health and parasite survey was carried out in parallel with the collection and processing of samples for the core program. The silver perch, *Bairdiella chrysura*, was the most widely distributed common fish species and was selected as the target host. Wherever possible, a maximum of 20 fish were collected at each site by otter trawl using the EMAP standard protocol (Hyland *et al.* 1996). Fish were held in aerated 50 gallon containers, returned to the lab, and processed within 48 hours.

Prior to necropsy, fish were examined for any obvious gross pathological features. External smears of skin and fins were made and evaluated for parasites. Basic fish morphometrics (length, weight) were measured and selected organs were evaluated for condition. Fish necropsies were conducted according to standard procedures (FDEP, 1994). Blood smears were made, gills from the left side of the fish were examined for parasites, gills from the right side were fixed for histopathology, and external examination of the body, fins, gills, buccal cavity, and mouth was completed. The fish were then frozen in liquid nitrogen for later examination. Freshly-thawed fish were examined for parasites in the following organs: stomach, ceca, anterior and posterior intestine, gall bladder, liver, spleen, swim bladder, anterior and posterior kidney, muscle, eye, and brain. Organosomatic indices of spleen (SSI) and liver (HSI) were determined. Parasites were enumerated using standard counts of 30 fields at 100 x for metazoa and 1000x for protists. Epitheliocystis, cysts (microsporea) and eggs (blood flukes) were counted either as present or absent. Metacercaria were discriminated based on larval characteristics, but these identifications were not always consistent.

2.2. ANALYSIS

Various measures were used to compare parasite assemblages at different sites. Ecological terms and measures used for parasite data analysis are adapted from those of Leong and Holmes, 1981; Margolis *et al.* 1982; Thoney 1991. Abundance refers to the mean number of individuals of a particular parasite species per host in a sample of n hosts that may include both infected and uninfected individuals. The following ecological parasite measures were used: the total number of parasite taxa in the sample at each site (S), which represents the parasite species richness; n = the number of fish in the sample; n_i = the number of parasite individuals of all parasite taxa per individual fish which represents the parasite infracommunity; N = the total number of individuals of all species of parasites in each sample of hosts (= Σn_1 per site); $\bar{N}$ = the mean number of all individual parasites in a sample of hosts (N/n); $\bar{S}$ = the mean number of parasite taxa per site (= S/n); $\bar{S}_1$ = the mean number of parasite taxa per fish (= $\Sigma S_1/n$) where S_1 = the number of parasite taxa per individual fish; the Shannon-Wiener diversity index, $H = \Sigma p_i \log_e p_i$, where $p_i = n_i/N$ and is the proportion of the sample belonging to the *i*th species; I/SI = the reciprocal of Simpson's index (an index of diversity) where $SI = 1 - \Sigma(p_i)^2$, and p_i is the proportion of individual parasites of species *i* in the infracommunity; and J' = the Brillouin evenness index. Jaccard's index of overlap (J), used as an index of qualitative similarity between pairs of samples was calculated as:

$$J = 100c/a + b - c,$$

where a is the number of taxa in the first sample, b is the number of taxa in the second sample and c is the number of taxa common to both. The higher the value of J, the greater the similarity between pairs of samples. Statistical analyses for comparisons of abundances by site were based on a one-factor ANOVA on log(x+1) transformed data. All possible pairwise comparisons among sites were conducted using Tukey's honest significant difference (Zar, 1984).

2.3. PARASITES AS RESPONSE INDICATORS

In order to study the response of parasites to environmental stress, phyletic groups of parasite taxa with similar life cycles were identified and sorted into 4 main groups according to the scheme outlined in Table I. The life cycle type, number of hosts required in the life cycle, habitat type of hosts, trophic level for parasite, mode of transmission, and stage represented in the fish are all factors that separate one group of parasites from another. The types of environmental stressors, and the parasite stage at which each stressor is likely to have an effect on parasite abundance and diversity, are also listed in Table I. Using parasite assemblages grouped in this way, the effect of a particular environmental stressor can be partitioned. Different factors will determine the abundance, prevalence, and diversity of each of these parasite groups. Some of these factors are water quality characteristics, sediment type, contaminant levels, silver perch host characteristics (age, length, weight, health), and the species diversity of the benthos, fish, and other organisms present at different trophic levels and involved in the parasite's life cycle.

2.4. PARASITE INDEX

For the development of the parasite index, a scoring procedure was used. This method was considerably modified from that used in other fish health indices (Novotny and Beeman, 1990; Goede and Barton, 1990; Adams *et al.*, 1993) and was developed to compare parasite responses to environmental stress. In this pilot study, the approach to understanding the differential parasite responses to environmental stress has been to analyze and interpret separately the grouped assemblages of representative parasite groups in relation to selected individual stressors. A parasite index can be developed using a scoring system that takes into account the prevalence of a particular parasite group (as defined in Table I) at each site as well as selected ecological parasite parameters that reflect the interaction of the parasites to environmental stress. For the development of the parasite stress index, the data can be scored according to the prevalences/values found at each of the sites. Fourteen criteria were chosen combining parasite responses and ecological parameters (Appendix 1). Individual prevalences/values can be compared between sites by setting a range of values using the expected parasite response values and scoring from 1 to 10 respectively (Appendix 1). The effects of environmental stressors such as contaminants can be tested for and included in the development of the parasite index. This system allows discrimination of the sensitivity of the biota to relatively low levels of environmental stress.

3. Results

3.1. FISH AS RESPONSE INDICATORS

A total of 122 silver perch were captured from 7 of the 19 EMAP sites sampled (Figure 1). Fish mean standard length by site ranged from 9.93 to 13.23 cm, mean weight from 20.46 to 45.97 g, and mean condition factor (cf) from 1.94 to 2.09 (Table II). Mean liver weights ranged from 0.179 to 0.383 g and spleen weights from 0.011 to 0.026 g. The hepatosomatic index (HSI) corrects for individual fish weight differences and means by site and ranged from 0.744 to 0.948; the splenosomatic index (SSI) ranged from 0.039 to 0.066 (Table II). The highest HSI's were found at sites 17 and 9 - the most heavily contaminated (Figure 2).

3.2. PARASITES AS RESPONSE INDICATORS

Parasite groups were fairly well distributed throughout the Florida portion of the Carolinian province. The relative abundance of parasite groups by site is shown in Figure 3. The species richness (*S*) of parasites ranged from 20 species at site 10 to 30 species at site 14 (Table III).

A total 18,461 parasites were found from all fish (n =122). Most parasites were located in the gills and intestine. Forty-four parasite taxa were identified from silver perch. Nine parasite taxa were present at every site: *Rhamnocercus* sp. 1 (group 1b), *Opecoeloides* spp. complex (group 4c), *Cryptobia* sp. (group 1a), *Spirocamallanus* sp. (group 4e), *Hysterothylacium* sp. (group 4f), *Scolex polymorphus*, *Scolex* large sp. (group 4g), a blood fluke (group 4b), and internal metacercaria (group 4c) (Table IV). Together these nine taxa comprised 85.7% of all the parasite individuals found. Ranking of the total number of parasites (as a dominance %) indicates that the 22 most abundant taxa comprised 99.16% of the individuals found. Rarer taxa comprised the last 1.0% of individuals found, yet these represented half of the total taxa identified (Table IV). The most dominant parasite species found at each site was the monogenean *Rhamnocercus* sp. 1 (group 1b) which accounted for over half of all parasites found (Table IV). The dominance of this species followed a south to north gradient (Figure 3).

TABLE I

Trophic interactions, life cycle type, and potential influential environmental stressors on the parasite assemblages of estuarine fish

Group	Life cycle type	Phyletic group and stage in fish	Habitat type for host	Mode of transmission	Hosts	Potential environmental stressors
1a	direct	protista - adult	pelagic	water	fish	water quality e.g. low D.O., NH_4,
1b	direct	monogenea - adult	pelagic	water	fish	water quality e.g. temperature
2	direct	crustacea - adult	pelagic	water	fish	water quality e.g. temperature, salinity, contaminants column e.g. pesticides
3	indirect	digenea - larva	benthic > pelagic > pelagic/terrestrial	water > predation > water	mollusc > fish > fish/bird	sediment contaminants for parasite stage in mollusc; water quality for transfer stage from mollusc to fish e.g. turbidity, pesticides. water quality for stage on fish e.g. salinity, low D.O. pesticides. Contaminants in tissues of fish/bird.
4a	indirect	protista - adult	benthic > pelagic	water > predation > water > predation	crustacea/ ?polychaete > fish	sediment contaminants and characteristics for parasite stage in polychaete/crustacean, water quality for crustacea e.g. lesticides, low D.O.
4b	indirect	digenea - adult	benthic > pelagic	water > water	mollusc > fish	sediment contaminants for parasite stage in mollusc; water quality for transfer stage from mollusc to fish e.g. turbidity pesticides; water quality for stage in fish e.g. PAH's, oil
4c	indirect	digenea - adult	benthic > benthic > pelagic	water > predation > water	mollusc > crustacea (amphipod, mysid)/polychaete > fish	sediment contaminants for parasite stage in mollusc; water quality for transfer stage from mollusc to crustacean; water quality and contaminants for parasite stage in crustacean
4d	indirect	digenea - larva metacercaria	benthic > pelagic > terrestrial?	water > predation > water	mollusc > fish > fish/?bird	sediment contaminants for parasite stage in mollusc; water contaminants for stage in fish e.g. PAH's, oil and transfer stages between hosts; contaminants in tissues of fish/bird
4e	indirect	nematoda - adult	bentho-pelagic > pelagic > water	water > predation	crustacean (amphipod, copepod, shrimp) > fish	contaminants in water e.g. PAH's, PCB's for stage in fish and for transfer stages between hosts; sediment and water contaminants for intermediate crustacean host; pesticides
4f	indirect	nematoda - larva	benthic > pelagic pelagic	water > predation > water > predation	crustacean (penaeid shrimp)/ fish > fish	sediment contaminants for parasite stage in crustacean; water quality and contaminants for parasite stage in fish and for transfer stages between hosts
4g	indirect	cestoda - larva, acanthocephala - larva	pelagic > pelagic	water > predation > predation > water	crustacean (copepod, amphipod) > fish > fish	water quality e.g. pesticides for stage in crustacean and for transfer stages between hosts; water contaminants for stage in fish e.g. PAHs

TABLE II
Fish health characterization

Site	Standard length (cm) Mean ±S.D. Range	Weight (g) Mean ±S.D. Range	Liver weight (g) Mean ±S.D. Range	HSI[1] Mean ±S.D. Range	Spleen weight (g) Mean ±S.D. Range	SSI[2] Mean ±S.D. Range	Condition factor[3] Mean ±S.D. Range
6	13.23 1.22 11.5 - 15.8 n = 20	45.97 12.34 27.52 - 68.99 n = 20	0.383 0.138 0.20 - 0.63 n = 19	0.824 0.173 0.59 - 1.25 n = 19	0.026 0.018 0.015 - 0.084 n = 19	0.058 0.031 0.040 - 0.15 n = 19	1.94 0.11 1.75 - 2.06 n = 20
7	12.93 0.85 12.1 - 15.2 n = 16	43.31 7.76 33.36 - 63.89 n = 16	0.326 0.131 0.18 - 0.67 n = 16	0.744 0.230 0.48 - 1.26 n = 16	0.021 0.007 0.015 - 0.044 n = 16	0.049 0.015 0.03 - 0.081 n = 16	1.99 0.15 1.77 - 2.24 n = 16
9	9.93 2.24 7.3 - 13.3 n = 14	20.46 11.71 6.8 - 32.31 n = 14	0.179 0.112 0.04 - 0.36 n = 14	0.850 0.216 0.56 - 1.24 n = 14	0.008 0.004 0.002 - 0.019 n = 14	0.042 0.018 0.023 - 0.08 n = 14	1.91 0.24 1.69 - 2.48 n = 14
10	10.67 0.56 9.8 - 11.7 n = 13	24.45 3.93 18.95 - 32.44 n = 13	0.186 0.046 0.10 - 0.26 n = 13	0.755 0.112 0.55 - 0.95 n = 13	0.013 0.003 0.008 - 0.020 n = 13	0.053 0.015 0.034 - 0.09 n = 13	2.00 0.16 1.61 - 2.25 n = 13
13	10.77 0.61 9.6 - 12.2 n = 19	25.44 4.20 17.12 - 34.87 n = 19	0.188 0.06 0.09 - 0.29 n = 19	0.726 0.15 0.46 - 1.05 n = 19	0.017 0.007 0.006 - 0.040 n = 19	0.066 0.022 0.031 - 0.14 n = 19	2.03 0.13 1.85 - 2.35 n = 19
14	12.27 1.63 9.0 - 15.1 n = 19	40.41 14.46 15.5 - 69.7 n = 19	0.316 0.131 0.08 - 0.48 n = 18	0.794 0.153 0.54 - 1.08 n = 18	0.016 0.136 0.007 - 0.021 n = 18	0.041 0.032 0.02 - 0.16 n = 18	2.09 0.17 1.66 - 2.39 n = 19
17	11.32 1.59 9.7 - 15.5 n = 19	30.85 17.13 18.57 - 85.09 n = 19	0.330 0.328 0.08 - 1.48 n = 18	0.948 0.386 0.34 - 1.74 n = 18	0.011 0.006 0.006 - 0.024 n = 18	0.039 0.016 0.021 - 0.09 n = 18	1.98 0.16 1.52 - 2.28 n = 19

hepatosomatic index[1] $= \frac{\text{liver weight X 100}}{\text{fish weight (g)}}$, splenosomatic index[2] $= \frac{\text{spleen weight x 100}}{\text{fish weight (g)}}$, condition factor[3] $= \frac{\text{fish weight (g) x 100}}{\text{standard length}^3}$

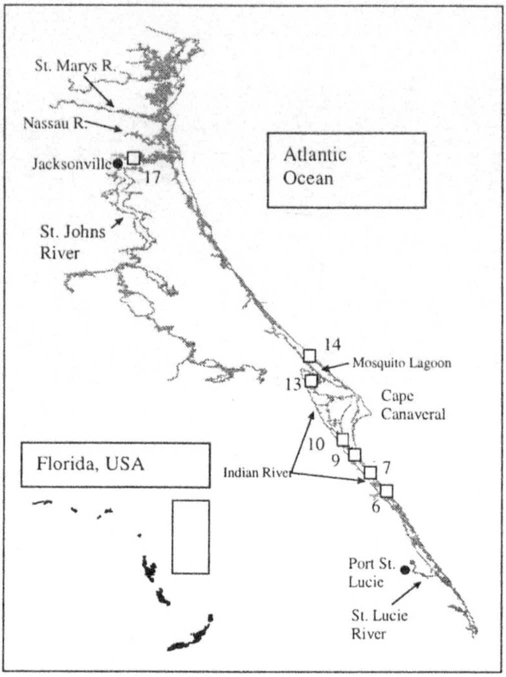

Fig 1. Sampling sites for fish parasites and health characterization.

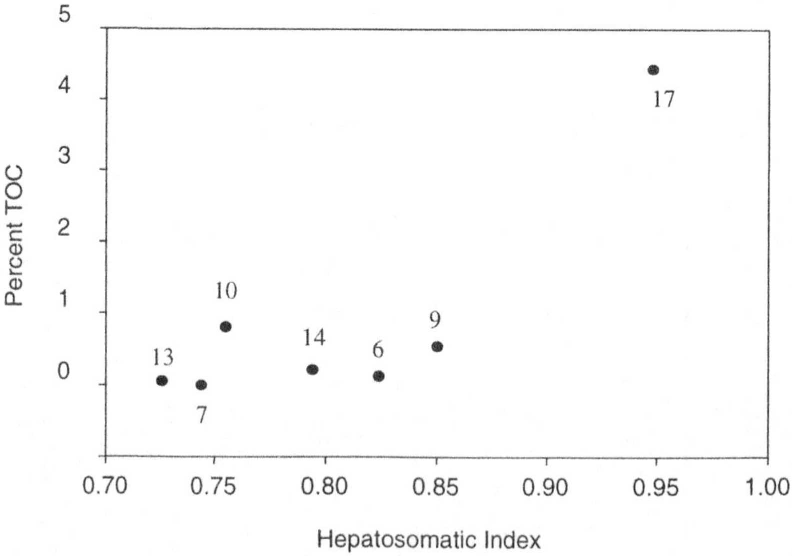

Fig 2. Relationship of hepatosomatic index to percent total organic carbon (TOC). Numbers represent sites in Figure 1.

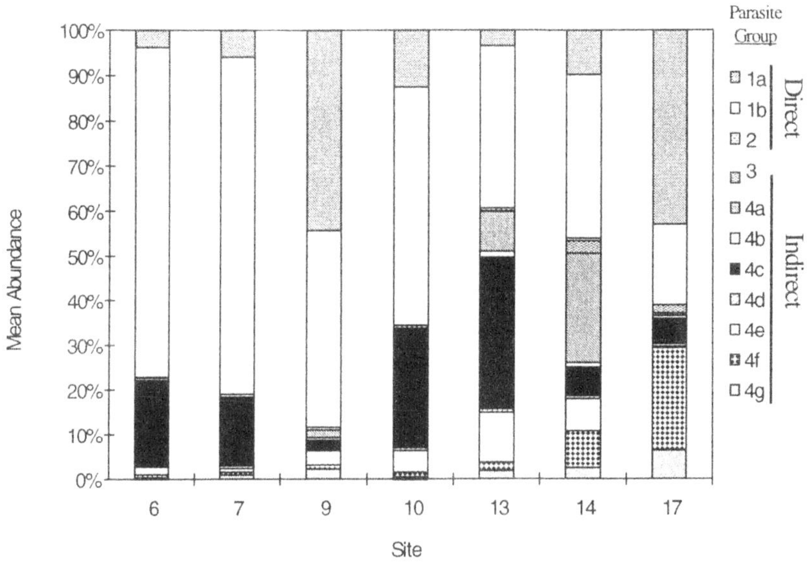

Fig 3. Relative abundance of parasite groups in silver perch by site

Table III
Ecological parasite data

Site	*n*	*S*	*N*	$\bar{N}$	$\bar{S}_1$	$\bar{S}$	H	1/*SI*	*J*
6	20	24	5866	293.30 A	9.90 AC	1.20	1.48	1.78	0.32
7	16	24	4707	294.19 A	9.37 A	1.50	1.49	1.73	0.33
9	14	24	2731	195.07 AB	11.71 C	1.71	2.56	3.21	0.56
10	13	20	1684	129.53 BC	8.54 A	1.54	2.29	3.22	0.53
13	20	23	1415	70.75 C	8.35 AB	1.15	2.82	4.76	0.62
14	19	30	1309	68.95 C	9.10 A	1.58	3.36	6.62	0.68
17	20	23	749	37.45 C	6.30 B	1.15	2.90	5.32	0.64

= $P < 0.05$. Values with the same letter are not significantly different

The species diversity ranged from 1.48 to 3.36 with site 14 having the highest species diversity and sites 6 and 7 having the lowest species diversity (Table III). The predominance of *Rhamnocercus* sp. 1 at these latter two sites strongly influenced the species diversity (*H*) and evenness (*J'*) measures (Table III). Some parasite taxa were found only at one site, with fish from five sites harboring unique taxa. Geographical proximity of sites was reflected in the close similarity of the parasite species assemblages. Some parasite genera were found only at pairs of sites ranked high in Jaccard's similarity (Table V): *Tergestia* (group 4c) at sites 6 and 7, *Tripartiella* (group 1a) at sites 9 and 10, and metacercaria types XI and XII (group 3) at sites 13 and 14.

TABLE IV
Dominance (%) of parasite taxa by sites, listed in descending order

Rank	Taxa	(%)	Sites	Rank	Taxa	(%)	Sites	Rank	Taxa	Dominance (%)	Sites
1	Rhamnocercus sp. 1	58.45	6,7,9,10,13, 14,17	16	blood fluke	0.51	6,7,9,10,13,14,17	29	Macrovalvitrematoides	0.04	17
2	Opecoeloides	10.25	6,7,9,10,13,14,17	17	Tergestia	0.47	6,7	32	metacercaria Type I	0.02	7,13,14
3	Cryptobia	9.53	6,7,9,10,13,14,17	18	microsporea	0.33	6,7,9,10,13,14	32	Amyloodinium	0.02	7,9,14,17
4	Spirocamallanus	3.28	6,7,9,10,13,14,17	19	Trichodina large	0.28	6,7,9,10	32	Lecithochirium	0.02	6,9,13,14
5	Siphodera	2.71	6,9,13,14,17	19	Diplomonorchis	0.28	6,7,9,13,17	32	metacercaria Type IV	0.02	17
6	Pseudopecoelus	2.33	6,7,13,14	21	sessile ciliate - body	0.21	7,10,14,17	32	Acanthocephala sp. 1	0.02	6
7	Hysterothylacium sp.	2.10	6,7,9,10,13,14,17	22	epitheliocystis	0.20	6,7,9,13,14,17	32	metacercaria Type XI	0.02	13,14
8	Eimeria sp. 1	1.70	6,9,13,14,17	23	metacercaria Type II	0.15	7,14,17	38	metacercaria Type V	0.01	7
9	Trichodinella epizootica	1.54	6,9,10,14,17	24	metacercaria Type XII	0.13	13,14	38	metacercaria Type XIII	0.01	14
10	Rhamnocercus sp. 2	1.35	9,10,13,14	24	internal metacercaria	0.13	6,7,9,10,13,14,17	40	Ceratomyxa	0.005	7
11	Scolex polymorphus	0.88	6,7,9,10,13,14,17	26	Capillaria	0.08	10,13,17	40	Myxobolus	0.005	17
12	Eimeria sp. 2	0.80	9,13,14	26	Lernanthropus	0.08	6,7,9,14	40	Argulus sp. 1	0.005	6
13	Tripartiella	0.73	9,10	28	Fabespora	0.07	14	40	Argulus sp.	0.005	14
14	Trichodina small	0.68	6,7,9,10, 14,17	29	sessile ciliate -buccal cavity	0.04	6,7,10,17	40	Acanthocephala sp. 2	0.005	10
15	Scolex large sp.	0.55	6,7,9,10,13,14,17	29	isopoda	0.04	9,10,13,14				

Table V
Jaccard's index of qualitative similarity among pairs of sites sampled for silver perch

	Site 6	Site 7	Site 9	Site 10	Site 13	Site 14
Site 7	60.0	-				
Site 9	60.0	54.8	-			
Site 10	51.7	41.9	57.1	-		
Site 13	51.6	41.9	51.6	48.3	-	
Site 14	50.0	50.0	63.6	42.9	65.6	-
Site 17	51.6	51.6	51.6	48.3	43.7	47.2

Other taxa were absent only at one particular site: all crustacea and microsporea from site 17, small *Trichodina* sp. from site 13 (possibly lost during handling), and the chlamydia-like epitheliocystis from site 10. All metacercarial types were absent from sites 6, 9, and 10. Individual silver perch harbored from 3 to 15 parasite taxa. The mean number of parasite species per fish ($\bar{S}_1$) ranged from 6.3 at site 17 to 11.7 at site 9. Site 17 also had the lowest mean number of individual parasites ($\bar{N}$ = 37.45) (Table III). Parasites may be positively or negatively influenced by different environmental stressors which can lead to dominance of particular species, taxa, or groups at individual sites. Trends in parasite abundance among sites can be studied by comparing the prevalence of the parasite groups at each site (Table VI).The predominance of *Rhamnocercus* sp. 1 (Group 1b) at sites 6 and 7 influences the measures of species diversity and evenness (Table III) but some general differences and trends between sites are still apparent (Figure 3).

3.3. ASSOCIATION OF PARASITE GROUPS WITH EXPOSURE INDICATORS

Two sites exhibited hypoxic conditions based on the instantaneous, surface-to-bottom profiles (Table VII).

Sites 9 and 17 had bottom dissolved oxygen levels of 1.6 and 1.4 ppm respectively. Group 1a parasites (protists) at site 9 (44.53%) and site 17 (43.39%) appeared to be strongly associated with low dissolved oxygen (Figure 4). The SSI appears to decline (Table II) with decreased dissolved oxygen at sites 9 and 17.

TABLE VI

Prevalence of parasite groups by site and apparent response of these groups to environmental stressors

Group	site 6	site 7	site 9	site 10	site 13	site 14	site 17	Stressor	Response
1a	3.8	5.95	44.53	12.71	3.46[2]	9.80	43.39	low dissolved oxygen	increase
1b	73.39	75.06	43.79	52.97	36.18	36.36	17.76	higher temperature	increase
2	0.10	0.04	0.33	0.12	0.14	0.38	0.0	pesticides/ salinity	decrease
3	0.0	0.21	0.0	0.0	0.21	2.98	1.87	contaminants/ salinity	decrease
4a	0.33	0.10	2.01	0.24	9.12	24.45	0.53	contaminants	[1]increase/ decrease
4b	0.32	0.32	0.48	0.77	1.34	0.99	0.40	contaminants	decrease
4c	19.43	15.02	2.56	26.19	33.85	6.42	5.87	contaminants	decrease
4d	0.02	0.06	0.04	0.24	0.71	0.23	0.13	contaminants	decrease
4e	1.64	1.72	3.19	5.41	11.31	7.56	0.93	contaminants	decrease
4f	0.46	0.15	0.88	1.00	1.84	8.25	22.83	contaminants	increase
4g	0.51	1.36	2.20	0.36	1.84	2.52	6.2	contaminants	increase

[1]The potential involvement for intermediate hosts in this parasite group has not been confirmed in this study. The response of intermediate hosts may be different, i.e. tubificids or shrimp may increase or decrease in response to contamination or low DO.

[2]Note that protist counts in Group 1a from site 13 were low because of handling procedures.

TABLE VII
Habitat and exposure indicators

Site	Bottom DO (ppm)*	pH	salinity	TOC (%)	% silt - clay	microtox corrected EC-50**	temp°C
6	7.6	8.3	20.0	0.13	5.4	1.550	33.2
7	7.8	8.0	25.5	BDL	1.6	4.376	29.7
9	1.6	8.1	23.0	0.55	17.3	0.420	30.6
10	6.1	7.9	21.8	0.81	14.5	0.172	28.5
13	6.7	7.8	24.8	0.06	2.5	0.774	29.2
14	5.5	7.9	27.9	0.22	7.2	0.612	28.3
17	1.4	7.5	16.4	4.43	88.1	0.446	28.4

Data means taken from the instantaneous, surface-to-bottom depth profiles.
*Data shown is the minimum range.**Sites with EC-50 below 0.5 are toxic (silt-clay <20%) or 0.2 (silt-clay > 20%) (Hyland *et al.* 1996)

The microtox assay showed that sediments at 2 out of the 7 sites sampled for fish parasites (sites 9 and 10) were toxic (Table VII). The contaminant data (Hyland *et al.*, 1996) can be examined with respect to potential effects on: changes in the abundance and diversity of the benthic infauna, the parasite fauna, and the associated effects on those benthic organisms that are intermediate hosts for silver perch. Comparison of criteria used to assess degraded sites (Long and Morgan, 1990; Long *et al.*, 1995) indicates that only Site 17 exceeded the ER-L (10%) values for arsenic, chromium, copper, nickel, lead, zinc, and mercury. ER-L values are defined as the concentration resulting in ecological effects 10% of the time (Long and Morgan 1990). Levels at 10% suggest that potential degradation may occur. Individual alkanes and isoprenoids ranged from 1 to 2588 ppb (Hyland *et al.* 1996).The highest total alkane concentration, at site 17 (14,362 ppb), was below the ER-L value. Total polynuclear aromatic hydrocarbons (PAHs) at site 17 (9179.2 ppb) exceeded the ER-L level. Twelve individual PAHs also exceeded the ER-L value at site 17 were naphthalene, acenaphthalene, acenaphthene, fluorene, phenanthrene, anthracene, fluoranthene, pyrene, benzo [a] anthracene, chrysene, benzo [a] pyrene and dibenzo [a,h] anthracene. All other sites were well below the currently accepted limits for PAHs. Individual polycyclic chlorinated biphenyl (PCB) congeners ranged from 0.01 to 25.65 ppb at all sites. Only at site 17 (282.83 ppb) were total PCB levels above both the ER-L and ER-M values (Long and Morgan, 1990; Long et al., 1995). Pesticide concentrations that exceeded the ER-L value were dieldrin (1.42 ppb), total chlordane (8.29 ppb), and total DDT (17.64 ppb) at site 17. Site 17 had a correspondingly poor parasite fauna that may have been a reflection of the high contaminant levels at this site. The low diversity in the parasite fauna at site 17 (Table III) is a reflection of a decline in the benthic infauna that act as intermediate hosts for many of the parasite groups. At site 17, only 4 species of the benthic infauna were present, whereas the most diverse infauna was found at site 13 with 75 species including annelids, arthropods, molluscs, and echinoderms. At this latter site, the parasite fauna was relatively diverse (Table III), again reflecting the presence of many intermediate benthic hosts for many of the parasite groups.

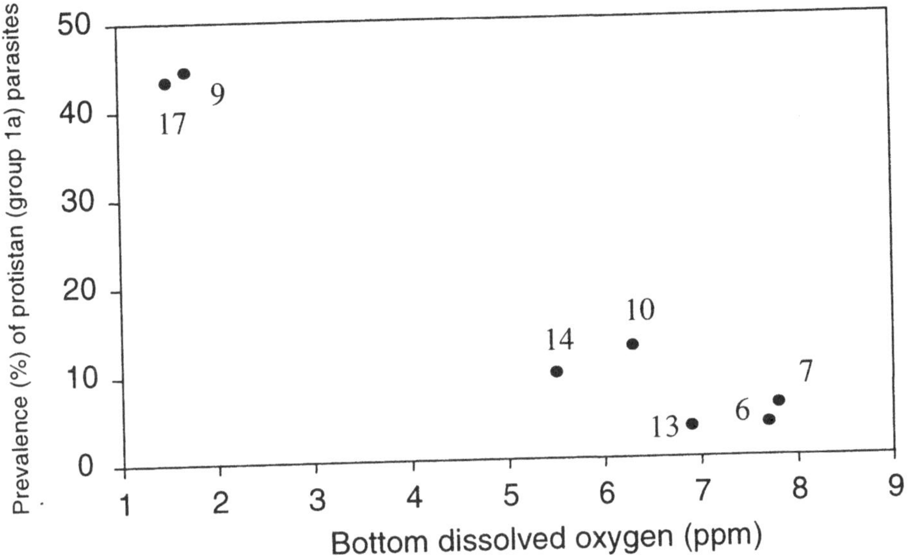

Fig 4. Prevalence (%) of protist (group 1a) parasites in relation to bottom dissolved oxygen at different sites. Numbers represent sites in Figure 1.

3.6. ASSOCIATION OF PARASITES WITH HABITAT INDICATORS

Of the sites selected for this study, site 17 had the lowest salinity (Table VII). The absence of crustacean and microsporean parasites may be associated with the low salinity at this site, but this was not tested. Some parasites are able to tolerate a wide range of salinities even as low as freshwater, whereas others can only survive at high salinities. The mean number of parasites at each site appeared to show a distinct geographical gradient with more parasites being found in the south (Table III). This trend appears to be associated with temperature (Table VII) and reflects the overall dominance of the monogenean parasites (group 1b), specifically *Rhamnocercus* sp. 1 (Figure 3). The high level of total organic carbon (TOC) at site 17 (4.43%) is indicative of a high % silt-clay sediment (Table VII), possible enrichment, and the capacity for this type of sediment to bind with contaminants. A positive association between the HSI and TOC was noted (Figure 2). This relationship would be anticipated because of the detoxifying function of the liver of fish in relation to external contaminant sources. The TOC concentration may best reflect the availability of all contaminants in the sediment and the fish response to such stressors.

3.7. PARASITE INDEX

The point score for the parasite index at each site is determined by adding the ranking values for each category of parasites (by prevalence) and for ecological parasite measures. The higher the score, the more unhealthy the site. Ranking criteria are not always applied from low to high values because the impact on the biota may differ. Groups 1a (protists), 1b (monogenea), 4f (larval nematodes) and 4g (larval cestodes and acanthocephala) are the tolerant indicator "opportunistic" groups that are expected to increase under adverse conditions (Table VI). For this reason, high prevalence of these groups is considered to reflect a relatively negative ecological condition. Ranking from high to low prevalence is then 10 > 1 (Table VIII). The groups 2 (crustacea), 3 (larval digenea), 4a (protists), 4b and 4c (adult digenea), 4d (larval digenea) and 4e (adult nematodes) are ranked (1 > 10) for a high to low prevalence, because these are the sensitive indicator groups that are expected to decline under adverse conditions. Ecological measures *H*, *S*, and $\bar{S}_1$, are also ranked this way (Table VIII). The parasite index rates sites 13 and 14 as being the healthiest and site 17 as being the unhealthiest site. Parasite richness and diversity was the lowest (Table III) and contamination was the highest at this site.

Table VIII
Parasite index

S	1a	1b	2	3	4a	4b	4c	4d	4e	4f	4g	H	S	$\bar{S}_1$	Points
6	1	8	8	10	10	9	6	10	9	1	1	8	6	6	93
7	2	8	10	10	1	9	7	9	9	1	2	8	6	6	97
9	9	5	4	10	10	8	10	10	7	1	3	5	6	4	92
10	10	6	8	10	10	7	4	6	5	1	1	6	6	7	87
13	1	4	8	10	7	4	2	1	1	1	2	5	6	7	59
14	2	4	3	5	1	6	9	6	3	4	3	4	4	6	60
17	9	2	10	7	10	8	9	8	10	9	7	5	6	9	109

Point scores by site for prevalence of parasite groups and ecological parasite measures. Scores are based on observed changes in abundance in response to particular environmental stressors as outlined in Table VII and ranked according to the scheme in Appendix 1. Symbols 1a-4g represent parasite groups, H, S, $\bar{S}_1$ ecological measures (see above). The higher the score, the more unhealthy the site.

4. Discussion

As fish grow and their diet becomes more diverse, the parasite fauna found in the fish also begins to become more diverse. Parasites acquired early are usually those with direct life cycles (protists, monogenea, and crustacea) while parasites with indirect life cycles are acquired later (such as larval tapeworms) through infection from prey such as copepods. The parasite fauna of silver perch essentially comprises two major groups. The first group, consisting primarily of the protists and monogeneans and, to a lesser extent, the crustacea, do not require other hosts to complete their life cycles and most of their life cycle is spent externally on the fish. For this reason, environmental factors such as temperature, dissolved oxygen, salinity and short-term pulses or fluctuations in contaminant loadings, are likely to determine short-term changes in the population dynamics of these parasites. Responses by the protistan and monogenean fauna to dissolved oxygen and temperature were particularly well demonstrated. The monogenean *Rhamnocercus* sp.1 was the most dominant species on silver perch. Changes in the abundance of this species are most likely to relate to increased temperature. The dominance of this species at sites 6 and 7 is reflected in the lower evenness and richness measures of the parasite assemblages at these sites when compared to those at other sites. Metabolism of this and many other parasite species with direct life cycles increases with temperature and hence their generation time decreases. Reproductive rates also increase and since this particular species lays eggs which hatch into larval stages directly on the fish, abundance can probably escalate even in one week. The low dissolved oxygen at sites 17 and 9 is reflected in a parasite response evidenced by the high abundance of protistan species such as *Cryptobia*, *Trichodinella epizootica*, and *Trichodina* spp. These opportunistic protists are commonly observed in warmwater aquaculture conditions where epizootics can erupt within a few days under poor water quality conditions (Landsberg, 1989). The adaptability of these parasite species to low dissolved oxygen and their ability to further compromise physiologically stressed fish will lead to an increase in the abundance of these groups under poor oxygen conditions. The increased respiratory stress of fish under hypoxic conditions will, in addition to predisposing the fish to protistan parasitemia, ultimately cause severe pathological changes to the gills. Fish in this condition may become moribund if they are unable to migrate to better conditions.

The second major component of the parasite fauna of silver perch is represented by those parasites that rely on intermediate hosts for the completion of their life history. Environmental impacts on intermediate hosts is a major factor in determining the presence or absence of these parasites. Habitat degradation or pollution should lead to a decline in intermediate hosts and to a subsequent decline in the diversity of fish endoparasites with indirect life cycles. In this situation, it is feasible that fish will switch to preying on the few resistant species of intermediate hosts. If certain parasite species are equally resistant, then a high abundance of these tolerant species in the fish host would be anticipated at degraded sites. Ultimately, highly degraded sites would be likely to contain few prey species. The parasite fauna of silver perch appear to mirror closely the availability of intermediate hosts in the benthos. Silver perch are predominantly bottom feeders, but their parasite fauna reflect a mixed composition of both benthic and pelagic hosts. Carr and Adams (1973) studied stomach contents of juvenile silver perch in the length range of 0.6 to 16.0 cm. With increasing age, silver perch prey changed from copepods to shrimp and mysids, some gammarid amphipods, ostracods, polychaetes, fish, and crabs. Present, but less dominant, were molluscs such as *Bittiolum* (=*Bittium*) and *Astyris* (=*Mitrella*). Typical fish found in the diet of older silver perch are pinfish, anchovy, goby, and silverside (Carr and Adams, 1973). The diverse number of parasite taxa (44) found in this study is partly a reflection of the varied diet of the silver perch.

Differences in the composition of the infauna in the Carolinian Province were apparent for the northern region in and around Jacksonville. For site 17 in the Trout River, for example, the infauna was very limited. Only 4 species were present in the sediment while mobile epibenthic shrimp and blue crabs were abundant. The predominance of the parasite nematode larvae *Hysterothylacium* sp. is linked to the polychaetes, penaeid shrimp, or amphipods at this site and some life cycle links between the parasite species and the intermediate host may be inferred. Site 17 had the highest TOC, metal contaminants, and silt-clay sediments, as well as the lowest salinity and dissolved oxygen levels. Indicator organisms for more polluted conditions are likely to be those that are able to tolerate these conditions and that can use the limited benthic fauna as intermediate hosts. The increased prevalence of *Hysterothylacium* sp. may be linked with the availability of tolerant benthic species found at site 17. In addition, it is presumed that the nematode is able to physiologically tolerate contaminated conditions. Site 17 was contaminated with metals, PAHs, low level alkanes, PCBs, and a few pesticides, in addition to having low dissolved oxygen. A few fish endoparasitic nematoda, acanthocephala, and cestoda are able to accumulate and tolerate higher levels of metal contaminants than their fish hosts (Sures *et al.*, 1994a, b, c). A positive aspect of the parasite-host relationship may be that certain fish are able to survive under contaminated conditions because their parasite fauna are able to store metals and thus aid in mobilizing contaminants away from the fish tissue. Contaminant analysis of parasites of fish may prove to be a very useful and extremely sensitive biomarker.

Digeneans such as *Diplomonorchis*, *Siphodera*, and *Tergestia* all require molluscs as first intermediate hosts. The proportionately high level of these species at sites 6 and 7 is a reflection of richness of the surrounding benthic fauna and the high diversity of molluscs. Low numbers of individuals of *Pseudopecoelus* and *Opecoeloides* spp. were present at site 17. The low abundance of these species at this site is likely due to the absence of molluscs. The presence of high numbers of *Opecoeloides* spp., species that require an amphipod or mysid shrimp as a second intermediate host, at sites 6, 7, and 13 suggests that such intermediate hosts are present. At these sites, amphipods, tanaids, and mysids were fairly common compared to sites 9, 10, and 17 where arthropod diversity was low and the number of *Opecoeloides* spp. was also low. The nematode *Spirocamallanus* sp. is typically associated with crustaceans as intermediate hosts. Adults of this nematode species were abundant in silver perch at all sites except site 17. The absence of most potential crustacean host species from that site would explain the low prevalence of this parasite. The absence of crustacean parasites at site 17 is associated with the high pesticide levels. Crustacean parasites are particularly sensitive to pesticides, e.g. organophosphates have been used in their control in fish aquaculture (Sarig, 1971; Nair and Nair, 1982). Salinity effects on crustacean parasites may also explain their absence at site 17. Salinity tolerances of each of the species found on silver perch needs clarification. Some crustacean parasites such as *Caligus elongatus* are unable to tolerate freshwater conditions whereas others such as *Lepeophtheirus salmonis* have a wide osmotic tolerance (Landsberg *et al.*, 1991).

A decline in parasite species richness, diversity, and abundance in degraded environments is expected. However, as shown in this study, the decline in diversity cannot be measured using standard ecological measures alone. The decline in species richness may be analogous to that seen in fish culture situations where many intermediate and final hosts are absent. An artificial culture environment results in an absence or lowered presence of parasite species with indirect life cycles, and a reduction in total parasite richness compared to wild fish (Landsberg, 1989). The proportion of parasites with direct life cycles increases significantly due to the inherently lower species diversity, the reduction in interspecific parasite competition, high fish densities, and ease of parasite transmission. At this point, parasite-associated disease and mortality may be observed in cultured fish.

Site 14 in the Mosquito Lagoon near the Merritt Island National Wildlife Refuge had the highest parasite species richness and diversity of all the sites studied. It might be anticipated that this site would have a high parasite diversity, given that the site is relatively undisturbed, is designated as a nature reserve, and has a relatively low-level tidal disturbance. However, the hydrological, geographical, and sediment characteristics of the site also predispose this area to act as a sink for contaminants. There is a lower species diversity of benthic infauna at this site than at site 13, and the contaminant levels are higher. Interpretation of a high diversity of parasite species present together with a lower benthic species diversity suggests that the parasite fauna of fish are only partially dependent upon the presence of a rich benthic fauna providing intermediate hosts. In addition the fish may have also migrated. The parasite assemblage also reflects other aspects of the site concerning water quality, other biota, trophic levels, and terrestrial influences. Site 13 was north of the dredge-spoil islands and, of all the sites studied in the Indian River area, was the furthest away from the Intracoastal Waterway. Seagrass habitat in the area of sites 13 and 14 is abundant whereas sites 6, 7, 9, and 10 have little seagrass coverage (Hart, 1994). This difference in seagrass abundance will affect the distribution of intermediate hosts, parasites with free-living mobile phases, and the fish host. This fact may explain the healthier rating of sites 13 and 14, and the more unhealthy rating of site 7 in the parasite index.

The parasite fauna of fish may also provide information on the trophic structure at the terrestrial/aquatic interface. Larval digenea in fish are likely to require piscivorous birds as final hosts. The infrastructure of the parasite communities may be used to conjecture about other aspects of habitat quality. In sites without surrounding terrestrial habitat for perching or nesting, or in highly urban areas, birds would less likely be present. The representative parasite community in fish at these sites would not be expected to include those that utilize birds as a final host. Generally, sites 13, and 14 had a higher representation of larval digenea. The presence of these digenean parasites may be related to the proximity of sites 13 and 14 to the Merritt Island National Wildlife Refuge with its associated bird fauna. Even though the parasite fauna at site 17 was lower in terms of abundance, the richness, evenness and species diversity measures of the parasite fauna in silver perch were not diminished compared to some other sites (Table III). These measures at site 17 reflect the number of parasite species with direct life cycles, other parasite species such as *Macrovalvitrematoides* and *Myxobolus* which are not found at other sites, and species such as *Scolex* spp. that are acquired from the water column. In this case, site degradation and a decline in parasite richness or species diversity do not necessarily follow for parasites of fish, at least at the level of degradation found in the current study.

Gross morphological or anatomical features, growth indicators, and calculations of condition factors or organosomatic indices of liver, gonad, viscera, and spleen can provide useful fish health indices (Goede and Barton, 1990). These fish health characteristics are relatively easy to determine, they are reliable, and they can provide additional data for the development of the parasite index. In many cases, studies of cellular changes in response to anthropogenic contaminants have led to definition of criteria to diagnose hepatic abnormalities, pathologies, or neoplastic conditions (Harshbarger *et al.*, 1993; Hinton, 1993). Early changes in liver structure are often manifested by an increase in liver weight due to hyperplasia or hypertrophy of the hepatocytes, both of which may be adaptive responses to increase the capacity of the liver to detoxify contaminants. A higher HSI reflects this change in liver capacity. This trend, although not significantly different, showed that liver HSI was increased at site 17, the site with the highest contaminant levels. The relationship between the HSI and the %TOC may be indicative of this effect (Figure 2), since a high silt-clay sediment (Table VII) will retain contaminants and be reflected in the higher TOC. A decrease in spleen weight would be anticipated in response to low dissolved oxygen (Heath, 1987). In fish, the spleen is linked with the storage and production of erythrocytes, as well as being the site of degradation of effete blood cells, accumulation of pigment and formation of melanomacrophage centers. Because low dissolved oxygen will require a metabolic response in the form of an increase in erythrocyte number to maximize oxygen uptake, it is likely that the spleen will compensate for the lack of oxygen by releasing as many erythrocytes as possible into the blood stream of the fish. The spleen weight would be likely to temporarily decrease as a consequence. This condition was noted at sites 9 and 17 where fish had the lowest spleen weights (Table II).

Both the parasite fauna and health of fish measured by selected criteria may be affected by synergistic interactions between different metal contaminants and/or dissolved oxygen levels. These interactions can affect parasite species individually, the intermediate hosts that they inhabit, the overall health of the fish, and the ability of the fish to mount an effective immune response. All these factors may interact to lead either to an increase or decrease in the abundance or diversity of particular parasites in their fish host. Synergistic effects of contaminants and parasites may also compound and confuse the interpretation of their individual effects (Möller, 1986). It is known that the presence of certain parasites may increase the susceptibility of fish to contaminants (Boyce and Yamada, 1977; Pascoe and Cram, 1977; Moles, 1980; Sakanari *et al.*, 1984) which fact, if not taken into account, could also lead to misinterpretations of data. Conversely, the effect of contaminants may weaken fish and hence increase their susceptibility to parasites. Parasites have different tolerance levels to contaminants or environmental stressors. For example, a variety of chemical bath treatments can be used in aquaculture to treat fish having certain ectoparasitic species that are more chemically sensitive than the fish (Landsberg, 1989, Thoney and Hargis, 1991). Conversely, some parasite species may be more tolerant than fish to particular pollutants (Pascoe and Mattey, 1977). This spectrum of tolerance will eventually be reflected in the abundance of the particular parasite species comprising the parasite community, and will vary depending on the type and duration of the contaminant exposure.

Silver perch appears to be an appropriate target fish species for the study of changes in the parasite fauna in relation to estuarine health. Within the geographic framework of the

Florida portion of the Carolinian Province, a small sample of silver perch provided data to test the concept of the use of parasites of fish as indicators of estuarine health. One of the major constraints of using nonbenthic biota such as fish within the framework of EMAP-E is the inherent problem associated with interpretation of data from mobile organisms. For this reason, the reliability of using fish as response indicators has often been questioned in environmental monitoring programs. Although silver perch were not obtained from all sites, apparent trends in the data obtained may show the adaptability and sensitivity of the fish parasite fauna to environmental stressors. Unless particular sites are severely degraded, gross pathology in fish traditionally sampled for such studies is usually present at such a low prevalence that it is often difficult to draw any meaningful conclusions regarding environmental effects. Large numbers of fish samples are often required to detect pathologies at a minimal level (Ziskowski *et al.* 1987), resulting in increased labor and time costs. Additionally, the detection of many pathological responses requires that fish must be exposed to contaminants for a long period. Parasites are built-in monitors of both fish and environmental stress and their responses in terms of population shifts can be quantified. Every individual fish has a parasite infracommunity that is sensitive to environmental stress at many trophic levels. The interaction of these parasite infracommunities and the health of the fish is closely related and can be correlated with abiotic and biotic components of the ecosystem.

Acknowledgments

This study was supported by the Environmental Monitoring and Assessment Program under Cooperative Agreement No. NA417OAO178 with the National Oceanic and Atmospheric Administration (NOAA) through the NOAA Carolinian Province Office, 217 Fort Johnson Road, P.O. Box 12559, Charleston, SC 29422-2559. Thanks to Jeff Hyland, Carolinian Province Officer, NOAA, Charleston, SC for advice during the field monitoring and comments on the report summaries. Many thanks to staff at the Florida Marine Research Institute, Florida Department of Environmental Protection, for field support during the collection of fish, especially Jim McKenna, Bob Heagey, Kim Amendola, Mike Wessel, Craig Norwicki (St. Petersburg), Rich Paperno, Derek Tremain, and Doug Adams (Melbourne). Lastly, a special acknowledgment to staff of the aquatic health group (St. Petersburg) for all their dedicated hours of fish and parasite collecting, preserving, and enumeration, especially Noretta Perry, James Pallias, Ann Forstchen, and Greg Vermeer; and for continued histological support, Pam Nagle and Iliana Quintero-Hunter. Special thanks to Bill Lyons for reviewing and editing the manuscript.

Appendix 1

Criteria and scoring method used to develop the parasite index

Parasite index	Rank 1	Rank 2	Rank 3	Rank 4	Rank 5	Rank 6	Rank 7	Rank 8	Rank 9	Rank 10
Group 1a (%)	0-4.99	5-9.99	10-14.99	15-19.99	20-24.99	25-29.99	30-34.99	35-39.99	40-44.99	45.0+
Group 1b (%)	0-9.99	10-19.99	20-29.99	30-39.99	40-49.99	50-59.99	60-69.99	70-79.99	80-89.99	90.0+
Group 2 (%)	0.45+	0.40-0.449	0.35-0.399	0.30-0.349	0.25-0.299	0.20-0.249	0.15-0.199	0.1-0.149	0.05-0.099	0.0-0.049
Group 3 (%)	4.5+	4.0-4.49	3.5-3.99	3.0-3.49	2.5-2.99	2.0-2.49	1.5-1.99	1.0-1.49	0.50-0.99	0.0-0.49
Group 4a (%)	22.5+	20.0-22.49	17.5-19.99	15.0-17.49	12.5-14.99	10.0-12.49	7.5-9.99	5.0-7.49	2.5-4.99	0.0-2.49
Group 4b (%)	1.8+	1.6-1.79	1.4-1.59	1.2-1.39	1.0-1.19	0.8-0.99	0.6-0.79	0.4-0.59	0.2-0.39	0.0-0.19
Group 4c (%)	36+	32-35.99	28-31.99	24-27.99	20-23.99	16-19.99	12-15.99	8-11.99	4-7.9	0.0-3.99
Group 4d (%)	0.45+	0.40-0.449	0.35-0.399	0.30-0.349	0.25-0.299	0.2-0.249	0.15-0.199	0.1-0.149	0.05-0.099	0.0-0.049
Group 4e (%)	9+	8-8.99	7-7.99	6-6.99	5-5.99	4-4.99	3-3.99	2-2.99	1-1.99	0-0.99
Group 4f (%)	0-2.49	2.5-4.99	5.0-7.49	7.5-9.9	10.0-12.49	12.5-14.99	15.0-17.49	17.5-19.99	20.0-24.99	25.0+
Group 4g (%)	0-0.99	1-1.99	2-2.99	3-3.99	4-4.99	5-5.99	6-6.99	7-7.99	8-8.99	9+
S	40+	36-39	35-39	30-34	25-29	20-24	15-19	10-1	5-9	0-4
H	4.50+	4.0-4.49	3.5-3.99	3.0-3.49	2.50-2.99	2.0-2.49	1.50-1.99	1.0-1.49	0.5-0.99	0-0.49
$\bar{S}_I$	14.0+	13.0-13.99	12.0-12.99	11.0-11.99	10.0-10.99	9.0-9.99	8.0-8.99	7.0-7.9	6.0-6.99	<5.0-5.99

Range values are based on a subjective ranking procedure for expected values appropriate to each response. Ranks are from 1 (excellent/healthy) to 10 (degraded/unhealthy).

References

Adams, S. M., Brown, A. M., Goede, R. W. :1993. *Trans. Am. Fish. Soc.* **122**, 63-73.

Boxrucker, J. C.:1979. *Parasitol.* 78:195-206.

Boyce, N. P., Yamada, S. B.: 1977. *J. Fish. Res. Bd. Can.* **34,** 706-709.

Carr, W.E.S. , Adams, C.A.: 1973. *Trans. Am. Fish. Soc.* **102**, 511-540.

Esch, G. W., Hazen, T. C., Dimock, R. V. Jr., Gibbons, J. W.:1976. *Trans. Am. Microsc. Soc.* 95:687-693.

FDEP 1994. Development of an environmental stress index using parasites of fish. Submitted research proposal.

Goede, R. W. , Barton, B. A.: 1990. In: *Biological indicators of stress in fish*, (ed. S. M. Adams), American Fisheries Society Symposium. **8**, 93-108.

Hart, A.W. 1994. Indian River Lagoon National Estuary Program. 85pp.

Harshbarger, J.C., Spero, P.M., Wolcott, N.M.: 1993. In: *Pathobiology of marine and estuarine organisms*, (eds. Couch, J.A. and Fournie, J.W.), pp. 157-176, CRC Press, Boca Raton, Florida.

Heath, A.G.: 1987. *Water pollution and fish physiology*. CRC Press, Boca Raton, Florida, 245pp.

Hinton, D.E.: 1993. In: *Pathobiology of marine and estuarine organisms*, (eds. Couch, J.A. and Fournie, J.W.), pp. 177-215, CRC Press, Boca Raton, Florida.

Hyland, J.L., Herringer, T.J., Snoots, T.R., Ringwood, A.H., Van Dolah, R.F., Hackney, C.T., Nelson, G.A., Rosen, J.S., Kokkanikis, S.A. 1996. *Annual statistical summary:EMAP-Estuaries Demonstration Project in the Carolinian Province - 1994.* NOAA Tech.Mem.NOS ORCA 97, pp. 102

Khan, R. A.: 1987. *Parasitol. Today* 3:99-100.

Khan, R. A.: 1990. *Bull. Environ. Contam. Toxicol.* 44:759-763

Khan, R. A., Thulin, J.:1991. *Adv. Parasitol.* 30:201-238.

Landsberg, J. H.: 1989. In: *Fish Culture in Warm Water Systems:Problems and Trends*, (eds. M. Shilo and S. Sarig), p. 195-252, CRC Press, Boca Raton, Florida.

Landsberg, J.H., Vermeer, G.K., Richards, S.A., Perry, N.: 1991. *J. Aquat. Anim. Health.* **3**, 206-209.

Leong, T.S., Holmes, J.C.: 1981. *J. Fish. Biol.* **18**, 693-713.

Long, E.R., MacDonald, D.D., Smith, S.L., Calder, F: 1995. *Environ. Management* **19**, 81-97.

Long, E. R., Morgan, L. G.: 1990. *NOAA Tech. Mem.* NOS OMA 52. 175pp.

Margolis, L., Esch, G.W., Holmes, J.C., Kuris, A.M., Schad, G.A.: 1982. *J. Parasitol.* **68**, 131-133.

McVicar, A. H., Bruno, D. W. and Fraser, C. O.:1988. *Mar. Pollution Bull.* 19:169-173.

Moles, A.: 1980. *Trans. Am. Fish. Soc.* **109,** 293-297.

Möller, H.: 1986. *Int. J. Parasitol.* **17**, 353-361.

Möller, H.: 1986. *Int. J. Parasitol.* **17**:353-361

Nair, A.G., Nair, N.B.: 1982. *J. Anim. Morphol. Physiol.* **29**, 265-271.

Novotny, J. F. , Beeman, J. W.: 1990. *Prog. Fish-Cult.* **52**, 162-170.

Overstreet, R. M. 1988. *Aquat. Toxicol.* 11:213-239.

Overstreet, R. M. 1993. In: *Pathobiology of marine and estuarine organisms*, (eds. Couch, J.A., Fournie, J.W.), pp. 111-156, CRC Press, Boca Raton, Florida.

Overstreet, R.M., Hawkins, W.E., Deardorff, T.L. 1996. In:*Environmental Fate and Effects of Pulp and Paper mill effluents.*

Pascoe, D., Cram, D.: 1977. *J. Fish Biol.* **10,** 467-472.

Pascoe, D., Mattey, D.: 1977. *Z. Parasitenkd.* **51**, 179-186.

Sakanari, J. A., Moser, M., Reilly, C. A., Yoshino, T. P.: 1984. *J. Fish Biol.* **24,** 553-563.

Sarig, S.: 1971. The prevention and treatment of diseases of warmwater fishes under subtropical conditions, with special emphasis on intensive fish farming. *Diseases of fishes*, Book 3, Snieszko, S.F. and Axelrod, H.R. (eds.), TFH Publications, Neptune City, New Jersey.

Summers, J.K., Macauley, J.M., Heitmuller, P.T., Engle, V.D., Adams, A.M., Brooks, G.T.: 1992. *Annual statistical summary:EMAP-Estuaries Louisianian Province - 1991*. U.S. EPA, ORD, Gulf Breeze FL, EPA/600/R-93-001.

Sures, B., Taraschewski, H., Jackwerth, E.: 1994a. *Bull. Environ. Contam. Toxicol.* **52**, 269-273

Sures, B., Taraschewski, H., Jackwerth, E.: 1994b. *J. Parasitol.* **80**, 355-357.

Sures, B., Taraschewski. H., Jackwerth, E.: 1994c. *Dis. aquat. org.* **19**, 105-107.

Thoney, D.A. :1991. *J. Fish Biol.* **39**:515-534.

Thoney, D.A. :1993. *J. Fish Biol.* **43**:781-804.

Thoney, D. A., Hargis, W. J., Jr.: 1991. *Ann. Rev. Fish Diseases.* **1,** 133-153.

Zar, J.H.: 1984. *Biostatistical analysis*. Prentice-Hall, New Jersey. 718pp.

Ziskowski, J.J., Despres-Patanjo, L., Murchelano, R.A., Howe, A.B., Ralph, D., Atran, S. 1987. *Mar. Poll. Bull.* 18, 496-504.

RELATING BENTHIC INFAUNAL COMMUNITY STRUCTURE TO ENVIRONMENTAL VARIABLES IN ESTUARIES USING NONMETRIC MULTIDIMENSIONAL SCALING AND SIMILARITY ANALYSIS

G. McRAE, D.K. CAMP, W.G. LYONS and T.L. DIX

Florida Department of Environmental Protection, Florida Marine Research Institute, 100 8th Ave. SE, St. Petersburg, FL 33701, USA

Abstract. In 1994, 19 stations were sampled (2 replicates/station) with Young grabs in association with the EMAP-Estuaries Carolinian Province Base Monitoring in Florida. A total of 295 unique benthic infauna taxa and 9647 individuals were identified and enumerated. Environmental data (bottom-water quality, sediment grain size, sediment metals, and organics) and benthic community data were analyzed using hierarchical agglomerative cluster analysis and ordination via nonmetric multidimensional scaling. Bray-Curtis similarities and Euclidean distance were used as the distance measures for biotic and abiotic data, respectively. Multivariate analyses were complemented by examining incremental contributions of benthic taxa to similarity values using a percentage similarity technique. A low-salinity site in a tributary to the St. Johns River had benthic communities uniquely different from those of moderate- to high-salinity sites. A diverse assemblage of polychaetes, gastropods, bivalves, amphipods, sipunculans, and phoronids was consistently associated with relatively unimpacted sites in the Indian River Lagoon. Infaunal community structure in the northern portion of the study area was influenced by the nearby Atlantic Ocean. Community shifts in association with latitudinal gradients and concentrations of sediment metals and organics were apparent. The nonparametric multivariate techniques used in this study were particularly effective at delineating and defining fine-scale community differences.

1. Introduction

Benthic infauna have long been used to assess environmental quality in estuaries (Rosenberg 1977; Harper *et al.* 1981; Engle *et al.* 1994). Researchers making these assessments often take a community-based approach and examine the relative numbers of taxa and their relationship to measures of water and sediment quality. These studies typically involve hundreds of taxa in each sample and many concurrent environmental measurements. Due to the multidimensional nature of the data, multivariate statistical methods such as clustering and ordination are preferred. Ordination techniques attempt to arrange sites along axes based on species composition, providing a summary of a multidimensional species array in two or three dimensions.

The ordination techniques that are most popular among ecologists are principal components analysis (PCA), correspondence analysis (CA), and techniques related to CA, such as weighted averaging and detrended correspondence analysis (Jongman *et al.* 1995). Canonical ordination techniques (e.g., canonical correspondence analysis; ter Braak 1986) combine ordination with regression analysis to detect the patterns of variation in the species data that can be explained "best" by the observed environmental variables. There are some practical concerns regarding the application of these techniques to biological data. PCA

Environmental Monitoring and Assessment **51**: 233–246, 1998.

uses Euclidean distance as the measure of similarity between samples and assumes that abundances are linearly related to environmental gradients. Some investigators have noted that the choice of similarity criterion should depend on the nature of the data rather than the method used, and that a linear response model is not always appropriate (e.g., Clarke 1993). Canonical correspondence analysis assumes a unimodal, rather than linear, response model, but this technique is mathematically quite complex, and it can be difficult to communicate the results to nontechnical audiences.

This study was conducted as part of the estuaries component of the Environmental Monitoring and Assessment Program (EMAP-E) initiated by the United States Environmental Protection Agency (EPA) in cooperation with the National Oceanic and Atmospheric Administration (NOAA). This program is designed to provide a quantitative assessment of the regional extent of potential environmental problems in estuaries by measuring status and change in selected ecological indicators. A summary of the EMAP-E program is presented in Hyland *et al.* 1996.

In the present work we use some alternative multivariate techniques based on nonparametric statistics to describe and quantify benthic community patterns and link these patterns to environmental factors. These techniques utilize the ranked biotic and abiotic similarity matrices as input into classification (hierarchical agglomerative or UPGMA clustering) and ordination (nonmetric multidimensional scaling) algorithms. Complementary analyses, which partition the factors most influential in site similarities and link biotic and abiotic data, are used to extend the inferential capabilities of the multivariate analyses. These techniques are advantageous in that minimal assumptions regarding the nature of the data are required, and they are comparatively easy to understand.

2. Material and Methods

The study was conducted in the Florida portion of the EMAP-E Carolinian Province in 1994. This area extends from the Florida-Georgia state boundary to St. Lucie inlet on the Atlantic coast (Figure 1). Sample sites were selected using a probability-based design, with spatial strata tailored to estuarine physiography (Hyland *et al.* 1996). The majority of sites were located in or near the Indian River Lagoon, although five sites were sampled in the Jacksonville area. A total of 17 sites were sampled in August 1994 with a 0.04-m^2 Young grab sampler. Two replicate samples were taken at each site, and 2 sites were sampled in replicate for a total of 19 sets of two replicates each. Abundance estimates used in this paper were means of the two replicates. Benthic samples were washed on a 0.5-mm-mesh screen, preserved in 10% formalin-rose bengal solution, and stored prior to processing. In the laboratory, macrobenthic samples were transferred from formalin to a 70% ethanol solution and sorted, identified to lowest practical taxonomic level, and counted. As part of the EMAP sampling protocol a large suite of water- and sediment-quality measurements were taken concurrent with the benthic samples (Table I). Methods and materials associated with these parameters are discussed in Hyland *et al.* (1996) and USEPA (1994).

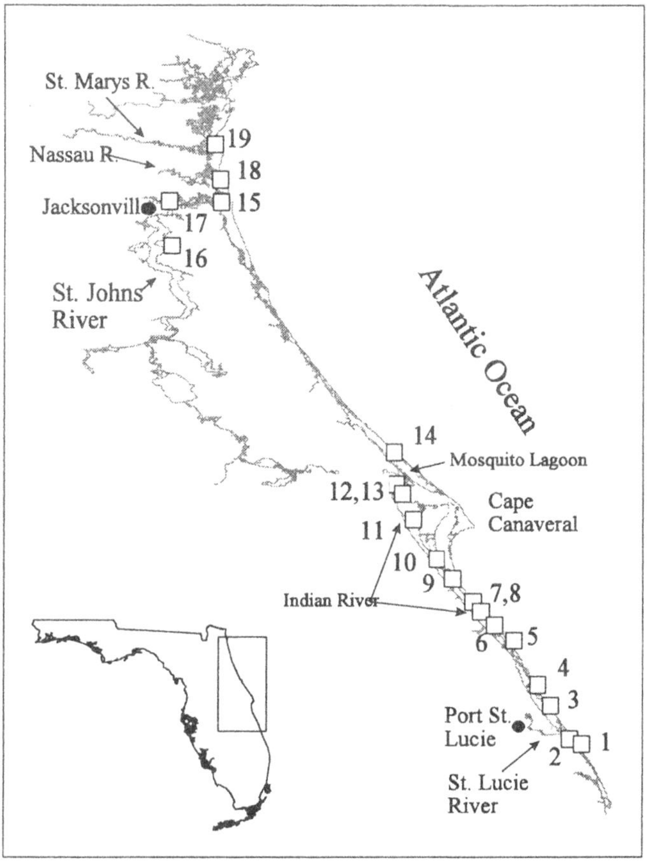

Fig. 1. Study site locations on the east coast of Florida.

Multivariate analyses of benthic community structure and its relationship to environmental parameters were undertaken using two complementary techniques. Hierarchical agglomerative cluster analysis using a group-average linkage method (UPGMA; Kaufman and Rousseeuw, 1990) based on Bray-Curtis similarities (Bray and Curtis, 1957) was used to search for groups among sample sites. Abundance data were double-square-root transformed prior to calculation of similarities in order to downweight the importance of the very abundant taxa and to allow taxa of intermediate and rarer abundance to contribute to the site similarities. Note that Bray-Curtis dissimilarity is simply the complement of Bray-Curtis similarity.

TABLE I
Environmental variables used in benthic infaunal community analyses

Water Quality/	Sediment Physical/Chemical	Sediment Contaminants	
Geographic		Metals (µg/g)	Organics
Depth (m)	Silt-Clay Content (%)	Silver (Ag)	Total PCBs
Bottom Dissolved Oxygen (mg/L)	Total Organic Carbon (%)	Aluminum (Al)	BHC
Salinity (ppt)		Arsenic (As)	Aldrin
Temperature (°C)		Cadmium (Cd)	Dieldrin
Latitude (decimal degrees)		Chromium (Cr)	Chlordane
		Copper (Cu)	Total Alkanes
		Iron (Fe)	Total PAHs
		Manganese (Mn)	Total DDT
		Nickel (Ni)	
		Lead (Pb)	
		Strontium (Sb)	
		Selenium (Se)	
		Tin (Sn)	
		Zinc (Zn)	
		Mercury (Hg)	

* PCBs = Polychlorinated biphenyls; BHC = Gamma Hexachlorocyclohexanes; PAHs = Polynuclear aromatic hydrocarbons; Total DDT = all six DDD, DDE, and DDT congeners

Analyses of each taxon's contribution to within- and among-group similarities was performed using the SIMPER procedure (Clarke and Warwick, 1994). For Bray-Curtis dissimilarity between two samples j and k, the contribution from the ith species, $d_{jk}(i)$, is defined as

$$d_{jk}(i) = \frac{100 * | y_{ij} - y_{ik} |}{\sum_{i=1}^{p} (y_{ij} + y_{ik})}$$

where p = the number of species, and y_{ij} and y_{ik} are the transformed abundances of taxon y at sites j and k, respectively. The values of $d_{jk}(i)$ are then averaged over all pairs (j,k) to give the average contribution $_i$ from the ith species to the overall dissimiliarity (d) between groups 1 and 2. If $_i$ is large and the standard deviation of the d_i values, $sd(d_{jk}(i))$, is small (and thus the ratio $_i / SD(d_{jk}(i))$ is large), then the ith species not only contributes much to the dissimilarity between groups 1 and 2, but it also does so consistently in inter-comparisons of all samples in the two groups; i.e., it is a good discriminating species.

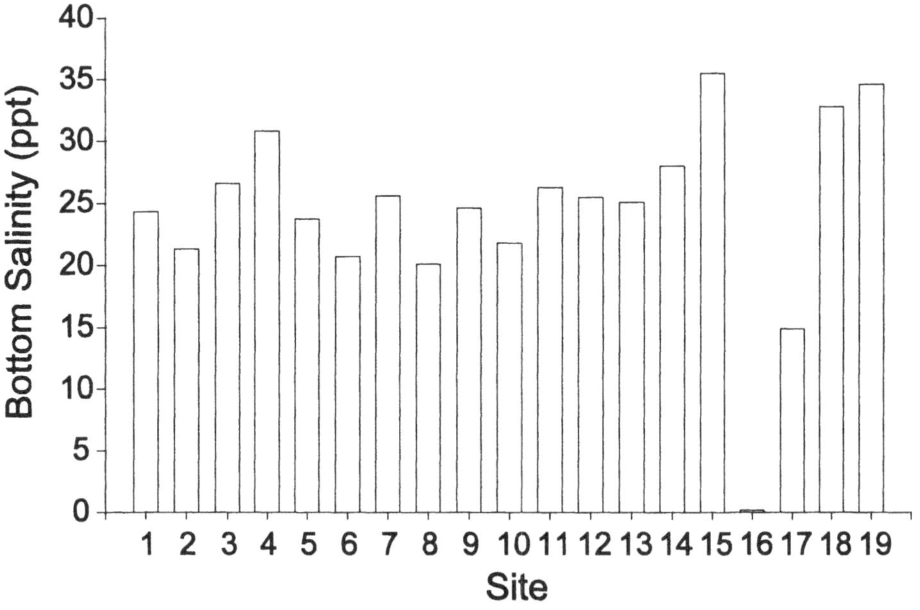

Fig. 2. Near-bottom salinities at the study sites

Likewise, similarity analyses can also identify which taxa are most influential in determining similarity within a group of sites. A nonparametric ordination technique, nonmetric multidimensional scaling (MDS; Kruskal 1964), was used to ordinate sites based on the Bray-Curtis biotic similarity matrices.

Patterns in benthic community structure were linked to environmental variables using the BIO-ENV procedure (Clarke and Ainsworth, 1993). This procedure is based on the weighted Spearman rank correlation coefficient (ρ_w, Conover 1980) between the ranked biotic and environmental similarity matrices. A suite of 30 environmental variables associated with sediment contaminants, sediment quality, and water quality was considered (Table I). For abiotic environmental data, the similarity matrices were constructed using Euclidean distance. Correlations were examined in a step-wise fashion; all combinations of 1 to 6 (drawn from the full set of 30) environmental variables were examined. The sets of environmental variables with the largest ρ_w were considered to provide the best match with the infaunal community data. Statistical analyses were performed using the PRIMER (Plymouth Routines in Multivariate Ecological Research) and the Statistical Analysis System software packages (SAS Institute 1990; Clarke and Warwick, 1994).

3. Results and Discussion

3.1 Site Characteristics

Bottom-water salinities at the study sites were generally in the range of 20-35 parts-per-thousand (ppt) (Figure 2). Two northern sites in the St. Johns River had bottom salinities that were significantly less than at the other sites (t-test, $p<0.05$); site 16 was essentially in fresh water, having a bottom salinity of 0.2 ppt.

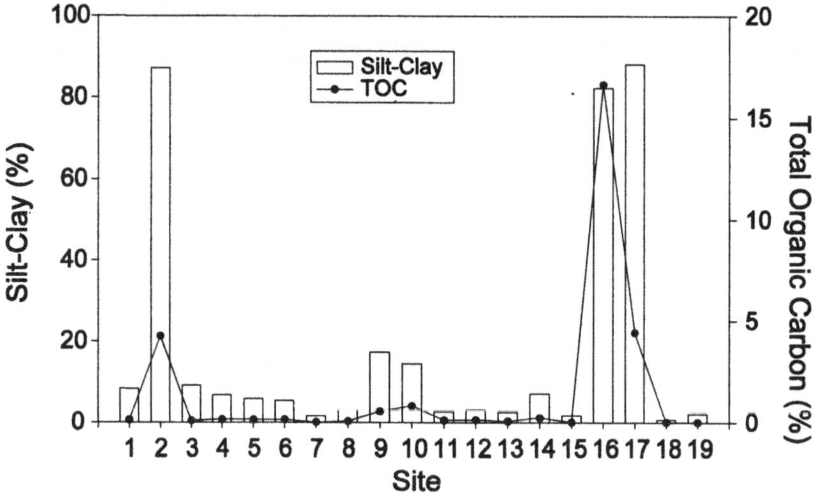

Fig. 3. Percent silt-clay and total organic carbon (TOC) content of sediments.

The sediments at the study sites were composed mostly of sand. The percent silt-clay content of sediments exceeded 20% at only three sites (2, 16, and 17; Figure 3). Site 2 was influenced by inflow from the nearby St. Lucie River, and sites 16 and 17 were in the St. Johns River near downtown Jacksonville (Figure 1). Likewise, total organic carbon (TOC) content of sediments was typically less than 1% at most sites. The three sites with relatively high silt-clay content in sediments also showed elevated levels of TOC.

Levels of sediment metals and organics at most sites were well below those expected to induce adverse effects (Long and Morgan, 1990; Long *et al.* 1995). Elevated metal and organic contaminant concentrations were present at sites 2, 16, and 17. Average sediment metal concentrations were seven times higher in this group of three sites than at the remaining sites. Concentrations of arsenic, chromium, copper, nickel, lead, zinc, and mercury exceeded Effects-Range Low values (Long and Morgan, 1990; Long *et al.* 1995) at site 17. Organic contaminants in sediment showed a similar pattern. Average levels of total PCBs, total alkanes, total PAHs, and total DDT were 47, 100, 108, and 198 times higher, respectively, at sites 2, 16, and 17 than at the remaining sites.

3.2 Community Structure

A total of 295 taxa and 9647 individuals were identified and enumerated from the infaunal samples. Sites 2, 15, 16, and 17 had substantially lower numbers of individuals than did the remaining sites (Figure 4). Polychaetes were the most abundant group in the samples, although other taxa were locally abundant. In particular, sites 3 and 4 were found to have relatively large numbers of gastropods, one bivalve taxon (*Mulinia lateralis*) dominated the samples from site 10, and barnacles of the order Thoracica were unique to site 19. Species diversity was generally lower in the northern group of sites than it was at those in the south (Figure 5), with site 19 as a notable exception. Only 4 taxa were found at site 2 and in very low numbers (average of 3 individuals per grab). Only six taxa were found at site 16, and all of those taxa were unique to that site, most likely because of very low salinity (see Figure 2). Sites 15 and 17 were also characterized by low numbers of taxa (mean = 5.7, sd = 2.1) and individuals (mean = 8.5, sd = 2.0).

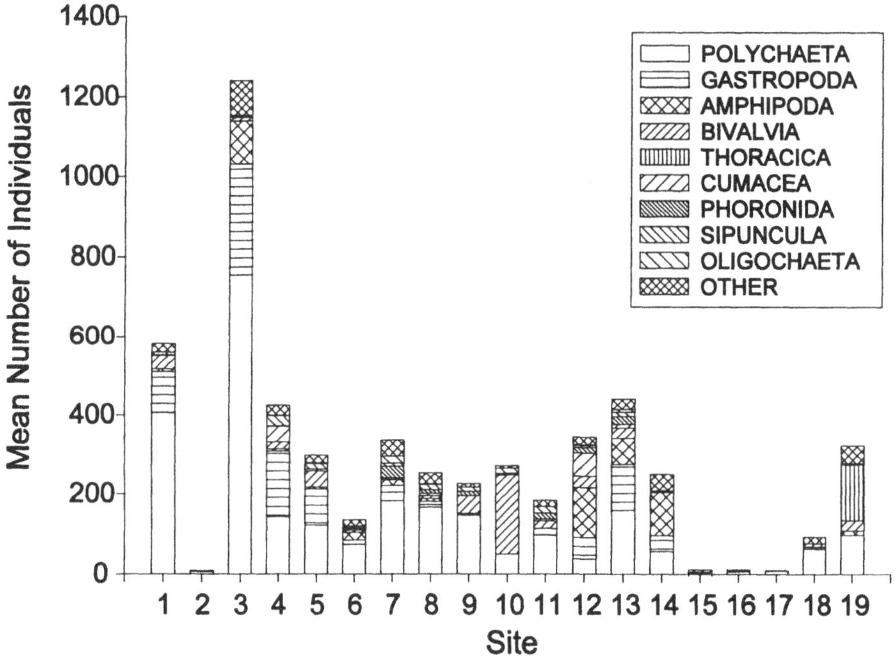

Fig.. 4. Abundances of benthic macroinvertebrate groups at the study sites

Cluster analysis of the infaunal community data identified six groups of sites (Figure 6a). The largest group was made up of sites 1 and 3-14; a second grouping consisted of sites 15 and 18; and sites 2, 16, 17, and 19 formed individual groups. As expected, because of the uniqueness of the fauna, site 16 was the last group to join the others in the clustering.

Biotic ordinations which included data from site 16 provided little useful information because the remaining sites occupied nearly coincident positions in the ordination space. Therefore, it was decided not to include site 16 in the ordination and similarity analyses. The groups identified by the cluster analysis were also apparent in the MDS ordination (Figure 6b). Because the simplicity of the classification in cluster analysis may force the data into artificially distinct classes when continua exist (Field *et al.* 1982), ordination often provides a more accurate representation of community similarities than does cluster

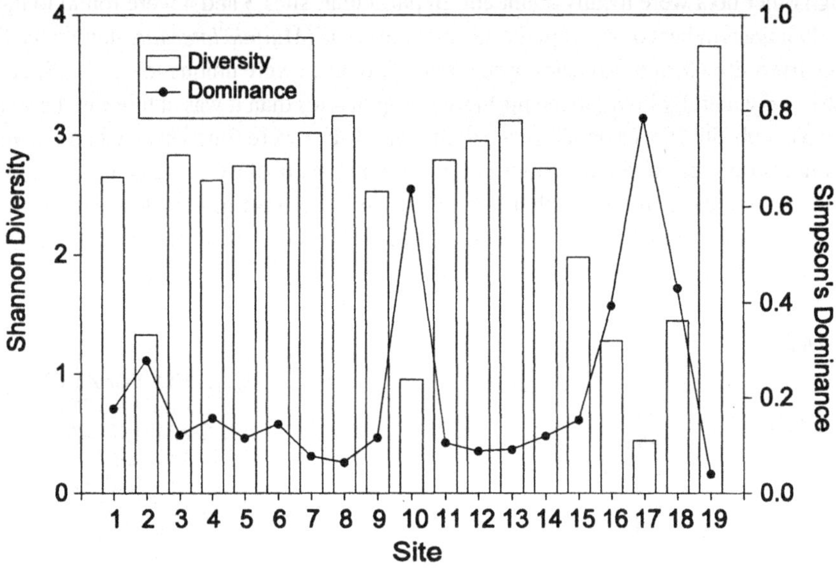

Fig. 5. Species diversity and dominance for benthic macroinvertebrate samples.

analysis. The groups identified by these analyses correspond to latitude-the northern sites 15 and 17-19 are all on the perimeter of the ordination plot, whereas the large group of similar sites in the center of the plot are within or very near the Indian River Lagoon system. The notable exception is the depauperate site 2, which, despite its geographic proximity to sites 1 and 3, had a quite dissimilar infaunal community.

The nature of the community groupings identified in the MDS ordination was explored further using similarity analyses (SIMPER). The assumption was made that the large group of sites in the center of the ordination (group 1, Figure 6b) were representive of a fairly typical, healthy community in the Indian River Lagoon. Therefore, similarity analysis was first used to identify which taxa were influential in determining the similarity among the sites in this group, and secondly, how the peripheral groups in the ordination differed from this central group. The average Bray-Curtis pairwise similarity among the 13 sites in group 1 was 39%. The average number of taxa at sites in this group was 52, with a range of 18 (site 10) to 100 (site 3). Overall, 109 different taxa occurred at these sites, but only 12 taxa accounted for more than 50% of the average similarity within the group (Table II). These

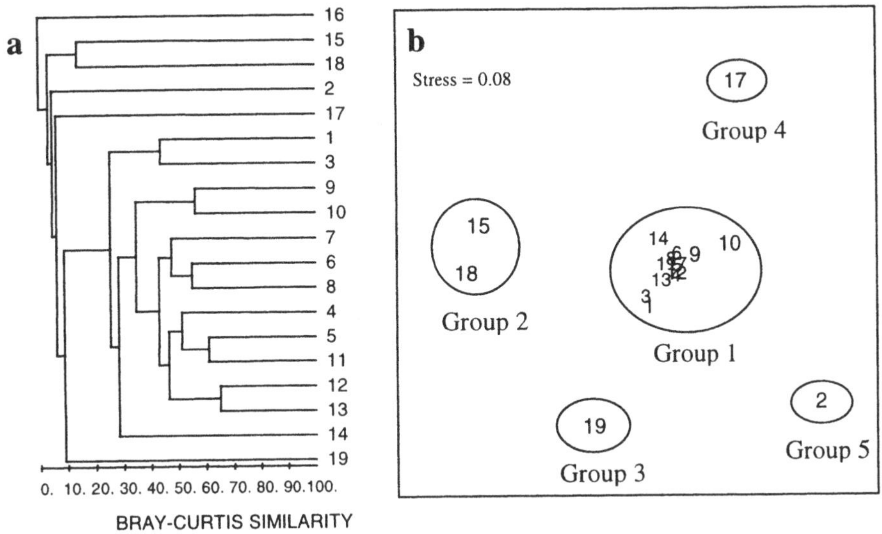

Fig. 6. Site dendrogram (a) and two dimensional nonmetric multidimensional scaling ordination (b) for the study sites based on invertebrate community data.

12 taxa represent a diverse assemblage of taxonomic groups and feeding types. Among the polychaetes, *Mediomastus ambiseta* is a deposit-feeder, *Spiochaetopterus oculatus* is a web-building filter-feeder, and *Diopatra cuprea* is an omnivore. The gastropod *Caecum pulchellum* is a detritivore (Ruppert and Fox, 1988).

The communities of group 1 and group 2 (sites 15 and 18) were very dissimilar, with only two taxa in common (Table II). Both of these taxa occurred at very low numbers (an average of less than 1 individual per station) in each group. Many of the taxa identified as exclusive to group 2 in Table II are characteristic of coastal areas rather than of sheltered estuarine systems. Sites 15, near the mouth of the St. Johns River, and 18, in Nassau Sound, were in areas directly connected to the Atlantic Ocean, unlike the sites of group 1, which are all behind barrier islands.

Community structure at site 19 (the sole site in group 3, Figure 6b) was also substantially different from that at the sites in group 1 (average dissimilarity = 88%). A total of 147 taxa occurred at the sites in groups 1 and 3, but only 22 were common to both groups (a total of 82 taxa were found at site 19). Like sites 15 and 18, site 19, which was located about 5 km upstream from the mouth of the St. Mary`s River (Figure 1), was subject to intrusion of Atlantic Ocean water. The infaunal community at site 19 was quite dissimilar from that at either site 15 ($d_{19,15}$ = 92%) or site 18 ($d_{19,18}$ = 94%), despite similar environmental conditions. This difference was likely due to the large amount of shell hash in the sediments at site 19. *Polydora barbilla*, a species unique to site 19, is known to occur in association with calcareous habitats (Blake 1980). *Spiophanes bombyx*, another species unique to site 19, has been found to be associated with deeper-water, coastal habitats (Blake 1983).

The most obvious community difference among sites in group 1 and those in group 4 (site 17) and group 5 (site 2) is the depauperate nature of the infauna at the latter sites. Only three taxa were found at site 17 (Table II). Two of these taxa, *Streblospio benedicti* and *Grandidierella bonnieroides*, were also found in group 1, but in higher abundances than at site 17. Four taxa, but only a total of six individuals (the fewest from any site), were found at site 2.

3.3 Linking Community Structure to Environmental Factors

Results of the stepwise nonparametric regression on the ranked biotic and abiotic similarity matrices indicated that a combination of natural variability in physical and geographic parameters and xenobiotic contaminants provided the best match between biotic and abiotic data (Table III). The highest correlation coefficient ($\rho_w = 0.79$) was associated with the 4-variable combination selenium, total alkanes, depth, and latitude. No five- or six-variable combination exceeded a ρ_w of 0.79. Selenium was highly correlated with other metals ($R^2 > 0.8$, $p< 0.001$), including silver, cadmium, chromium, copper, iron, and nickel, so its presence in Table III is indicative of general sediment contamination with heavy metals. Likewise, many of the organic sediment contaminants listed in Table I were highly correlated. Correlations were high ($R^2 > 0.8$, $p<0.001$) among the pairwise combinations of total alkanes, total PCBs, chlordane, total PAHs, and total DDT, so the appearance of total alkanes and total DDT in Table III is indicative of broad contamination with many organic compounds.

A useful visualization technique for extracting information about abiotic and biotic relationships is to overlay symbols, whose sizes correspond to the magnitude of environmental variables identified as important correlates with biotic community structure, on the biotic ordination. Figure 7 presents this information for the 1994 EMAP data. As mentioned previously, the stations on the periphery of the ordination are all in the northern portion of the sampling area, with the exception of site 2 (see Figure 6b). Three of the four northern sites, 15, 18, and 19, were also substantially deeper than the remaining sites, reaffirming the coastal nature of these areas. The patterns related to sediment metals and organics contamination in Figure 7 are striking. Community structure at site 17, the most estuarine-like site in the northern group, was very probably affected by high concentrations of both metal and organic contaminants in the sediments. Likewise, site 2, which had a dramatically different infaunal community structure from nearby stations, was apparently affected by elevated levels of metals and organics. Site 2 was near the mouth of the St. Lucie River, which is heavily impacted by nonpoint agricultural pollution. Interestingly, sites 9 and 10, in the Indian River just south of Merritt Island, lie on the edge of group 1 in the direction associated with increased sediment contamination, so the communities at these sites may be showing some initial signs of degradation due to sediment contamination.

TABLE II
Results of similarity analyses for benthic infaunal data

Average similarity (s_{avg}) or dissimilarity (d_{avg})				
$s_{avg} = 39\%$ Within Group 1	$d_{avg} = 96\%$ Group 1 vs. 2	$d_{avg} = 88\%$ Group 1 vs. 3	$d_{avg} = 93\%$ Group 1 vs. 4	$d_{avg} = 94\%$ Group 1 vs. 5
12 taxa account for >50% of S_{avg}	**Common taxa:**	**22 taxa in Common**	**3 taxa in Common***	**4 taxa in Common****
Polychaeta	Polychaeta		Polychaeta	Polychaeta
Mediomastus ambiseta	*Podarke obscura*		*Streblospio benedicti*	*Paraprionospio pinnata*
Glycinde solitaria	Bivalvia		*Eteone heteropoda*	*Scoloplos texana*
Spiochaetopterus oculatus	*Tellina versicolor*		Amphipoda	Nemertea
Diopatra cuprea			*Grandidierella*	Unidentified
Streblospio benedicti	**Taxa exlusive to Group 2:**	**Taxa exclusive to Group 3**	*bonnieroides*	Bivalvia
Gastropoda	Polychaeta	Polychaeta		*Mulinia lateralis*
Acteocina canaliculata	*Podarke* spp.	*Polycirrus* spp.		
Caecum pulchellum	*Hemipodus roseus*	*Neanthes micromma*		
Bivalvia	*Pseudeurythoe*	*Brania wellfleetensis*		
Mulinia lateralis	Bivalvia	*Caulleriella* (3 spp.)		
Amphipoda	*Pleuromeris tridentata*	*Eumida sanguinea*		
Grandidierella bonnieroides	*Crassinella lunata*	*Polydora barbilla*		
Sipuncula	*Caecum strigosum*	*Spiophanes bombyx*		
Phascolion strombus	Isopoda	*Sabellaria vulgaris*		
Phoronida	*Chiridotea stenops*	Amphipoda		
Phoronis architecta		*Unciola* spp.		
Other				
Unidentified Emplectonematidae				

*Only 3 taxa were found at site 17, the sole member of Group 4

** Only 4 taxa were found at site 2, the sole member of Group 5

TABLE III

Combinations of the 30 environmental variables, taken k at a time, that yielded the best matches of biotic and abiotic similarity matrices for each k, as measured by weighted Spearman rank correlation (ρ_w).

k	Best variable combinations (ρ_w)*			
1	Chlordane; (0.55)	Depth; (0.55)	Latitude; (0.55)	Total DDT (0.52)
2	Alkanes, Latitude; (0.74)	Se, Latitude; (0.74)	Total PAHs, Latitude (0.74)	
3	Alkanes, Depth, Latitude; (0.77)	Se, Depth, Latitude; (0.77)	DDT, Depth, Latitude (0.77)	
4	Se, Alkanes, Depth, Latitude; (0.79)	Dieldrin, DDT, Depth, Latitude; (0.77)	Alkanes, Depth, TOC, Latitude (0.77)	
5	Se, DDT, Alkanes, Depth, Latitude; (0.79)	Se, Dieldrin, Alkanes, Depth, Latitude (0.79)		

*Note: Alkanes = Total Alkanes, DDT = Total DDT, Se = Selenium, TOC = Total Organic Carbon

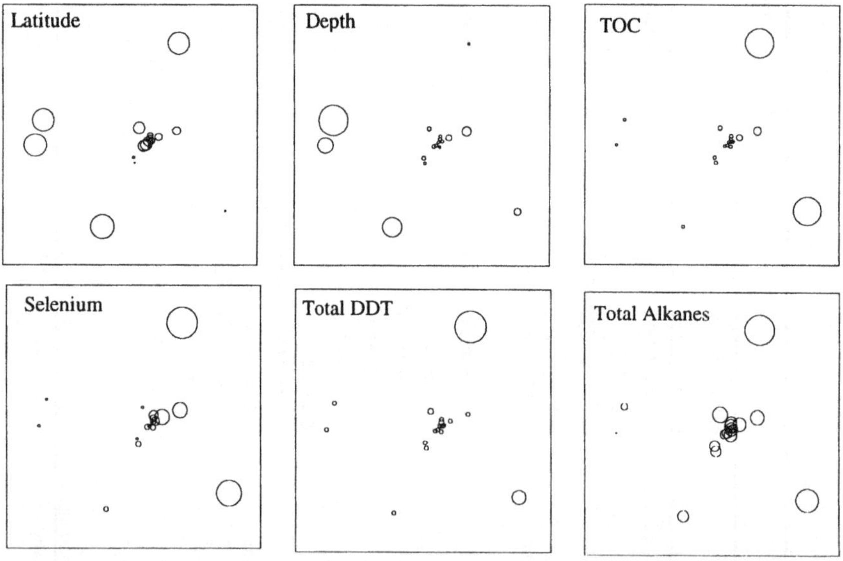

Fig. 7. Environmental data superimposed on the macroinvertebrate biotic ordination; diameter of circles is proportional to the magnitude of each environmental parameter at each site. The biotic ordination appears in Fig. 6b.

4. Conclusions

The analyses presented here indicate that most of the sites in and near the Indian River Lagoon have a relatively consistent, diverse infaunal community. There are predictable differences in benthic infaunal communities associated with estuarine geomorphology and the nature of the Atlantic Ocean connection. A combination of physical factors (salinity, silt-clay content of sediments, and TOC) and sediment contaminants accounted for community differences among sample sites.

Multivariate analyses based on nonparametric statistics and partitioning of pair-wise site similarities are useful tools in distinguishing which taxa are influential in determining community differences among sites. The BIO-ENV procedure, based on the weighted Spearman rank correlation, was useful in determining which environmental variables provided the best match between abiotic and biotic data.

Benthic infaunal community data are inherently complex in that numerous taxa, taxonomic uncertainty, numerous possible environmental correlates, and variability in space and time are all integrated into each analysis. Past attempts at developing environmental indices based on estuarine benthic community data (e.g., Engle *et al.* 1994) have proved applicable within a fairly limited geographic scope. The powerful analytical and data-reduction techniques presented here have the potential to augment such studies and improve the utility of benthic community data in an estuarine assessment context.

Acknowledgements

This study was funded through Cooperative Agreement #NA 47OA0178 from the National Oceanic and Atmospheric Administration to the Florida Department of Environmental Protection. The authors would like to thank Doug Adams, Kim Amendola, David Cook, Bob Heagey, James McKenna, Gary Nelson, Craig Norwicki, Richard Paperno, Jim Quinn, Derek Tremain, and Mike Wessel of the Florida Department of Environmental Protection (FDEP) for help with field sampling and processing. Jeffrey Hyland, Tim Herrlinger, Tim Snoots (NOAA), and Gary Nelson provided logistical and technical support, and Tom Perkins (FDEP) and Jim Quinn provided valuable taxonomic expertise. The authors would also like to thank Kevin Summers, Jeffrey Highland, Judy Leiby, Jim Quinn, and Llyn French for their insightful reviews.

References

Blake, J. A.: 1980, *Proc. Biol. Soc. Wash.* 93(4):947-962.

Blake, N. M.: 1983, "Systematics of Atlantic Spionidae (Annelida: Polychaeta) with special reference to deep-water species", *Ph.D. Dissertation* Boston University.

Bray, J. R. and Curtis, J. T.: 1957, *Ecol. Monogr.* **27**, 325-349.

Clarke, K. R.: 1993, *Aust. J. Ecol.* **18**, 117-143

Clarke, K. R. and Ainsworth, M.: 1993, *Mar. Ecol. Prog. Ser.* **92**, 205-219

Clarke, K. R. and Warwick, R. M.: 1994, "Change in marine communities: an approach to statistical analysis and interpretation", *National Environment Research Council*, UK. 144 pp.

Conover, W. J.: 1980, "Practical nonparametric statistics", Wiley, New York, 493 pp.

Engle, V. D., Summers, J.K., Gaston, G. R.: 1994, *Estuaries* **17**, 372-384.

Field, J. G., Clarke, K.R., Warwick, R. M.: 1982, *Mar. Ecol. Prog. Ser.* **8**, 37-52.

Harper, D. E.,McKinney, L. D., Salzer, R. R. and Case, R. J.: 1981, *Contrib. Mar. Sci.* **24**, 53-79.

Hyland, J. L., Herrlinger, T. L., Snoots, T. R, Ringwood, A. H., Van Dolah, R. F., Hackney, C. T., Nelson, G. A., Rosen, J. S. and Kokkinakis, S. A.: 1996, "Environmental quality of estuaries of the Carolinian Province: 1994. Annual statistical summary for the 1994 EMAP-Estuaries Demonstration Project in the Carolinian Province", *NOAA Technical Memorandum* NOS ORCA 97. NOAA/NOS, Office of Ocean Resources Conservation and Assessment, Silver Spring, MD. 102 pp.

Jongman, R. H. G., ter Braak, C. J. F., and Van Tongeren, O. F. R.: 1995, "Data analysis in community and landscape ecology", Cambridge University Press.

Kaufman, L. and Rousseeuw, P. J.: 1990, "Finding groups in data: an introduction to cluster anlaysis", Wiley, NY.

Kruskal, J. B.: 1964, *Psychometrica* **29**, 1-27.

Long, E. R., McDonald, D. D., Smith, S. L., Calder, F. D.: 1995, *Envionr. Manage.* **19**, 81-97.

Long, E. R. and Morgan, L. G.: 1990, "The potential for biological effects of sediment-sorbed contaminants tested in the National Status and Trends Program", *NOAA Technical Memorandum* NOS OMA 52.

Ruppert, E. E. and Fox, R. S.: 1988, "Seashore animals of the southeast", *Univ. South Carolina Press*, Columbia, SC, 429 pp.

Rosenberg, R.: 1977, *J. Exp. Mar. Biol. Ecol.* **26**, 107-133.

SAS Institute: 1990, *SAS/STAT users guide*. SAS Institute, Cary, NC, 1341 pp.

ter Braak, C. J. F.: 1986, *Ecology* **67**, 1167-1179.

U.S. Environmental Protection Agency (USEPA): 1994, "Environmental Monitoring and Assessment Program (EMAP): Laboratory methods manual - estuaries, Volume I: Biological and physical analyses", *Office of Research and Development, Environmental Monitoring and Systems Laboratory*, Cincinnati, OH.

SEED CLAM GROWTH: AN ALTERNATIVE SEDIMENT BIOASSAY DEVELOPED DURING EMAP IN THE CAROLINIAN PROVINCE

AMY H. RINGWOOD[1], CHARLES J. KEPPLER[2]

[1]*Marine Resources Research Institute, 217 Fort Johnson Rd, Charleston, SC 29412, USA,*
[2]*University of Charleston, Grice Marine Laboratory, 205 Fort Johnson Rd, Charleston, SC 29412, USA*

Abstract. A new sediment bioassay was developed in conjunction with EMAP studies conducted in the Carolinian Province using juvenile seed clams, *Mercenaria mercenaria*. This is a sublethal assay, based on growth (total dry weight) after a 7 day incubation period. Seed clam chronic growth assays were significantly more sensitive than amphipod acute toxicity assays. Optimization components include use of hatchery-reared juvenile clams in a rapid growth phase, and size-sieving to ensure a similar size range. Juvenile clam growth was not affected by sediment type, i.e., clams grew well in muddy and sandy sediments. Clams were slightly more sensitive to ammonia than amphipods (NOEC porewater total ammonia 14 - 16 mg/L for clams). Ammonia concentrations above these levels were more common in reference sites, so most of the false positives could be explained by ammonia toxicity. This assay possesses a number of other positive attributes that are desirable for a bioassay, including the requirement for a relatively small sample size (500 ml of sediments), balanced sensitivity, low incremental costs, and high information gained. The seed clam assay is believed to be a valuable tool for EMAP as well as other monitoring efforts for estimating potential chronic toxicity.

1. Introduction

Sediment contaminant analyses can document the presence of contaminants, but the potential for adverse effects is not readily predictable. The bioavailability of pollutants to organisms is a dynamic component that is the result of complex physical and chemical as well as biological interactions (Hamelink *et al.*, 1994; Ankley *et al.*, 1996). Since sediments are the primary sink for contaminants, sediment bioassays are conducted to identify environmental conditions that could affect biotic integrity. Laboratory toxicity tests are used as indicators of potential impacts on the biota and as indirect indicators of contaminant bioavailability (Luoma, 1995; Long *et al.*, 1996). Although most assessment strategies have used acute toxicity assays, sublethal endpoints represent a greater range of responses and provide continuous data that can be used to establish a variety of protective criteria. Sub-lethal bioassays are more likely to reflect the potential for chronic effects that could affect populations in subtle but serious ways. Criteria based on chronic responses are believed to be more effective for protecting early life history stages and sensitive species. The sustainability of populations and maintenance of ecosystem integrity requires successful growth, reproduction, and recruitment. Furthermore, it is important to recognize that no bioassay is ever likely to be perfect, so multiple assays and ancillary sediment characteristics (% silt-clays, ammonia, pH, salinity) are needed to minimize interpretive errors associated with false positives and false negatives.

Three sediment toxicity assays (amphipod acute toxicity, Microtox, and seed clam growth) were used for the EPA/NOAA-based Environmental Monitoring and Assessment

Environmental Monitoring and Assessment **51**: 247–257, 1998.

Program (EMAP) studies conducted in the Carolinian Province (the estuaries ranging from Cape Henry, VA to the southern end of Indian River Lagoon, FL) during 1995. The amphipod assay, which is a 10-day acute toxicity test with ampeliscid amphipods, has been the mainstay of EMAP estuaries and is widely used for other programs and assessments. However, exceptionally low levels of toxicity have been observed for tests with *Ampelisca abdita* during pilot year studies and 2 years of full-scale monitoring. During pilot year studies (summer, 1993) toxicity was observed at 2 of 24 stations, both of which were pristine reference sites characterized by very sandy sediments (Ringwood *et al.*, 1996). During the first year of full-scale assessment of EMAP (summer, 1994), 94 stations were tested with only 2 stations that exhibited toxicity, both reference stations with no evidence of contaminants (Hyland *et al.*, 1996). During 1995 only 1 station out of 105 exhibited toxicity, which was probably due to very high ammonia concentrations. The lack of toxicity was difficult to explain, because the stations tested included random and non-random sites, and many of the non-random sites were selected specifically because they were very heavily contaminated. Microtox is a sub-lethal assay based on the attenuation of light production by bioluminescent bacteria, *Vibrio fischeri*. It has been more sensitive than the amphipod assay, but the results must be very carefully interpreted because there is a strong sediment bias (Ringwood *et al.*, 1997). The seed clam assay is a chronic sub-lethal assay based on growth of juvenile clams, *Mercenaria mercenaria*, that was developed during the initial phases of EMAP, Carolinian Province, and tested on a full scale with sediments collected during Year 2 of the program, summer 1995. Seed clams possess a number of valuable attributes for bioassay organisms (Dewitt *et al.*, 1989). *Mercenaria mercenaria* is a widely distributed species, and juveniles (approximately 2-3 months from fertilization) are readily cultured and frequently available commercially. They are infaunal, crawling through the sediments and feeding at the surface-water interface. Encased in shells, they can be readily sieved and handled without harm. Newly metamorphosed bivalves exhibit rapid growth, so effects on growth can be detected in a relatively short time frame. The use of juveniles is also desirable to insure that risks to early life history stages and sensitive species are minimized. The methods, results, and guidelines for interpreting the data for this assay are presented.

2. Methods

2.1. Processing of Sediment Samples

Sediments were collected using a 1/25-m^2 stainless steel Young-modified Van Veen grab sampler using standard EMAP methods (Kokkinakis *et al.*, 1994) at 105 sites (85 random core stations and 20 non-random supplemental sites). The NC samples were collected by crews from the University of NC, Wilmington; the FL samples were collected by crews from FL Department of Natural Resources; the SC and GA samples were collected by crews from Marine Resources Research Institute (SC-MRRI), South Carolina Department of Natural Resources. Samples for the seed clam toxicity tests were taken as subsamples from the sediment composite (composed of the top 2 cm of approximately 8 to 10 grabs), and stored in new polypropylene jars at 4°C in the dark until the tests were conducted by SC-MRRI. Other subsamples were used for the amphipod toxicity and Microtox assays, sediment characterization studies, contaminant analyses (conducted by GERG, Texas A&M University), and benthic community studies.

The % silt-clay content of sediment samples was determined using standard EMAP methods (EMAP Laboratory Methods Manual, 1993). For silt-clay analyses, sediment samples were first dispersed with sodium hexametaphosphate and sieved through a 63-μm screen. Sediments retained on the screen were dried and weighed, and 3 replicate 40-ml subsamples of the filtrate were dried and used to estimate % silt-clays. The sediment porewater analyses (ammonia, salinity, pH) were conducted with porewaters extracted by centrifuging a 50 ml sediment subsample. The salicylate-cyanuarate method (Hach, 1994) was used to measure total ammonia-nitrogen (NH_3-N).

2.2. Seed Clam 7-Day Growth Assays

Juvenile clams (*Mercenaria mercenaria*) of approximately 1.0 mm in length (commonly referred to as seed clams) were exposed to sediments for 7 days and the effects on total dry weight were determined. Hatchery-reared seed clams were obtained from Atlantic Littleneck Clam Farms, Folly Beach, SC. On the day before initiation of an experiment, sediments were press sieved through a 500 μm screen and approximately 50 mls were added to 4 replicate 250 ml beakers. Control sediments (collected from Folly River, SC) were prepared in the same manner. Seawater was filtered through a 1 μm filter bag, adjusted to 25 ‰ with deionized water, and added to the replicate beakers for a total volume of 200 ml. The sediment suspension was allowed to settle overnight and clams (30 - 50 per replicate) were added the next day. Clams were size-selected prior to use with 500, 710 and 1000 μm sieves in series. Replicate subsets of clams were dried and weighed for initial weight estimates. All experiments were conducted at room temperature (23 - 25°C), with gentle aeration, and all replicates were fed 3 times during the course of the experiment (a phytoplankton mixture composed of equal volumes of *Isochrysis galbana* and *Chaetocerus gracilis*, cultured at MRRI and dialyzed against filtered seawater to remove excess nutrients and other components of the culture media).

All clam batches were evaluated for suitability and relative sensitivity using cadmium as the reference toxicant. Cadmium exposure experiments (water only, no sediments) were run at the same time as the sediment exposures, but were compared to their own water controls. Four Cd concentrations (25, 50, 100, 200 μg/L added as $CdCl_2$; 3 to 4 replicates of each) were typically used for each reference toxicant test. The effective Cd concentration that reduced growth by 50% (EC_{50}) relative to water controls was derived from regression analyses (exposure concentrations were log transformed; R^2s for the regression lines were generally > 0.9). A running mean and standard deviations over all batches were computed, and a control chart was established.

At the end of the 7 day exposure period, clams were sieved from the sediments (or water, in the case of the reference toxicant tests), placed in clean 25 ‰ seawater and allowed to depurate for approximately one hour. Clams were re-captured on a sieve, and rinsed briefly with distilled water to remove excess salt. Dead clams were removed before being processed for growth, although generally mortalities were less than 10%. The clams were dried overnight (60 - 70°C), counted, weighed on a micro-balance, and growth rates (μg/clam/day) were determined. The effects on growth rates were evaluated using a T-test or Mann-Whitney U test when variances were unequal. Sediments were defined as toxic when the mean growth rate was significantly different statistically from the control sediment growth rate ($p< 0.05$) and <80% of the control sediment growth rate. This dual criteria approach has been used throughout the EMAP program. The requirement for < 80% control rates is regarded as a biologically significant criterion which is used to avoid overestimation of toxicity when variances are small.

3. Results and Discussion

3.1. Data Validation Components

There are both laboratory components as well as field variables that are important to the evaluation of the performance of the assay. The laboratory components, growth rates of seed clams in control sediments and the results of reference toxicant tests with $CdCl_2$, are important QA/QC components that serve to validate the results of an assay series, and enable site-specific comparative statements based on different test series. The on-going laboratory control chart for growth of clams in the sediment controls (Figure 1) demonstrate that repeatable growth rates are obtained with these methods. This provides some assurance that the clams are healthy and growing as expected. Likewise, the control chart for Cd EC_{50}s (Figure 2) demonstrates generally good repeatability, and reference toxicant tests provide some confidence that the relative sensitivities of various clam batches are similar. When these parameters exceed or approach control limits, the results of a test series should be evaluated for suitablility, and sediments associated with a problematic series may require retesting.

3.2. Indicator Efficacy - Field Components

Critical issues regarding indicator performance include responses to sediment characteristics, porewater parameters, and contaminant mixtures (DeWitt *et al.*, 1988; Moore *et al.*, 1997; Ringwood *et al.*, 1997). How do the organisms used for the assay respond to a range of sediment types, from very sandy to muds and clays? Are they sensitive to ammonia or other porewater parameters? Are the organisms sensitive to mixtures of contaminants, characteristic of field settings? Ideally, a bioassay should demonstrate a high degree of balanced sensitivity, i.e. be sensitive enough to reflect toxicity due to anthropogenic perturbations without being hypersensitive to test conditions (sediment type, ammonia concentrations, etc.).

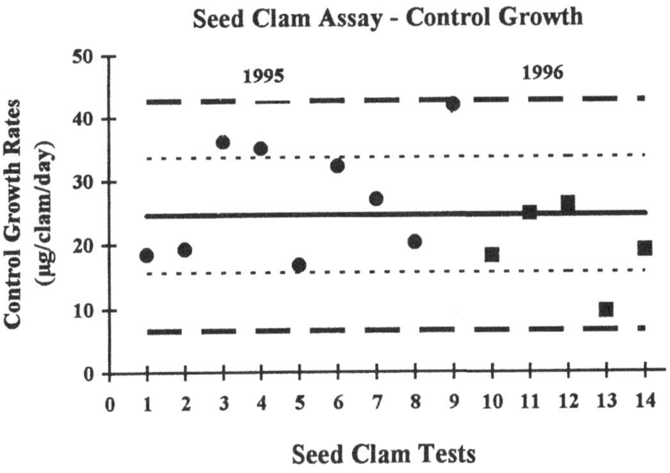

Fig. 1. Control chart for growth of clams in sediment controls (collected from Folly River, SC). The data for studies conducted during 1995 (●) and 1996 (■) tests are shown, although only the 1995 studies are discussed in this publication. Solid line indicates running mean, dotted lines are 1 standard deviation, and dashed lines are 2 standard deviations.

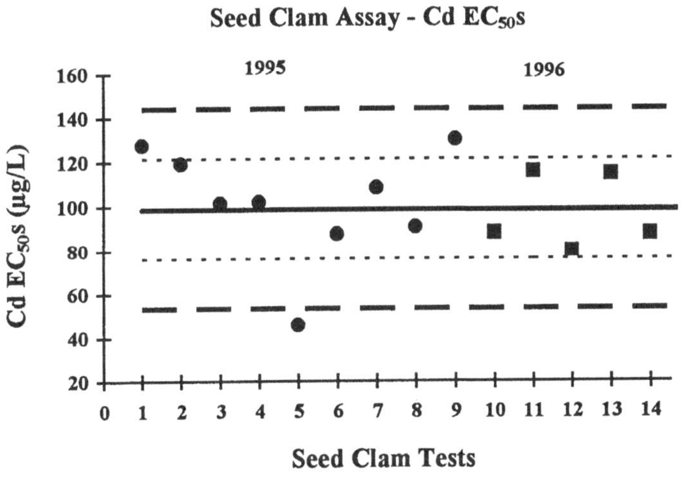

Fig. 2. Control chart for EC_{50}s of clams in reference toxicant studies with aqueous cadmium exposures. The data for studies conducted during 1995 (●) and 1996 (■) tests are shown, although only the 1995 studies are discussed in this publication. Solid line indicates running mean, dotted lines are 1 standard deviation, and dashed lines are 2 standard deviations.

One way of evaluating indicator performance is to determine the rates of false positives (i.e. reference sites that exhibit toxicity although toxicity would not be expected due to the apparent absence of contaminants) and false negatives (i.e. contaminated sites that do not exhibit toxicity although toxicity might be expected due to high levels of contaminants). This type of evaluation is of course affected by how the sites are classified. The site classification scheme used for these studies was based on ER-Ms, sediment contaminant guidelines defined by Long *et al.* (1995). Typically, habitats with one or more contaminants that exceed ER-M values are generally characterized as degraded. Although sediment guidelines have been established for individual contaminants, provisions for classifying stations with lower but enriched concentrations of multiple contaminants are less established. However, most polluted sites characteristically contain a variety of contaminants, and the combined effects of multiple pollutants may be as severe as those caused by high concentrations of a single contaminant, or may cause long-term chronic effects. Therefore, a quantitative approach used during pilot year studies based on the summed proportional contributions of contaminants was applied to the 1995 data (Ringwood *et al.*, 1996). A similar approach was also used by Carr *et al.* (1996). The measured concentrations of each contaminant were divided by the respective ER-M value to generate proportional concentrations (PC) and then the PCs were summed over a contaminant class or over all contaminants to yield the summed proportional concentrations (ΣPC). This exercise was conducted for only those analytes (9 metals, 14 PAHs, 3 pesticides, total PCBs) with ER-M concentrations (Long *et al.*, 1995). The ΣPC approach integrates over all contaminants by defining ER-M equivalents. When the ΣPC over all contaminants was greater than 1, a site was classified as degraded. It should be appreciated that this type of approach will also facilitate more graded classification schemes so that enriched sites (for example those with $\Sigma PC > 0.5$ but less than 1.0) may serve to identify sites in early or variable stages of decline. The general intent of this approach is to provide an estimate of the potential for interactions of multiple contaminants which may be sufficient to adversely affect biological integrity. Elevations above ER-M values for a single contaminant were rare in the Carolinian Province (10 out of 105 stations), so few stations would be classified as degraded based on this criteria. However, many sites had multiple contaminants that approached but did not quite exceed ER-M values, so there were numerous stations with $\Sigma PCs > 1$ ER-M equivalent (44 out of 105 stations). Additive or synergistic interactions can result in toxicity at lower concentrations of multiple contaminants than would be expected based on single contaminant criteria.

Significant effects on growth (< 80% of control growth) were observed at 41.9% of all 105 sites. However, there were false positives. Sediment composition and elevated ammonia levels are common causes of false positives. Generally there was good growth (> 80% of control growth) in sandy as well as muddy sediments (Figure 3), indicating that the assay is applicable over a wide range of sediment types.

Seed clams were found to be somewhat sensitive to ammonia. Generally when porewater total ammonia concentrations exceeded 16 mg/L, toxicity was observed (Figure 4). Notice that the majority of those sites in the lower right quadrant of Figure 4 (i.e., high

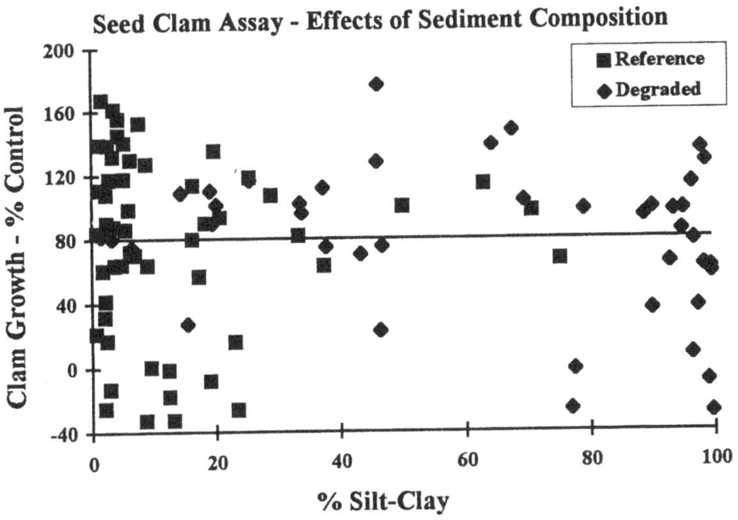

Fig. 3. The effects of sediment silt-clay content on seed clam growth.

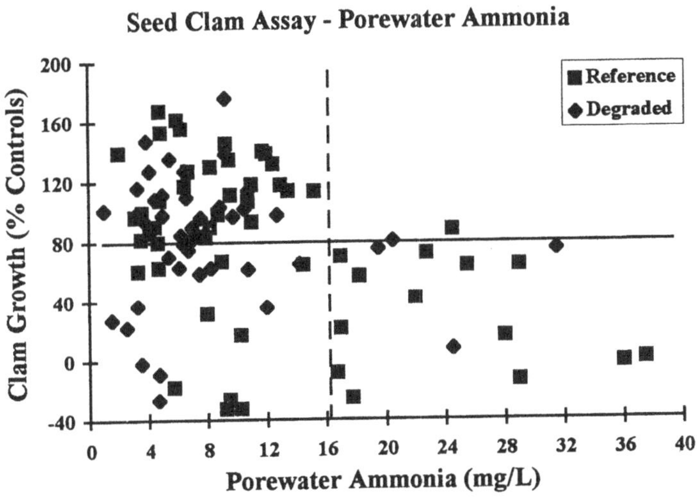

Fig. 4. The effects of porewater ammonia concentrations on seed clam growth.

ammonia, high toxicity sites) were reference sites. When the porewater ammonia concentrations were plotted against sediment composition (% silt-clays), it is apparent that high ammonia levels were more frequently observed in sandy sediments, < 20% silt-clays (Figure 5). Therefore, a large portion of the false positives could be explained by ammonia toxicity. However, since elevated ammonia concentrations occur more commonly in sandy, reference sites, the probability of confusing adverse effects due to ammonia from those due

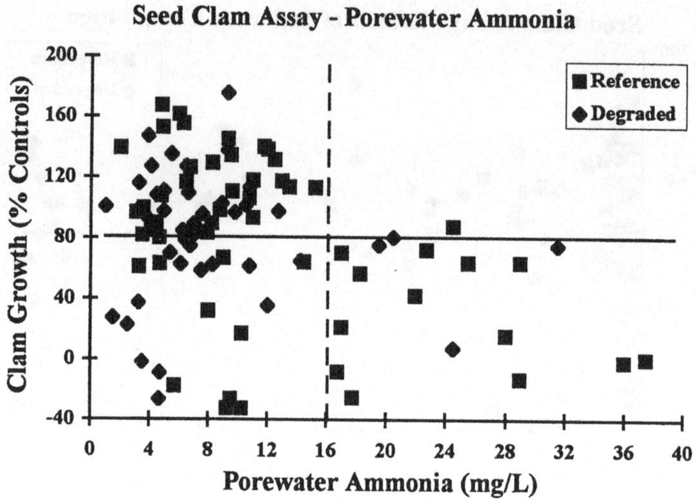

Fig. 5. The relationship between sediment silt-clay content and porewater ammonia concentrations.

to contaminants should be relatively low. Although higher ammonia concentrations might be expected in fine-grained sediments characterized by high organic content, high ammonia levels in sandy sediments were observed with sediments collected from a variety of southeastern estuaries (from shallow as well as deep systems, and from a range of salinity regimes). High ammonia concentrations in sandy sediments, and poor associations with grain size or organic content have been reported (Sims and Moore, 1995). High ammonia levels may be related to temperature, anoxia, transport and storage conditions, and the presence of resident organisms that excrete metabolic products, etc. (Carr *et al.*, 1996).

There was also some evidence that the assay may not be reliably interpretable with sediments characterized by low porewater salinities. Comparison of toxicity versus porewater salinity suggests that toxicity is observed when sediment porewater salinities are < 9 ‰ (Figure 6). However, it should be remembered that the assays are conducted with 25 ‰ seawater and the low porewater salinities are not maintained during the assay (porewater salinities fall within 5 ‰ of the overlying water due to mixing during sediment preparation steps). It may be significant that all of the samples that fall into the lower left quadrant of Figure 6 were collected from various estuaries over the same 2-4 day period and shipped together. Although a review of the sample tracking records revealed no evidence of sampling or shipping problems, the possibility that something happened with these samples that caused these anomalous results must be raised. Another plausible explanation is that other factors associated with low-salinity riverine sediments are inherently toxic to estuarine clams. We will continue to evaluate this issue. Porewater salinity data provide an important means for identifying these sites.

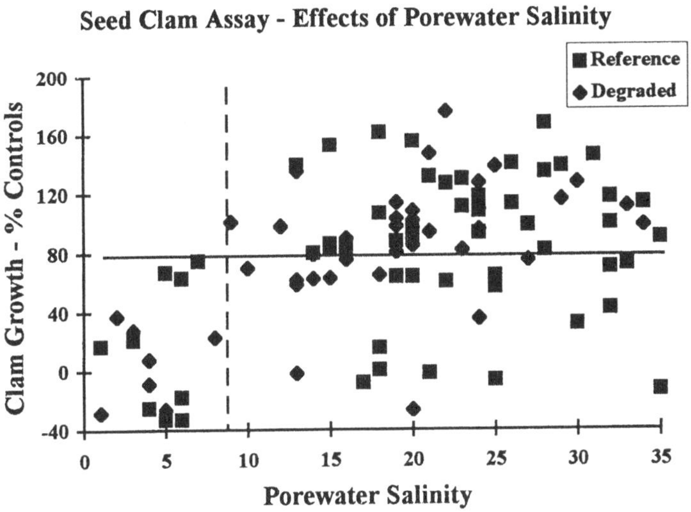

Fig. 6. The effects of porewater salinity on seed clam growth. All points that fall in the lower left quadrant were collected over a 2-4 day period and shipped together.

Therefore, when we consider the value of the assay as an indicator of sediment contaminants, we need to consider the false positive and false negative rates. How often do reference sites that have little evidence of contaminants cause toxicity and can we identify why, such as ammonia? How often do sites that have sediment contaminants not cause toxicity? These issues are summarized in the pie diagrams shown in Figure 7. For the 61 reference sites, 55.7% (34 sites) were non-toxic as expected, and the remaining pieces of the pie represent false positives. Most of the false positives could be attributed to ammonia. After accounting for ammonia (21.3% or 13 sites) and potential salinity or shipment effects (14.8% or 9 sites), there was only a small percentage that showed unexplained toxicity (8.2% or 5 sites). For the 44 degraded sites, approximately 23% (10 sites) showed toxicity that was believed to be due to contaminants. Ammonia was less of a problem with degraded sites, elevated at only 3 sites. The false negative rate was 54.5% (24 sites), which may seem high, but the false negative rate for the amphipod assay with these same sites was 100%. Lack of toxicity in the presence of contaminants is frequently assumed to reflect low bioavailability or toxicity due to unmeasured components.

4. Summary

Seed clam assays were conducted with sediments from 85 random core stations and 20 non-random supplemental stations collected over the range of the Carolinian Province. This is a sublethal assay based on growth of juvenile *Mercenaria mercenaria* as an indicator of potential sediment toxicity, and so may represent the potential for chronic as well as acute

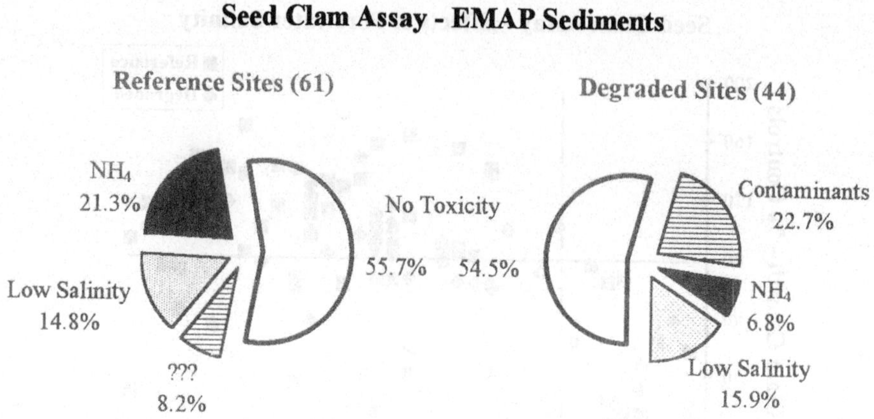

Fig. 7. Pie diagrams summarizing the results of seed clam tests in reference and degraded sites.

effects. The assay was more sensitive than the amphipod 10-day acute toxicity assay. There were 47 stations characterized as toxic (27 reference and 20 degraded stations). Porewater ammonia concentrations > 16 mg/L were measured for 16 of the toxic stations, 13 of which were reference sites. There were 16 stations (9 reference and 7 degraded) with sediments characterized by low salinity porewater or possible handling problems. There were 10 degraded stations for which toxicity was linked to contaminants.

In summary, the seed clam growth assay was more sensitive than the amphipod assay, and has a number of desirable attributes. A relatively small sediment sample size is required (approximately 300 - 500 ml sediment, compared to 3 L for the amphipod assay), juvenile organisms undergoing rapid growth enable identification of sublethal effects in a relatively short time frame (7 days), and test organisms can be readily acquired from hatchery facilities at a relatively small cost. Clam growth does not appear to be significantly biased by sediment type, so it is applicable across a wide range of habitat types, sandy as well as muddy. Seed clams do appear to be moderately sensitive to ammonia, which is primarily a problem with sandy reference sediments. A similar approach using *Mulinia lateralis* has recently been described, suggesting that the methodologies may be applicable with a variety of bivalve species (Burgess and Morrison, 1994). Assays based on growth of juvenile clams are believed to be a valuable tool for EMAP as well as other monitoring efforts for estimating potential chronic toxicity. When sublethal assays based on these methodologies are used in conjunction with other bioassays and ancillary environmental measurements to facilitate interpretations of the results, they can serve as important indicators of anthropogenic stress.

Acknowledgments

The contributions of the following people are gratefully acknowledged. Dr. J. Hyland (Director, EMAP Carolinian Province); Dr. A.F. Holland (Director, MRRI); Dr. R. Van

Dolah and J. Jones (MRRI) who conducted the ammonia and other porewater analyses; J. Scott and C. Mueller (SAIC) who conducted the amphipod studies; T. Wade, G. Denoux, J. Brooks, and P. Boothe (GERG, Texas A&M) who conducted the contaminant analyses; C. Batey and Atlantic Littleneck Clam Farm; and the Field Teams of UNC-W, MRRI/SCDNR, and FL DNR for sample collections. This work was funded by NOAA Cooperative Agreement No. NA470A0177. This document is South Carolina Marine Resources Research Institute Publication Number 397.

References

Ankley G.T., DiToro, D.M., Hansen, D.J. and Berry, W.J.: 1996, Assessing the Ecological Risk of Metals in Sediments. *Environ. Toxicol. Chem.* **15**, 2053-2055.

Burgess, R.M. and Morrison, G.E.: 1994, A Short-Exposure, Sublethal, Sediment Toxicity Test Using the Marine Bivalve *Mulinia Lateralis*: Statistical Design and Comparative Sensitivity. *Environ. Toxicol. Chem.* **13**, 571-580.

Carr, R.S., Long, E.R., Windom, H.L., Chapman, D.C., Thursby, G., Sloane, G.M. and Wolfe, D.A.: 1996, Sediment Quality Assessment Studies of Tampa Bay, Florida. *Environ. Toxicol. Chem.* **15**, 1218-1231.

DeWitt, T.H., Ditsworth, G.R. and Swartz, R.C.: 1988, Effects of Natural Sediment Features on Survival of the Phoxocephalid Amphipod, *Rhepoxynius Abronius. Mar. Environ. Res.* **25**, 99-124.

Hach: 1994, DR/700 Colorimeter Procedures Manual. Hach Chemical Company, Loveland, CO.

Hamelink, J.L., Landrum, P.F., Bergman, H.L. and Benson, W.H.: 1994, *Bioavailability: Physical, Chemical and Biological Interactions.* Lewis Publishers, Ann Arbor. 239 pp.

Hyland, J.L., Herrlinger, T.J., Snoots, T.R., Ringwood, A.H., Van Dolah, R.F., Hackney, C.T., Nelson, G.A., Rosen, J.S. and Kokkinakis, S.A.: 1996, Environmental Quality of Estuaries of the Carolinian Province: 1994. *NOAA Technical Memorandum NOS ORCA 97*, 102 pp.

Kokkinakis, S.A., Hyland, J.L., Mageau, C. and Robertson, A.: 1994, Carolinian Demonstration Project - 1994 Field Operations Manual. August 1994 Draft.

Long, E.R., MacDonald, D.D., Smith, S.L. and Calder, F.D: 1995, Incidence of Adverse Biological Effects Within Ranges of Chemical Concentrations in Marine and Estuarine Sediments. *Environ. Management.* **19**, 81-97.

Long, E.R., Robertson, A., Wolfe, D.A., Hameedi, J. and Sloane, G.: 1996, Estimates of the Spatial Extent of Sediment Toxicity in Major U.S. Estuaries. *Environ. Sci. Tech.* **30**, 3583-3592.

Luoma, S.N.: 1994. 'Prediction of Metal Toxicity in Nature From Bioassays: Limitations and Research Needs' in: Tessier, A. and Turner Metal, D.R. (eds.), *Speciation and Bioavailability in Aquatic Systems*, John Wiley & Sons, Chichester, England, pp. 609-659.

Moore, D.W., Bridges, T.S., Gray, B.R., and Duke, B.M.: 1997, Risk of Ammonia Toxicity During Sediment Bioassays with the Estuarine Amphipod *Leptocheirus plumulosus. Environ. Toxicol. Chem.* **16**, 1020-1027.

Ringwood, A.H., Holland, A.F., Kneib, R. and Ross, P.: 1996, EMAP/NS&T Pilot Studies in the Carolinian Province: Indicator Testing and Evaluation in Southeastern Estuaries. *NOAA Technical Memorandum NOS ORCA 102*, 115 pp.

Ringwood, A.H., DeLorenzo, M.E., Ross, P.E. and Holland, A.F.: 1997, Interpretation of Microtox® Solid-phase Toxicity Tests: The Effects of Sediment Composition. *Environ. Toxicol. Chem.* **16**, 1135-1140.

Sims, Jerre G. and Moore, David W.: 1995, Risk of Pore Water Ammonia Toxicity in Dredged Material Bioassays. *US Army Corps of Engineers, Miscellaneous paper D-95-3*, 66 pp.

BENTHIC BIOLOGICAL PROCESSES AND E_H AS A BASIS FOR A BENTHIC INDEX

WAYNE R. DAVIS[1], ANDREW F.J. DRAXLER[2], JOHN F. PAUL[1], and JOSEPH J. VITALIANO[2]

[1]*U.S. EPA, Atlantic Ecology Division, 27 Tarzwell Dr., Narragansett, RI 02882*
[2]*NOAA, NMFS, Northeast Fisheries Science Center, Highlands, NJ 07732*

Abstract. It is proposed that the common measures of benthic community condition can be augmented with a vertical E_H profile taken through the benthic bioturbation zone. Sediment E_H, an electrochemical measure of oxidized and reduced compounds in sediment porewater, measures the integrative consequences of all metabolic and transport processes of the benthic community. Biota, especially microbiota, metabolize carbon using a variety of electron acceptors, including O_2, SO_4 and some nitrogen and metal compounds. Motile benthic macrofauna ingest and transport particles, ventilate deep burrows and anoxic sediment with overlying seawater while sedentary suspension-feeding fauna deposit suspended organic matter onto the sediment surface. Collectively, these metabolic and behavioral processes advect particles and seawater between bottom water and deep sediment and define the overall structure of porewater chemistry. That structure creates a full spectrum of biogeochemical conditions of solubility, reactivity, and microbial metabolism which remineralizes excess organic carbon and most organic contaminants, defines solubility of trace metals, and pushes the vertical E_H profile toward oxidizing conditions. It is proposed that a standard E_H probe inserted downward through the bioturbation zone will provide a general measure of this resulting porewater chemistry and thus the impact of feeding, irrigation, and metabolism of the total macro, meio, and microbenthic community. If such a measure can be validated it will permit extended measurement of community function and reduced efforts in measuring community structure.

1. Introduction

Benthic condition or well-being is commonly expressed by counting the number and density of species and formulating some mathematical expression of those data, such as Sander's Rarifaction technique (1968) and is used to express species richness or diversity. To attain its full potential, assessment of biological condition requires further information on (a) biological structure (population structure and biomass), (b) models of behavior or natural history, and (c) rates of key functions (e.g., metabolism, irrigation, and sediment advection). Such an intensive research effort may be replaced with an impirical measure of major biological activities (feeding, metabolism, locomotion, and respiratory behaviors) estimated with a vertical profile of sediment porewater redox potential (E_H).

This possibility was explored using a recovery study of the New York Bight 12-Mile sewage sludge disposal site, where sediment E_H was found to correlate with the observed number of species (Figure 1; Draxler *et al.*, 1996). This positive exponential relationship ranges between a low of 5 to 6 species when E_H at 0.5 cm depth was -150mV, to 51 species when E_H was +300 mV. The increase in species was correlated with a reduction of labile organic matter (Draxler et al, 1996) and physical sediment erosion (Davis *et al.*, 1996). This work formed the basis for the following sequence of questions: Is the observed relationship of number species to sediment E_H causative and, if so, what is the driving variable? Does

Environmental Monitoring and Assessment **51**: 259–268, 1998.

E_H limit biology? Does biology determine E_H? And finally, could a E_H profile routinely and usefully monitor benthic condition?

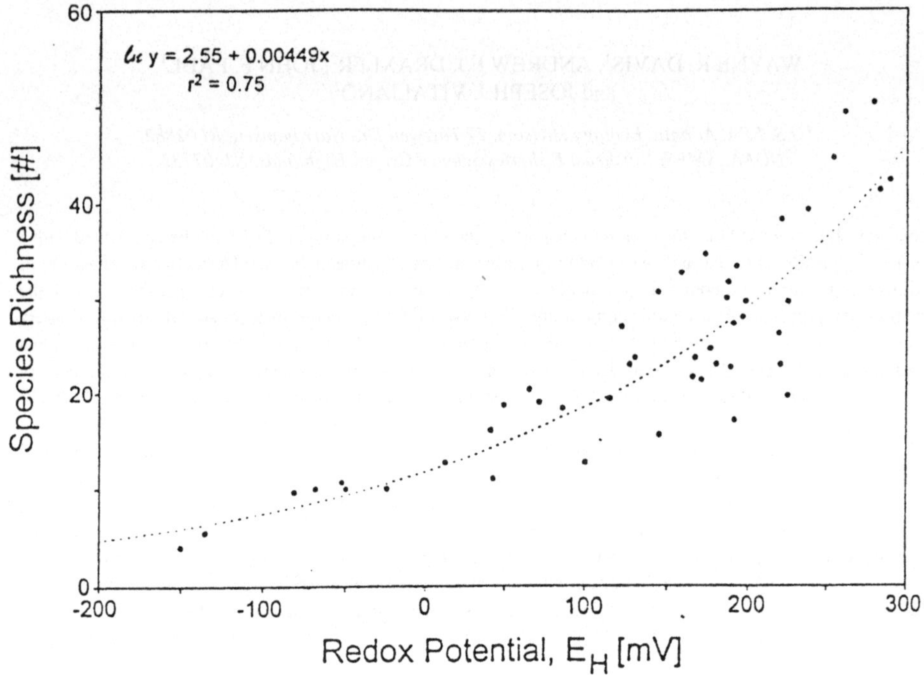

Fig. 1. From Draxler *et al.*, 1996: A corollation between number of species (Y-axis) and E_H (X-axis; mV @ .5 cm deep) at New York Bight station NY-6 (central Christiansen Basin), the station highly impacted by sewage sludge disposal measured over time.

The relationship between E_H and the presense of organisms has been explored by Rhoads and his colleagues (e.g., 1977). Aller (1982) demonstrated a mechanism for that relationship between biological presence and porewater E_H and showed that porewater diffusion occurs adjacent to irrigated animal burrows much like porewater diffusion at any sediment-water interface. While some biota like these burrow-irrigating vermiforms oxidize porewater, other biota, notably benthic filter feeders probably decrease porewater E_H through biodeposition of labile organic matter (Officer *et al.*, 1982). Thus, a vertical profile of E_H may integrate not only biological processes that raise E_H (metabolism and irrigation), but also biological processes that lower E_H (e.g., biodeposition and primary production). A general relationship is suggested by equations relating sediment mixing (bioturbation), sediment-water flux and sediment stratigraphy (Goreau, 1977). A highly used empirical measure of mixing functions utilizes the REMOTS® camera (Rhoad *et al.*, 1983).

The E_H measurement is the voltage potential between sediment porewater and overlying seawater and is caused by the relative concentration of porewater cations and anions, most of which reflect the kinds and quantity of metabolic activities of metazoa and microbes (Fenchel and Riedl, 1970). That metabolism is determined by the quantity and diversity of electron acceptors (compounds necessary to extract energy from organic carbon). Electron acceptor availability depends on adaptations of each biotic group or species and may include O_2, SO_4, Fe^{+3}, NO_3 and other compounds. Metazoa are generally limited by O_2 availability while microbes may use one (obligate) or more (facultative) electron acceptors. It may also be viewed as a relative measure of "reducing" verses "oxidizing" conditions. In the sense of a descriptor, it functions as an index of the chemical condition of the environment that can be compared with discrete measurements and biological community composition. Thus, porewater redox potential may indicate the recent history of biologically-mediated reactions.

The purpose of this paper is recast the collective findings of Rhoads, Aller and Fenchel, to lay out the procedures with which E_H profiles could be used as a measure of benthic condition and to use our data from a NY Bight sewage site recovery study to demonstrate this connection and test E_H as a economical reconnaissance tool.

2. Materials and Methods

The data on sediment E_H and benthic infauna were obtained during a study of the inner New York Bight, prior to and following cessation of sewage sludge disposal at the 12-Mile site. The focus of the present study was on staion NY-6, a highly impacted station in the Christiansen Basin (Figure 2). The E_H data was measured in a Smith-McIntyre sediment grab placed in a basin of seawater to prevent porewater drainage. Water above the sediment was baffled to minimize disturbance due to the ship's motion. The E_H probe was inserted vertically to measure E_H at various depth levels, using a Fisher Scientific model 640 portable pH/millivolt meter[1]. The instrument was calibrated to within 1 mV at 10, 100, and 1000 mV using a Cole-Palmer pH-mV calibrator (#5657-10). A platinum electrode (Thomas Scientific #4096-D20) with a band of platinum 6 mm dia. x 4 mm height was used as the sample electrode. The reference electrode was a Fisher Scientific (#13-639-62) sleeve junction calomel electrode. This system was calibrated using three $K_3Fe(CN)_6$-$K_4Fe(CN)_6$ solutions of differing redox potentials. Before measuring each profile, the electrode was first equilibrated with the sediment at 10 cm in the grab, then with the overlying water. Readings at each depth interval were accepted when the rate of change was less than 1 mV in 10 sec. (see Draxler *et al.*, 1996).

Biological assessment was made from standard sieving (1 mm), identification and counting (Reid *et al.*, 1996). The number of species and individuals were determined as a basis for comparing biological changes with concurrent E_H changes. These two data sets,

---

[1]The mention of commercial products is not to be interpreted as an endorsement by the U.S. Environmental Protection Agency.

sediment E_H and biological density, were regressed to show how each species related to E_H in each grab sample.

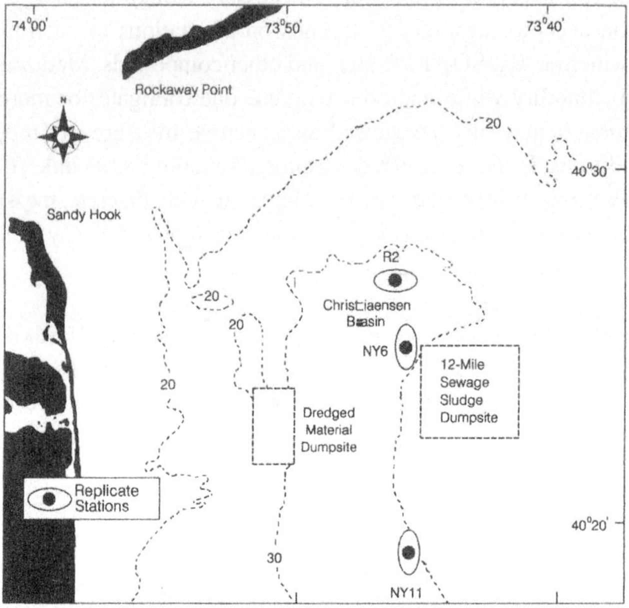

Fig. 2. Map of the inner New York Bight showing the 12-Mile sewage sludge disposal site and the study station, NY-6, in the nearby Christiansen Basin.

3. Results and Discussion

Following the cessation of sewage sludge disposal at the 12-Mile site there was an increase in both the biology and E_H at NY-6. The average sediment E_H increased throughout the study following cessation of disposal of sewage sludge from a low average of -125 mV to +25 mV (Figure 3). A likely interpretation of this linear increase in E_H (see Fig. 3 inset) is that the cessation of loading of highly labile organic carbon resulted in (a) fine particle erosion (Davis *et al.*, 1996), (b) increased dissolved oxygen in sediment and bottom water due to reduced biological oxygen demand (BOD), and (c) an increase in porewater E_H due to bioturbation or burrow ventilation.

Such an increase in E_H with biological recovery is predicted: fine-grained sediments (dominated by silt and clay) are typically populated by polychaete worms, filter and deposit-feeding bivalves, various crustacea, meiofauna (especially nematodes) and many genera of heterotrophic bacteria (e.g., Davis and Means, 1989). The presence of fauna and microflora suggest the possibility of raising porewater E_H since E_H is increased by the loss of organic matter and increased dissolved oxygen concentration. A prime biological mechanism in maximizing E_H, advection of aerobic seawater into deep sediment, is illustrated in photos of irrigated burrows of polychaete worms common to the New York

Bight (Figure 4). The irrigated burrows also aerate surrounding sediment and change sediment color from black to light brown (Fe^{+3} to Fe^{+2}). Worms like *Glycera* burrow as deep as 30 cm and thus aerate the sediment surrounding each portion of its burrow gallery.

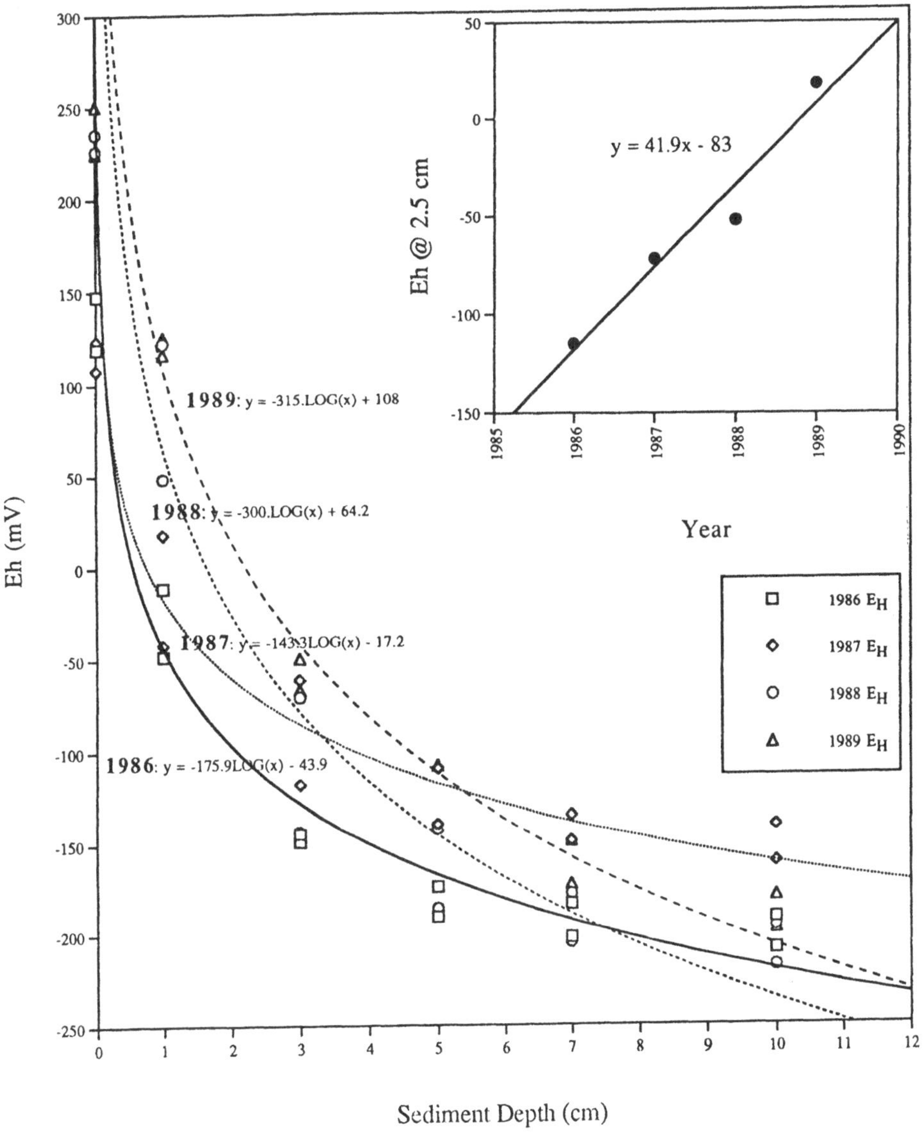

Fig. 3. Sediment E_H data on a four year shift in sediment E_H at NY-6 (central Christiansen Basin, New York Bight).

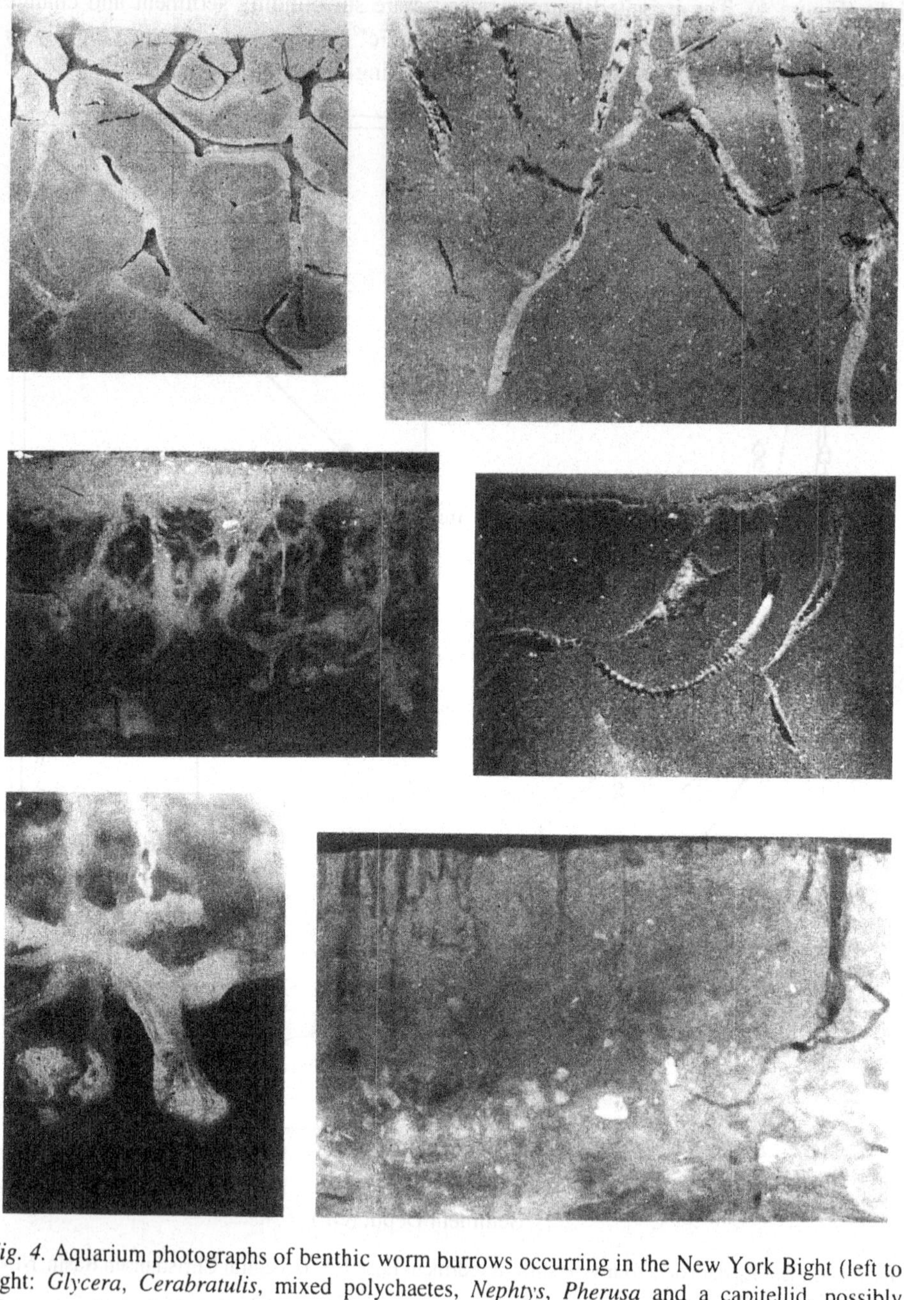

Fig. 4. Aquarium photographs of benthic worm burrows occurring in the New York Bight (left to right: *Glycera*, *Cerabratulis*, mixed polychaetes, *Nephtys*, *Pherusa* and a capitellid, possibly *Heteromastis*. Actively irrigated burrows turn from a black to light brown color due to Fe species and E_H relations.

As shown and discussed above (Figure 1), when the number of species is regressed against E_H a positive exponential relationship resulted (from Draxler *et al.*, 1996). An examination of the numbers of individuals of the 51 incountered species throughout the study indicates three general trends: a positive and negative relationship to E_H and a peaking in numbers at intermediate E_H levels (Figure 5). Of the species that could be categorized (80%), a positive relationship (a rise in numbers with rise in E_H) occurred with 19 species (48%). Species showing a steady rise in the population with E_H included all 4 bivalves (including *Arctica*, *Nucula*, *Pitar*), most polychaetes (e.g., *Ampharete*, *Glycera*, *Ninoe*, *Tharyx* and all 3 spionid polychaetes) and all 4 amphipods (Fig. 5). The interpretation made here is that a combination of higher dissolved oxygen, lower organic loading, and recruiting success resulted in the recovery of these species. Seven species (18%) were identified as peaking at the lowest E_H levels, including polychaetes (*Capitella*, *Harmothoe*, *Paranaitis*), a crab (*Cancer*), a Nemertinea (*Ceribratula sp.*?) and an Anthozoa (*Cerianthеopsis*). These species are presumed to be tolerant to low dissolved oxygen and/or high organic loading and/or are responding to a reduction in competitive processes such as carnivory. Six species (15%) peaked at intermediate E_H values and included 4 polychaetes (e.g., *Mediomastus*, *Nephtys*, *Pherusa*), a phoronid and an isopod (Figure 5). The key interpretation of this observation is that these species are sensitive to both very low E_H and to competitive biological conditions occurring under high E_H conditions. Data on 8 species (20%) could not be clearly related to E_H (e.g. *Asabellides*, *Crangon*, *Mytilus*, and *Tellina*), probably due to insufficient sampling density.

The evidence for a causal connection between biological presence and observed E_H, a necessary step to establish a mechanistic predictor, is limited to a few anecdotes. The number of species and individuals per species is rising in the four year trend in E_H (from Figures 3 and 4); the visual occurrance of oxidized sediment around irrigated burrows (Fig, 4); the ages of organisms, and thus their size and burrowing depth are increasing (sample bottle visual examination); and the reduction in sediment labile organic content (Draxler, in press). Other variables of grain size, organic content and time probably influenced biotic recovery. Grain size increased due to winnowing (Davis *et al.*, 1996) while labile carbon decreased (Draxler *et al.*, 1996) leaving the possibility that the biotic-E_H relationship is correlated but not causitive.

4. Conclusion

It is suggested that benthic condition can be expressed by total biological metabolism, including microbial metabolism and bioturbation-advective transport that collectively drives porewater redox condition, and can be empirically estimated by comparing a vertical E_H curve through the bioturbation zone, to deep sediment E_H values. It is recommended that a model of biotic community categorizing species on the basis of environmental impact or processing may be developed and applied to routine monitoring of benthic condition. It is possible that species mixes, while primarily representing how diverse species exploit

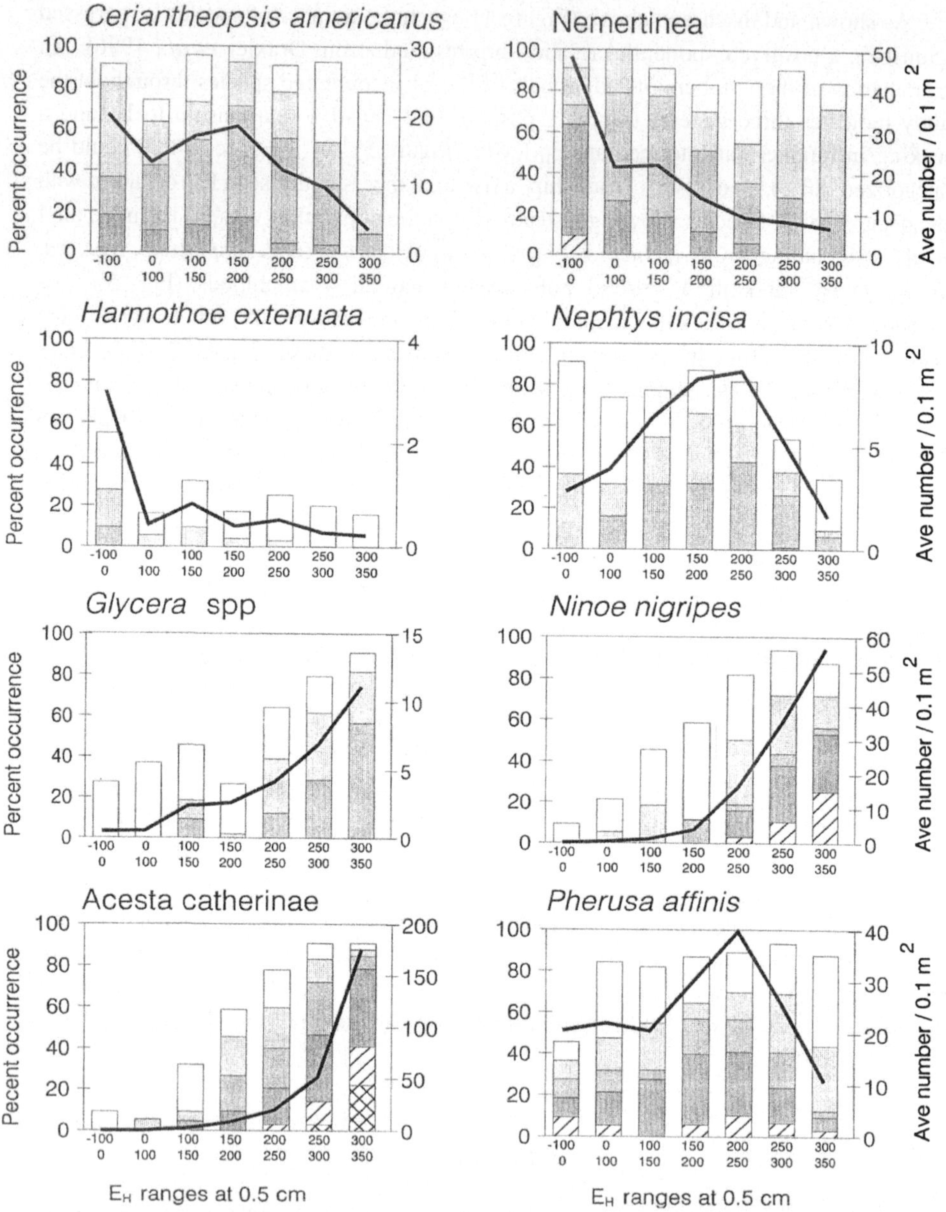

Fig. 5. Page 1. A variety of biotic changes in relation to observed changes in sediment E_H at NY6. Number of individuals of 8 out of 40 commonly occurring species indicating (1) a rise in numbers with E_H (48%), (2) a fall in numbers (18%) and (3) a peak in numbers at intermediate E_H levels (15%).

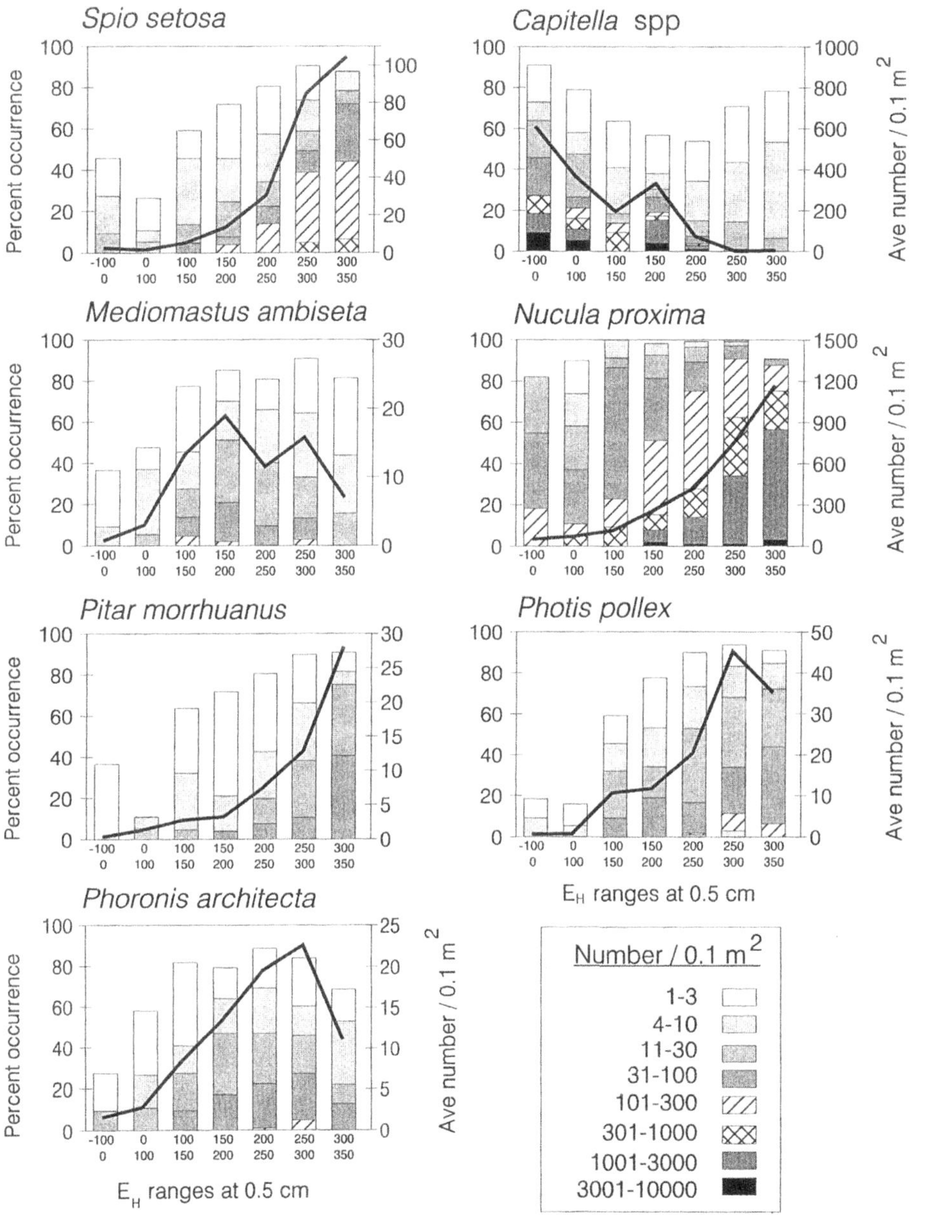

Fig. 5. Page 2. A variety of biotic changes in relation to observed changes in sediment E_H at NY6. Number of individuals of 8 out of 40 commonly occurring species indicating (1) a rise in numbers with E_H (48%), (2) a fall in numbers (18%), and (3) a peak in numbers at intermediate E_H levels (15%).

resources, also define the quality of processes, like irrigating sediment with overlying seawater. And it is thus asserted that total biological presence and its processes (e.g., feeding) are both adaptations to exploit available resources and result in predictable sediment chemical structure (i.e. porewater electrochemistry or E_H). In conclusion, it is asserted that porewater redox condition (E_H) should be evaluated as a reconnaissance tool to augment standard benthic biological assessment techniques.

Acknowledgements

We are grateful to Earl Davey, Richard Pruell, Brian Taplin, and Rob Burgess for editing assistance; to Richard Voyer for important encouragement, to Tom Pearson for his long-term inspiration and to EPA-ORD for funding. EPA contribution #1916.

References

Aller, R.C. 1982. "The effects of macrobenthos on chemical properties in marine sediment and overlying water", in P.L. McCall and M.J.S. Tevesz (Editors), *Animal-sediment Relations*. Plenum, New York, N.Y., pp.53-102.

Aller, R.C. 1994. "Bioturbation and remineralization of sedimentary organic matter: effects of redox oscillation", *Chem. Geol.* **114**, 331-345.

Bagander, L. E. and L. Niemisto. 1978. "An evaluation of the use of redox measurements for characterizing recent sediments", *Est. Coast. Mar. Sci.* **6**, 127-34.

Davis, W.R. 1993. "The role of bioturbation in sediment resuspension and its interaction with physical shearing", *J. Exp. Mar. Biol. Ecol.* **171**, 187-200.

Davis, W.R., R. McKinney and W. D. Watkins. 1996. "Response of the Hudson Shelf Valley sewage sludge-sediment reservoir to cessation of disposal at the 12-Mile site", in *U.S. Dep. Commer., NOAA Tech. Rep.* NMFS **124**, 49-60.

Davis, W.R. and J.C. Means. 1989. "A developing model of benthic-water contaminant transport in bioturbated sediment", *Proceedings of the 21st EMBS*, PL issn 0078-3234: 215-226.

Draxler, A.F.J. 1996. "Closure of the New York Bight 12-Mile sewage sludge dumpsite", *Northeastern Geology and Environmental Sciences* **128** (4): 1-11.

Fenchel, T.M. and R. J. Riedl. 1970. "The sulfide system: a new biotic community underneath the oxidized layer of marine san bottoms", *Marine Biology* **7**: 255-268.

Goreau, T.J. 1977. "Quantitative effects of sediment mixing on stratigraphy and biochemistry: a signal theory approach", *Nature* **265**, 525-526.

Pearson, T.H. and Rutger Rosenberg. 1978. "Macrobenthic succession in relation to organic enrichment and pollution of the marine environment", *Oceanogr. Mar. Biol. Ann. Rev.* **16**, 229-311.

Reid, R.N., S. Fromm, A. Frame, D. Jeffress, J. Vitaliano, D. Radosh & J. Finn. 1996. "Limited responses of benthic macrofauna and selected sewage sludge components to phaseout of sludge disposal in the inner New York Bight", in *U.S. Dep. Commer., NOAA Tech. Rep.* NMFS 124: 213-225.

Rhoads, D.C., R.C. Aller & M.B. Goldhaber. 1977. "The influence of colonizing benthos on physical properties and chemical diagenesis of the estuarine seafloor", in *Ecology of marine benthos*, ed. by B.C. Coull, 113-138. Belle Baruch, Univ. South Carolina Press.

Sanders, H.L. 1968. "Marine benthic diversity: a comparative study", *The American Naturalist* **102** (925): 243-282.

Whitfield, M. 1969. "E_H as an operational parameter in estuarine studies", *Limnol. Oceanogr.* **14**, 547-58.

STATE OF THE ESTUARIES IN THE MID-ATLANTIC REGION OF THE UNITED STATES

J.F. PAUL, C.J. STROBEL, B.D. MELZIAN, J.A. KIDDON, J.S. LATIMER, D.E. CAMPBELL, and D.J. COBB

Atlantic Ecology Division, National Health and Environmental Effects Research Laboratory, U.S. Environmental Protection Agency, Narragansett, RI 02882

Abstract. The U.S. EPA has prepared a State of the Region Report for Mid-Atlantic Estuaries to increase knowledge of environmental condition for improved environmental management. Sources of information included the National Estuary Programs, the Chesapeake Bay Program, the state monitoring programs in Delaware, Maryland, and Virginia, Federal programs such as National Status & Trends, National Shellfish Register, National Wetlands Inventory, the Environmental Monitoring and Assessment Program, and other primary literature sources. The state of the estuarine environment was summarized using indicators for water and sediment quality, habitat change, condition of living resources, and aesthetic quality. Each indicator was briefly discussed relative to its importance in understanding estuarine condition. Wherever possible, data from multiple programs were used to depict condition. Finally, an overall evaluation of estuarine condition in the region was determined. The usefulness of monitoring programs that collect consistent information with a well-defined sampling design cannot be overemphasized.

1. Introduction

At its inception in 1970, the U.S. Environmental Protection Agency (EPA) was given a broad charge to protect the nation's public health and the environment. This charge has been addressed by the Agency through the enforcement of numerous environmental statutes, such as the Clean Water Act, the Clean Air Act, the Safe Drinking Water Act, and Superfund laws. For many years, a command and control approach was taken to implement these environmental statutes, and enforcement under these Acts was maximized.

This approach to protecting the environment is changing. Over time, it was apparent that too much emphasis was being placed on the regulatory process and not enough on the end product--the condition of the environment. This change is consistent with the Agency's desire to increase the knowledge of environmental condition in making environmental decisions. The Agency supports the use of credible scientific information for making better environmental management decisions.

Beginning in 1990 in the Virginian Biogeographic Province, the EPA Office of Research and Development's Environmental Monitoring and Assessment Program (EMAP) tried to bridge the gap between the generation of scientific information and the use of such information by environmental managers (Holland, 1990; Weisberg *et al.*, 1993). Through workshops and seminars, a basic set of environmental managers' questions evolved that helped focus the analysis and synthesis of the ecological information (USEPA, 1991; Cochran, 1991; Queen *et al.*, 1992). These questions are:

Environmental Monitoring and Assessment **51**: 269–284, 1998.

- Are there problems with the ecological resources?
- What is the geographic distributions of the problem areas?
- What are the probable causes of the ecological problems?
- Have the ecological problems been changing with time?
- What is being done to address the ecological problems?

In an attempt to increase our knowledge of environmental condition and its use in management, EPA's Atlantic Ecology Division (AED) in Narragansett, Rhode Island, has been evaluating and assessing existing data and information to produce a state-of-the-estuaries report for the Mid-Atlantic Region of the United States that addresses the above series of questions (USEPA, 1997a). Our purpose here is to present a summary of the current status of estuaries in the Mid-Atlantic Region as reported in *Condition of the Mid-Atlantic Estuaries* (USEPA, 1997a).

2. Background for Preparing the *State of Estuaries*

Information used in preparing *Condition of the Mid-Atlantic Estuaries* was presented for estuaries in EPA Region 3 (states of Delaware, Maryland, Virginia, Pennsylvania, and West Virginia, and District of Columbia), which includes Chesapeake Bay, Delaware Bay, and the coastal bays of the Delmarva peninsula (the Atlantic coast lying in Delaware, Maryland, and Virginia). Portions of Delaware Bay in New Jersey (EPA Region 2) are also included. Conditions were presented for the entire region and for these major natural geographic subregions. No new data collection activities were undertaken, as the sources of information were published articles and program reports in the scientific literature. The intent was to integrate information from as many different sources as possible.

Actual sources of information for this included the National Estuary Programs, the Chesapeake Bay Program, the state monitoring programs in Delaware, Maryland, and Virginia, and Federal programs such as National Status & Trends, National Shellfish Register, National Wetlands Inventory, the Environmental Monitoring and Assessment Program, and other primary literature sources. We concentrated on these reports because of the broad geographic focus for the state of the estuaries and the availability of reports from these large-scale monitoring programs.

3. Description of Mid-Atlantic Estuarine Watersheds

The estuarine waters in the Mid-Atlantic Region consist of approximately 14,150 km^2, which includes 1.078 km^2 of Delaware Bay that lie in New Jersey. These estuaries contain a significant amount of the estuarine area in the United States; Chesapeake Bay is the single largest estuary in North America. The breakout of the estuarine area for the major natural subregions is 11,477 km^2 for Chesapeake Bay, 2,059 km^2 for Delaware Bay, and 613 km^2 for the Delmarva coastal bays. These estuaries drain 204,167 km^2 of land area, with

170,128 km² of these watersheds residing within the Region 3 state boundaries. Chesapeake Bay drains a watershed of 166,537 km², Delaware Bay drains 35,038 km², and Delmarva coastal bays drain 2,598 km².

Understanding patterns of land use within watersheds is basic to understanding the ecological condition of the Mid-Atlantic estuaries. Figure 1 summarizes the major land cover categories (urban, forests, agriculture, wetlands, and lakes and streams, as classified by the U.S. Geological Survey [Fegeas *et al.*, 1983]) for the region and subregions.

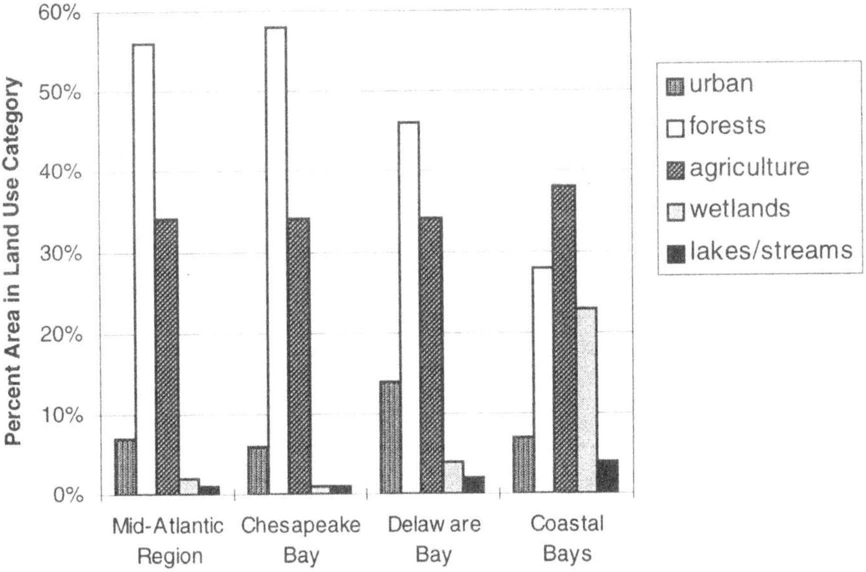

Fig. 1. Distribution of land cover across the Mid-Atlantic Region estuarine watersheds. Data from Fegeas *et al.* (1984).

For the region, forest is the dominant land cover, comprising 56% of the watershed area. Approximately one-third of the land in the region is agricultural. Agriculture is the dominant land cover in the Delmarva coastal bays watershed. Urban land (7% of the watershed) is generally close to the estuarine shoreline. It is expected that the amount of urban land will continue to increase across the region.

Human population growth is the single most important factor underlying various impacts on Mid-Atlantic estuaries. In 1950, the estuarine watersheds of the Mid-Atlantic Region contained 13 million residents. By 1990, this figure had grown to 21 million, and, by 2020, there will be an estimated 25 million people living in the estuarine watersheds of the Mid-Atlantic Region (Culliton *et al.*, 1990). Growing population requires land for homes, transportation, shops, jobs, and recreation.

The average depth of estuaries in the Mid-Atlantic Region is 6.2 m, with a maximum of 53.3 m in the mainstem of the Chesapeake Bay. The average depth is 6.2 m in Chesapeake Bay, 5.9 m in Delaware Bay, and 1.5 m in Delmarva coastal bays.

Oligohaline waters (salinity less than 5 ppt) represent a small fraction of the estuarine area of the region. Mesohaline waters (between 5 and 18 ppt) are dominant in Chesapeake Bay, while Delaware Bay and Delmarva coastal bays are mostly polyhaline/euhaline waters (greater than 18 ppt).

Most estuarine waters in the Mid-Atlantic Region exhibit little or no stratification. However, deeper portions of Chesapeake Bay do become highly stratified. Delaware Bay has areas of moderate stratification; Delmarva coastal bays are too shallow to stratify, since they are well mixed by the wind.

4. Estuarine Indicators in the Mid-Atlantic Region

A snapshot of estuarine condition is provided by summarizing what we know about a set of indicators for water and sediment quality, habitat change, and condition of living resources (USEPA, 1997a). Each indicator is briefly discussed relative to its importance for understanding estuarine condition. The current condition for each indicator is then summarized. A synopsis for each of the indicators is discussed in this section.

4.1 Water Quality: Nutrients

Nutrient concentrations are relatively high in many of the river tributaries and smaller bays of the Mid-Atlantic Region. High nutrient levels are not harmful in themselves; however, overenrichment can cause prolonged phytoplankton blooms, which can disrupt the normal estuary. Non-point sources, such as farming and atmospheric deposition, are the most common origins of excess nutrients, but municipal and industrial point sources are important near urban centers. Delaware Bay is one of the most enriched estuaries in the world, although harmful phytoplankton blooms are held in check by other factors, for example, by turbidity, which diminishes light necessary for plant growth (Sharp, 1984). Nutrient levels in Chesapeake Bay are declining in response to improved waste water management practices (nitrogen) and bans on certain types of detergents (phosphorus). However, there has been more success in controlling point sources than on controlling non-point sources of nutrients. While nutrient loading rates are generally low in the Maryland coastal bays, the Delaware coastal bays are overenriched.

4.2 Water Quality: Phytoplankton

An important component of estuarine systems, phytoplankton, is at the base of the estuarine food web. However, prolonged and excessive phytoplankton blooms can disrupt the growth of submerged aquatic vegetation and promote hypoxia (dissolved oxygen less than

5 mg/L) and anoxia (no dissolved oxygen) in poorly mixed or poorly flushed estuarine waters. Elevated concentrations of chlorophyll *a* (a measure of green pigment in phytoplankton) during the summer are evident in most tributaries in the region, but especially in the upper Chesapeake Bay. Extended blooms in Delaware Bay are uncommon despite high nutrient concentrations, in part because phytoplankton growth is limited by the high turbidity in this estuary. The high levels of chlorophyll in the upper Chesapeake Bay common during the 1970s have been reduced, in large part, by improved nutrient management practices. Increased levels of chlorophyll *a* are becoming increasingly noticeable, however, in the dead end canals along developed shorelines in the Delmarva coastal bays (Chaillou *et al.,* 1996; Maxted *et al.*, 1997).

4.3 WATER QUALITY: DISSOLVED OXYGEN

An adequate supply of dissolved oxygen (DO) is a fundamental requirement for estuarine organisms. During the mid summer (the critical time period for estuaries), 17% of the estuarine bottom waters of the region exhibit moderate hypoxia (DO between 2 and 5 mg/L) and 8% have severe hypoxia (DO less than 2 mg/L) (Figure 2). Hypoxia in Chesapeake Bay is associated with natural processes (stratification) and made worse by nutrient enrichment and eutrophication. Delaware Bay and the Delmarva coastal bays have small areas of hypoxia.

4.4 SEDIMENT CONTAMINATION

The public considers the contamination of sediments with trace metals, PAHs, PCBs, and pesticides, and the potential toxicity of these sediments, to be major threats to estuaries in the Mid-Atlantic Region. Sediment contaminant distribution[1] across estuaries of the region is shown in Figure 3. More than half of the estuarine sediments in the region have contaminant levels low enough to pose no potential risk. Only 6% of the sediments contain contaminant levels considered to pose a potential risk of effects to aquatic organisms.

---

[1]Informal guidelines for interpreting sediment contamination based on many field and laboratory studies have been developed. These guidelines attempt to relate observed chemical concentrations to concentrations known to either cause biological effect in laboratory spiked-sediments or spiked-water experiments or be associated with biological effects in field studies. Examples of these approaches are the Puget Sound (Malek, 1992) apparent effects thresholds (AETs), State of Washington (Phillips, et al., 1988) screening level concentrations (SLCs), Long and Morgan's (1990), as updated in Long et al. (1995), effects range median (ER-M) and effects range low (ER-L) concentrations and refinements to Long and Morgan by MacDonald (1994) for potential effects level (PEL) and threshold effects level (TEL). These approaches benefit from the weight of evidence afforded by large data sets associating sediment contaminant concentrations with biological effect, but suffer from a failure to incorporate the effects of multiple chemicals in complex mixtures, as the chemicals exist in the environment.

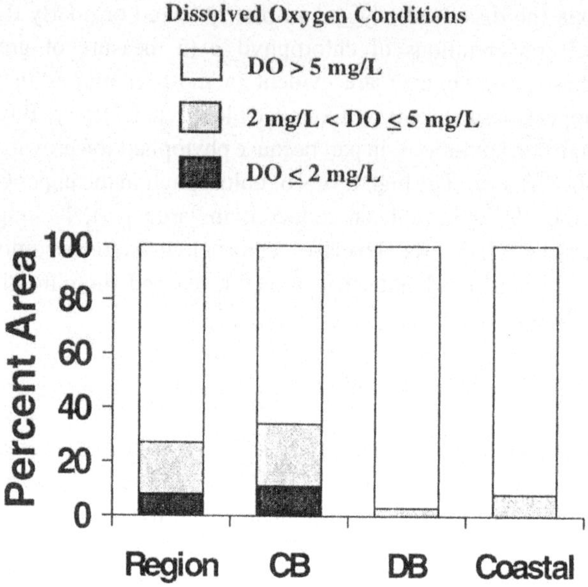

Fig. 2. Distribution of summer-time dissolved oxygen within one meter of bottom sediments across estuarine waters in the Mid-Atlantic Region. Percent area derived from EPA EMAP 1990-93 data (Strobel *et.al.*, 1995; Paul *et al.*, 1997).

The sources of toxic substances to estuaries include point sources (industrial and waste water treatment plant discharges), and non-point sources such as urban storm water runoff (from streets, parking lots, and grassy areas), atmospheric deposition (directly on the water surface and on the land which eventually runs off to the estuary), and agricultural runoff. Urban storm water runoff and point sources are the major sources for the metals entering the estuarine waters of the region (USEPA, 1994). The major sources for PAHs and PCBs are urban storm water runoff and atmospheric deposition (Greer and Terlizzi, 1997). Pesticide loadings to estuarine waters are primarily from agricultural runoff and atmospheric deposition. Recent studies by the National Oceanic and Atmospheric Administration National Status and Trends Program indicate that surficial sediment contaminant levels in the region have been generally decreasing over the last decade (O'Connor and Beliaeff, 1995).

4.5 Habitat Change

The historical loss of coastal wetlands in the Mid-Atlantic estuaries has largely been stabilized by state and federal conservation plans (Field *et al.*, 1991). The challenge now is to assure that the wetlands are healthy despite severe anthropogenic stresses. The precipitous loss of submerged aquatic vegetation (SAV) in Chesapeake Bay during the 1970s has been reversed. SAV beds are returning to Chesapeake Bay (Figure 4) in response to diminished eutrophication and improved nutrient management practices. SAV has historically been absent from Delaware Bay (Sharp, 1984) and the Delaware coastal bays (Weston, 1993) because of high natural turbidity in these estuaries.

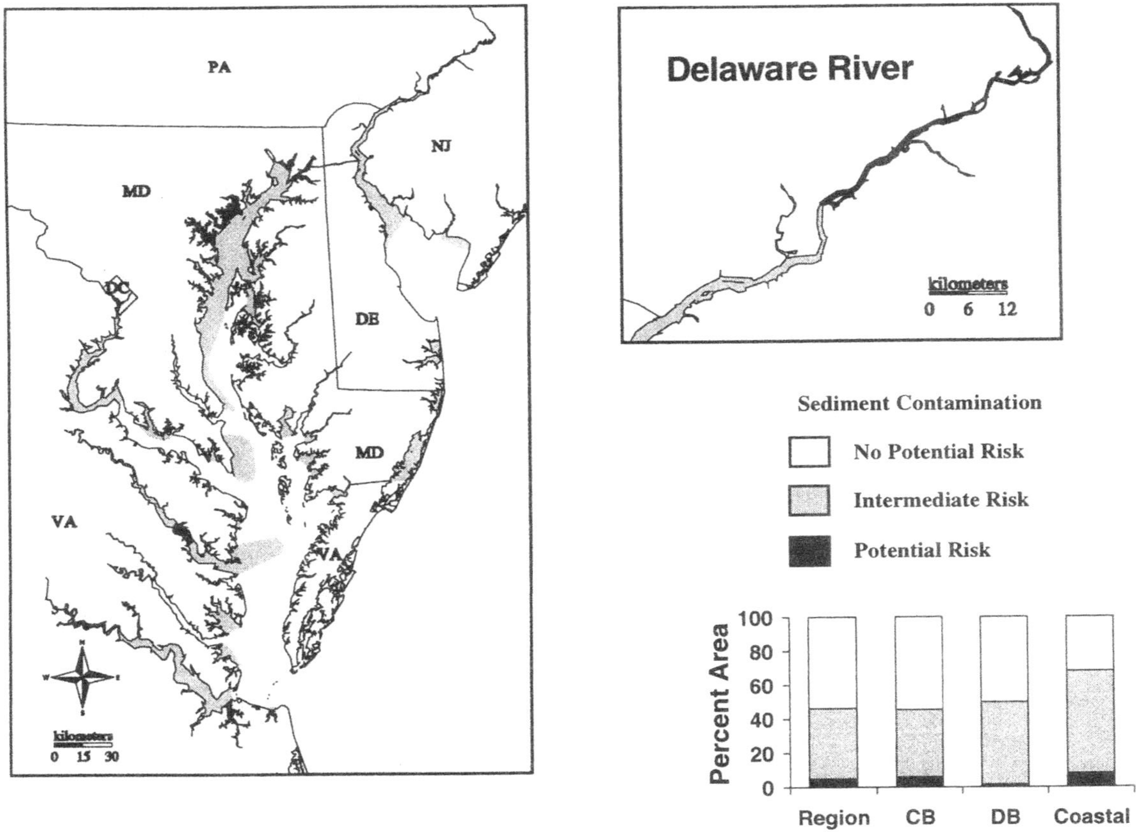

Figure 3. Sediment contaminant distribution across Mid-Atlantic estuaries expressed as risks to aquatic organisms, as discussed in text. Map depicts spatial distribution derived from multiple sources of information (USEPA, 1995; Chaillou *et al.*, 1996; Strobel *et al.*, 1995; Paul *et al.*, 1995). Bar graph shows percent area derived from EPA EMAP 1990-93 data (Strobel *et al.*, 1995; Paul *et al.*, 1997).

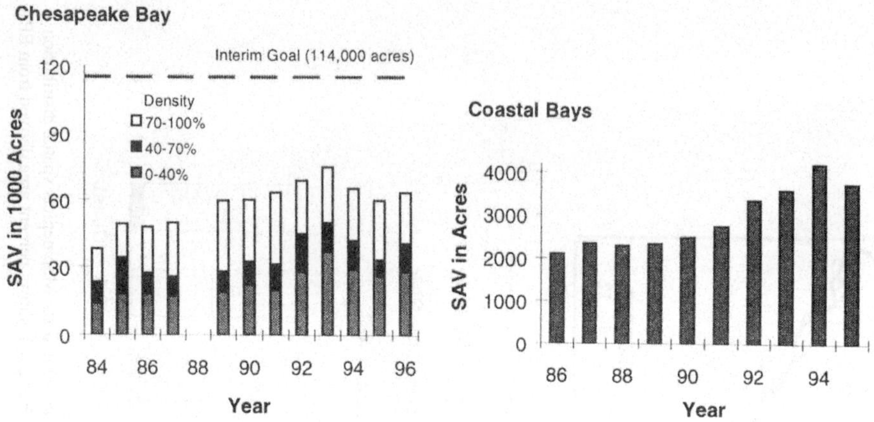

Fig. 4. Upper; SAV coverage in Chesapeake Bay (CPB, 1995). Lower: SAV coverage in Delmarva coastal bays (Bohlen and Boynton, 1996).

4.6 Condition of Living Resources: Benthic Communities

Benthic communities impacted by contaminants occur locally and in greater frequency near the urbanized centers of Philadelphia, Baltimore, Washington, and Norfolk, which have been major historical sources of industrial and municipal contaminants. Benthic communities impacted by eutrophication appear in Delaware Bay, the Delmarva coastal bays, and in the some small estuarine systems in Chesapeake Bay. However, the most prominent association of impacted benthic communities in Mid-Atlantic estuaries is with low dissolved oxygen concentrations in the bottom waters. This condition is primarily located in the mainstem portion of Chesapeake Bay. Similar types of impacts on the benthic community due to low dissolved oxygen, chemical contamination, eutrophication, or high velocity currents occur in each of the Mid-Atlantic geographic subregions; however, the relative importance and magnitude of the effects associated with each stressor is very different across estuaries.

4.7 Condition of Living Resources: Shellfish Harvest

The shellfish industry, and in particular the harvest of the American oyster, *Crassostrea virginica*, has traditionally been one of the major industries of the Mid-Atlantic states. It and is currently the one most seriously threatened. The decline in the oyster industry has been precipitous, from an annual catch of 133 million pounds in 1880 to about one million pounds today (Figure 5). The primary causes for the decline of the oyster fishery are the oyster diseases Dermo and MSX, although overfishing and pollution have contributed. Scientists are currently investigating the feasibility and ecological consequences of introducing disease-resistant strains of oysters to Mid-Atlantic estuaries to re-establish the fishery.

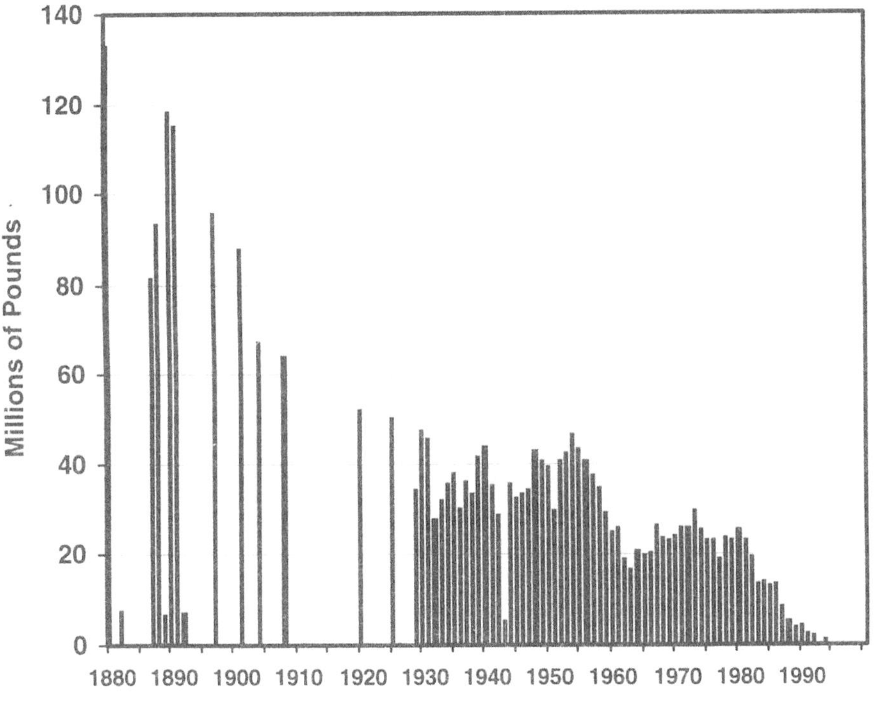

Fig. 5. Annual oyster harvest for Mid-Atlantic estuaries. Gaps in the late 1800s and early 1900s do not represent zeros, but rather missing data. Derived from Lyles (1967a, 1967b), NOAA (1997b), Chesapeake Bay Program (1997), Haskin Shellfish Lab (1997).

Another important component of the shellfish industry in the Mid-Atlantic Region is the blue crab, *Callinectes sapidus*. Annual harvest of blue crab has been highly variable over the past decades, especially for Delaware Bay. This variability is likely due to natural environmental factors, although it is thought to be compounded by fishing pressures. Although the annual catch of crabs has not decreased significantly, the catch per unit effort has. The current harvest is being kept up by increased effort, not stable populations (which are thought to be declining). Scientists are concerned that the increased fishing pressure would make it difficult for crab populations to recover if they were impacted by severe environmental conditions.

4.8 Condition of Living Resources: Shellfish Closures

Each state monitors its estuarine waters for coliform bacteria, and closes those waters to shellfishing where the concentration reaches a critical level. Of the 3,294,000 acres of potentially productive shellfish ground in the Mid-Atlantic estuaries, shellfishing is prohibited or restricted in some way in approximately 10% of the area (NOAA, 1991, 1997a). These closings can be attributed to contamination from a variety of sources, including sewage treatment plants, leaking septic systems, marinas, industry, wildlife, boating and runoff; or may be administrative in nature (e.g., inadequate monitoring to ensure the waters are safe). Considering the degree of urbanization in the area it is encouraging that such a small percentage of the total available area is closed to shellfishing. In addition, improvements in water treatment have led to a decrease in closed acreage, from 18% in 1985 to 10% in 1995.

4.9 Condition of Living Resources: Fish Stock Assessment

The abundance, distribution, and condition of fish are considered indicators of ecosystem health because fish integrate effects of environmental stress over space and time, and they are often the top carnivores. The Mid-Atlantic Region contains many diverse habitats, making region-wide judgements about fish populations difficult. In addition, market forces and environmental fluctuations obscure causes of fish population declines. Nevertheless, improvement has been chronicled for some species. For example, striped bass and American shad populations are improving after significant historical declines (Figure 6). These improvements are likely due to restrictions placed on fishing, but they may also be attributed to improved water quality. Species compositions of shore zone fish in the Delaware coastal bays indicate impacted environmental conditions (Chaillou *et al.,* 1996). In contrast, Maryland coastal bays species composition suggests a healthy habitat; however, recent evidence of initial degradation in northern areas of these bays has been observed (Chaillou *et al.,* 1996).

4.10 Condition of Living Resources: Contaminants in Fish and Shellfish

Fish and shellfish contaminant levels throughout the region appear to be decreasing over time due to bans and restrictions on the use of such chemicals as PCBs, DDT, and kepone, and stricter limits on point source discharges (NOAA, 1994). The contaminant concentrations in fish and shellfish are generally at or below the national mean levels; however, much higher levels may be present in organisms collected near urbanized areas, such as Baltimore Harbor. Generally, contaminant levels in fish and shellfish are higher in Delaware Bay than in Chesapeake Bay or the Delmarva coastal bays, perhaps reflective of the degree of urbanization in the estuarine watersheds (Figure 2), or the massive industrialization along the lower section of the Delaware River.

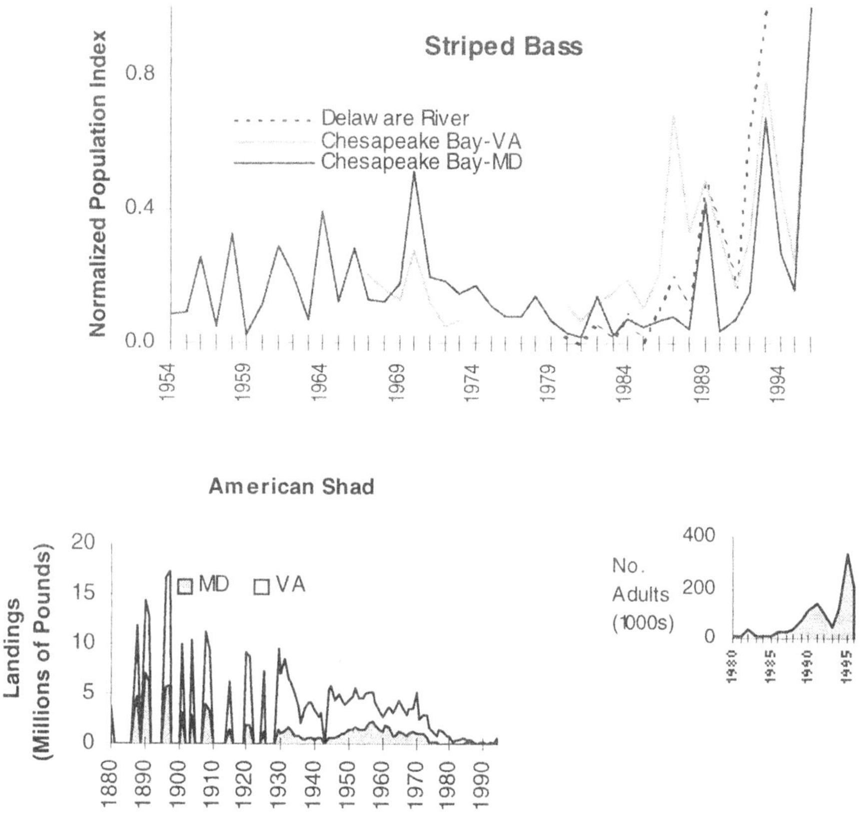

Fig. 6. Upper: Normalized indices for juvenile striped bass: seine index for Delaware River (Dove and Nyman, 1995) and juvenile index for VA and MD for Chesapeake Bay (NOAA, 1997b; USEPA, 1997b; CBP, 1997). Numerical comparisons between systems cannot be made due to the differences in the indices. Lower: American shad landings in Chesapeake Bay (USEPA, 1997b; CBP, 1997). Gaps in the late 1800s and early 1900s represent missing data. 1993-94 data are preliminary. Recent data on the abundance of adult fish in the upper Chesapeake Bay (right).

4.11 Condition of Living Resources: Incidence of Disease

EPA's EMAP included an examination of 13,467 fish from 177 stations in the Mid-Atlantic estuaries (Strobel *et al.*, 1995). Only three per thousand examined were observed with external pathological abnormalities (ulcers, lumps, growths, or fin erosion), indicating a low incidence of such abnormalities. This is lower than the 10 per thousand observed by EMAP for the estuaries of the Gulf of Mexico (Macauley *et al.*, 1996).

4.12 Condition of Living Resources: Waterfowl

In general, Mid-Atlantic waterfowl populations are in relatively good condition (USFWS, 1997). Although overall population levels have held relatively constant over the past 30 years, shifts have been seen in individual species (Sauer *et al.*, 1996). For example, Figure

7 shows a dramatic decrease in numbers of black ducks (which are less tolerant of habitat loss) in Chesapeake Bay, along with a concomitant increase in mallards (more tolerant of human activities). This trend has not been observed in either Delaware Bay or the Delmarva coastal bays. Of some concern is that favorable environmental conditions over the past few years in other parts of the country, where migratory birds nest during the summer, have resulted in increased populations of those birds, but that increase is not being seen in the Mid-Atlantic. Many of those birds are favoring other wintering sites, such as along the Gulf of Mexico. One possible explanation is habitat loss in the highly developed Mid-Atlantic Region.

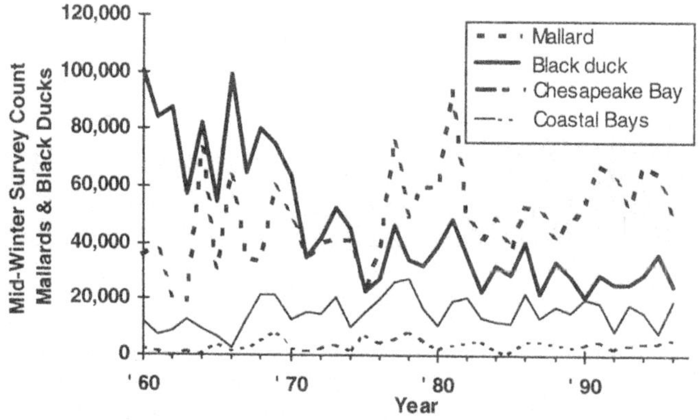

Fig. 7. Trends for mallards (dashed lines) and black ducks (solid lines) in Chesapeake Bay (heavy lines) and the Delmarva coastal bays (thin line) based on mid-winter survey counts, not actual population counts (USFWS, 1997).

5. State of Estuaries in Mid-Atlantic Region

In the prior section, we summarized our current understanding on how well estuaries as a whole are doing in the Mid-Atlantic Region by presenting information on individual measures or indicators. These individual measures tell us the condition of the Mid-Atlantic estuaries within the context of that particular indicator.

An attempt at putting together a "report card" on the condition of estuaries in the Mid-Atlantic Region is presented in Table 1. For the entire region and each of the major subregions, we assigned a shade representing the condition as evidenced by that particular indicator. Lighter shades represent better condition. These shades represent our best

Table 1. Summary of ecological conditions across the Mid-Atlantic estuaries. Shades represent best estimate of problems based upon information presented in USEPA (1997a): clear for minimal or no problem, grey for moderate problem, and dark for problem. Cross-hatching indicates inadequate information available. Where multiple shades are shown, best estimate is that condition ranges between the two categories.

	Mid-Atlantic Region	Chesapeake Bay		Delaware Bay		Coastal Bays		
		mainstem	tributaries	upper	lower	DE	MD	VA
water quality: nutrients								
water quality: phytoplankton								
water quality: dissolved oxygen								
sediment contamination								
habitat: coastal wetlands								
habitat: submerged aquatic vegetation								
living resources: benthos								
living resources: shellfish harvest (oyster)								
living resources: shellfish harvest (crab)								
living resources: shellfish closures								
living resources: fish stock								
living resources: contaminants in fish/shellfish								
living resources: disease (fish)								
living resources: disease (shellfish)								
living resources: waterfowl								

judgement summarization of the information. Where multiple shades are shown, our best estimate is that condition ranges between the two categories. Problem areas are determined by individual indicators. It should not inferred that problems are exclusively due to human activity.

The issues that are pervasive across the region include shellfish harvest for oysters and disease in shellfish. Chesapeake Bay, in addition to the regional issues, seems to be impacted by low dissolved oxygen, stemming in part from nutrient enrichment. Delaware Bay seems to be characterized by lack of submerged aquatic vegetation (which has been an historical problem), by water clarity, and lingering toxic contaminants associated with urbanization and industrialization of its major river. Delmarva coastal bays appear to have the least impact overall but are threatened by encroaching urbanization.

The mix of shades in Table 1 indicates that the Mid-Atlantic Region estuaries are being impacted to varying degrees. Therefore, they are at risk and in need of active management to restore and maintain environmental quality and sustainable resources. The states, in conjunction with the Chesapeake Bay Program and the National Estuary Programs, have instituted environmental management programs to address these concerns. We are seeing the results of these environmental programs, as evidenced by the generally improving ecological conditions in the Mid-Atlantic estuaries.

Acknowledgments

The work presented in this article was prepared with the collaboration of individuals from Delaware Department of Natural Resources and Environmental Control, Maryland Department of Natural Resources, Virginia Department of Environmental Quality, National Oceanic and Atmospheric Administration, U.S. Geological Survey, U.S. Fish and Wildlife Service, EPA Regions 2 and 3, AED Community-Based Assessment Team in Annapolis, EPA/ORD Gulf Ecology Division, EPA/ORD Western Ecology Division, Chesapeake Research Consortium Service, and EPA Office of Policy, Planning and Evaluation. Geographic information systems support was provided by Jane Copeland and George Morrison, OAO Corporation. The spatial displays were not prepared to meet EPA spatial locational guidelines; the displays were prepared from disparate data sets and represent a best attempt at approximating locations. Special thanks to Darryl Keith, Wayne Munns, Gerald Pesch, Steven Schimmel, and two anonymous reviewers for their critical reviews of this paper. The authors wish to recognize Tom DeMoss for his vision and persistence in ensuring that this project happened. This is contribution no. USEPA-NHEERL-NAR-1911 of the Atlantic Ecology Division.

References

Bohlen, C., and Boynton, W.: 1996, 'Maryland's Coastal Bays Status and Trends', Draft report prepared for Maryland Coastal Bays National Estuary Program.

Chaillou, J.C., Weisberg, S.B., Kutz, F.W., DeMoss, T.E., Mangiaracina, L., Magnien, R., Eskin, R., Maxted, J., Price, K., and Summers, J.K.: 1996, *Assessment of the Ecological Condition of the Delaware and Maryland Coastal Bays*, EPA/620/R-96/004, U.S. Environmental Protection Agency, Office of Research and Development, Washington, D.C.

Chesapeake Bay Program: 1995, *Trends in the Distribution, Abundance, and Habitat Quality of Submerged Aquatic Vegetation in Chesapeake Bay and its Tidal Tributaries: 1971-1991,* CBP/TRS 137/95, U.S. Environmental Protection Agency, Chesapeake Bay Program, Annapolis, MD.

Chesapeake Bay Program: 1997, *Chesapeake Bay Program Database.* U.S. Environmental Protection Agency, Chesapeake Bay Program, Annapolis, MD.

Cochran, J.K.: 1991, 'Forging a partnership between federal monitoring programs and the academic marine community', Summary of a workshop hosted by the Marine Sciences Research Center, State University of New York at Stony Brook, 29-30 April 1991, *Special Report 93 of the Marine Sciences Research Center*, State University of New York at Stony Brook.

Culliton, T.J., Warren, M.A., Goodspeed, T.R., Remer, D.G., Blackwell, C.M., and McDonough, III, J.J.: 1990, *50 Years of Population Change along the Nation's Coast. The Second Report of a Coastal Trends Series,* National Oceanic and Atmospheric Administration, National Ocean Service, Ocean Assessments Division, Strategic Assessment Branch, Rockville, MD.

Dove, L.E., and Nyman, R.M., Eds.: 1995, *Living Resources of the Delaware Estuary.* The Delaware Estuary Program.

Fegeas, R.C., Claire, R.W., Guptil, S.C., Anderson, K.E., and Hallam, C.A.: 1983, *Land use and land cover digital data,* Geological Survey Circular 895-E, U.S. Geological Survey, Washington, D.C.

Field, D.W., Reyer, A.J., Genovese, P.V., and Shearer, B.D.: 1991, *Coastal Wetlands of the United States. National Oceanic and Atmospheric Administration*, National Ocean Service, Rockville, MD.

Greer, J., and Terlizzi, D.: 1997, *Chemical Contamination of the Chesapeake Bay. A Synthesis of Research to Date and Future Research Directions. A Workshop Report,* Maryland Sea Grant, University of Maryland, College Park, MD.

Haskin Shellfish Laboratory: 1997, Unpublished data. Haskin Shellfish Laboratory, Rutgers University, Bivalve, NJ.

Holland, A.F., Ed..: 1990, *Near Coastal Program Plan for 1990: Estuaries,* EPA/600/4-90-033, U.S. Environmental Protection Agency, Environmental Research Laboratory, Office of Research and Development, Narragansett, RI.

Long, E.R., MacDonald, D.D., Smith, S.L., and Calder, F.D.: 1995, 'Incidence of adverse biological effects within ranges of chemical concentration in marine and estuarine sediments', *Environmental Management* 19(1), 81-97.

Long, E.R., and Morgan, L.G.: 1990, 'The potential for biological effects of sediment-sorbed contaminants tested in the National Status and Trends Program', NOAA Tech. Mem., NOS OMA 62, National Oceanic and Atmospheric Administration, Seattle, WA.

Lyles, C.H: 1967a, *Historical Catch Statistics (Middle Atlantic States),* U.S. Department of the Interior, Division of Economics, Branch of Fishery Statistics, Washington, DC.

Lyles, C.H.: 1967b, *Historical Catch Statistics (Chesapeake States),* U.S. Department of the Interior, Division of Economics, Branch of Fishery Statistics, Washington, DC.

Macauley, J.M., Summers, J.K., Heitmuller, P.T., Engle, V.D., and Adams, A.M.: 1996, *Statistical Summary - EMAP-Estuaries Louisianian Province 1993,* EPA/620/R-96/003. U.S. Environmental Protection Agency, Office of Research and Development, Environmental Research Laboratory, Gulf Breeze, FL.

MacDonald, D.D.: 1994, *Approach to the assessment of sediment quality in Florida coastal waters: Volume I - Development and evaluation of the sediment quality assessment guidelines*, Report prepared for Florida Department of Environmental Protection, Tallahassee, Florida.

Malek, J.: 1992, 'Apparent Effects Threshold Approach', in: *Sediment Classification Methods Compendium,* EPA 823-R-92-006, U.S. Environmental Protection Agency, Office of Water, Sediment Oversight Technical Committee, pp. 11-1 to 11-20.

Maxted, J.R., Weisberg, S.B., Chaillou, J.C., Eskin, R.A., and Kutz, F.W.: 1997, 'The ecological condition of dead-end canals of the Delaware and Maryland coastal bays', *Estuaries* 20(2), 319-327.

NOAA: 1991, *The 1990 National Shellfish Register of Classified Estuarine Waters,* National Oceanic and Atmospheric Administration, National Ocean Service, Ocean Assessments Division, Strategic Assessment Branch, Rockville, MD.

NOAA: 1994, *Assessment of Chemical Contaminants in the Chesapeake and Delaware Bays,* National Oceanic and Atmospheric Administration, National Ocean Service, National Status and Trends Program, Silver Spring, MD.

NOAA: 1997a, 'The 1995 National Shellfish Register of Classified Estuarine Waters', (prepublication data), Strategic Assessment Branch, National Ocean Service, National Oceanic and Atmospheric Administration, Silver Spring, MD.

NOAA: 1997b, *NMFS Database,* National Oceanic and Atmospheric Administration, National Marine Fisheries Service, Fishery Statistics Division, Silver Spring, MD.

O'Connor, T.P., and Beliaeff, B.: 1995, *Recent trends in coastal environmental quality: results from the Mussel Watch Project 1986 to 1993,* National Oceanic and Atmospheric Administration, National Status and Trends Program, Silver Spring, Maryland.

Paul, J.F., Gentile, J.H., Scott, K.J., Schimmel, S.C., Campbell, D.E., and Latimer, R.W.: 1997, *EMAP-Virginian Province Four-Year Assessment (1990-93),* Report in review, U.S. Environmental Protection Agency, Office of Research and Development, Narragansett, RI.

Phillips, K., Jamison, P., Malek, J., Ross, B., Krueger, C., Thornton, J., and Krull, J.: 1988, *Evaluation procedures technical appendix - Phase I (Central Puget Sound),* prepared for Puget Sound Dredged Disposal Analysis by the Evaluation Procedures Work Group, U.S. Army Corps of Engineers, Seattle, WA.

Queen, W.H., Copeland, B.J., and Schubel, J.R.: 1992, *Forging a partnership between federal monitoring programs and the academic marine community,* A report of the Board on Oceans and Atmosphere, National Association of State Universities and Land-Grant Colleges, Washington, DC.

Sauer, J.R., Schwartz, S., and Hoover, B.: 1996, 'The Christmas Bird Count Home Page, Version 95.1', Patuxent Wildlife Research Center, Laurel, MD. http://www.mbr.nbs.govbbs/cbc.html.

Sharp, J.H., Ed.: 1984, *The Delaware Estuary: Background for Estuarine Management and Development,* University of Delaware, Lewes, DE.

Strobel, C.J., Buffum, H.W., Benyi, S.J., Petrocelli, E.A., Reifsteck, D.R., and Keith, D.J.: 1995, *Statistical Summary: EMAP-Estuaries Virginian Province - 1990-1993,* EPA/620/R-94/026. U.S. Environmental Protection Agency, Office of Research and Development, National Health and Environmental Effects Research Laboratory, Atlantic Ecology Division, Narragansett, RI.

USEPA: 1991, *Summary, Environmental Monitoring and Assessment Program - Near Coastal, Virginian Province User Network Exchange,* A workshop held on 3-5 April 1991, Ocean City, Maryland.

USEPA: 1994, *Chesapeake Bay Basinwide Toxics Reduction Strategy Reevaluation Report,* CBP/TRS 117/9 ,U.S. Environmental Protection Agency, Chesapeake Bay Program, Annapolis, MD.

USEPA: 1995, *The State of the Chesapeake Bay, 1995,* U.S. Environmental Protection Agency, Chesapeake Bay Program, Annapolis, MD.

USEPA: 1997a, *Condition of the Mid-Atlantic Estuaries,* U.S. Environmental Protection Agency, Region 3, Philadelphia, PA. In review.

USEPA: 1997b, *Environmental Indicators: Measuring Our Progress,* Chesapeake Bay Program, Annapolis, MD. http://ww.epa.gov/r3chespk.

USFWS: 1997, *Mid-Winter Migratory Bird Survey,* United States Fish and Wildlife Service, Annapolis, MD.

Weisberg, S.B., Frithsen, J.B., Holland, A.F., Paul, J.F., Scott, K.J., Summers, J.K., Wilson, H.T., Heimbuch, D.G., Gerritsen, J., Schimmel, S.C., and Latimer, R.W.: 1993, *Virginian Province Demonstration Project Report, EMAP-Estuaries: 1990,* EPA/620/R-93/006. U.S. Environmental Protection Agency, Office of Research and Development, Washington, DC.

Weston, R.: 1993, *Characterization of the Inland Bays Estuary,* Report to the Delaware Inland Bays National Estuary Program, Delaware Department of Natural Resources and Environmental Control., Dover, DE.

A FRAMEWORK FOR A DELAWARE INLAND BAYS ENVIRONMENTAL CLASSIFICATION

KENT S. PRICE

University of Delaware, Lewes, DE 19958

Abstract. Since Delaware's coastal bays have been highly eutrophied for at least twenty years and Maryland's coastal bays are not nutrient stressed, dominance of the fish community in Delaware's coastal bays by *Fundulus* sp. may be an indicator of nutrient stress. Maryland's coastal bays are menhaden, spot, and anchovy dominated. The dominance of *Fundulus* sp. in a nutrient-stressed system relates to the hardy nature of these fishes, especially in low-oxygen conditions. Submerged aquatic vegetation as seagrasses (SAV) has been absent from the highly nutrient-stressed Delaware coastal bays for about twenty-five years. In contrast, SAV is still found in Maryland's coastal bays. The loss of SAV as a habitat for young fish may also be contributing to the apparent species shift in Delaware's coastal bays.

Indian River Bay is less hospitable to macroalgae (seaweeds) than Rehoboth Bay. Dominance of *Ulva* in Indian River Bay reflects its tolerance to varying salinities, higher nutrient levels, and increased turbidities, and indicates a stressed system. The total volume of macroalgae, especially in Rehoboth Bay, tends to follow the seasonal cycle for phosphorus.

Based on an assessment of the ecological condition of the Delaware and Maryland coastal bays conducted by EMAP in 1993 and other related studies, the author offers a conceptual framework for Delaware's inland bays environmental classification, considering the water quality parameters of turbidity, TSS, Chl_a, DIN, DIP, and O_2 as they relate to presence of SAV, seaweed abundance and diversity, benthic invertebrate diversity, and fish sensitivity to low oxygen.

1. Introduction

One challenge that estuarine scientists and pollution ecologists have is to develop a reliable predictor of environmental stressors in estuarine systems. Such predictors are often referred to as indicators. Although attempts have been made to relate individual stressors to individual species responses (Hinga *et al.*, 1995), few if any relations have been developed between stressors and multiple species responses in estuarine and coastal systems. This paper attempts to relate the stressors of plant nutrient enrichment (eutrophication) to multiple responses in a reasonably well studied coastal bay estuarine ecosystem. Based on an assessment of the ecological condition of the Delaware and Maryland coastal bays conducted by EMAP in 1993 and other related studies, the author offers a conceptual framework for Delaware's inland bays environmental classification considering the water quality parameters of turbidity, total suspended solids (TSS), Chlorophyll *a* concentration (Chl_a), dissolved inorganic nitrogen (DIN), dissolved inorganic phosphorus (DIP), and oxygen concentration (O_2) as they relate to the presence of submerged aquatic vegetation (SAV), seaweed abundance and diversity, benthic invertebrate diversity, and fish sensitivity to low oxygen.

Environmental Monitoring and Assessment **51:** 285–298, 1998.

THE SETTING IN DELAWARE

Delaware's inland bays consist of three interconnected water bodies—Rehoboth, Indian River, and Little Assawoman bays (Figure 1). The inland bays have a drainage area of about 300 square miles (777 square kilometers), a water surface area of thirty-two square miles (83 square kilometers), a marsh area of nine square miles, a mean-low-water volume of four billion cubic feet, and a freshwater discharge of 300 cubic feet per second. Almost thirty square miles of the inland bays are classified as shellfish waters, of which nineteen square miles presently are approved for shellfishing. There are about 126 people per square mile of the inland bay's watershed, and the land is about 10 percent urban, 44 percent forested, and 46 percent agriculture. The inland bays are tidally flushed, with estimates typically converging on 90-100 days for Indian River Bay and eighty days for Rehoboth Bay. No flushing estimates are available for Little Assawoman Bay (Weston, 1993).

The inland bays are suffering from plant nutrient enrichment (eutrophication) that causes unwanted phytoplankton blooms with resulting declines in light penetration and oxygen levels. These changes in environmental quality have led to eradication of submerged aquatic vegetation (sea grasses) and to declines in desirable finfish and shellfish. Major sources of these nutrients are land runoff from intensive agribusiness operations, intrusion of nutrient-contaminated groundwater from agricultural and domestic sources, and sewage treatment plant effluents.

Overall, the inland bays are highly nutrient enriched (eutrophic), especially in the tidal creeks. Characterization efforts in the Chesapeake Bay yielded a classification system for bay waters based upon total nitrogen and total phosphorous concentrations. Under that classification system, the inland bays' combination of ambient total nitrogen concentrations, generally in excess of one part per million (ppm), and total phosphorous concentrations, generally in the range of 0.1 to 0.2 ppm, would rank the inland bays among the most enriched of the thirty-two sub-estuarine systems of the Chesapeake Bay. Based upon the Chesapeake classification system, the middle and upper segments of the Indian River estuary are more enriched than any segment of the Chesapeake Bay. Significant increases in tidal flushing rates over the past twenty years may have mediated the progression of advancing eutrophic conditions, especially in the lower, higher salinity reaches of the system (Weston, 1993).

For Rehoboth Bay, agriculture is the principal source of nitrogen (33%), but point sources are the major source of phosphorus (57%), most of which originates from the Rehoboth wastewater treatment plant. For Indian River and Assawoman Bays, the principal source of both nitrogen (50%) and phosphorus (46%) is agriculture, through the application of inorganic fertilizers and manures (Weston, 1993). These practices, applied to the sandy, permeable soils of the watershed, have resulted in widespread contamination of the groundwater by nitrates (Andres, 1992).

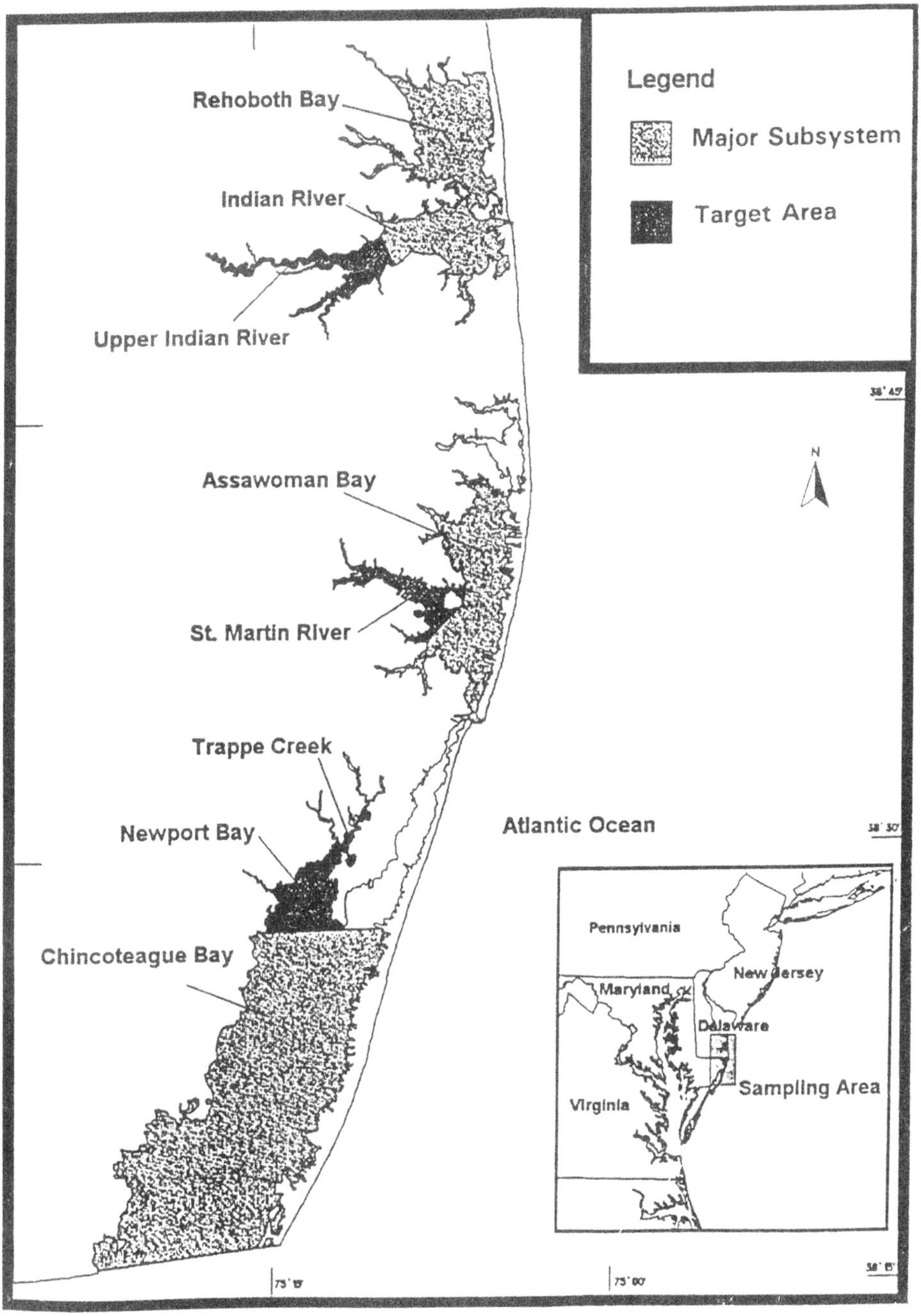

Fig. 1. Delaware and Maryland bays.

Groundwater is a highly significant component of freshwater flow into the bays. About 70 to 80 percent of total freshwater stream flow is composed of groundwater discharge. Groundwater also flows under the bay shores and discharges directly into the bays. Nearly all of this groundwater originates as precipitation in the inland bays watershed (Andres, 1992).

THE SETTING IN MARYLAND

Maryland's coastal bays (Figure 1) are contained within a single Maryland county and consist of six interconnected water bodies-St. Martin River and Assawoman, Isle of Wight, Sinepuxent, Newport, and Chincoteague Bays-as well as a number of smaller tributaries. Combined they have a total water surface area of 140.6 square miles. The watershed, however, is only about 205 square miles in size, primarily due to the proximity of the Pocomoke River to the west. The total length of the bays and watershed between the Virginia and Delaware lines is about 35 miles. The land is low, sandy, and generally poorly drained (EPA, 1996).

Nutrient inputs to the coastal bays are derived primarily from diffuse sources such as atmospheric deposition, agricultural sources, and urban runoff. Runoff accounts for an estimated 22% of the nitrogen and 34% of the phosphorous entering the coastal bays. Atmospheric deposition supplies another 32% nitrogen and 16% of phosphorous. Runoff associated with chicken and hog production facilities supplies about 32% of nitrogen and 32% of phosphorous. Only 4% of the nitrogen and 4% of the phosphorous entering the coastal bays comes from point sources. The remainder of the nutrients enter the coastal bays via the groundwater. Existing estimates of nutrient flows to the coastal bays, however, are preliminary (Bohlen and Boynton, 1997).

Absolute nutrient loading rates to the coastal bays as a whole are low or moderate in comparison to loading rates in other estuaries. For example, Bohlen and Boynton (1997) reported that the total nitrogen loading rate ($gNm^{-2}\ yr^{-1}$) is 2.4-3.1 (lower bays), 4.1-6.5 (upper bays), and 15.7-39.7 (tributaries) compared to 106.0 for the Delaware inland bays (Cerco, *et al.*, 1994). Therefore, Delaware inland bays have a total nitrogen loading rate that is an order of magnitude higher than the Maryland coastal bays. It is important to recall the great difference in watershed area and resulting nutrient impact on the two systems. The Delaware inland bays have a watershed to water ratio of 10 to 1, while the ratio for the Maryland bays is close to 1 to 1; which might go a long way in explaining the differences in nutrient loadings.

2. Materials and Methods

An assessment of the ecological condition of the Delaware and Maryland coastal bays was undertaken during 1993 by the Environmental Monitoring and Assessment Program (EMAP) (EPA, 1996). Two hundred sites were sampled in the summer of 1993 using a probability-based sampling design that was stratified to allow assessments of the coastal bays as a whole (Figure 1), each of the four major subsystems within coastal bays (Rehoboth Bay and Indian River Bay, DE and Assawoman Bay and Chincoteague Bay, MD) and four target areas of special interest to resource managers (upper Indian River, Delaware, St. Martin River and Trappe Creek, Maryland, and dead-end canals in both states). Measures of biological response, sediment contaminants, and eutrophication were collected at each site using the sampling methodologies and quality assurance/quality control procedures used by EPA's EMAP. An additional part of the study, trends in fish communities structure, were assessed by collecting monthly beach seine measurements during the summer at about 70 sites where historic measurements of fish communities have been made (EPA, 1996). In addition, Timmons and Price (1996) conducted a conventional study of the abundance and species composition of macroalgae for Rehoboth and Indian River Bays during 1992 and 1993 while Linder *et al.*, (1996) reported the distribution of submerged aquatic vegetation (SAV) in the Maryland coastal bays as mapped by Orth *et al.* (1992, 1993).

3. Results and Discussion

The results of the EMAP assessment for water quality parameters (SAV restoration criteria; Table 1) and the EMAP Benthic Index are summarized in Table II. Tables III and IV summarize regression analyzes conducted among parameters reported in Table II and EPA (1996), Timmons and Price (1996), Linder *et al.* (1996), and unpublished shore-zone fish data gathered by Linder and Price (1996).

TABLE I

Chesapeake Bay submerged aquatic vegetation habitat requirements for a polyhaline environment (Dennison *et al.*, 1993).

Parameter	Critical Value
Light attenuation coefficient (k_d; m^{-1})	1.5
Total suspended solid (mg/l)	15
Chlorophyll *a* (μg/l)	15
Dissolved inorganic nitrogen (μM)	10
Dissolved inorganic phosphorus (μM)	0.67

Table III indicates a significant positive correlation between the percent area with no SAV and the percent area not satisfying the chlorophyll *a* (Chl_a) and nutrient criteria for the restoration of SAV, but shows no significant relationship between SAV and Chl_a (percent area or concentration). Significant positive correlations were demonstrated among percent area not satisfying Chl_a and nutrient criteria and the percent area degraded according to the EMAP Benthic Index and percent of *Fundulus* sp. in the shore-zone fish population. Because of the very low percentage of *Fundulus* sp. in the Assawoman Bay survey, the inclusion of these data cause the relationship between percent area not satisfying Chl_a to be non-significant (Table III). However, the relationship is significant for the Chl_a criterion and concentration or for Chl_a and nutrient criteria with the Assawoman data omitted. Table IV demonstrates significant positive correlations among the percent area degraded according to the EMAP Benthic Index and the Chl_a concentration and percent *Fundulus* sp. in the shore-zone fish population. There is also an inverse correlation between the volume of macroalgae and the percent area degraded according to the EMAP Benthic Index for Indian River and Rehoboth Bays. No significant relationship was demonstrated among the percent area devoid of SAV and the percent area degraded according to the EMAP Benthic Index, although the inclusion of more data points could well provide a statistically significant relationship.

TABLE II

A summary of the results of the EMAP assessment of the ecological condition of the Delaware and Maryland coastal bays conducted in 1993 (EPA, 1996) augmented by fish and macroalgae studies by Timmons and Price (1996), Casey *et al.* (1996), and Price and Linder (1996).

Parameter	Entire Study Area	Major Subsystems				
		Rehoboth Bay	Indian River	Assawoman Bay	Chincoteague Bay	Upper Indian River
Percent area satisfying all 5 SAV criteria	22	10	5	0	45	0
Percent area not satisfying Chl_a and nutrient SAV criteria	38	62	76	65	16	92
Percent area exceeding SAV goal for Chl_a	23	37	52	41	7	93
Chl_a (μg/L)	12	13	21	16	6	35
Percent area devoid of SAV	96	100	100	99	96	100
Percent area degraded according to the EMAP Benthic Index	28	40	77	27	8	79
Percent *Fundulus* sp in shore-zone fish population	27	35	48	7	4	72
Mean volume of macroalgae (ml/m^2)		140	20			2

WATER QUALITY CONSIDERATIONS

The nutrient inputs to the Delaware inland bays affect the abundance and distribution of bay life. The microscopic floating plants (phytoplankton) are most prolific (as measured by chlorophyll concentrations) in the portions of the estuary closest to nutrient sources (e.g., in the upper and middle portions of Indian River Bay), while Rehoboth Bay generally represents an intermediate level of ambient nutrients and chlorophyll concentration, while the area nearest Indian River Inlet has the lowest concentrations of both. The same relationship is seen in the clarity (turbidity) of the water, with the upper portions of the tributaries having the most turbid water and the areas flushed near Indian River Inlet having the least turbid water. Turbidity also changes seasonally, with clarity of the water generally improving after Labor Day and lasting until about Memorial Day. The most turbid water in all three bays is seen during the summer season and probably results from a combination of biological effects (increased phytoplankton and microbial growth) and physical effects (boat traffic) (Ullman *et al.*, 1993).

TABLE III
Relationships among water quality parameters and living resource characteristics in Maryland/Delaware coastal bays.

	Percent Area Not Satisfying Chl_a and Nutrient Criteria		Percent Area Not Satisfying Chl_a Criteria		Mean Chl_a Concentration $\mu g/L$	
Percent area with no SAV	R^2 = 0.86 P = 0.008	(0.86) (0.023)	R^2 = 0.60 P = 0.072*	(0.61) (0.121)*	R^2 = 0.48 P = 0.126*	(0.50) (0.184)*
Percent area degraded according to EMAP Benthic Index	R^2 = 0.78 P = 0.019	(0.93) (0.008)	R^2 = 0.77 P = 0.022	(0.82) (0.034)	R^2 = 0.77 P = 0.021	(0.81) (0.038)
Percent *Fundulus* sp in shore-zone fish population	R^2 = 0.62 P = 0.062*	(0.95) (0.005)	R^2 = 0.72 P = 0.024	(0.97) (0.003)	R^2 = 0.77 P = 0.022	(0.94) (0.006)

Note: Data points involve: 1. aggregate data from entire area, 2. Chincoteague Bay, 3. Rehoboth Bay, 4. Assawoman Bay 5. Indian River Bay, and Upper Indian River Bay as reported in EPA 1996.

Figure in parenthesis represent the analysis without Assawoman Bay.

* = Not Significant

TABLE IV
Relationships among living resource characteristics in Maryland/Delaware coastal bays.

	Percent area degraded according to EMAP Benthic Index	
Chl_a ($\mu g/L$)	R^2 = 0.77 P = 0.021	(0.81) (0.038)
Volume of macroalgae**	R^2 = 0.99 P = 0.048	(0.99) (0.048)
Percent area with no SAV	R^2 = 0.63 P = 0.060*	(0.74) (0.062)*
Percent *Fundulus* sp in shore-zone fish population	R^2 = 0.86 P = 0.008	(0.89) (0.017)

Note: Data points involve: 1. aggregate data from entire area, 2. Chincoteague Bay, 3. Rehoboth Bay, 4. Assawoman Bay, 5. Indian River Bay, and 6. Upper Indian River Bay as reported in EPA 1996.

Figures in parenthesis represent the analysis without Assawoman Bay.

* = Not Significant

** = Indian River and Rehoboth Bay only

Secchi depths in upper Indian River now average about 50 cm year-round, but may be as low as 10 cm during summer months when extremely high chlorophyll concentrations (in excess of $100 \mu g L^{-1}$) occur in the mesohaline and tidal creek portions of the river (Ullman et al., 1993). Based upon the EPA Chesapeake Bay classification system, the middle and upper segments of Indian River estuary are more enriched than any segment of the Chesapeake Bay (Weston, 1993) and very likely any portion of the Maryland coastal bays (EPA, 1996 and Bohlen and Boynton, 1997). Key elements of water quality may be expressed as the five minimum water quality criteria necessary for the growth of SAV (Table I, Dennison *et al.*, 1993).

SUBMERGED AQUATIC VEGETATION

A major world-wide decline of seagrass beds occurred in the 1930s and affected the Chesapeake Bay and the Delmarva Peninsula (Delaware, Maryland, and Virginia). While many areas revived from the decline, the inland bays of Delaware never recovered. Eelgrass (*Zostera marina*) once present in the inland bays in the 1920s has been seen sporadically in small quantities, but has not been observed since 1970. Transplanting of seagrasses has been unsuccessful in Delaware, probably due to high levels of suspended chlorophyll, increased turbidity, and high levels of nutrients (Orth and Moore, 1988), as well as smothering by benthic attached algae (Timmons and Price, 1996).

The combination of excessive nutrient levels and high turbidity appears to eliminate the growth of submerged aquatic vegetation (SAV) such as eelgrass in the Delaware inland bays. This probably has significant ecological effects because SAV is desirable habitat for a variety of finfish and shellfish and is food for certain types of waterfowl, although the habitat function may be provided, to some extent, by attached benthic algae (seaweeds) (Timmons and Price, 1996). The seaweeds probably also play a role in sequestering excess nutrients during the summer, but we have evidence that extremely high levels of nutrients and turbidity have degrading effect on the seaweeds as well, especially in the upper portion of Indian River Bay (Timmons and Price, 1996).

SHORE-ZONE FISH

One way of attempting to examine trends in fish populations over time in the Delaware's inland coastal bays is to compare the composition for the earliest records in the area with current compositions. For White Creek, the earliest record (1957) and three representative studies conducted in 1968, 1973, and 1993, there seems to be a significant shift in the fish faunal dominance as shown in Table V.

During the past 36 years, it appears that dominance has shifted from juvenile menhaden, tidewater silversides, and bay anchovy to *Fundulus* sp. and sheepshead minnow. Basically, the general impression is that the Family Cyprinodontidae, which includes the killifish and sheepshead minnow, are becoming progressively more dominant with time, while menhaden, bay anchovy, and tidewater silversides are declining in dominance. Of these, the killifishes and silversides are year-round residents, while the anchovy and menhaden are warm-water migrants (Weston, 1993). Thornton (1975) reported that the killifish and sheepshead minnow have strong tolerances to low oxygen while menhaden and bay anchovy are quite sensitive to low oxygen. Based on the literature and his own research, Thornton (1975) constructed a classification of estuarine fish based on their sensitivity to low oxygen. Although *Anchoa mitchilli*, the bay anchovy, was not included in the original list by Thornton (1975), he updated the ranking to including the bay anchovy as shown in Table VI reported in Diaber *et al.* (1976).

TABLE V
Dominant fish species in Delaware's coastal bays.

Rank	1957	1968	1973	1993
1	Menhaden	Atlantic Silversides	Atlantic Silversides	Striped Killifish
2	Tidewater Silversides	Striped Killifish	Mummichog	Mummichog
3	Atlantic Silversides	Mummichog	Spot	Sheepshead Minnow
4	Mummichog	Winter Flounder	Tidewater Silversides	White Mullet
5	Bay Anchovy	Bay Anchovy	Striped Killifish	Spot
6	Rainwater Fish	Sheepshead Minnow	Striped Mullet	Atlantic Croaker
7	Silver Perch	Northern Puffer	Sheepshead Minnow	Striped Mullet
8	Striped Killifish	Rainwater Fish	Bay Anchovy	Atlantic Silversides
9	Sheepshead Minnow	Silver Perch	Banded Killifish	Winter Flounder
10	White Mullet	White Mullet	Top Minnow	Kingfish

Linder *et al.* (EPA, 1996) report on 20 years (1972-1993) of data collected from Maryland's coastal bays and tributaries, including Assawoman Bay, Isle of Wight Bay, Sinepuxent Bay, Newport Bay, St Martin's River, and Chincoteague Bay. The rankings of the dominant fish species in the Maryland coastal bays shore zone is shown in Table VII. The ranking of the top five dominants has essentially included the same five species for the past 20 years (EPA, 1996).

Orth and Heck (1980) found that the dominant fish species in Chesapeake Bay eelgrass meadows are *Leiostomus xanthurus* (1), *Syngnathus fuscus* (2), *Anchoa mitchelli* (3) *Bairdiella chrysoura* (4), and *Menidia menidia* (5). By contrast, *Fundulus heteroclitus* and *F. majalis* ranked 9th and 43rd in eelgrass meadows, respectively.

Table VI
Fish species ranked by low O_2 sensitivity.

	Scientific Name	Common Name
Most Sensitive	*Brevoortia tyrannus*	Atlantic Menhaden
↓	*Menidia Menidia*	Atlantic Silversides
	Anchoa mitchilli	Bay Anchovy
	Mugil cephalus	Striped Mullet
	Bairdiella chrysoura	Silver Perch
	Leiostomus xanthurus	Spot
	Cyprinodon variegatus	Sheepshead Minnow
	Fundulus heteroclitus	Mummichog
Least Sensitive	*Fundulus majalis*	Striped Killifish

Therefore, one can conclude that generally speaking the Maryland coastal bays are dominated primarily by Atlantic silversides, bay anchovy, Atlantic menhaden, and spot, not by *Fundulus majalis* and *Fundulus heteroclitus* which is the case in the Delaware coastal bays today. Indeed, if one compares the earliest available Delaware record for shore-zone fishes in Delaware Bay (1959) with the Maryland coastal bays fish fauna, they are strikingly similar. DeSylva *et al.* (1962) reported that the dominant shore-zone fish species for the Delaware Bay were *Menidia menidia* (53.0%), *Bairdiella chrysoura* (17.9%), *Anchoa mitchelli* (15.1%), *Brevoortia tyrannus* (2.3%), and *Fundulus majalis* (2.2%) for a total of 90.5 percent of the shore-zone fish community. Likewise, in 1957, the dominant species in White Creek, a tributary of Indian River Bay Delaware, were *Brevoortia tyrannus* (32.5%), *Menidia beryllina* (19.5%), *Menidia menidia* (18.2%), *Fundulus heteroclitus* (13.5%), and *Anchoa mitchelli* (5.9%) for a total of 89.6% of the shore-zone fish community (Table V; Pacheco and Grant, 1965). Therefore, if one goes back in history some 35 years, in Delaware's bays, the shore-zone fish community strongly resembled that of the less impacted Maryland coastal bays of today.

The fish community dominance in Delaware's coastal bays has shifted toward those species that are more tolerant to low-oxygen stress (Table VI) and which are also more tolerant to salinity and temperature extremes (Derickson and Price, 1973). There is also a strong possibility that *Fundulus* sp. and *Cyprinodon* sp. are more adaptable to eutrophication-mediated shifts in the food chain with its attendant increase in turbidity; i.e., under eutrophied conditions there would be a selective advantage for species that are omnivorous (Bigelow and Schroeder, 1953) and which do not feed primarily by sight. Grecay (1990) showed that weakfish juveniles (which are sight-feeding predators) were more successful at obtaining prey when light was not severely limited by turbidity. Vaas and Jordan (1991) also noticed a steady increase in *Fundulus* sp. in the Chesapeake Bay over the last 32 years, which they attributed to the effects of eutrophication. There might be some slight indication of an increase in *Fundulus* sp. in the Maryland coastal bays system as well, but it might be too early to judge if this is truly representing an impact of eutrophication (EPA, 1996).

TABLE VII
Dominant fish species in Maryland's coastal bays.

Rank	1972	1977	1987	1993
1	Menhaden	Menhaden	Atlantic Silversides	Atlantic Silversides
2	Atlantic Silversides	Atlantic Silversides	Bay Anchovy	Bay Anchovy
3	Spot	Spot	Striped Mullet	White Mullet
4	Mummichog	Bay Anchovy	Menhaden	Spot
5	Summer Flounder	Striped Mullet	Silver Perch	Silver Perch
6	Bluefish	Winter Flounder	Mummichog	Mummichog
7	Striped Killifish	Mummichog	Spot	Striped Killifish
8	Bay Anchovy	Summer Flounder	Striped Killifish	Rainwater Killifish
9	American Eel	Atlantic Needlefish	Atlantic Needlefish	Rough Silverside
10	Atlantic Needlefish	Striped Killifish	Summer Flounder	Menhaden

Table VIII shows the conceptual relationship of water quality and living resource variables in Maryland and Delaware coastal (inland) bays as derived from observations (Table II) and the regression analyses expressed in Tables III and IV. As eutrophication (as measured by chlorophyll and nutrient SAV restoration criteria) increases, phytoplankton-induced light extinction accelerates with accompanying declines in SAV (Table II and Dennison et al. 1993). Decline and eventual extinction of SAV are accompanied by increasing volumes of benthic attached algae and a decline in their species diversity with ultimately the eutrophication resistant sea lettuce, *Ulva lactuca*, surviving as a dominant. At the extreme, as in Upper Indian River Bay, where phytoplankton production is very high and light levels are low, even the resilient *Ulva lactuca* is scarce or absent (Timmons and Price, 1996). As eutroptrication increases, night-time oxygen levels go below 3.0 mg/ml (Schaffer, 1995) and probably decline close to zero under highly eutrophied conditions. Unfortunately, no recent validated diurnal studies of dissolved oxygen are available for the Maryland (Bohlen and Boynton, 1997) or the Delaware coastal bays (EPA, 1996). Increasing eutrophication has a debilitating effect on the benthic invertebrate community as measured by the EMAP benthic index (Table III) and shifts the shore-zone fish community from low-oxygen intolerant species (bay anchovy, menhaden,

TABLE VIII

A framework for Delaware inland bays environmental classification scheme.

% Area Chl_a & Nutrient Criteria Satisfied	Chlorophyll and Suspended Solids	Light Attenuation/ Turbidity	SAV Abundance % Coverage	Seaweed Abundance/ Diversity	Nighttime O_2 Levels	Benthic Invertebrate Diversity	Fish Sensitive to Low O_2		Equivalent Water Body
	Low	Low	High	Low/High Diversity	High	High	Most Sensitive		
84									South Eastern Chincogteague Bay
								Atlantic menhaden	
62								Atlantic silversides	Lower Indian River and Rehoboth Bays
35			▼ Low	▼ High/Medium Diversity	▼ Low			Bay anchovy	Assawoman Bay
								Striped mullet	
38			▼ None	▼ Low/Low Diversity				Silver perch	Upper Rehoboth Bay
								Spot	
24								Sheepshead minnow	Mid-Indian River Bay
				▼ None				Mummichog	Little Assawoman Bay
8								Striped killfish	Upper Indian River Bay
	▼ High	▼ High	▼ None	▼ None	▼ None	▼ Low	▼ Least Sensitive		

spot) to low-oxygen tolerant species (*Fundulus* sp. and *Cyprinodon* sp.) as seen in Tables II and V.

By using southeastern Chincoteague Bay as a target for nutrient reduction and as a restoration goal, one may infer that about a ten-fold reduction in the loadings of nitrogen and phosphorus would have to occur in the Delaware Inland Bays watershed in order to restore the ecology of these bays to that of Chincoteague Bay. Estimates using a hydrographic/water quality model (Cerco, C.F. et al., 1994) call for reductions of 10-15% total nitrogen and 60% total phosphorus from point sources and 30% total nitrogen and 70% total phosphorus from nonpoint sources in order to achieve satisfactory SAV chlorophyll and nutrient restoration criteria in Delaware's inland bays.

Acknowledgements

Special thanks go to Dr. Maryellen Timmons, Lexia Valdes, and Cecelia Linder, graduate students of the Graduate College of Marine Studies, University of Delaware, who assisted with field work and data analysis. I would also like to acknowledge the efforts of the Delaware/Maryland Coastal Bay Joint Assessment Steering Committee who designed the EMAP portion of this study under the leadership of Frederick W. Kutz of the U.S. Environmental Protection Agency, Region3 and, for the facilitation of the EMAP study by Janis Chaillou and Stephen Weisberg of Versar, Inc. Portions of the benthic algae and fish studies were sponsored through a grant from the Delaware Inland Bays National Estuary Program.

References

Andres, A.S.:1992. Delaware Geological Survey Open File Report #35, 36 pages.

Bigelow, H.B., and W.C. Schroeder.: 1953. *Fish Bull. Of Fish and Wildlife Ser.*, 53:577 pages.

Bolen, C and W. Boynton. 1997. Report to Maryland Coastal Bays Program. University of Maryland. 71 pages.

Cerco, C.F., B. Bunch, M.A. Cialone, and H. Wang.: 1994. U.S. Army Corps of Engineers, Technical Report EL-94-5. 246 pages.

Daiber, F.C., and multiple authors: 1976. *An Atlas of Delaware's Wetlands and Estuarine Resources.* Office of Coastal Zone Management, Technical Report 2. Dover, Delaware. 530 pages.

Dennison, W.C., R.J. Orth, K.A. Moore, J.C. Stevenson, V. Carter, S. Kollar, P.W. Bergstrom, and R. Batiuk.: 1993. *Bioscience* 43:86-94.

Derickson, W.K., and K.S. Price.: 1973. *Trans. Amer. Fish. Soc.* 102(3):552-562.

deSylva, D.P., F.A. Kalber, Jr., and C.N. Schuster.: 1962. University of Delaware, Marine Laboratory, Information Series, Publ. No. 5. 164 pages

Environmental Protection Agency.: 1996. EMAP.EPA/620/R-96/004. 78 pages.

Grecay, P.A.: 1990. Ph.D. Dissertation, University of Delaware. Newark, Delaware. 179 pages

Hinga, K.R., H. Jeon, and N.F. Lewis.: 1995. NOAA Coastal Ocean Program, DAS 4. 120 pages.

Linder, C.C., J. Casey, and S. Jordon.: 1996. A Report to Maryland Department of Natural Resources. 78 pages.

Linder, C.C. and K.S. Price.: 1996. Inland Bays Citizen Monitor. 5(2):3-6.

Orth, R.J., and K.L. Heck.: 1980. *Estuaries* 3(4): 278-288.

Orth, R.J. and K.A. Moore.: 1988. A report to Delaware Department of Natural Resources and Environmental Control. Dover, DE.

Orth, R.J., J.F. Nowak, G.F. Anderson, K.P. Kiley, and J.R. Whiting.: 1992. Final Report to U.S. Environmental Protection Agency Chesapeake Bay Program. Annapolis, Maryland. 268 pages.

Orth, R.J., J.F. Nowak, G.F. Anderson, and J.R. Whiting.: 1993. A Report for U.S. Environmental Protection Agency. Annapolis, Maryland.

Pacheco, A.L., and G.C. Grant: 1965. U.S. Fish and Wildlife Service. Spec. Sci. Rep. Fish. No. 504. 32 pages.

Schaffer, P.J.: 1995. Master's Thesis. College of Marine Studies, University of Delaware, Lewes, DE 101 pages.

Thornton, L.L.: 1975. Masters Thesis. University of Delaware. 82 pages.

Timmons, M. and K. Price.: 1996. *Bot. Mar.* 39: 231-238.

Ullman, W.J., R.J. Geider, S.A. Welch, L.M. Graziano, and B. Overman.: 1993. Report to DNREC and Delaware's Inland Bays Program. College of Marine Studies, University of Delaware. 43 pages.

Vaas, P.A. and S.J. Jordan.: 1991. In. J.A. Mihursky and A. Chaney (eds.). New Perspectives on the Chesapeake System: a Research and Management Partnership. Chesapeake Research Consortium., Inc. CRC Publ. No. 137. Solomons, Maryland.

Weston, Roy F., Inc.: 1993. Report to the Delaware Inland Bays National Estuary Program, DNREC., Dover,DE.

MARYLAND BIOLOGICAL STREAM SURVEY: A STATE AGENCY PROGRAM TO ASSESS THE IMPACT OF ANTHROPOGENIC STRESSES ON STREAM HABITAT QUALITY AND BIOTA

R. KLAUDA[1], P. KAZYAK[1], S. STRANKO[1], M. SOUTHERLAND[2], N. ROTH[2] and J. CHAILLOU[2]

[1]*Maryland Department of Natural Resources, 580 Taylor Ave., Annapolis, MD 21401, USA.* [2]*Versar, Inc., 9200 Rumsey Rd., Columbia, MD 21045, USA*

Abstract. The Maryland Department of Natural Resources is conducting the Maryland Biological Stream Survey, a probability-based sampling program, stratified by river basin and stream order, to assess water quality, physical habitat, and biological conditions in first through third order, non-tidal streams. These streams comprise about 90% of all lotic water miles in the state. About 300 sites (75 m segments) are being sampled during spring and summer each year. All basins in the state will be sampled over a three-year period, 1995-97. MBSS developments in 1995-96 included (1) an electrofishing capture efficiency correction method to improve the accuracy of fish population estimates, (2) two indices of biotic integrity (IBI) for fish assemblages to identify degraded streams, and (3) land use information for catchments upstream of sampled sites to investigate associations between stream condition and anthropogenic stresses. Based on fish IBI scores at 270 stream sites in six basins sampled in 1995, 11% of non-tidal stream miles in Maryland were classified as very poor, 15% as poor, 24% as fair, and 27% as good. IBIs have not yet been developed for stream sites with catchment areas less than 120 hectares (23% of non-tidal stream miles). IBI scores declined with stream acid neutralizing capacity (ANC) and pH, an association that was also evident for fish species richness, biomass, and density. Low IBI scores were associated with several measures of degraded stream habitat, but not with local riparian buffer width. There was a significant negative association between IBI scores and urban land use upstream of sampled sites in the only extensively urbanized basin assessed in 1995. Future plans for the MBSS include (1) identifying all benthic macroinvertebrate samples to genus, (2) developing benthic macroinvertebrate, herpetofaunal, and physical habitat indicators, and (3) enhancing the analysis of stream condition-stressor associations by refining landscape metrics and using multi-variate techniques.

1. Introduction

Because of its location to the east and downwind from the heavily-industrialized Ohio River valley, Maryland receives precipitation that is as acidic as any state east of the Mississippi River (Padmanabah and Olem, 1991). A spring 1987 water chemistry survey of 585 randomly-selected stream reaches with watershed areas $\leq$ 100 km^2 showed that almost a third of Maryland's headwater streams (about 4200 km) were either chronically acidic or vulnerable to episodic acidification (Knapp *et al.*, 1988). The Appalachian Plateau in western Maryland and the southern Coastal Plain of eastern Maryland (where 52% and 74% of the streams were acidic or acid-sensitive) were the two most sensitive regions. These survey results for Maryland are consistent with results from the U.S. Environmental Protection Agency's (U.S. EPA) National Stream Survey in Maryland also conducted in the mid-1980's (Herlihy *et al.*, 1991; Kaufman *et al.*, 1992).

Environmental Monitoring and Assessment **51: 299–316, 1998.**

Chronic and episodic acidification of surface waters from acidic deposition, with associated depressions of pH and increases in concentrations of inorganic aluminum, can have serious impacts on stream biota, especially fish (Baker and Christensen, 1991; Wigington, *et al.* 1996; Gerritsen *et al.*, 1996). The sensitivity of Maryland streams to acid deposition led to the hypothesis that biological resources are being adversely affected in headwater streams. Quantitative information on species composition and abundance of fish and other stream biota needed to test this hypothesis are spatially limited. Differences in sampling methods among monitoring programs also make comparisons among watersheds, basins, or regions very difficult.

To better understand how biological resources in Maryland streams are affected by acidic deposition, the Department of Natural Resources initiated the statewide, probability-based Maryland Biological Stream Survey (MBSS) as a pilot study in 1993 and a demonstration project in 1994. The objectives of the MBSS are to assess the current status of biological resources and acidification in non-tidal streams; quantify the extent to which acidic deposition has affected or may be affecting the biota; and examine other water quality, physical habitat, and land use factors that may also be affecting stream biota. MBSS results will also be used to prioritize watershed protection and restoration actions to restore degraded streams; establish benchmarks for long-term monitoring of water quality, physical habitat, and biological resource trends; contribute to an inventory of the state's aquatic biota; and document biodiversity "hot spots." The MBSS has two assessment factors (or endpoints): biological integrity and fishability. The purpose of this paper is to describe the MBSS sampling design and data collection methods, present some results from 1995 sampling, and discuss future directions for the survey.

2. Methods

2.1. Survey Design

The MBSS uses a probability-based or random design called lattice sampling or multi-stratification (Cochran, 1977; Jessen, 1978) to ensure that all first through third order (Strahler, 1957) non-tidal streams have a non-zero and known probability of being sampled. All river basins (Figure 1) will be sampled over a three-year period, 1995-97. Random sampling allows the estimation of unbiased summary statistics (e.g., means, proportions, variances) for the entire state, a particular basin, or for sub-populations of particular interest (e.g., all streams with pH less than 5).

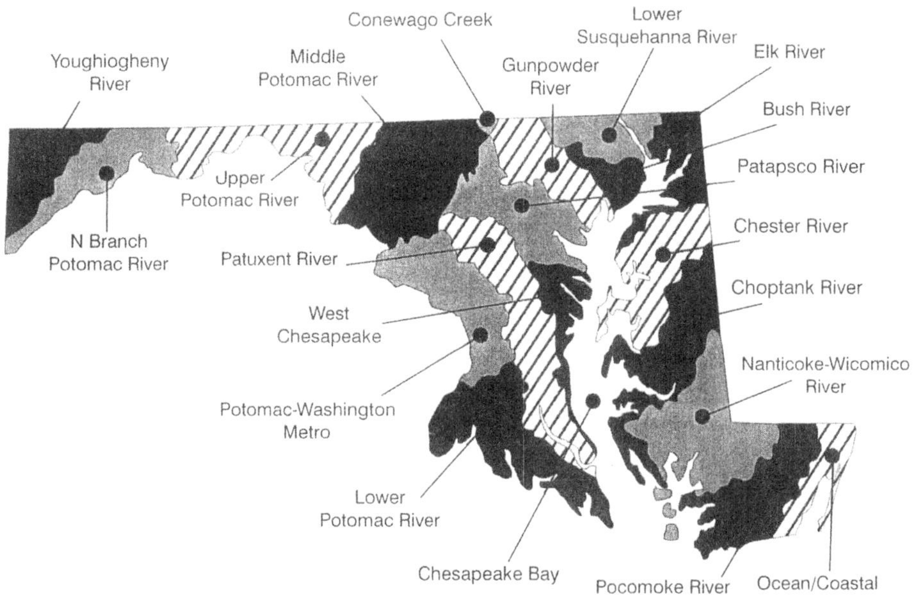

Fig. 1. Major drainage basins in Maryland.

For logistical reasons, the MBSS study area was divided into three regions (western, central, eastern) with five to seven basins each. Two basins are randomly selected (without replacement) from each region for sampling each year. Six basins were sampled in 1995, seven in 1996, and eight in 1997. One randomly-selected basin in each region will be sampled twice during the three-year survey period to quantify inter-annual variability.

The sampling frame for the MBSS was constructed by overlaying basin boundaries on a map of blue line stream reaches in the state as digitized on a U.S. Geological Survey 1:250,000 scale topographic map. This sample frame was also used in the 1987 Maryland synoptic stream chemistry survey (Knapp *et al.* 1988). Sampling in the MBSS is restricted to non-tidal, third order and smaller streams, excluding impoundments that are non-wadeable or substantially alter the riverine nature of the stream. Stream reaches are divided into non-overlapping 75-m segments (sites), the elementary sampling units from which biological, water chemistry, and physical habitat data are collected.

The MBSS sampling design used in 1995-97 is based on a stratified random selection of stream sites located within each basin. An approximately equal number of sites are selected from each of the three stream orders within each basin. This method weights third order streams most heavily (to better assess fishability) because the number of third order stream miles is smallest, with progressively more second and first order streams (Table I). Within each basin, the total number of sites selected from each stream order is roughly proportional to the total number of stream miles.

TABLE I
Kilometer of first, second and third order, non-tidal streams in Maryland based on a 1:250,000 scale U.S. Geological Survey topographic map

Region	Drainage Basin	First Order	Second Order	Third Order	Total
West	Youghiogheny	390	139	69	598
	North Branch Potomac	619	208	123	950
	Upper Potomac	742	259	69	1070
	Middle Potomac	1187	368	208	1763
	Potomac-Washington Metro	786	192	125	1103
Central	Patuxent	1117	251	85	1453
	Lower Potomac	805	160	77	1042
	West Chesapeake	288	46	18	352
	Patapsco	677	214	96	987
	Gunpowder	557	120	69	746
	Bush	210	50	38	298
East	Lower Susquehanna	333	67	40	440
	Elk	261	61	18	340
	Chester	347	102	16	465
	Choptank	334	51	26	411
	Nanticoke-Wicomico	309	46	10	365
	Pocomoke	350	61	22	433
	Total	9312	2395	1109	12816

To maximize the number of randomly-selected sites that can be accessed and sampled, several hundred landowners are contacted by letter and telephone each year to obtain permission to cross privately-owned lands. Mean success rate for obtaining landowner permission to sample in 1993-96 was 92.0% (range 75-100%). In 1995 and 1996, 284 and 357 stream sites were sampled. A total of 341 sites were sampled in 1997. Each stream site is visited twice during the sampling year, for the reasons described below.

2.2. Data Collection

The MBSS is conducted during two index periods. Benthic macroinvertebrate and water chemistry sampling is conducted during spring baseflow (March-May) when benthos are reliable indicators of environmental stress (Plafkin *et al.*, 1989). Benthos are collected with a "D" net from all available habitats including riffles, root wads, woody debris, leaf packs, macrophytes, and undercut banks. Benthos are preserved in 70% ethanol and identified to family and eventually to genus in the laboratory. Water samples are analyzed for a suite of chemical parameters associated with acidic deposition: closed pH, specific conductance,

ANC, dissolved organic carbon, sulfate, and nitrate using methods described in U.S. EPA (1987).

During summer (June-September), *in situ* measurements of dissolved oxygen, pH, temperature, and conductivity are measured at each site. Fish, herpetofauna, aquatic macrophyte, and freshwater mussel sampling, along with physical habitat evaluations, are also conducted during the low flow summer period. This period was selected to avoid the effects of spring and fall spawning movements on fish assemblages and to maximize electrofishing catch efficiencies.

Fish are sampled with double-pass electrofishing using direct current backpack units. Block nets are placed at each end of the 75 m segments and all available habitats are thoroughly sampled. For each pass, all captured fish are identified to species and counted. Up to 100 individuals of each species are examined for external anomalies. Up to 50 individuals of each game fish species (trouts, black basses, walleye, pikes, striped bass) are measured for total length; all game fish are weighed in aggregate. All non-game species are also weighed in aggregate but not measured for total length. Voucher specimens of each species collected in each basin are preserved and retained at the Smithsonian Institution (Washington, DC) and at Frostburg State University (Frostburg, MD). All remaining fish are released.

Amphibians and reptiles observed during electrofishing and a 15-minute search of the riparian zone within 5 m of the stream channel on both sides of the segment are recorded as present. Voucher specimens are retained. Each site is also searched for the presence of freshwater mussels and aquatic vegetation.

Physical habitat assessments are conducted to evaluate habitat effects on biological integrity and fishability. MBSS habitat assessment procedures were derived from two methods: EPA's Rapid Bioassessment Protocols (Plafkin *et al.* 1989), as modified by Barbour and Stribling (1991), and Ohio EPA's Qualitative Habitat Evaluation Index (Ohio EPA, 1987). Several parameters (instream habitat, epifaunal substrate, velocity/depth diversity, pool/glide/eddy quality, riffle quality, channel alteration, bank stability, embeddedness, channel flow status, shading) are scored based on visual observations. Minimum riparian buffer width (up to 50 m from the stream) and vegetation type in the buffer are recorded. Scores for aesthetic value and site remoteness are also recorded. Beaver ponds, point sources of pollution, quantity of root wads and other woody debris, and local land usage visible from the site are noted. Quantitative measurements at each site include maximum depth, stream gradient, wetted width, sinuosity, over bank flood height, a velocity/depth profile, and discharge.

2.3. Quality Assurance

A Quality Assurance Officer experienced in all aspects of the MBSS administers the quality asssurance program. Specific activities include written sampling protocols, annual field crew training and proficiency examinations, standard forms for recording field data, double entry of all data, and field audits to assess field crew performance. For water sample analyses, routine daily quality control checks (5%) are performed that include duplicate, blank, and calibration samples processed according to U.S. EPA (1987) guidelines for each analyte. Quality control checks for benthos include duplicate samples collected at 5% of the sites and laboratory reprocessing of 5% of the samples. An independent assessment of habitat quality is performed at 5-10% of the sites by the Quality Assurance Officer.

2.4. Statistical Methods

The stratified random sample of stream sites collected by the MBSS allows for stream order and basin-specific estimates of chemical, physical habitat, and biological parameters of interest. Statewide estimates will also be available after the first round of the MBSS is completed in fall 1997. Because samples are independent and identically distributed within strata, the survey design also allows for regression and correlation analyses.

Observations at individual sites are used to estimate totals, means, medians, proportions, and percentiles. The mean for all sites in a basin (across stream orders) can be estimated as a weighted mean of the sample values. The proportion of all stream miles in a basin that falls into a given category (e.g., percentage of stream miles in the Upper Potomac basin with ANC less than 0 ueq/L) can be estimated by introducing an indicator variable that takes the value "1" if the observation falls into the specified category, and "0" otherwise. Within a particular stream order in a basin, various means (e.g., mean number of blacknose dace per segment) can be estimated as the ordinary mean derived from the simple random sample of sites in that basin. If 100% capture efficiency is assumed for electrofishing, total abundance and biomass, by individual species or for all species, can be obtained by extrapolating the mean numbers of fish collected per site (density) or biomass to total stream miles, by stream order or for the entire basin. In section 2.5, the method used to adjust these abundance and biomass estimates for actual capture efficiency achieved by double-pass electrofishing (with block nets) is described.

To estimate several biological characteristics of game fish populations in a basin (e.g., the size composition of brook trout in the Youghiogheny River basin), the proportion of fish falling into various size categories can be estimated. The sampling unit for the electrofishing effort in the MBSS is the individual stream site, not individual fish (Pennington and Volstad, 1994). Therefore, a combined ratio estimator can be used to estimate the proportion of fish within a specific size group (Cochran, 1977). This same method can be used to estimate the proportion of fish with a specific type of external anomaly. The variance of each proportion can be estimated by jack knifing (Saerndal *et al.*, 1992).

2.5. Capture Efficiency Adjustment for Fish Population Estimates

Estimates of fish density, abundance, and biomass are corrected for capture efficiency using a new analytical method developed with 1995 MBSS data (Heimbuch *et al.*, 1997). The method uses electrofishing catch date to estimate actual density and abundance based on the rate of decline in catch per unit effort between the first and second collection passes. It is difficult to estimate capture efficiency with only two passes at a single stream site because equal or even greater numbers of fish are frequently collected during the second pass compared to the first pass. To address this problem, samples were pooled over several stream sites within the same order and basin. Using a modified Seber-LeCren estimator (Seber and LeCren, 1967), the methods described in Heimbuch *et al.* (1997) analytically corrected for the bias introduced by a variable probability of capture and minimized the bias typically resulting from small sample sizes. An average capture probability of 59.9% was estimated for 26 stream fish species collected in 1995, with a range of 28.3% for largemouth bass (*Micropterus salmoides*) to 82.7% for rainbow trout (*Oncorhynkus mykiss*).

2.6. Development of an Index of Biotic Integrity for Fish

A provisional index of biotic integrity, or IBI (Karr *et al.*, 1986), was developed for stream fish assemblages sampled by the MBSS in 1994 and 1995. This multi-metric ecological indicator will be used in the biological assessment of Maryland streams and will be validated with 1996 and 1997 data. The IBI compares the condition of fish assemblages at sampled stream sites to regional reference sites sampled by the MBSS in 1994-95 that represent minimally-disturbed habitat conditions. Individual metrics that quantitatively describe attributes of the fish assemblages are scored and combined into a single index, the IBI.

Procedures used to develop the fish IBI for Maryland streams are fully described in Roth *et al.* (1997 a,b). Briefly, expectations for minimally degraded streams (or reference sites) were first established. The large 1994 and 1995 MBSS data sets (419 sites) were then used to identify reference and degraded sites based on physical habitat, water quality, and land use. After establishing appropriate geographic strata, the ability of candidate metrics to distinguish between reference and degraded stream sites was evaluated. Scoring thresholds were established and adjustments made for catchment area. Finally, various combinations of metrics were used to formulate an eight-metric IBI for each of two geographic regions in Maryland: coastal and non-coastal plain (Table II). Because of relatively low species richness and total abundance of fishes in very small headwater streams, IBIs have not yet been developed for MBSS sites with catchment areas less than 300 acres (120 hectares).

2.7. LANDSCAPE ANALYSIS

Land use and cover within the catchments upstream of the MBSS sites are derived with a geographic information system (GIS) that uses the Map Information Processing System (MIPS) and PC Arc Info software. Catchments are digitized using topographic lines from digital county topographic maps (1:62,500 scale). The catchment file is intersected with 1990 digital land use and land cover maps obtained from the Maryland Office of Planning (Fisher, 1991). The land use file (1:63,360 scale) has a one acre grid cell accuracy and identifies land uses according to the Anderson Level II Land Use Classification System. To simplify the analyses, land uses are collapsed to Anderson Level I classifications: urban, agriculture, forest, water, wetland, and barren. Total area and percentage of the area in each Level I land use are calculated for each MBSS site catchment.

TABLE II
Eight metric index of biotic integrity (IBI) developed for stream fish assemblages in Maryland with 1994 and 1995 MBSS data

Coastal Plain Streams		**Non-Coastal Plain Streams**
	Number of Native Fish Species	
	Number of Benthic Species	
	Percent Tolerant Species	
	Percent Generalists, Omnivores and Invertirores	
	Percent Abundance of Dominant Species	
	Percent Lithophilic Spawners	
	Total Abundance per Square Meter	
+		+
Total Biomass per Square Meter		Percent Insectivores

3. Results and Discussion

Examples of results obtained from 1995 MBSS sampling in six basins are presented below. Additional results were reported in Volstad *et al.* (1995, 1996). A more detailed discussion of fish IBIs is presented in Roth *et al.* (1997 a,b).

3.1. CHARACTERIZATION OF BIOLOGICAL RESOURCES

FISH

A total of 67 fish species was collected in the 1995 MBSS. More species were collected in eastern and central Maryland streams (average of 6.1-7.1 species/site) than in the two most western basins (averages of 2.9 and 3.7 species/site). As expected, species richness increased with stream order: a mean of 3.9 species per site for first order streams, 7.9 species for second order sites, and 10.9 species for third order sites. Over all six basins, the most abundant species were the mottled sculpin (*Cottus bairdi*) at 1663/stream mile and blacknose dace (*Rhinichthys atratulus*) at 1501/stream mile. The most abundant game fish was the largemouth bass at 43/stream mile. Brook trout was the most abundant catchable-size game fish across the six basins (14,639), followed by brown trout (11,470). Fish densities (all species) were about equal in second and third order streams (15,400 and 15,692/stream mile), but much lower in first order streams (5,379/stream mile). Across all basins, mean fish biomass was 50 kg/stream mile, ranging from 35 kg/stream mile in the Lower Potomac basin to 68 kg/stream mile in the Patapsco basin.

External anomalies were infrequently observed (range of 1.2% in the Nanticoke-Wicomico basin to 20.3% in the Upper Potomac basin), but tended to be more common in larger streams. Excluding black spot, other parasites, ich, and injuries, the frequency of pathological anomalies was even lower (range of 0.6-1.2%). Most pathological anomalies were skin lesions which comprised 62.5% of all observed anomalies in game fish and 81.5% in non-game fish.

BENTHIC MACROINVERTEBRATES

Ninety-seven benthic macroinvertebrate taxa (family level) were collected (genus level information is not yet available). Eight taxa were very common and occurred at more that 50% of the sites: Chironomidae (98%), Simuliidae (73%), Tipulidae (61%), Hydropsychidae (70%), Ephemerellidae (64%), Heptageniidae (56%), Nemouridae (56%), and Elmidae (56%). In contrast, 47 taxa were collected at less than 5% of the sites. Mean taxa richness for all sites was 12.7, with little variation among basins (range 10.9-16.7) and stream orders (range 12.5-13.6). Identification of benthos below the family level might reveal greater spatial differences, if any differences exist.

HERPETOFAUNA

Thirty-nine species of reptiles and amphibians were observed. Average species richness across all six basins varied from 1.9/site in the Nanticoke-Wicomico basin to 4.1/site in the Lower Potomac basin. Frogs, toads, and salamanders were most commonly observed and occurred at more than 50% of the sites. Turtles, snakes, and lizards were found at only 16%, 14%, and 3% of the sites. No strong pattern of herpetofauna species richness among stream orders was observed, except that salamanders were more common at first order stream sites (63%) than at larger sites (47%). The species richness of salamanders in first order streams suggests they may be useful indicators of biological integrity in small headwater streams with few or no fish, at least in western Maryland. Few salamanders were observed at eastern shore sites, presumably because of habitat differences in and around the relatively low-gradient, predominantly sand and silt substrate streams in this topographically less diverse region of the state.

FRESHWATER MUSSELS

Six species of unionid mussels were collected in 1995, as represented by live specimens, shells of recently dead individuals, and relic shell: the alewife floater (*Anodonta implicata*), Carolina lance (*Elliptio angustata*), eastern elliptio (*E. complanata*), eastern floater (*Pyganodon cataracta*), northern lance (*E. fisheriana*), and yellow lance (*E. lanceolata*). Overall, these species were collected at 15% of the sites, but in only three of the six basins sampled: Chester, Lower Potomac, and Nanticoke-Wicomico. The most commonly occurring species was the eastern elliptio, found at 14% of the sites. Mussels were collected at 1%, 15%, and 28% of the first, second, and third order stream sites. In Maryland, 20 species of unionid mussels have been reported, including three that are listed as state endangered and one (dwarf wedge mussel, *Alasmidonta heterodon*) that is listed as state and federally endangered. Three freshwater mussel species have been introduced to Maryland: plain pocketbook (*Lampsilis cardiuim*), pocketbook (*L. ovata*), and Asian clam (*Corbicula fluminae*).

AQUATIC MACROPHYTES

Eighteen species of aquatic macrophytes were observed, including ten species of emergents, five submerged species, two species that can be either submerged or floating, and one floating species. The greatest diversity of macrophytes was observed in low gradient, coastal plain streams of the two eastern shore basins (averages of 0.8 and 1.2 species/site) and the Lower Potomac basin on the lower western shore (average of 0.6 species/site). By comparison, aquatic macrophyte diversity was low in central and western Maryland streams (average of 0.1 species/site).

3.2. ASSESSMENT OF BIOLOGICAL CONDITIONS

FISH IBI

The IBI used to assess the condition of headwater stream sites sampled by the MBSS in 1995 compared the fish assemblage at each site to a reference condition representative of streams sampled in 1994-95 that appear to be minimally impacted by anthropogenic disturbance (Karr, 1991). Site scores for the fish IBI were calculated as the mean of eight metric scores and could range from 1.0 (worst) to 5.0 (best). In general, an IBI score of 3.0 or greater is comparable to reference conditions. Fish IBI scores were defined as follows: 4.0-5.0 (good), 3.0-3.9 (fair), 2.0-2.9 (poor), and 1.0-1.9 (very poor). For more details on the fish IBI, see Roth *et al.* (1997 a,b).

Over all six basins, 27% of the stream sites were classified as good, based on fish IBI scores; 24% were fair, 15% were poor, and 11% were very poor. IBIs have not yet been developed for stream sites with catchment areas less than 120 hectares (23% of the sites in 1995). The Youghiogheny basin had the highest percentage of good and fair stream miles (67). The Nanticoke-Wicomico basin had the highest percentage of poor and very poor stream miles (44). First order streams had a lower percentage of stream miles in the good and fair categories (43) and a higher percentage of poor and very poor stream miles (28), compared to second and third order streams. These results reflect stream order differences for sites where no fish were collected (17%, 4%, and 2% for first, second, and third order streams). Those sites where no fish were collected received the lowest possible IBI score (1.0).

BENTHIC MACROINVERTEBRATE INDICATORS
Benthic macroinvertebrate assemblages are commonly used to evaluate the condition of freshwater streams (Southerland and Stribling 1995). Because a multi-metric benthic indicator similar to the fish IBI has not yet been developed for Maryland streams, two commonly used metrics of benthic community condition were used: Ephemeroptera, Plecoptera, and Trichoptera (EPT) taxa richness (Plafkin *et al.,* 1989) and a modified Hilsenhoff biotic index (Hilsenhoff, 1987; 1988). The biotic index incorporated tolerance values for Maryland benthos derived from research results from the mid-western United States (Hilsenhoff, 1987), New York (Bode 1988), and North Carolina (Lenat, 1993). This index has not been calibrated specifically for Maryland stream benthos. These two benthic community measures were applied to the family-level data collected by the 1995 MBSS. Genus-level data will be available in 1988 and used for future analyses.

Across all six basins, EPT taxa richness ranged from 0 to 14 taxa/site. Based on the modified Hilsenhoff biotic index, 36% of the stream miles in these basins were categorized as being in good condition, 55% fair, and 9% poor. Based on the biotic index, fair and poor stream conditions were least common in the Youghiogheny basin (46% of the stream miles) but widespread in the Chester basin (99%). The Nanticoke-Wicomico basin had the highest percentage of stream miles rated as poor (26%). Mean number of EPT taxa was highest in the Youghiogheny basin (8.6/site) and lowest in the two eastern shore basins—Chester (3.6/site) and Nanticoke-Wicomico (3.9/site). EPT taxa richness did not vary greatly among stream orders.

3.3. ANTHROPOGENIC EFFECTS ON BIOLOGICAL RESOURCES

This section explores associations, based on graphical analyses, between the biological conditions at 1995 MBSS sites and three factors known to adversely affect stream biota: acidification, physical habitat degradation, and land use.

ACIDIFICATION
One of the primary objectives of the MBSS is to assess the effects of acidic deposition on Maryland streams. MBSS data can be used to quantify the extent of stream acidification in the mid to late 1990's and compare these results to the 1987 survey. During spring 1995, 2% of the stream miles in the six basins had pH values less than 5; 10% had pH between 5 and 6. Low spring pH values were most common in the Lower Potomac basin where 5% of the stream miles had pH less than 5; 39% had pH between 5 and 6, due primarily to acidic deposition. Small streams were most susceptible to acidification. Across all six sampled basins, 84% and 89% of the stream miles with spring pH below 5 or between 5 and 6 were first order streams.

Although pH is a commonly used measure of stream acidity, ANC is a better overall measure of acid sensitivity. Across all basins, only about 1% of the stream miles were acidic (ANC < 0 ueq/L) in spring 1995, only 7% were highly sensitive (ANC 0-50), but 32% were fairly sensitive (ANC 50-200), especially to episodic acidification. In four of the six basins, 40% of the stream miles had ANC below 200, a threshold that has been used to identify acid-sensitive waters (Knapp *et al., 1988*). The Lower Potomac basin had the highest percentage of stream miles with ANCs below 200 (78), reflecting the relatively low buffering capabilities of surface waters in this coastal plain basin. ANC did not vary significantly among stream orders. Percentages of stream miles with ANC below 200 were 40% for first order, 31% for second, and 45% for third order streams.

Although the MBSS will not complete the entire state until fall 1997, spatial trends in low ANC streams during spring 1995 were consistent with results from the 1987 synoptic stream chemistry survey conducted from early March through early May (Table III). The most acid sensitive regions of Maryland in both years were in the Appalachian Plateau, Blue Ridge, and Southern Coastal Plain provinces. Furthermore, there is no clear evidence that the percentages of acidic and acid-sensitive stream miles in these regions changed between 1987 and 1995. These comparisons of water quality trends will be repeated after the 1997 MBSS data are collected.

Among all MBSS sites sampled in 1995, very few fish were captured where spring pH was less than 5; no fish were captured at sites with summer pH less than 5. Fish species richness, density, biomass, and IBI scores decreased at low pH and ANC stream sites. Fish species that exhibited the greatest declines at sites with ANCs below 200 were mottled sculpin, blacknose dace, rosyside dace (*Clinostomus elongatus*), white sucker (*Catostomus commersoni*), and brown bullhead (*Ameiurus nebulosus*).

The mean number of EPT taxa (benthos) was lowest at sites with ANC less than 50. The modified Hilsenhoff biotic index, which increases as the number of pollution-tolerant taxa increases, was highest at sites with ANC less than 0, but did not show any clear differences among the other ANC classes. The occurrence of freshwater musssels decreased in low ANC streams, but herpetofauna did not exhibit any diversity or abundance trends with ANC.

PHYSICAL HABITAT DEGRADATION

A number of physical habitat characteristics were qualitatively assessed within the channel at each stream site. The instream habitat parameter represents the amount of stable habitat such as cobble, boulders, logs, undercut banks, root wads, aquatic plants, and other structures that provide cover for fish. Over all six basins sampled in 1995, 48% of the stream miles had either poor or marginal instream habitat; 18% of the stream miles were rated in the highest category, optimal. Poor and marginal instream habitat scores were most common in the Upper Potomac, Lower Potomac, Chester, and Nanticoke-Wicomico basins (59-73% of the stream miles). The Youghiogheny basin had the highest percentage of stream miles with optimal instream habitat (47).

Physical habitat scores varied with stream order for many metrics. First order sites generally received lower scores for instream habitat, epifaunal substrate, velocity/depth

TABLE III

Percentage of stream miles in Maryland that were acidic and acid-sensitive in 1987 and 1995 based on acid neutralizing capacity

1987						
	Physiographic Region					
ANC (μeq/L)	Appalachian Plateau	Valley and Ridge	Blue Ridge	Piedmont	Northern Coastal Plain	Southern Coastal Plain
< 0	10.7	0	0	0	2.1	7.6
< 50	15.7	0	58.0	0.9	4.7	29.3
<200	53.3	1.5	26.0	8.9	28.3	74.4

1995						
	Basins					
	Youghiogheny	Upper Potomac	Lower Potomac	Patapsco	Chester	Nanticoke-Wicomico
	Physiographic Regions Represented					
ANC (μeq/L)	Appalachian Plateau	Appalachian Plateau, Valley & Ridge	Southern Coastal Plain	Piedmont, Northern Coastal Plain	Northern Coastal Plain	Southern Coastal Plain
< 0	0	0.1	4.7	0	0	0
< 50	3.3	6.1	20.2	0	5.0	7.9
<200	45.8	45.9	78.1	0	14.9	41.7

diversity, and pool/glide/eddy quality, compared to second and third order sites. These results are consistent with the river continuum theory (Vannote *et al., 1980)* and suggest that physical habitat ratings should be adjusted to account for different habitat expectations for the smallest headwater streams. Over all six basins, 66% of the stream miles had forested buffers, 11% had other kinds of vegetated buffers (old field, tall grass, lawn), and 21% had no vegetated buffers. Forty-six percent of the stream miles had at least a 50 m wide vegetated buffer on both sides of the stream channel.

Thirteen physical habitat metrics were graphically compared with fish IBI scores to determine which measures were most strongly associated with biotic integrity in headwater streams. Instream habitat showed the strongest positive association with IBI scores in all six basins. IBI scores showed fairly strong positive associations with pool/glide/eddy quality and velocity/depth diversity, and slight positive trends with epifaunal substrate, riffle quality, channel alteration, and aesthetic rating. IBI scores did not show any clear associations with maximum depth, vegetated riparian width at the site, bank stability, amount of woody debris, discharge/catchment area, and site remoteness. A combination of seven physical habitat parameters that were qualitatively assessed within the stream channel (instream habitat, epifaunal substrate, velocity/depth diversity, pool/glide/eddy quality, riffle quality, channel alteration, and bank stability) and measured habitat features of direct importance to stream biota showed a strong positive association with fish IBI scores. The median number of EPT taxa (benthos) increased slightly with increasing epifaunal substrate scores.

LAND USE IMPACTS

Rivers and streams are by nature hierarchical systems, so their chemical, physical, and biological characters are controlled to some degree by the larger-scale systems to which they belong (Frissell et al., 1986). To fully understand multiple, cumulative impacts on Maryland stream systems, conditions at a broad landscape scale, as well as the local or site-specific scale, must be assessed. One measure of anthropogenic influence at the landscape scale is watershed land use. The associations between land use and three biological indicators of stream condition (fish IBI, number of EPT taxa, modified Hilsenhoff biotic index) were investigated using 1995 MBSS data. Because the MBSS uses a probability-based sampling design, examining land use-biological indicator associations for individual stream sites also allows inferences about conditions at stream order, whole basin, and statewide scales.

Across all stream site catchments sampled in 1995, the dominant land use was forest (52%), followed by agriculture (35%), and urban (13%). Forest cover was most extensive for site catchments in the Lower Potomac (67%), Upper Potomac (65%), and Youghiogheny (59%) basins. Agricultural land use was highest on the eastern shore at sites in the Nanticoke-Wicomico (56%) and Chester (54%) basins. Urban land use was greatest for stream site catchments in the Patapsco basin (37%).

When stream sites were combined across all six basins, fish IBI scores generally decreased with increasing urban land use; however, this relationship was variable (r=0.22, p <0.001). Most stream sites with greater than 50% urban land use in their catchments had IBI scores in the poor and very poor categories. But among sites with a lower percentage of urban land use (below 25%), a wider range of IBI scores was observed from good to very poor. Factors other than land use also influenced biological conditions at many of these sites.

The association between urban land use and fish IBI score varied across the six basins. A strong negative association between urban land use and IBI score (r=0.82, p < 0.001) occurred in the Patapsco basin (Figure 2), where urban land use in the catchments of 61 stream sites ranged from 0-95%. No other basins sampled in 1995 showed significant associations between fish IBIs and urban land use, probably because of the paucity of urban stream sites. Comparison of fish IBI scores and other land use categories were less informative. For all six basins combined, there were no significant positive associations between forest land cover and IBI score, and no significant negative associations between agricultural land use and IBI scores, either for all six basins combined or within individual basins. The agricultural land use classification used in these analyses included pasture, orchards, and crop land (Fisher *et al.*, 1991). Less intensive forms of agriculture may have lower impacts on stream systems.

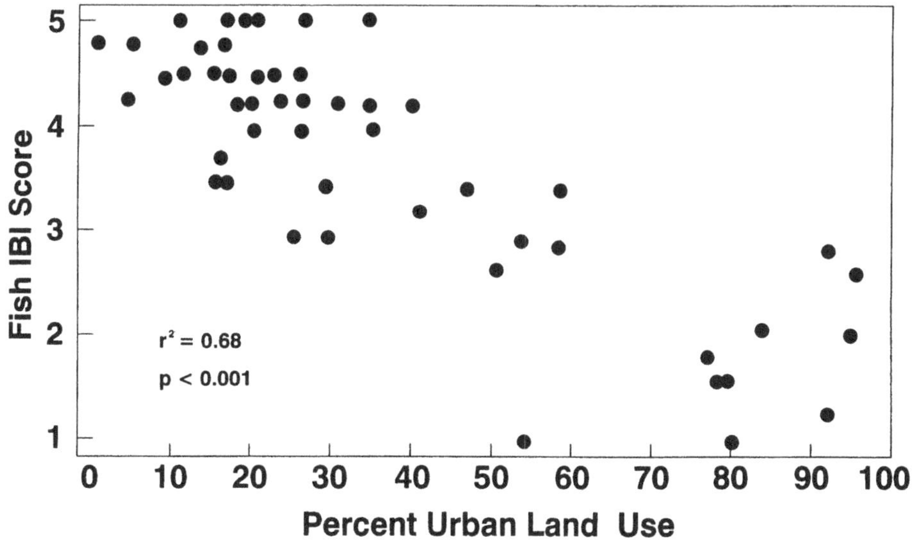

Fig. 2. Associations between fish IBI scores and percent of catchment areas in urban land use at MBSS sites, Patapsco River basin, 1995.

The interactions of combinations of factors were examined graphically to evaluate the cumulative effects of three anthropogenic stressors (acidification, physical habitat degradation, land use) on stream biota. Extensive urban development was associated with low IBI scores that were indicative of degraded fish communities. However, at many stream sites with low IBI scores and low urban land use (below 25%), ANCs were below 200 ueq/L, suggesting that acidification was the more important stressor.

4. Future Directions

The characterizations, assessments, and resource-stressor associations developed from 1995 MBSS data that are highlighted in this paper will be further refined when the three-year sampling cycle in all Maryland basins is completed in fall 1997. The applicability of MBSS results will be evaluated statewide and in terms of interannual variability among basins. These statewide asssessments will provide a framework for targeting those basins that need further stream assessments at more local scales, for targeting reforestation in riparian stream buffers, for prioritizing areas for mitigation of identified impacts, for monitoring habitat restoration success, and for mapping biodiversity "hotspots" that contain spatially-restricted or rare species and communities.

The fish IBI will be validated with 1996-1997 data and refined if necessary. Additional spatial stratification for Piedmont, very small, and cold water streams will be investigated. Multi-variate analyses will be used to determine habitat preferences for various fish species. Multimetric indicators will also be developed for benthic macroinvertebrates, herpetofauna, and physical habitat. Future MBSS activities will characterize the extent of stream acidification statewide and compare the status of streams in the mid to late 1990's with 1987 statewide stream chemistry survey results to evaluate success of the 1990 amendments to the Clean Air Act in improving water quality in Maryland's headwater streams. A comparative risk assessment will also be conducted to distinguish the relative impacts of acidic deposition, acid mine drainage, and natural organic sources of acidity on stream water quality and biota. Geographic differences in the distribution and abundance of acid-adapted fish communities will be investigated. Aquatic biodiversity in Maryland's streams will be quantified and "hot spots" identified.

Land use information will be improved by including data for those portions of Maryland watersheds that extend upstream into neighboring states. Indices of watershed imperviousness will be developed from Normalized Difference Vegetation Index (NDVI) information available from Landsat records. Land uses within riparian corridors will be isolated to evaluate their effectiveness at a basin-wide scale. Other landscape metrics such as habitat abundance, habitat proportion, habitat patchiness, patch size, and amount of edge will be developed. Several multi-variate analytical approaches will be explored for analyzing the cumulative effects of anthropogenic stresses on stream biota.

Acknowledgements

The 1995 MBSS was a cooperative effort that involved many individuals, too numerous to list, from the Maryland Department of Natural Resources, Versar, the University of Maryland's Appalachian Laboratory and Wye Research and Education Center, and Coastal Environmental Services. These organizations selected and mapped sampling sites, obtained landowners' permissions, collected and managed the data, conducted analyses, and prepared survey reports. Assistance was also provided by several seasonal interns and volunteers. We appreciate the comments on the draft manuscript from Paul Angermeier, Paul Massicot, and William Sharpe. We are grateful to the Maryland Power Plant Research Program for providing the major source of funding for the MBSS, and to NOAA's Coastal Zone Management Program and U.S. EPA for their contributions.

References

Baker, J.P. and Christensen, S.W.: 1991, *Acidic Deposition and Aquatic Ecosystems*, Springer-Verlag, New York, pp. 83-106.

Baker, J.P., VanSickle, J., Gagen, C.J., DeWalle, D.R., Sharpe, W.E., Carline, R.F., Baldigo, B.P., Murdock, P.S., Bath, D.W., Kretser, W.E., Simonin, H.A., Wigington, R.J., Jr.: 1996, *Ecol. Applications* **6**, 422-437.

Barbour, M.T. and Stribling, J.B.: 1991, *Biological Criteria: Research and Regulation*, U.S. Environmental Protection Agency, Washington, D.C.

Bode, R.W.: 1988, *Methods for Rapid Bioassessment of Streams*, New York Department of Environmental Conservation, Albany, New York.

Cochran, W.G.: 1977, *Sampling Techniques*, John Wiley and Sons, New York.

Fisher, G.T.: 1991. *Preparation of 1990 Land Use/Land Cover Maps and ARC/INFO Digital Data Base*, Daft-McCune-Walker, Towson, Maryland.

Frissel, C.A., Liss, W.J., Warren, C.E., Hurley, M.D.: 1986, *Environ. Mgmt.* **10**, 199-214.

Gerritsen, J., Dietz, J.M., Wilson, H.T., Jr.: 1996, *Ecol. Applications* **6**, 438-448.

Heimbuch, D.G., Wilson, H.T., Weisberg, S.B., Volstad, J.H., Kazyak, P.F.: 1997, *Trans. Amer. Fish. Soc.* (In press).

Herlihy, A.T., Kaufman, P.R., Mitch, M.E.: 1991, *Water Resources Res.* **27**: 629-642.

Hilsenhoff, W.L.: 1987, *Great Lakes Entomol.* **20**, 31-39.

Hilsenhoff, W.L.: 1988, *Great Lakes Entomol.* **21**, 9-13.

Jessen, R.J.: 1978, *Statistical Survey Techniques*, John Wiley and Sons, New York.

Karr, J.R.: 1991, *Ecol. Applications*, **1**, 66-84.

Karr, J.R., Fausch, K.D., Angermeier, P.L., Yant, P.R., Schlosser, I.J.: 1986, *Nat. History Survey Spec. Publ.* **5**: 1-28.

Kaufman, P.R., Herlihy, A.T., Baker, L.A.: 1992, *Environ. Pollut.* **77**, 115-122.

Knapp, C.M., Saunders, W.P., Heimbuch, D.G., Greening, H.S., Filbin, G.J.: 1988, *Maryland Sympotic Stream Chemistry Survey*, International Science and Technology, Reston, Virginia.

Lenat, D.R.: 1993, *N. Amer. Benthological Soc.* **12**, 279-290.

Ohio EPA: 1987, *Biological Criteria for the Protection of Aquatic Life*, Ohio Environmental Protection Agency, Columbus, Ohio.

Padmanabah, A.P. and Olem, H.: 1991, *Water Environ. Tech.* 3, 5, 40.

Pennington, M. and Volstad, J.H.: 1994, *Biometrics* **50**, 725-732.

Plafkin, J.L., Barbour, M.T., Porter, K.D., Gross, S.K., Hughes, R.M.: 1989, *Rapid Bioassessment Protocols for Use in Streams and Rivers*, U.S. Environmental Protection Agency, Washington, D.C.

Rankin, E.T.: 1989, *The Qualitative Habitat Evaluation Index (QHEI)*, Ohio Environmental Protection Agency, Columbus, Ohio.

Roth, N.E., Southerland, M.T., Chaillou, J.C., Volstad, J.H., Neisberg, S.B., Wilson, H.T., Heimbuch, D.G., Seibel, J.C.: 1977a, *Maryland Biological Stream Survey*, Versar, Inc., Columbia, Maryland.

Roth, N.E., Southerland, M.T., Chaillou, J.C., Klauda, R.J., Kazyak, P.F., Stranko, S.A., Hall, L.W., Jr., Morgan, R.P., II: 1977b, *Environ. Monit. Assess.* (this volume).

Saerndal, C.E., Swensson,B., Wretman, J.: 1992, *Model Assisted Survey Sampling*, Springer-Verlag, New York.

Seber, G.A.F. and LeCren, E.D.: 1967, *Jour. Animal Ecol.* **36**: 631-643.

Southerland, M.T. and Stribling, J.B.: 1995, *Biological Assessment and Criteria: Tools for Water Resources Planning and Decision Making*, Lewis Publishers, Boca Raton, Florida.

Strahler, A.N.: 1957, *Trans. Amer. Geophysical Union*, **38**, 913-920.

U.S. EPA: 1987, *Handbook of Methods for Acid Deposition Studies,* U.S. Environmental Protection Agency, Washington, D.C.

VanSickle, J., Baker, J.P., Simonin, H.A., Baldigo, B.P., Kretser, W.A., Sharpe, W.E.: 1996, *Ecol. Applications* **6**, 408-421.

Vannote, R.L., Minshall, G.W., Cummins, K.W., Sedell, J.R., Cushing, C.E.: 1980, *Can. J. Fish. Aquat. Sci.* **37**, 130-137.

Volstad, J.H., Southerland, M., Chaillou, J., Wilson, H., Heimbuch, D., Jacobson, P., Weisberg, S.: 1995, *The Maryland Biological Stream Survey,* Versar, Inc., Columbia, Maryland.

Volstad, J.H., Southerland, M.T., Weisberg, S.B., Wilson, H.T., Heimbuch, D.G., Seibel, J.C.: 1996, *Maryland Biological Stream Survey*, Versar, Inc., Columbia, Maryland.

Wigington, P.J., Jr., DeWalle, D.R., Murdoch, P.S., Simonin, H.A., VanSickle, J., Baker, J.P.: 1996, *Ecol. Applications* **6**, 389-407.

ASSESSMENT OF THE CONDITION OF AGRICULTURAL LANDS IN FIVE MID-ATLANTIC STATES

A. S. HELLKAMP[1], S. R. SHAFER[1,2], C. L. CAMPBELL[1], J. M. BAY[1], D. A. FISCUS[1], G. R. HESS[1], B. F. MCQUAID[1,3], M. J. MUNSTER[1], G. L. OLSON[1], S. L. PECK[1], K. N. EASTERLING[1], K. SIDIK[1], and M. B. TOOLEY[1]

[1]*EMAP-Agricultural Lands Resource Group, North Carolina State University, 1509 Varsity Dr., Raleigh, NC 27606, USA*
[2]*U.S. Dept. of Agriculture, Agricultural Research Service*
[3]*U.S. Dept. of Agriculture, Natural Resources Conservation Service*

Abstract. Indicators of the condition and sustainability of agricultural lands in five mid-Atlantic states were measured in 1994. Indicators were selected to reflect crop productivity and land stewardship on annually harvested herbaceous crop (AHHC) land, which covers almost 10% of the land area in this region. Overall, condition of agricultural lands in the region is good. Crops generally yielded more than those grown in the 1980s, with a mean observed/expected yield index greater than 1. The mean soil quality index was slightly better than a "moderate" rating for crop growth. Almost 2/3 of the AHHC land is covered by crop rotation plans, with the remaining land mostly in hay fields. Insecticides were applied to less than 20% of AHHC land, and less than 20% of the land where pesticides were applied has high to moderately high potential for pesticides leaching into groundwater. However, integrated pest management (IPM) is practiced on less than 20% of AHHC land. Hay showed more efficient use of nitrogen than seed crops, and non-tilled sites, which are mostly hay, had more microbial biomass (suggesting more nutrient cycling) than tilled sites. This information could provide a baseline for a long-term monitoring program for agroecosystems in the region.

1. Introduction

The Environmental Monitoring and Assessment Program (EMAP) was initiated in 1989 by the U.S. Environmental Protection Agency (US-EPA) to help determine whether current national environmental policies are meeting their objectives and to provide information that can be used to improve those policies (Kutz and Linthurst, 1990). This interdisciplinary project involved multiple federal agencies and universities. Whereas many monitoring efforts are site- or problem-specific and do not allow assessment of condition with statistical confidence over large regions, EMAP attempted to monitor and evaluate condition on a regional basis.

The mid-Atlantic states comprise an ecologically diverse and densely populated part of the United States. The states of Delaware, Pennsylvania, Maryland, Virginia, and West Virginia cover approximately 31 million hectares (120,000 square miles) along the eastern seaboard. The climate and terrain vary considerably from the Appalachian Mountains, through the Piedmont plateau, to the coastal plain. Agriculture is an important presence throughout these five states. Of the 16.6 million hectares in the Chesapeake watershed alone, for example, crops are planted on over 3 million hectares (Horton, 1993; USDA-SCS, 1981). Thus, the condition of agricultural lands and their interactions with other resources are important to environmental managers and policy makers in the region.

Environmental Monitoring and Assessment **51:** 317–324, 1998.

The objectives of our work were to develop indicators of the condition of agricultural lands within an ecological framework, and to monitor and evaluate this condition within the mid-Atlantic region. Data from 1994 (the first of two years) are summarized here.

2. Materials and Methods

A probability sampling frame based on the area frame developed by the National Agricultural Statistics Service (NASS) of the U.S. Department of Agriculture (USDA) (Cotter and Nealon, 1987) was used to select sites for data collection. Based on this frame, the number of samples collected in a particular area was proportional to the extent of agriculture there. Across the five-state area, 150 sampling sites were selected from land planted to annually harvested herbaceous crops (AHHCs; i.e., herbaceous plants that are harvested annually, regardless of whether the plants are annuals or perennials). For each site, data were collected as part of the annual June Enumerative Survey conducted by NASS. Furthermore, NASS enumerators attempted to revisit each site in the fall after the harvest to collect surface soil samples and information about crop yields and management practices. Enumerators gathered complete information on 122 sites, and partial information was gathered from three others; 25 sites were unavailable for the fall collection. Indicators of agroecosystem condition were developed from these data.

Agricultural systems are managed to produce food and fiber. Therefore, one measure of the condition of agricultural lands is agricultural productivity. For each sampled field, agricultural productivity was expressed as the ratio of reported yield for the crop in that field in 1994 relative to the county-wide average yield for the same crop during 1980-89.

Many factors contribute to overall soil quality. Soil samples were analyzed for physical and chemical properties by the USDA Natural Resources Conservation Service in Lincoln, NE (USDA-SCS, 1992). Soil microbial biomass was determined elsewhere by the substrate-induced respiration method (Smith *et al.*, 1985). Eight properties from three categories were considered in determining soil quality: percent clay (physical category); cation exchange capacity, base saturation, pH, sodium (Na) absorption ratio, total nitrogen (N), and total carbon (C) (chemical category); and microbial biomass (biological category). For each site, each factor was rated as low (=1), moderate (=2), or high (=3) quality for supporting crop plant growth, and the eight numerical ratings were averaged to provide an overall index of soil quality for crops on the site.

Management of farmland affects sustainability (i.e., ability to maintain desired condition over time). Growers' plans for crop rotation, which increases the spatial and temporal diversity of crops, were characterized for each site. The use of practices involved in integrated pest management (IPM) was also recorded for each site; these practices involve, among other pest management approaches, application of pesticides according to observed pest populations and damage rather than on spray schedules.

Contamination of groundwater by pesticides applied on farms is most likely when the most leachable pesticides are applied over soils that are most conducive for movement of solutions through the profile. Pesticide leaching risk on each site was classified according to a system developed by the North Carolina Cooperative Extension Service (McLaughlin *et al.*, 1994). Soil leaching potential was determined by assigning rankings to organic matter content, clay content, and acidity, which are the three most important characteristics that control movement of pesticides through soil. Pesticide leaching potential was determined by pesticide persistence, strength of adsorption to clays and organic matter, and application rate and method. Relative risk at a particular site was determined by the combination of soil leaching potential and pesticide leaching potential.

Nitrogen use efficiency at each site was defined as the amount of N applied in commercial fertilizer per unit of harvested commodity (kg N/Mg crop).

3. Results

In 1994, annually harvested herbaceous crops (AHHCs) covered approximately 3.67 x 10^6 ha (95% confidence interval [CI] of 3.48 to 3.85 x 10^6 ha), or slightly more than 10% of the land area of the five states. The most extensively grown crops were hay (alfalfa and other forage), corn, and soybean. Barley, oat, Irish potato, rye, sorghum, tobacco, watermelon, and wheat also were common. In general, the randomly selected fields were small: only 6% were 39 ha (100 acres) or larger in size. Crops were irrigated on only 3% of the land planted in AHHCs.

Crops in the mid-Atlantic area exhibited yields in 1994 that were greater than expected from the county averages for 1980-89 (Figure 1). The mean observed/expected yield index for the region was 1.41 (CI = 1.25 to 1.57, n=91 fields). Only 29.7% (CI = 20.3% to 39.1%) of sites had an index value less than 1.00.

Soil quality was generally moderate for crop production throughout the region. For the five-state area, the mean soil quality index was 2.23 (CI = 2.17 to 2.29), which was significantly greater than the value of 2.00 that indicates moderate condition for all eight factors in the index. Where indices were lower than 2.00, low cation exchange capacity and low total soil carbon were the most frequent causes (data not shown). Tillage had a negative effect on soil quality index. Non-tilled sites such as hay fields, pastures, and idle land (n=48 fields) had a mean soil quality index of 2.39, whereas conventionally tilled fields (n=56 fields) had a significantly ($p<0.001$) lower mean index of 2.11. Fields subjected to reduced tillage (no-till or mulch till, n=14 fields) had a mean index of 2.17, which was not significantly different from the mean index for conventionally tilled fields but was significantly ($p<0.05$) less than the mean index for non-tilled fields (Figure 2A).

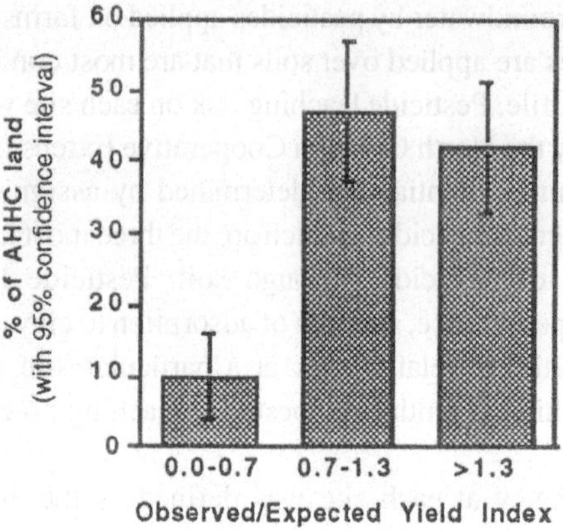

Fig. 1. Yield index for annually-harvested herbaceous crops (AHHCs) in the five-state region.

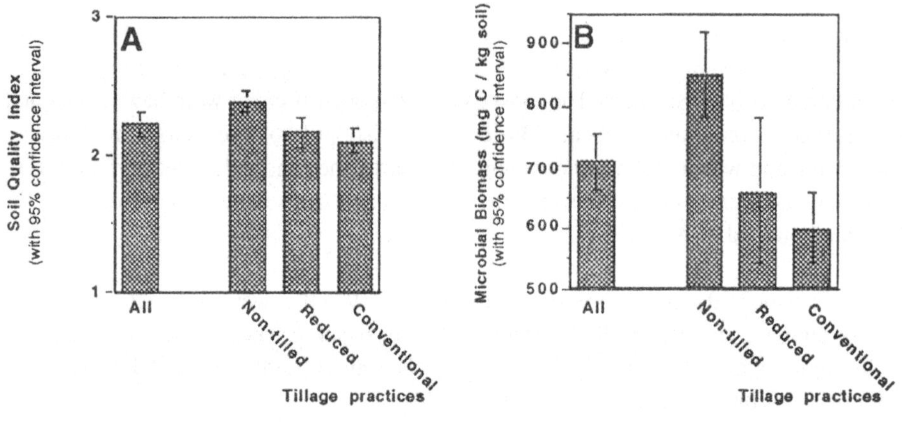

Fig. 2. Soil characteristics for the five-state region ("All") and segregated by tillage practices. A. Soil quality index. Values of 1.00, 2.00, and 3.00 correspond to low, moderate, or high quality, respectively, for characteristics important to crop production. B. Soil microbial biomass.

Tillage type also affected estimates of microbial biomass in soil. Microbial biomass averaged slightly greater than 700 mg C/kg soil across the entire region, but non-tilled soils had significantly ($p<0.05$) greater values than either conventionally tilled or reduced-tillage soils (Figure 2B).

Growers reported crop rotation plans for 61% (CI = 52% to 70%) of the sampled fields. Plans included rotations of 2-3 years (30% of AHHC land in the region), 4 or more years (16%), or unknown duration (15%). Approximately two-thirds of the 39% of AHHC land not in rotation was planted to perennial hay crops, which are not rotated.

Herbicides were the most frequently applied pesticides, but half the farmers who responded to the survey used no pesticides. Farmers in the region reported practicing IPM on approximately 16% (CI = 9% to 22%) of AHHC land. Insecticides were applied to approximately 16 % (CI = 9% to 20%) of the AHHC land, and insecticides were applied in one-third of the fields in which IPM is practiced (i.e., 6 of 18 fields). Coleopteran insects were the most frequently targeted insect pests; no respondent reported targeting lepidopteran insects, which are a major pest group on crops.

A large proportion of the AHHC land in 1994 had soils with properties conducive to pesticide leaching, but application of highly leachable pesticides was uncommon. Based on data from the 61 sample sites on which pesticides were applied in 1994, an estimated 80% (CI = 70% to 90%) of AHHC land in the mid-Atlantic region has soil with high to very high leaching potential. About 77% (CI = 65% to 89%) of the AHHC land where pesticides were applied received pesticides with low to moderately low leaching potentials, however. Of the AHHC land where pesticides were applied, 16% (CI = 7% to 25%) involved application of pesticides with moderately high to high leaching potentials to cropland with highly to very highly leachable soils (Figure 3).

Hay crops showed more efficient use of N than did seed crops. The mean nitrogen use efficiency index for seed crops (corn, soybean, grain sorghum, wheat, oat, barley) was 25 kg N/Mg (CI = 10 to 39) of harvested commodity, which was significantly (p<0.05) greater than the value of 5 kg N/Mg for hay (alfalfa or other forages).

		Pesticide Leaching Potential	
		Low to moderately low	Moderately high to high
Soil Leaching Potential	Low to moderate	13%	7%
	High to very high	64%	16%

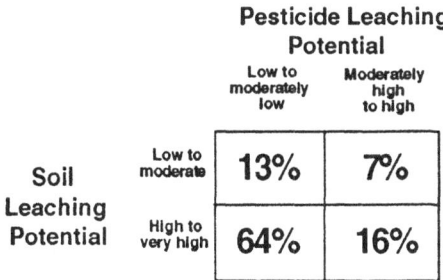

Fig.3. Proportion of annually harvested herbaceous crop lands in the five-state region to which leachable pesticides were applied, on which soils conducive to leaching occur, and both; estimates based on 61 fields in which pesticides were applied.

4. Discussion

The condition of agricultural lands in the mid-Atlantic region in 1994 was good, based on our indicators. Much of this information would be most valuable, however, as benchmarks for comparative evaluations in subsequent years.

We realized the importance of more than just agronomic yield in even a preliminary assessment of the condition of agricultural lands (specifically, land planted to AHHCs). Of obvious importance was the growth and health of the plants themselves (which are the focal point of agroecosystems) and the condition of their supporting medium (the soil). We also recognized the need to quantify the occurrence of management practices that confer sustainability (the ability to maintain a productive condition in the future); thus, we assessed the use of crop rotations and integrated pest management. Furthermore, we selected two characteristics of agricultural lands that reflect stewardship (the maintenance of good condition of both the agricultural resource and the larger surrounding resources): the potential for applied nutrients (fertilizer N) and pesticides to remain on the target site, thus maximizing the effectiveness of scarce energy resources and minimizing their movement into non-agricultural resources.

On a regional average, crop yields were 40% greater than expected, based on yields during the previous decade. The general condition of soil in agronomic systems in the region was slightly better than "moderate." Whether the condition of soils in the region is stable or the soils are degraded or improved from some past condition is difficult to ascertain. Furthermore, an index calculated as an aggregate of measurements of eight different soil properties could obscure the importance of any single factor that in fact had great effect on plant growth. Many environmental, technological, or even societal factors could have contributed to the increased yields. Continued monitoring of condition and comparison to 1994 data could help answer this question and identify trends in soil condition that affect yields. Understanding trends in soil condition and their relationships to trends in crop yields, rather than immediate focus on broad characterizations such as low or high quality, may help determine the role of soil characteristics and other factors in the agricultural productivity of the region.

Soil microbial biomass may be an indicator of important microbial activities such as nutrient cycling. The use of microbial biomass to help describe soil quality is a fairly recent innovation. In our study, microbial biomass data varied within the region, suggesting microbial biomass was greatest in West Virginia and least in Delaware and Maryland. Crops vary through the region, however, and crops are also confounded with tillage to some extent. For example, most of the non-tilled sites were planted to hay; microbial biomass tended to be greatest in hay fields, and a high proportion of the fields sampled in West Virginia were planted to hay. Thus, extrapolating results from microbial biomass analyses to the entire region is unjustified. The utility of microbial biomass estimates in soil quality assessments requires further study.

The spatial and temporal diversity introduced by crop rotation can affect plant health. Continuous monocropping can be conducive to pest outbreaks and depletion of soil nutrients (Bullock, 1992). This practice was not extensive in the mid-Atlantic region in 1994, with only 13% of the AHHC land neither included in a crop rotation nor planted to a perennial forage.

In contrast to crop rotation plans, integrated pest management was not practiced widely, although interpretation of IPM information is problematic. IPM is perceived widely as "environmentally friendly" because a major goal is to limit pesticide applications to those justified by biological and economic considerations. Of the 12 respondents who said they used insecticides but not according to IPM practices, half claimed that they sprayed only when they saw the pests or their damage, not according to a schedule. This approach is central to IPM. This information suggests that IPM is still understood poorly by many growers but that aspects of IPM practice are in widespread use nevertheless. Thus, the small sample size reported for IPM use (18 sites) and grower misperceptions of IPM make these data difficult to interpret.

Even if current crop yields are being maintained or increasing, they may be doing so only because of increasing subsidies to the agroecosystem from non-renewable inputs such as fertilizers and pesticides. A deteriorating system will require increasing inputs to maintain production (NRC, 1993). The N use efficiency index is intended to indicate whether increasing inputs are required to sustain yields. It may also provide insight to how well nutrients are retained on farms for their intended use rather than contaminating off-farm resources such as groundwater or local surface waters. As expected, seed crops required more fertilizer N per unit of harvested commodity than did hay crops. Alfalfa and clovers are leguminous plants that often are major components of hay crops and, in association with bacterial symbionts in the roots, fix atmospheric N. Thus, their requirement for fertilizer N is low, and little N fertilizer is applied to them; accordingly, these crops exhibit a low index value (high efficiency). The combination of N fixation and no tillage of perennial forages may have contributed to the greater microbial biomass that occurred in hay fields than in fields planted to other AHHCs. There is no basis for delimiting N use efficiency index values as good or poor, but a long-term record of this index potentially would be a useful indicator of sustainability.

The pesticide leaching risk evaluation also is intended to assess the extent to which non-renewable resources can be expected to be retained in the target area. Retention at the target maximizes pesticide efficacy and minimizes off-site contamination. The highest-risk combination (the most leachable pesticides applied to the soils most conducive to their movement) represented less than one-sixth of the AHHC land on which pesticides were applied in 1994. Because an estimated half of the AHHC land in the region received no pesticide application, less than one-tenth of the land was at moderately high to high risk for loss of pesticides from the on-farm target area.

5. Conclusion

Our assessment of the condition of agricultural lands in the mid-Atlantic region in 1994 is neither all good nor all bad. Informed use of integrated pest management is not as widespread as might be desired, and pesticide leaching may pose a threat from a small proportion of farms, but neither of these should be considered an insurmountable problem. Regionally, crop plants in 1994 met or exceeded the average yields during the 1980s, soil quality is moderate for crop production, extensive monocropping was not observed, and pesticide applications were not as widespread as might be anticipated given the intensity of agriculture in the region. This information will provide a baseline for future evaluations of the condition of agricultural lands in the region and for considering the role of agroecosystems in the health of other components of the landscape.

Acknowledgments

Cooperative investigations of the USDA-ARS, USDA-NRCS, USDA-NASS, US-EPA, and North Carolina State University. The use of trade names in this publication does not imply endorsement by any U.S. Government agency or the North Carolina Agricultural Research Service of the products named, nor criticism of similar ones not mentioned. The U.S. Government retains the right in and to a non-exclusive, royalty-free copyright.

References

Bullock, D. G. 1992. *Crit. Rev. Plant Sci.* **11,** 309-326.

Cotter, J. and Nealon, J.: 1987, *Area Frame Design for Agricultural Surveys.* USDA, National Agricultural Statistics Service, Research and Applications Division, Area Frame Section. Washington, DC.

Horton, T.: 1993, *Natl. Geo.* **185,** 2-35.

Kutz, F. W. and Linthurst, R. A.: 1990, *Toxicol. Environ. Chem.* **28,** 105-114.

McLaughlin, R. A., Weber, J. B. and Warren, R. L.: 1994. *Protecting Groundwater in North Carolina - A Pesticide and Soil Ranking System.* SoilFacts Series, AG-439-31. North Carolina Cooperative Extension Service, Raleigh, NC. 6 p.

Smith J. L., McNeal, B. L. and Cheng, H. H.: 1985, *Soil Biol. Biochem.* **17,** 11-16.

NRC (Committee on Long-Range Soil and Water Conservation, Board on Agriculture, National Research Council): 1993, *Soil and Water Quality: An Agenda for Agriculture.* National Academy Press, Washington, DC.

USDA-SCS: 1981, *Land Resource Regions and Major Land Resource Areas of the United States.* U.S. Dept. of Agriculture, Agriculture Handbook 296, Washington, DC.

USDA-SCS: 1992. Soil survey laboratory methods manual. Soil Survey Investigations Report No. 42, Version 2.0. USDA-Soil Conservation Service, Lincoln, Nebraska.

AN INTERACTIVE, SPATIAL INVENTORY OF ENVIRONMENTAL DATA IN THE MID-ATLANTIC REGION

L.E. JACKSON[1] and M.P. GANT[2]

[1]*Office of Research and Development, U.S. Environmental Protection Agency, Research Triangle Park, NC 27711,* [2]*Office of Research and Development, U.S. Environmental Protection Agency, Annapolis, MD 21401*

Abstract. The U.S. Environmental Protection Agency (EPA) is working with federal, state, local, and non-governmental partners to produce an interactive, spatial inventory of environmental data in the mid-Atlantic region. The inventory will include maps of sampling locations, lists of measurements, and design information for hundreds of research sites and monitoring programs. It will also feature user-defined queries, resulting in customized maps that satisfy search criteria. (For example, "Display the probability-based surveys that measure dry deposition and nutrient availability in soils"). The inventory will be used in an interagency pilot study, instigated by the National Science and Technology Council's Committee on the Environment and Natural Resources, to integrate environmental monitoring and research activities. The inventory will also provide information for a regional ecological assessment led by EPA Region 3 and the Office of Research and Development. In addition, an interagency consortium will use the inventory to identify suitable field data for assessing the accuracy of satellite imagery. In each of these three applications, the inventory will be tested and evaluated as a potential prototype for completing additional regions of the U.S. Maintained as an Oracle database, the inventory is accessible on the internet at http://www.epa.gov/monitor/. Currently, ten inventory records are on-line for demonstration. The complete federal inventory of approximately 180 records will be accessible on-line by October, 1997; approximately 200 state, local and non-governmental records are scheduled for on-line access by April, 1998.

1. Introduction

As we continue to uncover the complexity and scope of environmental issues, we recognize the necessity of integrating diverse datasets to evaluate environmental conditions beyond narrow specialties and management jurisdictions. Hundreds of public and private organizations in the U.S. collect environmental data for research and management. Our federal government spends approximately $600 million annually on environmental data collection within 30 separate national programs (National Science and Technology Council, 1997). This estimate does not include the large and costly programs that collect our weather data and satellite imagery, which are critical to supplement and understand field research and monitoring. Compelled by increasing challenges and shrinking budgets, environmental organizations are beginning the process to maximize our tremendous and disparate data collection efforts by encouraging broader utilization of our data archives and compatibility of our data management systems.

A leader in this process is the National Science and Technology Council (NSTC), established by President Clinton in 1993 to coordinate science policy across the federal government. The NSTC established the Committee on the Environment and Natural Resources (CENR) specifically to improve coordination among federal agencies involved

Environmental Monitoring and Assessment **51:** 325–329, 1998.

in environmental research and monitoring. One of the first products of the CENR was a conceptual framework with which to begin integrating the multiple environmental research networks and monitoring programs of these agencies. This framework, along with the objectives, participants, and other achievements to date of the CENR, is described on-line at http://www.epa.gov/monitor/.

The CENR framework established three tiers with which to characterize environmental data collection strategies: 1) complete censuses, including remote sensing, 2) surveys of discrete resources, and 3) intensive field research of complex systems. Interagency research staff must now explore methods to nest, aggregate, or otherwise associate relevant data from multiple programs within each tier, as well as to link data across tiers. To address this enormous challenge in research coordination, the CENR is testing its framework on a regional scale. The CENR selected the mid-Atlantic as its pilot region because of the breadth of available environmental data in all three tiers, over time, and across multiple resources and systems. This paper describes an electronic inventory we developed to catalogue the hundreds of environmental data collection activities in the region by a standard, manageable, and descriptive set of attributes. The mid-Atlantic inventory represents the first step in testing the CENR framework.

2. Three Applications of the Mid-Atlantic Inventory

Figure 1 depicts the CENR pilot region, which is centered on Standard Federal Region 3 (Delaware, Maryland, Pennsylvania, Virginia, West Virginia, and the District of Columbia). The CENR pilot region encompasses the entire watersheds of the Chesapeake and Delaware Bays and the Albemarle-Pamlico Sounds, which add portions of New Jersey, New York, and North Carolina to the pilot. The pilot region also includes the coastal ocean to the edge of the continental shelf.

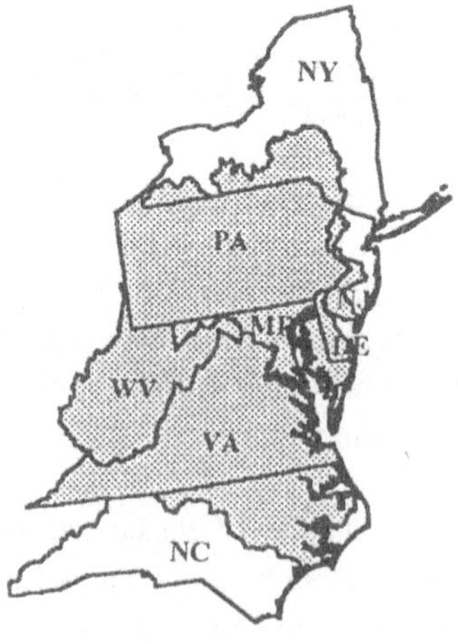

Fig. 1. The mid-Atlantic pilot region.

This area is currently the focus of an extensive interagency assessment of ecological condition (Holland and DeMoss, 1996). Known as the Mid-Atlantic Integrated Assessment (MAIA), this initiative was instigated in 1992 by the U.S. Environmental Protection Agency's (EPA's) Office of Research and Development and EPA Region 3, as part of the Environmental Monitoring and Assessment Program (EMAP). The core of this assessment is region-wide monitoring and reporting of environmental status and trends across all major systems. In addition to collecting new data, the research partners in MAIA are exploring and incorporating existing data of known quality, where possible, to fortify the assessment. Clearly, the CENR pilot to integrate environmental data collection activities in the mid-Atlantic region is particularly timely for the MAIA initiative. In order to increase the utility of the inventory for both the CENR and MAIA, we extended its original federal scope to include state, local, and non-governmental data collection efforts.

We also incorporated into the inventory structure the information needs of a third interagency user community. The Multi-Resolution Land Characteristics Consortium (MRLC) jointly purchases and processes satellite imagery for numerous environmental research and management programs (Loveland and Shaw, 1996). The first MRLC product is a land-cover map of the mid-Atlantic region derived from Landsat Thematic Mapper imagery. To enable accuracy assessment of this large regional dataset, the MRLC is exploring the use of existing field data that meet specific criteria concerning vegetation measures and known accuracy of geographic coordinates. Aware of the MRLC's coincident need for an inventory of data collection activities in the mid-Atlantic region, we worked with Consortium members to customize an additional set of inventory attributes. The final structure of the inventory is designed to serve three major user communities—the CENR, MAIA, and the MRLC.

3. The Structure of the Inventory

The inventory is maintained in an Oracle database and ARC/INFO files; these are converted to user-friendly charts and maps for internet display. When complete, the inventory will provide metadata for approximately 180 federal research sites and monitoring programs, and for an additional 200 state, local, and non-governmental data collection activities. The inventory will not include raw environmental data; users can obtain datasets through links to other websites and from listed contacts.

Each database record contains a general information section (goals and objectives, funding agencies, points of contact, data format and availability, etc.), a design section (the CENR tier, the number, size, and configuration of sites, geographic extent, site location method, etc.), and a measurements section, which is presented as a detailed hierarchy of commonly-measured variables for all resources (i.e., atmosphere, land cover, fresh surface water, groundwater, etc.). The database contains fields for several hundred attributes. While each record has an entry for all fields in the sections on general information and design, most records contain entries for only a fraction of the possible fields in the measurements section.

The inventory also features location maps and user-defined database querying capability. Accompanying each chart on the websites is a mid-Atlantic regional map, which depicts the location of the data collection area(s) for the research site or monitoring program described. The querying capability involves both the spatial and tabular features of the inventory.

We designed the database to contain standardized entries so that users can conduct comprehensive queries by accessing a search menu that displays all database fields. A query may contain any combination of fields that conform to the user's interests; the results are returned in the form of a customized map. An example query might ask, "Display the probability-based surveys funded by the U.S. Forest Service that collect soil nutrient data and have been in operation since 1950." Hypertext map legends take the user to the complete chart for any program of interest, and from there to the program's websites, if one is available. Through a link with EPA's on-line spatial data library, users will also be able to query the inventory by geographic subregions such as watersheds and counties.

4. Methods for Developing and Populating the Inventory

The several hundred fields in the inventory were selected through extensive consultation with interagency researchers to ensure that resulting information would adequately address the three primary applications: 1) to identify opportunities for linking programs within and across CENR tiers, 2) to identify existing environmental data that may be incorporated into a planned research or assessment study, and 3) to identify existing field data that may be used to interpret and assess the accuracy of satellite imagery.

We used a variety of sources to identify research sites and monitoring programs for inclusion. The CENR had already identified 35 major federal programs in a preliminary national inventory (also accessible on the websites). We searched for additional activities in published compilations, on the internet, and through ongoing interagency discussions.

We collected information for the inventory using a questionnaire, administered during a site visit or telephone interview, in order to ensure full respondent comprehension and completion. This process took approximately one hour per record. A completed questionnaire was keyed into the Oracle database and a printout was checked against the handwritten form. Concurrently, an ARC/INFO map of the data collection location(s) was developed from coordinates or a file provided by the respondent's spatial data manager. The printout and map were then returned to the respondent to ensure accuracy.

5. Status of the Inventory and Future Plans

Tabular and spatial information for 180 federal environmental data collection activities will be accessible on the internet by October 1, 1997. Information for ten inventory records has been posted on the websites for demonstration purposes. The next phase will focus on state, local, and non-governmental data collection activities. We anticipate administering approximately 200 questionnaires and posting complete information for these activities by April 1, 1998. Database querying capability will become available in stages. By October 1, 1997, we expect to provide attribute searches of major inventory subject areas. Unlimited searches of all database fields should be possible by April 1, 1998.

We plan to keep the inventory current by adding records as new activities are initiated, and by requesting annual updates on existing entries. Where possible, we will automate the update process, sending electronic notification and providing an on-line form. We will work with hardtop for those contacts without internet access. Both electronic and hardtop updates will be quality-checked and added to the inventory using the same process as the initial entries.

As various efforts to coordinate and interpret diverse environmental databases mature, we will doubtless require additional and alternative information than that developed for the mid-Atlantic pilot inventory. We recognize that this is an iterative process, and that we have not accounted for every information need. If necessary, we will modify the database fields as information needs become more apparent over time. Primarily, we have developed the mid-Atlantic inventory as a prototype to facilitate an integrated approach—a new paradigm—to address the enormity and complexity of environmental research and assessment.

References

DeMoss, T.B. and Holland, M.M.: 1996, 'The Mid-Atlantic Integrated Assessment: Focus on process', in *North American Workshop on Monitoring for Ecological Assessment of Terrestrial and Aquatic Ecosystems*, Bravo, C.A., ed., USDA Forest Service General Technical Report RM-GER.-284, pp. 194-211.

Loveland, T.R. and Shaw, D.M.: 1996, 'Multi-Resolution Land Characteristics: Building collaborative partnerships', in *Gap Analysis: A landscape approach to biodiversity planning*.

Scott, J.M., Tear, T., and Davis, F., eds., *Proceedings of the ASPRS/GAP Symposium*, Charlotte, NC. (National Biological Survey: Moscow, ID.) pp. 83-89.

National Science and Technology Council: 1997, 'Integrating the Nation's Environmental Monitoring and Research Networks and Programs: A Proposed Framework,' Available on-line at http://www.epa.gov/monitor/pubs.html.

6. Status of the Inventory and Future Plans

Tabular and spatial information for 180 federal environmental data collection activities will be accessible on the Internet by October 1, 1998. Information for all inventory records has been posted on the websites for demonstration purposes. The next phase will focus on state, local and non-governmental data collection activities. We anticipate submission of approximately 250 questionnaires and posting complete information for these activities by April 1, 1999. Database querying capability will become available in stages. By October 1, 1998, we expect to provide attribute searches of tabular inventory-based [illegible]. [illegible]

[illegible]

SEDIMENT QUALITY OF ESTUARIES IN THE SOUTHEASTERN U.S.

JEFFREY L. HYLAND[1], TIMOTHY R. SNOOTS[2], AND W. LEONARD BALTHIS[3]

[1] *NOAA Carolinian Province Office, 217 Fort Johnson Road, P.O. Box 12559, Charleston, South Carolina, 29422-2559, USA.* [2] *South Carolina Department of Natural Resources, 217 Fort Johnson Road, P.O. Box 12559, Charleston, South Carolina, 29422-2559, USA.* [3] *University of Charleston South Carolina, Grice Marine Biological Laboratory, 205 Fort Johnson Road, Charleston, South Carolina, 29412, USA*

Abstract. A study was conducted to assess the condition of estuaries in the EMAP Carolinian Province (Cape Henry, VA – St. Lucie Inlet, FL). Synoptic measures of sediment contamination, toxicity, and macroinfaunal condition were made at 82 and 86 stations in 1994 and 1995, respectively, in accordance with a probabilistic sampling design. These data were used to estimate percentages of degraded vs. undegraded estuarine area from the perspective of sediment quality. Each year a sizable portion of the province (36% in 1994, 51% in 1995) showed some evidence of either degraded benthic assemblages, contaminated sediments in excess of bioeffect guidelines, or significant sediment toxicity (based on *Ampelisca abdita* and Microtox® assays). However, co-occurrences of a degraded benthos and adverse exposure conditions (sediment contamination and/or toxicity) were much less extensive — 17% of the province in 1994 and 25% in 1995. Each year only four sites, representing 5% of the province in 1994 and 8% in 1995, had degraded infauna accompanied by both sediment contamination and toxicity, suggesting that strong contaminant-induced effects on the benthos (based on such combined weight-of-evidence) were limited to a fairly small percentage of estuarine area province-wide. PCBs and pesticides (lindane, dieldrin, DDT and derivatives) were the most dominant contaminants over the two-year period. The broad-scale sampling design of EMAP was not intended to support detailed characterizations of potential pollutant impacts within individual estuaries. Thus, some estuaries classified as undegraded may include additional degraded portions outside the immediate vicinity of randomly sampled sites. Such localized impacts (not accounted for in the above estimates) were detected in this study at additional nonrandom supplemental sites near potential contaminant sources.

1. Introduction

In 1993, NOAA and EPA formalized an agreement to conduct a joint study of the quality of estuaries of the Carolinian Province, one of 12 coastal regions established under the nationwide Environmental Monitoring and Assessment Program (EMAP). The study combined resources and methods of EMAP and NOAA's National Status and Trends (NS&T) Program.

The Carolinian Province extends from Cape Henry, Virginia through the southern end of Indian River Lagoon along the east coast of Florida. Estuaries of this region, covering an estimated 11,622 km^2 (based on EMAP boundaries and exclusive of coastal wetlands), provide extensive habitat and breeding grounds for diverse assemblages of terrestrial and aquatic species. There is an increasing need for effective management of these resources given a predicted influx of people and businesses to southeastern coastal states over the next decade and the ensuing pressures of human activities in these areas. Culliton *et al.* (1990) estimated that the coastal population of the southeastern United States will increase by 181% over the 50-year period from 1960–2010 (the largest increase in the country).

Environmental Monitoring and Assessment **51:** 331–343, 1998.

The EMAP Carolinian study was initiated to help support such management needs by providing year-to-year estimates of the condition of southeastern estuaries based on a variety of synoptically measured indicators of environmental quality. This information is intended to provide scientists and resource managers with a better understanding of the types of environmental problems that exist, how living resources will respond to various types of anthropogenic alterations, and how to prioritize management strategies for dealing with existing problems as well as potential impacts of further coastal development. The Carolinian study also has provided opportunities to refine methods for use in future monitoring and assessment studies in this and other regions.

An initial pilot study was conducted in the Carolinian Province in 1993 to collect background information on ranges of key environmental variables and to determine appropriate indicators of environmental quality to include in subsequent monitoring efforts. Results of the pilot study are presented by Ringwood *et al.* (1996). Full province-wide assessments of Carolinian Province estuaries (using a variety of biological, chemical, toxicological, and aesthetic indicators) were conducted during the summers of 1994 and 1995. A detailed report on the overall quality of these estuaries, based on the 1994 survey, was presented by Hyland *et al.* (1996). A similar report on results of the 1995 survey is in preparation. In the present paper, we use selected data on sediment contamination, sediment toxicity, and macroinfaunal composition from both years to summarize salient points about the conditions of these estuaries from the perspective of sediment quality. Combining measures of sediment chemistry, toxicity, and *in situ* benthic condition has been shown to be very effective as a weight-of-evidence approach to assessing pollution-induced degradation of the benthos (Chapman, 1990; Chapman *et al.*, 1991).

2. Materials and Methods

An overall goal of EMAP is to make statistically unbiased estimates of ecological condition with known confidence. To approach this goal, a probabilistic sampling framework was established among the overall population of estuaries comprising the Carolinian Province. Under this design, each sampling point is a statistically valid probability sample. Thus, percentages of total estuarine area with values of selected indicators above or below suggested environmental guidelines can be estimated based on the conditions observed at individual sampling points. Statistical confidence intervals around these estimates also can be calculated. Further details on sampling design and statistical methods are given in Hyland *et al.* (1996).

Figure 1 shows the province-wide distribution of the 82 and 86 random base stations sampled during the summers of 1994 and 1995, respectively. These stations formed the probability-based monitoring design. Data from these sites were used to estimate condition based on a variety of environmental indicators. Each year, samples also were taken from an additional series of "supplemental stations" selected non-randomly in areas for which there was some prior knowledge of ambient environmental conditions. These sites, which

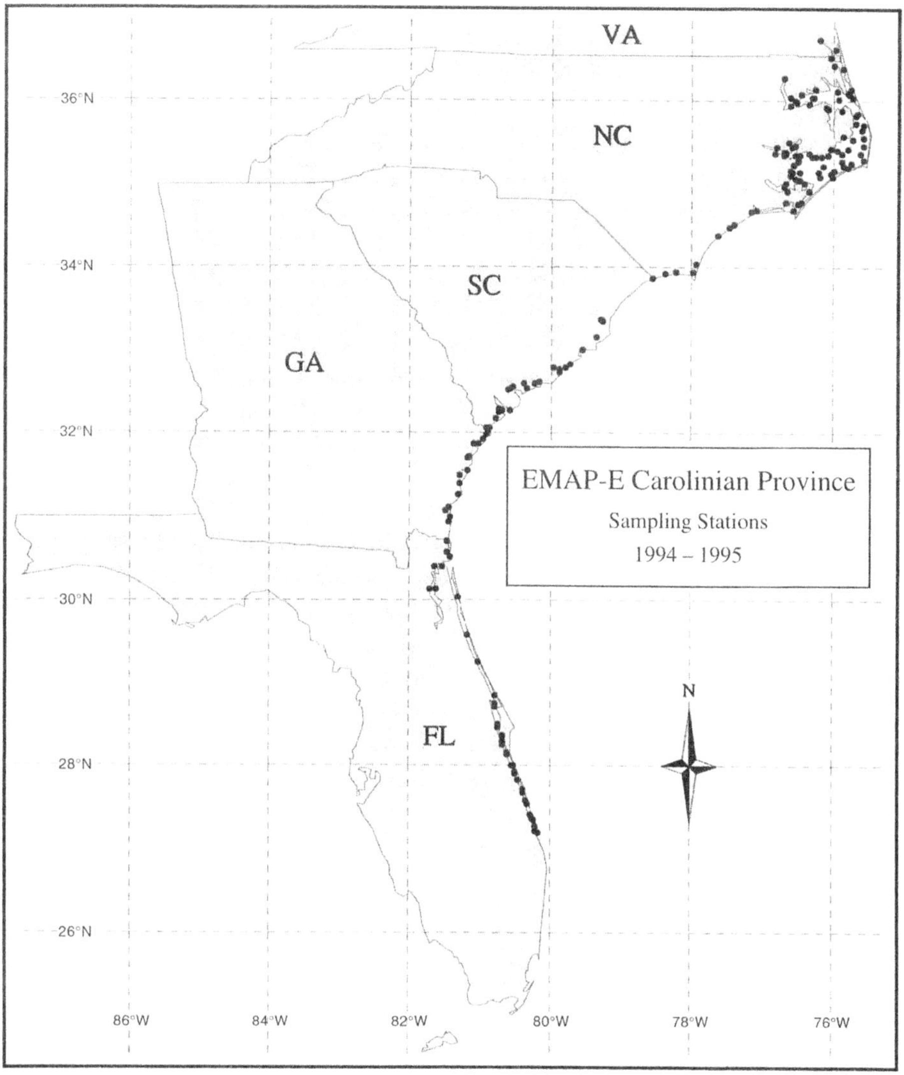

Fig 1. Carolinian Province random sampling sites (summer 1994 and 1995).

represented a combination of both reference areas and places with histories of anthropogenic disturbance, were used to test the discriminatory power of new ecological indicators being developed in the program.

At each station, samples and *in situ* measurements were obtained for characterization of: (1) general habitat conditions, including depth, physical properties of water (dissolved oxygen, salinity, pH, and temperature), sediment silt-clay and moisture content, and organic carbon content of sediment; (2) potential pollution exposure (sediment contaminants, sediment toxicity, ammonia and sulfide in sediment porewater, and low dissolved-oxygen conditions in the water column); (3) biotic conditions (diversity and abundances of macroinfauna and demersal fishes and invertebrates); and (4) aesthetic quality (presence of anthropogenic debris, visible oil, noxious sediment odor, and water clarity based on secchi depths). The present discussion of the quality of southeastern estuaries focuses on evidence of sediment quality based on the combined measures of sediment contamination, toxicity, and macroinfaunal conditions.

Organic and metal contaminants were measured in subsamples of composited surface sediment (upper 2 cm) collected from each station with a 0.04-m^2 Young grab sampler. These subsamples were taken from the same sediment composite used for toxicity testing and the analysis of other supportive physical/chemical characteristics. Stations were represented usually by unreplicated samples, with the exception of duplicates run for ~ 10% of the stations as a quality-control procedure. A total of 16 inorganic metals, four butyl-tins, 27 aliphatic hydrocarbons, 44 polynuclear aromatic hydrocarbons (PAHs), 20 polychlorinated biphenyls (PCBs), and 24 pesticides were measured at each station. Table I summarizes the measurement units, detection limits, analytical methods, and protocol references for various analyte groups.

Sediment toxicity was measured using up to four different assays: (1) the Microtox® solid-phase assay (Bulich, 1979; Microbics, 1992a,b); (2) the 10-day, solid-phase test for survival of the marine amphipod *Ampelisca abdita* (ASTM, 1991); (3) a similar amphipod test with the congeneric species *Ampelisca verrilli* (Ringwood *et al.*, 1997); and (4) a one-week, solid-phase test for sublethal effects of sediment exposure on growth of juvenile clams *Mercenaria mercenaria* (Ringwood and Keppler, In Press). The *A. verrilli* and *Mercenaria* assays, both developed on this program, were not applied on a province-wide basis until the second year of the study (summer 1995). Thus, in order to maximize comparability between years, toxicity interpretations in this discussion are based largely on results of the Microtox® and *A. abdita* assays. Both assays were conducted with subsamples of the same sediment on which analyses of contaminants and other sediment characteristics were performed.

TABLE I.
Summary of analytical methods for the analyses of contaminants in sediments.

Analyte	Min. Detection Limits [a]	Units (dry wgt.)	Method [b]	Reference
Si	10,000	μg/g	FAA	Taylor and Presley 1993
Al	1500	μg/g	FAA	Taylor and Presley 1993
Fe	500	μg/g	FAA	Taylor and Presley 1993
Cr	5.0	μg/g	FAA	Taylor and Presley 1993
Zn	2.0	μg/g	FAA	Taylor and Presley 1993
Mn	1.0	μg/g	FAA	Taylor and Presley 1993
Cu	5.0	μg/g	GFAA	Taylor and Presley 1993
As	1.5	μg/g	GFAA	Taylor and Presley 1993
Ni	1.0	μg/g	GFAA	Taylor and Presley 1993
Pb	1.0	μg/g	GFAA	Taylor and Presley 1993
Sb	0.2	μg/g	GFAA	Taylor and Presley 1993
Se, Sn	0.1	μg/g	GFAA	Taylor and Presley 1993
Cd	0.05	μg/g	GFAA	Taylor and Presley 1993
Ag	0.01	μg/g	GFAA	Taylor and Presley 1993
Hg	0.01	μg/g	CVAA	Taylor and Presley 1993
Butyltins [c]	1.0	ng Sn/g	GC/FPD	Wade et al. 1990
PAHs [d]	5.0	ng/g	GC/MS-SIM	Wade et al. 1993
Aliphatics [e]	25	ng/g	GC/FID	Wade et al. 1994
Pesticides [f]	0.1	ng/g	GC/ECD	Wade et al. 1993
PCBS [g]	0.1	ng/g	GC/ECD	Wade et al. 1993

[a]Based on sample size of 0.2 g for metals and 15 g for organics.
[b]GC/ECD = Gas Chromatography/Electron Capture Detection; GC/MS-SIM = GC/Mass Spectroscopy - Selective Ion Monitoring Mode; GC/FID = GC/Flame Ionization Detection; FAA = Flame Atomic Absorption; GC/FPD = GC/Flame Photo Detection.
[c]Butyltins: mono-, di-, tri-, tetra-
[d]PAHs: 44 parent compounds & alkylated homologues, Tot. PAHs
[e]Aliphatics: C10–C34 alkanes, Tot. Alk., pristane, phytane
[f]Pesticides: DDD (2,4′& 4, 4′), DDE (2,4′ & 4,4′), DDT(2,4′ & 4,4′), Total DDD/DDE/DDT, aldrin, chlordane (alpha-, gamma-, oxy-), dieldrin, heptachlor, heptachlor epoxide, hexachlorobenzene, BHC (alpha-, beta-, gamma-, delta-), mirex, trans- & cis-nonachlor, endrin, endosulfan, toxaphene
[g]PCBs: Congener Nos. 8, 18, 28, 44, 52, 66, 77/110, 101, 105, 188/108/149, 126, 128, 138, 153, 170, 180, 187/182/159, 195, 206, 209, Tot. PCBs

The *Ampelisca abdita* assays were conducted under static conditions at a temperature of 20 ± 1 °C and salinity range of 25–35‰. Field samples were considered to be toxic if mean survival relative to the corresponding negative control (sediment from a reference site) was < 80% and statistically different at $\alpha = 0.05$.

The Microtox® assay provides a sublethal measure of sediment toxicity based on attenuation of light production by the photoluminescent bacterium *Vibrio fischerii*. An aliquot of each sediment sample was used to make a dilution series ranging from 0.01 to 10% sediment in a 2% saline diluent. A reagent solution containing the bacteria was then added to each sediment suspension. After a 20-min incubation period, a column filter was used to separate the liquid phase and bacterial cells from the sediment. Post-exposure light output in each of the filtrates was measured on a Microtox® Model 500 Analyzer. A log-linear regression model was used to determine an EC_{50} — the sediment concentration that reduced light production by 50% relative to a control (nontoxic reagent blank). EC_{50} values were corrected for percent-water content and reported as dry-weight concentrations. Ringwood *et al.* (1995, 1997) have shown a strong inverse relationship between Microtox® EC_{50} values and the percent silt-clay content of sediment. Lower EC_{50} values in muddier sediments are believed to be caused by physical adsorption of the bacteria to the finer sediment particles. Thus, criteria for interpreting Microtox® results were established for two separate silt-clay classes (*sensu* Ringwood *et al.*, 1995): (1) sediments with ≥ 20% silt-clays (muddy sands to muds) were classified as being toxic if EC_{50} values were ≤ 0.2%, and (2) sediments with < 20% silt-clays (sands) were classified as being toxic if EC_{50} values were ≤ 0.5%.

Four replicate macroinfaunal samples were collected from each station with a 0.04-m^2 Young grab sampler. Contents of the grabs were live-sieved in the field with a 0.5-mm mesh screen. Material retained on the screen was fixed in 10% buffered formalin with rose bengal (to facilitate subsequent sorting) and transferred to the laboratory for further processing. Two of the four samples from each station were further processed to characterize the infaunal assemblages and the remaining two samples were archived (for possible future analysis). Animals were sorted from sample debris under a dissecting microscope and identified to the lowest possible taxon (usually to species). The data were used to compute: numbers of species and individuals; the Shannon information function, H′ (Shannon and Weaver, 1949); densities of dominant species; percent abundance of key taxonomic groups; and a combined benthic index of biotic integrity modeled after a similar index developed for benthic assemblages in the Chesapeake Bay (Weisberg *et al.*, 1997). The index for the Carolinian Province is a combined score computed from four component metrics (number of species, total faunal abundance, dominance, and % pollution-sensitive taxa) that showed, in comparison to other metric combinations, the closest agreement with predictions of sediment bioeffects based on sediment chemistry and toxicity data. Stations were evaluated as having a degraded benthos if any one of the following conditions existed: (1) mean species richness ≤ 3 species/sample, (2) mean abundance (all fauna combined) ≤ 25/sample, (3) mean H′ ≤ 1/sample, or (4) IBI score ≤ 1.5.

3. Results and Discussion

Figure 2 provides a summary of the spatial extent of sediment contamination relative to reported bioeffect guidelines—either the Effects Range-Low (ER-L) and Effects Range-Median (ER-M) guidelines of Long *et al.* (1995), and Long and Morgan (1990), or the comparable Threshold Effects Level (TEL) and Probable Effects Level (PEL) guidelines of MacDonald (1994). ER-M and PEL values both represent higher-end probable effect levels above which adverse effects on a wide variety of benthic organisms are likely to occur. ER-L and TEL values represent lower threshold levels below which bioeffects are rarely expected.

TABLE II

Summary of contaminant concentrations and sediment quality guideline exceedances at EMAP sites in the Carolinian Province during summer 1994 and 1995. ER-L and ER-M values are in Long (1995) and Long and Morgan (1990); TEL and PEL values are in MacDonald (1994).

	1994				1995			
			# Sites Exceeding				# Sites Exceeding	
Contaminant	Median Conc.	Range (Min – Max)	ER-L/ TEL[c]	ER-M/ PEL	Median Conc.	Range (Min – Max)	ER-L/ TEL[c]	ER-M/ PEL
Metals (μg/g)								
Antimony	0.2	0.0 – 3.4	4	0	0.0	0.0 – 0.9	0	0
Arsenic	2.9	0.0 – 20.5	13	0	3.0	0.0 – 22.3	18	0
Cadmium	0.1	0.0 – 1.1	0	0	< 0.1	0.0 – 1.3	1	0
Chromium	25.2	3.8 – 130.9	8	0	25.7	0.8 – 98.1	7	0
Copper	2.4	0.5 – 36.3	1	0	2.5	0.5 – 35.4	1	0
Lead	8.3	1.8 – 52.7	2	0	8.9	0.9 – 45.6	0	0
Mercury	< 0.1	0.0 – 0.3	5	0	< 0.1	0.0 – 0.2	2	0
Nickel	3.8	0.5 – 34.3	14	0	3.8	0.5 – 40.3	12	0
Silver	< 0.1	0.0 – 0.4	0	0	< 0.1	0.0 – 0.5	0	0
Zinc	23.3	< 0.1 – 182.8	3	0	25.7	5.8 – 156.7	1	0
PAHs (ng/g)								
Acenaphthene	0.2	< 0.1 – 33.6	1	0	0.3	0.0 – 53.2	1	0
Acenaphthylene	0.2	< 0.1 – 74.2	1	0	0.4	0.0 – 56.3	1	0
Anthracene	0.5	< 0.1 – 136.4	1	0	0.5	0.0 – 142.4	1	0
Benzo[a]anthracene	1.1	< 0.1 – 426.9	2	0	1.3	0.0 – 333.2	2	0
Benzo[a]pyrene	1.4	< 0.1 – 431.3	1	0	1.8	0.0 – 685.9	1	0
Chrysene	1.5	< 0.1 – 469.9	1	0	1.9	0.0 – 620.5	1	0
Dibenz[a,h]anthracene	0.3	< 0.1 – 79.8	1	0	0.3	0.0 – 71.4	1	0
Fluoranthene	2.8	< 0.1 – 802.1	2	0	3.0	0.1 – 701.6	1	0
Fluorene	0.3	< 0.1 – 46.3	1	0	0.5	0.1 – 45.6	1	0
2-Methylnaphthalene	0.6	0.1 – 56.1	0	0	0.8	0.1 – 12.0	0	0
Naphthalene	1.4	< 0.1 – 167.0	1	0	2.9	1.1 – 39.9	0	0
Phenanthrene	1.3	< 0.1 – 263.1	1	0	1.2	0.2 – 114.6	0	0
Pyrene	2.5	0.1 – 867.7	1	0	3.5	0.3 – 3855.4	1	1
Total PAHs [a]	36.1	1.7 – 9179.2	2	0	50.7	9.1 – 12307.9	2	0
PCBs (ng/g)								
Total PCBs	3.7	2.4 – 311.5	8	3	4.1	2.2 – 80.9	5	0
Pesticides (ng/g)								
Chlordane [b]	0.0	0.0 – 5.2	1	1	0.1	0.0 – 3.1	1	0
4,4'-DDD (p,p'-DDD)	< 0.1	0.0 – 6.6	8	0	< 0.1	0.0 – 150.9	13	5
4,4'-DDE (p,p'-DDE)	< 0.1	0.0 – 10.1	8	0	< 0.1	0.0 – 34.2	10	2
4,4'-DDT (p,p'-DDT)	0.0	0.0 – 3.6	3	0	0.0	0.0 – 35.0	10	6
Dieldrin	0.0	0.0 – 1.4	2	0	0.0	0.0 – 38.5	11	5
Lindane [c]	0.0	0.0 – 0.8	2	0	0.0	0.0 – 30.5	15	10
Total DDT [d]	0.1	0.0 – 18.8	17	0	0.3	0.0 – 213.2	22	4

[a]without Perylene
[b]alpha-, gamma-, and oxychlordane
[c]gamma BHC
[d]all six DDD, DDE, and DDT congeners
[e]note that ER-M/PEL exceedances are included in the ER-L/TEL exceedance counts.

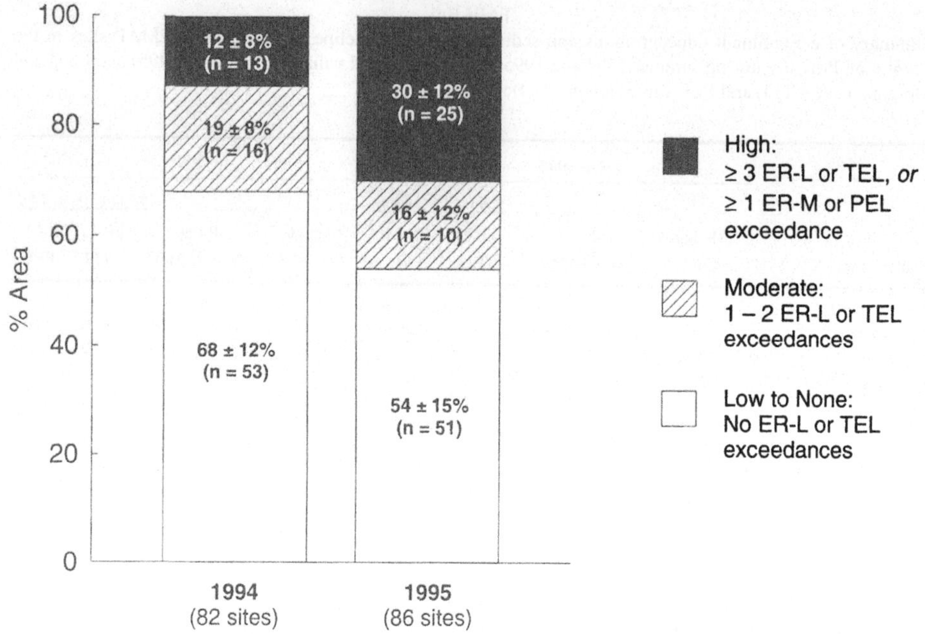

Fig. 2. Spatial extent of sediment contamination (note that sum of percentages given for 1994 do not equal 100% due to rounding).

Over half of the province in both years (68 ± 12% in 1994 and 54 ± 15% in 1995) showed low levels of sediment contamination with all of the measured contaminants falling below corresponding threshold ER-L or TEL bioeffect guidelines. Estuaries with high sediment contamination — defined here by the presence of three or more contaminants in excess of the lower ER-L/TEL values, or one or more contaminants in excess of the higher ER-M/PEL values — represented 12 ± 8% of the province in 1994 and 30 ± 12% in 1995. Though not depicted in Figure 2, the majority of the contaminated area was represented by North Carolina stations.

Dominant contaminants in 1994 were antimony, arsenic, chromium, mercury, nickel, zinc, total PCBs, total chlordane, and DDT and derivatives (Table II). These contaminants were found either at concentrations in excess of ER-M/PEL values in at least one estuary (chlordane and total PCBs) or at concentrations in excess of the lower ER-L/TEL values in three or more estuaries. PCBs were the most pronounced contaminants in 1994. Total PCBs were found at three stations in excess of the ER-M value of 180 ng/g and at five

The range for arsenic (0–20.5 μg/g) included moderately high concentrations, above the ER-L value of 8.2 μg/g but below the ER-M value of 70 μg/g, at 13 of 82 stations in 1994 and at 18 of 86 stations in 1995. Windom *et al.* (1989) reported that southeastern estuarine and coastal sediments are enriched with arsenic relative to concentrations expected from average continental crustal rocks and soils and that these higher concentrations may be related to phosphate deposits that occur commonly throughout the region.

Overall sediment quality was assessed by evaluating the combined data on sediment contamination, toxicity, and macroinfaunal conditions (Figure 3). A sizable portion of the province in both years — 36% in 1994 and 51% in 1995 — showed some evidence of stress (combined shaded areas in Figure 3). Included, however, were sites with no signs of degraded biological conditions. Co-occurrences of a degraded benthos and adverse exposure conditions (high sediment contamination in excess of bioeffect guidelines and/or significant sediment toxicity based on the *Ampelisca abdita* and Microtox® assays) were much less extensive. Such conditions were found at 16 of the 82 random stations in 1994 (representing 17% of the province) and 18 of the 86 random stations in 1995 (representing 25% of the province).

Each year only four sites, representing 5 ± 5% of the province in 1994 and 8 ± 10% in 1995, had degraded infauna accompanied by both sediment contamination and toxicity. These sites were places that showed the strongest evidence of degraded sediment quality associated with anthropogenic contaminant influences. The data suggest that such places represented a fairly small percentage of estuarine area province-wide. Note that a similar conclusion would be reached if results of the two additional bioassays run in 1995 (*Mercenaria mercenaria* and *Ampelisca verrilli* assays) were included in the evaluation of sediment toxicity. For example, the 1995 estimate for percent area of estuaries with a degraded benthos, sediment contamination, and sediment toxicity would shift only from 8 to 13% (due to addition of five more stations) if the determination of toxicity were based on a significant hit in any one of the four assays performed on 1995 samples.

Data presented here suggest that contaminant-induced degradation of the benthos is perhaps limited to a fairly small percentage of estuarine area province-wide. However, it must be understood that the low percentage of degraded estuarine area reported here is based on estimates resulting from the broad-scale probabilistic sampling framework of EMAP-Estuaries. This design was not intended to support detailed characterizations of pollutant distributions and sources within individual estuarine systems. In fact, only one

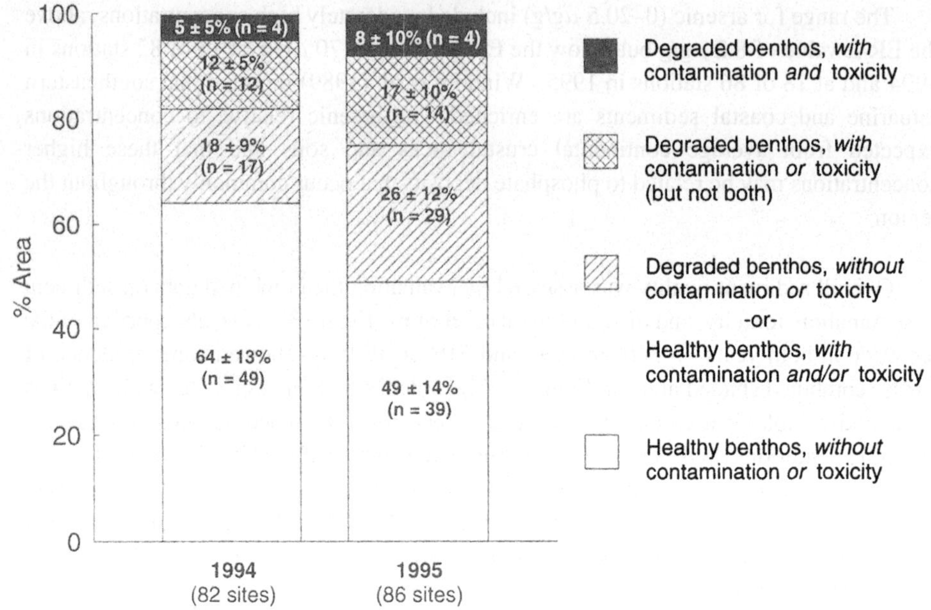

Fig. 3. Summary of sediment quality based on combined measures of sediment contamination, toxicity, and *in situ* macroinfaunal conditions (note that sum of percentages given for 1994 do not equal 100% due to rounding.)

station was sampled in many of these estuaries. Thus, some estuaries classified as undegraded may include additional contaminated portions outside of the immediate vicinity of randomly selected sites. Such localized impacts were detected in this study at some nonrandom supplemental sites near suspected contaminant sources (Ringwood *et al.*, 1995, 1996, 1997).

For example, significant chromium contamination was observed in sediments at Shipyard Creek, a supplemental site in Charleston Harbor, South Carolina. The chromium concentration at this site was 1,911–20,660 μg/g in 1994–95, respectively. This concentration range exceeds the ER-M bioeffect value for chromium (370 μg/g; Long *et al.*, 1995) by up to a factor of 56. Several other PAHs, pesticides, and metals exceeded threshold ER-L/TEL bioeffect values, but none (other than chromium) were found at concentrations in excess of the higher ER-M/PEL values. The high chromium concentrations found at this site also were accompanied by degraded infaunal assemblages and high sediment toxicity (e.g., Ringwood *et al.*, 1997). Impacts at this and other supplemental sites were not accounted for in the above estimates of sediment quality.

Coastal managers often are in need of information to support various policy decisions for individual estuaries or watersheds. To address such needs, additional finer sampling scales (relative to the present EMAP design) should be included to support the characterization of pollutant distributions, sources, and impacts within the specific systems of concern.

4. Conclusion

Results of synoptic sampling of chemical, toxicological, and biological variables at 82 and 86 random sites in 1994 and 1995, respectively, indicated that 49% – 60% of the Carolinian Province has estuaries of relatively high sediment quality. PCBs and pesticides (lindane, dieldrin, DDT and derivatives) were the most dominant contaminants over the two-year period, being present at concentrations in excess of higher probable bioeffect guidelines at multiple sites. While at least some evidence of stress was detectable over broad areas, co-occurrences of degraded benthic fauna, sediment contamination in excess of reported bioeffect guidelines, and sediment toxicity based on *Ampelisca abdita* and Microtox® assays were found in similarly low proportions each year (5% in 1994 and 8% in 1995). These results suggest that contaminant-induced effects on the benthos are perhaps limited to a fairly small percentage of estuarine area province-wide. The broad-scale sampling design of EMAP-Estuaries, however, was not intended to support detailed characterizations of environmental conditions within individual estuaries or watersheds. Thus, some estuaries classified as undegraded may include additional degraded portions outside the immediate vicinity of randomly sampled sites. Such localized impacts (not accounted for in the above estimates) were detected in this study at additional nonrandom supplemental sites near suspected contaminant sources.

Acknowledgments

This program was jointly sponsored by EPA and the NOAA National Ocean Service (NOAA/NOS). EPA funds were provided through Interagency Agreement #DW13936394-01 from EPA's National Health and Environmental Effects Research Laboratory (NHEERL), Gulf Ecology Division. NOAA/NOS funds were provided by the Coastal Monitoring and Bioeffects Assessment Division (CMBAD) of the Office of Ocean Resources Conservation and Assessment (ORCA), and the Coastal Services Center in Charleston, SC. Special recognition is extended to the following contributors who played key roles in the portions of research discussed herein:

Program Design and Coordination—Andrew Robertson and Steve Kokkinakis (NOAA-ORCA/CMBAD); Kevin Summers (EPA-Gulf Ecology Division); Amy Ringwood, Robert Van Dolah, and Fred Holland (S.C. Department of Natural Resources); Courtney Hackney, Steve Ross, and Tracy Wheeler (University of North Carolina-Wilmington); Gil McRae, Gary Nelson, Jan Landsberg, and James McKenna (Florida Department of Environmental Protection).

Contaminant Analyses—Terry Wade, Guy Denoux, Paul Boothe, and James Brooks (Texas A&M University).

Benthic Analyses—Martin Posey and Mary Smith (University of North Carolina-Wilmington); Robert Van Dolah and David Goldman (S.C. Department of Natural Resources); David Camp and Tom Perkins (Florida Department of Environmental Protection).

Toxicity Testing—Amy Ringwood, Robert Van Dolah, and Charles Keppler (S.C. Department of Natural Resources); Philip Ross (Citadel); Marie DeLorenzo (Clemson University); John Scott, Cornelia Mueller, and Glen Thursby (Science Applications International Corporation).

Information Management—Jeff Rosen (Technology Planning and Management Corporation) and Timothy Herrlinger (University of Charleston South Carolina).

Sample Collections—Field teams of the University of North Carolina-Wilmington, S.C. Department of Natural Resources, and Florida Department of Environmental Protection.

References

American Society for Testing and Materials (ASTM): 1991, *Guide for conducting 10-day static sediment toxicity tests with marine and estuarine amphipods, ASTM Standard Methods, 11.04, Method Number E-1367-90*, ASTM, Philadelphia, PA.

Bulich, A.A.: 1979, *Use of luminescent bacteria for determining toxicity in aquatic environments*, In L. L. Marking and R. A. Kimerle (eds.), *Aquatic Toxicology*, 98–106, ASTM STP 667, American Society for Testing and Materials, Philadelphia, PA.

Chapman, P.M.: 1990, *Sci. Total Environment*, **97/98**, 815-825.

Chapman, P.M., E.A. Power, R.N. Dexter, and G.A. Burton Jr: 1991, *Integrative assessments in aquatic ecosystems*, In G.A. Burton Jr. (ed.), *Contaminated Sediment Toxicity Assessment*, 313-340, Lewis Publishers, Chelsea, MI.

Culliton, T.J., M.A. Warren, T.R. Goodspeed, D.G. Remer, C.M. Blackwell and J.J. McDonough III: 1990, *50 years of population change along the Nation's coast, 1960–2010*, National Ocean Service, NOAA, U.S. Department of Commerce, Rockville, MD. 41 p.

Hyland, J.L., T.J. Herrlinger, T.R. Snoots, A.H. Ringwood, R.F. Van Dolah, C.T. Hackney, G.A. Nelson, J.S. Rosen, and S.A. Kokkinakis: 1996, *Environmental Quality of Estuaries of the Carolinian Province: 1994. Annual Statistical Summary for the 1994 EMAP-Estuaries Demonstration Project in the Carolinian Province*, NOAA Technical Memorandum NOS ORCA 97, NOAA/NOS, Office of Ocean Resources Conservation and Assessment, Silver Spring, MD.

Long, E.R. and L. G. Morgan: 1990, *The potential for biological effects of sediment-sorbed contaminants tested in the National Status and Trends Program*, NOAA Technical Memorandum NOS OMA 52, U.S. Department of Commerce, National Oceanic and Atmospheric Administration, National Ocean Service, Rockville, MD.

Long, E.R., D.D. MacDonald, S.L. Smith, and F. D. Calder: 1995, *Envir. Man.*, **19**, 81–97.

MacDonald, D.D.: 1994. *Approach to the assessment of sediment quality in Florida coastal waters. Vols. I–IV*, Report prepared for Florida Department of Environmental Protection, Tallahassee, FL.

Microbics Corporation: 1992a, *Microtox Manual (5 volume set)*, Carlsbad, CA.

Microbics Corporation: 1992b, *Microtox Update Manual*, Carlsbad, CA, 128 p.

Ringwood, A.H., A.F. Holland, R. Kneib, and P. Ross: 1996, *EMAP/NS&T pilot studies in the Carolinian Province: Indicator testing and evaluation in southeastern estuaries.* Final Report under Grant NA90AA-D-SG790 through S.C. Sea Grant College Program, S.C. Dept. of Natural Resources, Marine Resources Research Institute, Charleston, S.C. NOAA Technical Memorandum NOS ORCA 102.

Ringwood, A.H. and C. Keppler: In Press, *Seed clam growth: A sediment bioassay developed in the EMAP Carolinian Province*, Environ. Monitor. & Assess, Submitted May 1997.

Ringwood, A.H., R. Van Dolah, A.F. Holland, and M.G. Delorenzo: 1995, *Year one demonstration project studies conducted in the Carolinian Province by Marine Resources Research Institute: Results and summaries*, Year 1 Final Report under NOAA Cooperative Agreement No. NA470A0177, South Carolina Department of Natural Resources, Marine Resources Research Institute, Charleston, S.C.

Ringwood, A.H., R. Van Dolah, A.F. Holland, M.G. Delorenzo, C. Keppler, P. Maier, J. Jones, and M. Armstrong-Taylor: 1997, *Year two demonstration project studies in the Carolinian Province by Marine Resources Research Institute: Results and summaries*, Year 2 Final Report under NOAA Cooperative Agreement No. NA470A0177, South Carolina Department of Natural Resources, Marine Resources Research Institute, Charleston, S.C.

Shannon, C.E. and W. Weaver: 1949, *The mathematical theory of communication*, Univ. of Illinois Press, Urbana, 117 p.

Taylor, B.J. and B.J. Presley: 1993, *GERG trace element quantification techniques*, In: G.G. Lauenstein and A.Y. Cantillo (eds.), *Sampling and analytical methods of the National Status and Trends Programs, National Benthic Surveillance and Mussel Watch Projects, 1984-1992*, NOAA Technical Memorandum, NOS ORCA 71.

Wade, T.L., J.M. Brooks, M.C. Kennicutt II, T.J. McDonald, J.L. Sericano, and T.J. Jackson: 1993, *GERG trace organics contaminant analytical techniques*, In: G.G. Lauenstein and A.Y. Cantillo (eds.), *Sampling and analytical methods of the National Status and Trends Programs, National Benthic Surveillance and Mussel Watch Projects, 1984-1992*, NOAA Technical Memorandum, NOS ORCA 71.

Wade, T.L., B. Garcia-Romero, and J.M. Brooks: 1990, *Chemosphere*, **20**, 647-662.

Wade, T.L., D.J. Velinsky, E. Reinharz, and C.E. Schekat: 1994, *Estuaries*, **17**, 321-333.

Weisberg, S.B., J.A. Ranasinghe, D.M. Dauer, L.C. Shaffner, R.J. Diaz, and J.B.Frithsen: 1997, *Estuaries*, **20**, 149-158.

Windom, H.L., S.J. Schropp, F.D. Calder, J.D. Ryan, R.D. Smith Jr., L.C. Burney, F.G. Lewis, and C.H. Rawlinson: 1989, *Environ. Sci. Technol.*, **23**, 314-320.

EVALUATION OF R-EMAP TECHNIQUES FOR THE MEASUREMENT OF ECOLOGICAL INTEGRITY OF STREAMS IN WASHINGTON STATE'S COAST RANGE ECOREGION

J. WHITE AND G. MERRITT

Washington State Department of Ecology, P.O. Box 47710, Olympia, WA, USA

Abstract: We used methods from EPA's Environmental Monitoring and Assessment Program (EMAP) to assess the regional status of streams within the Coast Range ecoregion of Washington State. Study objectives were: to determine the ecological condition of wadable, 1st-order through 3rd-order streams; to provide information for the development of water quality biological criteria; and to determine the applicability of EMAP-derived methods in Washington. Stream condition was assessed using EMAP indicators for habitat (chemical and physical) and biology (invertebrate and vertebrate assemblages). EMAP's probability survey was used to select 75 1st through 3rd-order stream sites from the USGS 1:100,000 series hydrographic layer. Of these, 45 sites were sampled. Multivariate techniques were used to identify community types and related physical and chemical habitat. Overall, about 25% of the sites were rated least-impacted. Most impacts were associated with non-point source pollution, mainly forestry practices. The R-EMAP method was a successful tool for assessment of regional status and ecological integrity; however, in order to use it for biological criteria development in Washington State, the method would require some modification to complement the current state protocols.

1. Introduction

The United States Environmental Protection Agency (EPA) developed the Environmental Monitoring and Assessment Program (EMAP) to provide broad scale information on the current status and trends of the United States major ecological resources (EPA 1993a). EMAP's objectives are to use a multi-indicator approach to answer questions about broad-scale ecological conditions. Regional EMAP (R-EMAP) was later initiated by EPA to test the applicability of the EMAP approach at smaller (state or regional) scales (EPA 1993b).

During 1994-1995 the Oregon Department of Environmental Quality and the Washington State Department of Ecology (Ecology) conducted sampling for a R-EMAP project in the Coast Range ecoregion of the Pacific Northwest. Here we discuss Ecology's results of the sampling effort in Washington's portion of the Coast Range. The project goals were to evaluate the usefulness of applying EMAP indicators and sampling design to regional and statewide programs in Oregon and Washington and to assist each state in building biological assessment programs. The specific objectives were to 1) determine the ecological condition of wadable, 1st-order through 3rd-order streams in Washington's Coast Range ecoregion; 2) provide information for the development of water quality biological criteria; and 3) determine the applicability of EMAP-derived methods in Washington State's surface waters.

Environmental Monitoring and Assessment **51**: 345–355, 1998.

2. Materials and Methods

The Coast Range Ecoregion of Washington is located on the western edge of the continental United States bounded by the Strait of Juan de Fuca to the north and the Columbia River to the south. A geologically diverse region of volcanic and sedimentary parent material, the region is divided into 12 sub-ecoregions (Thiele *et al.*, 1992). Coniferous forests of Sitka spruce, western red cedar, western hemlock and Douglas fir uniformly cover the region (Franklin and Dyrness, 1984). Average yearly rainfall ranges from 80-200 cm. The region was last glaciated about 15,000 years ago (McPhail and Lindsey , 1986); however, active glaciers are still present in the Olympic Mountains. The dominant land-use is commercial forestry.

A systematic-random technique was used to locate sites according to EMAP protocols (EPA, 1991). The technique targeted 1st through 3rd-order streams (Strahler, 1957) from a USGS 1:100,000 hydrographic data layer. Once sites were selected, reconnaissance was used to determine the sampling status and accessibility of each site. A site was sampled if it had a definable channel, flowing water, and was wadable across more than half of its reach. Each sampled site also had to be safely accessible.

Field collection occurred during the low-flow period of July-October. This index period was chosen to minimize seasonal variability from changing biological assemblages and the chemical and habitat conditions (EPA-ORD, 1994). At each site, sampling was done over a stream reach length equal to 40 times the channel wetted width (but no shorter than 150 m). Eleven equidistant transects were located within the stream reach. All physical habitat and biological data were collected in relation to these transects. General water chemistry parameters included *in situ* measurements and grab samples which were analyzed at the Ecology/EPA Region 10 Manchester Environmental Laboratory. For this discussion, we limited our analyses to parameters related to watershed impact (e.g. nutrient enrichment): total phosphorus, total nitrogen, and dissolved organic carbon.

The R-EMAP physical habitat measurements were fairly extensive and are described in detail in EPA-ORD (1994). We used several of them to calculate metrics for describing physical habitat quality: channel morphology was evaluated by measuring cross-sectional depths, thalweg profiles, compass bearing between transects, a pebble count was used to estimate reach substrate (bedrock, boulder, cobbles, coarse gravel, fine gravel, sand, fines), and reach gradient was measured with a hand held clinometer. Habitat metrics used in the analysis include:

- Large woody debris greater than 10cm were tallied across the reach.
- Small woody debris was visually estimated at each transect.
- Natural instream cover was visually estimated at each transect.
- Human disturbance index (HDI) - the presence and proximity of evidence of human activity - was visually estimated along each bank to calculate an index.
- Percent riparian canopy cover estimated with a spherical densiometer at midstream.

- Residual pool depth was calculated from thalweg profile according to Robison and Kaufmann (1994).
- Percent coarse, sand and fine substrate from pebble count.

Invertebrates, fish, and amphibians were sampled during each visit. Invertebrates were sampled using D-frame nets with 500mm mesh. One grab was located randomly at each of the 11 transects and designated as either erosional (swiftly flowing water) or depositional (slow-moving or pooled water). Two composites were created, one each for erosional and depositional grabs. Samples were preserved with 70% ethanol and taken back to the Ecology laboratory. Once in the lab, samples were placed into a gridded tray and subsampled until 300 organisms were counted. A minimum of 2 out of the total 30 square grids were subsampled regardless of the number of invertebrates. Identifications were made to a standardized taxonomic level defined in Plotnikoff and White (1996). Aquatic vertebrate populations were sampled using a backpack electroshocker. One pass was made moving upstream for 5000 seconds or the length of the reach. Fish and amphibians were identified to species and released. Those specimens which were not identifiable in the field were preserved in 10% formalin and sent to the University of Washington (Seattle, WA) fish museum for verification.

Without designated reference sites, impact was inferred from spatial patterns of benthic invertebrates as described in Green (1979). Reference sites were not identified due to the lack of available water or habitat quality data in Washington's Coast Range Ecoregion. Invertebrate community types were used to classify sites into like groups by assuming that communities of least-impacted or reference conditions would be more similar to each other than those at sites which were more impacted. The Bray-Curtis similarity coefficient was used as a multivariate description of each community, and site groups were identified by a hierarchical, agglomerative, group-averaged cluster routine. Discriminant analysis was used to test whether site groups identified by the invertebrate community classification differed by their habitat characteristics. Variables chosen from the habitat survey were those found to be affected by forest practices. These variables included substrate measures, woody debris, HDI, canopy cover, and residual pool depth. All site scores from the analysis were plotted on the first two discriminant function axes.

Each site cluster was evaluated to determine which community represented reference or least-impacted conditions. Because of the nature of forestry practices, determining least-impacted conditions was difficult. Within each watershed, the type, extent, and duration of forestry activities has been changing for over 100 years. Each watershed was in a different state of recovery and measuring the actual amount of disturbance was not possible. Forest practices influence streams differently depending on condition of the riparian zone and type of roads. Obvious least-impacted sites were located in the Olympic National Park; however, sites located in the rest of the Coast Range were surrounded by commercial forest lands. To determine healthy stream reaches, several assumptions were made, with each site needing to have more than one of them to qualify:

1) Potentially least-impacted sites would be located in the Olympic National Forest
2) Most of the watershed was in non-commercial lands
3) Riparian zones were intact
4) Instream habitat was diverse

To generate a single ecological integrity score for each site, a multi-metric analysis was used. A scoring system for the Washington Coast Range has not been identified but several studies have shown reliable metrics for use in an index of biotic integrity for Puget Sound and southwestern Oregon (Fore *et al.*, 1996, Rick Hafele, OR DEQ, personal communication). These metrics were used in a principle components analysis (PCA) to calculate factor 1 site scores (Table I). Hughes *et al.* (in press) has shown that for fish communities, the PCA factor 1 site scores were a gross indicator of the IBI score. The invertebrate classification groupings were used to verify the estimated IBI scores for each site.

Table I
Macroinvertebrate metrics used in factor analysis.

Invertebrate Metrics
Total Taxa Richness
Ephemeroptera Taxa
Plecoptera Taxa
Trichoptera Taxa
Intolerant Taxa
Percent Tolerant Individuals
Community Tolerance Index (modified HBI)

3. Results and Discussion

Of the 73 targeted sites, 48 sites were actually sampled. Twenty-five sites did not meet the identified criteria, or access permission was not obtained from the landowner. Table II illustrates the range of site conditions. Both erosional and depositional invertebrate communities were used in the analysis, but no patterns that related to habitat quality were identified from samples from depositional habitats. Using the invertebrate communities from the erosional habitat, classification of the sites by cluster analysis showed three distinct clusters of sites with several smaller, intermediate clusters (Figure 1). Fish and amphibians were not used in the ecological integrity analysis because few species were caught, and no tolerant or non-native species were identified. All sites in one cluster were identified as least impacted according to the previous definition. A gradient of conditions were visually identified from moderate to severe impact depending on the clusters' nearness in similarity to the identified least least-impacted group. Four sites from two clusters were removed from further analysis because of anomalous conditions (Figure 1). One cluster included two higher elevation sites in the

Olympic National Park with bedrock substrate, and the other cluster represented wetlands. Sites with anomalous characteristics will add variability to the date set and decrease the power to detect differences between impacted and non-impacted sites (Reynoldson and Rosenberg, 1996).

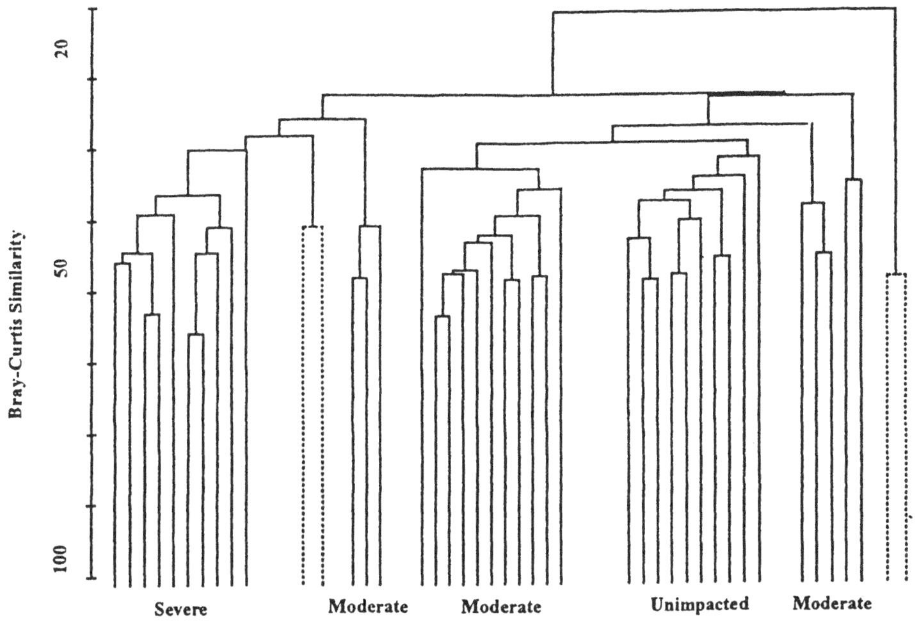

Fig. 1. Cluster analysis of Bray-Curtis similarity for benthic macroinvertebrates. Sites are classified according to human disturbance categories of unimpacted to severe. Groups removed from the analysis are indicated by dashed lines.

The discriminant analysis separated all five identified groups by their habitat characteristics (Wilkes Lambda < 0.001). All variables were significant in the separation of groups except for percent canopy. A biplot of the site scores from the first two discriminant functions shows a definite separation between the least-impacted and severely-impacted sites (Figure 2). The habitat variables that best separated the groups on the first canonical axis size were substrate size and amount of woody debris. Residual pool depth, a measure of habitat complexity, best separated the groups for the second axis.

The average values for several habitat and chemical variables were plotted by disturbance category (Figures 3 and 4). All indicators of watershed disturbance specific to forestry increased over the disturbance categories (McDonald *et al.*, 1993). Decreased particle sizes, increased nutrients, and visual disturbance were all related with the higher levels of impact.

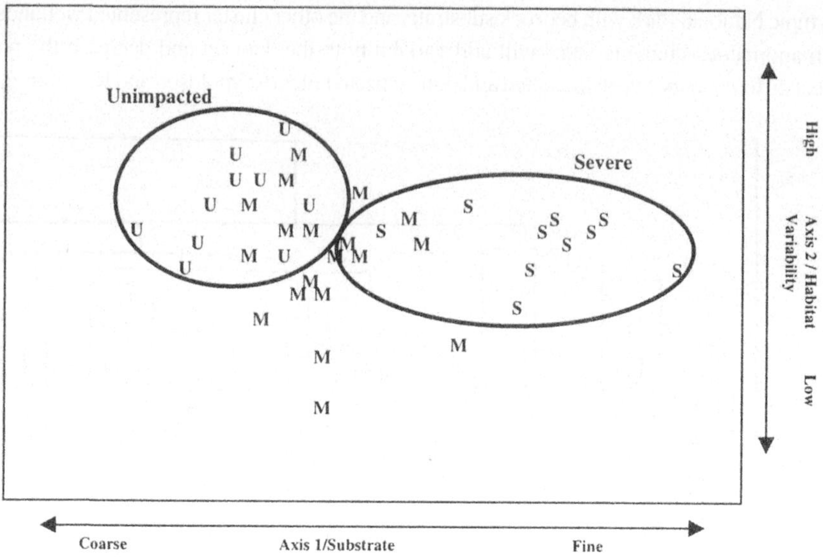

Fig. 2. Plot of habitat data from the first two discriminant functions. Axis 1 is measuring substrate size and axis 2 habitat variability. Circles illustrate the major disturbance clusters from the macroinvertebrate classification and sites are designated by letters (U = unimpacted; M = moderate impact; S = severe impact).

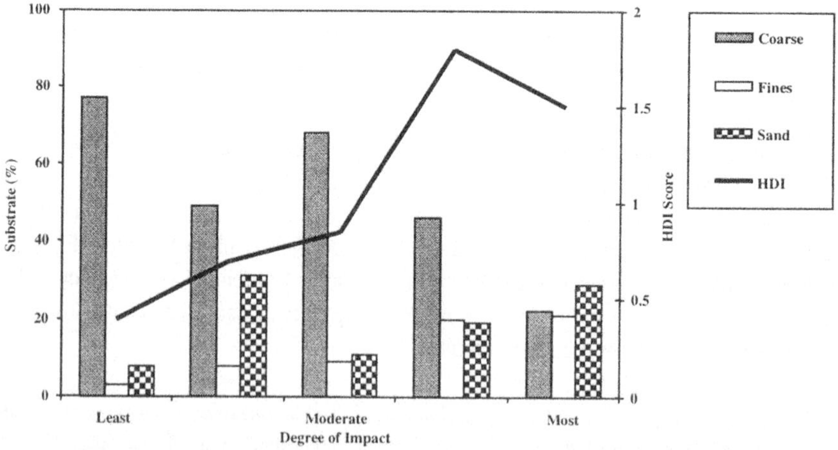

Fig. 3. Human disturbance measures including substrate measures (% coarse particles, % fine particles, and % sand) and the human disturbance index (HDI) for identified impact groups from the macroinvertebrate classification.

The multi-metric PCA factor 1 described 43% of the variation of the macroinvertebrate metric data. A cumulative distribution frequency was plotted using the factor 1 site scores and the upper quartile of the sites were assumed to be least impacted within the sampled population of sites (Figure 5). To verify this assumption, factor 1 site scores were grouped into the impact clusters described earlier and plotted. Of the 48 sites sampled, only four were misclassified, two as least-impacted and two as impacted (Figure 6).

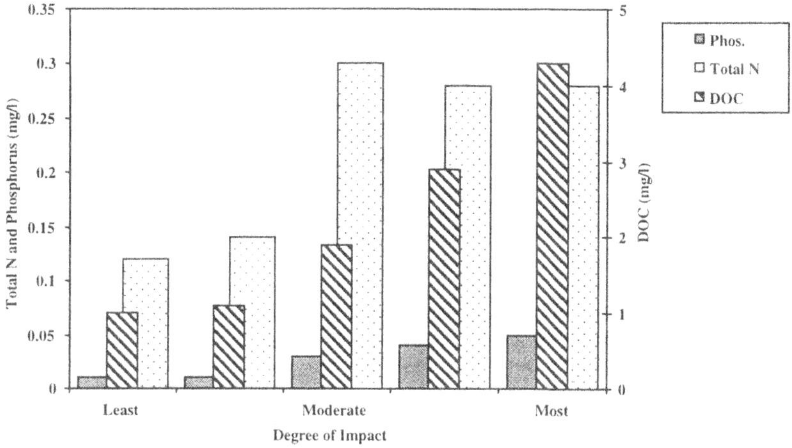

Fig. 4. Measured site levels (mg/l) of total phosphorus (Phos.), total nitrogen (Total N), and dissolved organic carbon (DOC) for identified impact groups from the macroinvertebrate classification.

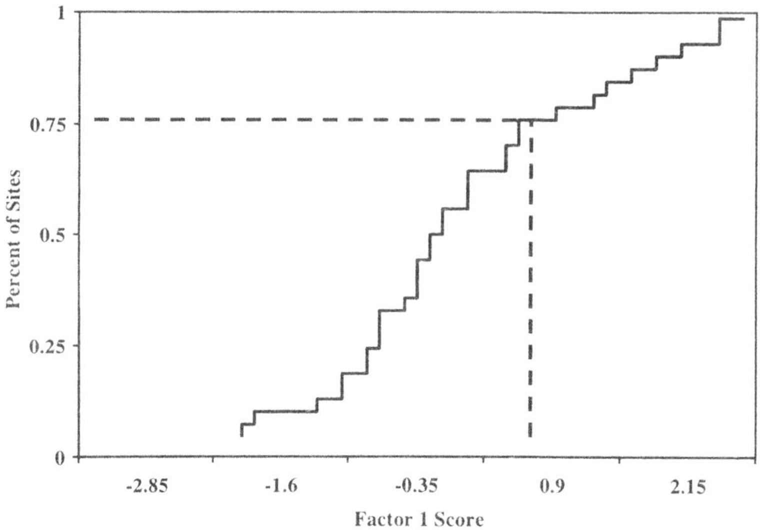

Fig. 5. Cumulative distribution function of site scores from PCA Factor 1 of macroinvertebrate metrics. Dashed line marks the upper 25th percentile of site scores.

Several reasons for misclassification of sites include lack of identifiable reference conditions, sampling design, and analysis techniques. Lack of a suitable tool for *a priori* determination of human disturbance levels can cause some consternation in the analysis.

However, using several methods, both multi-metric and multivariate, we were able to demonstrate a strong relationship between reach quality and invertebrate communities. One aspect of the sampling design, unequal sampling effort for invertebrates, had confounding effects on site response, because each site could potentially have 1 to 11 kicks depending on the distribution of transects across habitat types (e.g. erosional versus depositional). Taxa richness, an important attribute of the community, increases as sampling effort increases (Rosenberg and Resh, 1993).

Table II

Range of site conditions found in Washington's Coast Range ecoregion (median value in parentheses, n=48).

Watershed Area (Ha)	9 - 14,855 (2950)
Elevation (m)	6 – 672 (140)
Stream Gradient (%)	1 - 22 (2)

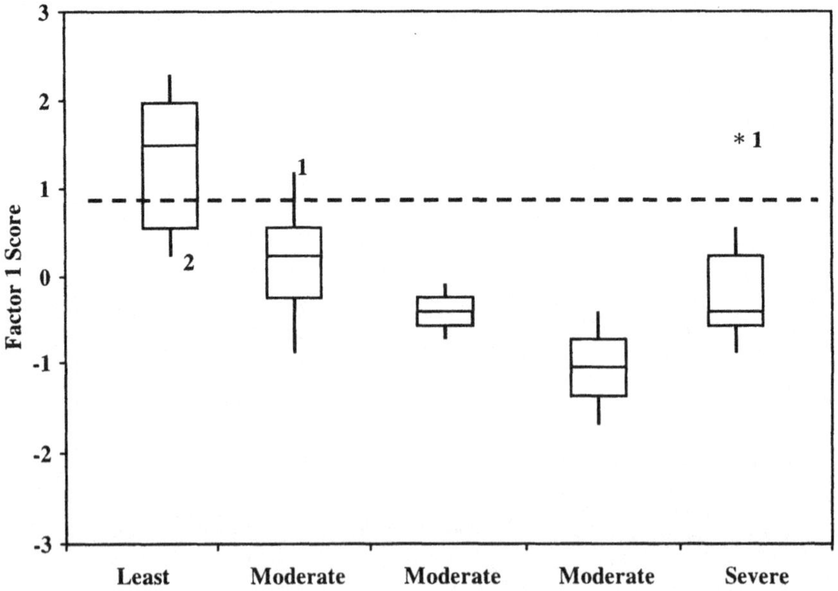

Fig. 6. Box plots of PCA Factor 1 site scores of invertebrate metrics for each disturbance group. Dashed line indicates factor scores that are in the upper 25th percentile. The number of misclassified sites per disturbance group are indicated next to the box plots.

In our study, the correlation between the number of kicks at each site and the total number of taxa collected equaled 0.28 r^2.

We used the upper 25% of the site scores as an arbitrary designator for least-impacted sites. This assumed level may misrepresent the actual conditions, especially with varying stream type and size. Relationships between stream size or type to invertebrate response could also be confounding the analysis. Least-impacted large streams or lower gradient sites might have inherently lower IBI sites scores than the mid-order, higher gradient streams.

Fish and amphibian data were not used in the ecological assessment of the Coast Range data but we present several important findings from this data. Of the 17 fish species collected, all were native, none were identified as pollution tolerant, seven were anadromous, and none showed obvious lesions or deformities. Fifty percent of the sites had less than three species. The low species richness was attributable to high gradient streams or extremely small streams for most sites. For multi-metric analysis, the fish metrics did not show any differences between sites due to impact groups. The metric analysis was probably measuring response due to elevation or watershed size as well as habitat quality. We found a high correlation between taxa richness and watershed size and elevation. However, watershed size was found to have little correlation to elevation and most anadromous species were found in the lower reaches of watersheds.

Anadromous species pose a unique problem to ecological integrity studies. These fish not only represent watershed conditions at a site but also oceanic conditions, harvest rates, and migratory corridor health. Since population levels or presence/absence information of anadromous fish can be influenced greatly by factors other than watershed conditions, it is difficult to use the data as an indicator of watershed health. According to Washington State Fish and Wildlife Department collection permit rules, electroshocking is not allowed when adult salmon or steelhead are present in a stream. Also, with the potential listing of several anadromous stocks in Washington, electroshocking for juveniles in these streams may also be prohibited. Limitations placed on fish community sampling could affect representative sampling under EMAP's probability-based design.

Several additional factors limit the use of EPA's R-EMAP design for ecological assessments and biocriteria development in the State of Washington. Most notable was that the invertebrate sampling methods did not complement the state's current methods of collecting one composite of four samples from a discrete riffle reach (Plotnikoff, 1994). The R-EMAP data would not be comparable to the state's invertebrate database due to differences in sampling effort at each site. Also, the R-EMAP sampling design can target transitional habitats (random locations within the reach) which is not consistent with the "discrete" habitat sampling method. The time and budget required for R-EMAP sampling also limits its use by the state. For example, the more remote sites took from 2 to 4 days per site to complete sampling. Also, fish and amphibian sampling occupied more than 50% of the overall effort, yet the resultant data were not very useful for assessing ecological integrity within the Coast Range. With budgets for environmental sampling decreasing, this level of effort for sampling vertebrate populations is difficult to justify.

4. Conclusions

Use of the EPA R-EMAP sampling design for ecological studies within the Coast Range of Washington had mixed results depending on the stated objective. The design showed promise in identifying stream ecological integrity over varying ranges of impacts. However, to better complement state invertebrate community sampling, R-EMAP methods need to be modified to reflect habitat-level assessments. To be directly comparable to the state invertebrate protocol, four, 0.1 m^2 kick samples should be sampled from discrete riffle units. The probability sampling design has been shown by many to achieve excellent spatial representation of sites. However, because of the remoteness of many of the streams in Washington, time and budgetary constraints would drastically reduce the target population of sites. This, in turn, affects the statistical assumptions underlying the probability design and methods are needed that deal with the issue of non-random site selection.

R-EMAP methods and probability sampling are better adapted to the statewide water quality assessment required under Section 305b of the Federal Clean Water (EPA, 1995). Present state sampling for the 305b assessment is often conducted on sites of known impairment, and thus may skew results toward greater impairment than actually occurs. If the sampling protocols are altered to complement the current state protocols, the use of random or stratified-random sampling techniques should provide a more representative assessment of water quality impairments statewide. Nonetheless, if the field sampling methods are not altered to complement current state invertebrate protocols as well as dropping sampling stream vertebrate populations from the field sampling methods, access issues and sampling costs associated with the R-EMAP design may continue to limit its usefulness in Washington.

References

Environmental Protection Agency (EPA): 1991, *Design Report for EMAP, the Environmental Monitoring and Assessment Program*, EPA/600/3-91/053.

Environmental Protection Agency (EPA): 1993a, *Surface Waters 1991 Pilot Report*, EPA/620/R-93/003, 201 pp.

Environmental Protection Agency (EPA): 1993b, *Regional Environmental Monitoring and Assessment Program*, EPA/625/R-93/012, 82 pp.

Environmental Protection Agency - Office of Research and Development (EPA-ORD): 1994, *1994 Field Operations and Methods Manual for Streams in the Coast Range Ecoregion of Oregon and Washington and the Yakima River Basin of Washington*, 63pp.

Environmental Protection Agency (EPA): 1995, *Guidelines for Preparation of the 1996 State Water Quality Assessments (305b reports)*, EPA/841/B95/001, 239 pp.

Fore, L.S., J.R. Karr, R.W. Wisseman: 1996, *J. N. Am. Benth. Soc.* **15**, 212-231.

Franklin, J. F. and C. T. Dyrness: 1984, *Natural Vegetation of Oregon and Washington*, Oregon State University Press, 452 pp.

Green, R. H.: 1979, *Sampling Design and Statistical Methods for Environmental Biologists*, John Wiley & Sons, New York, 257 pp.

Hughes, R. M., P. R. Kaufmann, A. T. Herlihy, T. M. Kincaid, L. Reynolds, D. P. Larsen: (in press), *Can. J. of Fish. Aquat. Sci.*

McPhail, J. D. and C.C. Lindsey: 1986, *Zoogeography of the freshwater Fishes of Cascadia*, in C. H. Hocutt (ed.), *The Zoogeography of North American Freshwater Fishes*, E.O. Wiley, NY, 866 pp.

McDonald, L. H., A. W. Smart, R. C. Wissmar: 1993, *Monitoring Guidelines to Evaluate Effects of Forestry Activities on Streams in the Pacific Northwest and Alaska*, EPA 910/9-91-001, 166 pp.

Thiele, S.A., C. W. Kiilsgaard, J. M. Omernik: 1992, *The Subdivision of the Coast Range Ecoregion of Oregon and Washington*, EPA-ORD (Corvallis, OR) Report, 20 pp.

Plotnikoff, R. W.: 1994, *Instream Biological Assessment Monitoring Protocols: Benthic Macroinvertebrates*, Washington Dept. of Ecology, Olympia, WA, 27 pp.

Plotnikoff, R. W. and J. S. White: 1996, *Taxonomic Laboratory Protocol for Stream Macroinvertebrates Collected by the Washington State Department of Ecology*, Washington Dept. of Ecology, Olympia, WA, 32 pp.

Reynoldson , T. B. and D.M. Rosenberg: 1996, *Study Design and Data Analysis in Benthic Macroinvertebrate Assessments of Freshwater Ecosystems Using a Reference Site Approach*, North American Benthological Society Technical Bulletin, Lawrence, Kansas, 69 pp.

Robison, E. G. and P. R. Kaufmann: 1994, *Evaluating Two Objective Techniques to Define Pools in Small Stream*, in R. A. Marston and V. R. Hasfurther (eds.), *Effects of Human-induced Changes on Hydrologic Systems*, Amer. Water Res. Ass., Summer Symposium Proceedings, Jackson Hole, WY, 1182 pp.

Rosenberg, D. M. and V. H. Resh: 1993, *Freshwater Biomonitoring and Benthic Macroinvertebrates*, Chapman & Hall, NY, 488 pp.

Strahler, A. N.: 1957, *Am. Geophy. Union Tran.* **38**, 913-920.

SITE ACCESS AND SAMPLE FRAME ISSUES FOR R-EMAP CENTRAL VALLEY, CALIFORNIA, STREAM ASSESSMENT

ROBERT K HALL, PETER HUSBY, GARY WOLINSKY, OLOF HANSEN, and MICHIKO MARES

Region 9, U. S. Environmental Protection Agency, 75 Hawthorne St., San Francisco, CA 94105

Abstract. The Central Valley of California contains critical habitat for many aquatic and terrestrial biological resources. The purpose of this R-EMAP project was to assess the effects from a highly modified agriculturally dominated landuse area on the aquatic resources of the lower portion of the Central Valley watersheds. The study area is 24,346 mi^2 and comprises the Sacramento Valley and San Joaquin Valley watersheds to the 1,000 ft. elevation contour. Populations of interest are man-made conveyances and wadeable natural streams. There are 40,756 miles of streams and constructed conveyances within the Central Valley as designated by RF3 database. Sample sites were selected to represent 14,399 miles of streams and sloughs, and 16,697 miles of constructed conveyances.

In an arid ecosystem, the presence or absence of water and the residence time of that water are primary factors for determining if a water body will support aquatic life. The flow regime in natural streams in this ecoregion varies both with the timing and intensity of the previous rainy season and human flow alterations, resulting in the optimal index period varying from year to year.

Site access to constructed conveyances and wadeable natural streams is affected by local ownership. Conveyances are generally owned and/or maintained by irrigation districts or water agencies, or in some cases maintained by large corporate farms.

This study assesses 31,096 (76%) of the total reach miles, as designated by RF3, of streams and constructed conveyances within the Central Valley. Overall, the R-EMAP probabilistic design provides a snap shot in time of how water is distributed throughout the Central Valley, and the potential for aquatic life in constructed conveyances and wadeable natural streams. The major obstacle to implementing and analyzing the R-EMAP design is obtaining landowner permission for site access.

INTRODUCTION

The Central Valley of California is one of the nation's most productive agricultural areas with approximately 31,000,000 acres in production. California agriculture uses approximately 80% of the State's water supply. The State Water Resources Control Board (SWRCB) identifies metals and pesticides contained in agricultural drainage as a major cause of aquatic impairment in Central Valley rivers and streams. Studies by Moyle *et al.*, (1986a and 1986b) and Saiki (1984) also indicate human-related activities such as water

Environmental Monitoring and Assessment **51:** 357–367, 1998.

withdrawals, contamination by agricultural wastes, and hydro-modifications as contributing to the decline of environmental conditions of aquatic biota.

The purpose of this R-EMAP Surface Water study is to assess the current status of aquatic resources in constructed conveyances and wadeable natural streams within the Central Valley, California. The purpose of this paper is to evaluate the success of the EMAP statistical design for assessing a highly modified and managed environment with large private land ownership. This paper discusses how private land ownership affects implementation of the statistical design, and lessons learned on identifying land ownership. This paper also discusses the usefulness of the US EPA River Reach File version 3 (RF3) as a representation of aquatic resources in the Central Valley.

PROJECT DESIGN

The ecosystem within the Central Valley has been highly modified and dominated by agricultural uses. Constructed conveyances and streams provide important habitat for biological communities. The study area is 24,346 mi^2 and comprises the Sacramento River and San Joaquin River watersheds to the 1,000 ft. elevation contour (Figure 1). According to the US EPA River Reach File version 3 (RF3) database, there are 40,756 miles of streams and constructed conveyances within the Central Valley. Sample sites are selected to represent 14,399 miles of the RF3 designated as streams and sloughs, and 16,697 miles of the RF3 designated as constructed conveyances within the Central California Valley and Southern and Central California Plains and Hills ecoregions. This study assessed 31,096 miles (76%) of the total reach miles, as designated by RF3, of streams and constructed conveyances within the Central Valley.

The areal extent of the study area is from just below Shasta Dam in the north southward to the Tehachapi Mountains. The study area is bounded by the Sierran foothills on the east and Coast Ranges on the west. The study area includes, completely or partially, 24 counties.

The objectives of the Central Valley R-EMAP are:

- Assess the current biotic condition of surface waters in the Central Valley
- Establish baseline conditions for different water body types in the Central Valley
- Correlate biological index measures with other available data (e.g., bio-toxicity monitoring of the Central Valley Regional Water Quality Control Board)
- Modify existing metric schemes/indices to better assess the current biotic condition of surface waters for this ecosystem
- Demonstrate to the EPA Region 9 and the state of California utility of EMAP indicators and sampling design for various environmental programs.

EMAP Stream Protocols are used for the study (US EPA, 1994). The index period for this study is from mid-July through the end of September, and the sampling was done in 1994 and 1995. The indicators measured are water chemistry, sediment chemistry (1995 only), physical habitat, benthic invertebrates, fish, and periphyton.

Fig. 1. Location map of the US EPA Region 9 Regional Environmental Assessment and Monitoring Program (R-EMAP) study area, Central Valley, California. Image is courtesy of the USGS, Flagstaff, Arizona.

SAMPLE DESIGN

Two populations of interest, constructed conveyances and streams, were identified according to the River Reach File version 3 (RF3) 1:100,000 scale Digital Line graph (DLG). The monitoring network was established by overlaying the national EMAP 40 km^2 hexagonal frame (Stevens, 1994) over the Central Valley. Hexagons were selected in the geographic target area of the Sacramento and San Joaquin river valleys according to the following RF3 water body types: Regular reach (R), Isolated reach (N) and Unknown (U). Annual rainfall for the Central Valley is heaviest from November to April with very little to no precipitation from May through October. Line segments designated as Intermittent (P) and Start (S) reaches were not used because of the high probability these reaches would be dry during the index period of August and September. A reconnaissance survey of the Central valley prior to sample selection found that approximately 95% of Start reaches were dry in April.

Sites were selected to represent the two main populations of interest with natural streams identified as "R" and "N", and constructed conveyances as "U." The site selection requirements were:

- Equal area sampling representation of the Sacramento and San Joaquin River watersheds
- Equal representation of natural stream courses (R, N) and constructed conveyances (U) line segment miles
- Equal representation by year for the two study years of 1994 and 1995
- Detection of trends in a set of indicators by revisiting at least 10% of the sites sampled the previous year (Stevens and Olsen, 1991).

Site selection was done by using RF3 as the database framework. Optimal statistical representation of aquatic resources in the Central Valley would be best achieved, with the available financial and personnel resources, if there were 40 sample sites from each population type (natural streams and constructed conveyances). It was difficult to discern from RF3 whether line segments would in fact contain water, be accessible, and be wadeable. In addition, it was anticipated that some landowners would refuse permission to enter sampling locations. Therefore, the number of prospective sampling sites selected were increased to compensate for these discrepancies. As a result, in 1994, 160 sites were initially selected to reach the statistical target of 40 sampled sites for each population - 77 (48%) in natural streams and 83 (52%) in constructed conveyances. However, even with the over selection of sampling sites, the target of 40 sampled sites of each population of interest was not reached in 1994. In 1995, using site access and sampling information from 1994, the number of sites selected was increased to - 122 (51%) in natural streams and 118 (49%) in constructed conveyances. In addition, to assess interseasonal variability, 10 sites from 1994 were randomly selected for revisits in 1995 - 4 in natural streams and 6 in constructed conveyances. In 1995 two sites were sampled twice to assess intra-seasonal calibration.

RECONNAISSANCE

Reconnaissance for this study was done in a two-step process. Step 1 was an office evaluation, and Step 2 was a field evaluation of the selected sites. For the two-year study, 400 sites were randomly selected throughout the Central Valley. Sample sites were selected from RF3 and mapped onto transparent overheads at 1:62,500 map scale, which is equivalent to a 7.5 min USGS quadrangle map sheet. Each overhead showed the R-EMAP hexagon, sampling site location, latitude, longitude, county, and 7.5 min. USGS topographical map name. Each overhead was overlaid onto the appropriate map sheet to determine altitude and directions to the site.

For 1994, sites with the highest probability of having water during the index period were identified for further reconnaissance. Sites on the topographic sheets which had a dashed blue line and were located on the south-western side of the valley were eliminated from further consideration because of the high probability of the stream being dry. Sites were also discarded if they were in steep and inaccessible terrain. Sites were selected for further field reconnaissance to verify the presence or absence of water, wadeability, accessibility, safety, or if ownership could be determined. A major portion of the sites field visited could be viewed without prior landowner permission. Reconnaissance was done to assess the site and to identify land ownership.

Site reconnaissance was accomplished by regional EPA staff. Upon reaching a site, it was determined whether it was sampleable or non-sampleable. The site was considered non-sampleable if it was:

- wider than 30 meters
- potentially dangerous
- dry
- no longer existing
- difficult to access, or having no visible access point

If the site was determined to be sampleable, then the approximate width, depth, directions, possible site contacts, and any additional comments were noted.

SITE ACCESS

The U.S. Department of Agriculture (USDA), Consolidated Farm Service Agency (CFSA) was contacted for assistance in identifying land ownership to receive permission to access property where the sample sites were located. CFSA agreed to provide property maps, ownership information (name and mailing address) and aerial photographs for R-EMAP sites. EPA regional staff supplied photo-copies of the sample sites located on the topographical maps along with the township, range, section, and latitude and longitude of the sites by county to the CFSA. The aerial photographs supplied by the USDA provided important information on the accessibility, physical habitat, and availability of water for the proposed sampling sites.

Land ownership was also determined by gathering information from the outdated topographical maps. Several sites were located within Fish and Wildlife National Refuges, Bureau of Reclamation, military installations, and state and city parks. Reconnaissance of the sites provided a few potential addresses and names to determine actual ownership of the sites. The State Department of Water Resources also provided assistance.

Following office and field reconnaissance, EPA staff contacted by mail landowners whose sites were deemed sampleable. To reduce any site selection bias, EPA staff also contacted landowners whose sites had no reconnaissance information. A form letter was mailed as initial correspondence to landowners, irrigation districts, federal agencies, etc. The letter stated a brief summary of the program describing the actual protocol and goals of the R-EMAP project. It also sought to determine ownership along with a request for access to the property. Included along with the form letter was a summary sheet about EMAP, distributed by EPA Office of Research and Development. An access permission form requesting approval or denial was attached with a copy of a topographical map showing the exact location of the site. Follow-up with landowners who did not respond to the form letter was done by telephone. In some cases initial contact was via the telephone followed up by the form letter. Many of the contacts expressed concern regarding whether R-EMAP data would be used for regulatory purposes against agricultural interests.

SITE ACCESS RESULTS

An effort was made to identify land ownership and garner some assessment information for all 400 selected sites. For sampling year 1994, 160 sites were randomly selected from the 40 km^2 grid. Of the 160 selected sites, 140 (88%) of the sites were assessed. Of the remaining 20 (12%) sites, access was denied to 10 sites, and the remaining 10 sites have no landowner or reconnaissance information. In 1995, 240 sites were randomly selected, of which 195 (81%) sites were assessed. Of the 45 sites not visited, access was denied to 10 sites, and the remaining 35 sites had no landowner or reconnaissance information regarding sampleability. Table 1 summarizes the response of landowners to requests for permission to sample. In 1994, permission was given by landowners at 34% of the sites;

18% of the landowners denied access. At 33% of the sites the landowner(s) did not respond, was unwilling to respond, or was unknown. Site selection errors in the RF3 framework database occurred at 15% of the sites.

Table 1. Landowner response to request for permission to sample.

Comments	1994 n = 160	1995 n = 240	Total n = 400
Permission given by landowners	55 (34%)	62 (26%)	117 (29%)
Denied access	29 (18%)	35 (15%)	64 (16%)
No response from landowner	52 (33%)	110 (46%)	162 (41%)
Error in RF3 framework	24 (15%)	33 (13%)	58 (14%)

For 1995, Table 1 shows permission was given by 26% of the landowners; 15% of the landowners denied access; 46% of the landowners did not respond; and site selection errors occurred in the RF3 database at 13% of the sites. Of the revisit sites from 1994, 8 sites were sampled, 1 site was denied access, and 1 site was unsampleable because of safety issues (i.e., water was too deep).

Table 2 summarizes the results of field sampling team visits to the 117 sites for which access permission was received. In 1994, of the 55 sites given access permission, 45 (82%) were sampleable, 4 (7%) were dry and 6 (11%) were not sampleable. Sites not sampled were determined to be non-target for safety reasons.

Table 2. Site visit results.

Year	Sampled	Dry	Not Sampled
1994 Total # of Sites	**45**	**4**	**6**
Conveyances	28	2	3
Natural streams	17	2	3
1995 Total # of Sites	**44**	**8**	**8**
Conveyances	33	2	6
Natural Streams	11	6	2
1994/1995 Total # of Sites	**89**	**12**	**14**
Conveyances	61	4	9
Natural Streams	28	8	5

In 1995, 72 sites were given access permission, 62 sites were new for 1995 and 8 sites were revisits from 1994, and 2 sites were intra-season revisits. Of the 62 new sites, 44 (73%) were sampleable, 8 (13%) were dry and 8 (13%) were not sampleable (Table 2).

Table 2 also summarizes the sites sampled by the populations originally identified as constructed conveyances and wadeable natural streams. In 1994, constructed conveyances made up 60% of the sites given access permission, and 62% of the sampleable sites (Table 2). In 1995, constructed conveyances made up 70% of the sites given access permission, and 75% of the sampleable sites (Table 2).

DISCUSSION

In an arid ecosystem, the presence or absence of water and the residence time of that water are primary factors for determining if a waterbody will support aquatic life. In the Central Valley, a highly modified and managed system, residence time of water in a channel is the main component for determining the presence or absence, and type of aquatic community which will be supported. Central Valley Water Districts, which maintain the constructed conveyances, indicate residence time of water ranges from 2 weeks to 6 months, depending on water demand. Under these conditions, the selection of the index period is critical. The index period should correspond to the time when the target assemblages are at their highest taxonomic diversity and water and flow conditions are appropriate for sampling protocols (Gibson *et al.,* 1994). For the conveyances in the study, the optimal index period would be as late in the irrigation season as possible to allow for the biological communities to mature. However, given the range of residence times for water in channels, dry conditions should be expected to be common. As such, information on the presence or absence of water and the time water has been present is valuable.

The flow regime in natural streams in this ecoregion varies both with the timing and intensity of the previous rainy season and human flow alterations. As a result, the biologically optimal index period for streams will vary from year to year, and indeed the presence or absence of water in the streams at a certain index period may change annually. Information on the flow status of these streams and constructed conveyances then becomes a valuable indicator of the biotic condition of the overall resource.

Consolidating information on the number of dry sites with sampled sites increases the total number of sites and total area evaluated for this study (Table 3). For the two sampling years, 44% of the total sites can be assessed for aquatic resources (53% for 1994 and 38% for 1995). Statistical characterization of the study area is further strengthened by eliminating non-target sites (errors in the RF3 framework database; sites altered by urbanization, agricultural use, or some other land management practice; unsampleable sites because of safety, wetland, or submerged by dam construction).

Table 3. Assessed sites.

Comments	1994 n = 160	1995 n = 240	Total n = 400
Target - Wet sampled	45 (28%)	44 (18%)	89 (22%)
Target - Dry	40 (25%)	47 (20%)	87 (22%)
Target - Access denied	13 (8%)	19 (8%)	32 (8%)
Target - No response from landowner	9 (6%)	46 (19%)	55 (14%)
Non Target	33 (21%)	39 (16%)	72 (18%)
Not Assessed - Access denied	10 (6%)	10 (4%)	20 (5%)
Not Assessed - No response from landowner	10 (6%)	35 (15%)	45 (11%)

The RF3 framework upon which the site selection was based does contain map errors. In addition, the RF3 reach types selected, especially the U type (constructed conveyances), contained a variety of sizes of channels, some of which were unsampleable. As a result, a number of sites selected were found to be outside the target population for the study and methods used. A total of 18% of the selected sites (21% for 1994 and 16% for 1995) were eliminated because water or a channel were not present at the index site (due to map error, altered, moved as a result of urbanization, agricultural activity, etc.), or could not be sampled because of unsafe conditions (too large, high stream flow, cement lining, difficult access, etc.). If unsampleable sites are removed from the study as non-target, the total number (n) of sites is reduced to 328 (n = 328), n = 127 for 1994, and n = 201 for 1995. Using sampled and dry sites as the key components for assessing the Central Valley, the area evaluated improves to 54% overall (67% in 1994 and 45% in 1995).

As illustrated in Table 3, the largest obstacle to the implementation of the study was landowner permission for access. From the reconnaissance information gathered, sites were determined to be sampleable if they:

- were within water bodies containing flowing water, or that would contain flowing water during the sampling index period;
- were easily accessible by the field crews;
- had some definable habitat within the reach, if dry.

It was not possible to access a number of sites, even for reconnaissance. Therefore, not all of the 400 sites were pre-visited. Additionally, "no response" from the landowners resulted in having no information of any kind to make a minimal assessment. In total, 16% of the sites had no assessment information, 12% in 1994 and 19% in 1995.

For sites that were assessed from the reconnaissance survey and found to be sampleable overall, 20% were denied access, or there was no response from the landowner. By combining Target (Access denied/No response from landowner) with Not assessed (Access denied/no response from landowner) from Table 3 it can be seen that 38% of all sites, 26% in 1994 and 46% in 1995, could not be completely assessed because of private ownership/access questions.

The two populations of interest, conveyances and wadeable natural streams, were affected differently by ownership. Conveyances were generally owned/maintained by irrigation districts or water agencies, or in some cases maintained by large corporate farms. Natural streams tended to flow across private lands, making permission critical to arriving at the designated latitude and longitude, or center point (X point) of the reach. The public or quasi-public agencies and large corporate farms provided access to the resource more often than the smaller private farmer and landowner.

As seen in Table 2, the over-selection of constructed conveyances for 1995, did not overly increase the number of sampleable and accessible sites. With the additional over-selection the target number of 40 sites for constructed conveyances was almost achieved for 1995. However, the assessment of natural streams did not improve by over-selection of sites. The low number of natural stream sites will make it difficult to reach statistically valid assessment conclusions for this population. The natural streams for which permission was received may not represent all streams in the valley, and a closer examination of the stream orders and geographical distribution of the sites assessed will be necessary to evaluate the success of the characterization.

Overall, the EMAP probabilistic design did what was asked of it. It has given the authors a snapshot in time of how water is distributed throughout the Central Valley, and the potential for aquatic life in these waterways.

CONCLUSION

- The RF3 database framework, for this study area, was not an accurate representation of the stream resources in the Central Valley. The RF3 Start Reach Type, which was not used in this study, and which represented 20% of the stream miles in the Central Valley, was over 95% dry or nonexistent.
- In an agricultural area with extensive private land ownership, site access is the single biggest impediment to achieving the projected goals of the statistical design. To best implement the EMAP statistical design, a major increase in personnel time and resources needs to be expended on site access.
- Site access for conveyances was more successful when the sites were owned by agencies or public/water districts. Coordination with the local Regional Water Quality Control Board (RWQCB), Natural Resources Conservation Services (NRCS), the local

Resource Conservation Districts (RCD), Agricultural Stabilization and Conservation Service (ASCS), Agricultural Industry and the community is critical in getting permission for site access and cooperation in the monitoring effort.

- For arid environments, where flows are often regulated, sampling during more than one index period may provide a better assessment of the status of the biological resource, and how the biological community responds to the presence or absence of water.
- The design is probably adequate for assessing constructed conveyances of the RF3 U Reach/line segment type.

ACKNOWLEDGEMENT

The authors would like to acknowledge Dr. Brian Hill, Dr. Anthony Olsen and Dr. Don Stevens for their efforts in guiding the project and their critical review of this article. We would also like to acknowledge Mary Dunne and Curtis Hagan, California Department of Fish and Game, for leading the field sampling crews and performing the fish identification, and John Smythe and Robert Molleur of the Agricultural Stabilization and Conservation Service (ASCS) for their assistance in obtaining land ownership information. A special thanks to Ms. Janet Hashimoto for her program knowledge and guidance, and to Ms. Diana Woods and Ms. Janet Parrish for their review and editorial comments of this article.

REFERENCES

Gibson, G.R., Barbour, M.T., Stribling, J.B., Gerritsen, J., and Karr, J.R.: 1994, *Biological Criteria: Technical guidance for streams and small rivers*, U.S. Environmental Protection Agency, Office of Science and Technology.

Moyle, P.B., *et al.*: 1986a, *Final Report on Development and Preliminary Tests of Indices of Biotic Integrity for California,* Final Report to the US EPA Environmental Research Laboratory, Corvallis, OR.

Moyle, P.B., *et al.*: 1986b, 'Evaluating the Condition of California's Streams using Indices of Biotic Integrity: Evidence for Continuing Decline', *Technical Completion Report W-659,* Water Resources Center, UC-Davis, Davis, CA.

Saiki, M.K.: 1984, *Environmental Conditions and Fish Faunas in Low Elevation Rivers on the Irrigated San Joaquin Valley Floor, California;* California Department of Fish & Game 70(3), 145-157.

Stevens, Don L., Jr., and Olsen, Anthony R.: 1991, 'Statistical Issues in Environmental Monitoring and Assessment', *Proceedings of the Section on Statistics and the Environment, 1991*, American Statistical Association, 10.

Stevens, Don L., Jr.: 1994, 'Implementation of a National Environmental Monitoring Program', *Journal of Environmental Management* **42**, 1-29.

US EPA: 1994, *Environmental Monitoring and Assessment Program Surface Waters and Region 3 Regional Environmental Monitoring and Assessment Program: 1994 Pilot Field Operations and Methods Manual for Streams;* Klemm, Donald J., and Lazorchak, James M., editors, EPA/620/R-94/004, March.

LINKING MONITORING AND EFFECTS RESEARCH: EMAP'S INTENSIVE SITE NETWORK PROGRAM

J. KEVIN SUMMERS[1] and KATHY E.TONNESSEN[2]

[1]*U.S. Environmental Protection Agency, National Health and Environmental Effects Research Laboratory, Gulf Ecology Division, 1 Sabine Island Drive, Gulf Breeze, FL 32561-5299,* [2] *U.S. National Park Service, Air Resources Division, P. O. Box 25287, Denver, CO 80225.*

Abstract. The EMAP program has been organized into three primary elements: Multi-Tier Design, Indicators, and Index Sites. The Index Sites program (DISPro - Demonstration Intensive Site Project) is the primary activity within the Index Sites element of EMAP. This project represents an inter-agency effort between EPA/ORD and DOI/NPS to develop a demonstration of an intensive site network of monitoring and research locations throughout the United States, utilizing the Nation's parklands as "outdoor laboratories." Twelve parks were selected to establish this demonstration. These 12 parks were selected because they are readily accessible, have a history of monitoring environmental information, and represent a broad spectrum of ecological communities. EMAP, through DISPro, is examining whether a "network" of sites existing within the parks can be used to address monitoring issues for global-scale environmental stressors (e.g., air pollution) as well as locale-specific stressors (e.g., air deposition, water-borne) and coordinated with cause-effect, issue-based research related to these environmental stressors. As a first activity, EPA will provide each of the sites with the instrumentation to monitor UV-B. The intent of the program is to initiate a consistent air monitoring program at each site to be followed by consistent monitoring within other media. The project will initiate research projects at all the sites (eventually) to examine the effects of environmental stressors of importance at each of the sites.

1. Introduction

The development and demonstration of the utility of a network of intensively monitored index sites is one of the four major components of U.S. EPA's next phase of the Environmental Monitoring and Assessment Program (EMAP) (Paulsen 1997). The initial phase of EMAP focused on the demonstration of utilizing probabilistic sampling designs to assess the condition of the Nation's resources at a regional scale (Messer *et al.* 1991). EMAP-Phase II is comprised of four major elements: Indicator Research and Development; Regional Geographic Initiatives; Intensive Site Network Development; and Special Research for Monitoring and Assessment Problems (REMAP). This incorporation of the intensive site component directly relates to criticisms leveled at the earlier stages of the program by a review by the National Research Council (NRC 1995). This review suggested that the EMAP approach could benefit from the strategic placement of long-term sites that were intensively monitored to establish linkages between observed changes in environmental stressors and concomitant changes in ecological resources. This approach was incorporated into EMAP-Phase II planning in 1994 and into the planning of other entities addressing the development of environmental monitoring networks; namely, the Committee for the Environment and Natural Resources (CENR 1997). In its strategic plan, the CENR also describes the use of a local monitoring tier representing defined index locations where intensive monitoring is used to assess cause-effect relationships. As a

Environmental Monitoring and Assessment **51:** 369–380, 1998.

result of these two activities outside EMAP, the concept of a demonstration network of intensively monitored sites has been established within EMAP-Phase II.

2. The Concept of an Intensive Site Network

The development of a intensive site network for examining the short-term variability in long-term trend behavior was initiated as an investigation into the questions, "Can you develop a set of index sites that is representative of national trends in ecological behavior?" and "Can you use a set of index sites to realistically examine the relationships between changes in environmental stressors and ecosystem behavior?" Investigators within EMAP quickly came to the conclusion that the first question would not produce fruitful results. The results from early EMAP monitoring, while having some difficulties, clearly demonstrated that *a-priori* selection of sites could not guarantee "representativeness." Representativeness resulted from the probabilitistic nature of the monitoring sample set, not from the selection of a single site for monitoring. Given that cost criteria would likely limit the number of intensive sites to less than 20 to "represent" all terrestrial and freshwater environments, *a-priori* selection of the sites were unlikely to produce representative behavior of "all" of these ecosystems. Also, EMAP-Phase I had not completed its demonstrations in these ecosystems so that the results could not be used to identify representative locations.

As a result, representativeness of all terrestrial/freshwater environments was not an initial constraint placed on the development of an intensive site monitoring network. The network would be developed as a set of long-term trend sites where research would be supported to examine the interactions of environmental stressors and environmental effects. This research would focus on the "why" questions concerning observed ecological effects. Thus, much of the emphasis for the development of an intensive site network was placed on the effects research question. In order to address this question, active ecological research programs would be required at the intensive sites that are based on the research and monitoring issues known or determined for the sites (i.e., specific relationships between stressors and effects), as well as, the broader regional- and national-scale research issues (e.g., air deposition patterns and their effects of terrestrial/aquatic ecosystems).

Two primary approaches to the development of an intensive site network were examined: (1) utilize existing site networks such as NSF's Long-Term Ecological Research (LTER) network (NSF 1993) or the U.S. Forest Service's Forest Health Monitoring Program site network (Forest Health Monitoring 1994) and (2) the development of an EPA network that could be combined with other networks at a later date. Examination of existing networks rapidly showed that they were created to address specific objectives and were incompatible with the EMAP objectives of long-term monitoring of environmental stressors and associated issue-driven effects research including all major ecosystem types. The LTER sites included all ecosystem types but focused almost exclusively on cause-and-effect research, virtually excluding monitoring activities. The Forest Health Monitoring Network examined forest condition primarily from an economic perspective and supported little ecological monitoring and research. As a result, the EMAP Intensive Site Network was determined to be developed outside existing networks with a core set of monitoring indicators. The network would consist of two elements representing terrestrial ecosystems and coastal ecosystems. However, it would be constructed in a way that other networks would be complementary for the assessment of specific issues. Strategically, the EMAP Intensive Site Network needed to meet the following criteria:

1) The selected sites would be of interest to multiple federal agencies and at least one agency was interested in participating in the development of the network.;

(2) The selected sites would represent all major ecosystem types.

(3) The selected sites would have long-term accessibility (i.e., no potential for property rights issues).

(4) The selected sites would have some degree of environmental monitoring and/or ecological effects research already in place.

Because federal resource agencies, other than EPA, tend to be focused on either terrestrial or coastal ecosystems (but rarely both), EMAP decided that the two elements of the Intensive Site Network would be developed separately but consistently with the above criteria. The initial Federal agencies that should be contacted to co-develop these two network elements seemed obvious: Department of Interior (DOI) for the terrestrial systems and the National Oceanic and Atmospheric Administration for the coastal systems. The remainder of this paper will describe our interactions with DOI to develop a 12-site terrestrial/aquatic intensive site monitoring/research network.

3. DISPro - A National Park Network of "Outdoor Laboratories"

Examination of existing networks within the Department of Interior (e.g., NAWQA in USGS) showed the same difficulties that were experienced when trying to force the EPA criteria on the LTER Program or the Forest Service Programs. In short, the DOI networks were constructed with specific objectives that addressed some, but not all, of the EPA criteria. For example, NAWQA examines primarily only one ecosystem type—rivers and streams. While NAWQA is watershed-based, it focuses almost exclusively on stream condition with regard to chemistry with some elements dedicated to other biota. Relatively little effects research is associated with the program. With the merger of the National Biological Service into USGS (its new Biological Resources Division), perhaps NAWQA will incorporate more biological aspects. However, in the short term, EMAP continued to examine other network possibilities.

The National Park system seems to provide at least the potential for all terrestrial/aquatic ecosystem types but the existence of ongoing monitoring/research programs and an interest to participate in the construction of the network needed to be established. Knowing that the success of the network construction was dependent on "buy-in" at the grassroots as well as organizational level, EMAP contacted several park superintendents to establish interest before interacting with mid- and upper-level park personnel. These initial contacts were greeted with some suspicion, and a great number of questions, but a general acceptance that the concept of an intensive monitoring/research network centered in the National Parks of the United States seemed intriguing or, more importantly, plausible. These early discussions also provided several significant inroads into EMAP's question regarding ongoing activities and possibilities for collaboration. At the time of these discussions, the National Park Service was developing its Inventory and Monitoring Program at selected parks and several environmental stressors, particularly air stressors, were being monitored at many of the park sites. In fact, many of the goals and objectives of the I&M Program were similar to those of EMAP.

At this time, EMAP entered into discussions with the National Park Service's Air Resources Division and Inventory and Monitoring Program to co-develop a 10-15 site terrestrial intensive monitoring/research network. In short, both agencies would contribute funds and efforts toward this development with the intent to invite other DOI groups (e.g., Fish and Wildlife Service, U.S. Geological Survey) as well as other federal agencies (e.g., Department of Agriculture, U.S. Forest Service) to participate in the longer term. In 1996, EPA and NPS created a formal interagency agreement to create DISPro, the Demonstration of Intensive Sites Project. This project represents an inter-agency effort between EPA/ORD and DOI/NPS to develop a demonstration of an intensive site network of monitoring and research locations throughout the United States utilizing the Nation's parklands as "outdoor laboratories." Twelve parks were selected to establish this demonstration. Table I lists the parks and the ecosystem types that they represent. All 12 parks are readily accessible, have a history of monitoring environmental information, and represent a broad, sometimes unique, spectrum of ecological communities. EMAP, through DISPro, is examining whether a "network" of sites existing within the parks can be used to address monitoring issues for global-scale or regional-scale environmental stressors (e.g., air pollution) as well as locale-specific stressors (e.g., air deposition, water-borne) and coordinated with cause-effect, issue-based research related to these environmental stressors. The intent of the program is to initiate a consistent air monitoring program at each site to be followed by consistent monitoring within other media. Part of EPA's contribution to the interagency agreement is to provide some instrumentation to improve this consistency and to initiate some monitoring activities at all the DISPro sites. An example of the existing, ongoing monitoring activities at the DISPro sites is shown in Table II for air-related stressor monitoring. In order to demonstrate the relevancy of this monitoring, the project will initiate research projects at all of the sites (eventually) to examine the effects of environmental stressors of importance at each of the sites.

TABLE I
National Parks included in DISPro with ecosystem types.

National Park	Ecosystem Type
Big Bend National Park, Texas	Arid, Multiple Elevations
Everglades National Park, Florida	Tropical Wetlands, Lagoon, Coral Reefs
Virgin Islands National Park, Virgin Islands	Coral Reefs, Tropical Estuaries, Tropical Forests
Sequoia National Park, California	Multiple Elevation Forests, Unique Species, High Elevation Lakes
Rocky Mountain National Park, Colorado	High Elevation Forests and Lakes, Tundra
Great Smoky Mountains National Park, North Carolina	Multiple Elevation Forests, Lakes, Streams, Unique Species
Shenandoah National Park, Virginia	Multiple Elevation Forests and Streams
Acadia National Park, Maine	Rocky Fjord Estuaries, Northeastern Coastlines, High Elevation Vegetation, Lakes and Streams
Denali National Park, Alaska	Subarctic Ecosystems, High Elevation Forests, Glaciers, Tundra
Olympic National Park, Washington	Pacific Northwest Humid Ecosystems, Precipitation Gradients, Multiple Elevation Forests, Lakes, and Streams
Glacier National Park, Montana	High Elevation Forests, Lakes and Streams, Glaciers, East-West Gradients
Canyonlands National Park, Utah	Arid Ecosystems, Cryptogamic Species

4. Organizational Development of DISPro

Once the interagency agreement was in place, EPA and NPS developed an organizational structure to manage DISPro. This management included the determination of monitoring and research needs at individual sites and across sites, arrangement for those needs to be addressed, and management of the information to come from these sites. In late 1996, EPA and NPS initiated the DISPro Oversight Committee with membership from both agencies and, through its members, ties to CENR, NPS' I&M Program, and EPA's Offices of Air and Water.

The primary charge of the DISPro oversight committee is the responsibility for successful execution of the DISPro program. The activities of the Oversight Committee consist of five elements: (1) Organization of DISPro into a functional program; (2) Provision of guidance to the organizational units; (3) Responsibility for ensuring the execution of DISPro activities; (4) Selection of monitoring and research projects to be funded each annual cycle, and (5) Development of a long-term (5-10 years) Inter-Agency Agreement between EPA, NPS, and other interested parties to establish a ecological monitoring-research network throughout the United States.

Table II
Current(X) and Added (*) Air Quality Monitoring at DISPro Locations

National Park	O_3	SO^2	NO_x	VOC	NDDN	MET	IMPROVE	VIS(T)	VIS(N)	WET	UV-B	Hg	Organics	Metals	EDCs
Acadia	X	*	X	X	*	X	X		X	X	*	*	*	*	*
Big Bend	X	*	*		X	X	X	X		X	*				
Canyonlands	X	*	*		X	X	X	X			*	*	*	*	*
Denali	X	X[1]	*		*	X	X			X	*				
Everglades	X	*	*		*	X	X			X	*	*	*	*	*
Glacier	X	*	*		X	X	X	X		X	*				
Great Smoky Mountains[2]	X	X	X	X	X	X	X		X	X	*				
Olympic	X	X	*		*	X	*			X[3]	*				
Rocky Mountain	X	*	*		X	X	X	X		X	*				
Sequoia/ Kings Canyon	X	*	*		X	X	X			X	*	*	*	*	*
Shenandoah NP	X	X	X	X	X	X	X	X		X	*				
Virgin Islands	*	*	*		*	*	X[4]			*	*				

NOTES:

[1] Integrated sampling using filter pack on IMPROVE sampler; 2 24h samples/week

[2] There are multiple ozone monitoring locations at the park; not all parameters measured at same location

[3] Wet deposition sampler located on west side; AQ station located on north side near pulp mills

[4] IMPROVE module A only ($PM_{2.5}$ plus elements, B_{abs}, H)

Unless otherwise indicated measurements are made on a continuous basis

ABBREVIATIONS:

VOC	Speciated volatile organic compounds using SS canisters; 1-3h samples on event basis
NDDN	National Dry Deposition Network Filter Pack (SO_2, HNO_3, NH_4+, SO_4=); weekly samples
MET	Most NPS stations measure WS, WD, RH or dew pt., T, solar radiation (pyranometer), delta T at NDDN sites, Precip
IMPROVE	National visibility monitoring network sampler, 2 24h samples/week
VIS(T)	Transmissometer used for optical measurements (b_{ext})
VIS(N)	Nephelometer used for optical measurements (b_{scat})
WET	National Atmospheric Deposition Program wet deposition sampler; weekly samples

The Oversight Committee created twelve committees, one corresponding to each park unit, to establish the monitoring and research needs at each site. Each site-needs committee is comprised of three to five members representing the park, academic researchers involved in park research and, when available, the EPA Region where the park is located. Each site-needs committee developed a prioritized list of monitoring and research needs for their site to characterize the long-term exposure of parklands to environmental stressors, including anthropogenic stressors. However, anthropogenic stressors do not include human use of the parklands and its facilities as a public land. This list could include anthropogenically induced and natural stressors such as air deposition of contaminants and/or nutrients, visibility issues as they relate to ecological concerns (not as they relate to human visual experience), water contamination and eutrophication, groundwater contamination, habitat destruction, disease, and ozone/UV exposure. Natural "stressors" might include temperature, water flow, rainfall, habitat distribution patterns, or lightning fires as examples. This list was based on the long-term environmental monitoring and research needs of their individual park sites rather than short-term activities related to park operation. In parallel, the prioritized needs list could request field and laboratory research efforts to ascertain the biological/ecological effects of these stressors, particularly with regard to their effects on population, community, and/or ecosystem dynamics. An example of this emphasis would be a monitoring contaminant may have a biochemical or physiological effect on a fish that is measurable. This occurrence represents an effect but the broader question is whether this effect has any biological or ecological meaning. For example, does the measured effect result in changes in reproductive success, growth, population-level mortality, or ecological function. These broader effects are the primary interests of the research activities in DISPro.

Based on these site reports and requests from either EPA or NPS for broader-scale monitoring or research activities, the Oversight Committee developed an Annual Monitoring and Research Plan for the DISPro including not only site-specific needs but also multi-site monitoring and research activities needed to address important EPA, DOI, or NPS issues (e.g., broad-scale nitrogen deposition effects, broad-scale biodiversity changes). On the basis of this plan, specific topics were selected for which EPA/NPS then solicited research proposals from external academic sources and other federal agencies. The selected topics are listed in Table III. The process used to define the research and monitoring needs is depicted in Figure 1.

5. Merging DISPro and Other Networks to Address National Issues

It is now widely documented that reduced ozone results in increased levels of ultraviolet (UV) radiation, especially UV-B (280-320nm), incident at the surface of the earth (Watson 1988; Anderson *et al*,. 1991; Schoeberl and Hartmann 1991; Frederick and Alberts 1991; WMO 1991; Madronich 1993; Kerr and McElroy 1993). There is considerable and increasing evidence that these higher levels of UV-B radiation may be detrimental to various forms of biota ranging from freshwater amphibians, to arid-based reptiles, to

TABLE III
Recommended Project Areas for DISPro Monitoring/Research Competition

1.	Complete basic air monitoring stations for all 12 parks, to include visibility, wet and dry deposition, and gaseous air pollutants. Candidate Park Sites: All
2.	Set up of demonstration sites in 2-4 parks to monitor atmospheric deposition of organic contaminants. Candidate Park Sites: Everglades, Acadia, Sequoia or Canyonlands.
3.	Inventory of amphibian species for all 12 parks, with surveys to include incidence of deformities and numbers of species. Interagency activity with USGS/BRD.
4.	Effects Research in four major topic areas will be proposed in the form of Requests for Assistance (RFAs) or Announcements of Opportunity (AOs) for clusters of parks: UV-B effects on aquatic and terrestrial resources: Most likely resources are coral ecosystems, near-coastal plankton communities and amphibian/reptile populations. Candidate parks include Virgin Islands NP, Everglades NP, Big Bend NP, and Denali NP. Nitrogen effects on upland aquatic and terrestrial systems: Examination of the effects of air deposition of nitrogen on the biogeochemical cycles in forest, lakes, streams, and coastal waters. Candidate parks are Great Smoky Mountains NP, Shenandoah NP, Canyonlands NP, Acadia NP, Rocky Mountain NP and Olympic NP. Ozone effects on forests: This research would primarily focus on the effects of concentration gradients on ozone effects. Examples of these gradients might be age-structure in forests, elevation gradients, or longitudinal gradients. Candidate parks are Sequoia NP, Olympic NP, Acadia NP, Rocky Mountain NP, Denali NP, Shenandoah NP, and Great Smoky Mountains NP. Problems of temporal and spatial variability in environmental measurements: This research is needed but many on the committee felt that this topic would cost too much for DISPro to make much progress. One suggestion was to approach The National Center for Exploratory Research and Quality Assurance (NCERQA) and the Science To Achieve Results program (STAR) with regard to spatial variability as a research topic in FY98. However, the committee felt a focused RFA concerning spatial variability of wet/dry deposition or ozone exposure mean be successful. Candidate Parks: Sequoia NP, Olympic NP, Glacier NP, Great Smoky Mountains NP and Shenandoah NP.
5.	Routine monitoring of other media, e.g. water quality. The first task will be to determine the quality and quantity of water data available at the DISPro parks and then establish a standard monitoring protocol for water resources.

marine life inhabiting the upper layers of the oceans. With respect to aquatic ecosystems, we know this biologically-damaging mid-ultraviolet radiation can penetrate to ecologically-significant depths in marine and freshwater systems (Jerlov 1950, Lenoble 1956, Smith and Baker 1979, 1980, 1981; Kirt *et al.* 1994). Several books have been recently published reviewing various aspects of environmental UV photobiology (Young *et al.* 1993), UV effects on humans, animals and plants (Tevani 1993), and UV research in freshwater ecosystems (Williamson and Zagarese 1994). In short UV-B radiation in aquatic systems: 1) affects adaptive strategies (e.g., motility, orientation; 2) impairs important physiological functions (e.g., photosynthesis and enzymatic reactions); and 3) threatens marine organisms during their developmental stages (e.g., young-of-year finfish, shrimp larvae, crab larvae).

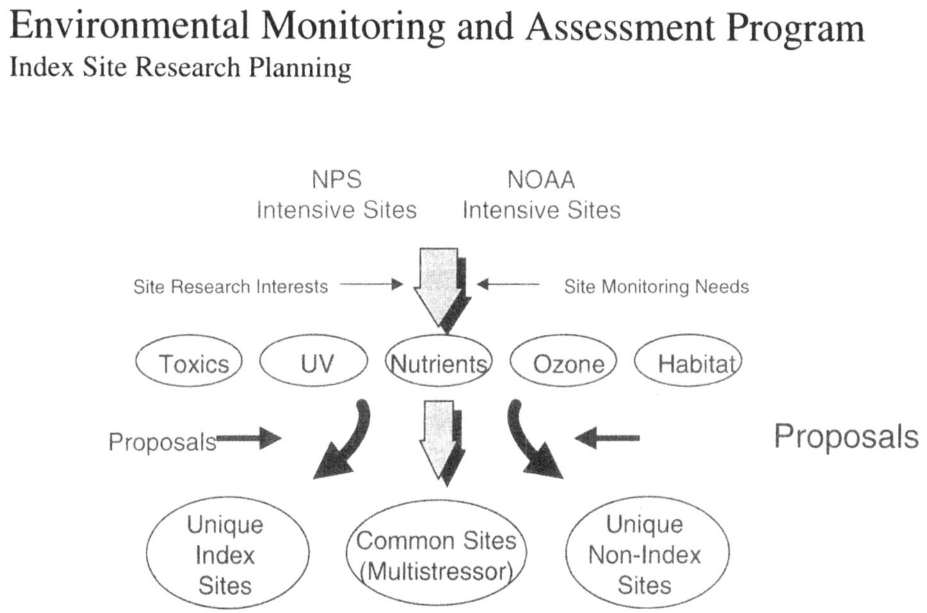

Fig. 1. Process Used to Determine Monitoring/Research Needs.

Possible consequences to aquatic systems include: reduced biomass production; changes in species composition and biodiversity; and alteration of aquatic ecosystems and biogeochemical cycles associated with the above changes. Consensus is building toward the view that current levels of UV play a major role as an ecological determinant, influencing both survival and distribution of organisms.

The EPA's Office of Research and Development is presently constructing a UV radiation monitoring network throughout the United States to document changes in land- and water surface exposure and to document the ecological and human health related effects of these changes. The network, constructed in concert with the National Park Service, includes 12 sites located in National Parks to represent areas often termed "pristine" with regard to other environmental stressors and 10 sites located in or near urban centers (Figure 2). However, the ubiquitous nature of UV exposure renders remote locations just as susceptible to UV effects as other locations. The parks include a variety of ecosystem types: tropical lagoons, coral reefs, and wetlands in Everglades National Park; arid systems in Big Bend NP, subarctic polar ecosystems in Denali NP, high elevation lakes in Rocky Mountain NP, and old growth forests in Sequoia NP are examples. The human health sites include urban areas where the incidence of melanoma and non-melanoma skin cancers are documented (Table IV). The network of UV monitors will be in place by July 1997 measuring UV exposure daily. As described earlier, to complement the monitored levels of UV exposure, ORD, through its Environmental Monitoring and Assessment Program, will sponsor ecological effects research examining the potential biotic damage resulting from the measured changes. This research will be accomplished in the form of grants for original investigator-initiated research, cooperative agreements between EPA research staff and academic researchers, and interagency agreements between EPA and other federal researchers, particularly NPS, and USGS researchers.

6. Conclusions

The Demonstration Intensive Site Project (DISPro) represents one of the first attempts to co-locate monitoring and research activities by multiple federal agencies in the National Parks to examine the long-term trends and effects of global-level and regional-level environmental stresses on terrestrial ecosystems. This program, initiated by EPA/ORD's Environmental Monitoring and Assessment program (EMAP) and the DOI/NPS's Air Resources Division in 1996, is designed to demonstrate that planned interaction by multiple federal agencies: (1) can provide useful monitoring and research data to all participating

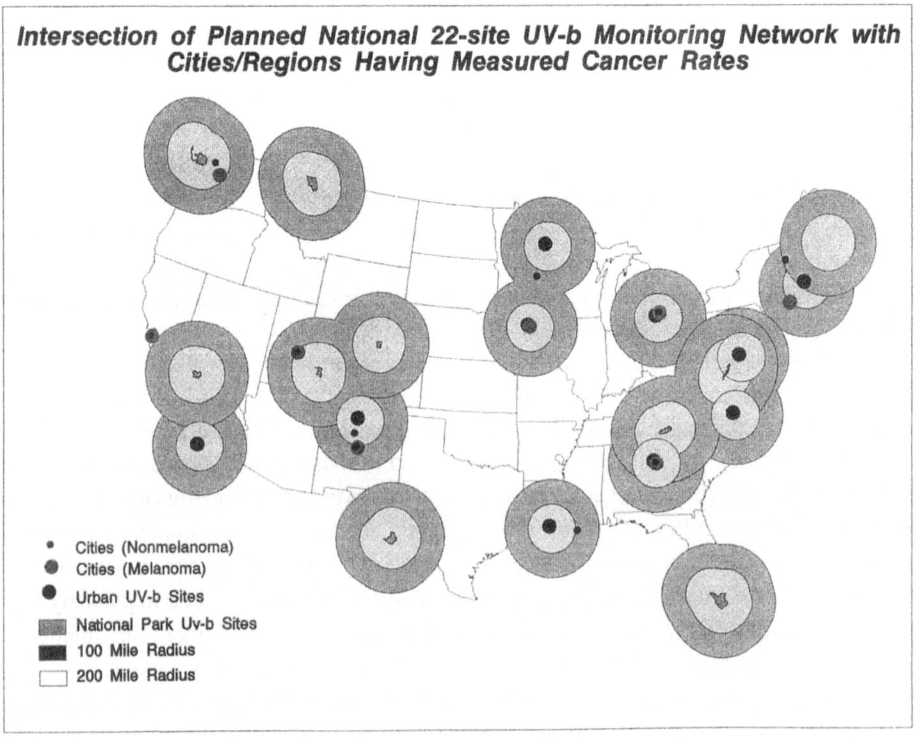

Fig. 2. Locations of UV-b Sampling.

agencies, (2) can incorporate research ideas and activities of academia to broaden the knowledge base applied to the global environmental issues of today, (3) can use the National Park System as an "outdoor laboratory" to examine global environmental stressors without the interference of the local environmental stressors generally encountered in unprotected areas, and (4) can work.

TABLE IV
UV-b sampling sites corresponding to cities with cancer registries

City or State	Type of Registry	Proposed Collection Site
Seattle, WA	Non-melanoma	Olympic NP
West Washington State	Melanoma	Olympic NP
New Hampshire/Vermont	Non-melanoma	Acadia NP/Boston, MA
Connecticut	Melanoma	Boston, MA
Detroit, MI	Non-Melanoma	Grosse Ile, MI
	Melanoma	Grosse Ile, MI
Minneapolis/St. Paul, MN	Non-Melanoma	Duluth, MN
Iowa	Melanoma	Ames, IO
Utah	Non-Melanoma	Canyonlands NP
	Melanoma	Canyonlands NP
San Francisco, CA	Non-Melanoma	San Francisco, CA
	Melanoma	San Francisco, CA
Atlanta, GA	Non-Melanoma	Atlanta, GA
	Melanoma	Atlanta, GA
Albuquerque, NM	Non-Melanoma	Las Cruces, NM
	Melanoma	Las Cruces, NM
San Diego, CA	Non-Melanoma	San Diego, CA
New Orleans, LA	Melanoma	Lafayette, LA

Acknowledgments

The authors would like to thank the numerous people involved in trying to make DISPro a successful effort including Miguel Flores, John Ray, Bill Barnard, Tom Heitmuller, John Karish, Cat Hoffman, Gary Williams, Bill Hogsett, Hal Walker, Jay Messer, Larry Cupitt, Steve Hedtke, Gilman Veith, Rick Linthurst, and Foster Mayer. The authors acknowledge the untimely passing of John Christiano, Director of the NPS Air Resources Division, an important link in developing the original interagency agreement between EPA and NPS.

References

Anderson, J.G., Toohey, D.W. and Brune, W.H.:. 1991, *Science* **251**, 39-46. Committee on Environment and Natural Resources (CENR): 1997. *Integrating the Nation's Environmental Monitoring and Research Networks and Programs; A Proposed Framework*, National Research and Technology Council, Washington, D.C.

Forest Health Monitoring: 1994, *Forest Health Monitoring 1992 Annual Statistical Summary*, EPA/620/R-94/010.

Frederick, J.E. and Alberts, A.D.: 1991, *Geophys. Res. Let.* **18**, 1869-1871.

Kerr, J.B. and McElroy, C.T.: 1993, *Science* **262**, 1032-1034.

Madronich, S.: 1993, *The atmosphere and UV-B radiation at ground level*, In: Environmental UV Photobiology. Plenum Press, New York.

Messer, J.J., Linthurst, R.A. and Overton, S.W.: 1991, *Environ. Monitor. Assess.* **17**, 67-78. National Research Council (NRC): 1995. *Review of EPA's Environmental Monitoring and Assessment Program*, National Research Council, Washington, D.C.

National Science Foundation (NSF): 1993. *Ten-Year Review of the National Science Foundation Long Term Ecological Research (LTER) Program*, National Science Foundation, Washington, D.C.

Paulsen, S.: 1997, *A Research Strategy for the Environmental Monitoring and Research Program, Phase II.* U.S. Environmental Protection Agency, Office of Research and Development, National Health and Environmental Effects Research Laboratory, Research Triangle Park, NC.

Schoeberl, M.R. and Hartmann, D.L.: 1991, *Science* **251**, 46-52.

Tevani, M. (Ed.): 1993, *UV-B Radiation and Ozone Depletion: Effects on Humans, Animals, Plants, Microorganisms and Materials*, Lewis Publications, Boca Raton, FL.

WMO.: 1991, *Scientific Assessment of Ozone Depletion: 1991*, World Meteorological Organization, Geneva, 1991. Global Ozone Research and Monitoring Project - Report No. 25.

Watson, R.: 1988, *Ozone Trends Panel. Executive Summary*, NASA, Washington, DC.

Weiler, C.S. and Penhale. P.A. (Eds.): 1994, *Antarctica Research Series* **62**.

Williamson, C.E. and Zagarese, H.E.: 1994, *Impacts of Solar UVB on Pelagic Freshwater Ecosystems*. Wiley & Sons, New York. 226 pp.

Young, A.R., Bjorn, L.O., Moan, J. and Nultsch, W. (Eds.): 1993, *Environmental UV Photobiology*. Plenum Press, New York. 479 pp.

DETERMINING THE CAUSES OF BENTHIC CONDITION

VIRGINIA D. ENGLE[1] and J. KEVIN SUMMERS[2]

[1]*Gulf Breeze Project Office, Biological Resources Division, National Wetlands Research Center, U.S. Geological Survey,* [2]*Gulf Ecology Division, National Health and Environmental Effects Research Lab, U.S. Environmental Protection Agency, 1 Sabine Island Drive, Gulf Breeze, FL 32561-5299*

Abstract. A benthic index for northern Gulf of Mexico estuaries has been developed and successfully validated by the Environmental Monitoring and Assessment Program for Estuaries (EMAP-E) in the Louisianian Province. The benthic index is a useful and valid indicator of estuarine condition that is intended to provide environmental managers with a simple tool for assessing the ecological condition of benthic macroinvertebrate communities. Associations between the benthic index and indicators of hypoxia, sediment contamination, and sediment toxicity were investigated to determine the most probable cause(s) of degraded benthic condition. The results showed that, on a local scale, the associations between the benthic index and potential environmental causes differed among estuaries. In Pensacola Bay, FL, for example, there was a significant association between the levels of toxic chemicals (e.g. DDT, silver, and TBT) in the sediment and the benthic index, especially in the bayous which have known sediment contamination problems. In Mobile Bay, however, degraded benthic communities were more closely associated with eutrophication and hypoxia. Nevertheless, a benthic index is a valuable tool for identifying areas that could be already degraded and tracking the status of environmental condition in large geographical regions.

1. Introduction

Monitoring benthic macroinvertebrate communities has been widely used to measure the status of and trends in the ecological condition of estuaries. Environmental managers and policy makers desire straightforward, manageable methods of identifying the extent of potentially degraded areas and the means with which to associate biotic responses with environmental exposures (Summers *et al.*, 1995). Multimetric biotic indices, especially, have been implemented as a standardized means of tracking the ecological condition of estuaries (Engle *et al.*, 1994; Ranasinghe *et al.*, 1994). While ecological indicators such as biotic indices have been developed to serve as tools for the preliminary assessment of ecological condition, they are not intended to replace a complete analysis of the benthic community dynamics.

A benthic index was developed by the U.S. Environmental Protection Agency's (USEPA) Environmental Monitoring and Assessment Program for Estuaries (EMAP-E) in the Louisianian Province (Engle *et al.*, 1994) and subsequently refined as more data became available. This index characterizes the environmental quality of estuaries in the northern Gulf of Mexico by summarizing the composition and diversity of benthic infaunal macroinvertebrate communities. The benthic index combines the Shannon-Wiener index (H' - adjusted for salinity), the mean abundance of tubificids (Family: Tubificidae), and the percentages of total abundance represented by each of bivalves (Class: Bivalvia), capitellids (Family: Capitellidae), and amphipods (Order: Amphipoda). This index successfully discriminates between reference sites and sites that are degraded with respect to sediment

Environmental Monitoring and Assessment **51**: 381–397, 1998.

contaminants, sediment toxicity, and hypoxia, and has been validated successfully using independent data.

The benthic index is intended to provide a means for comparing benthic conditions across a large geographical area. However, the benthic index can also be used to examine small-scale, local variations in benthic conditions such as those in a single estuarine system. Once the benthic index has been used to indicate areas of degradation within an estuary, the next logical step would be to identify potential causes of the observed degradation. Although a true cause-and-effect relationship can only be determined in an experimental setting, several statistical methods can be used to show associations of probable causal agents (e.g., sediment contaminants, hypoxia) with degraded benthic condition. Information about both the locations of and the environmental factors related to degraded benthic communities can greatly benefit environmental managers in focusing their future monitoring or research efforts.

Five estuarine systems have been intensively sampled as a direct part of or in association with the EMAP-E monitoring effort in the Louisianian Province (Figure 1). The sampling design and protocols used in each estuary are identical to those employed by EMAP-E. Each estuary was sampled during only one season between 1992 and 1993 (Pensacola Bay was sampled in April and the rest were sampled during the summer). Using the benthic index as an indicator of benthic condition, we have explored the spatial distribution of degraded benthic communities in each estuary. We have also investigated statistical associations between various environmental indicators and the benthic index in these estuaries.

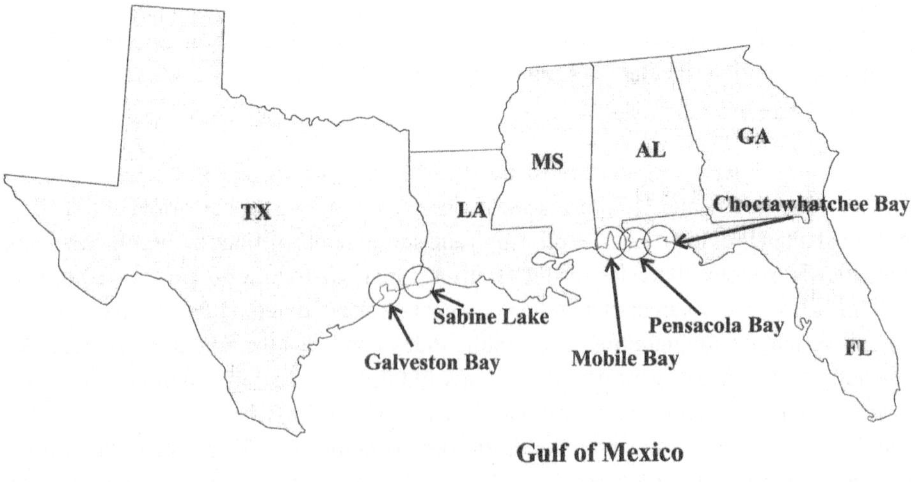

Fig. 1. Estuaries of the Gulf of Mexico.

2. Methods

Table I lists the five estuarine systems discussed in this paper and the program under which each was sampled. The sample design used to determine the number and locations of sampling sites was based on the probabilistic sampling design for large estuaries developed by EMAP-E (Summers *et al.*, 1991; USEPA-ORD, 1993). A hexagonal grid was randomly placed over an estuary and a station location was randomly selected as a latitude/longitude coordinate from within each hexagon cell resulting in a double randomization of each site. The size of each hexagon in the sampling grid for these estuaries was reduced from the standard 280 km^2 used by EMAP-E to a size more appropriate to the given estuary. This smaller-scale grid provided a more intensive sampling effort for each estuary than would have been obtained by the base EMAP-E design. In addition, small tributaries and bayous entering the estuaries were sampled at random locations within linear segments.

TABLE I
Estuaries that were intensively sampled in association with EMAP-E.

Estuary	Year	Program	Number of Stations
Choctawhatchee Bay, FL	1993	USEPA EMAP-E LA Province	12
Pensacola Bay, FL	1992	USEPA NHEERL - GED	40
Mobile Bay, AL	1993	AL Dept. Env. Mgmt. AL-EMAP	62
Sabine Lake, LA/TX	1993	USEPA EMAP-E LA Province	18
Galveston Bay, TX	1993	USEPA Region VI R-EMAP-TX	48

The field sampling methods used in almost all cases were identical to those used by EMAP-E in the Louisianian Province (Macauley, 1992). Benthic macroinvertebrates, fish, and sediment samples were collected according to EMAP-E sampling protocols and sent to laboratories for processing. Water quality parameters were measured on site using a Hydrolab Surveyor 2, DataSonde 3, or Licor LI 1000. All field and laboratory procedures were subjected to rigorous quality control requirements (Heitmuller and Valente, 1991). All organic contaminants were normalized to total organic carbon (TOC) and metals were normalized to aluminum for use in statistical analyses except when compared to established guidelines. All contaminants were compared to guidelines established by Long *et al.* (1995) that were developed from a biological effects database (BEDS) containing the concentrations of contaminants at which adverse biological effects occurred (i.e., altered benthic communities, sediment toxicity, and histopathological disorders in demersal fish). The guidelines delineate concentrations at which adverse biological effects occur rarely (< ER-L), occasionally (ER-L - ER-M), or frequently (> ER-M). The benthic index was calculated for each site by applying updated versions of the formulas and methods detailed in Engle *et al.* (1994). The index was then compared to environmental indicators (Appendix I) to identify any associations between degraded benthic communities and indicators of degraded habitat.

Although the benthic index has a univariate normal distribution and is suitable for use in regression analysis, many of the associative parameters (i.e., contaminants, dissolved oxygen, fish abundance) are not normally distributed. The nature of chemical data, in particular, makes it difficult to transform in order to approximate normality because the distribution of values is often highly skewed toward "non-detect" concentrations (below laboratory method detection limits [MDL]). Traditional stepwise regression analysis using continuous variables was not useful in identifying those causal parameters that explain the greatest amount of variance in the benthic index.

The sites in each estuary were divided into degraded or reference sites based on the value of the benthic index (degraded <3; reference >5; transition 3-5). Two common statistical methods, Wilcoxon rank-sum test and logistic regression, were used to identify parameters that were associated with differences between degraded and reference sites as indicated by the benthic index. The Wilcoxon rank-sum test is used to test the null hypothesis (H_0) that the populations from which two data sets (degraded and reference sites) have been drawn have the same mean. The advantages of this test over traditional means testing are that the two data sets need not be drawn from normally distributed populations, and because this test is based on ranks, it can handle a moderate number of "non-detect" values by treating them as ties (Gilbert, 1987). Rejecting H_0 indicates that the values of a given parameter are significantly different between degraded and reference sites.

Logistic regression is often used when the dependent variable is a binary response variable (e.g., Yes/No, True/False, Good/Bad). This technique is especially useful when the important covariates or explanatory variables are not known and it is desirable to screen a large number of possible covariates for significant associations (Hosmer and Lemeshow, 1989). Logistic regression was applied in this context to determine a set of parameters that were important in explaining the differences between degraded and reference sites in each estuary. SAS/STAT® was used to conduct all statistical analyses (SAS Institute, 1990).

Maps of all of the estuaries were delineated from U.S. Geological Survey (USGS) 1:100,000 digital line graph hydrography files. Point maps were created to provide the location and data value for each sampling station. A continuous interpolated surface map was derived from the data values by utilizing an interpolation technique available within a geographic information system (GIS).

3. Results and Discussion

3.1 CHOCTAWHATCHEE BAY, FL

Choctawhatchee Bay is the third largest estuary on the gulf coast of Florida and is characterized by its limited exchange with the Gulf of Mexico through a narrow pass at the western end (Blaylock, 1983). Although historically water quality has been good, sedimentation and hypoxia due to poor circulation have become problems in the bay (Rabalais, 1992). Recently, priority ecological issues in Choctawhatchee Bay have included eutrophication, fish kills, and localized metals contamination from point and nonpoint source pollution (Rabalais, 1992). Benthic index values < 3 were calculated for five sites in Choctawhatchee Bay (Figure 2). The most striking difference between the degraded sites and the three reference sites identified by the benthic index was in sediment characteristics. The degraded sites were predominantly muddy sites with high silt-clay content (>80% silt-clay for four out of five degraded sites), high TOC, and high aluminum (Al) concentrations (Al 3.75 - 10.30%). The reference sites, on the other hand, occurred in sandy sediments (<5% silt-clay), with low TOC, and low Al concentrations (<0.5%). Metal concentrations were very high at the degraded sites, with as many as six metals with concentrations that were enriched with respect to Al concentrations (Summers *et al.*, 1995) and up to four metals in excess of ER-L guidelines. Chromium (Cr), in particular, exceeded the ER-L guideline of 51 ppm at four out of five of the degraded sites, whereas the maximum concentration of Cr at the reference sites was 5.8 ppm.

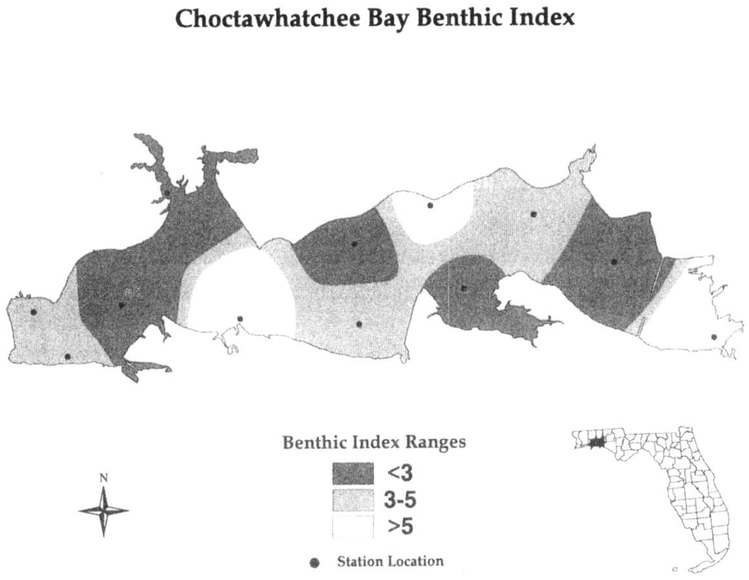

Fig. 2. Spatial distribution of benthic index values in Choctawhatchee Bay, Florida.

The question arises whether the difference in metal concentrations is simply an artifact of the differences in sediment characteristics at the two groups of sites. Higher metal concentrations are expected in sediments with high Al concentrations, high silt-clay, and high TOC, but these metals may not be bioavailable (Luoma, 1989). Recent research has indicated that the ratio of simultaneously extracted metals (SEM) to acid-volatile sulfides (AVS) may determine whether metal concentrations are potentially toxic to biota or whether they are bound in the sediment as sulfides (Bryan and Langston, 1992; Pesch *et al.*, 1995). If the SEM:AVS ratio exceeds 1, then the metals are bioavailable (Di Toro *et al.*, 1990). In contrast to what would be expected, in Choctawhatchee Bay, SEM:AVS was greater than 1 at only one of the degraded sites and at all of the reference sites.

The Wilcoxon rank-sum test did not identify any parameters that differed significantly between these two groups. The results of stepwise logistic regression indicated that only Cr was important as an explanatory variable in distinguishing between degraded and reference sites (Table II).

TABLE II
Results of stepwise logistic regression.

Estuary	Parameter	Score χ^2	Prob.
Choctawhatchee Bay, FL	Chromium	6.68	0.0098
Pensacola Bay, FL	Silver	4.12	0.0423
	Lead	5.18	0.0230
	Number of contaminants > ER-L	13.41	0.0003
Mobile Bay, AL	Dissolved oxygen	9.23	0.0024
	Chlorophyll a	10.00	0.0016
	Ammonia (NH_3-N)	4.47	0.0345
	Nitrate (NO_3-N)	1.95	0.1623
	Tin	5.82	0.0158
	Mercury	2.91	0.0881
Sabine Lake, LA/TX	Salinity	6.65	0.0099
	Selenium	7.87	0.0057
Galveston Bay, TX	Number of contaminants > ER-L	11.66	0.0006
	DDT	2.97	0.0850
	PCB 66	8.77	0.0031
	Secchi depth	9.97	0.0016
	Chromium	2.01	0.1566

3.2 PENSACOLA BAY, FL

Pensacola Bay, an estuary in northwest Florida, has a history of sedimentation problems due to poor flushing and locally high inputs of suspended sediments which are generally retained within the system. The priority problems in the bay have been evidenced by low benthic diversity, decline of seagrasses and oyster populations, and contaminated sediments, especially in the bayous (Collard, 1991). The components of the benthic index indicated that Pensacola Bay's benthic community was dominated by polychaete and nemertean worms. Although the silt-clay content of sediments was not measured during this sampling period, Pensacola Bay has predominantly muddy (fine-grained) sediments (Seal *et al.*, 1994). In general, sites that were degraded exhibited a low proportion of expected diversity and dominance by capitellid polychaetes. Some degraded sites exhibited extremely low abundance and low species richness. Most reference sites exhibited diversity close to expected and low proportions of capitellids.

The benthic index identified 12 degraded sites (Figure 3) that were located primarily in the mainstem of Pensacola Bay and in the three bayous proximal to the city of Pensacola (Bayous Chico, Grande, and Texar). Historically, the bayous, especially Chico and Grande, have exhibited very high concentrations of trace metals (cadmium [Cd], Cr, lead [Pb], and zinc [Zn]) as well as polycyclic aromatic hydrocarbons (PAHs) and polychlorinated biphenyls (PCBs) (Seal *et al.*, 1994) and these areas of high concentrations were confirmed by our sediment contaminant analyses. Bayou Grande, part of the Pensacola Naval Air Station complex, exhibits toxic levels of sediment chemicals, and is currently on the USEPA's National Priority list of Superfund sites. Both Bayou Chico and Bayou Texar have high levels of toxic sediment contaminants as a result of urban runoff or extensive industrial activities.

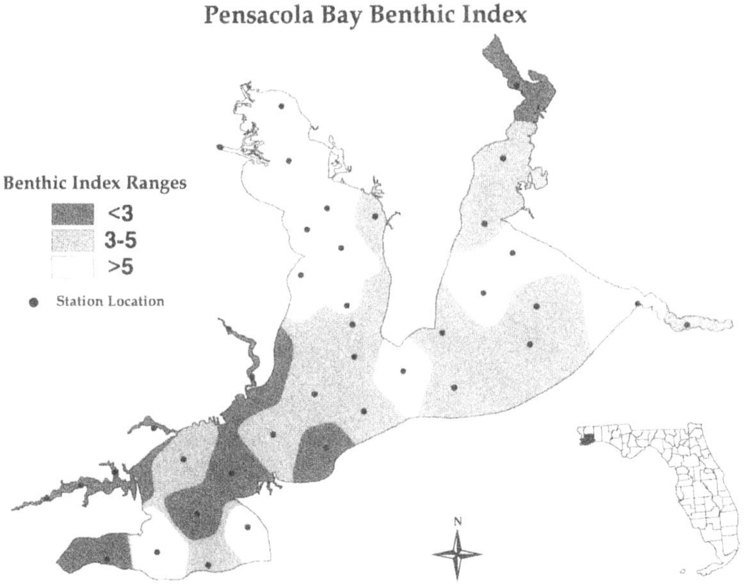

Fig. 3. Spatial distribution of benthic index values in Pensacola Bay, Florida.

Although contaminant levels were high at almost all sites in Pensacola Bay, there were some differences between degraded and reference sites. Concentrations of sediment contaminants were compared to ER-L and ER-M guidelines to determine if any contaminants were at levels that could potentially affect benthic invertebrates. At least one contaminant exceeded the ER-L guideline at every site in Pensacola Bay with a maximum of 44 contaminants exceeding the ER-L guidelines at a degraded site in Bayou Chico. Half of the degraded sites had high concentrations of Pb and tributyltin (TBT), whereas none of the reference sites exceeded the ER-L for Pb and only one site showed TBT > 1 ppb.

The Wilcoxon rank-sum test indicated that concentrations of silver (Ag), TBT, and several PCBs, PAHs, and pesticides, including DDT, were significantly higher at degraded sites than at reference sites. Stepwise logistic regression analysis indicated that concentrations of Pb and Ag and the number of contaminants with concentrations greater than ER-L guidelines were the most important parameters in explaining the differences between degraded and reference sites (Table II). Dissolved oxygen stress was not a factor in this survey because all samples were taken in April, 1992, before the onset of seasonal hypoxia. Because silt-clay content of sediments and TOC were not measured for this study, salinity was the only habitat parameter available for comparison with the benthic index. There was no significant relationship ($p>0.1$) between salinity at the bottom and the benthic index even though salinity ranged from 3.5 to 35 ppm throughout the bay and its tributaries.

3.3 MOBILE BAY, AL

Mobile Bay, the largest estuary in Alabama, is shallow (avg. depth 3 m) with some deep holes and a dredged shipping channel (Turner *et al.*, 1987). Historically, Mobile Bay has endured seasonal hypoxic events that often culminate in "jubilees," where fish and invertebrates migrate onto the shore to escape waters with low dissolved oxygen (May, 1973). Twenty-two degraded and twenty-two reference sites in Mobile Bay were identified by the benthic index (Figure 4). Most of the degraded sites occurred in the northeast sub-region of Mobile Bay and in Bon Secour Bay. The benthic communities at degraded sites,

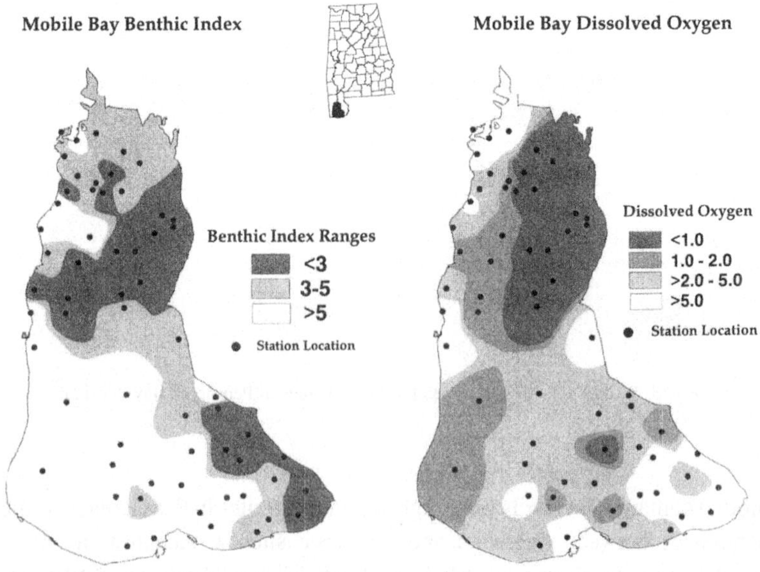

Fig. 4. Spatial distribution of (a) benthic index values and (b) dissolved oxygen concentrations in Mobile Bay, AL.

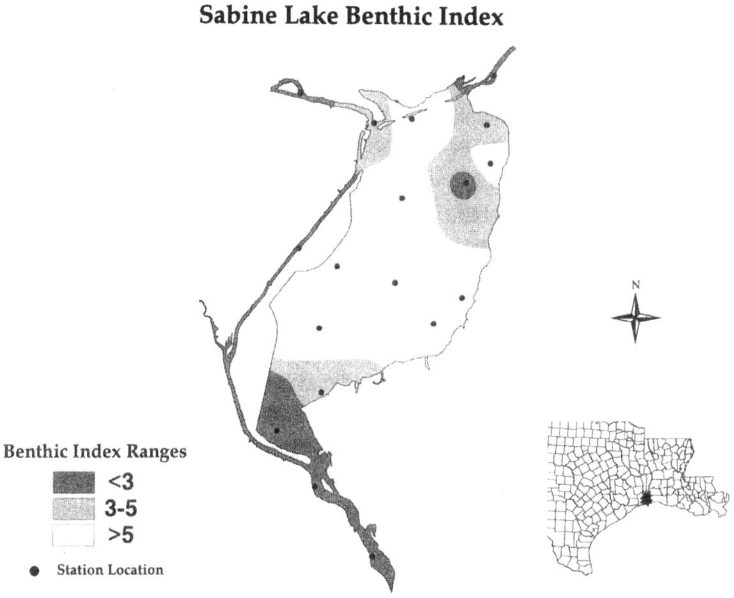

Fig. 5. Spatial distribution of benthic index values in Sabine Lake, Texas.

in general, had less than 50% of the diversity expected based on salinity and were dominated by capitellids (19 sites had >70% capitellids), primarily *Mediomastus ambiseta* and *Streblospio benedicti*. The percent of total abundance represented by capitellids averaged 19% at reference sites.

Fourteen degraded sites had hypoxic conditions (DO < 2.0 mg/L; Figure 4). At ten of the hypoxic sites, dissolved oxygen was near zero (anoxia). In contrast, only five reference sites were hypoxic. Nutrients and chlorophyll *a* were higher at degraded sites than at reference sites. These elevated concentrations, combined with low dissolved oxygen, indicate that the degraded sites were eutrophic. Recent research also suggests that both nutrient enrichment and strong vertical stratification lead to seasonal oxygen depletion in the bottom waters of Mobile Bay (Rabalais, 1992). Eutrophication is further evidenced by the high proportions of capitellids at degraded sites; capitellids often respond favorably to nutrient enrichment (Pearson and Rosenberg, 1978) and may be more resistant to hypoxia.

The Wilcoxon rank-sum test identified significant differences between degraded and reference sites in Ag, mercury (Hg), dissolved oxygen, chlorophyll *a*, nitrate (NO_3-N), phosphate (PO_4-P), and turbidity. In general, however, concentrations of heavy metals, particularly Hg, nickel (Ni), arsenic (As), and Cr, were high (> ER-L guidelines) throughout Mobile Bay. Dissolved oxygen, chlorophyll *a*, nitrogen (N) in the forms of NO_3-N and ammonia (NH_3-N), and metals (tin [Sn] and Hg) were the most important explanatory variables in identifying the differences between degraded and reference sites (Table II).

3.4 SABINE LAKE, LA/TX

The Sabine Lake estuarine system lies on the border between Louisiana and Texas and includes the Sabine and Neches Rivers and a dredged shipping canal. Land use in the Sabine Lake watershed is primarily urban and industrial (Rabalais, 1992). Of all estuaries in Texas, Sabine Lake has the largest area of surrounding marshland, receives the largest areal nutrient load, and may be the Texas estuary most affected by human activity (McFarlane, 1996). Hypoxia has been regularly documented in the deeper rivers and in Sabine Pass but rarely in the open waters of the lake (Rabalais, 1992). Benthic index values < 3 were calculated for seven sites: four in the southern end of the lake, two in the rivers, and one in the Sabine-Neches canal (Figure 5). All seven reference sites occurred in the open area of the lake. The most striking difference between sites was that the reference sites were all oligohaline (bottom water salinity ≤ 5 ppt), whereas almost all of the degraded sites were mesohaline or polyhaline. The expectation that higher benthic diversity would occur at the sites with higher salinity is supported by the literature (Berger *et al.*, 1995). In Sabine Lake, however, the reference sites had low diversity but were close to the expected diversity given the low salinity. Four of the degraded sites had diversity values that were < 50% of expected diversity values given the higher salinities. The benthic communities at degraded sites were also characterized by high tubificid and capitellid abundances, low bivalve abundance, and amphipods present at only one site.

Four out of seven of the degraded sites had more than one contaminant with concentrations greater than ER-L guidelines. Although six out of eight reference sites had one contaminant with concentrations greater than ER-L, in all cases this contaminant was dieldrin, which has an ER-L of 0.02 ppb but a laboratory MDL of 0.16 ppb. At all reference sites where dieldrin concentrations exceeded >0.02, the levels were less than 0.16, making these concentrations suspect.

Several heavy metals (As, Cd, and copper [Cu]) and mono-butyltin (MBT) were identified by the Wilcoxon rank-sum test as having significantly greater concentrations at degraded sites than at reference sites. Bottom salinity and bottom DO were also significantly different between sites. Stepwise logistic regression indicated that only bottom salinity and selenium (Se) were important explanatory variables (Table II). There was no significant difference in silt-clay or TOC between the two groups of sites.

3.5 GALVESTON BAY, TX

Region VI of the USEPA began the regional EMAP (R-EMAP-TX) project to focus on potential environmental problems in Louisiana and Texas estuaries that had been identified by EMAP-E's full-scale monitoring of the Louisianian Province. EMAP-E had discovered highly contaminated sediments in Galveston Bay (particularly by TBT) and high rates of fish pathology and sediment toxicity in East Bay Bayou. Galveston Bay has a history of environmental problems that led to the creation of the Galveston Bay National Estuary Program (GBNEP) in 1988 to identify and assess the priority issues that affect water quality in this estuary. The greatest threats to the health o the Galveston Bay system are habitat destruction (wetland and seagrass losses), freshwater inflow that impacts the hydrology of the bay and carries pollutants, and bacterial contamination that leads to shellfish closures (GBNEP, 1996).

Galveston Bay is a polyhaline estuary with predominantly medium to fine-grained sediments. The benthic communities of Galveston Bay are dominated by polychaetes but many sites support good benthic diversity, with a total of 136 taxa identified from this sampling program. There was no difference in sediment characteristics or salinity between degraded or reference sites. Twelve degraded sites were characterized by lower than expected diversity and low abundance (three sites had no benthic organisms present), dominance by either tubificids or capitellids, and no amphipods present at any sites. Eighteen reference sites were characterized by benthic diversity close to expected based on salinity, fewer tubificids and capitellids, and presence of amphipods. Only one site in the main portion of Galveston Bay was degraded; the rest of the degraded sites occurred in East Bay Bayou or Trinity Bay, at the marinas, or in the small lakes (Figure 6). All of the reference sites had fewer than two contaminants with concentrations greater than ER-L guidelines. In contrast, the degraded sites had as many as seven contaminants exceeding ER-L guidelines. In East Bay Bayou, several PAHs, including fluorene and phenanthrene, exceeded ER-L guidelines. At the marinas and in Offats Bayou, Cu and chlordanes exceeded ER-L values. Offats Bayou also had high levels of Pb, Zn, and DDT.

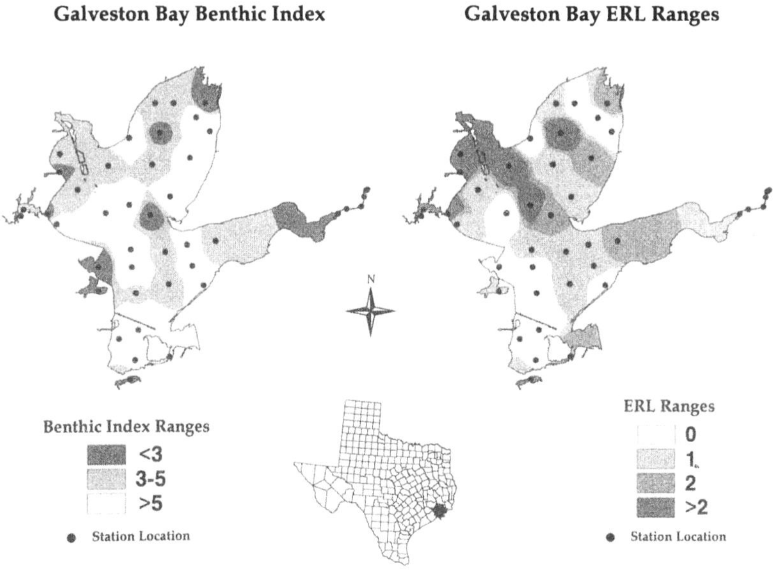

Fig. 6. Spatial distribution of (a) benthic index values and (b) number of contaminants > ER-L in Galveston Bay, Texas.

The Wilcoxon rank-sum test identified significantly higher levels of PAHs (benzo(a)anthracene, benzo(a)pyrene, benzo(e)pyrene, chrysene, 2,6-dimethylnaphthalene, fluoranthene, pyrene, 2,3,5-trimethylnaphthalene, total high-molecular weight PAHs, total PAHs), metals (Al, As, Cd, Cr, Cu, Pb, antimony [Sb], Se, Sn, Zn), and pesticides (alpha-chlordane, trans-nonachlor, total chlordanes) at degraded sites. The number of contaminants with concentrations greater than ER-L guidelines was determined by stepwise logistic regression to have the most important association with differences between degraded and reference sites (Table II). Other important explanatory variables included DDT, PCB 66, secchi depth (turbidity), and Cr.

4. Conclusion

The benthic index is intended to be used as an indicator of the ecological health of estuaries by ranking and classifying the conditions of benthic invertebrate communities over large geographical areas. It can also be used successfully to classify specific areas of a single estuary as degraded or reference with respect to benthos. We can then try to identify possible stressors that may exist in the degraded areas but not in the reference areas. This provides insight to determine environmental impacts that may be affecting the benthic communities at the degraded areas. However, the techniques used here to determine associations between degraded benthic communities and possible stressors are not intended to imply a cause-and-effect relationship. The stressors that may impact benthic communities differ among estuaries in the Gulf of Mexico. Intensive site monitoring, like the sample designs used here, provides more detailed information about the condition of a single estuary than does a large-scale monitoring project like EMAP-E. Although each estuary investigated here has some similarities in hydrography, dominant benthic communities, and stressors that affect them, none of these estuaries can be considered typical of Gulf of Mexico estuaries.

Choctawhatchee Bay has a strong sediment gradient, from sand in the shallow, nearshore areas to mud from the mouths of its tributaries to the deeper channel. The greatest impacts to Choctawhatchee Bay stem from rapid population growth rate due to tourism and urbanization (Blaylock, 1983) and from its proximity to one of the largest military bases in the nation. The degraded areas in Choctawhatchee Bay occur in the high silt-clay sediments near the tributaries and bayous, which are also the sources of discharges from urban outfalls, stormwater systems, and poultry processing plants. The hydrography of the bay contributes to limited flushing, sedimentation, and stratification, leading to seasonal hypoxia.

Pensacola Bay also has a sedimentation problem due to poor flushing and high suspended sediment input from its tributaries. However, the sediment and biological quality of Pensacola Bay have deteriorated since the 1950s and recovery is improbable without substantial intervention (Collard, 1991). Pensacola Bay has severely contaminated sediments with as many as 40 chemicals at concentrations greater than ER-L guidelines, especially in the bayous. The benthic community is impoverished throughout the bay, with the worse conditions occurring in the areas with low sediment quality.

The benthic communities of Mobile Bay are more affected by hypoxia and nutrient enrichment than by toxic sediments. Although hypoxia in Mobile Bay is primarily driven by salinity stratification and the timing and duration of wind events, the severity and extent of hypoxic bottom waters may be exacerbated by nutrient enrichment (Turner, *et al.*, 1987). In this case the dominant benthic taxa at degraded sites are small, tube-dwelling polychaetes, indicating a stressed environment.

The hydrography of Sabine Lake is unlike that of any other estuary presented here in that high salinities are found in the rivers and low salinities occur in the open waters of the lake. Benthic communities in this highly urbanized and industrialized lake were impacted primarily by salinity and dissolved oxygen, as well as by heavy metals.

Galveston Bay is similar to Pensacola Bay in its prevalence of sediments contaminated by mixtures of organics and inorganics. Like Pensacola Bay, most of the degraded benthic communities occur in the bayous or at specific locations (i.e., marinas and industrialized areas). Galveston Bay has a well-documented history of environmental problems stemming, primarily, from urban and industrial development (GBNEP, 1996; Carr, *et al*, 1996).

In conclusion, the benthic index, in association with additional monitoring data, is a successful indicator of stressed or impacted benthic communities in estuaries. It is also possible to identify potential causal agents using the techniques demonstrated here. The results from this type of analysis may aid environmental managers in the decision-making processes that direct future research and/or monitoring efforts in these estuaries.

This study also demonstrates the benefits of intensive monitoring when the goal is to reveal local-scale patterns and associations that are difficult to detect with broader sampling scales (e.g., how associations between benthic condition and environmental exposures can differ among estuaries from the same region).

Acknowledgements

The authors would like to thank the entire staff at the Gulf Breeze Project Office (USGS/BRD/NWRC and Johnson Controls World Services) for their assistance with the preparation of this manuscript. We also thank J. Macauley (USEPA) and D. McGrath (JCWS) for their helpful reviews. Any mention of trade names does not imply endorsement or recommendation for use by the U.S. Government.

APPENDIX I.

Mean and range of values of parameters used to test for associations between benthic condition and environmental exposures for each estuary.

PARAMETER	UNITS	Choctawhatchee Bay Mean (Range)	Sabine Lake Mean (Range)	Galveston Bay Mean (Range)	Pensacola Bay Mean (Range)	Mobile Bay Mean (Range)
Bottom DO	mg/L	5.55 (3.5 - 6.65)	5.64 (3.45 - 6.8)	6.88 (3.45 - 10.2)	6.64 (3.5 - 9.1)	3.10 (0.10 - 9.10)
Bottom pH	s.u.	8.01 (7.1 - 8.2)	7.68 (7.2 - 8.3)	7.97 (7.2 - 8.55)	7.96 (6.5 - 8.6)	
Bottom Salinity	ppt	25.41 (9.7 - 36.6)	10.91 (3.85 - 29.45)	20.07 (3.85 - 32.3)	20.51 (3.55 - 35.15)	18.78 (7.00-31.80)
Bottom Temperature	C	29.63 (28.3 - 30.9)	31.38 (29.9 - 32.55)	27.38 (29.9 - 30.3)	22.73 (20.4 - 25.8)	
Aluminum	%	4.75 (0.06 - 10.3)	4.5 (1.44 - 6.81)	5.43 (1.44 - 7.91)	0 (0 - 0)	3.78 (0,10-9.49)
Antimony	µg/g dwt	0.51 (0 - 1.1)	0.43 (0 - 1)	0.43 (0 - 0.86)	0.5 (0.03 - 0.98)	
Arsenic	µg/g dwt	14.9 (0.42 - 32.3)	5.74 (2.64 - 13.3)	5.98 (2.64 - 11.09)	13.82 (0.29 - 26.4)	12.23 (1.00-28.00)
Cadmium	µg/g dwt	0.14 (0 - 0.61)	0.12 (0.06 - 0.22)	0.14 (0.06 - 0.78)	0.42 (0.02 - 6.6)	0.21 (0.12-0.38)
Chromium	µg/g dwt	56.63 (0.9 - 115)	7.8 (14.6 - 65)	51.03 (14.6 - 79.2)	71.25 (2.9 - 222)	62.38 (6.00-108.00)
Copper	µg/g dwt	10.52 (0 - 24.7)	7.35 (2.9 - 13)	15.71 (2.9 - 57.8)	21.85 (0.3 - 241)	14.49 (1.40-26.00)
Iron	%	3.29 (0.03 - 7.76)	1.89 (0.46 - 3.81)	2.55 (0.46 - 4)	3.32 (0.1 - 6.72)	
Lead	µg/g dwt	20.73 (1.2 - 59)	16.12 (9.1 - 33.8)	18.77 (9.1 - 50.94)	46.48 (0.19 - 311)	18.83 (1.00-34.00)
Manganese	µg/g dwt	387 (0 - 981)	429.11 (79 - 1320)	399.95 (79 - 1194)	251.79 (2 - 679)	
Mercury	µg/g dwt	0.07 (0.02 - 0.16)	0.06 (0.03 - 0.09)	0.05 (0.03 - 0.1)	0.15 (0.01 - 2.23)	0.44 (0.10-2.12)
Nickel	µg/g dwt	15.49 (0 - 34.6)	14.06 (4.2 - 25.5)	20.92 (4.2 - 33.8)	19.63 (0.9 - 34)	20.52 (1.00-36.00)
Selenium	µg/g dwt	0.69 (0 - 1.5)	0.22 (0 - 0.42)	0.32 (0 - 0.69)	0.26 (0.14 - 0.54)	
Silver	µg/g dwt	0.1 (0 - 0.5)	0.15 (0.12 - 0.18)	0.15 (0.12 - 0.35)	0.14 (0.02 - 0.88)	0.41 (0.01-0.90)
Tin	µg/g dwt	1.02 (0 - 2.26)	0.86 (0.22 - 1.7)	1.57 (0.22 - 3.4)	1.9 (0.11 - 5.6)	1.61 (1.00-5.00)
Zinc	µg/g dwt	55.97 (5 - 109)	63.01 (23.4 - 111)	86.26 (23.4 - 216.6)	118.88 (5.9 - 790)	85.14 (5.00-165.00)
(i)1,2,3 - c,d - Pyrene	µg/g TOC	0.41 (0 - 1.99)	0.36 (0.06 - 1.03)	0.95 (0.06 - 6.17)	83.87 (0.23 - 1925.14)	
1 - Methylnaphthalene	µg/g TOC	0.24 (0.01 - 0.97)	0.22 (0.03 - 0.63)	0.17 (0.03 - 0.97)	167.27 (0.16 - 588.64)	
1 - Methylphenanthrene	µg/g TOC	0.15 (0 - 0.65)	0.21 (0 - 0.54)	0.28 (0 - 1.25)	25.5 (2.95 - 100.46)	
2,3,5 - Trimethylnaphthalene	µg/g TOC	0.08 (0 - 0.32)	0.13 (0.02 - 0.27)	0.17 (0.02 - 1.35)	374.71 (1.02 - 1094.68)	
2,6 - Dimethylnaphthalene	µg/g TOC	0.14 (0 - 0.65)	0.22 (0.03 - 0.5)	0.14 (0.03 - 1.53)	420.31 (0.25 - 1223.17)	
2 - Methylnaphthalene	µg/g TOC	0.28 (0.01 - 0.97)	0.45 (0.05 - 1.93)	0.23 (0.05 - 1.25)	287.52 (0.19 - 947.85)	

PARAMETER	UNITS	Choctawhatchee Bay Mean (Range)	Sabine Lake Mean (Range)	Galveston Bay Mean (Range)	Pensacola Bay Mean (Range)	Mobile Bay Mean (Range)
Acenaphthene	μg/g TOC	0.11 (0 - 0.65)	0.16 (0.01 - 1)	0.1 (0.01 - 0.36)	12.25 (0.08 - 74.15)	
Acenaphthylene	μg/g TOC	0.08 (0 - 0.32)	0.86 (0 - 3.13)	0.35 (0 - 3.69)	8.17 (0.04 - 101.9)	
Anthracene	μg/g TOC	0.1 (0 - 0.32)	0.38 (0.05 - 1.13)	0.47 (0.05 - 5.1)	13.16 (0.49 - 94.04)	
Benzo(a)anthracene	μg/g TOC	0.29 (0 - 1.18)	0.74 (0.09 - 2.01)	1.37 (0.09 - 9.7)	79.27 (0.32 - 1464.51)	
Benzo(a)pyrene	μg/g TOC	0.48 (0 - 1.92)	0.8 (0.1 - 2.56)	1.5 (0.1 - 11.06)	75.15 (0.18 - 1517.13)	
Benzo(b)fluoranthene	μg/g TOC	0.51 (0 - 2.32)	0.63 (0.09 - 1.88)	1.42 (0.09 - 11.57)	97.88 (0.3 - 1957.74)	
Benzo(e)pyrene	μg/g TOC	0.44 (0 - 1.85)	0.78 (0.11 - 2.44)	1.4 (0.11 - 8.01)	76.49 (0.22 - 1472.23)	
Benzo(g,h,i)perylene	μg/g TOC	0.41 (0 - 1.72)	0.5 (0.09 - 1.37)	1.24 (0.09 - 6.9)	64.5 (0.25 - 1366.16)	
Benzo(k)fluoranthene	μg/g TOC	0.54 (0 - 2.38)	0.65 (0.09 - 1.94)	1.5 (0.09 - 12.27)	94.54 (0.13 - 1893.97)	
Chrysene	μg/g TOC	0.43 (0.01 - 1.72)	0.86 (0.11 - 2.46)	1.9 (0.11 - 14.9)	86.12 (0.3 - 1895.8)	
Dibenzo(a,h)anthracene	μg/g TOC	0.07 (0 - 0.33)	0.18 (0.03 - 0.48)	0.3 (0.03 - 1.46)	9.19 (0.03 - 173.13)	
Fluoranthene	μg/g TOC	0.66 (0.03 - 2.25)	1.13 (0.1 - 3.26)	2.42 (0.1 - 11.43)	137.09 (0.84 - 3418.37)	
Fluorene	μg/g TOC	0.12 (0 - 0.65)	0.26 (0.03 - 1.13)	0.14 (0.03 - 0.67)	55.64 (0.42 - 155.09)	
Naphthalene	μg/g TOC	0.64 (0.04 - 2.9)	0.58 (0.11 - 2.19)	0.3 (0.11 - 1.32)	74.44 (0.73 - 408.16)	
Perylene	μg/g TOC	0.23 (0 - 0.54)	3.11 (0.29 - 30.62)	1.02 (0.29 - 4.12)	70.53 (0.34 - 1173.79)	
Phenanthrene	μg/g TOC	0.47 (0.02 - 2.58)	0.68 (0.06 - 2.46)	0.76 (0.06 - 4.16)	94.2 (5.16 - 832.75)	
Pyrene	μg/g TOC	0.62 (0.04 - 1.99)	1.64 (0.17 - 5.19)	2.95 (0.17 - 13.97)	123.67 (2.45 - 2666.76)	
Total PCBs	μg/g TOC	0.39 (0.01 - 2.77)	0.4 (0.05 - 0.93)	0.45 (0.05 - 3.24)	38.16 (0.79 - 249.5)	
alpha - Chlordane	μg/g TOC	0.01 (0 - 0.03)	0 (0 - 0.03)	0 (0 - 0.03)	1.22 (0 - 42.52)	
Hexachlorobenzene	μg/g TOC	0.05 (0 - 0.45)	0.03 (0 - 0.07)	0.02 (0 - 0.08)	0.05 (0 - 0.97)	
o,p'DDD	μg/g TOC	0 (0 - 0)	0 (0 - 0.01)	0 (0 - 0.01)	0.21 (0 - 3.71)	
o,p'DDE	μg/g TOC	0.06 (0 - 0.24)	0 (0 - 0.01)	0 (0 - 0.03)	0.03 (0 - 1.35)	
o,p'DDT	μg/g TOC	0 (0 - 0)	0 (0 - 0.01)	0 (0 - 0.01)	0.29 (0 - 2.57)	
p,p'DDD	μg/g TOC	0.02 (0 - 0.08)	0.01 (0 - 0.03)	0.01 (0 - 0.08)	1.05 (0 - 24.82)	
p,p'DDE	μg/g TOC	0.04 (0.01 - 0.1)	0.01 (0 - 0.03)	0.03 (0 - 0.16)	3.64 (0 - 54.61)	
p,p'DDT	μg/g TOC	0 (0 - 0.02)	0 (0 - 0.01)	0.01 (0 - 0.12)	1.16 (0 - 29.67)	
Tributyl Tin	ng/g dwt	0.44 (0.07 - 1.31)	0.94 (0 - 7.53)	4.12 (0 - 40.75)	10.73 (0 - 392)	
Chlorophylla	mb/m^3					12.11 (1.10-34.70)
Ammonia	mg/L					0.08 (0.01-0.60)

PARAMETER	UNITS	Choctawhatchee Bay Mean (Range)	Sabine Lake Mean (Range)	Galveston Bay Mean (Range)	Pensacola Bay Mean (Range)	Mobile Bay Mean (Range)
Nitrate	mg/L					0.18 (0.00-1.36)
Phosphate	mg/L					1.14 (0.30-2.00)
Secchi Depth	m					16046 (6680-27080)
Total Dissolved Solids	mg/L					0.60 (0.18-3.97)
Total Kjeldahl Nitrogen	mg/L					29.00 (4.00-65.00)
Total Suspended Solids	mg/L					7.58 (3.00-28.00)

References

Berger, V. Ya., Naumov, A. D., and Babkov, A. I.: 1995, *Russian J. of Mar. Biol.* **21**, 41-46.

Blaylock, D. A.: 1983, *Choctawhatchee Bay: Analysis and interpretation of baseline environmental data*, Technical Paper No. 29, Florida Sea Grant College, 237 pages.

Bryan, G. W. and Langston, W. J.: 1992, *Environ. Poll.* **76**, 89-131.

Carr, R. S., Chapman, D. C., Howard, C. L., and Biedenbach, J. M.: 1996, *Ecotoxicology* **5**, 341-364.

Collard, S. B.: 1991, *Surface water improvement and management (S.W.I.M.) program. The Pensacola Bay System: Biological Trends and Current Status.* Water Resources Special Report 91-3, Northwest Florida Water Management District, Havana, Florida.

Di Toro, D. M., Mahony, J. D., Hansen, D. J., Scott, K. J., Hicks, M. B., Mayr, S. M., and Redmond, M. S.: 1990, *Environ. Toxicol. Chem.* **9**, 1487-1502.

Engle, V.D., Summers, J.K., and Gaston, G.R.: 1994, *Estuaries* **17**, 372-384.

Galveston Bay National Estuary Program (GBNEP): 1996, *The State of the Bay - a Characterization of the Galveston Bay Ecosystem*, GBNEP-44, 232 pages.

Gilbert, R. O.: 1987, *Statistical Methods for Environmental Pollution Monitoring*, Van Nostrand Reinhold, 320 pages.

Heitmuller, P. T. and Valente, R.: 1991, *Environmental Monitoring and Assessment Program: EMAP- Estuaries Louisianian Province: 1991 quality assurance project plan*, EPA/ERL-GB No. SR-120, U.S. Environmental Protection Agency - Office of Research and Development (USEPA-ORD).

Hosmer, D W. and Lemeshow, S.: 1989, *Applied Logistic Regression*, John Wiley & Sons, 307 pages.

Long, E. R., MacDonald, D. D., Smith, S. L., and Calder, F. D.: 1995, *Environ. Mgmt.* **19**, 81-97.

Luoma, S.: 1989, *Hydrobiologia* **176/177**, 379-396.

Macauley, J. M.: 1992, *Environmental Monitoring and Assessment Program: Louisianian Province: 1992 Sampling: Field Operations Manual*, EPA/ERL-GB No. SR-119, U.S. Environmental Protection Agency -Office of Research and Development (EPA-ORD).

McFarlane, R.W.: 1996, *A conceptual ecosystem model for Sabine Lake*, In: Proceedings of the Sabine Lake Conference, Sept. 13-14, 1996, Beaumont Texas, TAMU-SG-97-101, National Oceanic Administration, National Sea Grant Office, 66 pages.

May, E.B.: 1973, *Limnol. Oceanogr.* **18**, 353-366.

Pearson, T. H., and Rosenberg, R.: 1978, *Oceanogr. Mar. Biol. Ann. Rev.* **16**, 229-311.

Pesch, C. E., Hansen, D. J., Boothman, W. S., Berry, W. J., and Mahony, J. D.: 1995, *Environ. Toxicol. and Chem.* **14**, 129-141.

Rabalais, N. N.: 1992, *An updated summary of status and trends in indicators of nutrient enrichment in the Gulf of Mexico*, EPA/800-R-92-004, U.S. Environmental Protection Agency, Office of Water, 421 pages.

Ranasinghe, J.A., Weisberg, S.D., Frithsen, J.B., Dauer, D.M., Schaffner, L.C., and Diza, R.J.: 1994, *Chesapeake Bay benthic community restoration goals*, Report CBP/TRS 107/94, U.S. Environmental Protection Agency, Chesapeake Bay Program, Annapolis, Maryland.

SAS Institute, Inc.: 1990, *SAS/STAT® User's Guide*, 1686 pages.

Seal, T. L., Calder, F. D., Sloane, G. M., Schropp, S. J., and Windom, H. L.: 1994, *Florida coastal sediment contaminants atlas*. Florida Department of Environmental Protection, 75 pages.

Summers, J. K., Macauley, J. M., and Heitmuller, P. T.: 1991, *Environmental Monitoring and Assessment Program, Implementation Plan for Monitoring the Estuarine Waters of the Louisianian Province - 1991 Demonstration*, EPA/600/5-91/228, U.S. Environmental Protection Agency - Office of Research and Development (USEPA-ORD).

Summers, J. K., Paul, J. F., and Robertson, A.: 1995, *Toxicol. and Environ. Chem.* **49**, 93-108.

Turner, R. E., Schroeder, W. W., and Wiseman, W. J.: 1987, *Estuaries* **10**, 13-19.

U.S. Environmental Protection Agency - Office of Research and Development (USEPA-ORD): 1993, *R-EMAP Regional Environmental Monitoring and Assessment Program*, EPA/625/R-93/012, 82 pages.

A REGIONAL ANALYSIS OF LAKE ACIDIFICATION TRENDS FOR THE NORTHEASTERN U.S., 1982-1994

JOHN L. STODDARD[1], CHARLES T. DRISCOLL[2], JEFFREY S. KAHL[3], and JAMES H. KELLOGG[4]

[1] *Western Ecology Division, U.S. Environmental Protection Agency, 200 S.W. 35th Street, Corvallis, OR 97333,* [2] *Dept. Civil & Environ. Engineering, Syracuse Univ., Syracuse, NY 13244,* [3] *Water Resources Institute, Univ. Maine, Orono, ME 04469,* [4] *Vermont DEC, Water Quality Division, 103 South Main Street, Waterbury, Vermont 05676*

Abstract. Acidic deposition is a regional phenomenon, but its effects have traditionally been studied using site-specific, intensive monitoring. We present trends information for 36 lakes of high-to-moderate acid sensitivity (defined as acid neutralizing capacity [ANC] < 100 μeq L^{-1}), and 15 deposition monitoring stations, in the northeastern U.S. for the period 1982-1994. Trends at each site were assessed through use of the Seasonal Kendall tau test; the resulting statistics were combined, through a technique analogous to analysis of variance, to produce quasi-regional estimates of change for key chemical variables. Rates of sulfate deposition declined significantly across all of the northeastern region during this time period, while rates of nitrate and ammonium deposition were unchanged. All lakes exhibited strong decreases in sulfate concentrations (ΔSO_4^{2-} = -1.7 μeq $L^{-1}yr^{-1}$, $p<0.001$) in response to declining sulfate deposition, but there was a strong contrast in the response of acid/base status between lakes in New England and lakes in the Adirondacks. As a group, the New England lakes exhibited recovery (ΔANC = +0.8 μeq $L^{-1}yr^{-1}$, $p<0.001$), while the Adirondack lakes exhibited either no trend or further acidification (as a group, ΔANC = -0.5 μeq $L^{-1}yr^{-1}$, $p<0.01$). This contrast can be attributed to changes in base cation concentrations: New England lakes exhibited base cations declines that were smaller in magnitude than declines in sulfate, producing the observed recovery in ANC; Adirondack lakes showed base cation declines that were very similar to those of sulfate, and no recovery was evident.

1. Introduction

One of the most anticipated changes that is expected to result from the 1990 amendments to the Clean Air Act (CAAA) is the recovery of lakes in the northeastern U.S. that have been acidified by atmospheric deposition. Changes in deposition, particularly of sulfate (SO_4^{2-}), are expected to occur over all of the region (Lynch *et al.*, 1996), and to be of sufficient magnitude (ca. 40-50%) to reverse the trends in acidification that have characterized the northeastern region for several decades (e.g., Charles, 1991; Driscoll *et al.*, 1995). Model projections suggest that declines of this magnitude will be sufficient to produce recovery in a large proportion of the acid-sensitive lakes in the region (Church *et al.*, 1989), but there are few empirical examples of lake recovery from acidification on which to test these projections.

Most of the effects of the CAAA emissions reductions are not expected to be measurable until sometime early in the next century. In the meantime, sulfate deposition has been declining over much of the eastern U.S. since ca. 1970 (Husar *et al.*, 1991), and monitoring data for many Northeast lakes have been collected since early in the 1980s (Driscoll and Van Dreason, 1993; Kahl *et al.*, 1993; Stoddard and Kellogg, 1993). While

Environmental Monitoring and Assessment **51**: 399–413, 1998.

deposition declines during this period (1980-1995) are likely to be smaller in magnitude than those resulting from the CAAA, they none-the-less provide the best available evidence with which to test our hypotheses about lake recovery from acidification.

In this paper we present trend results for 36 low acid neutralizing capacity (ANC) lakes in the Northeast, along with results for 15 stations collecting wet deposition data. By using a meta-analytical technique, we are able to combine these trends results to produce subregional estimates of change in acid/base chemistry in the Northeast.

2. Methods

2.1. Data Collection

The lake data presented here were collected as part of the U.S. EPA Long-Term Monitoring (LTM) project (Ford *et al.*, 1993; U.S. Environmental Protection Agency, 1991). We present trends from chemistry data collected from 36 lakes (Table I) in two subregions of the northeastern U.S.: the Adirondack mountains and New England (represented by lakes in Vermont and Maine). For the analyses presented here, we have excluded relatively insensitive lakes (e.g., those with ANC>100 $\mu eq\ L^{-1}$), because they are not expected to exhibit (nor do they exhibit) trends similar to the low ANC lakes included here (Stoddard *et al.*, in press). While the data are collected by three separate groups of investigators, each group participates in a common program of quality assurance (Morrison, 1991), including regular analysis of samples from an independent audit program. Most of the lakes have been sampled since 1982. Samples are collected quarterly in Maine and Vermont and monthly in the Adirondacks; in order to maintain comparability, and the ability to combine trend results, the Adirondack data set used for the analyses presented here was modified to include only one sample per quarter by taking the mean of the three monthly observations made in each quarter. Trends (through 1989) from individual lakes in the LTM project, as well as descriptions of sampling and analytical methods, have been reported previously (Driscoll and Van Dreason, 1993; Kahl *et al.*, 1993; Stoddard and Kellogg, 1993); trend results through 1994 have recently been updated for the Adirondack lakes (Driscoll *et al.*, 1995).

Deposition trends were calculated from 15 sites in eastern New York and New England (Table II). We used quarterly wet deposition data collected as part of the National Atmospheric Deposition Program/National Trends Network (NADP/NTN), as reported in annual reports (National Atmospheric Deposition Program/National Trends Network, 1980-94), and available through the World Wide Web (http://nadp.nrel.colostate.edu/NADP/). At each site, all of the available time series was utilized for the period 1982 through the end of 1994 (the period when nearly all of the LTM sites were operating); this results in some variability in the beginning dates for the sites, as a small number of NADP/NTN sites did not begin data collection until 1984.

Table I

Locations and median acid neutralizing capacity (ANC) values of 36 low ANC (< 100 μeq L^{-1}) Long-Term Monitoring lakes

Subregion	Lake Name	Latitude	Longitude	Median ANC (μeq/L)
Adirondacks	ARBUTUS LAKE	43.9875	74.2417	64
	BIG MOOSE LAKE	43.8292	74.8500	0
	BUBB LAKE	43.7708	74.8542	39
	CASCADE LAKE	43.7911	74.8041	82
	CONSTABLE POND	43.8333	74.7958	8
	DART LAKE	43.7972	74.8583	5
	HEART LAKE	44.1822	73.9694	39
	LAKE RONDAXE	43.7639	74.9055	47
	MOSS LAKE	43.7861	74.8500	68
	OTTER LAKE	43.1880	74.5000	12
	SQUASH POND	43.8264	74.8897	-32
	WEST POND	43.8111	74.8792	10
	WINDFALL POND	43.8110	74.8500	59
New England (Maine)	ANDERSON POND	44.6475	68.0593	12
	LITTLE LONG POND	44.6375	68.0781	12
	MUD POND	44.6306	65.0939	-19
	SALMON POND	44.6294	68.0871	57
	TILDEN POND	44.6347	68.0722	45
New England (Vermont)	BIG MUD	43.3139	72.9305	1
	BIG MUDDY	44.7556	72.6000	65
	BOURN	43.1055	73.0028	-1
	BRANCH	43.0811	73.0186	-15
	GRIFFITH	43.3022	72.9597	27
	GROUT	43.0455	72.9458	26
	HARDWOOD	44.4680	72.5000	43
	HAYSTACK	42.9167	72.9167	-9
	HOWE	42.7856	72.9875	53
	LILY	43.2342	72.7514	34
	LITTLE-WINHALL	43.1236	72.9417	21
	LITTLE-WOODFORD	42.9250	73.0653	-3
	SOUTH-MARLBORO	42.8439	72.7125	38
	STAMFORD	42.8222	73.0653	15
	STRATTON	43.1042	72.9694	23
	SUCKER	42.8250	73.1292	86
	SUNSET	42.9194	72.6833	17
	UNKNOWN	44.9097	71.8444	20

Table II
Locations of 15 NADP/NTN desposition monitoring stations in the Northeast

Site ID	Site Name	State	Latitude	Longitude
MA01	North Atlantic Lab	MA	41.9758	70.025
MA08	Quabbin Reservoir	MA	42.3925	72.344
MA13	East	MA	42.3839	71.215
ME02	Bridgton	ME	44.1075	70.729
ME09	Greenville Station	ME	45.4892	69.665
ME98	Acadia Nat'l Park	ME	44.3739	68.261
NH02	Hubbard Brook	NH	43.9431	71.703
NY08	Aurora Research Farm	NY	42.7339	76.660
NY20	Huntington Wildlife	NY	43.9731	74.223
NY52	Bennett Bridge	NY	43.5261	75.947
NY68	Biscuit Brook	NY	41.9936	74.504
NY98	Whiteface Mountain	NY	44.3933	73.859
NY99	West Point	NY	41.3508	74.049
VT01	Bennington	VT	42.8761	73.163
VT99	Underhill	VT	44.5283	72.869

2.2. Statistical Methods

For the lake data, trends in ANC, base cations ($C_B = \Sigma(Ca^{2+} + Mg^{2+} + Na^+ + K^+)$), SO_4^{2-} and nitrate (NO_3^-) were analyzed using the nonparametric Seasonal Kendall test (SKT; Hirsch *et al.*, 1982). The SKT has become somewhat of a standard for detecting site-specific trends in water quality data, largely because it can accommodate the non-normality, missing and censored data, and seasonality that are common in data of this type, but it is nevertheless a powerful (in a statistical sense) trend test (Loftis and Taylor, 1989). Because the SKT is sensitive to serial correlation, we use a modified form of the test described by Hirsch and Slack (1984), which uses a covariance term to correct the seasonal Z scores for serial dependence. One limitation of the SKT is that it detects only monotonic trends; trends need not be linear, but they must proceed in only one direction (increasing or decreasing) to be detectable. In brief, the SKT analyzes the data within seasonal blocks, and compares the rank value for a single seasonal observation to the rank values of subsequent data from that same season. The signs (indicating whether the second observation in each pairwise comparison is higher or lower than the first) for all pairwise comparisons within each block are summed, and a Z statistic calculated as the ratio of the sum of signs divided by the standard deviation in the signs. Z statistics are produced for each season, and if the trends are homogenous among the seasons, the seasonal Z statistics can be combined to produce an overall trend result. The SKT does not estimate the slopes of trends, but it has become customary to associate slopes calculated according to the method of Sen (1968), which estimates the slope by calculating the median of all between-year differences in the variable of interest. As with Z statistics, the slope values can be produced separately for each season, or combined to create an overall trend magnitude.

The deposition data were also analyzed by SKT; trends were calculated for SO_4^{2-}, NO_3^-, nitrogen ($NO_3^- + NH_4^+$), sea salt-corrected C_B, acid anions ($SO_4^{2-} + NO_3^-$), and hydrogen ion. We excluded deposition data that did not meet the completeness criteria suggested by the NADP/NTN program office (National Atmospheric Deposition Program/National Trends Network, 1980-94).

In order to infer regional trends from the separate trend tests performed by SKT, we employ a variation of meta-analysis (Hedges and Olkin, 1985); the term meta-analysis describes a family of techniques for analyzing the summary statistics of a series of individual tests or studies, in order to determine whether the combined results have more significance than the individual tests. The methodology for combining test results has its roots in the work of Fisher (1932), who recognized that p values from independent tests could be combined and tested against a chi-square distribution.

The procedure for combining SKT results was first presented by van Belle and Hughes (1984). The general concept is to perform a test on the combined Z statistics from individual SKT tests, with the goal of determining whether the distribution of individual trend tests deviates from the distribution that would be expected from chance. In its simplest form, meta-analysis involves squaring the individual Z statistics, taking a sum, and testing the result against a chi-square distribution with degrees of freedom equal to the number of sites subjected to the trend test.

Just as homogeneity of trend results among seasons is a concern when combining seasonal Z statistics from a single site into an overall trend result, the homogeneity of trends among sites and seasons is a concern in meta-analysis. If sites do not exhibit the same or similar trends, then it would be misleading to combine them. We test for homogeneity by attributing the variance among the Z statistics (in this case, from 36 lakes and 4 seasons, or a total of 144 Z values, and for 15 NADP/NTN sites and 4 seasons, for a total of 60 Z values) to each of the possible sources of interest (site, season, and site*season interaction). In practice, this is accomplished by submitting the Z statistics to an analysis of variance (ANOVA), and testing the resulting sums of squares against a chi-square distribution (Mattson *et al.*, 1997; van Belle and Hughes, 1984). This procedure has been termed Analysis of Chi-Squares (ANOCHIS) by Mattson *et al.* (1997).

There is no established rule for determining when heterogeneity among trends (either among sites or among seasons) is sufficient to invalidate combining trend tests from individual sites. van Belle and Hughes (1984) propose setting a relatively high significance level (e.g., $p = 0.01$), provided that all of the Z statistics are of similar sign and magnitude. We find that in some cases where there is little seasonal variability in Z scores (e.g., for lake SO_4^{2-} concentrations) all of the Z statistics may be very similar (all of the same sign, and within a factor of 3 of one another), yet between-lake differences are estimated to be highly significant ($p < 0.01$). For this reason we use the threshold suggested by van Belle and Hughes (1984) as a "red flag," requiring more in-depth analysis of the Z scores, rather than as an absolute limit. For each case where homogeneity tests yield p values less than

0.01, we assess the consistency in the magnitude and signs of Z values, and determine qualitatively whether the homogeneity is ecologically significant (e.g., whether the trend is the overwhelmingly dominant pattern in the data, despite some heterogeneity in absolute Z values).

3. Results and Discussion

3.1. Deposition Trends

Time series plots of SO_4^{2-} deposition at four northeastern NADP/NTN sites (Figure 1) demonstrate the value of combining trend tests from single sites. While all of the sites show a tendency for SO_4^{2-} deposition to decrease over time, none of the single-site trends is significant (i.e., all $p > 0.05$). In fact, all 15 of the deposition sites exhibited negative slopes for SO_4^{2-}, and the meta-analytical technique of combining trend results from individual sites allows us to capture statistically this tendency for all of the sites to behave similarly.

There was strong homogeneity among the trends in deposition chemistry from 1982-94 (Table III) especially for variables that exhibited large changes (e.g., SO_4^{2-} and C_B). Across the Northeast, SO_4^{2-} deposition declined at a rate of -1.8 eq $ha^{-1} \cdot yr^{-1}$; there was very little variation among the sites, and no evidence that deposition rates were different among the two subregions (eastern New York vs. New England) examined ($\text{Chi-Square}_{subregion}$ was not significant for this or any other deposition variable, indicating no significant differences in deposition trends between the two subregions). Base cations deposition (corrected for sea salt effects, Baker *et al.*, 1990) also declined (slope = -0.5 eq $ha^{-1} \cdot yr^{-1}$), with little variation across the region. There were no significant region-wide trends in NO_3^-, NH_4^+, nitrogen ($NO_3^- + NH_4^+$), or hydrogen ion.

3.2. Lake Trends

One of the primary variables of interest in the recovery of lakes from acidification is ANC. Acid neutralizing capacity is a measure of the capacity of water to neutralize strong acid inputs, and is inversely related to acid sensitivity (a low ANC value indicates high sensitivity). At its simplest, ANC can be defined as:

$$ANC = \Sigma(\text{cations}) - \Sigma(\text{anions}), \text{ or}$$

$$ANC = 2[Ca^{2+}] + 2[Mg^{2+}] + [Na^+] + [K^+] - 2[SO_4^{2-}] - [NO_3^-] - [Cl^-]$$

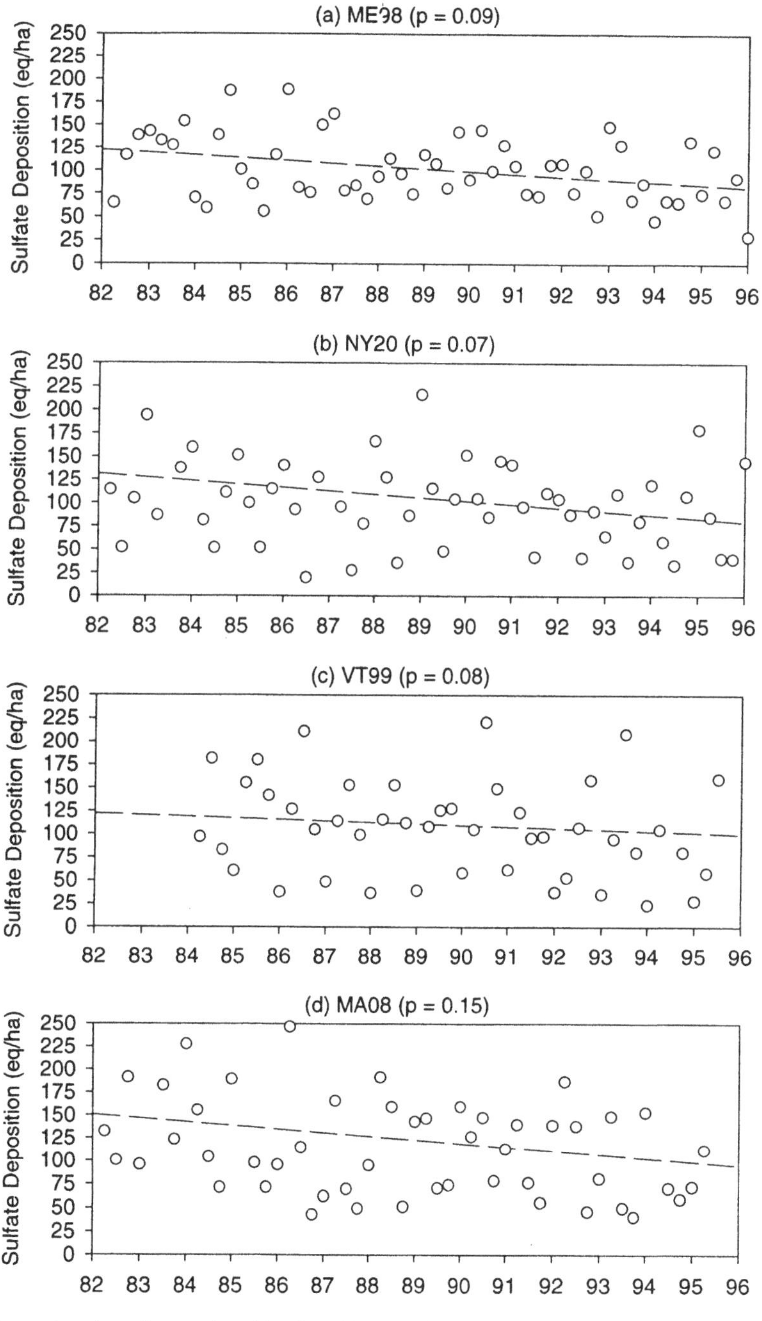

Fig. 1: Time series plots of quarterly sulfur deposition data from four NADP/NTN sites in the Northeast: (a) Acadia National Park (Maine); (b) Huntington Wildlife Refuge (Adirondacks, New York); (c) Underhill (Vermont); and (d) Quabbin Reservoir (Massachusetts). Significance values for each panel are the results of Seasonal Kendall tau trends tests on each site. While all of the sites show a tendency for SO_4^{2-} deposition to decline, none of the trends for individual sites is significant (i.e., all $p > 0.05$).

Table III
Results of meta-analysis (ANOVA on SKT results) of deposition trends, 1982-1994. Slopes and trend statistics are for the combined trends in 15 NADP/NTN sites in the Northeast; significant trends ($p<0.05$) are shown in bold (d.f.=1 for both subregions and trend).

Variable	$\text{Chi-Square}_{\text{subregion}}$	p	$\text{Chi-Square}_{\text{trend}}$	p	Slope (eq $ha^{-1} \cdot yr^{-1}$)
Sulfate*	1.147	n.s.	**23.76**	**<0.001**	**-1.77**
Nitrate	0.01	n.s.	0.08	n.s.	+0.06
Nitrogen	0.02	n.s.	0.40	n.s.	+0.30
Acid Anions	0.67	n.s.	**4.86**	**<0.05**	**-1.53**
Base Cations*	0.74	n.s.	**59.43**	**<0.001**	**-0.52**
Hydrogen	0.90	n.s.	2.63	n.s.	-0.00

* sea-salt corrected

where all ions are in molar concentrations. In order to maintain electrical neutrality over time, changes in ANC will reflect the differences between changes in strong acid anions (SO_4^{2-} and NO_3^-) and changes in base cations. At ANC values near or below zero, changes in other ions (especially H^+ and aluminum) may also occur. At ANC values above zero, most of the changes in acidity that occur will be reflected in changing ANC values, rather than changes in pH. This is one of the most important reasons for our focus on ANC rather than pH.

In a previous paper, we have presented lake trend results for a number of lake subpopulations in the Northeast (Stoddard *et al.*, in press), demonstrating that the magnitudes of trends in acid/base variables such as ANC and C_B are a function of the acid-sensitivity of the lake classes. Within each subregion (Adirondacks vs. New England) lakes with high ANC values tend to exhibit smaller (and non-significant) trends in ANC and larger C_B trends; at higher ANC values (e.g., >100 μeq L^{-1}), most or all of the change in SO_4^{2-} that results from a decrease in deposition is expected to be matched by a stoichiometric decrease in C_B, and that is exactly what the data show. The result is that, while SO_4^{2-} and C_B may change markedly in high ANC lakes, they show little or no response in their acidity or ANC. In order to focus on lakes that are likely to show recovery, we excluded all lakes with ANC values greater than 100 μeq L^{-1} from the current analysis.

One of the primary concerns in combining trend test results is that we do not obscure the patterns present in the data by ignoring important differences among the lakes. With this concern in mind, we conducted an analysis of variance on the seasonal Z scores from the 36 LTM lakes, to identify the important sources of variation. The results of this analysis are shown in Table IV, and indicate that significant differences are present among lakes (for all of the variables) and seasons (for base cations). This heterogeneity makes it unwise

Table IV

Results of meta-analysis (ANOVA on individual SKT results) of trends in northeastern lakes, 1982-1994. Significant sources of variability in the results are shown in bold. These results are provided primarily to demonstrate that significant subregional (most, if not all, of the lake effect can be removed by analyzing separately for the two subregions) and seasonal differences exist in the trends. Slopes for trends are included for completeness, but are not valid at this level (all of the Northeast) due to the lack of homogeneity.

Variable	Source of Variability	Chi-Square	d.f.	p
Sulfate*	Lakes	**94.57**	**35**	**<<0.001**
	Seasons	9.94	3	0.02
	Lake*Season Interaction	47.26	104	0.99
	Trend (slope = -1.72 μeq $L^{-1}yr^{-1}$)	**621.49**	**1**	**<<0.001**
Nitrate	Lakes	53.70	35	0.02
	Seasons	**18.71**	**3**	**<0.001**
	Lake*Season Interaction	83.34	104	0.85
	Trend (slope = -0.03 μeq $L^{-1}yr^{-1}$)	0.24	1	0.62
ANC	Lakes	**174.82**	**35**	**<<0.001**
	Seasons	7.83	3	0.05
	Lake*Season Interaction	119.77	104	0.09
	Trend (slope = +0.32 μeq $L^{-1}yr^{-1}$)	**34.38**	**1**	**<<0.001**
Base Cations*	Lakes	**81.55**	**35**	**<<0.001**
	Seasons	**19.61**	**3**	**<0.001**
	Lake*Season Interaction	56.87	104	0.99
	Trend (slope = -1.06 μeq $L^{-1}yr^{-1}$)	**100.11**	**1**	**<<0.001**

* sea-salt corrected

to summarize the trends for any of the variables across all of the Northeast, and the trend statistics are provided in Table IV only for the sake of completeness. A further examination of the trends results suggests that most of the heterogeneity is due to the different responses of lakes in the two subregions covered by LTM (Adirondacks vs. New England), and for this reason we conducted further analyses on the subregions separately.

When the trend results are separated by subregion (Table V), important patterns emerge. As was the case with sulfur deposition, lake SO_4^{2-} concentrations decrease significantly (Table V), and at similar rates in both the Adirondacks and New England (slopes = -1.8 and -1.6 μeq $L^{-1}yr^{-1}$, respectively). Deposition is the source of nearly all of the SO_4^{2-} in Northeast lakes (Baker *et al.*, 1991), and the size and consistency of lake SO_4^{2-} trends is reflective of this. According to the threshold suggested by van Belle and Hughes (1984), there was significant heterogeneity in the SO_4^{2-} trend results among the lakes in New England (i.e., χ^2_{lakes} had p<0.01). We regard this as a good example of a case where the homogeneity test breaks down for lack of a good rule to determine when to reject the null hypothesis; of the 92 seasonal Z statistics calculated for New England lakes, all but 3 where negative (indicating a declining trend in SO_4^{2-}). There was some apparent difference in the rates of decline between lakes in Vermont and lakes in Maine, but the clear tendency among all of the lakes was for lake SO_4^{2-} to decrease over time. We feel justified in using

Table V

Results of meta-analysis (ANOVA on individual SKT results) of lake trends in two subregions of the northeastern U.S., 1982-1994. Significant trends, and significant sources of heterogeneity in trends, are shown in bold. Heterogeneity was considered significant if $p < 0.01$; trend results were considered significant at $p < 0.05$. See text for discussion of heterogeneity and its implications.

Variable	Subregion	Lakes			Seasons			Lake*season Interaction			Trend			
		Chi-Square*	d.f.	p	Chi-Square*	d.f.	p	Chi-Square*	d.f.	p	Chi-square*	d.f.	p	Slope (μeq/L/yr)
Sulfate**	Adirondacks	23.73	12	0.02	6.99	3	0.07	15.47	36	0.99	**319.0**	**1**	**<<0.001**	**-1.8**
	New England	**39.66**	**22**	**0.006**	7.03	3	0.07	25.93	56	0.99	**332.38**	**1**	**<<0.001**	**-1.6**
Nitrate	Adirondacks	13.50	12	0.33	**35.20**	**3**	**<<0.001**	13.73	36	0.99	**4.43**	**1**	**0.04**	**+0.1**†
	New England	31.0	22	0.10	1.2	3	0.75	51.90	61	0.79	**5.0**	**1**	**0.02**	**-0.0**†
ANC	Adirondacks	22.1	12	0.04	5.59	3	0.13	12.51	36	0.99	**26.89**	**1**	**<<0.001**	**-0.5**
	New England	36.90	22	0.02	9.40	3	0.02	29.58	56	0.99	**147.36**	**1**	**<<0.001**	**+0.8**
C_B**	Adirondacks	8.63	12	0.73	6.07	3	0.11	6.70	36	0.99	**97.42**	**1**	**<<0.001**	**-2.0**
	New England††	**53.9**	**22**	**<0.001**	6.4	3	0.09	36.84	61	0.99	**31.8**	**1**	**<<0.001**	**-0.7**

* Chi-square values are the Type I sums of squares resulting from an Analysis of Variance performed on the results (Z scores) of Seasonal Kendall tau trend tests

** Sea-salt corrected.

† NO_3^- trends showed significant heterogeneity that could not be resolved and should be interpreted with caution - see text for further discussion.

†† C_B trends in New England exhibited significant heterogeneity among lakes, which can be resolved by splitting the dataset into western New England (Vermont) and Eastern New England (Maine). When analyzed this way, the trends were homogenous, with $\chi^{**}_{trend} = 46.37$, $p << 0.001$ and slope = -.8 μeq $L^{-1}yr^{-1}$ (Vermont), and $\chi^2_{trend} = 1.77$, $p = 0.18$ and slope = -0.1 μeq $L^{-1}yr^{-1}$ (Maine).

a more qualitative test of homogeneity (the vast majority of Z scores with the same sign and within the same order of magnitude) in this case, and present an overall trend for SO_4^{2-} in New England lakes (Table V). The reader should consider the potential for between-lake heterogeneity in drawing his or her own conclusion, however.

Nitrate trends differ in the two subregions, but in both cases are small in magnitude (slopes = +0.1 μeq $L^{-1}yr^{-1}$ in the Adirondacks, -0.0 μeq $L^{-1}yr^{-1}$ in New England). Nitrate trends exhibited significant between-season heterogeneity in the Adirondacks; this seems to result primarily from the stronger trends typical of spring values in this subregion (Figure 2b; Stoddard, 1994). Given clear evidence that NO_3^- increases at different rates in different seasons in the Adirondacks, we cannot establish a reliable overall rate of change for this variable. Because NO_3^- plays such a minor role in the interpretation of other acid/base trends in both the Adirondacks and New England, we have not gone the extra (but difficult to interpret) step of reporting separate trends for each season, and instead use the overall rates of change in Table V with the caveat that the Adirondack number may be an overestimate of overall change.

In combination, SO_4^{2-} and NO_3^- trends in the Adirondacks and New England suggest very similar rates of change in acidic anions across the entire Northeast for the period 1982-1994 (slopes = -1.7 vs. -1.6 μeq $L^{-1}yr^{-1}$, respectively). Based solely on these changes in the driving variables for acidification, we would expect the two subregions to exhibit very similar levels of recovery.

In fact, trends in ANC (Table V) differ markedly between the two subregions, with strong recovery evident in the New England lakes (ANC slope = +0.8 μeq $L^{-1}yr^{-1}$), and continued acidification suggested for the Adirondacks (ANC slope = -0.5 μeq $L^{-1}yr^{-1}$). As mentioned earlier, supply rates of base cations from watershed soils largely determine whether lakes or watersheds will respond to acidic deposition (Galloway *et al.*, 1983). Recovery in ANC will only occur if the difference between trends in base cations and acid anions is positive (i.e., acid anions decline more markedly than base cation concentrations). Declines in base cations are expected, because a portion of the base cation supply to each lake results from the movement of mobile anions through the watershed, and so declining SO_4^{2-} would produce, in most cases, a decline in base cations (e.g., Kirchner and Lydersen, 1995). The trend results for base cations (Table V) indicate widespread declines in base cation concentrations, but with much stronger declines in the Adirondacks (slope = -2.0 μeq $L^{-1}yr^{-1}$) than in New England (-0.7 μeq $L^{-1}yr^{-1}$). The heterogeneity observed in base cation trends among lakes in New England results from lower rates of base cation decline in Maine (-0.1 μeq $L^{-1}yr^{-1}$) than in Vermont (-0.8 μeq $L^{-1}yr^{-1}$), suggesting a west-to-east gradient in lake base cation decline across the Northeast. If analyzed by three separate subregions (Adirondacks, western New England, eastern New England), there was no significant heterogeneity in base cation trends (Table V, footnote 4) among lakes in the subregions.

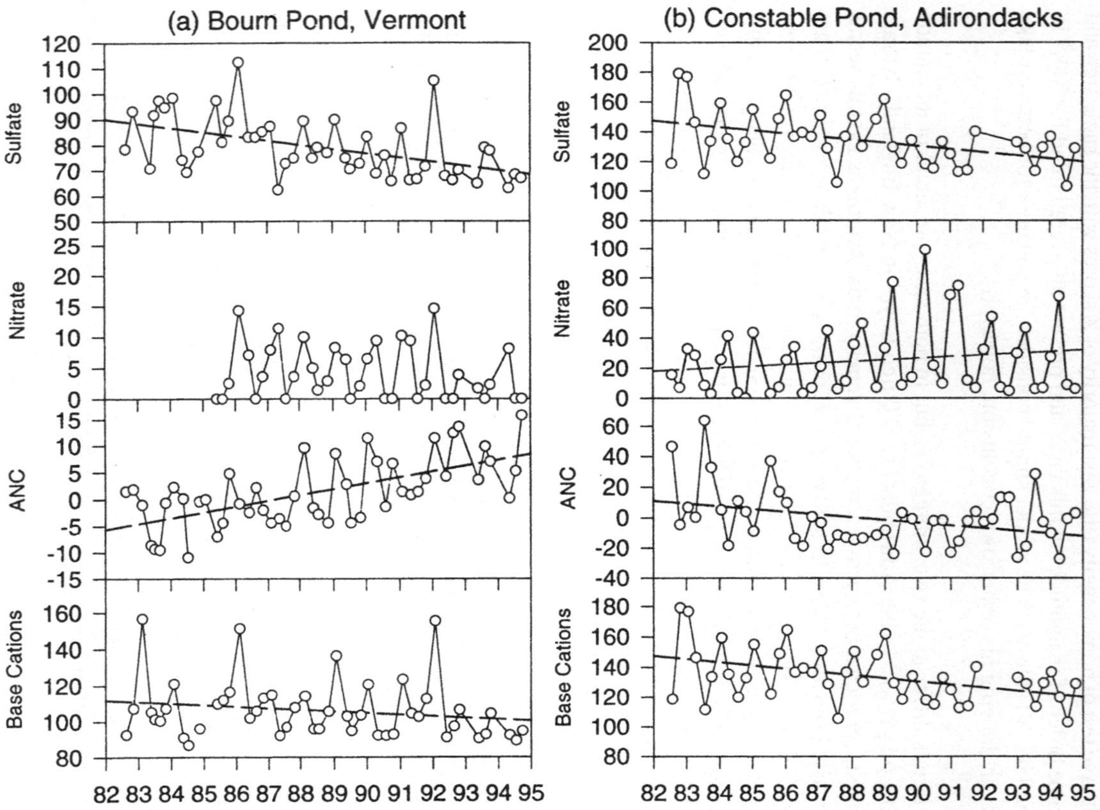

Fig. 2. Time series plots of typical lakes from each subregion. (a) SO_4^{2-}, NO_3^-, ANC and base cation trends in Bourn Pond, Vermont. (b) SO_4^{2-}, NO_3^-, ANC and base cation trends in Constable Pond, Adirondacks. All concentrations are in μeq L^{-1}. Presence of trend line in each panel indicates a significant trend ($p < 0.05$) for this variable.

Time series plots for two typical lakes illustrate the subregional differences revealed by the trend analyses (Figure 2). In New England lakes (e.g., Figure 2a), decreases in SO_4^{2-} concentrations are accompanied by little or no change in NO_3^-, and small decreases in base cation concentrations. The sum effect of these changes is a measurable recovery in ANC (Δ[ANC] » Δ[base cations] - Δ[sulfate]); the calculated change in ANC (based on trends in SO_4^{2-}, NO_3^- and C_B) is +0.9 and the measured change is +0.8 μeq $L^{-1}yr^{-1}$. In the Adirondack lakes (e.g., Figure 2b) decreases in SO_4^{2-} (or in acidic anions, if both SO_4^{2-} and NO_3^- are considered) are matched or exceeded by decreases in base cations. The difference between the two trend magnitudes (» Δ[ANC]) is near zero in some lakes, and negative in others; overall the calculated change in ANC, based on trends in SO_4^{2-}, NO_3^- and C_B, is -0.1, while the measured change is -0.5 μeq $L^{-1}yr^{-1}$. The trends are remarkably consistent internally, and the magnitude of change in C_B appears to control the likelihood of recovery in each subregion.

Several possible mechanisms could produce the observed difference in C_B response in the two subregions. One hypothesis that has received substantial attention of late is that of soil cation depletion. Several authors have proposed that, in watersheds of very low native buffering capacity, chronic levels of acidic deposition have depleted soil C_B stores to the point where soils no longer have sufficient cation exchange capacity to neutralize even lowered levels of acidity (Lawrence *et al.*, 1997; Likens *et al.*, 1996). It is certainly suggestive that the greatest declines in C_B occur in the subregion (the Adirondacks) with the highest historical rates of acidic deposition (Driscoll *et al.*, 1991) in this study. The evidence for a west-to-east gradient in base cation decline (-2.0 μeq $L^{-1}yr^{-1}$ in the Adirondacks, -0.8 μeq $L^{-1}yr^{-1}$ in Vermont, and -0.1 μeq $L^{-1}yr^{-1}$ in Maine) is further support, as the historical gradient in acidic deposition follows the same pattern. Other authors have suggested that declining C_B concentrations in deposition may play a role in delaying surface water recovery from acidification (Driscoll *et al.*, 1989; Hedin *et al.*, 1994); a difference in rates of C_B decline in deposition could produce subregional differences in lake trends. And finally, rapid decreases in NO_3^- that occurred in the Adirondacks in the 1990s (Driscoll *et al.*, 1995; Mitchell *et al.*, 1996) may drive C_B concentrations downward more steeply in this subregion than in the rest of the Northeast. While a detailed examination of these hypotheses is beyond the scope of the current analysis, we can probably eliminate the C_B deposition theory on the basis of the deposition trends presented in Table III. There is no evidence of a difference in the magnitude of C_B deposition trends in the Adirondacks vs. New England, and so little likelihood that lake trends would be driven by such differences.

4. Conclusions

We analyzed trends in the chemistry of deposition and of low ANC lakes in two subregions (Adirondacks and New England) of the northeastern U.S. for the period 1982-1994. Both the deposition and lake data suggest region-wide decreases in SO_4^{2-} of the magnitude -1.8 eq $ha^{-1}yr^{-1}$ (deposition) and -1.8 and -1.6 μeq $L^{-1}yr^{-1}$ (lakes in the Adirondacks and New England, respectively). Rates of NO_3^- , NH_4^+, and nitrogen (NO_3^- + NH_4^+) deposition were unchanged, while small (but significant) differences in the NO_3^- concentrations of lakes were evident (slopes were +0.1 μeq $L^{-1}yr^{-1}$ in the Adirondacks and -0.0 μeq $L^{-1}yr^{-1}$ in New England).

Surprisingly, lakes in the two subregions showed very different responses to declining acid anion inputs. In New England, values of ANC increased at a rate of +0.8 μeq $L^{-1}yr^{-1}$, while lakes in the Adirondacks continued to acidify (ANC slope = -0.5 μeq $L^{-1}yr^{-1}$). This dichotomous behavior appears to result from differences in base cation supply rates in watersheds of the two subregions. While decreases in base cation concentrations are expected to result from declining SO_4^{2-}, C_B declines (slope = -0.7 μeq $L^{-1}yr^{-1}$) in New England were smaller in magnitude than SO_4^{2-} declines, and recovery of ANC resulted. Base cation declines in the Adirondacks (slope = -2.0 μeq $L^{-1}yr^{-1}$) were very similar in magnitude to acid anion (NO_3^- + SO_4^{2-}) declines, and recovery was not evident.

Acknowledgments

The research described in this paper has been funded by the U.S. Environmental Protection Agency, under Contract #68-C8-005 to Dynamac International Corp., cooperative agreements CR818842 and CR824135 to the University of Maine, CR815196 and CR822726 to the Vermont DEC, CR815733 to Syracuse University, and CR822788 to the Adirondack Lake Survey Corporation. It has been subjected to the Agency's peer and administrative review, and it has been approved for publication. Mention of trade names or commercial products does not constitute endorsement or recommendation for use. The authors are grateful for the comments of P.J. Wigington, M. Mattson, and D. DeWalle; they improved the quality of the final product significantly.

References

Baker, L. A., Herlihy, A. T., Kaufmann, P. R., and Eilers, J. M.: 1991, *Science*, **252**, 1151-1154.

Baker, L. A., Kaufmann, P. R., Herlihy, A. T., and Eilers, J. M.: 1990, *Current Status of Surface Water Acid-Base Chemistry*, NAPAP State-of-Science/Technology Report No. 9, National Acid Precipitation Assessment Program, Washington, DC.

Charles, D. R.: 1991. *Acidic Deposition and Aquatic Ecosystems. Regional Case Studies*, Springer-Verlag, New York, 747 pages.

Church, M. R., Thornton, K. W., Shaffer, P. W., Stevens, D. L., Rochelle, B. P., Holdren, G. R., Johnson, M. G., Lee, J. L., Turner, R. S., Cassell, D. L., Lammers, D. A., Campbell, W. G., Liff, C. I., Brandt, C. C., Liegel, L. H., Bishop, G. D., Mortenson, D. C., Pierson, S. M., and Schomoyer, D. D.: 1989, *Direct/Delayed Response Project: Future Effects of Long-Term Sulfur Deposition on Surface Water Chemistry in the Northeast and Southern Blue Ridge Province. Volume III. Level III Analyses and Summary of Results.*, U.S. Environmental Protection Agency, Washington, D.C.

Driscoll, C. T., Likens, G. E., Hedin, L. O., Eaton, J. S., and Bormann, F. H.: 1989, *Environ. Sci. Technol.*, **23**, 137-143.

Driscoll, C. T., Newton, R. M., Gubala, C. P., Baker, J. P., and Christensen, S. W.: 1991, Adirondack Mountains, in *Acidic Deposition and Aquatic Ecosystems: Regional Case Studies*, D. F. Charles, ed., Springer-Verlag, New York, NY, 133-202.

Driscoll, C. T., Postek, K. M., Kretser, W., and Raynal, D. J.: 1995, *Water, Air, and Soil Pollut.*, **85**, 583-588.

Driscoll, C. T., and Van Dreason, R.: 1993, *Water, Air, and Soil Pollut.*, **67**, 319-344.

Fisher, R. A.: 1932, *Statistical Methods for Research Workers*, Oliver and Boyd, London.

Ford, J., Stoddard, J. L., and Powers, C. F.: 1993, *Water, Air, and Soil Pollut.*, **67**, 247-255.

Galloway, J. N., Norton, S. A., and Church, M. R.: 1983, *Environ. Sci. Technol.*, **17**, 541-545.

Hedges, L. V., and Olkin, I.: 1985, *Statistical Methods for Meta-Analysis*, Academic Press, New York.

Hedin, L. O., Granat, L., Likens, G. E., Buishand, T. A., Galloway, J. N., Butler, T. J., and Rodhe, H.: 1994, *Nature*, **367**, 351-354.

Hirsch, R. M., and Slack, J. R.: 1984, *Water Resour. Res.*, **20**, 727-732.

Hirsch, R. M., Slack, J. R., and Smith, R. A.: 1982, *Water Resour. Res.*, **18**, 107-121.

Husar, R. B., Sullivan, T. J., and Charles, D. F.: 1991, Historical trends in atmospheric sulfur deposition and methods for assessing long-term trends in surface water chemistry, in *Acidic Deposition and Aquatic Ecosystems: Regional Case Studies*, D. F. Charles, ed., Springer-Verlag, New York, 65-82.

Kahl, J. S., Norton, S. A., Haines, T. A., and Davis, R. B.: 1993, *Water, Air, and Soil Pollut.*, **67**, 281-300.

Kirchner, J. W., and Lydersen, E.: 1995, *Environ. Sci. Technol.*, **29**, 1953-1960.

Lawrence, G. B., David, M. B., Bailey, S. W., and Shortle, W. C.: 1997, *Biogeochemistry*, **38**, 19-39.

Likens, G. E., Driscoll, C. T., and Buso, D. C.: 1996, *Science*, **272**, 244-246.

Loftis, J. C., and Taylor, C. H.: 1989, *Environ. Manag.*, **13**, 529-539.

Lynch, J. A., Bowersox, V. C., and Grimm, J. W.: 1996, *Trends in Precipitation Chemistry in the United States, 1983-94: An Analysis of the Effects in 1995 of Phase I of the Clean Air Act Amendments of 1990, Title IV*, Open File Report 96-0346, U.S. Geological Survey.

Mattson, M. D., Godfrey, P. J., Walk, M., Kerr, P. A., and Zajicek, O. T.: 1997, *Water, Air, and Soil Pollut.*, **96**, 211-232.

Mitchell, M. J., Driscoll, C. T., Kahl, J. S., Likens, G. E., Murdoch, P. S., and Pardo, L. H.: 1996, *Environ. Sci. Technol.*, **30**, 2609-2612.

Morrison, M.: 1991, Quality Assurance Plan for the Long-Term Monitoring Project, in *Data User's Guide to the United States Environmental Protection Agency's Long-Term Monitoring Project: Quality Assurance Plan and Data Dictionary*, U.S. Environmental Protection Agency, Corvallis, OR, 1.1-B.1.

National Atmospheric Deposition Program/National Trends Network: 1980-94, *NADP/NTN Annual Data Summary. Precipitation Chemistry in the United States*, , NADP/NTN Coordination Office, Natural Resource Ecology Laboratory, Colorado State University, Fort Collins, CO.

Sen, P. K.: 1968, *Annals of Mathematics and Statistics*, **39**, 1115-1124.

Stoddard, J. L.: 1994, Long-term changes in watershed retention of nitrogen: its causes and aquatic consequences, in *Environmental Chemistry of Lakes and Reservoirs*, L. A. Baker, ed., American Chemical Society, Washington, DC, 223-284.

Stoddard, J. L., Driscoll, C. T., Kahl, S., and Kellogg, J.: in press, *Ecol. Appl.*

Stoddard, J. L., and Kellogg, J. H.: 1993, *Water, Air, and Soil Pollut.*, **67**, 301-317.

U.S. Environmental Protection Agency: 1991. *Data User's Guide to the United States Environmental Protection Agency's Long-Term Monitoring Project: Quality Assurance Plan and Data Dictionary*, U.S. Environmental Protection Agency, Corvallis, OR.

van Belle, G., and Hughes, J. P.: 1984, *Water Resour. Res.*, **20**, 127-136.

REGIONAL LAND COVER CHARACTERIZATION USING LANDSAT THEMATIC MAPPER DATA AND ANCILLARY DATA SOURCES

J.E. VOGELMANN[1], T.L. SOHL[1], P.V. CAMPBELL[2], and D.M. SHAW[3]

[1]*EROS Data Center, Hughes-STX Corporation, USGS, Sioux Falls, SD 57198,* [2]*MRLC, U.S. EPA, MD-75A, Research Triangle Park, NC 27711,* [3]*Landscape Characterization, U.S. EPA, MD-75A, Research Triangle Park, NC 27711*

Abstract. As part of the activities of the Multi-Resolution Land Characteristics (MRLC) Interagency Consortium, an intermediate-scale land cover data set is being generated for the conterminous United States. This effort is being conducted on a region-by-region basis using U.S. Standard Federal Regions. To date, land cover data sets have been generated for Federal Regions 3 (Pennsylvania, West Virginia, Virginia, Maryland, and Delaware) and 2 (New York and New Jersey). Classification work is currently under way in Federal Region 4 (the southeastern United States), and land cover mapping activities have been started in Federal Regions 5 (the Great Lakes region) and 1 (New England). It is anticipated that a land cover data set for the conterminous United States will be completed by the end of 1999. A standard land cover classification legend is used, which is analogous to and compatible with other classification schemes. The primary MRLC regional classification scheme contains 23 land cover classes.

The primary source of data for the project is the Landsat thematic mapper (TM) sensor. For each region, TM scenes representing both leaf-on and leaf-off conditions are acquired, preprocessed, and georeferenced to MRLC specifications. Mosaicked data are clustered using unsupervised classification, and individual clusters are labeled using aerial photographs. Individual clusters that represent more than one land cover unit are split using spatial modeling with multiple ancillary spatial data layers (most notably, digital elevation model, population, land use and land cover, and wetlands information). This approach yields regional land cover information suitable for a wide array of applications, including landscape metric analyses, land management, land cover change studies, and nutrient and pesticide runoff modeling.

1. Introduction

Many organizations require accurate intermediate-scale land cover information for a variety of applications. As an example, the National Oceanic and Atmospheric Administration's (NOAA) Coastal Change Analysis Program (C-CAP; Dobson *et al.*, 1995) has strong requirements for such information for assessing changes in coastal areas. In this case, the effects of land cover changes are being investigated with special emphasis on determining long-term effects on estuarine systems. Similarly, the U.S. Geological Survey (USGS) Water Resources Division National Water-Quality Assessment Program (Leahy *et al.*, 1993; National Research Council, 1990) is using medium-scale land cover data as input for nutrient and pesticide runoff models. This is a concerted effort involving the major watershed drainage units within the United States. Additionally, the USGS Biological Resources Division's Gap Analysis Program (Scott *et al.*, 1996) uses intermediate-scale land cover data to generate detailed data sets mapping natural and semi-natural plant assemblages. This information is linked with modeled vertebrate habitat preference distribution data to map (and ultimately manage) biodiversity on a national scale. The field

Environmental Monitoring and Assessment **51: 415–428, 1998.**

base 30-m resolution to facilitate maximum use. The MRLC national land cover data can then be incorporated with spatial data, such as the AVHRR, at other scales and resolutions to provide a true multi-resolution land characteristics data base. This data set will have the following specifications: (1) there will be a nationally consistent hierarchical legend; (2) final data will be maintained at a minimum spatial resolution of 30 meters; (3) the data set will be produced and stored in a generic raster data format; (4) the data set will include the classified and labeled land cover data, appropriate ancillary data, and metadata documentation; and (5) data will comply with Federal Geographic Data Committee standards. In addition to these characteristics, all MRLC data will be easily accessible to the user community. To this end, the MRLC is developing Internet access as well as conventional delivery routes, such as Compact Disc or tape media.

The MRLC elected to execute its national land cover initiative using a template of 10 Standard Federal Regions as defined by the April 4, 1974, Executive Order OMB Circular A-105. The first "pilot" regional data set was completed for Federal Region 3, which includes the States of Pennsylvania, West Virginia, Virginia, Maryland, and Delaware. Federal Region 2 (New York and New Jersey) has also been completed, and classification work is currently under way in Federal Region 4 (the southeastern United States). Initial land cover mapping activities are under way in Federal Regions 5 (the Great Lakes region) and 1 (New England), and it is projected that land cover generation for the eastern United States will be completed during 1998. It is anticipated that a land cover data set for the entire conterminous United States will be completed by the end of 1999.

3. Data Sources

The primary source of data for this effort is Landsat TM data acquired in 1991, 1992, and 1993 for the MRLC (Loveland and Shaw, 1996). As part of this effort, data sets have been destriped, terrain-corrected using the 3-arc-second digital terrain elevation data (DTED), and georegistered using ground control points, resulting in a root mean square registration error of less than 1 pixel (30 m). Both leaf-on and leaf-off TM data sets are being analyzed.

Other intermediate-scale spatial data are being used as ancillary information in the analysis, including DTED (U.S. Geological Survey, 1993) and derivative DTED products (slope, aspect, and shaded relief), population density data at the census block level (Bureau of the Census, 1991a and b; 1992), Land Use and Land Cover (LUDA) data, and National Wetlands Inventory (NWI; U.S. Fish and Wildlife Service, 1996) data. Additionally, available water capacity and organic carbon (0-40 cm depth) data from the State Soil Geographic (STATSGO) Data Base (U.S. Department of Agriculture, 1994) are being used. Land cover information from various state or national programs, such as the USGS Biological Resources Division Gap Analysis Program (Scott *et al.*, 1996), are being incorporated when appropriate.

4. Classification System

The MRLC classification system (Table I) provides a consistent hierarchical approach to defining 23 classes of land cover across the lower 48 United States. Because of landscape differences found among the ten federal regions, all classes may not appear in the legend of a particular regional data set. The land cover classification approach is a merging of the C-CAP classification protocol (Dobson *et al.,* 1995) and draft Federal Geographic Data Committee standards. The intent behind creating this hybrid classification approach is to provide a linkage between existing generalized land cover data sets, such as the C-CAP system, with the more detailed natural vegetation data, such as that provided by the USGS Gap Analysis Program. In addition, the C-CAP classification protocol is based on the Anderson system (Anderson, 1976), and thus land cover data generated by the MRLC system can be easily related to legacy data sets, such as the LUDA data set.

5. Classification Procedure

The general procedure for creating the national land cover data set includes (1) generating mosaics of leaf-on and leaf-off TM scenes and clustering each using an unsupervised classification algorithm, (2) interpreting and labeling clusters into MRLC classes using aerial photographs, (3) resolving confused clusters by constructing models that make use of the appropriate ancillary data sources, and (4) incorporating information from onscreen digitizing (e.g., quarries and transitional bare areas, such as clear cuts) and additional available land cover data sets to refine and augment the basic classification developed above. Depending on the region being analyzed, either leaf-on or leaf-off mosaics are used as the primary source of land cover information. The other mosaic is then used as an ancillary data layer to aid in class-splitting operations.

The smallest federal regions are analyzed as single units, whereas large federal regions are broken down into smaller areas for analyses. In general, current software and hardware limitations make it difficult to work with files that exceed 2 gigabytes. Assuming that a three-band composite is required at some stage of the analysis procedure (e.g., for onscreen digitizing steps, or for overlaying classifications onto imagery), then the upper limit of the land area that can be analyzed as a single unit using TM data is approximately 700,000 km^2. Once the individual units of a given federal region have been classified, the pieces are edge-matched with the goal of obtaining a consistent and seamless land cover product for the region. Land cover results from adjacent federal regions are similarly edge-matched, which will ultimately result in a consistent land cover data set for the conterminous United States.

Table I
Multi-Resolution Land Characterization regional land cover classification system.
Definitions of classes available by request.

1.0 Water
- 1.1 Open Water
- 1.2 Perennial Ice/Snow

2.0 Developed
- 2.1 High Intensity
  - 2.11 Residential
  - 2.12 Commercial/Industrial/Transportation
- 2.2 Low Intensity
  - 2.21 Residential

3.0 Bare
- 3.1 Transitional
- 3.2 Quarries/Strip Mines/Gravel Pits
- 3.3 Bare Rock/Sand

4.0 Vegetated
- 4.1 Woody Upland Vegetation
  - 4.11 Natural Forested
    - 4.111 Deciduous Forest
    - 4.112 Evergreen Forest
    - 4.113 Mixed Forest
  - 4.12 Natural Shrubland
    - 4.121 Deciduous Shrubland
    - 4.122 Evergreen Shrubland
    - 4.123 Mixed Shrubland
  - 4.13 Planted/Cultivated *
    (orchards, vineyards, groves)
- 4.2 Herbaceous Upland Vegetation
  - 4.21 Natural/Semi-natural Herbaceous
    - 4.211 Grasslands
  - 4.22 Planted/Cultivated Herbaceous
    - 4.221 Bare Soil
    - 4.222 Small Grains
    - 4.223 Row Crops
    - 4.224 Grasses
      - 4.2241 Pasture/Hay
      - 4.2242 Other (parks, lawns, golf courses)
- 4.3 Wetlands
  - 4.31 Woody Wetlands
  - 4.32 Emergent Herbaceous Wetlands

* Classification of woody planted/cultivated vegetation subject to availability of sufficient ancillary data to differentiate from natural woody vegetation.

For mosaicking purposes, an attempt is made during the scene selection process to minimize the unwanted effects of interscene phenological variability by choosing scenes acquired at approximately the same time of year. After scenes have been selected, a "master" scene (Homer *et al.*, 1997) is selected, and regions of spatial overlap with adjacent "slave" scenes are used to normalize digital data. From these zones of overlap, histograms of digital values from the slave scenes are adjusted to match the histogram brightness values of the master image on a band by band basis. Prior to normalization, areas with clouds and water are masked out so that normalization is performed using only digital data from areas dominated by land cover. Once a slave image is radiometrically matched to the master, it, in turn, becomes a master for its adjacent scenes.

Mosaicked scenes are clustered into 100 spectrally distinct classes using the CLUSTER algorithm developed at Los Alamos National Laboratory (Kelly and White, 1993; Benjamin *et al.*, 1996). Classification is accomplished using TM bands 3 (0.63-0.69 micrometers), 4 (0.76-0.90 micrometers), 5 (1.55-1.75 micrometers), and 7 (2.08-2.35 micrometers). Previous work has indicated that relatively little unique land cover information is derived by using greater numbers of clusters (Vogelmann *et al.*, 1997a), and it was decided that 100 clusters capture most of the regional land cover variability that could be derived from the TM data. Clusters are assigned into MRLC classes (Table I) using National High Altitude Photography (NHAP) Program aerial photographs as reference information.

Almost invariably, the individual spectral clusters derived from classification represent two or more of the targeted land cover classes. These clusters (hereafter designated "multi-class clusters") are split into more meaningful land cover units using ancillary raster data that have the same pixel size (30 m) and the same projection parameters as the imagery. Slope, aspect, and shaded-relief data sets are derived from the DTED data using standard raster-based image processing software, whereas the NWI, LUDA, STATSGO, and population census block group data layers are obtained by rasterizing and combining available vector-based coverages.

Briefly, for each multi-class cluster, digital ancillary values are obtained for a sample of individual pixels representing the suite of land cover classes represented by that cluster. The digital values of the various ancillary data layers are then compared to (1) determine which data layers are the most effective for splitting the multi-class clusters into the appropriate land cover units, and (2) derive the appropriate thresholds for splitting the clusters. Models are developed using one to several ancillary data sets to split each multi-class cluster into the desired land cover categories.

Each model consists of a series of conditional statements that split the cluster in question into two or more classes. The statements that are the most effective in splitting the clusters into the appropriate land cover classes are placed in the beginning of the models, whereas those that are less effective (but are still useful) are placed towards the end of the models. As an example, consider a multi-class cluster that includes both deciduous forest and low intensity developed land cover classes. Logically, regions of especially high elevation are not likely to be urbanized, and thus the first step of a model for this hypothetical cluster might be to assign pixels of the cluster that occur at relatively high elevations (e.g., 700 meters or greater) into the deciduous forest class. Those pixels from that cluster located in regions of high population density will most likely be low-intensity developed rather than deciduous pixels, and thus the next statement of the model might be to assign all pixels of the cluster that are located in areas with population densities greater than a particular threshold into the low intensity developed class. The last statement in the model might be to use LUDA data to assign the classes for the remaining pixels. The model development procedure is very empirical, and it generally takes several trials and modifications of model parameters using the subset of the mosaicked data set before the class-splitting models are considered refined enough to apply to the entire region (determined by visual inspection of model runs).

Most spectral clusters require development of class-splitting models. After all of these models are run, data are recombined into first-order classification products. It should be noted that there are advantages to conducting separate analyses of clustered leaf-on and leaf-off data sets rather than clustering and analyzing leaf-on and leaf-off data sets together. The analyst can make effective use of seasonally specific phenological information during the labeling process when data sets represent distinct time periods (e.g., leaf-on versus leaf-off). Such information is more difficult to use when the two dates are clustered as a single unit. It should also be noted that previous work (Vogelmann *et al.*, 1997a) has indicated that minimal gain in class discrimination is achieved after multi-temporal clustering of two TM data sets as opposed to clustering of the two data sets separately. However, it should also be noted that other investigators (Slaymaker *et al.*, 1996) have achieved excellent results after multi-temporal clustering of two seasonally distinct TM data sets.

Many bare areas (especially clearcuts and quarries) and the "other grass" category (i.e., parks, golf courses, large lawns) are spectrally similar to other land cover classes and consequently are difficult to accurately classify using spectral data alone. However, when spatial characteristics are combined with their spectral properties, these areas can often be readily discerned in the TM imagery. Such classes are obtained through onscreen digitizing of the TM images. These digitized data sets are rasterized and recoded into the appropriate land cover categories and are incorporated into the land cover mosaics. The resulting product is then compared with the raw imagery, and obvious errors are corrected on a case by case basis.

6. Sample Land Cover Products

A mosaic of leaf-off Landsat TM images is shown for Federal Region 2 (Figure 1). This mosaic was produced using TM bands 5, 4, and 3 in red, green, and blue color planes. In this image, bright green areas correspond with areas of hay and pasture, pink areas are mostly deciduous forest, dark green areas are evergreen forest, purple areas are urban centers, and turquoise areas relate to snow cover. Note the spruce-fir forests associated with the high peaks region of the Adirondack Mountains in the northeastern part of the image, and the Pine Barrens in southern New Jersey (both dark green). Because of the normalization process, the mosaic is mostly seamless; this data set was produced using 14 TM scenes.

Comparison of the land cover classification data set for Federal Region 2 (Figure 2) with the imagery indicates good general agreement between the two products. Although

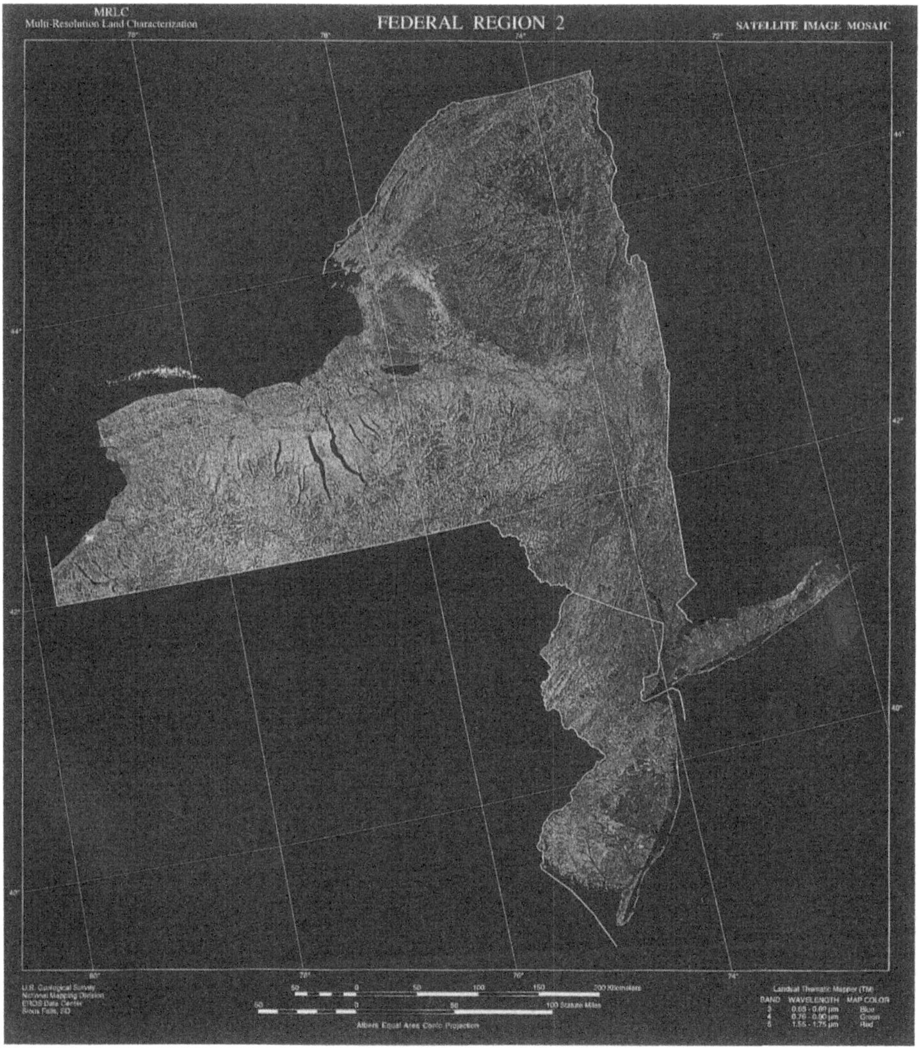

Fig. 1. Landsat thematic mapper mosaic of Federal Region 2 produced using bands 5, 4, and 3 in the order of red, green, and blue. Data represent leaf-off conditions.

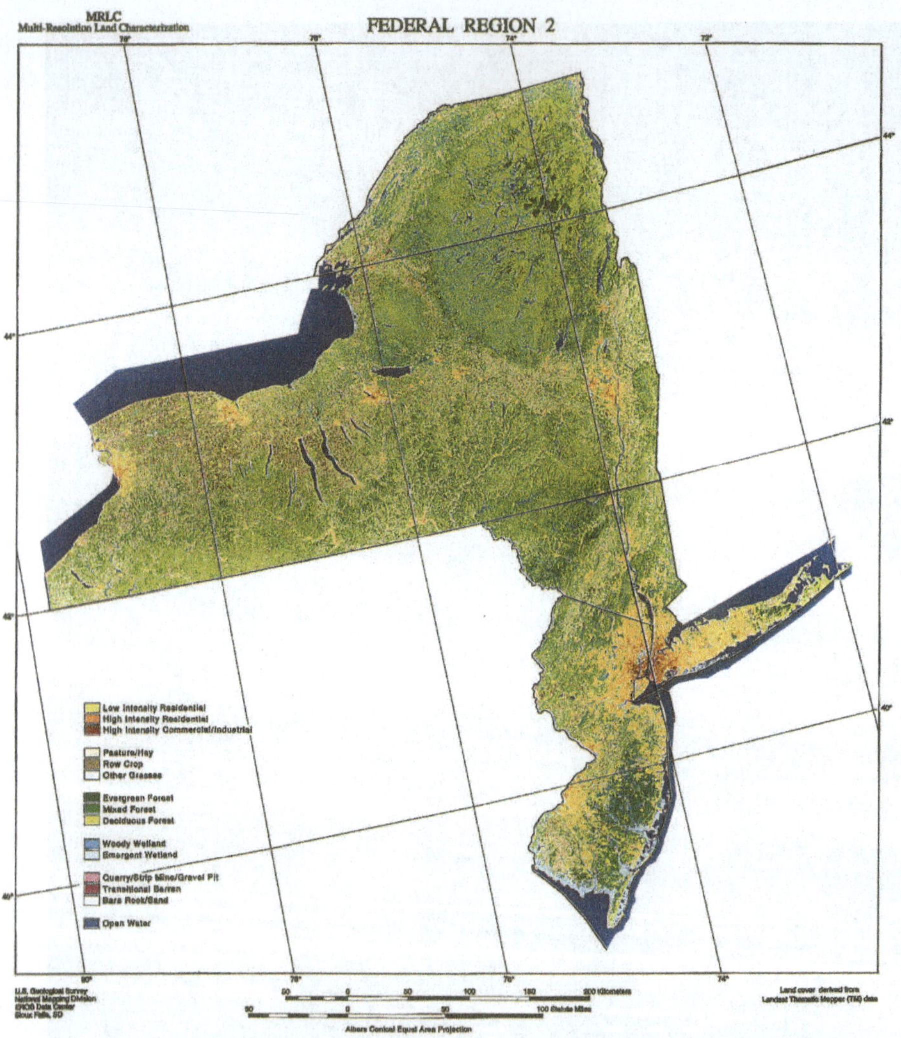
MRLC
Multi-Resolution Land Characterization
FEDERAL REGION 2
Low Intensity Residential
High Intensity Residential
High Intensity Commercial/Industrial
Pasture/Hay
Row Crop
Other Grasses
Evergreen Forest
Mixed Forest
Deciduous Forest
Woody Wetland
Emergent Wetland
Quarry/Strip Mine/Gravel Pit
Transitional Barren
Bare Rock/Sand
Open Water
U.S. Geological Survey
National Mapping Division
EROS Data Center
Sioux Falls, SD
Kilometers
Statute Miles
Albers Conical Equal Area Projection
Land cover derived from
Landsat Thematic Mapper (TM) data

Fig. 2. Land cover data set developed for Federal Region 2.

the classification product was derived using 14 leaf-off and 14 leaf-on TM scenes, the methods used produced a nearly seamless classification product. It should be noted that both Figures 1 and 2 have been resampled for the purposes of presentation, and that substantial additional spatial detail is contained within the actual digital data files. An enlarged sample of the classified Region 2 data set (Figure 3) provides an example of the full-resolution characteristics and quality of the regional data sets being produced. Class area estimates (Table II) indicate that about 54 percent of the Federal Region 2 is forested, about 22 percent is in agriculture, and about 6.5 percent is urban/residential. A product of similar quality was produced for the mid-Atlantic region (Federal Region 3; Vogelmann *et al.*, 1997b).

7. Error Assessment and Consistency Checks

Once a land cover data set for a given region has been generated, three general stages of error assessment are conducted. Phase 1 includes initial checks, in which the classification product is compared with the imagery and obvious errors are fixed. During this phase, the preliminary land cover product is released to selected groups, especially those most familiar with the area. Feedback from these individuals is encouraged, and misclassifications are fixed when warranted. During phase 2, the land cover data set is compared with other sources of data from the region. Possible data sources for comparison include, but are not limited to, other classifications (often done for smaller parts of the regions), Census of Agriculture (Bureau of the Census, 1993) information, and aerial photograph point observation data. Comparisons with such data sources do not provide users with absolute values of accuracy, but do provide general information regarding the degree of consistency between the data sets being compared. This approach was found to be useful in Federal Region 3 (Vogelmann *et al.*, 1997b), where the combined assessments of comparisons with several sources of data provided information regarding which classes were the most trustworthy. During phase 3, a formal, statistically designed accuracy assessment (Congalton, 1991) will be done. The most appropriate methods for conducting phase 3 are being explored.

8. Conclusions

The approach being implemented appears to have provided users with very good general land cover classification products for large regions. Although there are some classification errors within the data sets, the products appear to have many desirable characteristics (e.g., mostly seamless, and reasonable in terms of accuracy on the basis of visual inspection and consistency checks). The data sets produced to date are being used by researchers in the Environmental Protection Agency's Landscape Ecology Program, Mid-Atlantic Integrated Assessment, and Regional Vulnerability Assessment Program. In addition, some GAP principal investigators in the mid-Atlantic and southeastern United States are using MRLC-derived land cover data to help map their state's natural and semi-natural vegetation. Additionally, USGS National Water Quality Assessment personnel are using the data to

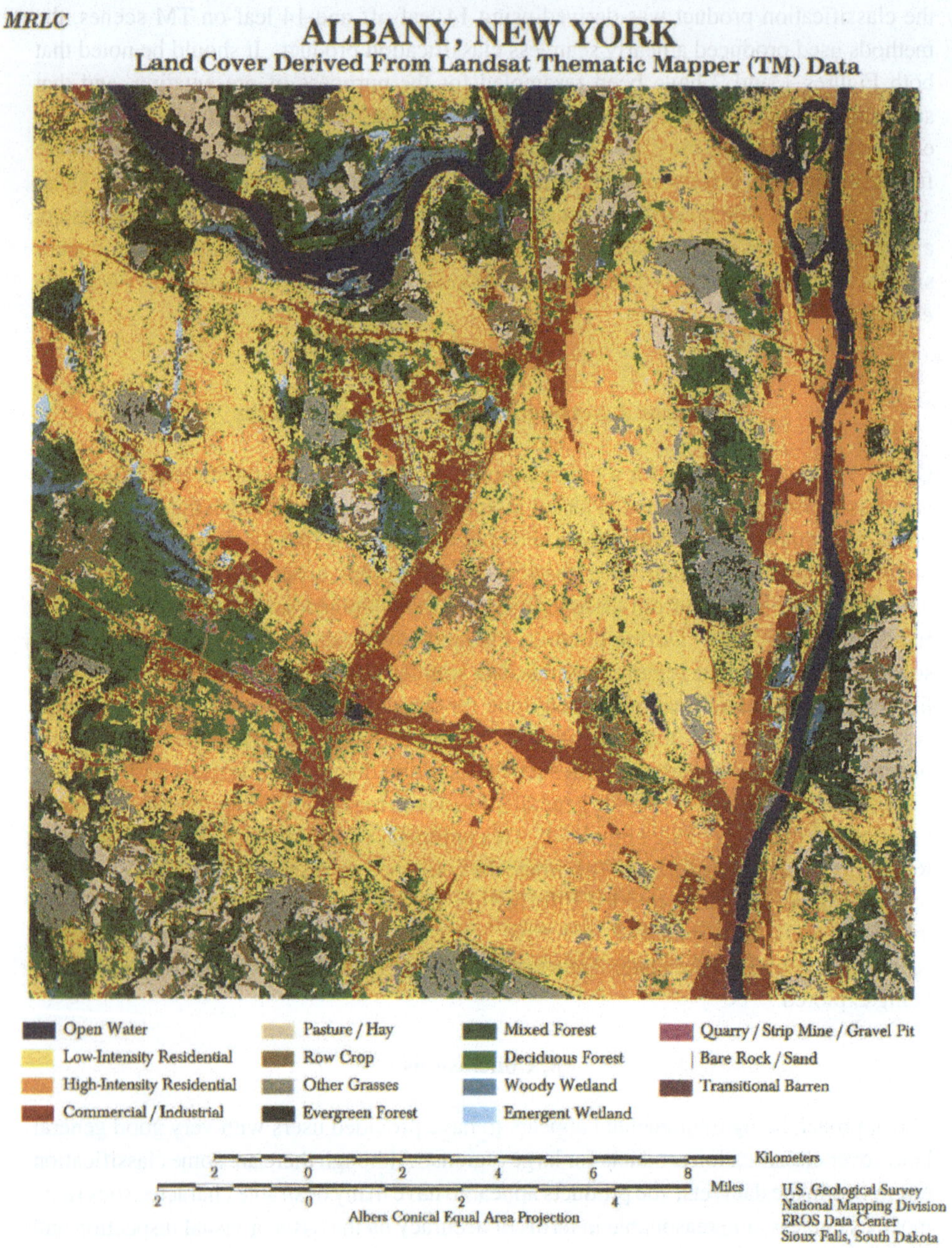

Fig. 3. Full resolution land cover data developed for the Albany, New York region.

of landscape ecology (Formon and Godron, 1986) also has strong requirements for accurate and consistent land cover data. A series of landscape metrics using intermediate-scale land cover data (Riiters *et al.*, 1995) has been developed for assessing ecologically significant landscape patterns and processes, including forest contiguity and fragmentation, wildlife corridors, and patch size variables.

Despite the demand for land cover in these applications, many of the intermediate-scale spatial land cover data sets now available for the United States are outdated and of questionable accuracy. The only intermediate-scale land cover data set currently available for the conterminous United States is the Land Use and Land Cover (LUDA) data set (USGS, 1990). This data set, which was developed in the 1970's by interpreting and digitizing high-altitude aerial photographs, is probably still adequate for some applications. However, many land cover changes have occurred since the data set was compiled, and a more up-to-date national data set is needed. Recently, a land cover classification for the conterminous United States using 1-km advanced very high resolution radiometer (AVHRR) data (Loveland *et al.*, 1991; Brown *et al.*, 1993) was developed. Although it meets the needs of many researchers within the global change research community (Reed *et al.*, 1994), this data set is spatially too coarse for dealing with the problems and questions being addressed by other groups.

The Multi-Resolution Land Characteristics (MRLC) Interagency Consortium project was established as a partnership among Federal programs responsible for producing or using land cover data (Loveland and Shaw, 1996). Current partners include the U.S. Environmental Protection Agency, the U.S. Forest Service, NOAA, and the USGS. Initial priorities of the MRLC were concentrated on acquiring a common set of Landsat thematic mapper (TM) data sets for the conterminous United States from 1991 to 1993, and processing and georeferencing them to a set of standard specifications for a multitude of agency-specific purposes. More recently, MRLC activities have focused on using these processed data sets to help develop an intermediate-scale (30 m) land cover data set for the conterminous United States. One of the goals of this work is to produce a thematically consistent, seamless, and reasonably accurate land cover data set for the United States for multiple applications. This effort is the primary focus of this paper.

2. National Land Cover Initiative

The genesis of the MRLC's National Land Cover Initiative (NLCI) can be found in the March 1995 Memorandum of Understanding (MOU) signed at the director's level of the program's agencies. The MOU identified several long-term goals, including the development of a flexible and functional land characteristics data base for use by the MRLC and other federal, state, and local organizations. Although the overall vision of the MRLC is to provide a multi-resolution data base, current efforts are focused on generating an intermediate-scale national land cover data set that is based on remotely sensed satellite data acquired by the Landsat TM. The classified land cover data will be maintained at the develop pesticide and herbicide runoff models. Because of the scope of the analyses, it needs to be emphasized that the data sets are especially appropriate for regional analyses and applications. It should be cautioned, however, that many local-scale phenomena may be missed in such efforts, and that there is no surrogate for more in-depth analyses to obtain more detailed and precise information relating to localized conditions. For these latter purposes, however, we believe that the data sets being produced may be useful for providing a first-order overview.

Table II
Class area estimates for Federal Region 2.

Class	Area (km^2)	Percentage Area
Low Intensity Residential	6,881	4.2
High Intensity Residential	2,013	1.2
Commercial/Industrial	1,770	1.1
Pasture/Hay	15,034	9.2
Row Crop	20,001	12.3
Other Grasses	1,222	0.8
Evergreen Forest	8,931	5.5
Mixed Forest	26,814	16.4
Deciduous Forest	52,111	31.9
Woody Wetland	4,750	2.9
Herbaceous Emergent Wetland	1,170	0.7
Quarry/Mines/Gravel Pits	210	0.1
Bare Rock/Sand	114	0.1
Transitional Bare	78	0.1
Open Water	22,326	13.7

Acknowledgements

This work was performed in part by the Hughes STX Corporation under U.S. Geological Survey Contract 1434-CR-97-CN-40274. We thank the following people for their many contributions throughout the course of this study: Tom Loveland, Jess Brown, Paul Seevers, Brenda Jones, Bruce Wylie, Rachel Clement, Chuck Larson, Brian Wardlow, Don Ohlen, Kent Hegge, Brad Kontz, Gayla Evans, and Steve Howard.

References

Anderson, J.F., Hardy, E.E., Roach, J.T., Witmer, R.E.: 1976, 'A land use and land cover classification system for use with remote sensor data,' U.S. Geological Survey Professional Paper 964, 28 pp.

Benjamin, S., White, J.M., Argiro, D., Lowell, K.: 1996, 'Land cover mapping with Spectrum', in *Gap Analysis: A Landscape Approach to Biodiversity Planning* (eds. J.M. Scott, T. Tear, and F. Davis), Proceedings of the ASPRS/GAP Symposium (National Biological Service, Moscow, ID), pp. 279-288.

Brown, J.F., Loveland, T.R., Merchant, J.W., Reed, B.C., and Ohlen, D.O.: 1993, 'Using multisource data in global land cover characterization: concepts, requirements, and methods,' *Photogrammetric Engineering and Remote Sensing* **59**, 977-987.

Bureau of the Census: 1991a, 'Census of population and housing', 1990, public law 94-171 data (United States) (machine readable data files), The U.S. Bureau of the Census (producer and distributor), Washington, D.C.

Bureau of the Census: 1991b, 'Census of population and housing,' 1990, public law 94-171 data, on-line documentation (United States), The U.S. Bureau of the Census, Washington, D.C.

Bureau of the Census: 1992, 'TIGER/Line Files', (machine readable data files), The Bureau of the Census (producer and distributor), Washington, D.C.

Bureau of the Census: 1993, 'Census of the Agriculture, Final county files (machine-readable data file)', 1992, The Bureau of the Census (producer and distributer), Washington, D.C.

Congalton, R.G.: 1991, 'A review of assessing the accuracy of classifications of remotely sensed data', *Remote Sensing of Environment* **37**, 35-46.

Dobson, J.E., Bright, E.A., Ferguson, R.L., Field, D.W., Wood, L.L., Haddad, K.D., Iredale, H., Jensen, J.R., Klemas, V.V., Orth, R.J., and Thomas, J.P.: 1995, 'NOAA Coastal Change Analysis Program (C-CAP): guidance for regional implementation', NOAA Technical Report NMFS 123, U.S. Department of Commerce, Seattle, Washington.

Forman, R.T.T. and Godron, M.: 1986, *Landscape Ecology*, John Wiley and Sons, New York, NY.

Homer, C.G., Ramsey, D., Edwards, T.C., Jr., and Falconer, A.: 1997, 'Landscape cover-type mapping and modeling using a multi-scene Thematic Mapper mosaic', *Photogrammetric Engineering and Remote Sensing*, **63**, 59-67.

Kelly, P.M. and White, J.M.: 1993, 'Preprocessing remotely sensed data for efficient analysis and classification, Knowledge-Based Systems in Aerospace and Industry', Proceedings of SPIE 1993, 24-30.

Leahy, P.P., Ryan, B.J. and Johnson, A.: 1993, 'An introduction to the U.S. Geological Survey's National Water-Quality Assessment Program', *Water Resources Bulletin* **29**, 529-532.

Loveland, T.R., Merchant, J.W., Ohlen, D.O., and Brown, J.F.: 1991, 'Development of a landcover characteristics database for the conterminous U.S.', *Photogrammetric Engineering and Remote Sensing* **57**, 1,453-1,463.

Loveland, T.R. and Shaw, D.M.: 1996, 'Multiresolution land characterization: building collaborative partnerships', in *Gap Analysis: A Landscape Approach to Biodiversity Planning* (eds. J.M. Scott, T. Tear, and F. Davis), Proceedings of the ASPRS/GAP Symposium, Charlotte, NC, (National Biological Service, Moscow, ID), pp. 83-89.

National Research Council: 1990, *A review of the USGS National Water-Quality Assessment Pilot Program*, Washington, D.C., National Academy Press, 153 pp.

Reed, B.C., Loveland, T.R., Steyaert, L.T., Brown, J.F., Merchant, J.W., Ohlen, D.O.: 1994, 'Designing global land cover databases to maximize utility', in *Environmental Information Management and Analysis: Ecosystem to Global Scales*, W.K. Michener, J.W. Brunt, and S.G. Stafford, editors: London, Francis and Taylor, pp. 299-314.

Riiters, K.H., O'Neill, R.V., Hunsaker, C.T., Wickham, J.D., Yankee, D.H., Timmins, S.P., Jones, K.B., and Jackson, B.L.: 1995, 'A factor analysis of landscape pattern and structure metrics', *Landscape Ecology* **10,** 23-39.

Scott, J.M., Tear, T.H., and Davis, F.W. (eds.),: 1996, *Gap Analysis. A Landscape Approach to Biodiversity Planning*, American Society for Photogrammetry and Remote Sensing, Bethesda, Maryland, 320 pp.

Slaymaker, D.M., Jones, K.M.L., Griffin, C.R., and Finn, J.T.: 1996, 'Mapping deciduous forests in southern New England using aerial videography and hyperclustered multi-temporal Landsat TM imager', in *Gap Analysis: A Landscape Approach to Biodiversity Planning* (eds. J. M. Scott, T. Tear, and F. Davis), Proceedings of the ASPRS/GAP Symposium (National Biological Service, Moscow, ID), pp. 87-101.

U.S. Department of Agriculture: 1994, 'State Soil Geographic (STATSGO) Data Base', *Data Use Information,* United States Department of Agriculture Miscellaneous Publication Number 1492.

U.S. Geological Survey: 1990, 'Land use and land cover digital data from 1:250,000- and 1:1,000,000-scale maps', *Data User's Guide 4*, Reston, Va: Department of the Interior, U.S. Geological Survey, 33 pp.

U.S. Geological Survey: 1993, 'US GeoData digital elevation models', *Data User's Guide 5*, Reston, Va: Department of the Interior, U.S. Geological Survey, 51 pp.

U.S. Fish and Wildlife Service: 1996, 'National Wetlands Inventory (NWI) metadata,' U.S. Fish and Wildlife Service, National Wetlands Inventory, St. Petersburg, Florida.

Vogelmann, J.E., Seevers, P.M., and Oimoen, M.: 1997a, 'Effects of selected variables for discriminating land cover: multiseasonal data, different clustering algorithms, and varying numbers of clusters', *Proceedings of the Pecora 13 Symposium*, Sioux Falls, South Dakota, August 20-22, 1996, in press.

Vogelmann, J.E., Sohl, T., and Howard, S.M.: 1997b, 'Regional characterization of land cover using multiple sources of data,' *Photogrammetric Engineering and Remote Sensing*, in press.

MANAGING SCIENTIFIC DATA: THE EMAP APPROACH

STEPHEN S. HALE[1], MELISSA M. HUGHES[2], JOHN F. PAUL[1], R. SCOTT MCASKILL[2], STEVEN A. REGO[1], DAVID R. BENDER[2], NANCY J. DODGE[2], THOMAS L. RICHTER[2], and JANE L. COPELAND[2]

[1] *U.S. Environmental Protection Agency, Atlantic Ecology Division*
[2] *OAO Corp., 27 Tarzwell Drive, Narragansett, RI, USA 02882*

Abstract. Many data sets used by EPA's Environmental Monitoring and Assessment Program (EMAP) will be collected and managed by groups other than EPA as the Committee on Environment and Natural Resources develops the inter-agency National Environmental Monitoring Initiative. Managing these data requires a change from a database managed solely by EPA to a model where there is, in addition to distributed databases, truly distributed ownership and responsibility. Common standards, data directories, and data descriptions allow data of interest to be located, understood, and downloaded. The level of EMAP data management practices applied to a data set is based on the degree of EMAP responsibility for the data. The EMAP Data Directory is an Oracle database that tracks data sets of interest and contains sufficient information about a data set for a user to determine if the data are of interest. Some of the data sets listed in the Directory are in the possession of EMAP and are accessible on the EMAP WWW site (*http://www.epa.gov/emap*). Other data sets in the Directory are managed, documented, and made available by other organizations. The Data Catalog contains metadata about data sets in the possession of EMAP and provides the user with information about methods, assumptions, and data quality.

1. Introduction

Assessment of ecological condition often requires data from several different databases, particularly when terrestrial, atmospheric, and aquatic influences need to be considered. It is rare for any one organization to be able to conduct sufficient studies to obtain a complete picture or to maintain a comprehensive, consistent, well-described database management system for monitoring ecological condition over a large geographic region. The need for ecological assessments is increasing as the federal Committee on Environment and Natural Resources (CENR) develops its National Environmental Monitoring Initiative (NEMI) to coordinate the individual monitoring programs of several agencies. Research conducted by EPA's Environmental Monitoring and Assessment Program (EMAP), a NEMI cooperator, is increasingly dependent on data sets that are collected and managed by other organizations. A major challenge in conducting ecological assessments is the acquisition and integration of data of varying ownership, format, quality, and degree of documentation (ESA, 1995; NRC, 1995). The difficulty can be reduced by data sources moving toward common standards, data directories, and data descriptions (Chinn and Bledsoe, 1997; Williams, 1997; Barton, 1996, 1997; LTER, 1995; CENR, 1994; FGDC, 1994). Many useful guidelines and standards have evolved from the U.S. Global Change Research Program (GCRP, 1995a, 1995b). These developments allow data of interest to a researcher to be more easily found, understood, and downloaded. Another part of the solution is the increasing use of World Wide Web (WWW) browsers as the common user interface to distributed databases that use a variety of software packages and database structures.

Environmental Monitoring and Assessment **51**: 429–440, 1998.

1.1 ENVIRONMENTAL MONITORING AND ASSESSMENT PROGRAM

The goal of EMAP is to "Monitor the condition of the Nations's ecological resources to evaluate the cumulative success of current policies and programs and to identify emerging problems before they become widespread or irreversible" (U.S. EPA, 1997a). The EMAP strategy is centered on three principles: (1) pursue all tiers in the CENR monitoring framework, i.e., from index sites to regional ecological assessments to landscape characterization, (2) provide the research necessary for scientific credibility for the monitoring network, and (3) build the national monitoring network by starting with existing networks and filling in gaps. EMAP is one component of the national monitoring network (U.S. EPA, 1997b).

EMAP has occurred in two stages. In its early years, EMAP focused on monitoring of individual resources, such as streams, lakes, estuaries, forests, agro-ecosystems, and rangelands. This involved extensive data collection on a randomly chosen grid. The current phase of EMAP recognizes the availability of extensive data from existing regional and local monitoring projects. The role of EMAP is to determine where data gaps exist and to help fill those gaps. Where existing data are available, EMAP will be cataloging these data and making them available to researchers. These researchers will attempt to aggregate the existing data along with data gathered by EMAP and other large monitoring programs. The current phase is focused on research and will involve a number of research/monitoring efforts. These include: regional-scale assessments (with the pilot study located in the U.S. mid-Atlantic region); long-term fixed index sites, such as the demonstration project that monitors UV-B radiation in several National Parks; and EMAP-like studies done by EPA regional offices (REMAP). EMAP is also conducting research on monitoring network designs, the development of ecological indicators, and landscape characterization and landscape ecology methods. Data collected from the aforementioned projects will be managed by a variety of groups with primary data management responsibilities being maintained by the principal investigator or associated organization.

1.2 EMAP INFORMATION MANAGEMENT

The EMAP information management approach supports a research program that is developing methods for conducting different levels of environmental assessments across resources and across geographic scales. The data being used in the assessments come from a variety of sources including EMAP databases and other data systems. These data systems are physically located at different sites and have varying degrees of ownership, quality assurance, and documentation. The objectives are to (1) provide an EMAP Data Directory so that data of interest can be identified; (2) ensure a distributed data structure that allows responsibility for the data to reside with the owners; (3) provide an information management system to support EMAP research; (4) make EMAP data sets and metadata files available on the World Wide Web; and (5) maximize interoperability with other environmental monitoring data systems under the framework of the Committee on Environment and Natural Resources.

This paper describes the approach used by EMAP to manage data where environmental assessments require EMAP-collected data to be integrated with data collected by other entities. This is a challenge faced by many organizations participating in the framework of the CENR National Environmental Monitoring Initiative.

2. Methods

There are three basic components to the EMAP information management system: a Data Directory, a Data Catalog (metadata), and the actual data sets. The Data Directory keeps track of all data sets of interest. Some of these data sets are in the possession of EMAP and are accessible on the EMAP WWW site (*http://www.epa.gov/emap*); other data sets are managed and documented by other organizations. Data at the different data sources are managed using a variety of software and hardware. The Data Catalog contains metadata about data sets in the possession of EMAP so that a user can understand enough about the methods, assumptions, and quality to use the data appropriately. With a WWW browser, users locate data sets of interest by querying the Data Directory and then download the selected files (Figure 1). Metadata transferred with each downloaded data set provide the user with the context and assumptions under which the data were collected. Subsequent manipulation and analysis of data sets are under control of the user, using tools of their choice.

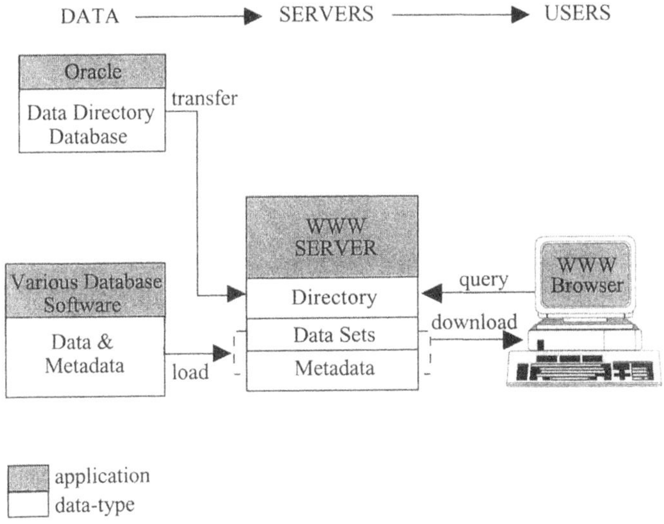

Fig. 1. Structure for EMAP data access

The level of EMAP data management practices applied to a data set is based on the degree of EMAP responsibility for the data (U. S. EPA, 1996a). For example, remote sensing data collected by other than EMAP sources may only be referenced in the Directory. The Directory reference would allow users interested in these types of data to locate the data sets. However, the actual acquisition of the data and metadata files would not occur via the EMAP information system; rather, the request would be routed to the appropriate data center, such as the EROS Data Center. On the other end of the spectrum, data collected by EMAP in the estuaries of the mid-Atlantic U.S. will be included in the Directory and have a full set of accompanying metadata written to EMAP standards. These data will be stored and made available via the EMAP information system. An example of the middle of this spectrum might include data sets and metadata files that are frequently used by several EMAP investigators, such as stream discharge data collected and maintained by the USGS. A subset of these data may be copied to the EMAP system to facilitate regular access, particularly if EMAP further processes the data (such as clipping a GIS coverage to the boundaries of an EMAP study area). The authoritative database for these data would remain the USGS database. Regular updates of the EMAP held data would be made to ensure compatibility with the authoritative database.

This approach was driven by the need to keep the system simple during a transition time (from the original intent of EMAP to develop a national environmental monitoring program to the current objectives, which are more focused on an environmental monitoring research program), while retaining the ability to expand when necessary. Another driving force was the need for compatibility with other federal agency data systems in support of the CENR National Environmental Monitoring Initiative (*http://www.epa.gov/monitor*). These needs are met by: (1) adopting and updating existing EMAP information management guidelines and standards to emerging inter-agency standards; (2) updating the existing Data Directory Oracle database to be compatible with federal standards; (3) developing procedures for producing metadata, including a Web directory entry form; and (4) making the directory, data, metadata, and publications accessible through the EMAP WWW site.

Each organization involved in EMAP manages the data they collect or acquire and is responsible for quality assurance, documentation, and transfer to the EMAP web site, in accordance with established standards and formats. Researchers who collect a particular type of data are also likely to be experienced users of that type of data and, therefore, in the best position to determine how those data should be formatted and described for use by other researchers. The organizations use a number of data management systems based on different software packages, although SAS, Oracle, and Arc/Info are commonly used in EPA. The Atlantic Ecology Division (AED) in Narragansett, RI functions as the network site coordinator, providing guidance, standards, and formats, with the assistance of the national EMAP Information Management Working Group (U.S. EPA, 1996a). AED operates the EMAP Data Directory and maintains the EMAP home page, while the individual organizations are responsible for populating the databases.

2.1 EMAP DATA DIRECTORY

The EMAP Data Directory (Frithsen and Strebel, 1995; updated by U.S. EPA, 1996d) is the primary means for users to find out what data are available and where the data files can be found. The Directory is an Oracle database with a searchable listing of sufficient information about a data set for a user to determine if the data are of interest. The Directory provides (1) a means for users to find out what EMAP data have been collected and where those data can be accessed and (2) interoperability with other federal agency data directories, particularly those associated with the CENR National Environmental Monitoring Initiative.

The original EMAP Data Directory design, based on the NASA Directory Interchange Format (NASA, 1991), has been updated (Frithsen, 1996a) to emerging standards such as those of the Federal Geographic Data Committee (FGDC, 1994; *http://fgdc.er.usgs.gov*). Use of the restricted keyword vocabulary of the Global Change Master Directory (*http://gcmd.gsfc.nasa.gov*), operated by NASA, helps provide compatibility among agencies within the CENR monitoring framework and allows users to acquire common expectations of what they will retrieve with a query (Frithsen, 1996b).

Entries are loaded into the Data Directory through an Oracle client application (Oracle Forms software). Users who do not have Oracle software can enter Directory information into a Directory Interchange Format (DIF) Web entry template (adapted from a NASA application). This produces an ASCII file that is subsequently loaded into a temporary database table. It can then be validated by the database administrator with the assistance of a semi-automated program, and inserted into the Oracle Data Directory database.

The Oracle Data Directory database can be queried from a Web browser through use of a form (Figure 2) developed using Oracle WebServer Option software. This direct Web browser to Oracle database simplifies database administration and gives Web users access to the most recent information.

2.2 EMAP DATA CATALOG

The Data Catalog contains information about data sets (metadata) so that a user can understand the methods, assumptions, and quality assurance procedures used to create the data set. The Data Catalog provides (1) information about the data files so that the data can be correctly interpreted and used and (2) interoperability with other federal agency data catalog standards, particularly those associated with the CENR. Good quality metadata are essential in a multi-investigator research program, such as EMAP or the NSF Long-Term Ecological Research Program (Ingersoll et al., 1997). EMAP uses a common text format for catalog information (Figure 3) that is then converted to HTML (hypertext markup language) format for access by Web browsers.

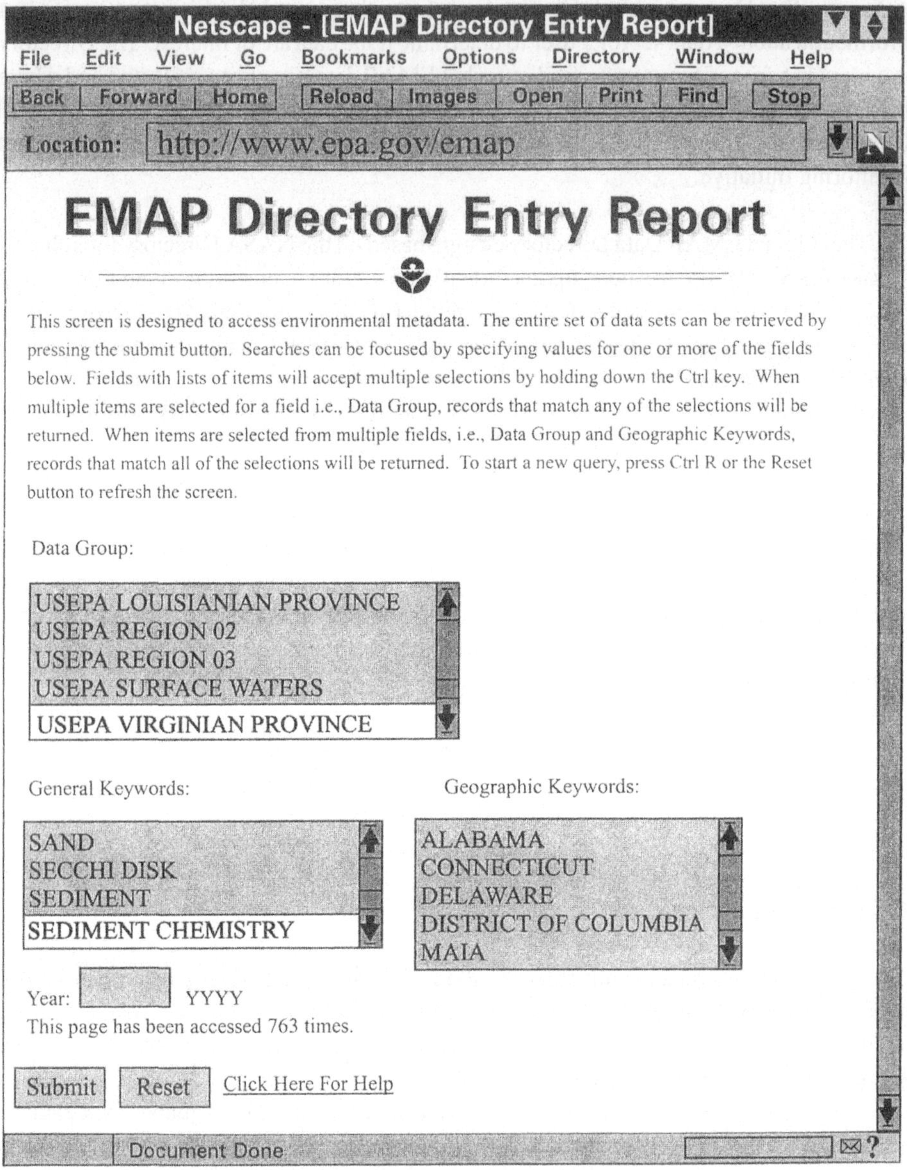

Fig. 2. Oracle WebServer Option query form for the EMAP Data Directory. Source code adapted from program written by TPMC.

The EMAP Data Catalog is based on guidelines derived from NASA and other sources (Strebel and Frithsen, 1995b; updated by U.S. EPA, 1996e). Similarities with other metadata formats, such as that of the Global Change Master Directory (GCMD) are useful for compatibility with other agencies.

EMAP policy requires that metadata (Data Catalog) files accompany all data files (U.S. EPA, 1996a). Metadata files are kept for all data sets maintained by EMAP—whether collected by EPA or another organization. Metadata for those data sets that have entries in the EMAP Data Directory, but are not in EMAP possession, reside at the original source.

2.3 EMAP WORLD WIDE WEB SITE

EPA information management policies encourage sharing of data with other agencies and organizations and the public (U. S. EPA, 1997d, 1996b, 1996c, 1995). Data sharing is a necessity if the CENR National Environmental Monitoring Initiative is to succeed. The primary method for distributing EMAP data is through the EMAP web site on the EPA public access WWW server at Research Triangle Park, NC (Figure 4).

Primary objectives of the EMAP web site are to (1) provide information about the EMAP program, (2) provide access to a Directory of EMAP data sets, with links to data sets maintained by other organizations, and (3) provide downloadable EMAP data sets and metadata files. Guidelines for data collection sources to add data and metadata files to the EMAP home page (Strebel and Frithsen, 1995a; as updated by U.S. EPA, 1997c) provide a base level of uniformity.

Contents (and data formats) of the EMAP web site include:

- Data Directory (Oracle, HTML)
- Data and metadata files (ASCII, SAS, Oracle, Arc/Info, HTML, PDF)
- Publications (WordPerfect, with ASCII summaries; PDF)
- List of contacts (HTML)
- EMAP Bibliography (Oracle, HTML)
- EMAP Information: EMAP Research Plan, EMAP Information Management Plan, EMAP Newsletter (WordPerfect, with ASCII summaries; PDF)
- Links to other environmental monitoring research sites (HTML)
- Geographic Reference Database for GIS coverages; links to EPA GIS library; links to the EROS Data Center for Multi-Resolution Landscape Characterization data

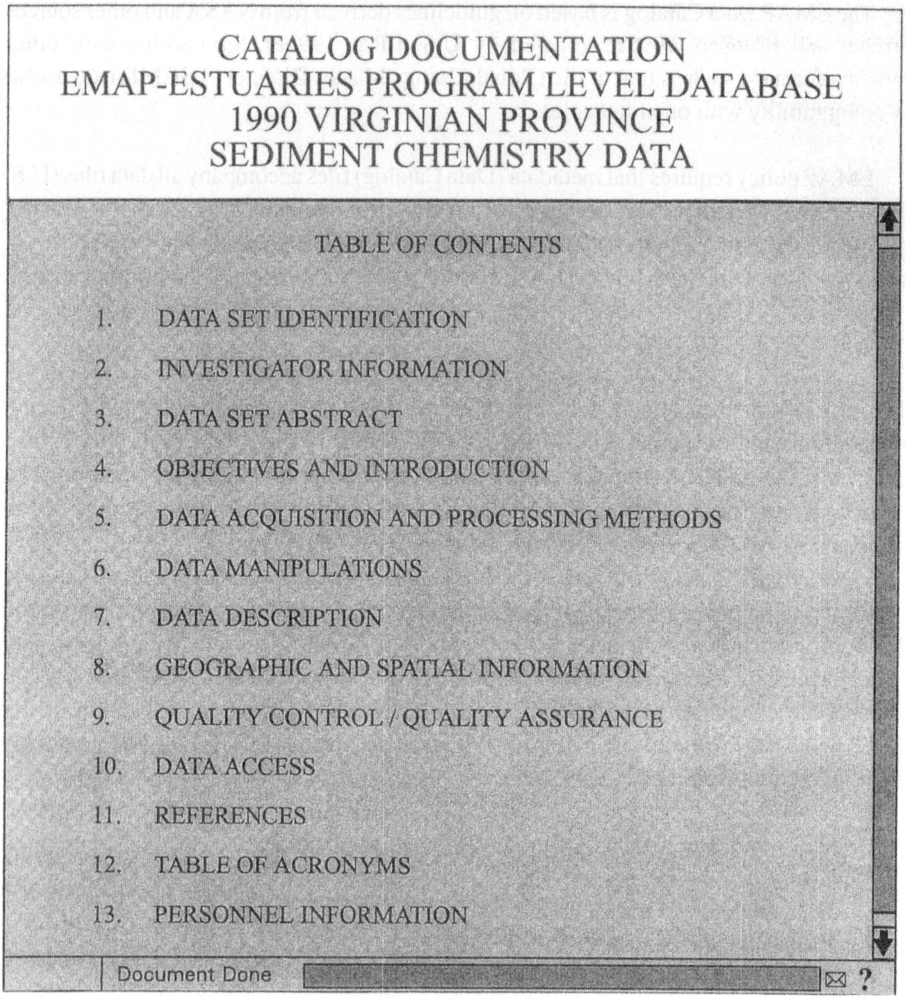

CATALOG DOCUMENTATION
EMAP-ESTUARIES PROGRAM LEVEL DATABASE
1990 VIRGINIAN PROVINCE
SEDIMENT CHEMISTRY DATA

TABLE OF CONTENTS

1. DATA SET IDENTIFICATION
2. INVESTIGATOR INFORMATION
3. DATA SET ABSTRACT
4. OBJECTIVES AND INTRODUCTION
5. DATA ACQUISITION AND PROCESSING METHODS
6. DATA MANIPULATIONS
7. DATA DESCRIPTION
8. GEOGRAPHIC AND SPATIAL INFORMATION
9. QUALITY CONTROL / QUALITY ASSURANCE
10. DATA ACCESS
11. REFERENCES
12. TABLE OF ACRONYMS
13. PERSONNEL INFORMATION

Document Done

Fig. 3. Structure of metadata fields for the EMAP Data Catalog

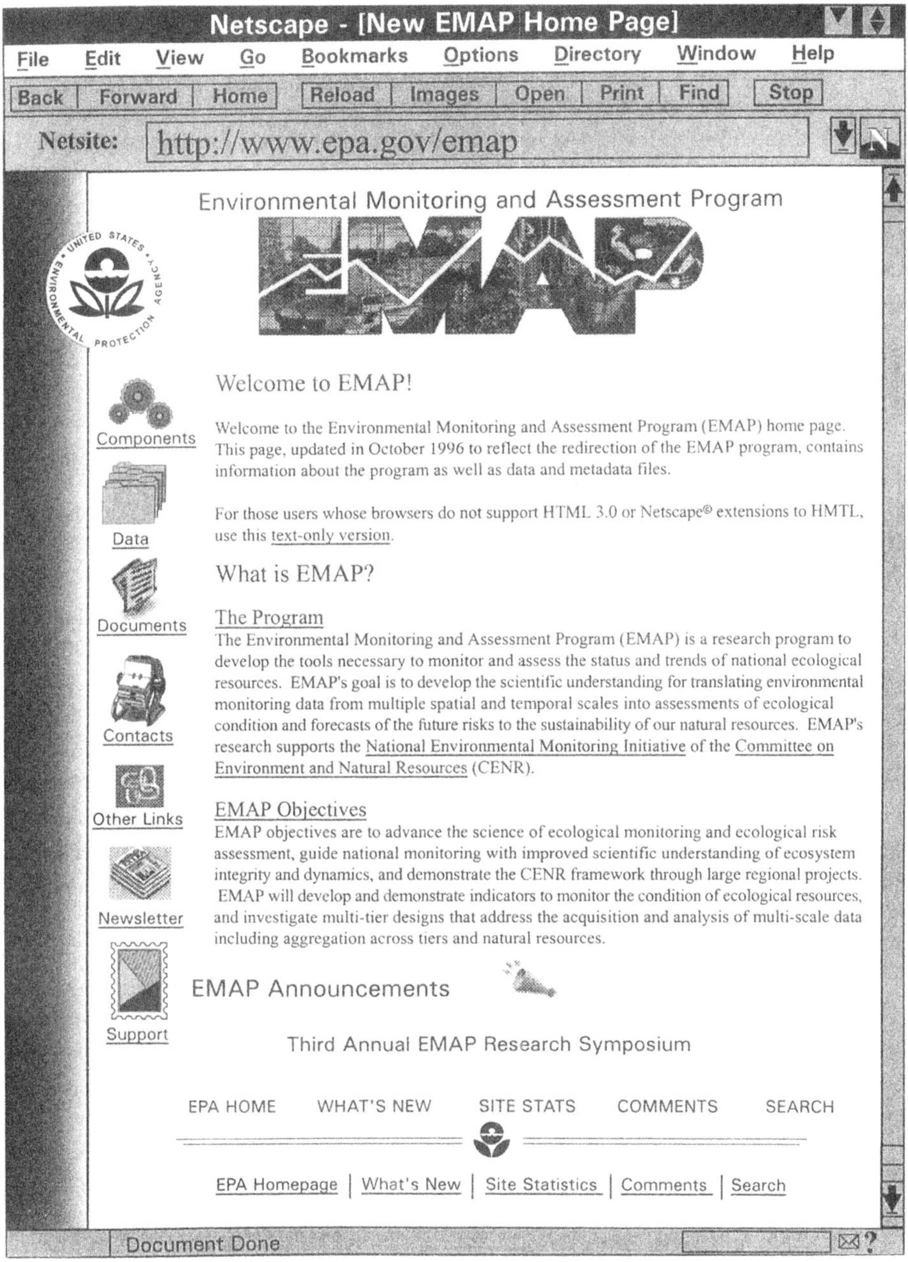

Fig. 4. The EMAP home page on the World Wide Web.

WAIS (Wide-Area Information Server) software is used to (1) search the bibliography, and (2) search the entire Web collection for any text string. An Oracle WebServer option form is used to search the Data Directory database.

Some data collected for EMAP are managed at other Web sites. For example, GIS (Geographic Information System) data from the Mid-Atlantic Integrated Assessment are kept on the Geographic Reference Database Web page (*http://www.epa.gov/docs/grd/grd_home.html*). Remote sensing data from the Multi-Resolution Landscape Characterization (MRLC) consortium are distributed by the USGS EROS Data Center (*http://edcwww.cr.usgs.gov/eros-home.html*).

3. Conclusions

EMAP is one component of the national environmental monitoring network envisioned by the federal Committee on Environment and Natural Resources (CENR). For effective environmental assessments to be made, there must be effective sharing of data and information among all components. Managing the data from this inter-agency environmental monitoring effort requires EMAP to change from a database managed by EPA laboratories to a new model where data management responsibilities are shared by numerous organizations.

The EMAP approach to information management uses a simple, but expandable system, consisting of data and metadata files on the World Wide Web, tracked by an Oracle database that includes directory searching. Common standards and formats allow a degree of interoperability with other federal environmental monitoring databases. This is becoming of increasing importance, as no one agency can collect all of the data needed for analyses of ecological condition across large geographic regions.

The future will bring easier access to databases through the Internet by more powerful Web browsers. EMAP has implemented Web browser to Oracle database searching and is exploring Web browser access to Arc/Info GIS coverages and SAS data sets. Software that delivers map products to a Web browser on demand would enhance spatial queries and the user's understanding of monitoring data.

The future will require increasing coordination among federal agencies and other organizations involved in environmental monitoring. Complex environmental assessments require data from numerous databases. Common directories and metadata will simplify the acquisition and use of these data, regardless of what data management software is used or where the data reside.

Acknowledgments

Development of EMAP information management system policies, standards, and databases have been conducted by numerous individuals and groups since 1989. Prominent among developers at the early stages were: Bob Shepanek, Jeff Frithsen, Jeff Rosen, many people at EPA and other labs throughout the country, and programmer/analysts at Technology Planning and Management Corporation and Lockheed-Martin. We are grateful for the vision and order they have brought to a complex problem and for the software applications they developed. Larry Rossner, Brian Melzian, Darryl Keith, Barbara Brown, and members of the national EMAP Information Management Working Group have contributed to recent progress. We thank Jerry Pesch, Dan Campbell, Jim Latimer, and anonymous reviewers for editorial improvements and Tricia Bussiere for help with the figures.

The information in this paper has been funded by the U.S. Environmental Protection Agency. It does not necessarily reflect the views of the Agency, and no official endorsement should be inferred. Mention of trade names or commercial products does not constitute endorsement or recommendation for use. This is contribution number 1906 of the Atlantic Ecology Division, National Health and Environmental Effects Research Laboratory, U.S. Environmental Protection Agency.

References

Barton, G. 1997. NOAA and the Federal Geographic Data Committee: Earth System Monitor 7(3), March 1997.

Barton, G. 1996. NOAA Environmental Services Data Directory. Earth System Monitor, December 1996. 6-8.

CENR. 1994." The U.S. Global Change Data and Information System Implementation Plan", *Report, Committee on Environment and Natural Resources*, National Science and Technology Council, Washington, D.C.

Chinn, H. and Bledsoe, C. 1997. "Internet access to ecological information-the US LTER All-Site Bibliography Project", *BioScience* **47**(1), :50-57.

ESA. 1995. Report of the Ecological Society of America Committee on the Future of Long-Term Ecological Data. Vol. 1. *http://www.sdsc.edu/~ESA*

FGDC. 1994. "Content standards for digital geospatial metadata, June 8, 1994", *Federal Geographic Data Committee*, Washington, DC.

Frithsen, J. B. 1996a. "Suggested modifications to the EMAP data set directory and catalog for implementation in US EPA Region 10. Draft, June 10, 1996", *Report prepared for the U.S. Environmental Protection Agency, National Center for Environmental Assessment*, Washington, DC., by Versar, Inc., Columbia, MD.

Frithsen, J. B. 1996b. "Directory Keywords: Restricted vs. unrestricted vocabulary. Draft, May 21, 1996", *Report prepared for the U.S. Environmental Protection Agency, National Center for Environmental Assessment*, Washington, DC., by Versar, Inc., Columbia, MD.

Frithsen, J. B. and Strebel, D. E. 1995. "Summary documentation for EMAP data: Guidelines for the Information Management Directory. 30 April 1995", *Report prepared for U.S. Environmental Protection Agency, Environmental Monitoring and Assessment Program (EMAP)*, Washington, DC. Prepared by Versar, Inc., Columbia, MD.

GCRP. 1995a. GCDIS Implementation 1995. Vol. **I**-*Interagency Implementation*. U.S. Global Change Research Program. Committee on Environment and Natural Resources, National Science and Technology Council, Washington, D.C.

GCRP. 1995b. GCDIS Implementation 1995. Vol. **II**-*Agency Implementation*. U.S. Global Change Research Program. Committee on Environment and Natural Resources, National Science and Technology Council, Washington, D.C.

Ingersoll, R. C., Seastedt, T. R. and Hartman, M. 1997. "A model information management system for ecological research", *BioScience* **47**(5):310-316.

LTER. 1995. Draft proceedings of the 1995 Long-Term Ecological Research Data Management Workshop, July 27-29, 1995, Snowbird, Colorado.

NASA. 1991. "Directory Interchange Format Manual; Version 4.0", *NASA, National Space Science Data Center*, Greenbelt, MD. December 1991.

NRC. 1995. "Finding the forest in the trees: The challenge of combining diverse environmental data", *National Academy Press*, Washington, DC. 129 pp.

Strebel, D. E., and Frithsen, J. B. 1995a. "Guidelines for distributing EMAP data and information via the Internet. April 30, 1995", *Prepared for U.S. Environmental Protection Agency, Environmental Monitoring and Assessment Program (EMAP)*, Washington, DC. Prepared by Versar, Inc., Columbia, MD.

Strebel, D. E., and Frithsen, J. B. 1995b. "Scientific documentation for EMAP data: Guidelines for the information management catalog. Draft: April 30, 1995", *Prepared for U.S. Environmental Protection Agency, Office of Modeling, Monitoring Systems and Quality Assurance*, Washington, DC. Prepared by Versar, Inc., Columbia, MD.

U.S. EPA. 1997a (in prep). "EMAP Research Strategy" (January 1997 draft). *U.S. Environmental Protection Agency, ORD, NHEERL*, Research Triangle Park, NC.

U.S. EPA. 1997b (in prep). "EMAP Research Plan" (March 1997 draft). *U.S. Environmental Protection Agency, ORD, NHEERL*, Research Triangle Park, NC.

U. S. EPA. 1997c. Update to: "Guidelines for distributing EMAP data and information via the Internet", *U. S. EPA, NHEERL*, Atlantic Ecology Division, Narragansett, RI.

U.S. EPA. 1997d. 1997 update to ORD's strategic plan. EPA/600/R-97/015. *Office of Research and Development, U.S. Environmental Protection Agency*, Washington, DC.

U. S. EPA. 1996a. "EMAP information management plan. Draft, Oct 30, 1996", *U. S. EPA, NHEERL*, Narragansett, RI.

U.S. EPA. 1996b. (in prep.). "ORD Information Management Strategic Plan", *U.S. Environmental Protection Agency, Office of Research and Development*, Washington, DC.

U.S. EPA. 1996c. (in prep.). "Providing information to decision makers to protect human health and the environment. Information Resources Management Five-Year IRM Implementation Plan, February 1996 draft", *U.S. Environmental Protection Agency, Administration and Resources Management*, Washington, DC.

U. S. EPA. 1996d. Addendum to: "Guidelines for the information management directory", *U. S. EPA, NHEERL*, Atlantic Ecology Division, Narragansett, RI.

U. S. EPA. 1996e. Addendum to: "Guidelines for the information management catalog", *U. S. EPA, NHEERL, Atlantic Ecology Division*, Narragansett, RI.

U.S. EPA. 1995. "Providing information to decision makers to protect human health and the environment", *Information Resources Management Strategic Plan. EPA-220-B-95-002.* April 1995. *U.S. Environmental Protection Agency, Administration and Resources Management*, Washington, DC.

Williams, N. 1997. "How to get databases talking the same language", *Science* **275**, 301-302.

EXPLORING ENVIRONMENTAL DATA IN A HIGHLY IMMERSIVE VIRTUAL REALITY ENVIRONMENT

DIANNE COOK[1], CAROLINA CRUZ-NEIRA[2], BRADLEY D. KOHLMEYER[2], ULI LECHNER[3], NICHOLAS LEWIN[4], LAURA NELSON[2], ANTHONY OLSEN[5], SUE PIERSON[6], and JÜRGEN SYMANZIK[1]

[1]Department of Statistics, Iowa State University, Ames, IA 50011, [2]Iowa Center for Emerging Manufacturing Technology, Iowa State University, Ames, IA 50011, [3]German National Research Center for Information Technology, Institute for Media Communications, Schloß~Birlinghoven, 53574 Sankt Augustin, Germany, [4]Geographic Information Systems Support and Research Facility, Iowa State University, Ames, IA 50011, [5]US EPA National Health and Environmental Effects Research Laboratory, Western Ecology Division, Corvallis, OR 97333, [6]OAO, c/o USEPA NHEERL Western Ecology Division, Corvallis, OR 97333.

Abstract. Geography inherently fills a 3D space and yet we struggle with displaying geography using, primarily, 2D display devices. Virtual environments offer a more realistically-dimensioned display space and this is being realized in the expanding area of research on 3D Geographic Information Systems (GISs). Traditionally, a GIS has only limited tools for statistical analysis, and 3D GIS research has concentrated on the visualization of the geographical terrain. Here we discuss linking multivariate statistical graphics to geography in the highly immersive C2 virtual reality environment at Iowa State University using mid-Atlantic streams data.

1. Introduction

In this paper we discuss exploring multivariate spatial data in a highly immersive virtual reality environment. There is considerable excitement about rendering geography in 3D, but there (as yet) are few efforts to build in analytical tools. Entwined to make good data analysis are visualization tools and analytical tools. This work explores developing these tools in the highly immersive C2 virtual reality environment.

We have data on 501 sampling sites on streams in the mid-Atlantic states of Delaware, Maryland, New York, Pennsylvania, Virginia, and West Virginia. It is a probability sample of wadeable streams in the period 1993-1995 by the U.S. E.P.A.'s Environmental Monitoring and Assessment Program (Klemm and Lazorchak, 1995). At each site the subset of measurements we use are Closed System pH, Calcium (µeq/L), Sodium (ueq/L), Ammonium (µeq/L), Chloride (µeq/L), Nitrate (µeq/L), Sulfate (µeq/L), Dissolved Organic Carbon (mg/L), Total Suspended Solids (mg/L), and Total Phosphorous (µg/L). In addition to these measurements we have the latitude and longitude, elevation, and the Aggregated Omernik Level 4 ecoregion membership of each sampling site.

In order, the sections of the paper discuss geographic information systems, dynamic statistical graphics for multivariate data, immersive virtual reality technology, linking the geography with the multivariate graphics in the virtual environment and applied to the streams data.

Environmental Monitoring and Assessment **51:** 441–450, 1998.

2. Geographic Information Systems

Geographic Information Systems (GISs) have played and do play an enormous role in the analysis of environmental data. They perform the task of storing and displaying spatial data and concomitant geographic variables. Critical to any good GIS are database storage and retrieval and solid map drawing capabilities.

The existing software for GIS concentrates on 2D display. Figure 1 shows the mid-Atlantic streams data (described in the Introduction) displayed in ArcView 3.0™[1]. There is growing interest in constructing and designing appropriate data structures for 3D GIS, which is quite natural given that geography is inherently 3D. GRASS is an example of software that is developing tools for 3D GIS (Brown *et al.*, 1995; Mitasova *et al.*, 1995). Also, see http://www.esri.com/base/products/arcinfo/3dtin/tin.html for ESRI developments. Raper (1989) provides more introduction into this field.

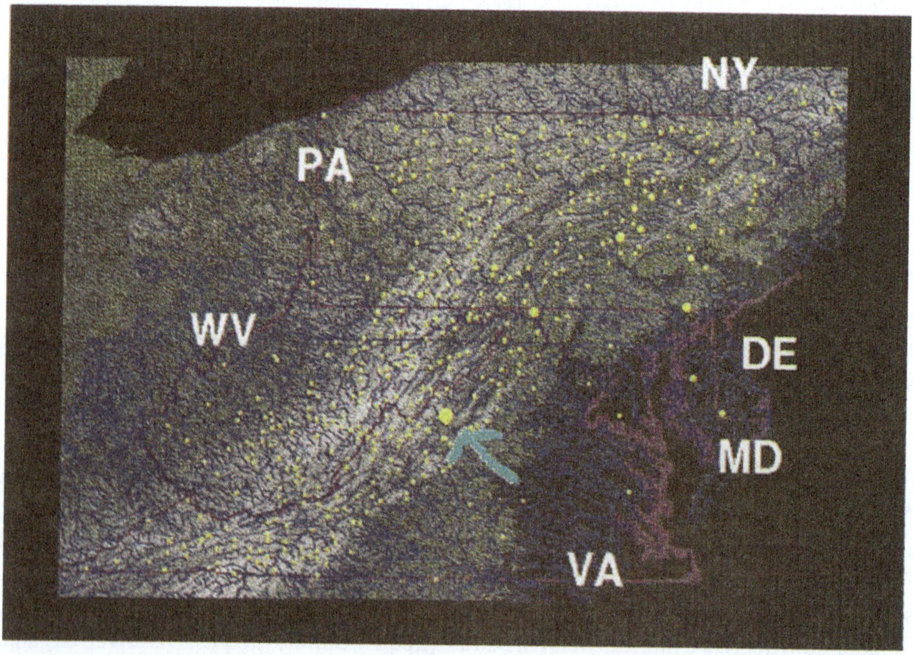

Fig. 1. ArcView 3.0™ view of the elevation and sampling sites in the mid-Atlantic states with Nitrate value coded proportional to circle diameter.

[1] ™ArcView 3.0 is a trademark of Environmental Systems Research Institute, Inc.

In addition, traditionally, GISs lack tools for visualizing multivariate spatial data. In Figure 1, the diameter of the spot at each sampling site is a function of the nitrate concentration. One site has a very high nitrate concentration (highlighted in Figure 1). Does this site also have high ammonium concentration? To answer this, we could code ammonium similarly. But keeping track of the relationships between these variables, and more, becomes increasingly difficult. A nicer approach to exploring the multivariate nature of the measurements is to use dynamic statistical graphics linked to the geographic location. An example of this approach can be found in Cook *et al.* (1996) and Cook *et al.* (1997) (shown in Figures 2,3).

In short, a GIS provides the geographic context but lacks the tools for multivariate statistical visualization and analysis.

3. Dynamic Statistical Graphics

Dynamic statistical graphics enables data analysts in all fields to carry out visual investigations leading to insights into relationships in complex data. Graphics for displaying one or two variables simultaneously are well-known and familiar but plotting three, four, or five variables simultaneously and in a manner that allows easy insight into the data is a tricky proposition. This has been the domain and challenge of statistical graphics research for over 30 years. Through an evolution of ideas, we have scatterplot matrices, parallel coordinate plots, icons (for example, Chernoff faces), and animated plots for viewing many variables. (Videos demonstrating these methods can be borrowed from the American Statistical Association Statistical Graphics Section Video Lending Library. See http://orion.oac.uci.edu/~rnewcomb/statistics/graphics/graphics.html.)

Using the animation approach to build directly from the familiar 2D scatterplot, we can animate plots by choosing different linear combinations of several variables. The most familiar technique is to rotate three variables by looking at a continuous sequence of 2D linear combinations of the three variables. This method extends naturally to linear combinations of four, five, or more variables, and the extension is called a grand tour (Asimov, 1985; Buja *et al.*, 1997). With a grand tour we can see multivariate structure such as clusters, multicollinearity, or outliers, that may not be visible in plots of one or two variables.

Using multivariate graphical methods we might take two windows to view the data and in one look at the measured variables in a grand tour and in the other plot the geographic location of the sampling sites. As a simple example, Figure 2 shows two views of the mid-Atlantic streams data. The software used is XGobi (Swayne *et al.*, 1991) and the geographical locations are shown in ArcView 3.0™. In the XGobi window (top), several observations that are extreme in comparison to the majority of points are identified. These

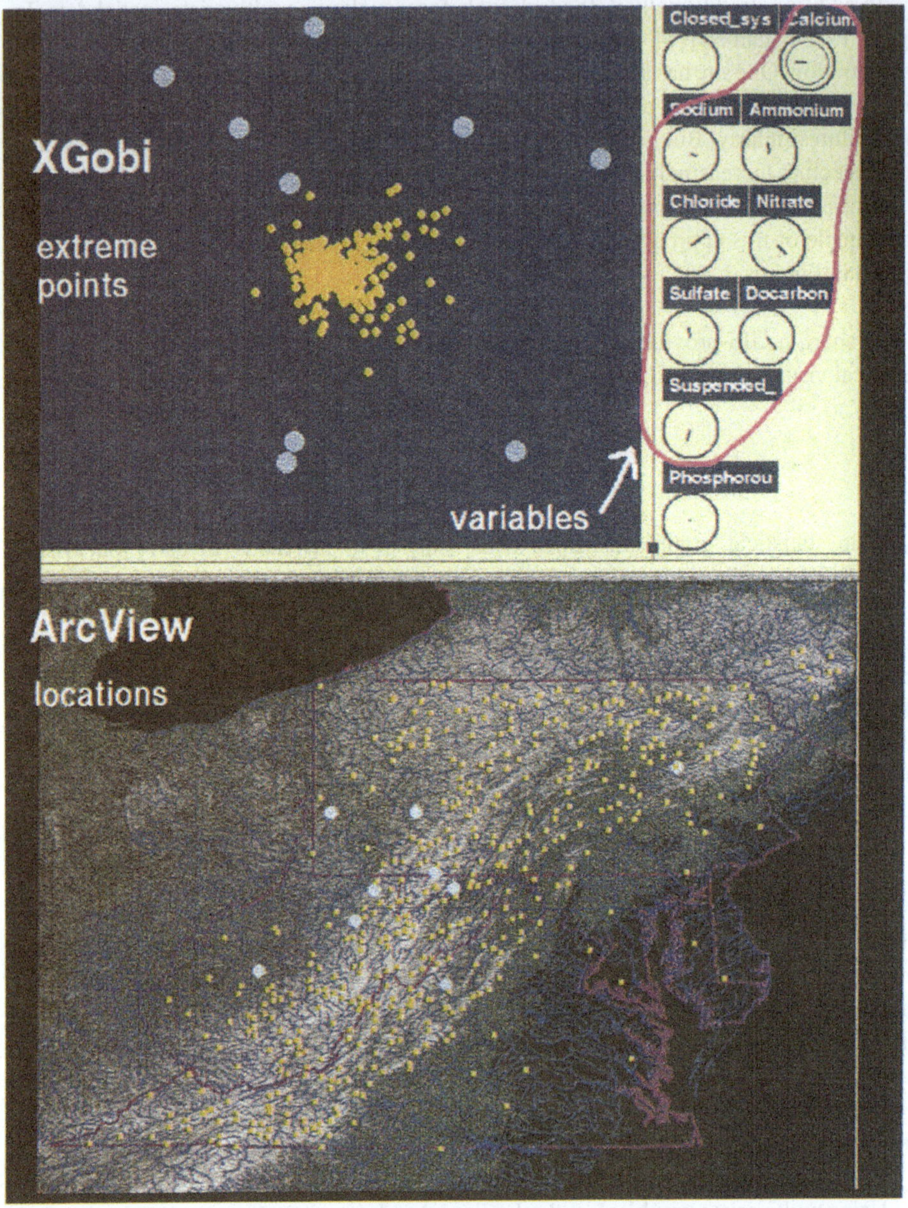

Fig. 2. Using XGobi linked to ArcView 3.0™ to explore the streams data. Several observations that are extreme in comparison to the majority of points are identified (top). These have outlying values on one or more of several variables: Calcium, Sodium, Ammonium, Chloride, Nitrate, Sulfate, Dissolved Organic Carbon, Total Suspended Solids. These extreme values are spread throughout the geographic region (bottom).

have outlying values on one or more of several variables: Calcium, Sodium, Ammonium, Chloride, Nitrate, Sulfate, Dissolved Organic Carbon, Total Suspended Solids. These extreme values are spread throughout the geographic region as the ArcView window (bottom) shows. In another view, a dotplot of Closed System pH has the more acidic samples brushed (Figure 3, left) and the locations appear mostly at higher altitudes (Figure 3, right).

Connecting multivariate graphics with a geographic information system allows for a more comprehensive analysis because available concomitant information, such as elevation, ecoregion, or census information, can be overlaid onto the map.

4. Immersive Virtual Reality Technology

Virtual reality (VR) is a rapidly developing technology that involves many aspects of computer-augmented visualization. The technology dates back to 1965 when Sutherland (1965) proposed the Ultimate Display and built the first head-mounted display, the Sword of Damocles, with cathode ray tubes and a ceiling suspension system in 1968. A force-feedback system was developed in 1971 by Frederick Brooks, and the Data Glove which measures finger angle was developed in 1985 by Thomas Zimmerman. A brief chronology of events that influenced the development of VR can be found in Cruz-Neira (1993), and a more complete overview can be found in Pimentel and Teixeira (1995).

The C2 at Iowa State University is a highly immersive VR environment in which images are projected onto the walls and floor of a small "room" to create the illusion of 3D when CrystalEyes Stereographics' LCD shutter glasses are worn. It uses position tracking and auditory feedback through multiple speakers to immerse users in a 3D environment. The position and orientation of the user's hands and head are determined through the use of a magnetic based tracker, a cyberglove, and a hand-held wand. It is possible for multiple viewers to enter the C2 and view the same scene with minimal equipment. Technical details and applications of a VR environment similar to the C2 can be found in Cruz-Neira (1995) and Roy *et al.* (1995).

5. Linking Multivariate Graphics and Geography in the C2

In the C2 environment, we have two viewing areas: a viewing box which contains the scatterplot view of the variables allowing us to conduct a grand tour and an elevation surface for the region of the study on which locations of the sampling sites are shown. In addition, we have several controls/interaction tools: a speed pole for changing the speed of the tour, a color palette for selecting a brush color, a symbol type, and a way to resize/reshape the brush from a sphere to a rectangle (Figure 4). There is sound feedback on interaction activities which makes the interaction more efficient. More details on dynamic statistical graphics in the C2 (at earlier stages) can be found in Symanzik *et al.*

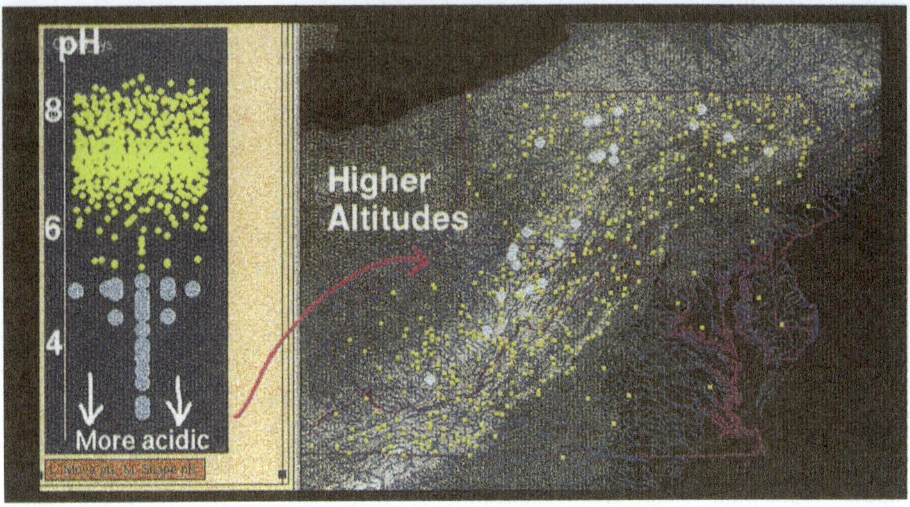

Fig. 3. Using XGobi linked to ArcView 3.0™ to explore the streams data. A dotplot of Closed System pH has the more acidic samples brushed (left) and the locations appear all to be at the higher altitudes (right). The dotplot is read like a vertical histogram. Closed System pH can be seen to be skewed with most measurements at the top around the slightly alkaline values 7-8, and less in the acidic range at the bottom of the dotplot. Values less than 5 have been brushed light blue.

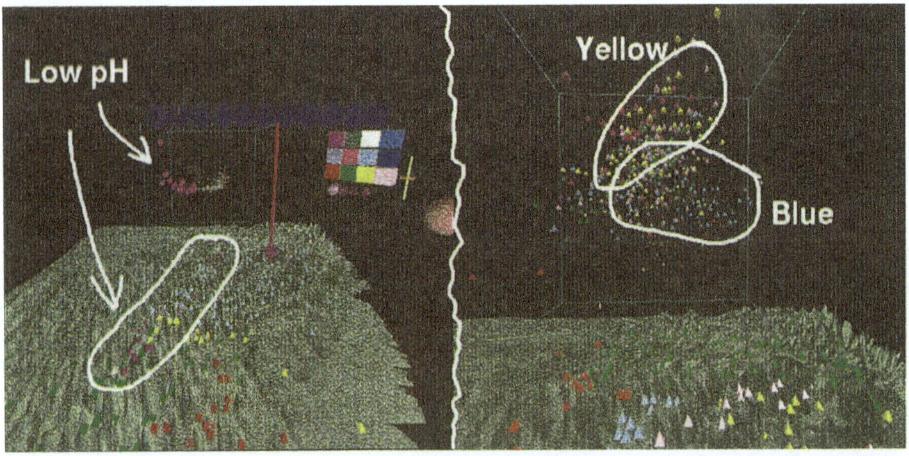

Fig. 4. In simulator mode: (left) Low pH values are brushed purple in the scatterplot, and these sampling sites which are more acidic fall higher up in the mountains. This corroborates what was seen in Figure 3 but in the C2 it is much easier to see the sites are at high altitude because the mountains are rendered in 3D. (right) Grand tour view over the elevation indicating differences in combination of chemical contaminants over ecoregion.

(1997). (Other examples of work on building a general multivariate data visualization system in a virtual reality environment are discussed in van Teylingen *et al.* (1997) and Carr *et al.* (1996).

Sampling sites are linked one-to-one between the scatterplot and the elevation map. So painting a point (or group of points) in the scatterplot highlights it (them) on the elevation map. In this way, locations of interesting features can be discovered and assessed in relation to other interesting features or elevation. And most importantly, the multivariate nature of the data can be explored more extensively than is possible with a univariate variable-by-variable analysis.

6. Mid-Atlantic Streams Application

We have displayed the chemical information in the viewing box of the C2, and pulled out geographic location of the sampling site and drawn these with regard to the elevation of the site. In addition we have colored the observations according to their ecoregion identity. The elevation map is drawn on the floor that we walk on (Figure 5). The effect is stunning: acting like Gulliver we can walk on the mountaintops and look along valleys.

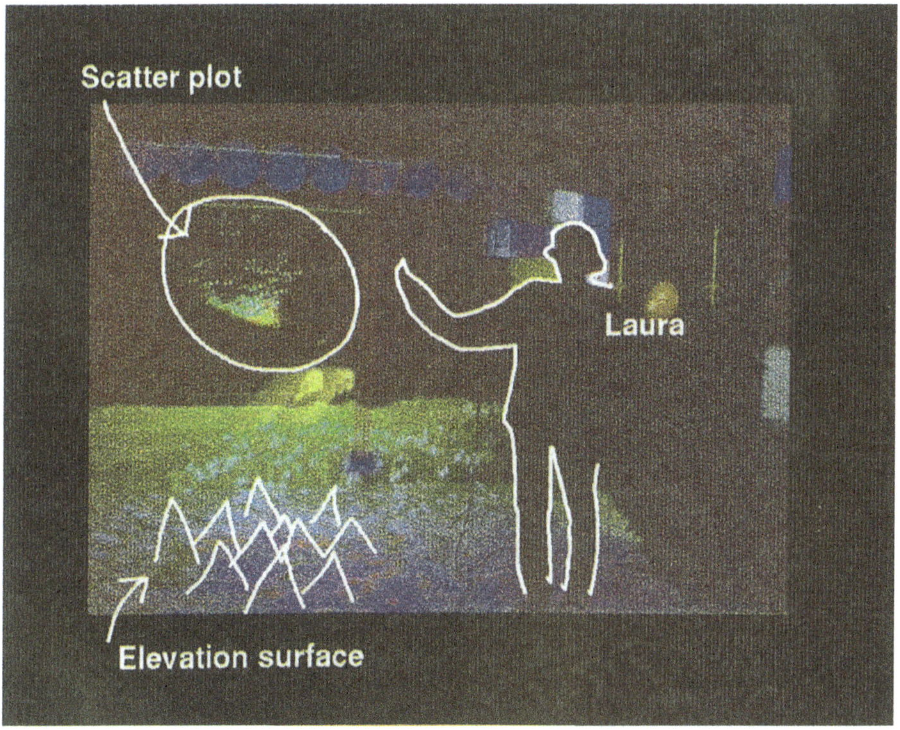

Fig. 5. The measurements on samples in streams of the mid-Atlantic states in the C2.

Some initial observations of the data show that there are some isolated sampling sites that have extreme values on several chemistry-related variables. In the viewing box, these points are seen as outliers from the main clustering of points. These individual points are brushed to see their locations in the study region. You might expect that these sampling sites are located in a similar geographic region, but they appear to be spread throughout the entire sampling region.

Low pH values are concentrated at higher elevations (Figures 3,4) and the different level 4 ecoregions have different combinations of the measured variables (Figure 4). The relationship between pH and elevation has been noticed in other studies and is understood: acid deposition is neutralized as water flows downstream. The few sites which do not correspond to high elevation occur in the western part of Pennsylvania and are most likely associated with mine drainage.

The data are heavily skewed towards small quantities of contaminants, so we looked at both the raw data and power transformations of the variables. (Transformations of skewed data are commonly used in data analysis to make the data more symmetric. Once transformed it is easier to see other informative patterns in the data which may have been hidden by the skewness.) Amongst the transformed variables strong linear dependencies exist which are not immediately obvious from the pairwise plots.

7. Conclusions

Motivating this work is the question of "How this technology might enhance environmental monitoring and assessment?" The virtual environment does offer a lot of scope for visualizing environmental data. It is possible to represent geography to its full 3D extent. A virtual space allows for enormous flexibility in zooming or panning into global or local views. It has potential for communicating information to an uninformed audience by placing it in the context of a familiar physical environment, appropriate to the problem being addressed. But virtual reality is still far from being a desktop tool and the highly immersive tools that we have discussed are available at only a few (but expanding) locations worldwide.

The C2 is a highly immersive and very expensive environment. The benefit is that it provides a very realistic sense of being surrounded by the scene. It is a developing technology, especially at Iowa State University, and so it is constantly being enhanced by new software libraries and undergoing hardware updates and additions. Modeling to make a truly believable physical environment would require enormous efforts and probably considerably more computational power than even this state-of-the-art setup has, to track the user's movements and update the scene sufficiently quickly. The C2 uses the OpenGL graphics library. So to draw the elevation surface we had to resort to the first principles of computer graphics: start with a regular and ordered grid, and calculate the normals to the individual polygons to determine the appropriate light and shade.

It would be appropriate to collaborate with a 3D GIS research group who could model the terrain of the study region quickly and in a more sophisticated manner. Incorporating multivariate visualization tools with terrain visualization shifts 3D GIS "up a gear" into the domain of data analysis.

Acknowledgements

Helpful comments and information about spatial data analysis methods were provided by Professor Noel Cressie.

The research reported in this article has been funded by National Science Foundation Grant DMS–96–32662 and the U.S. Environmental Protection Agency through Cooperative Agreement CR822919-01-0 with Iowa State University. This paper has not been subjected to the Agency's peer and administrative review. No endorsement of the contents by the Agency should be inferred.

Additional Information

Further information and developments to the work can be found at:
http://www.public.iastate.edu/~dicook/research/C2/statistic.html.

References

Asimov, D.: 1985, "SIAM" *J. of Sci. and Stat. Comp.*, **6** (1):128–143.

Brown, W. M., Astley, M., Baker, T., and Mitasova, H.: 1995, "Twelfth International Symposium on Computer-Assisted Cartography", pages 89--99, Charlotte, NC.

Buja, A., Cook, D., Asimov, D., and Hurley, C.: 1997, J. of Computational and Graphical Statistics. Submitted.

Carr, D. B., Wegman, E. J., and Luo, Q.: 1996, "Technical Report 129", Center for Computational Statistics, George Mason University.

Cook, D., Majure, J. J., Symanzik, J., and Cressie, N.: 1996, "Computational Statistics: Special Issue on Computer Aided Analyses of Spatial Data", **11** (4):467–480.

Cook, D., Symanzik, J., Majure, J. J., and Cressie, N.: 1997, Computers and Geosciences: Special Issue on Exploratory Cartographic Visualization. **4** (1); 371-385, web material at www.elsevier.nl/locate/cgvis.

Cruz-Neira, C.: 1993, SIGGRAPH '93 Course Notes 23, 18 pages.

Cruz-Neira, C.: 1995, PhD thesis, University of Illinois at Chicago.

Klemm, D. J. and Lazorchak, J. M. editors.: 1995, "Technical Report EPA/620/R-94/004", U.S. Environ. Protection Agency, Office of Res. and Dev., Environ. Monitoring Systems Laboratory, Cinncinnati, Ohio.

Mitasova, H., Mitas, L., Brown, W. M., Gerdes, D. P., Kosinovsky, I., and Baker, T.: 1995, *Intl. J. of Geographical Info. Systems*, **9** (4):433–446.

Pimentel, K. and Teixeira, K.: 1995, "Virtual Reality through the new Looking Glass (Second Edition)", McGraw-Hill, New York, NY.

Raper, J.: 1989, "Three Dimensional Applications in Geographic Information Systems", Taylor Francis, London, UK.

Roy, T., Cruz-Neira, C., and DeFanti, T. A.: 1995, "Presence: Teleoperators and Virtual Environments", **4** (2):121–129.

Sutherland, I. E.: 1965, Proc. IFIP 65, 2, pages 506--508, 582–583.

Swayne, D. F., Cook, D., and Buja, A.: 1991, "ASA Proceedings of the Section on Statistical Graphics", pages 1–8, Alexandria, VA. American Statistical Association.

Symanzik, J., Cook, D., Kohlmeyer, B. D., Lechner, U., and Cruz-Neira, C.: 1997, IASC Proceedings, Forthcoming.

Van Teylingen, R., Ribarsky, W., and Van Der Mast, C.: 1997, *IEEE* "Transactions on Visualization and Computer Graphics", **3** (1):65–74.

PROTOTYPING A VISION FOR INTER-AGENCY TERRESTRIAL INVENTORY AND MONITORING: A STATISTICAL PERSPECTIVE

CAROL C. HOUSE[1], J. JEFFERY GOEBEL[1], HANS T. SCHREUDER[1], PAUL H. GEISSLER[2], WILLIAM R. WILLIAMS[2], and ANTHONY R. OLSEN[3]

[1]*U.S. Department of Agriculture,* [2]*U. S. Department of Interior,* [3]*Environmental Protection Agency*

Abstract. A demonstration project in Oregon examined the feasibility of combining Federal environmental monitoring surveys. An integrated approach should remove duplication of effort and reduce the possibility of providing apparently conflicting information to policy makers and the public. Data collection teams made photo interpretation measurements and on-site soil/vegetation/animal observations at locations that were selected from the Forest Inventory and Analysis (FIA), National Forest System (NFS) Region 6, and National Resource Inventory (NRI) surveys in a six-county area in Northern Oregon. The project demonstrated the feasibility of conducting a combined FIA/NFS/NRI survey and suggests an approach that will preserve the utility of the critical historical information from these surveys. We suggest a framework for estimating the extent of forest and range land that explains FIA/NRI differences and provides a common basis for both surveys. We suggest indicator and protocol criteria that will allow consistent national and regional estimates over all vegetation types, and stress the importance of including measurement repeatability in the design of the combined survey.

1. Introduction

The Federal Government currently funds several inventory efforts based on probability surveys to measure status and trends of the Nation's natural resources (Olsen, *et al.,* 1997). Although each has a unique focus, there exists considerable potential for duplication of effort and for providing apparently conflicting information to policy makers. This paper examines the feasibility of integrating national inventories, emphasizing those from the U.S. Forest Service (USFS) and the Natural Resources Conservation Service (NRCS).

An integrated approach has a number of advantages. A common database of environmental information would facilitate interdisciplinary and interagency studies that could address the Nation's major environmental issues. Uniformity of definitions, sample design, and measurements throughout the United States would permit data collected by different agencies to be meaningfully combined, leveraging their investments and allowing agencies to address broader issues. The use of common definitions and estimators would reduce the occurrence of apparent conflicts in related estimates. The pooling of resources and the elimination of duplicate efforts would improve efficiency in data collection. Greater opportunity for state involvement and funding could be available through participation in interagency efforts. A broad-based interagency survey would increase flexibility to accommodate multiple conditions and objectives, including the ability to cut across jurisdictional boundaries when defining domains of interest.

Environmental Monitoring and Assessment **51**: 451–463, 1998.

This paper reports on a demonstration project conducted in Oregon, which combined the sampling frames from the NRCS National Resources Inventory (NRI) (Nusser and Goebel, 1997), the USFS Forest Inventory and Analysis (FIA) (USFS, 1992), and the USFS National Forest System (NFS) Region 6 (Max *et al.*, 1996.). Additionally, it focused on the data collection needs of the Forest Health Monitoring Survey (FHM) (FHM, 1994) of the USFS, and the National Biological Service (now part of the U.S. Geological Survey). The project used the combined sampling frame to field a three-phase integrated inventory that included many of the important classification and measurement variables currently collected on the independent inventories, and experimented with a number of developmental protocols.

2. Methods

The Oregon demonstration project was conducted in six counties in northern Oregon, (which included Mt. Hood): Clackamas, Multnomah, Hood River, Wasco, Jefferson, and Sherman (Fig. 1). The area encompassed a diversity of land cover and land use, including forest land, range land, transition areas, urban boundary, juniper woodlands, cropland, orchards, tree farms, and major riparian zones. Considerable inter-agency activity was ongoing in that region because of the Northwest Forest Plan (USDA/FS and DOI/BLM, 1994). The area was fairly compact for data collection efficiency and convenient to management staff in Portland.

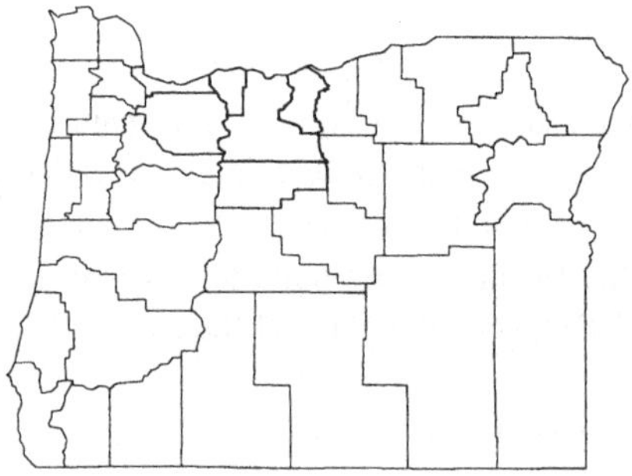

Fig.1. Study Area.

The demonstration project included three phases of data collection. Phase I was the photo interpretation of 613 photo points based on 337 FIA and NFS photo points and 276 NRI second stage points. In each case the photo points were a random subsample of existing FIA/NFS and NRI sample points. From these Phase I sample points, 78 Phase II sample points were selected systematically for ground data collection of classification, vegetation and soil related variables. The Phase I and II subsamples from FIA/NFS and

NRI were selected independently using their respective survey designs. Each Phase II sample point was visited by two different teams, to assess the repeatability of the measurement process. Phase II also included the laboratory analysis of soil samples. Phase III, which focused on a selection of animal relative abundance measurements, was carried out by an independent field crew on 14 Phase II sample points located on Federal land.

The project did not attempt to collect all the information currently collected by the NRI, FIA, NSF Region 6, and FHM surveys. Instead, it included a selection of important current measurements and experimented with several measures associated with soil quality, range and forest health, wildlife habitat, and animal relative abundance. Combined NRCS and USFS field crews did the photo interpretation and collected vegetation and soil measurements on the ground plots. A National Biological Service (NBS) field crew made Phase III measurements. The National Agricultural Statistics Service (NASS) produced a combined data base and conducted the initial data analyses. A more detailed description of data collection and analysis is provided in Goebel *et al.,* 1997.

PHASE I

Phase I data collection was carried out in the office from photos, GIS data layers, and hard copy ancillary materials. The photos were 1994 stereo 1:40,000 scale black and white National Aerial Photography Program film positives. Photo-interpreted data items were: earth cover class based primarily on a draft Federal Geographic Data Committee (FGDC) vegetation classification; evidence of disturbance associated with earth cover; land class; land use context; degree of urbanization; and wildlife habitat diversity. The project required four trained photo interpreters for approximately three months.

PHASE II

Phase II ground data were collected by field crews, which observed site characteristics, vegetation structure, ground cover, herbaceous vegetation, species frequency, shrub canopy cover, shrub density, woody debris, tree tallies, and soil quality. The plot design was laid out in a pattern of four subplots (Figure 2) (Scott and Bechtold, 1995). At each subplot, data were collected within three concentric fixed-radius plots and along transects extending out from the subplot center. The center of subplot 1 was the grid point used in the photo interpretation.

Field crews were usually composed of two technical specialists trained in soils and/or vegetation. Crews were in the field for approximately four months; each site visit required approximately one field day. Each Phase II site was enumerated by two independent crews. The first crew enumerated subplots 1, 2, and 3, and the second crew enumerated subplots 1, 2, and 4. Thus, one half the measurements were independently replicated. Soil samples were collected from two specified locations within each plot. Samples were taken from the A Horizon, and separately from the O horizon and crust when they were present. Samples were analyzed for organic C, total N, pH, aggregate stability, and particle size.

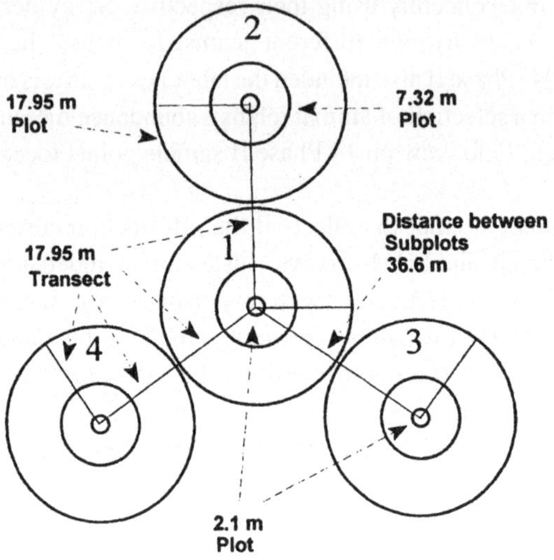

Fig. 2. Phase II Plot Design.

PHASE III

Phase III focused on animal relative abundance measures. A specialized field crew visited 14 of the Phase II sample points located in the Douglas Fir and Ponderosa Pine areas in the Ochoco, Mt. Hood, and Deschutes National Forests. The protocols required three visits to each site. Insect traps were set out on the first visit and measurements taken on the second and third visits. Bird counts were conducted at each site on a visit between June 7 and July 13, with a second count on a subsequent visit at 5 sites. Field crews also searched for amphibians under logs, stumps, and rocks located along transects connecting the plot center with the bird count locations. The use of mammal traps would have required daily field visits, and, therefore, were not included in this project.

3. Results

FEASIBILITY OF A COMBINED SURVEY

SURVEY DESIGN

The NRI and FIA/NFS sampling frames were defined and had complete coverage of the study area. The NFS frame was used on National Forests, and the FIA frame was used on other lands. The NRI frame covered all land, although the NRI historically had not sampled ground points on public land. We were able to locate frame materials and determine appropriate sampling weights for the respective frame units. We examined only the study area, and we emphasize that these findings might not apply to other areas.

FIA and NRI used permanent sample points, which have important historical data. The FIA began in 1930, and there have been three to six repetitive measurements on some sites since the early 1960's. NRI dated from 1982, with revisits once every five years per site. It was critical to preserve the continuity of this information. The most direct way was to incorporate the existing permanent sample points into the new design. This study used standard multiple-frame estimation methodology to combine two independent sub-samples from existing sampling frames into a single estimation procedure. In the analysis, the existing frames were equally weighted; however, more research is needed to determine optimal frame weights, and to determine the sample sizes that would be optimal from each existing frame.

There was concern about the continued use of permanent plots. If an integrated survey design meant a higher visitation rate and visibility, there were possible implications for their use. It might be difficult to keep plot locations confidential and representative. Confidentiality was necessary to prevent the plots from being treated differently from other areas, biasing the estimates. Increased visits might cause more landowners to refuse access. There might be disturbance and change caused by the crews. An integrated survey must address these issues.

FIELD EXPERIENCES

The use of joint photo interpretation teams and joint field crews proved to be very successful. Combined crews brought together the varied expertise and experiences developed on current inventories. The use of joint crews helped to produce common interpretations of standard protocols. The experience with joint field crews demonstrated the blending of cultures necessary to support an integrated survey.

Integration resulted in a substantial increase in the complexity of the field measurements because a broader range of measurements was needed. It also highlighted several needs of the field crews: 1) Each field crew needed specialists in vegetation and soils. 2) Continuous training throughout the season was important for updating skills for identifying soils, plants, wildlife, insects, and diseases. 3) Field crews needed access to experts for plant species identification and for interpretation of the protocols in unusual situations. 4) A reconnaissance person was needed to obtain landowner permission and to determine the best access route prior to the arrival of the field crew. 5) It was important for the crews to understand the objectives and rationales for the field procedures to increase their ability to interpret the instructions and to increase their commitment.

The field crews provided the following feedback concerning field protocols: 1) Field measurements of vegetation height structure were complicated by the difficulty in positioning the quadrat measuring device in heavy brush. 2) All observations along a transect should be done in a single pass on steep terrain to minimize trampling. 3) Both the fixed quadrat measuring device and the nested frequency device could use some engineering design improvements. 4) Crews felt it was important to be able to qualitatively describe soil disturbance and observed erosion.

A unified database was established, including all the photo interpretation, soils, vegetation, and animal relative abundance data, as well as the weighting factors required for estimation. A common set of six-county estimates was developed. Documentation was developed, discussing observation identification, file format, variable names, missing data conventions, the weighting factors required for estimation, and other information that would allow data users to perform analyses of their own in a statistically valid manner.

ESTIMATES OF FOREST LAND AND RANGE LAND EXTENT

The differences in estimates of resource extent between NRI and FIA had two causes. First, although the two agencies agreed to a definition of forest land as 10% stocking of trees, it was difficult to implement this definition. This difficulty led to different operational definitions in the field. Additionally, inconsistencies existed about which species were classified as "trees." For example, the NRI might classify "oak woodland" and "pinyon-juniper woodland" as range land, while the FIA might classify this same area as forest land. Also, FIA equated 10% crown cover in trees to 10% stocking while the NRI used 25% crown cover. To avoid these differences, we recommended that the field crews record what was actually observed (using "land class" definitions consistent with FGDC standards), rather than classify land as forest or range. For example, a crew member would record "pinyon-juniper woodland" with 15% crown cover. Later, the inventory data could be classified into many different groupings, including the historical FIA and NRI definitions, while illuminating the differences between these classifications.

Second, some differences might result from the way that the estimation procedures utilized photo interpretation and on-site classification. Such differences in procedures could create substantial differences in the final estimates. FIA employs a standard double sampling approach with photo and ground data to provide ratio or regression estimates of forest land. NRI was unable to use a similar classical double sampling technique because of operational practices. All NRI plots were initially field-visited. Subsequent surveys at five-year intervals relied heavily on aerial photography to update condition and classification, with field visits used mostly where photography was inadequate. This area of difference between the two operational procedures needed additional investigation.

Table I displays a subset of land class categories used for data collection (Phase I) during this project. Four of these categories are further sub-divided by "percent crown cover". The table shows the total number of hectares (in the six county study area) classified to each category. Simultaneously, and separately, it shows whether the USFS and NRCS would have classified that land as "forest" or "range," according to their operational definitions. Table I also shows the difference between total land estimates (and percent of total land) based on Phase I photo interpretation alone and those adjusted by Phase II data through a regression estimate.

Table I
Forest and Range Land Estimates with USFS and NRCS Definitions, in Hectares.

Land Class[1]	Crown Cover %	Forest Land		Range Land	
		USFS	NRCS	USFS	NRCS
Timberland	10-25	36,517			36,517
	≥ 25	706,972	706,972		
Oak Woodland	10-25	3,036			3,036
	≥ 25	30,358	30,358		
Unclassified Woodland	10-25				
	≥ 25	6,361	6,361		
Juniper Woodland	10-25	98,403			98,403
	≥ 25	43,912			43,912
Chaparral		3,036			3,036
Desert Shrub				169,548	169,548
Grass/herbaceous				392,820	392,820
Phase I Total		928,595 45%	743,691 36%	562,368 27%	747,272 37%
Regression estimates using Phase I (photo) totals, adjusted with a regression based on the Phase II (ground) observations.					
		793,246 39%	700,043 34%	613,710 30%	706,913 35%

[1] Other lands not included are cropland, pasture and hayland, permanent snow, barren land, wetlands, developed land, transportation and utilities, and water which total 554,906 ha.

INDICATOR AND PROTOCOL CRITERIA

Plot Design

We identified some design characteristics that were valuable and effective.

1. All measurement protocols were centered at the sampled grid point, but the definitions of the subplots or transects might have been different. This flexibility allowed the plot design to accommodate both protocols that required measurements within fixed areas, and protocols that required measurements along a transect. There were areas around and between subplots where destructive protocols could be accommodated.

2. Multiple subplots were used to increase efficiency (considering travel costs to reach the point) and to provide a better characterization of the plot. An important survey design issue was determining the optimal number of subplots for each indicator.

3. The number of repetitions of the same measurement required for a given precision was often dependent on the vegetation density. Protocols and the plot design should have accommodated these differences, basing the decision on information available prior to ground data collection (i.e., photo interpretation, historical ground information, etc.). This rule would allow the Phase I points to be poststratified and ensures objectivity.

4. Field crews should not decide on the type or number of measurements based on any land use classification or other field observation.

5. Measurement locations within plots should not be changed due to field conditions encountered. When impossible to make the prescribed measurement in the prescribed location, record the measurement as inaccessible or not applicable. Any movement biases the estimates and reduces the credibility of the survey.

COMMON DEFINITIONS AND MEASUREMENT PROTOCOLS

The demonstration project prompted discussions among the agency representatives about protocol selection and provided field experience in measuring them. We identified and refined common indicators that measured selected qualities important to the cooperating agencies. The goal was to complete all field measurements at a single site in one field day. This goal was based on the need for efficiency and the need to reduce the political impact of multiple visits on private land. We considered three options to reduce workload to achieve the one-day site visit: increase the size of crews, reduce the number of indicators, or modify field protocols for specific indicators to reduce the number of repetitions. In general, we chose to reduce the number of repetitions to achieve this goal.

We rejected the option of changing protocols based on field plot conditions. In a national survey, it would be important to select indicators and develop protocols which would work for all vegetation conditions and terrain. At the time of the survey, forest and range indicators were very different. Rangeland assessments emphasized soil erosion factors, species frequency, and canopy cover, all measured on relatively small plots. The proportion, or the contribution, of each species toward total annual biomass production was often estimated. In forest inventories, species frequency estimates required larger plots; size and age distributions were as important as the total number of individuals. Traditionally, biomass measurement only addressed periodic production of wood fiber with much less concern for biomass production in tree branches and foliage and in non-tree vegetation. Forest land assessments usually placed less emphasis on soil factors.

We have recommended developing indicators that can measure compatibly in all areas so that estimates from forest and range lands can be aggregated to obtain regional and national estimates. If different protocols were used in forest and range lands, the results could not be meaningfully combined or compared. For example, in forests, the O soil horizon could be easily measured, but it was absent in range soils. The depth of that horizon can be compatibly measured by recording a zero depth when it was not present. We found that the number of repetitions of a measurement required for a given level of precision might be dependent on the type of terrain. Therefore, it might be useful to vary the number of repetitions and/or the number of subplots, depending on the density of the vegetation or other factors, while maintaining the same protocols so that compatible estimates could be obtained.

More work on protocols and optimum plot design was found necessary. Three issues were: (1) What variables or attributes were important to address resource health or condition? (2) What was the best way to measure the attributes? and (3) What was the most efficient protocol under varying landscape conditions of topography, vegetation life form, and vegetation density for an attribute?

ANIMAL OBSERVATIONS

Animal relative abundance estimates greatly improved our understanding of forest ecosystems. Separate crews were necessary for conducting animal counts because bird observations must be made at dawn to be effective, because there was often a short season when animals were active and observable, and because at least two visits were required to set and collect pit-fall traps for arthropods. Although separate crews were necessary, close coordination with the vegetation/soils crews was essential.

Making a single visit to a plot would increase the number of plots that could be visited and would reduce the impact on land owners. However, we found that substantially more than twice the information was obtained from two visits to the plot, because of the information obtained from the arthropod traps. Whereas bird populations were migratory and reflected conditions on both their winter and summer ranges and on their migration route, arthropod populations reflected conditions on the plot. Taxonomically diverse arthropod populations represented several trophic levels and strategies, providing a much better representation of ecological conditions on the plot.

Field crews visited 14 of 16 targeted plots in the Ochoco, Mt. Hood, and Deschutes National Forests during a 5-week period. Bird and amphibian sampling for each site was completed in a single day, but a second visit was required to collect arthropods from pitfall traps. Eighty-three bird species were detected at the 14 plots. Survey results were compared to lists of species predicted by GAP (Scott, Tear, and Davis, 1996) to occur in study areas. Fewer than 25 percent of predicted species were detected. Eleven species were detected that were not predicted to occur. Amphibian surveys were unsuccessful, because observations occurred too late in the season, when amphibians were inactive. Arthropod sampling was successful. Numerous species and individuals were collected. However, sorting and identification of specimens was time consuming and required the assistance of a qualified taxonomist. (Availability of taxonomic expertise is likely to be a limiting factor for monitoring programs which include arthropods as an indicator group.) The effects of phenology were also a major concern, because there often was a short peak period of activity, with few captures before and after the peak.

MEASUREMENT REPEATABILITY

Measurements had to be repeatable in order to be useful, and this was especially true for measurements used in status and trend analysis. Each Phase II site was enumerated by two different crews. Both crews visited a site on the same day to eliminate temporal variability. The first crew enumerated subplots 1, 2, and 3 and the second crew enumerated subplots 1, 2, and 4. Thus, one half of all Phase II measurements were independently replicated.

The correlation (r) between the paired measurements of different crews was estimated for selected variables (Table II). Tree counts, DBH, and the number of distinct tree species were believed to be accurate and repeatable ($0.90<r<1.00$). Other variables were subject to substantial measurement error: ground cover ($r= 0.44$) and shrub seedlings ($r=0.27$).

The difference between the measurements of the two crews, expressed as a percent of the mean, also indicated the repeatability of the measurements: $(\Sigma\ w_i\ |y_{i1}-y_{i2}|)\ /\ (\Sigma\ w_i\ \bar{y})$ where w_i was the weight and y_{i1} y_{i2} were the measurements on the ith plot. Another indicator was the percent of the plot variance that was due to measurement error: $s_m^2 / (2s_p^2+s_m^2)$. The mean differences between the measurements ranged from 6% to 100%, while measurement error ranged from 0.4% to 73% of the plot variance (Table II).

To determine if measurements were sufficiently repeatable, one needed to determine whether or not important differences can be detected. If measurements were not repeatable, there would be a large measurement error component to the variance, which would prevent the survey from detecting even large changes. If important changes could not be reliably detected within a feasible budget, the survey would not achieve its objective. Estimates based on variables measured with serious measurement errors could be quite misleading.

Making independent replicate observations with different crews allowed us to separate measurement error S^2 from the other components of variance. This estimate allowed us to identify measurements that would benefit from refinements to make them more repeatable. Estimates of variance components and cost (time) estimates for the various measurement activities allowed us to determine the optimal number of subplots and replicates to maximize power.

Table II
Repeatability of Selected Measurements.

Attribute	Correlation	Difference Percent	Measurement Error %
Nested Frequency			
Average number of species per location	0.89	22.5	6.1
Total number of species per plot	0.96	19.5	2.0
Average to Total species ratio	0.39	60.1	44.0
Shrub Density			
Percent seedlings	0.27	100.0	73.0
Percent saplings	0.39	53.4	44.2
Percent mature	0.52	50.4	32.4
Percent decadent	0.74	57.6	15.6
Percent dead	0.87	47.2	7.2
Total count	0.93	24.6	3.8
Tree Tally For Woodlands			
DBH	0.90	8.0	5.6
Number of trees	1.00	5.8	0.4
Number of species	0.96	5.8	2.1

4. Recommendations

We recommend a phased transition toward a national integrated inter-agency natural resource status and trends inventory. It would begin with a core program based on the integration of the NRI and the FIA surveys. This would create a successful, cost effective and flexible monitoring framework, and a comprehensive information package that should encourage other agencies and programs into the partnership. For example, the Forest Service is encouraging closer ties between the FIA, the FHM program, and the monitoring programs within the National Forest Systems. Integration of these programs with the core may be an appropriate second phase of integration. Agencies such as the Bureau of Land Management, the Department of Defense, and the National Park Service could expand the coverage of the core monitoring effort by bringing additional Federal land into the framework. Other agencies such as the National Agricultural Statistics Service, the Economic Research Service, the Environmental Protection Agency and Geological Survey could expand the type of information collected. Inclusion of the Fish and Wildlife Service could lead to more comprehensive and consistent information on wetland resources. The framework for this transition is outlined in Goebel *et al.,* (1997).

This framework has been approved by USDA policy makers and the implementation process is beginning. The recommendations are consistent with the intent of the National Science and Technology Council's Committee on the Environment and National Resources' ongoing efforts to "develop a national framework for integration and coordination of environmental monitoring and related research through collaboration and building upon existing networks and programs."

Acknowledgements

We particularly acknowledge five individuals whose commitment and leadership contributed substantially to the successful outcome of the Oregon Demonstration Project: Mark Tilton (NRCS) and Dale Baer (USFS) provided the local project leadership, managing training, data gathering, and innumerable daily details; John Amrhein (NASS) developed the database, data documentation, and conducted the analysis of that data; Carol Chambers (USGS) managed the Phase III data gathering and analysis; and Glen Miller (BLM) stepped in to manage a multitude of different critical activities in support of the project.

Many other scientists and technicians provided their time, experiences, and expertise in various aspects of the design and implementation of this project: They include: Jim Alegria (BLM), Carol Franks (USFS), Leon Liegel (USFS), David Pike (USGS), Ed Starkey (USGS), John Teply (USFS), Marty Stapanian (BLM), and Al Winward (USFS).

Our thanks to the Oregon Agricultural Statistics Service who provided data entry support for the project.

Finally our special thanks for a job well done to those individuals who collected data on one or more of the three phases of this project. Field Crew Coordinator: Walter Grabowiecki (NRCS). From NRCS: Catherine Darby, Steven Fedje, Eileen Larkin, Sara Zimmerman, Rich Edlund, Monte Graham, Ron Myhrum, Matt Ricketts, Lorna Stolen, Randall Wilson. From USFS: Dale Baer, Sarah Butler, Erica Hanson, Dan Kenitz, Sarah Butler, Paul Dunham. From BLM: Glen Miller. Field Crew (Animal) Coordinator: Carol Chambers (USGS). From USGS: Becky Fasth, Rachel Johnson.

References

Birdsey, R. A., and Schreuder, H.T.: 1992, 'An Overview of Forest Inventory and Analysis Estimation in the Eastern United States - with an Emphasis on Components of Change', *USDA Forest Service RM Technical Report* RM-214.

FHM.: 1994, 'National Forest Health Monitoring Program.', U. S. Department of Agriculture, Forest Service, Forest Health Monitoring Program, Research Triangle Park.

Goebel, J. Jeffery, Schreuder, Hans T., Olsen, Anthony. R., House, Carol C., Geissler, Paul H., and Williams, William R.: 1997, 'The Oregon Demonstration Project: A Study on Integrating Surveys of Terrestrial Natural Resources', *Inventory and Monitoring Institute Report No. 2*, Ft. Collins, Colorado, U.S. Department of Agriculture, Forest Service, Rocky Mountain Research Station.

Max, T.A., Schreuder, H.T., Hazard, J.W., Teply, J., and Alegria, J.: 1996, 'The Region 6 Vegetation Inventory and Monitoring System', *USDA Forest Service Research Paper* PNW-R8-493.

Mandel, J.: 1972. 'Repeatability and Reproducibility', *J Qual Techn* **4**, 74-85.

Nusser, Sarah M., and Goebel, J. Jeffery.: 1997, 'The National Resources Inventory: A Long Term Multi-Resource Monitoring Program', *Environmental and Ecological Statistics* **4**, 181-204.

Olsen, A. R., Sedransk, J., Edwards, D., Gotway C. A., Liggett, W., Rathbun, S. L., Reckhow, K. H., and Young, L. J.: 1997, 'Statistical Issues for Monitoring Ecological and Natural Resources in the United States', *Environmental Monitoring and Assessment,* (This volume).

Schreuder, H.T., Gregoire, T.G., and Wood, G.B.: 1993, *Sampling Methods for Multiresource Forest Inventory*, J. Wiley and Sons, New York.

Scott, C. T., and Bechtold, W. A.: 1995, 'Techniques and Computations for Mapping Plot Clusters That Straddle Stand Boundaries', *Forest Science Monograph* **31 (41) 3,** 46-41.

Scott, J. M., Tear, T. H., and Davis, F. W.: 1996, 'Gap Analysis: A Landscape Approach to Biodiversity Planning', American Society for Photogrammetry and Remote Sensing, Bethesda, Maryland, ISBN-1-57083-03603. http://www.gap.uidaho.edu/gap/

USDA/FS and DOI/BLM.: 1994, 'Record of Decision for Amendments to Forest Service and Bureau of Land Management Planning Documents Within the Range of the Northern Spotted Owl and Standards and Guidelines for Managment of Habitat for Late-Successional and Old-Growth Forest Related Species Within the Range of the Northern Spotted Owl.'

USFS.: 1992, *Forest Service Resource Inventories: An Overview*, USGPO 1992-341-350/60861, U. S. Department of Agriculture, Forest Service, Forest Inventory, Economics, and Recreation Research, Washington, DC. 39 p.

GENETIC PATTERNS AS A TOOL FOR MONITORING AND ASSESSMENT OF ENVIRONMENTAL IMPACTS: THE EXAMPLE OF GENETIC ECOTOXICOLOGY

NATALIE M. BELFIORE[1,2] and SUSAN L. ANDERSON[2]

[1]*University of California, Davis (California) 95616, USA, and* [2]*Lawrence Berkeley National Laboratory, 1 Cyclotron Road MS 70-193A Berkeley (California) 94720, USA*

Abstract. Genetic techniques are widely applied to assess the effects of environmental variation or exogenous impacts on populations. Many studies fail to provide convincing evidence that genetic patterns are attributable to the factors proposed. We assert that a rigorous approach must be followed to distinguish patterns of natural genetic variation from genetic change. We review the principles of natural genetic variation and population structure and present them in the context of their interaction with biological and stochastic sources of genetic change. Key steps are articulated which are often overlooked when applying genetic techniques. These are consideration of population structure when comparing populations, developing a specific test against a model of genetic change, and testing for evidence of direct effects and mechanisms of impact. Use of these steps in genetic ecotoxicology is described in detail and includes three primary methods of linking genetic patterns to the effects of contaminants. We propose that this combined approach is critical to the use of genetic techniques to assess and predict long-term effects of environmental impacts on populations or ecosystems.

1. Introduction

In recent years, genetic techniques have been used in new ways to address environmental questions. The genetic characteristics of organisms are described as a step in understanding population processes and this information is employed to address ecosystem-level questions. These efforts take advantage of powerful new molecular genetic techniques that permit a more rapid and larger-scale assessment of genetic patterns. Genetic patterns, because they are heritable, may be thought of as a record of past biological and environmental processes. Such processes include breeding, migration, and selection, for example. This paper summarizes key considerations when using genetic patterns to assess and monitor environmental impacts in natural systems, and considers applications of this approach in genetic ecotoxicology.

We propose that the primary challenge of the genetic approach is the careful distinction between genetic variation and genetic change. Genetic variation is used here to indicate the level of genetic polymorphism ("variability") and the patterns of genotype frequencies that result from natural population processes and natural spatial and temporal fluctuation in environmental conditions. Genetic change refers here to significant alterations to genetic patterns due to the acute or chronic imposition of anthropogenic or exogenous forces. The second important challenge of the genetic approach is to link the change to the correct source(s), if it is determined that genetic change has occurred.

Environmental Monitoring and Assessment **51:** 465–479, 1998.

What can be done to distinguish natural genetic variation from genetic change due to exogenous forces and to link genetic change to probable causes? A reliable assessment of environmental impacts on genetic patterns must combine a rigorous evaluation of natural genetic variation, a confirmation that genetic change has occurred, and a variety of short-term and direct indicators. All steps are emphasized because they characterize several common pitfalls of studies that attempt to understand genetic patterns. Specifically, we propose that natural genetic variation be characterized by testing for correlation between genetic patterns and an objectively-selected suite of population and environmental factors, genetic change be measured by comparing affected to unaffected populations as a surrogate for monitoring populations over time, and, mechanisms of potential direct or short-term effects be tested directly.

The goal of this paper is to address several questions which relate directly to studying genetic patterns for assessing long-term impacts on populations. These questions are:

1) What is the role of ecological, environmental, and life-history parameters in creating patterns of natural genetic variation and how might natural genetic variation be taken into consideration when testing for genetic change?

2) What are the models of genetic and stochastic sources of genetic variation in populations that might interact with exogenous population pressures to cause genetic change?

3) What tools are available to assist in discriminating among the exogenous factors that affect genetic patterns?

We then discuss this approach in the context of genetic ecotoxicology to demonstrate how a comprehensive approach would enhance current efforts. Ecotoxicologists have studied tolerance and resistance in relation to contaminant exposure. However, they have been slow to exploit the potential of molecular genetic markers to understand how contaminants may be interacting with genetic and population processes, and what impact these interactions might have on organisms in the field. Understanding these interactions may help characterize large-scale and potential long-term effects of contaminants in natural systems. Several genetic ecotoxicologists have suggested studying genetic variation and genetic change as a means to understand long-term impacts in the wild (e.g., Anderson *et al.*, 1994; Bickham, 1994; Shugart and Theodorakis, 1996), but few studies have undertaken this approach (see Anderson and Belfiore, submitted, for review). Fewer studies have provided evidence convincingly attributing changes in genetic patterns to contaminants (e.g., Theodorakis *et al.*, 1997; Theodorakis and Shugart, 1997).

We conclude by summarizing a generic flow chart for monitoring and assessment of long-term environmental impacts, combining tools of both population genetics and traditional environmental monitoring programs.

2. Understanding Genetic Patterns

2.1 NATURAL VARIATION AND POPULATION STRUCTURE

In general terms, natural genetic variation arises from a combination of natural environmental variation and population structure. Gradients of environmental conditions are often reflected in gradual shifts in genotype frequency, for example, because certain genotypes are favored in different conditions. Population structure, on the other hand, describes the relationships among individuals, subpopulations, and the total population. Typical population parameters such as inbreeding coefficients, effective population size, and neighborhood size are included when qualitatively describing population structure (e.g., Hartl, 1988). Population structure is generally quantified genetically by measuring and interpreting genotype frequencies. Ocean species, for example, are expected to show little geographic population structure because the neighborhood size, therefore the rate of gene flow, is extremely high (e.g., Palumbi, 1996; Garcia de Leon, *et al.*, 1997; but see Pogson *et al.*, 1995). By contrast, stream species are generally restricted to smaller geographic areas where the rate of outcrossing is severely reduced compared to ocean species, and local environmental and population effects may have comparatively greater influences on genetic patterns.

A thorough characterization of the genetic structure of the population and the covariation due to environmental variables is needed to ascertain whether genetic change has occurred. For example, three environmental variables, flow, altitude, and temperature, were found to covary with allozyme genotype variation in mosquitofish (*Gambusia affinis*) (Smith *et al.*, 1983). However, it is not possible to determine whether authors tested for covariance with other possible environmental or population causes of genetic variation in this study. If these authors were going to test for genetic change due to some imposed factor, it would be important to characterize initial covariance with a wider range of factors, such as the presence of potential competitors or predators, or available food items. In another study, mitochondrial DNA (mtDNA) sequence variation at two conserved loci in loggerhead shrikes (*Lanius ludovicianus*) confirmed strong population subdivision among subspecies, but indicated a small amount of gene flow among the subspecies populations and a recent northward range expansion (Mundy *et al.*, 1997). This latter study evaluated genetic patterns to estimate population parameters, specifically degree of population subdivision, and the rates and patterns of migration. Understanding initial population structure in these ways helps researchers to predict and to observe the possible genetic effects of exogenous factors.

If natural genetic variation due to population parameters and environmental conditions may be confidently estimated, then these patterns of variation or covariation may be factored out in order to test for genetic change. Although it is never feasible to test for every possible environmental covariant or all possible genetic patterns of population structure, it is crucial to consider these factors thoroughly. Without a basic understanding of natural genetic variation, genetic differences may be incorrectly called genetic change.

2.2 MODELS OF GENETIC CHANGE

QUESTION DEFINITION

What are the models of sources of natural genetic variation and genetic change in populations? An important step in testing for genetic change due to exogenous forces is to formulate specific questions and predictions based on models proposed to be causing the change. These questions are best tested by appropriate genetic marker selection.

Evolutionary models of genetic patterns include mechanisms that play a role in natural genetic variation as well as potential genetic change due to exogenous forces. Armbruster and Schwaegerle (1996) discuss five possible mechanisms that lead to genetic patterns. These are:

1) Genetic adaptation or selection, resulting in adaptive character complexes, selective covariation, or coadapted gene complexes.

2) pleiotropy (in which single genes affect several traits) or linkage (in which genes do not randomly recombine).

3) Mutation (which could be highly deleterious, highly advantageous, or very slightly disadvantageous direct genetic alterations).

4) Genetic drift (random changes in gene frequencies).

5) Gene flow (clinal covariance among nearby subpopulations).

Any of these mechanisms may be predominantly responsible for covariance or directional shift in genotype frequencies, or, there may be strong interactions producing very different results. For example, the mechanism of genetic drift would be expected to produce a predictable outcome in the face of large population size fluctuations. Populations with dramatic size fluctuations over time are expected to have very low effective population sizes (Lande and Barrowclough, 1987). This may result in a strong effect of genetic drift whereby every time a population is severely reduced in size, the "founding" gene pool for subsequent generations is very small, and by chance, population genotype frequencies will be dramatically altered. By contrast, subpopulation genotype frequencies established by genetic drift (or population subsampling) may result in several different adaptive character complexes if each subpopulation is subject to strong selective pressure because of the initial range of genetic variation.

It is important to be able to evaluate genetic change that may be due to population structure alterations unrelated to exogenous impacts or to indirect effects of exogenous impacts. Background patterns of natural genetic variation must be characterized so that genetic change that is a result of complex interactions between natural genetic variation and exogenous forces might be separated from the background. Indirect processes, for example, might include a significant reduction of the population size which could in turn change density-dependent population breeding strategies and cause a shift in population structure. Migration barriers, created by an imposed exogenous factor, could alter the rate of gene flow among subpopulations, increasing the effective population density. This change could then alter the population structure and increase the level of inbreeding.

Exogenous factors of interest might be expected to affect genetic covariation in very specific ways. Identifying which models of pattern development are most likely to be acting in your system will lead to the most precise question definition and thus the most specific test of predicted effects.

TESTING THE QUESTION

Simple levels of genetic variability are not actual indicators of adaptation or fitness, but are merely assumed to be contributors. Measuring genetic variability in isolated populations is, therefore, inadequate as a specific test of genetic change. Rather, Milligan *et al.*, 1994 suggest constructing a "genealogical-based" analysis. A genealogical-based analysis is one that characterizes patterns of genotype frequency and interprets them relative to relatedness among individuals and populations. Molecular genetic loci should be selected to provide the most informative data for the question you are testing.

The term "genetic locus" refers, somewhat arbitrarily, to the portion of the molecule being analyzed by genetic techniques. Genetic loci include regions of DNA (or RNA) that are actively transcribed into functional proteins or that act as control regions for transcription of other segments of DNA. These are generally highly conserved, because small mutations would be expected to cause damaging effects on gene transcription. Other loci are thought to play little or no role in coding or controlling, including regions that are excised from coding sequences in the process of making proteins, and short repetitive sequences located throughout the genome. These loci are much more variable, because mutations are more likely to be neutral with respect to selection.

A genealogical-based analysis incorporates locus selection with known population parameters to distinguish natural genetic variation from genetic change, a process which has been called "retrospective population genetics" (Milligan *et al.*, 1994). The goal of retrospective population genetics in this context is to carefully evaluate tests of genetic change in light of all possible contributors.

We propose that selection of the appropriate genetic locus to test for effects of exogenous factors follow the scheme presented in Figure 1. The first major issue to consider is the availability of unaffected or reference (control) sites to compare to affected populations. If reference sites are not available, then it will be very difficult to distinguish between natural genetic variation and genetic change. Only a couple of approaches are possible, including genetic adaptation experiments and single-locus studies designed to directly test a model of genetic change based on a proposed mechanism. These are often not possible. An alternative approach is to survey for genetic patterns, possibly at two or

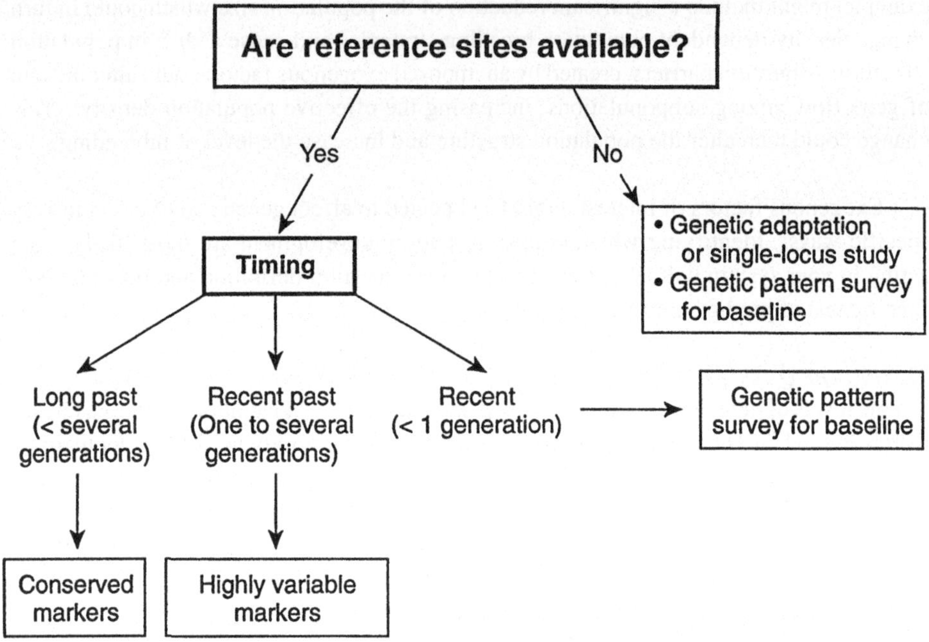

Fig. 1. Selecting genetic loci to test for the effects of exogenous impacts on populations.

more loci that differ in level of variation, to establish baseline patterns for comparison over time.

If reference sites are available, it is important to consider the length of time (relative to the generation time) since the exogenous impact began. If the impact occurred fewer than one generation ago, then the only way to estimate long-term impacts is to characterize genetic patterns as described above. If the impact occurred recently, within several generations, then highly variable loci, such as microsatellites or mtDNA, should be used, to ensure sufficient variation to be informative. If the impact occurred many generations ago, more conserved markers should be selected to reduce the possibility that loci will have accumulated new, obscuring mutations since the event.

Important steps of retrospective population genetics include the following:

1) Estimate the proportions of variance and covariance caused by different biological and environmental factors. For example, genetic patterns may be tested for covariation with geographic distance, local effective population size, and other population parameters expected to contribute.

2) Determine the possible relationship of loci or traits examined to fitness measures to evaluate selection. For example, genetic patterns may be tested for covariation with adaptive measures such as tolerance to chemicals or ability to thermoregulate in extreme temperatures.

3) Evaluate multiple loci or traits for correlation with each other. For example, patterns at highly variable loci may be compared to patterns at conserved loci in order to better estimate the timing of the observed genetic change.

4) Evaluate the probability of mutation, based on the taxon and genetic locus, and on the putative mutagen.

5) Incorporate as much historical evidence as is available into understanding background genetic structure.

6) Calculate the power of your test to discern among the causes based among other factors, on locus choice, actual locus variability, sample size, and population structure (power analysis).

(Brewer *et al.*, 1990; Hale and Singh, 1991; Frankham, 1995; Armbruster and Schwaegerle, 1996; Taylor and Dizon, 1996). Many genetic, mathematical, and statistical analytical techniques are available to perform each of these tests (see Hedrick *et al.*, 1976; D'Angelo *et al.*, 1994; and Rand, 1996 for examples).

In summary, we advocate meticulous question definition that results from consideration of the models of genetic change. Strategic experimental design should include consideration of retrospective population genetic steps in order to include the most thorough understanding of the effects of population structure on genetic patterns and to ultimately increase confidence that genetic change has occurred. These steps are necessary to avoid the first two of the common pitfalls described earlier.

2.3 TESTS OF DIRECT EFFECTS

Genetic change, if identified, should be linked to direct evidence for particular mechanisms. This step may be considered to be the difference between correlation and attribution. Hedrick *et al.* (1976) advocated that experimenters provide "supportive evidence" to distinguish among possible direct effects, and, in particular to demonstrate the possibility of selective pressures. Several examples from the literature demonstrate different approaches to accumulating direct evidence. For example, a genetic basis for thermal optima in fish (*Catostomus clarkii*) have been examined using a combination of genetic data and experimental testing. In these studies, the allele frequency of an enzyme varied clinally with climate (Hedrick *et al.*, 1976). The activity of each allele was subsequently found to be optimal at temperatures that correlated to climatic temperatures where it prevailed. In another example, various factors were identified as contributors to population genetic structure. Endler (1995) summarized several decades of studies on guppies (*Poecilia reticulata*) in northeastern South America. Genetic variation at allozyme loci followed a pattern consistent with gene flow (geographic), and relatedly, with stream geometry (higher stream order means more tributaries and therefore more gene flow), but not with predation intensity, which was shown experimentally to vary among locations.

Many different potential environmental and ecological contributors to genetic patterns should be considered when designing tests of direct effects. Some important factors include gradients of contaminant exposure, environmental gradients such as geomorphology or salinity, or significant barriers to gene flow. Such broad considerations are in line with Endler's (1995) emphasis on the use of data on a geographic scale large enough to distinguish gene flow effects from more localized selection effects. As factors are identified, they should be controlled for by ensuring they are represented equally in experimental and reference populations. Alternatively, hypothesized mechanisms by which these factors might effect change should be tested experimentally or by direct indicators, such as the enzyme activity found at different temperatures as described above. The extent to which these indicators show the same patterns differentiating experimental from reference populations represents the link between the genetic patterns and their proposed cause.

A variety of specific tests of direct effects might be designed for different categories of exogenous factors. Often, such collateral experiments or measures are standard for researchers involved in laboratory studies. To find appropriate direct tests, we emphasize that communication and collaboration across disciplines is vital.

Thorough cross-examination that compares genetic patterns to models of genetic change and emphasizes the linkage with tests of direct effects will often lead to comprehensive understanding of the factors involved in genetic change. In the case of industrial melanism in the moth, *Biston betularia*, researchers showed experimentally that light-colored moths were more readily preyed-upon after trees turned dark. But, Hedrick *et al.* (1976) point out that even in relatively "obvious" cases of selection, strong evidence such as this does not always explain all the patterns observed. Mathematical predictions of spatial distribution of the trait based on selection alone did not correlate to its actual distribution. This discrepancy indicated that selection was not the only factor causing the gene frequencies observed. However, studies that make this mechanistic link are significantly more convincing than mere correlations, and go far in explaining genetic patterns observed.

3. The Genetic Ecotoxicology Example

3.1 OVERVIEW OF GENETIC ECOTOXICOLOGY

A recent conference on Genetic and Molecular Ecotoxicology defined the overall goal of the field as: "To assess, predict, and prevent significant radiation- or chemical-induced genetic and epigenetic damage in populations" (Anderson *et al.*, 1994). We maintain here that in order to measure, much less characterize, genetic damage in this context, two main processes must be elucidated. As indicated in earlier sections, these are the interaction of population structure with environmental factors to create background genetic patterns and rates of change, and the mechanisms by which contaminants may cause direct effects that lead to alterations of genetic patterns. Genetic patterns must therefore be understood before any changes in genetic patterns may be attributed to contaminants.

Contaminant exposure is proposed to effect genetic change by three main processes: mutation, genetic drift, and genetic adaptation. Direct effects include highly deleterious mutations, and very slightly deleterious mutations which are predicted to accumulate and result in increased mutational loads (Kondrashov, 1995; Lynch *et al.*, 1995). Indirect effects include genetic drift and genetic adaptation. Genetic drift may be defined as "stochastic variation in allele frequencies" that can differentiate rapidly as a result of population subdivision (Selander and Whittam, 1983). This may occur as a result of contaminant-induced barriers to gene flow or severe fluctuations in effective population size. Genetic adaptation may be defined as a shift in allele frequency due to selective forces, and could result from a response to environmental change (Selander and Whittam, 1983). In response to contaminants, sensitive individuals might experience a variety of life-history costs. Examples include reproductive impairment, slow growth, reduced longevity, low survival of young, and increased disease susceptibility (e.g., Kurelec, 1993) due to effects of mutational load (Lynch *et al.*, 1995), induction of detoxication processes (e.g., Sanders, 1990), or immune suppression (Weeks, 1992).

Each mechanism for genetic change could result in alterations in gene frequency or in genetic variation. Mutational load is predicted to increase overall variation at first, leading to a slow selection process against mutation-susceptible individuals ultimately resulting in reduced population genetic variation. Genetic drift could result in reduced genetic variation and in gene frequency changes due to founder effects. Genetic adaptation may result in gene frequency alterations due to founder effects, selection for resistant genotypes or against susceptible genotypes, or selection for specific resistant genes. It may also result in reduced overall genetic variation if selection for certain genotypes results in reduced heterozygosity overall due to linkage.

Several examples have shown some correlation between contaminant exposure and gene frequencies (e.g., Kopp *et al.*, 1994) or levels of genetic variability (e.g., Murdoch and Hebert, 1994) that appear to support predictions of genotoxic effect. Studies vary widely, however, in their efforts to critically examine whether these correlations are actually sufficiently robust to attribute observed genetic changes to contaminant effects. Some studies fail to report basic chemical site characterization as evidence of contaminant exposure. Several investigations, however, do provide various kinds of corroborating evidence for attribution of changes to contaminants, for example, indices of community effects (e.g., Fore *et al.*, 1995). In another example, multiple molecular genetic loci were compared as evidence that selection must explain patterns observed at a particular locus (e.g., Wirgin *et al.*, 1990). In a final example, population structure, in particular, gene flow and drift, was characterized in order to test for any effects of environmental factors by first factoring out effects of population structure (e.g., Woodward *et al.*, 1996). Each of these efforts is an appropriate step to understanding long-term effects. However, we believe the best approach to such understanding is a more systematic and rigorous one that combines more traditional toxicology with these approaches.

3.2 ESTABLISHING LINKS WITH GENETIC CHANGE

Linking Changes in Gene Frequency with Genetic Adaptation

The most direct means of establishing that genetic change may occur as a result of contaminant exposure is through measuring genetic adaptation. In laboratory experiments or field collections, organisms (especially plants and invertebrates with short generation times) may be exposed to contaminants at various doses. Tolerance levels to the contaminant are determined in progeny (Andreasen, 1985; Klerks and Levinton, 1989). Over one to many subsequent generations, tolerance to contaminant exposure may increase, and such tolerance is termed resistance. By establishing the evolution of resistance in the organisms, one may conclude that genetic change may be possible in the population by this mechanism.

The primary limitation to this approach is that few species may be bred in the laboratory and exposed in tolerance assays. In addition, genetic adaptation *per se* may or may not be considered significant genetic damage if it results in greater tolerance to environmental conditions, including contaminants, and does not result in any deleterious effects. Links to deleterious effects may be found in the agricultural literature which has outlined approaches for measuring life history and other potential long-term costs associated with genetic adaptation (e.g., Baker, 1982; Carriere and Roff, 1995). Furthermore, the population and conservation genetics literature has proposed models of potential deleterious effects of an overall reduction of genetic variation that could result from genetic adaptation (e.g., Turelli and Ginzburg, 1983). Debate has centered on mechanisms such as inbreeding depression, originally emphasized in response to long-term understanding of its effects by animal breeders (e.g., Soule, 1986), and the loss of adaptive potential associated with loss of heterozygosity (e.g., Turelli and Ginzburg, 1983; Guttman and Dykhuizen, 1994). Combining genetic adaptation studies with an assessment of gene frequency changes will permit insight into the potential for significant genetic damage. This conclusion may only be confidently drawn relative to background levels of genetic variation for the population, rates of gene flow, and rates of natural accumulation of variation predicted from population parameters.

Linking Changes in Gene Frequency with Single-locus Genetic Markers

Some genes are potentially directly involved with contaminant detoxication and response. Many of these genes have been characterized, especially in agricultural species in which resistance is well-studied. One way to establish that population genetic changes may occur due to selection at specific gene loci is to study variation at those loci. Patterns of variation in unexposed populations are then compared to gene frequencies in exposed (i.e., post-selection) populations. Extreme shifts in the frequency of these critical genes may provide evidence that selection or mutation is occurring at this locus. Concomitant shifts in overall genetic variation, or in gene frequencies at other loci may indicate founder-effects, or the hitch-hiking effect (e.g., Baker, 1982) and support the hypothesis that selection on single loci may cause significant genetic change in populations.

Several studies have pursued this means of linking contaminant exposure to genetic change. Dramatic shifts in the frequency of the c-abl oncogene (Wirgin *et al.*, 1990) and the cytochrome P4501A detoxication gene (Roy *et al.*, 1995) in Hudson River tomcod (*Microgadus tomcod*) have been demonstrated when they were compared to tomcod from cleaner rivers in the northeast. Other workers have demonstrated multiple alleles in the p53 tumor suppressor gene in fish (Kusser, 1995) and in the metabolic enzyme family, the esterases (e.g., Mouches *et al.*, 1986). Agricultural research has provided evidence linking laboratory resistance experiments with shifts in multilocus gene frequency in crop plants. Genetic markers for resistance identified in the multilocus tests are being traced to specific genes for resistance (e.g., Michelmore, 1995; Anderson *et al.*, 1996). This kind of evidence, if properly coupled with power analysis, good reference comparison, and laboratory tolerance tests, where possible, is an extremely robust connection between contaminant exposure and population genetic change.

LINKING CHANGES IN GENE FREQUENCY WITH DIRECT AND SHORT-TERM INDICATORS

Biomarkers are biochemical or physiological measures of biological response (Huggett *et al.*, 1992) and may be used to ascertain that exposure to a given contaminant has occurred and to assess absorbed dose. If properly applied, such markers can be used to discriminate contaminant-exposed populations from relatively unexposed counterparts and to assess physiological damages. These responses, therefore, can be used to link changes in gene frequency with contaminant exposure. In situations where multiple contaminants are present, specific biomarkers help to discriminate among contaminants that may be having an effect. For example, serum levels of acetylcholinesterase (AChE) may indicate exposure to organophosphates or carbamates, AChE inhibitors (e.g., Dell'Omo *et al.*, 1997); thymine dimers indicate exposure to ultraviolet light (e.g., Applegate and Ley, 1988), and DNA adducts have been shown to form from exposure to polycyclic aromatic hydrocarbons (PAHs) (e.g., Stein, 1994; Sadinski *et al.*, 1995).

Genotoxic chemicals, of primary interest to genetic ecotoxicologists, cause damage directly to genetic material, and responses quantifying such damage provide additional evidence that can be used to determine whether changes in gene frequency are linked to exposure to a mutagenic chemical. Measures of the effects of such chemicals include DNA adducts, DNA strand breaks, micronuclei, and mitotic aberrations. Genotoxic damage, if not repaired, may result in genetic change in the form of mutations. In addition, many of these responses have been shown to have long-term effects such as reduced reproductive success and reduced longevity due to factors such as gamete loss, developmental delay, and tumorigenesis (Anderson and Harrison, 1990; Hemminki, 1993; Anderson and Wild, 1994). DNA adducts are now also recognized as valuable measures of absorbed dose and of chronic exposure (Stein, 1994). Careful selection of biomarkers is an important step in establishing that the contaminant in question produces damaging effects in organisms. Although a large number of techniques are currently available, a variety of logistical parameters such as the size and accessibility of organisms dictate which measures might be useful in a given study.

Despite their great value, the short-term measures commonly measured by genetic ecotoxicologists are not sufficient indicators of long-term effects, for several reasons. DNA repair and cellular regeneration mechanisms, for example, may reverse the effects of contaminants (Anderson and Harrison, 1990; Shugart *et al.*, 1992). Even acute doses causing some lethality in a population may not increase the population mortality rate significantly to cause long-term damage from population fluctuation or from selective pressure.

4. Conclusions

Using retrospective population genetics, combined with careful experimental design, we believe that genetic patterns resulting from stochastic processes, the influence of geographic and geophysical characteristics, and natural genetic variation, may be characterized during genetic analysis and distinguished from possible effects of contaminants or other exogenous factors. This has the potential for a determination of the relative contribution of the selective or mutation effects of anthropogenic environmental impacts on a population.

Figure 2 represents the steps proposed, schematically. The scheme presented here is formulated for genetic ecotoxicological studies, but applies to more general cases. The principal substitution into the model, for examination of other stressors such as species introductions or global warming, is at the level of linking genetic differences to mechanisms of change. In this scheme, genetic patterns, which may be gene frequencies or levels of genetic variability, are derived from genetic analysis of appropriate loci. These patterns are compared to each factor identified that might effect genetic change, following the steps described in section 2.2. If genetic patterns are found to covary with any feature of both experimental and reference locations, this covariation may be factored out of the genetic pattern. If these effects are factored out, then genetic patterns of exposed populations may be compared to those of unexposed populations to test for differences. If genetic patterns do not covary with any feature considered, then population genetic indices may be compared directly. Simultaneously, or following genetic comparison, supportive evidence of the presence and effects of the contaminants being tested is gathered. Thorough experimental design consideration should result in informative genetic locus(i) selection, chemical site characterization, selection of appropriate experimental and reference locations, and good indicator species. Biomarker selection could discern among contaminants, assess absorbed dose, or estimate direct damage by contaminants. Where possible, genetic adaptation experiments on progeny of field-collected organisms, or examination of relevant single-locus genetic change would provide robust support for mechanisms of genetic change.

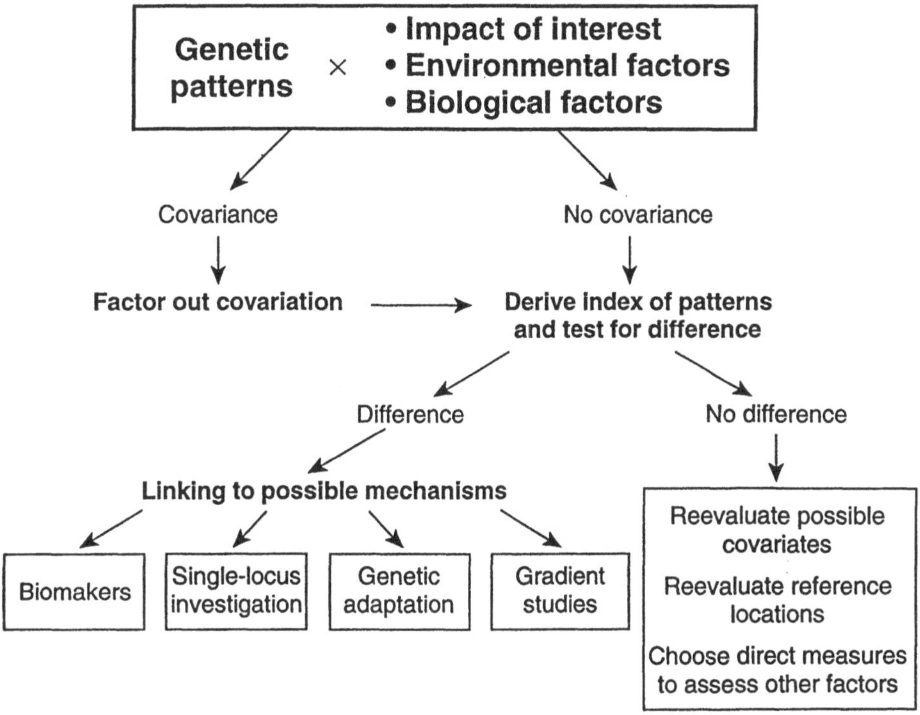

Fig. 2. Testing for genetic changes due to contamination through considering covariation, and testing for mechanisms and direct effects.

Direct and short-term measures in conjunction with genetic studies may lend supportive evidence that genetic patterns are not solely attributable to population processes or natural ariation. If significant genetic differences are found between experimental and reference populations, and supportive evidence indicates that there is a direct effect of the contaminant on genetic processes or biochemical responses, then a link has been made between the change in genetic pattern and the purported source.

We have presented background on genetic variation, and the basic steps required to evaluate population genetic data. We contend that monitoring and assessing environmental impacts on populations (and by extension, communities and ecosystems) must begin with understanding the basic principles of the population dynamics and inherent variation of the system(s) under study. We propose, therefore, that long-term monitoring of the impacts of environmental change may benefit greatly from studies of genetic patterns, but not without an understanding of genetic processes and background genetic variation, combined with a careful selection of short-term or direct indicators and careful experimental design.

Acknowledgements

This work was supported by the Pew Scholars in Conservation and the Environment award to S. Anderson, through the U.S. Department of Energy under contract DE-AC03-76SF00098.

References

Anderson, S.L., Harrison, F.L.: 1990, *In:* S. S. S. *et al.* (ed.) In Situ *Evaluations of Biological Hazards of Environmental Pollutants. Plenum Press*, New York, 81-93.

Anderson, I.A., Okubura, P.A., Arroyo-Garcia, R., Myers, B.C., Michelmore, R.W.: 1996, Molecular and General Genetics **251,** 316-325.

Anderson, S., Sadinski, W., Shugart, L., Brussard, P., Depledge, M., Ford, T., Hose, J., Stegeman, J., Suk, W., Wirgin, I., Wogan, G.: 1994, Environmental Health Perspectives **102,** 3-8.

Anderson, S., Wild, G.: 1994, Environmental Health Perspectives **102,** 9-12.

Anderson, S.L., Belfiore, N.M.: submitted, Mutation Research.

Andreasen, J.K.: 1985, Archives of Environmental Contamination and Toxicology **14,** 573-577.

Applegate, L.A. and Ley, R.D.: 1988, Mutation Research **198**, 85-92.

Armbruster, W.S., K.E. Schwaegerle: 1996, Journal of Evol. Biol. **9,** 261-276.

Baker, J.: 1982, Agriculture and Environment **7,** 187-198.

Bickham, J.W., Smolen, M.J.: 1994, Environmental Health Perspectives **102,** 25-28.

Brewer, B.A., Lacy, R.C., Foster, M.L., Alaks, G.: 1990, Journal of Heredity **81,** 257-266.

Carriere, Y., Roff, D.A.: 1995, Heredity **75,** 618-629.

D'Angelo, D.J., Howard, L.M., Meyer, J.L., Gregory, S.V., Ashkenas, L.R.: 1995, Canadian Journal of Fisheries and Aquatic Science **52,** 1893-1908.

Dell'Omo, G., Bryenton, R., Shore, R.F.: 1997, Environmental Toxicology and Chemistry **16,** 272-276.

Endler, J.A.: 1995, Trends in Evolution and Ecology **10,** 22-29.

Fore, S.A., Guttman, S.I., Bailer, A.J., Altfater, D.J., Counts, B.V.: 1995, Ecotoxicology and Environmental Safety **30,** 24-35.

Frankham, R.: 1995, Conservation Biology **9,** 792-799.

Garcia de Leon, F.J., Chikhi, L., Bonhomme, F.: 1997, Molecular Ecology **6,** 51-62.

Guttman, D.S., Dykhuizen, D.E.: 1994, Genetics **138,** 993-1003.

Hale, L.R., Singh, R.S.: 1991, Genetics **129**, 103-117.

Hartl, D.L., 1988 *A Primer of Population Genetics*, 2 ed. Sinauer Associates, Inc., Sunderland, 305 pp.

Hedrick, P.W., Ginevan, M.E., Ewing, E.P.: 1976, Annual Review of Ecology and Systematics **7,** 1-32.

Hemminki, K.: 1993, Carcinogenesis **14,** 2007-2012.

Huggett, R.J., Kimerle, R.A., Mehrie, P.M., *et al.*: 1992, *In:* R. J. Huggett, R. A. Kimerle, P. M. J. Mehrie and H. L. Bergman (eds.) *Biomarkers Biochemical, Physiological, and Histological Markers of Anthropogenic Stress*. Lewis Publishers, Ann Arbor, Michigan, 1-3.

Klerks, P.L., Levinton, J.S.: 1989, Biol. Bull **176,** 135-141.

Kondrashov, A.S.: 1995, Journal of Theoretical Biology **175,** 583-594.

Kopp, R.L., Guttman, S.I., Wissing, T.I.: 1994, Environmental Toxicology and Chemistry **11,** 665-676.

Kurelec, B.: 1993, Marine Environmental Research **35,** 141-348.

Kusser, W.C., Brand, D., Cretney, W., Glickman, B.W.: 1995, Society of Environmental Toxicology and Chemistry, Vancouver, British Colombia.

Lande, R., Barrowclough, G.F.: 1987, *In:* M. E. Soule (ed.) *Viable Populations for Conservation*. Cambridge University Press, Cambridge, 69-86.

Lynch, M., Conery, J., Burger, R.: 1995, The American Naturalist **146,** 489-518.

Michelmore, R.W.: 1995, Current Opinion in Biotechnology **6,** 145-152.

Milligan, B.G., Leebens-Mack, J., Strand, A.E.: 1994, Molecular Ecology **3,** 423-435.

Mouches, C., Pasteur, N., Berge, J.B., *et al.*: 1986, Science **233,** 778-780.

Mundy, N.I., Winchell, C.S., and Woodruff, D.S.: 1997, Molecular Ecology **6**, 29-37.

Murdoch, M.H., Hebert, V.: 1994, Environmental Toxicology and Chemistry **13,** 1281-1289.

Palumbi, S.R.: 1996, *In:* J. D. Ferraris and S. R. Palumbi (eds.) *Molecular Zoology: Advances, Strategies, and Protocols*. Wiley-Liss, New York, 101-117.

Piertney, S.B., Carvalho, G.R.: 1995, Journal of Experimental Marine Biology and Ecology **188,** 277-288.

Pogson, G.H., Mesa, K.A., Boutilier, R.G.: 1995, Genetics **139,** 375-385.

Rand, D.M.: 1996, Conservation Biology **10,** 665-671.

Roy, N.K., Kreamer, G.L., Konkle, B., Grunwald, C., Wirgin, I.: 1995, Archives of Biochemistry and Biophysics **322,** 204-213.

Sadinski, W.J., Levay, G., Wilson, M.C., Hoffman, J.R., Bodell, W.J., Anderson, S.L.: 1995, Aquatic Toxicology **32,** 333-352.

Sanders, B.: 1990, *In:* J. F. McCarthy and L. R. Shugart (eds.) *Biomarkers of Environmental Contamination*. Lewis Publishers, Boca Raton, 165-192.

Selander, R.K. and Whittam, T.S.: 1983, In: M. Nei and R.K. Koehn (eds.) *Evolution of Genes and Proteins*. Sinauer Associates, Inc., Sunderland, 89-114.

Shugart, L., Bickham, J., Jackim, G., McMahon, G., Ridley, W., Stein, J., Steinert, S.: 1992, *In:* R. J. Huggett, R. A. Kimerie, P. M. J. Mehrie and H. L. Bergman (eds.) *Biomarkers: Biochemical, Physiological, and Histological Markers of Anthropogenic Stress*. Lewis Publishers, Ann Arbor, Michigan, 125-153.

Shugart, L.R., Theodorakis, C.: 1996, Comparative Biochemistry and Physiology **113C,** 273-276.

Stein, J.E., Reichert, W.L., Varanasi, : 1994, Environmental Health Perspectives **102,** 19-23.

Smith, M.W., Smith, M.H., Chesser, R.K.: 1983, Copeia **1983,** 182-193.

Soule, M.E.: 1986, *In:* M. E. Soule (ed.) *Conservation Biology: The Science of Scarcity and Diversity*. Sinauer Associates, Inc., Sunderland, 13-18.

Taylor, B.L., Dizon, A.E.: 1996, Conservation Biology **10,** 661-664.

Theodorakis, C.W., Blaylock, B.G., Shugart, L.R.: 1997, Ecotoxicology **5,** 1-15.

Theodorakis, C.W., Shugart, L.R.: 1997, Ecotoxicology **in press,**

Turelli, M., Ginzburg, L.R.: 1983, Genetics **104,** 191-209.

Weeks, B.A., Anderson, D.P., Dufour, A.P., Fairbrother, A., Goven, A.J., Lahvis, G.P., Peters, G.: 1992, *In:* R. J. Huggett, R. A. Kimerle, P. M. Mehrle and H. L. Bergman (eds.) *Biomarkers: Biochemical, Physiological, and Histological Markers of Anthropogenic Stress*. Lewis Publishers, Chelsea, 211-234.

Wirgin, I.I., D'Amore, M., Grunwald, C., Goldman, A., Garte, S.J.: 1990, Biochemical Genetics **28,**

Woodward, L.A., Mulvey, M., Newman, M.C.: 1996, Environmental Toxicology and Chemistry **15,** 1309-1316.

PRELIMINARY STUDIES ON THE POPULATION GENETICS OF THE CENTRAL STONEROLLER (*Campostoma anomalum*) FROM THE GREAT MIAMI RIVER BASIN, OHIO

R. N. SILBIGER[1], S. A. CHRIST[2], A. C. LEONARD[3], M. GARG[3], D. L. LATTIER[1], S. DAWES[1], P. DIMSOSKI[3], F. McCORMICK[4], T. WESSENDARP[2], D. A. GORDON[2], A.C. ROTH[2], M. K. SMITH[2] and G. P. TOTH[2]

[1]Pathology Associates International, [2]Molecular Ecology Research Branch, [3]Oak Ridge Institute for Science and Education, [4]Ecosystems Research Branch, Ecological Exposure Research Division, National Exposure Research Laboratory, Office of Research and Development, United States Environmental Protection Agency, 26 W. Martin Luther King Dr., Cincinnati, Ohio 45268

Abstract. Molecular approaches are particularly useful for measuring genetic diversity and were applied to samples of central stonerollers obtained from sites along tributaries to the Great Miami River in Ohio. We used Random Amplified Polymorphic DNA (RAPD) analysis to assess the level of genetic diversity within and among these populations. RAPD analysis generates genetic profiles that were used to develop indices of genetic similarity. The RAPD method provides a cost effective means of generating an arbitrary sample of anonymous loci across the genome and generate a virtually unlimited set of loci for use in genetic analysis in the absence of specific sequence information. These attributes make RAPDs well suited for use in evaluating the diversity and assessing the potential vulnerability to exposure of populations across multiple spatial scales. The results demonstrate that a significant amount of structuring exists among populations analyzed to date and that a trend exists towards genetic diversity being an inverse function of site distance from the main stem as well as a being directly related to stream order. This indicates that populations farthest from main conduits or in lower order streams, and thereby most isolated, may be the most vulnerable populations to stressor exposure. It is hoped that information pertaining to genetic diversity, when integrated with other metrics of resource condition, will aid in making scientifically grounded decisions on resource management that enhance the probability of population survival and preserve natural evolutionary processes.

1. Introduction

1.1. IMPORTANCE OF GENETIC DIVERSITY

Changes in the hereditary makeup of a species provide the natural substrate which leads to evolutionary change (Dobzhansky, 1950). Resource managers are increasingly concerned about the maintenance of genetic diversity in populations of wild species exposed to stressors. It is feared that these populations may become more genetically monomorphic. Extensive genetic monomorphism is considered to be undesirable, being equated with decreased fitness and increased vulnerability to additional stressors, with loss of evolutionary potential foreshadowing population extinction (Soulé, 1986). For example, native Pacific northwest salmon and steelhead are subdivided into local populations with genetic differences which are adaptive (Nehlsen *et al.*,1991; Maclean and Evans, 1981). Human activities have led to the extinction of some and declines in many of these populations which in turn have resulted in a reduction of the genetic resources available (Nehlsen *et al*, 1991). The response to stress by a species depends on the genetic variability possessed by the species (Thorpe, 1981). It has been recognized that decreases

Environmental Monitoring and Assessment **51:** 481–495, 1998.

in complexity, as evidenced by negative changes respectively in genetic composition and genetic diversity of salmonid populations, make these species more vulnerable to the effects of environmental perturbations and to eventual collapse (Nehlsen *et al.,* 1991; Caswell, 1976).

Evaluation of the levels of intrapopulation genetic diversity is especially important for those species exposed to environmental stressors (e.g. chemical, biological, physical). A stressor can cause mortality in the population forcing it through a "bottleneck" and causing a decrease in the population's variability and fitness. Stressors can act indirectly by decreasing reproductive success and by acting as barriers to gene flow. The stressor can act on the population through mortality of, or decreased reproductive success for, some phenotypes, leading to inbreeding depression.

As a population loses heterogeneity, a depression in fitness traits, such as growth, survival, age to sexual maturity, resistance to disease, and fecundity, may occur making the population more vulnerable to stressor exposure (Allendorf and Leary, 1986; Quatro and Vrijenhoek, 1989). Leberg (1990), demonstrated that a 25% loss of heterozygosity due to inbreeding led to a 56% reduction in brood population size for the eastern mosquitofish. In gentically identical populations of California pocket gophers (as determined by DNA fingerprint analysis [Sanjayan and Crooks, 1996; Soulé and Zegers, 1996; Patton and Smith, 1990]), Sanjayan and Crooks (1996) showed that skin tissue grafts among individuals from within naturally inbred populations were not rejected. However, skin tissue grafts were rejected by members of inbred populations when provided from donors outside of these populations, indicating that these individuals remained immunocompetent. Individuals from genetically diverse populations rejected both inter- and intra-population skin tissue grafts. This extreme loss of heterozygosity may lead to population wide anergy and a consequent increased susceptibility to disease. Both the above studies point to the interplay between diversity within a population and a population's fitness. We have developed diagnostic indicators of genetic diversity which can be used in making critical decisions concerning landscape use and management (Angermeier and Karr, 1994; Soulé and Wilcox, 1980). We have used RAPD analysis to assess the levels of genetic diversity for wild populations of the central stoneroller (*C. anomalum*) found in the Eastern Cornbelt Plains (ECBP).

1.2. DNA PROFILING ANALYSIS

We have applied the tools of molecular biology towards the assessment of genetic diversity in wild populations. The ability to estimate the level and distribution of genetic variation in wild species is an integral part of determining species' vulnerability to exposure. Genetic data can provide an understanding of the pattern of relationships among populations of a species and their degree of isolation. This information, when integrated with population size and reproductive dynamics, can be used to define significant management units and formulate best management practices that would preserve genetic integrity and ensure sustainability. Towards this goal we have developed random amplified polymorphic DNA [RAPD] (Welsh and McClelland, 1990; Williams *et al.*, 1990) analysis as an indicator for species genetic diversity. The process is amenable to robust assessment of the genetic diversity for populations across multiple spatial scales of resolution (demes to metapopulations) in order to measure emergent properties for these populations. DNA profile analysis using the RAPD method produces highly variable and genetically distinct markers at loci, which are for the most part thought to be neutral. This provides the opportunity to sample across the genome revealing the genetic variation that exists in wild populations, for which specific gene sequence information is lacking. An implicit assumption is that the polymorphism in the DNA architectures assayable by profiling techniques represents an acceptable surrogate of overall genomic variation. These types of analyses have been applied to studies of the genetic structure of wild populations, and are reviewed in (Smith and Wayne, 1996; Caetano-Anolles *et al.*, 1991).

The RAPD method of DNA profiling can detect differences among individuals and populations. Variation in the form of polymorphism are typically neither homogeneous nor fixed, but rather exist as differences in the frequency of genotypes. DNA profiling allows for the quantification of levels of genetic variability within and between populations. The diversity indices of populations over a region would comprise a distribution curve with sites at the extremes (especially the low diversity tail of the curve) being theorized to be vulnerable to exposure. These types of measurements can identify trends across populations on a regional basis. This provides a measure to assess a specific population's vulnerability to the impact of stressors. Similar factors (i.e. exposure to environmental stressors) should push the genetic structures of independently evolved species in a concordant fashion (Avise *et al.*, 1987).

1.3. STUDY REGION

The United States Environmental Protection Agency (USEPA) has designed programs to address large scale, long term environmental problems which occur at regional and national levels by adopting comprehensive, multimedia perspectives of the environment to answer questions about overall ecological condition. A study, which stretches across the states of Michigan, Ohio, and Indiana, within the Eastern Cornbelt Plains (Omernik, 1995; Figure 1, inset), was initiated as a USEPA Region 5, Regional Environmental Monitoring and Assessment Program (R-EMAP) project to assess stream resource status (Ohio Environmental Protection Agency, 1987), and to develop indicators for the ECBP because this ecoregion has complex sources of agricultural and industrial stressors which are similar to those found in many areas of the United States. The ECBP was once composed of tall grass prairie, eastern deciduous forest, and wetlands that have since been converted to agriculture. Industrialized cities developed along major rivers. The stressors associated with development of the region include point-source (e.g. industrial discharge, waste water treatment plants, combined sewer overflows) and non-point source (e.g. nutrient enrichment, toxics, sedimentation) stressors. These alterations to the landscape alter ecosystem services which impinge on the sustainablity of valued resources.

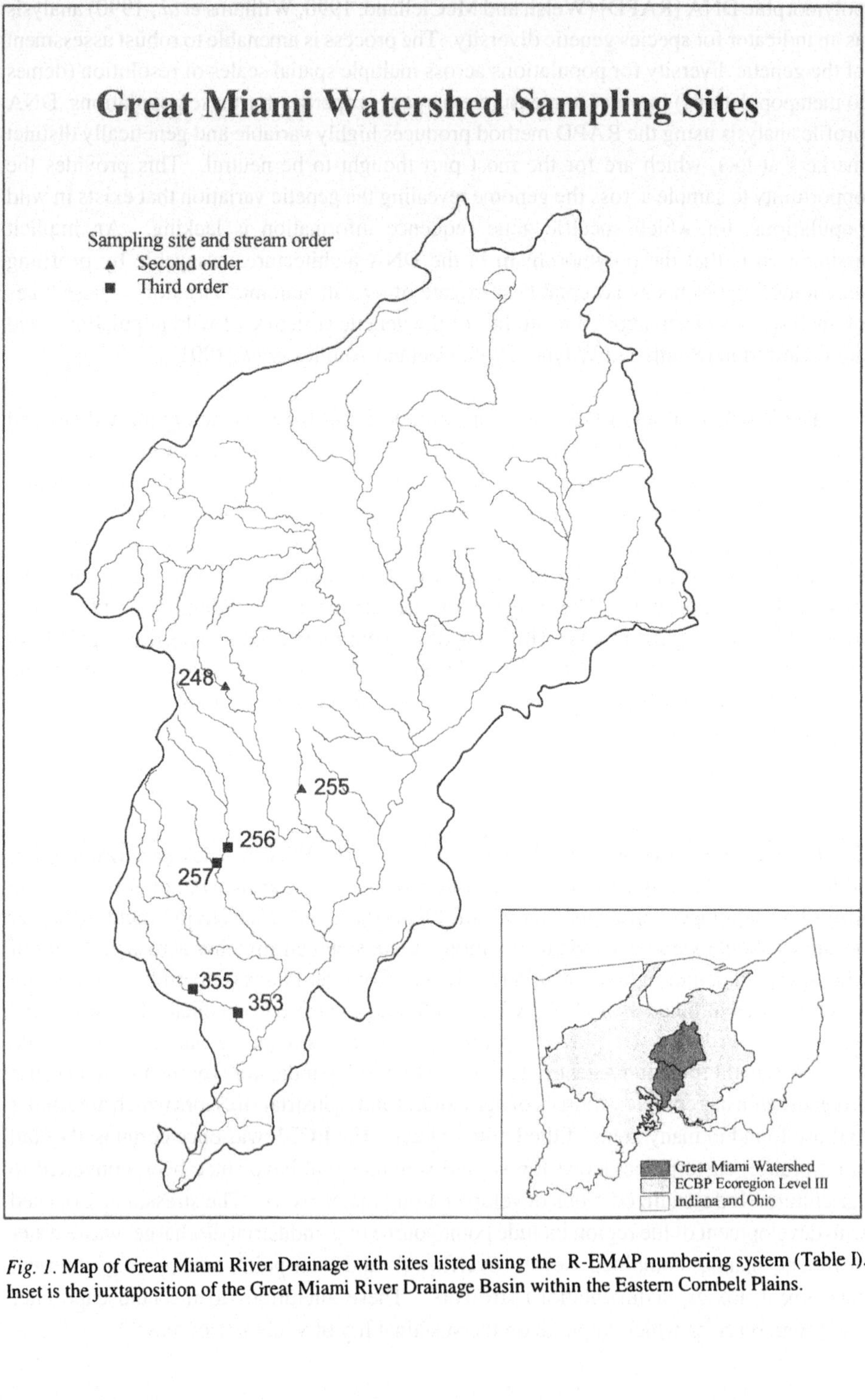

Fig. 1. Map of Great Miami River Drainage with sites listed using the R-EMAP numbering system (Table I). Inset is the juxtaposition of the Great Miami River Drainage Basin within the Eastern Cornbelt Plains.

This program combined a statistical sampling protocol with indicators for biological integrity. In so doing, species threatened with impact by multiple stressors can be assessed as to their vulnerability to stressor exposure. Indicators of exposure are measurements that can be used to quantitatively estimate the condition of ecological resources, the magnitude of stress, the exposure to a stressor(s), or the amount of change in condition. Indicators can provide correlations that point to patterns and trends in biological integrity that can be used to further investigate causation.

1.4. STUDY GOALS

Populations are generally more responsive to stressors than are ecosystems and, therefore, can act as indicators of environmental change (Root and Schneider, 1995). These indicators can be followed through time in monitoring programs and help construct models for the linkages between populations and ecosystems. These changes can occur, therefore, in populations at stressor levels which have little effect on the functioning of the ecosystem, and at a point when the impact by stressor(s) is reversible (Stone, 1995).

We have used the RAPD method to measure intra- and inter-populational genetic diversity for populations of central stoneroller in order to develop indicators of exposure to environmental stressor(s). We expect that a representative sample of individuals from stressor impacted sites would display a level of diversity significantly different from the distribution of diversity levels observed for other populations in the region. If the diversity found in impacted populations is lower than the observed distribution, this would indicate that the stressor exerts a selective pressure or has led to population declines. If the diversity is found to be higher, then increased mutational or migration rates, as well as patchiness in response to stressor impact, could be at play. If no significant differences are found in the levels of diversity between impacted and the level of diversity from the distribution of diversity levels observed for other populations in the region, this would indicate that the stressor either had multiple effects on diversity which balanced out, had no effect on genetic diversity in the sampled species, or that the diversity within the species has experienced some event(s) that reduced variability below a threshold where further impacts from stressors can be detected.

2. Materials and Methods

2.1 STUDY AREA

In 1995, a study of the fish communities of the ECBP was initiated by the USEPA Region 5, state agencies of Ohio, Indiana and Michigan and the USEPA National Exposure Laboratory in Cincinnati. Four hundred sites were selected based on EMAP's probability design (United States Environmental Protection Agency, 1993). Of these 232 sites were sampled. Fish were collected at 59 randomly selected sites in the Great Miami River Basin using the EMAP protocols for wadeable streams (USEPA, 1993). We obtained stonerollers from 20 sites. Six of these sites have been analyzed to date (Table I).

Table I
Study Sites and Coordinates

Stream Name	R-EMAP #	River Mile	Latitude (N)	Longitude (W)
Twin Creek [TC]	248	43.09	39.9048	84.594
Tom's Run [TR]	255	6.89	39.7246	84.455
Sevenmile Creek [7M]	256	17.55	39.6438	84.631
Paint Creek [PC]	257	0.73	39.6206	84.631
Indian Creek [IC(l)]	353	6.01	39.3781	84.651
Indian Creek [IC(u)]	355	13.60	39.4252	84.742

2.2 SAMPLE PREPARATION AND RAPD PROCEDURE

Fish were maintained in a live well until identification, after which up to ten individuals were selected per site and a postanal resection was performed. The caudal peduncle and caudal fin were placed into a labeled cryovial and immediately frozen in a liquid nitrogen dry shipper. Resection was performed using utility razor blades. Razor blades were exchanged and the dissection boards rinsed with water and 70% ethanol between individuals. Upon arrival from the field, samples were transferred to -80°C freezers.

For DNA purification, a portion of the frozen tail section was removed to a 1.5 ml microfuge tube containing 100µl PBSET (1.5 mM KH_2PO_4: 2.7 mM Na_2HPO_4 [pH 7.2], 154 mM NaCl, 100 mM EDTA, 0.1% Triton -X 100) and homogenized. To the homogenate an additional 400µl of PBSET was added as well as 27µl Proteinase K (20mg ml^{-1}) and 30µl of 20% SDS. This was mixed by gentle inversion and then incubated overnight at 65°C. Following digestion, the preparation was centrifuged at 12,000 rpm for 15 minutes, to remove particulate and undigested material. The supernatant was removed to a clean microfuge tube and 200µl of 10M ammonium acetate added and mixed by gentle inversion. The DNA was precipitated by the addition of an equal volume of 100% ice cold isopropanol, mixed by gentle inversion and placed at -20°C for at least 20 minutes. The precipitated DNA was pelleted for 15 minutes at 15,000 rpm in a refrigerated centrifuge set to 0°C. After pelleting, the supernatant was removed and the DNA washed with ice cold 70% ethanol. The DNA was again pelleted, the supernatant removed and the pellet washed with 95% ethanol. The DNA was pelleted, the supernatant removed and the DNA dried in a 65°C dry block. After drying the pellet was resuspended in 200µl of dilution buffer (10mM Tris-Cl [pH 8.5, 25°C], 50mM KCL).

RAPD analysis used was modified from that of Williams *et al.* (1990), Welsh and McClelland (1990) and Lattier *et al.* (1996). Sample was serially diluted in dilution buffer using large orifice pipette tips. A 5 minute, 65°C warming period was allowed prior to initial sample removal and between dilutions. A 20µl aliquot of the diluted sample was placed into a PCR tube and 5µl of master mix (10mM Tris-Cl [pH 8.5, 25°C], 50mM KCL,

15mM $MgCl_2$, 25 pmoles μl^{-1} RAPD primer, 1.25 Units of Native *Taq* polymerase, 25% acetamide, 1mM dNTP mix) was added (the same results were achieved when the reaction volume was cut in half). Amplification reactions carried out over a series of cycles demonstrated that increased cycle counts led to a change in primer template stoichiometry (plateau effects (Sardelli, 1993)). These stoichiometric changes led to the evolution of product hybrids occurring, due to lowered primer competition, allowing band homologies to cause templated product bands to fade with increased times and non-templated bands to evolve (data not shown). This prompted a change to lower cycle numbers. After the reaction constituents were added, the mix was kept on ice up to fifteen minutes prior to initiating thermal amplification (retention of reaction mixes beyond 15 minutes resulted in a degradation of sample quality possibly due to the 5'-3' exonuclease activity of the polymerase). The sample was amplified by 34 cycles of 45 sec. at 94°C, 1 min. at 41°C, 1 min. at 72°C.

After completion of the amplification steps a 15% Ficoll 400/ 0.25% bromophenol blue solution was added to the RAPD products, as well as to 500/100 bp ladder (GenSura, CA) along with a buffer solution that equalized salt concentrations across samples and molecular weight markers. Samples were arranged on the gel using a randomization protocol. Each sample lane was placed adjacent to a molecular weight marker lane. These were electrophoresed through 1.65% (w v^{-1}) agarose gels containing 1X TBE (45mM Tris-borate, 2mM EDTA) at 7°C, 6.8 V cm^{-1} for 4.5 hours with continuous buffer recirculation. Upon completion of electrophoresis the gels were stained in 1000ml of 1X TBE with 4μg ml^{-1} ethidium bromide for 30 minutes with constant agitation and then destained in water for 15 minutes. The gels were then visualized with a Molecular Dynamics Fluorimager.

Bands were declared using FragmeNT Analysis software (Molecular Dynamics) and then further edited visually to eliminate both false bands and those bands below a level of optical density that could be reliably scored. Sample band molecular weights were calculated by a point to point logarithmic interpolation method within FragmeNT using the standard lane nearest the unknown.

2.3 PRIMER SELECTION

Primers used in RAPD analysis were optimized for PCR temperature by thermal amplification of HeLa cell DNA at a concentration of 1 ng μl^{-1} in the reaction mix. Amplification reactions were carried out at five annealing temperatures (37, 38, 39, 40, and 41°C) and the temperature that gave the strongest and most reproducible bands was chosen as the optimal temperature. This procedure was carried out for the 40 primers of the Operon F and L kits.

Primers for this study were selected based solely upon their ability to elaborate bands for DNA of stoneroller populations from the Mill Creek in Butler and Hamilton Counties in Ohio and Tanner's Creek, Dearborn County Indiana. These populations lie outside the sampled area for the ECBP study and were selected to assure that no bias towards polymorphism was being introduced by primer selection using the study populations. This will allow for the use of these primers for studies of other stoneroller populations. Genomic DNA samples from within a sample site along the Mill and Tanner's Creeks were pooled and a bulk segregant analysis was run using RAPD primers. From the pool of 40 optimized primers, 21 primers met the annealing temperature criteria of 41 °C. Of these 21 primers, 10 primers were tested on stoneroller DNA. Eight of these primers elaborated bands with stoneroller DNA and four of this pool were selected at random for use in RAPD analysis of study populations (Table II).

Table II
RAPD Primers

Name	Sequences (5'-3')
OPF-04	GGTGATCAGG
OPL-02	TGGGCGTCAA
OPL-05	ACGCAGGCAC
OPL-07	AGGCGGGAAC

2.4 STATISTICAL ANALYSIS

For the RAPD datasets, tests of population subdivision and comparisons of within population diversity were based on the Similarity Index (Lynch, 1991; Lynch and Milligan, 1994; Lynch, 1990; Lynch, 1988) computed for each possible pairing of individuals, both within and across populations. The similarity index (S) between two individuals is twice the number of matching pairs of bands they share, divided by the sum of the total number of bands for each; S thus ranges from zero for a pair of individuals sharing no bands, to unity (1) in the case of identical profiles. The total set of similarity indices (S's) can be partitioned into two parts, (1) within population (denoted, S_w) and (2) across population S's (denoted, S_a), each set with its own mean S ($\bar{S}$). For tests comparing within population mean S's (tests of within population diversity, where diversity is equal to 1-S or the dissimilarity between individuals), populations were compared on a pairwise basis using an approximate t-test. Variances of the two within mean S's being compared were estimated in a fashion similar to that proposed by Lynch (Lynch, 1991; Lynch and Milligan, 1994; Lynch, 1990) for estimating variance due to sampling individuals, but modified so as to make it more precisely defined and more efficient (manuscript in preparation).

Tests of population subdivision which test whether the populations show evidence of being genetically distinct were conducted by permuting the individuals randomly into populations of the same size as in the original data. The test statistic was the ratio of the overall $\bar{S}$ within populations to overall $\bar{S}$ across populations (a high ratio indicating population subdivision). The test p-value was based on the ranking of the observed ratio relative to the set of ratios generated from the permuted data.

3. Results and Discussion

3.1 GENETIC SIMILARITY WITHIN AND ACROSS POPULATIONS

Analysis of DNA profiles for fish produced 42 nominal bands, between 200 and 1200 bp in size. Of these bands, 40 bands were found to be polymorphic. An approximate F-test comparing variation among all six within population mean S's ($\bar{S}_w$) (Table III, bold diagonal) to the variation due to sampling within populations showed no more than chance variation ($p=0.42$) where the test statistic $F'=1.01$, $df=5, 52$; however, a pairwise comparison of the two most extreme populations [Tom's Run and Indian Creek (lower), $\bar{S}_w=0.816$ and 0.720 respectively], without adjusting for multiple comparisons, shows very significant differences in $\bar{S}_w$s [$t(17)=2.96$, $p=0.0088$].

A test for overall population subdivision using the $\bar{S}_a$s (Table III, above diagonal) indicates that there is evidence of highly significant population structure across the sites tested, $p<0.01$. These $\bar{S}_a$ values were taken into NTSYS ver. 1.80 (Applied Biostatistics Inc., Setauket N.Y.) and used to construct phenograms which show nested patterns of population similarity (Figure 2). From this analysis two populations (Sevenmile Creek and Indian Creek) stand out as being different from the rest of the populations as well as from each other. When the average number of bands exhibited per individual in a population (Table IV) are compared across the populations, two populations fall out as being significantly different ($p=0.05$); the Sevenmile Creek population sample which had an average of 25.0 bands and the Indian Creek (lower) population with 27.2 bands. The other populations averaged between 29.2-30.9 bands per individual.

Table III

Values calculated for $\bar{S}_w$ (bold diagonal) and $\bar{S}_a$(above diagonal)[1]

	TC	PC	7M	IC(l)	TR	IC(u)
TC	**0.782**	0.789	0.694	0.739	0.796	0.772
PC		**0.779**	0.680	0.733	0.803	0.771
7M	<0.01	<0.01	**0.772**	0.734	0.650	0.707
IC(l)	0.13	0.13	0.09	**0.720**	0.715	0.734
TR	0.24		<0.01	<0.01	**0.816**	0.780
IC(u)			<0.01	0.27	0.27	**0.752**

1. Cells below the diagonal are p values for pairwise $\bar{S}_w/\bar{S}_a$ ratios. Empty cells indicate that the test statistic ratio was <1.0.

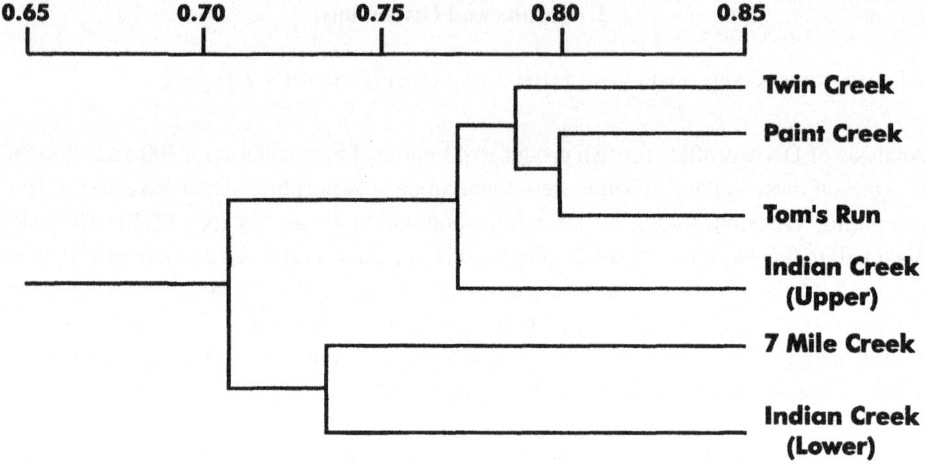

Fig. 2. Genetic relationship for six populations of stonerollers based on across population similarity indices developed for some tributaries to the Great Miami River.

Table IV
Average number of bands exhibited per individual in the study populations

Population	Number of individuals	Mean Total Bands per Individual	Standard Deviation of individuals from mean	Range for band number per individual in population
TC	9	30.5	4.5	23-36
PC	10	30.5	2.8	27-35
7M	10	25.0	2.6	21-28
IC(l)	10	27.2	3.7	21-33
TR	9	30.9	2.2	27-34
IC(u)	10	29.2	3.4	24-35

Using these preliminary data, one can view emerging relationships. For instance a relationship appears between the distance of the sample population from the mainstem of the Great Miami River and the genetic similarity within the population ($\bar{S}_w$)where the change in genetic diversity is related to the inverse of the distance from the mainstem (Figure 3). Additionally, a relationship can be discerned between stream order and $\bar{S}_w$ (Figure 4). These observations are in keeping with the pioneering nature of the stoneroller in that those populations that have penetrated farthest would establish themselves with fewer individuals and therefore be expected to naturally have lower diversity values. These observations provide two hypotheses which require further testing. Further, those populations in lower ordered streams with a limited carrying capacity would have limited effective population sizes and therefore limited levels of diversity. These observations fit the with theory of island biogeography (MacArthur and Wilson, 1967). These data, though limited, do indicate that a significant amount of population structure exists in lowered ordered streams which may provide the nucleus for species diversity. Further, it is these streams which may be the most vulnerable to exposure.

Our data set was part of a larger R-EMAP project for which other measures of resource condition were made (Table V). Data for genetic diversity will allow our correlation of S values with other exposure indicators as well as provide insights for assessing resource condition for which other indicators may not be sensitive. An example for this is in the consideration of the IBI score obtained for Sevenmile and Twin Creeks in relation to bile benzo[a]pyrene (BAP) metabolite levels (an indicator of combustion by-products) measured in white sucker (*Catostomus commersoni*) from these sites (Table V).

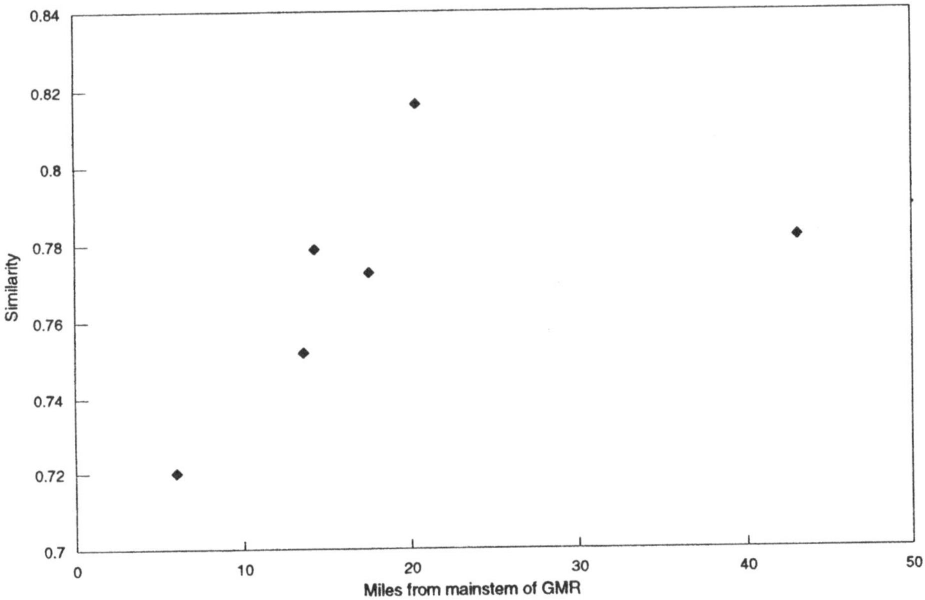

Fig. 3. Scatter plot of the mean within S vs. distance from mainstem of the Great Miami River.

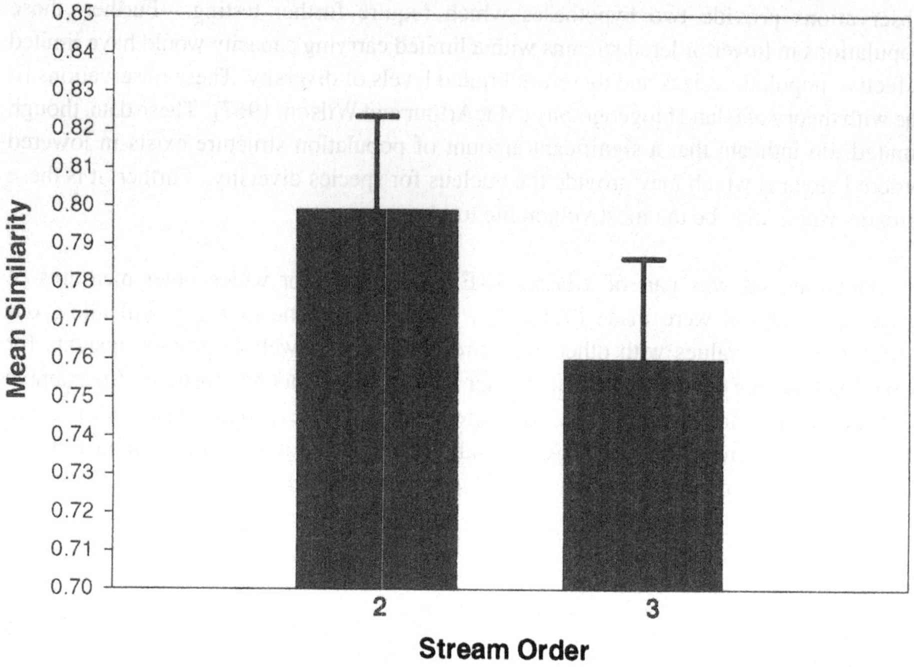

Fig. 4. Bar graph of the mean within S vs. stream order.

Here, both the fish community index and BAP levels for the sites are relatively high, and therefore seemingly contradictory. This could be interpreted to mean that the exposure to BAP precursors is "transparent" to the community; however, the trend in $\bar{S}_w$ for these sites is towards the high end of the index scale developed, perhaps suggesting that the populations are responding to the stressor exposure in advance of effects being observed at the community level. Therefore, genetic diversity when used in concert with other indicators would identify valued resources which may warrant special consideration by land managers. The validity of the above inferences will be more stringently tested upon completion of the remaining sample sites across the ECBP.

4. Conclusion

We have begun a study of the population genetic diversity of central stonerollers in the Great Miami River Basin, one of ten basins in the ECBP ecoregion. We hypothesize that there will be significant differences in both genetic variation and population subdivision within and across basins given the existence of point and non-point source stressors. Early indications of trends in levels of genetic diversity relative to exposure to environmental stressors support this hypothesis.

Table V
Site characteristics

Stream Name	Stream Order	Stressor[1]	QHEI[2]	IBI[3]	BAP[4]	NAPH[5]
Twin Creek	2	agriculture	59.0	48	145.2(±39.3)[6] 99.8(±40.6)[7]	26.3(±2.3)[6] 49.0(±11.5)[7]
Tom's Run	2	agriculture	62.0	N/A[9]	130.8(±140.2)[6]	11.6("4.0)[6]
Sevenmile Creek	3	agriculture	73.5	46	176.4(±32.0)[6] 85.8(±25.0)[7]	26.4(±1.1)[6] 31.4(±4.3)[7]
Paint Creek	3	agriculture	74.0	42	37.2(±27.8)[8]	16.4(±4.1)[8]
Indian Creek	3	WWTP agriculture	71.5	46	N/A	N/A
Indian Creek	3	none	80.5	42	39.1(±40.4)[7]	23.0(±4.8)[7]

1 Primary stressor identified, followed by secondary stressor, if any.
2 Qualitative Habitat Evaluation Index (score range 0-100).
3 Index of Biotic Integrity (score range 12-60).
4 benzo[a]pyrene metabolites (μg mg^{-1} protein).
5 naphthalene metabolites (μg mg^{-1} protein).
6 white sucker.
7 common carp.
8 creek chub.
9 data not available

The RAPD technique is a powerful tool by which to evaluate the genetic diversity across a drainage basin, as well as provide for the assessment of status and trends in genetic diversity which occur across populations at multiple spatial scales. Anthropogenic stressors can affect genetic structuring and dynamics of populations. Stressors can restrict or eliminate gene flow between populations and may lead to genetic divergence. Gene frequencies may shift under the selective pressures exerted by stressors. These selective pressures may lead to an intensification of the effects by exposure and increase the levels of divergence among populations owing to differential adaptive mechanisms for each stressor regime. The combination of lack of gene flow and divergence between populations can leave these populations genetically impoverished and vulnerable to exposure, due to the lack of genetic repertoire from which to muster a response to new stressor regimes. These populations, therefore, may be more vulnerable to extirpation with the introduction of new stressors. RAPDs as a means of measuring genetic diversity within and across populations when coupled to other measures of resource condition can be used to assess the vulnerability of valued resources to exposure and thus provide resource managers with critical scientific data upon which to base management decisions. Methods developed for the ECBP would be portable to other regions in the central United States to aid in sound environmental decision making.

Acknowledgments

Special thanks to the field crews, from DynCorp: Dan Williams, Scott Jacobs, Joseph Loucek, Steve Harmon, Jason Jannot, Dennis McMullen as well as Kelly Capuzzi from Ohio EPA. Susan Cormier (USEPA), Bhagya Subramanian, Sharon Detmer, Dan Williams of PAI for bile metabolite data, Rona Fan of PAI for GIS support and Melanie Bell of OAO for graphic support. A.C.L. and M.G. were supported in part by appointments to the Postgraduate Participation research Program at the National Exposure Research Laboratory administered by the Oak Ridge Institute for Science and Education through an interagency agreement between the U.S. Department of Energy and the U.S. Environmental Protection Agency. This document has been reviewed in accordance with U.S. Environmental Protection Agency policy and approved for publication. Approval does not signify that the contents necessarily reflect the views or policies of the Agency nor does mention of trade names or commercial products constitute endorsement or recommendation for use.

References

Allendorf, F.W., Leary, R.F.: 1986, Conservation Biology, Sinauer Associates, Sunderland, p 57.

Angermeier, P.L., Karr, J.R.: 1994, *BioScience* **44**, 690.

Avise, J.C., Arnold, J., Ball, R.M., Bermingham, E., Lamb, T., Neigel, J.E., Reeb, C.A., Saunders, N.C.: 1987, *Annual Review of Ecology and Systematics* **18**, 489.

Caetano-Anolles, G., Bassam, B.J., Gresshoff, P.M.: 1991, *Bio/Technology* **9,** 553.

Caswell, H.: 1976, *Ecol. Monogr.* **46**, 327.

Dobzhansky, T.: 1950, *Sci. Amer.* **182(1),** 32.

Lattier, D.L., Gordon, D.A., Silbiger, R.N., McCormick, F., Smith, M.K.: 1996, Techniques in Aquatic Toxicology, Lewis Publishers, Boca Raton, p. 569.

Leberg, P.L.: 1990, *Journal of Fish Biology* **37,** 193.

Lynch, M.: 1988, *Mol. Biol. Evol.* **5(5),** 584.

Lynch, M.: 1990, *Molecular. Biology. &. Evolution* **7,** 478.

Lynch, M.: 1991, DNA Fingerprinting: Approaches and Applications, Birkhauser Verlag, Switzerland.

Lynch, M., Milligan, B.G.: 1994, *Molecular. Ecology.* **3,** 91.

MacArthur, R.H., Wilson, E.O.: 1967, *The Theory of Island Biogeography,* Princeton University Press, Princeton.

MacLean, J.A., Evans D.O.: 1981, *Can J. Fish. Aquat. Sci.* **38**, 1889.

Nehlsen, W., Williams, J.E., Lichatowich, J.A.: 1991, *Fisheries* **16**, 4.

Ohio Environmental Protection Agency: 1987, Biological Criteria for the Protection of Aquatic Life, Vol II, Division of Water Quality Monitoring and Assessment, Surface Water section, Columbus.

Omernik, J.M.: 1995, Biological Assessment and Criteria: Tools for water resource planning and decision making, Lewis Publishers, Boca Raton, p. 49.

Patton, J.L., Smith, M.F.: 1990, *The Evolutionary dynamics of the pocket gopher Thomomys bottae, emphasis on California populations,* University of California Press, Berkley.

Quatro, J.M., Vrijenhoek, R.C.: 1989, *Science* **245,** 976.

Root, T.L., Schneider, S.H.: 1995, *Science* **269,** 334.

Sanjayan, M.A., Crooks, K.: 1996, *Nature* **381,** 566.

Sardelli, A.D.: 1993, *Amplifications* 1.

Smith, T.B. and Wayne, R.K.: 1996, *Molecular Genetic Approaches in Conservation*, Oxford University Press, New York.

Soulé, M.E. and Wilcox, B.A.: 1980, *Conservation Biology: an evolutionary-ecological perspective*, Sinauer Associates, Sunderland.

Soulé, M.E.: 1986, *Conservation Biology: the science of scarcity and diversity*, Sinauer, Sunderland.

Soulé, M.E., Zegers, G.P.: 1996, *Journal of Heredity* **87,** 341.

Stone, R.: 1995, *Science* **269,** 316.

Thorpe, J.E., *et al.*: 1981, *Can J. Fish. Aquat. Sci.* **38**, 1899.

United States Environmental Protection Agency: Klemm, D.J. and Lazorchak, J.M.: 1993, *Environmental Monitoring and Assessment Program. Surface Waters and Region 3 Regional Environmental Monitoring and Assessment Program. 1993 Pilot Field Operation and Methods Manual. Streams*, USEPA, ORD, EMSL-Ci, Washington, D.C.

Welsh, J., McClelland, M.: 1990, *Nucl. Acids Res.* **18,** 7213.

Williams, J.G., Kubelik, A.R., Livak, K.J., Rafalski, J.A., Tingey, S.V.: 1990, *Nucleic. Acids. Research.* **18,** 6531.

GENETIC IMPACT OF LOW-DOSE RADIATION ON HUMAN AND NON-HUMAN BIOTA IN CHERNOBYL, UKRAINE

NATASHA A. MAZNIK

Kharkiv Research Institute of Medical Radiology, Pushkinskaya st., 82, Kharkiv, 310024, Ukraine

1. Introduction

The accident at the Chernobyl Nuclear Power Plant (NPP) on 26 April 1986 resulted in extensive contamination of the environment from the Unit 4 reactor. A cloud consisting of a complex mixture of radionuclides, with different radiological properties, was released into the atmosphere for at least 10 days. The Chernobyl NPP is located in Ukrainian territory close to the borders of Russia and Belorus. The scale of the radiation emergency was so large that, by 14 May 1986, the entire human population had been evacuated from the 30-km radius exclusion zone around the reactor. The largest city in the 30-km zone was Prip'at, with a population of 49,360. This population consisted mainly of the staff of the Chernobyl NPP and members of their families. In addition, many thousands of workers, called liquidators, were drafted in the ensuing months and years to undertake a variety of clean-up and other tasks inside the affected regions.

The human and non-human biota in contaminated areas were affected by external and internal irradiation from many radionuclides with different radiological properties. The ratio of external and internal dose mainly depends on (1) the radionuclide composition of the cloud and fallout from it in each particular region, (2) on meteorological conditions of fallout, and (3) on soil type, agricultural practices, and the countermeasures applied (Balonov et al., 1996). These radioactive contaminants have caused both short-term and long-term genetic effects in the human population and other living organisms that still need to be assessed.

2. Materials and methods

2.1. Composition of control and Chernobyl evacuee cohorts

In the present study, 40 people (21 males and 19 females, in the age range from 15 to 66 years) were investigated. They had been evacuated from the city of Prip'at and adjacent villages between 27 April and 5 May 1986, i.e. about one week after the accident. The evacuees were moved by rail to the "clean" city Kharkiv, more than 500 km from the Chernobyl NPP. Clinical examinations were performed on all of them in the Kharkiv Research Institute of Medical Radiology from 1 day to 1 year after their evacuation. Blood samples for cytogenetic analysis also were taken at the same time.

Environmental Monitoring and Assessment **51:** 497–506, 1998.

Another cohort studied consists of 145 liquidators (including 8 females, in the age range from 20 to 60 years) who lived in the Kharkiv region and worked in Chernobyl for 0.5 to 3 months in the period 1986-1987. The majority of them were mobilized from the military reserve and undertook a wide range of duties at Chernobyl. They were firefighters, decontamination technicians, radiation monitors, physicians, drivers, builders, and guards. Some of them were issued personal dose meters. This cohort only included persons who undertook only one period of duty in the zone. The other point of selection was that their medical and occupational histories gave no indication that they would have received any radiation, in excess of background, apart from their work as liquidators.

Blood samples also were taken during clinical examinations from an additional 107 persons from 1 week to 1 year after moving from the Chernobyl zone, and within 8 to 9 years after the accident in 38 other persons.

The control cohort was drawn from persons recruited to work in Kharkiv medical institutions and from those who had no previous occupational and medical overexposure to ionizing radiation. These blood samples were taken at the time of their pre-employment medical examination. The control cohort was comprised of 24 males and 6 females in the age range of 26 to 52 years.

2.2. Preparation of blood samples for cytogenetic analysis

Heparinized blood samples were cultured to metaphase by a standardized protocol (IAEA., 1986). In brief, PHA stimulated lymphocytes were grown in MEM with 20% fetal calf serum supplemented with 2mM L-glutamine. The cultures were incubated at 37°C for 50 h with colcemid added for the final 3 h. Metaphases were harvested from the cultures by hypotonic treatment in 0.075M KCl at 37°C followed by the changes of 3:1 methanol:acetic acid fixative. The slides were Giemsa-stained for conventional cytogenetic analysis. Blind scoring, using coded slides, of the first division metaphases was performed to detect all the kinds of chromosome aberrations: acentric fragments, dicentrics, centric rings and translocations; and wherever possible; chromatid type breaks and exchanges; as well as hyperploid and polyploid cells. 100 cells or more were scored from each person. The criteria for scoring were as described in IAEA (1986). For biological dosimetry only the frequency of dicentric chromosome aberrations were used.

Stable chromosome aberration analysis, using the fluorescence *in situ* hybridization (FISH) technique, was performed in 9 liquidators in the period 1994-1995. Sequential hybridizations of fluorescein isothiocyanate (FITC)-labeled probes for chromosomes 6, 9, 15, and 21 (Cambio), and a biotin-labeled pan-centromeric probe (Oncor) were carried out according to the protocol designed at the National Radiological Protection Board of UK (Finnon et al., 1995). The pan-centromeric probe was detected by alternating layers of avidin-texas red and biotinylated goat anti-avidin (both Vector). The chromosome paint signal was amplified by successive layers of rabbit anti-FITC IgG (Dako) and FITC-conjugated goat anti-rabbit IgG (Sigma). The slides were mounted with Vectashield antifade solution (Vector) containing 4', 6-diamidino-2-phenylindole (DAPI). The slides were analyzed using a Zeiss Axioscope fluorescence microscope equipped with separate filters for DAPI, FITC and a triple-band pass filter. Under the last filter the painted chromosomes appeared green-yellow, the unpainted chromosomes blue and the centromeres red. A chromosome was deemed to contain a translocation if it exhibited a single centromeric signal and possessed a bicolored (FITC/DAPI) junction. Since approximately 15% of the genome was painted, the translocation yield in the whole genome was derived using the formula of Lucas *et al.* (1992).

2.3. Statistical analysis

The Student's t-test was used for all statistical analyses.

3. Results and Discussion

3.1. Investigation of Early Cytogenetic Effects in the Chernobyl Cohorts

The total level of aberrations in those evacuees and liquidators examined during the first year after the accident was increased 3 to 4 times higher in comparison with the control. The frequency of chromosomal aberrations in control adults is generally observed to be about 10 to 20 aberrations (including 1 dicentric) per 1000 cells. The basic results of our scoring of chromosome and chromatid aberrations are presented in Table I.

Table I. Early Cytogenetic Effects in Chernobyl Evacuees and Liquidators

Cohort	Aberrations per 100 cells				Documented
	Cytogenetic dose	Dicentrics and rings	Acentric breaks	Chromatid exchanges	dose range (mGy)
Controls	0.10±0.05	0.88±0.13	0.66±0.10	0.08±0.05	0
Evacuees	1.57±0.28	2.12±0.34	1.67±0.25	0.71±0.17	330-420
Liquidators	1.73±0.23	2.38±0.30	1.83±0.22	0.84±0.22	17-940

These data demonstrate a significant increase ($P < 0.05$) for all chromosome type aberrations in exposed persons. Chromosome aberrations (dicentrics, centric rings, acentric fragments), which are well-known to be induced by ionizing radiations, were observed in both liquidators and evacuee cohorts. Dicentrics and rings were almost always accompanied by an acentric fragment but only excess acentric fragment frequencies are listed in Table I. These cohorts are characterized by a significant increase in the yields of dicentrics.

An increased chromosome aberration level in Chernobyl victims protractedly exposed to low-dose radiation was also shown by Sevan'kaev and Zhloba (1991), Pilinskaya *et al.* (1992), Shishmarev *et al.* (1992).

Dose estimation in liquidators and evacuees was performed soon after the accident by scoring for the frequency of dicentrics. Dicentrics are the aberration of choice for biological dosimetry because of its inherent low background frequency, near exclusivity to radiation, and a well-documented dose-response relationship (Bender *et al.*, 1988). Because dicentrics seem to appear at random and the total number of chromosome aberrations have a normal distribution among the sampled persons, the conclusion could be reached that the evacuees and liquidators in the present study are normally distributed according to biological dose. The cytogenetic dose estimates listed in Table I were obtained by solving for dose (D) in equation (1) derived protracted irradiation from the basic equation (2) for in vitro acute dose response data from a Cobalt 60 gamma source (Lloyd *et al.*, 1996).

Equation (1): $Y = C + (\alpha) \times D$

Equation (2): $Y = C + (\alpha) \times D + (\beta) \times D^2$

In these two equations, Y is the number of dicentrics per cell measured in Chernobyl victims, D is dose in Gy, C is the background frequency that is typically about 0.001 dicentric per cell (Lloyd et al., 1983), (α) is 3.08×10^{-2}) dicentric per cell per Gy and (β) is 6.27×10^{-2} dicentric per cell per Gy^2.

The use of such coefficients from *in vitro* calibration curves to estimate the dose to exposed individuals is justified by the observation that identical aberration frequencies are induced in human lymphocytes *in vitro* and *in vivo* (Dolphin *et al.*,1973; Sasaki, 1983; IAEA, 1986).

The cytogenetic doses in liquidators were 1.5 to 2 times higher than that listed officially in their documents. The uncertainties in physical dosimetry data obtained in conditions of such a large scale accident as that in Chernobyl are not surprising. Since the biological doses estimated for each of our cohorts did not exceed 1 Gy, this information was extremely helpful for determining the health treatment strategy of choice for Chernobyl victims immediately after the accident.

It should be noted that cytogenetic retrospective dosimetry can be use successfully where there is a lack of physical dosimetry data and where one suspects non-uniform exposure. It was of particularly great value in 1986 for the acute radiation syndrome medical treatment of firefighters as to whether or not bone marrow transplantation should be provided (Gus'kova *et al.*, 1987). The unexpectedly high level of chromatid aberrations, which are usually induced only by exposure to chemical mutagens, was found in both evacuee and liquidator cohorts. The possible source of these chromatid aberrations could be either the pesticides used in this agricultural region, or chemicals that were ejected into the reactor zone from the helicopters for fire fighting.

3.2. Investigation of Late Cytogenetic Effects in the Chernobyl Cohorts

The total aberration levels in the liquidators cohort determined during 1994-95 was still higher than in the control cohort. Those liquidators, with an officially registered dose, were selected for further analysis to determine whether there was any correlation between cytogenetic effects and physical radiation dose. It should be noted that recorded radiation dose distribution was rather non-uniform, with a number of persons with official doses as large as 250 mGy. The basic results of scoring chromosome and chromatid aberrations in liquidators subdivided into 3 sub-cohorts with registered doses of (1) less than 250 mGy, (2) exactly 250 mGy, or (3) greater than 250 mGy, are presented in Table II.

Table II. Late Cytogenetic Effects in Chernobyl Liquidators

Cohort and estimated dose	No. of persons	Cytogenetic abnormalities per 100 cells			
		Chromosome aberrations	Chromatid aberrations	Total aberrations	Aneuploidy
Control (0 mGy)	30	0.99±0.15	0.75±0.10	1.78±0.20	0.03±0.02
Liquidators (Σ)	38	1.64±0.31	1.40±0.29[a]	3.03±0.42[a]	0.40±0.17[a]
<250 mGy	12	0.74±0.28	0.70±0.25	1.43±0.37	0.16±0.13
250 mGy	16	2.12±0.63	1.35±0.50	3.47±0.81[a]	0.26±0.13[a]
>250 mGy	10	1.92±0.50[a, b]	2.30±0.68[a, b]	4.22±0.84[a, b]	0.98±0.54[a, b]

[a] Statistically significant differences (P<0.005) in comparison with the control
[b] Statistically significant differences (P<0.005) in comparison with the subcohort <250 mGy

The total aberration frequency in the whole cohort of 38 liquidators (3.03 per 100 cells), which was higher than the control cohort frequency, was more than two times lower than 6.78 per 100 cell determined in the period 1986-1987.

The radiation-induced chromosome aberration frequency for the whole cohort of liquidators did not differ significantly from the control frequency, but an inter-cohort comparison shows a positive dose trend for total aberration frequencies as was found for chromosome aberration frequencies. For example, chromosome aberration frequency in the liquidator sub-cohort >250 mGy was significantly increased ($P < 0.05$) in comparison not only with the control cohort but also with liquidator sub-cohort <250 mGy (Table II). The same effect also was shown for chromatid aberrations.

3.3. Utility of Different Types of Chromosome Aberrations for Bio-Dosimetry

The chromosome aberration frequency that demonstrates the most rapid decrease with time was that for unstable chromosome exchange aberrations. The dicentric and centric ring group frequency in liquidators reached 2.9±0.8 per 1000 cells, that was higher than control cohort frequency but 6 times lower than the frequency found early after the accident. These data are in a good agreement with the studies of Sevan'kaev *et al.* (1995).

Dicentric chromosome aberrations in human peripheral blood lymphocytes have been the mainstay in radiation bio-dosimetry. Numerous studies have demonstrated the utility of the dicentric chromosome aberrations for dose assessment within a few weeks after acute or subacute exposure. However, it is well understood that dicentrics are much less useful for evaluation of exposures that occurred a long time ago (Bender *et al.*,1988; Lloyd and Sevan'kaev, 1996; Ramalho *et al.*, 1995).

But, there are so-called "stable" chromosome aberrations that can pass successfully through cell division and which should persist in the body for many years. Stable chromosome aberrations observed in peripheral blood lymphocytes a number of years after radiation exposure are probably the products of the division of the stem cells in which the initial stable chromosome lesions were induced (Awa, 1983; Buckton, 1983; Lucas et al., 1992).

3.4. Comparison of Methods for Measurement of Stable Chromosome Aberrations

There are two methods to measure the frequencies of stable chromosome aberrations. The well established G-banding method enables a skilled microscopist to identify most translocations in metaphase preparations. However, this method is extraordinarily time consuming and not suitable for the study of large numbers of cells from a large number of subjects. It is useful for accidents with a few victims but it is less reliable for the large number of subjects in each of the Chernobyl cohorts investigated.

The alternative method is fluorescence *in situ* hybridization (FISH) which allows more rapid screening of cells. In addition, discrimination between translocations and dicentrics can be done accurately using centromeric staining with pan-centromeric probes.

A pilot study on 9 liquidators was made soon after their exposure in 1986 using conventional analysis. In 1994, both conventional and FISH technique also were carried out to compare the biological dose estimation based on the dicentric and translocation methods. The translocation dose estimation was made in the same manner as described for dicentrics (see Section 3.1), i.e. using the linear alpha-term obtained from the *in vitro* linear-quadratic equation (2), where Y is the number of translocations per cell measured in Chernobyl liquidators, D is dose in Gy, C is the background frequency 0.0055 translocations per cell, α is 0.0563 translocation per cell per Gy, and β is 0.1465 translocation per cell per Gy^2. This equation was derived for whole genome equivalent using formula of Lucas *et al.*, (1992).

The mean dicentric frequency obtained in 1986 for the group investigated was 1.702 per 100 cells; corresponding to a cytogenetic dose of 510 mGy. As expected, the dicentric frequency was lowered with time to 0.342 per 100 cells in 1994-1995. But the translocation level, measured at the same time, was 3.250 per 100 cells (whole-genome equivalent); corresponding to cytogenetic dose of 480 mGy. We concluded the FISH-calculated dose was in a good agreement with earlier dicentric dose estimation. In spite of some limitations (i.e. large number of cells to be analyzed, strong statistics required, etc.) the FISH technique opens exciting prospects for the study of late genetic effects and for use as a biological dosimeter, as well as for genetic and cancer risk assessment.

3.5. Genetic Instability as General Late Effect in Human and Non-Human Biota after the Chernobyl Accident

Apart from the detectable level of stable aberrations, we also found some late cytogenetic effects in groups examined by conventional methods: the trend to dose-dependent increases of chromatid aberrations and such mitotic damage as aneuploidy (Table II). It was unpredictable because chromatid aberrations are induced by chemical mutagens preferably, and are non-specific to radiation. We assumed that the chromatid breakage effect might be explained by a dose-dependent decrease of DNA-repair system, that could be caused by radiation-induced gene mutations affecting DNA-repair enzymes. Such damage would make somatic cells more sensitive to common environmental chemical and physical mutagens.

The genomic instability displayed in the high spontaneous level of chromosomal abnormalities and chromosomal fragility under *in vitro* mutagenic exposure is well known in patients with inherited DNA-repair defects, e.g. ataxia-telangiectasia, Bloom's syndrome, or Fanconi's anemia. For these patients, the increased level of chromosome breakage is associated with high cancer susceptibility (Higurashi and Conen, 1973; Ray and German, 1981; Taylor, 1983). Thus, the indices of mutagen-induced elevated chromosomal sensitivity could be used as the cytogenetic criteria for cancer risk assessment in addition to the usual dose-related risk calculations.

3.6. Enhanced Sensitivity of Human and Non-Human Biota to Radiation

Several genetic endpoints indicating enhanced chromosome instability were detected in human and non-human biota in the post-Chernobyl period. For example, increased chromosome sensitivity to additional irradiation could be seen in both in plants, animals, and humans. Using the chromosome aberrations in seedling root meristem test (Dmitrieva, 1996), it was shown that the seeds developed in the Chernobyl contaminated zone were more radiosensitive than those from the "clean" zone. As reported by Goncharova *et al.*,

(1996), the increased level of the mutations in somatic cells and gametes have been demonstrated in many generations (1 to 18) of natural populations of small mammals that inhabited the Chernobyl contaminated zone in the period 1986-1991.

The same authors detected a high frequency of chromosome aberrations in somatic cells of young carps from contaminated ponds in the period 1988-1992. Radiosensitivity of the somatic and germ cells of animals was increased in subsequent generations as compared with generations that lived in the Chernobyl contaminated zone in the period 1986-1988. (Goncharova *et al.*, 1996).

The analysis of the character of cytogenetic variability, and the correlative interactions between them, demonstrates the group specificity of karyotype instability in bone marrow cells in some lines of laboratory and wild mice that lived in the Chernobyl zone (Glazko et al., 1996; Pelevina et al, 1996).

Among the adult and child populations from contaminated areas, the frequency of individuals with definite adaptive response, measured by chromosome aberrations and micronuclei tests after low level irradiation of lymphocytes, was decreased and there also were individuals with elevated radiosensitivity (Pelevina et al., 1996; Ryabchenko et al., 1996). The investigators concluded that the chronic exposure to radionuclides while living in Chernobyl fallout contaminated areas did not induce the adaptive response

It is our opinion that the genetic instability detected as increased sensitivity to radiation exposure and environmental mutagens and decreased adaptive response was a general effect for both human and non-human biota after the Chernobyl disaster. These effects might be explained by induced lesions at the molecular, gene and/or genomic level, as, for example, gene mutations in DNA-repair systems, the persistence of stable translocation in somatic cells years after exposure. Undoubtedly, the early and late genetic impact induced by the mutagenic factors of the Chernobyl accident stipulate the necessity of further long-time investigations. Aftermath assessment is considered essential to elaborate the strategy of environmental monitoring and protection of genetic diversity in areas suffered by any technological disaster.

4. Conclusion

The post-Chernobyl genetic effects measured by chromosomal analysis in human lymphocytes were displayed as significantly increased level of chromosome and chromatid aberrations. In the evacuee and liquidator cohorts, the cytogenetic data were used for biological dosimetry, and a good agreement was shown between early dose estimation from the assay for dicentric chromosome aberrations and by stable chromosome translocations using the FISH assay.

The genomic instability indices, such as dose-dependent chromosomal breakage, adaptive response decrease, and abnormal chromosome sensitivity to mutagenic exposure seem to be the common end-point for human and non-human biota late after the Chernobyl accident. The conventional and molecular cytogenetic techniques were proven to be definite indicators and bio-dosimeters of accidental radiation exposure in humans and can be employed in non-human investigations for environmental monitoring and genetic diversity protection.

Acknowledgments

We wish to thank Drs. D. C. Lloyd and A. A. Edwards (National Radiological Protection Board, United Kingdom) for their collaboration and helpful comments on the biological dosimetry data discussion, and to Mr. Vladimir Vinnikov (Kharkiv Research Institute of Medical Radiology, Ukraine) for technical assistance.

References

Awa, A.A.: 1983, in: Ishihara, T. and Sasaki, M.(Eds.), "Radiation Induced Chromosome Damage in Man", Liss, New York, 433-453.

Balonov, M., Jacob, P., Likhtarev, I., Minenko, V.: 1996, in: Karaoglou, A., Desmet, G., Kelly, G.N., Menzel, H.G. (Eds.), "The radiological consequences of the Chernobyl accident", European Commission, EUR 16544, Luxembourg, 235-249.

Bender, M.A., Awa, A.A., Brooks, A.L., Evans, H.J., Croer, P.C., Littlefield, L.C., Pereira, C., Preston, R.J., Wachholz, W.: 1988, *Mutation. Res.* **196**, 103-159.

Buckton, K.E.: 1983, in: Ishihara, T. and Sasaki, M.(Eds.), "Radiation Induced Chromosome Damage in Man", Liss, New York, 491-511.

Dmitriyeva, S.: 1996, *Tsitologia i Genetika* **4**, 3-8.

Dolphin, G.W., Lloyd, D.C., Purrot, R.J.: 1973, *Health Physics* **25**, 7-15.

Finnon, P., Lloyd, D.C., Edwards, A.A.: 1995, *Int. J. Radiat. Biol.* **68**, 429-435.

Glazko, T.T., Safonova, N.A., Buntova, E.G., Glazko, G.V., Sozinov, A.A.: 1996, *Tsitologia i Genetika* **4**, 25-34.

Goncharova, R.I., Ryabokon', N.I., Slukvin, A.M.: 1996, *Tsitologia i Genetika* **4**, 35-41.

Gus'kova, A.K., Baranov, A.E., Barabanova, A.V., Gruzdev, G.P., Pyatkin, E.K., Nadezhina, N.M., Metlyaeva, N.A., Selidovkin, G.D., Gusev, I.A., Moiseev, A.A., Dorofeeva, E.M., Zykova, I.E., Konchalovsky, M.V.: 1987, *Meditsinskaya Radiologia* **12**, 3-18.

Higurashi, M., Conen, P.E.: 1973, *Cancer* **32**, 380-383.

International Atomic Energy Agency (IAEA): 1986, "Biological Dosimetry: Chromosomal Aberrations Analysis for Dose Assessment", Technical Report Series No.260, IAEA, Vienna.

Lloyd, D.C., Purrot, R.J., Reeder, E.J.: 1980, *Mutation Res.* **72**, 523-532.

Lloyd, D.C., Edwards, A.A., Sevan'kaev, A.V., Bauchinger, M., Braselman, H., Georgiadou-Schumacher, V., Salassidis, K., Darroudi, F., Natarajan, A.T., Van der Berg, M., Fedortseva, R., Fomina, Z., Maznik, N.A., Melnov, S., Palitti, F., Pantelias, G., Pilinskaya, M., Vorobtsova, I.E.: 1996, in: Karaoglou, A., Desmet, G., Kelly, G.N., Menzel, H.G. (Eds.), "The radiological consequences of the Chernobyl accident", European Commission, EUR 16544, Luxembourg, 965-974.

Lloyd, D.C. and Sevan'kaev, A.V.: 1996, "Biological Dosimetry for Persons Irradiated by the Chernobyl Accident", European Commission, EUR 16532, Luxembourg, 83 pages.

Lucas, J.N., Awa, A.A., Straume, T.: 1992, *Int. J. Radiat. Biol.* **62**, 53-63.

Pelevina, I.I., Gotlib, V.Ya., Kudryashova, O.V., Serebryani, A.M.: 1996, *Raditsionnaya Biologia. Radioecologia* **36**(4), 546-560.

Pilinskaya, M.A., Shemetun, A.M., Dybsky, S.S., Redko, D.V., Eremeyeva, M.N.:1992, *Radiobiologia* **32**(5), 632-639.

Ramalho, A.T., Currado, M.P. and Natarajan, A.T.: 1995, *Mutation Res.* **331**, 47-54.

Ray, J.H. and German, J.: 1981, in: Arrighi, F.E., Rao, P.H., Stubblefield, E.(Eds.), Genes, Chromosomes and Neoplasia, Raven Press, New York, 351-358.

Ryabchenko, N.I., Antoschina, M.M., Nasonova, V.A., Fesenko, E.V.: 1996, *Radiatsionnaya Biologia. Radioecologia* **35**(5), 670-675.

Sasaki, M.S.: 1983, in: Ishihara, T. and Sasaki, M.(Eds.), "Radiation Induced Chromosome Damage in Man", Liss, New York, 585-604.

Sevan'kayev, A.V. and Zhloba, A.A.: 1991, *Acta Oncologica* **12**, 201-204.

Sevan'kayev, A.V., Lloyd, D.C., Braselmann, H., Edwards, A.A., Moiseenko, V.V. and Zhloba, A.A.: 1995, *Radiation Protection Dosimetry* **2**, 85-91.

Shishmarev, Yu.N., Alekseev, G.I., Nikiforov, A.M., Larchenko, G.K., Krivoruchko, A.A., Pronin, M.A., Ivanov, I.A.: 1992, *Radiobiologia* **32**(3), 323-332.

Taylor, A.M.R.: 1983, in: Ishihara, T. and Sasaki, M.(Eds.), "Radiation Induced Chromosome Damage in Man", Liss, New York, 167-199.

HUMAN CARRYING CAPACITY AS AN INDICATOR OF REGIONAL SUSTAINABILITY

J. DAVID YOUNT

Mid-Continent Ecology Division, National Health and Environmental Effects Research Laboratory, U. S. Environmental Protection Agency, 6201 Congdon Blvd., Duluth, MN 55804, USA, tel.: 218-720-5752, fax: 218-720-5539, e-mail: yount.david@epamail.epa.gov

Section overview

The theme of this symposium, "Developing the Tools to Meet the Nation's Monitoring Needs: The Evolution of EMAP," implies a view both of how EMAP has evolved up to the present and in what way it might evolve in the future. The objectives of the 1997 EMAP research strategy include the formulation of "policies and programs that promote the preservation of ecosystem integrity and sustainable use of natural resources" (USEPA, 1997). Among the information needs for such programs are "approaches to monitor important ecosystem characteristics and the human perturbations which alter them over space and time." EPA's Science Advisory Board, according to the strategy, has recommended research on "techniques that can be used to help anticipate environmental problems" before they become widespread or irreversible.

The development of effective indicators of ecological condition is central to the goal of EMAP. According to the 1997 EMAP research strategy, "we are concerned about whether or not our human activities are having an adverse effect on the ability of ecological resources to continue providing a variety of goods and services into the future. Have our actions somehow limited the options available to future generations by impacting certain ecological processes or systems?" One term which has been applied to this attribute of ecological systems, the strategy points out, is "sustainability." Thus "EMAP research must contribute to an understanding of the conceptual basis for defining sustainability."

Is the development of indicators of ecological condition, as currently practiced by EMAP, sufficient to anticipate environmental problems before they become widespread or irreversible? Consider the following idealized thought-experiment. Imagine a wealthy island nation in which most of the citizenry lives in a few dense urban areas. With an internationally competitive service and value-adding economy, they are able to supply most of their resource needs and to export their wastes through international trade, and are thus able to maintain most of the island nation in a state of near-wilderness. Suppose further that they have instituted an Ecological Monitoring and Assessment Program consisting of a suite of indicators

Environmental Monitoring and Assessment **51**: 507–509, 1998.

of ecological integrity that demonstrate that the terrestrial and aquatic ecosystems on the island are in excellent condition. The Environmental Protection Agency of that nation therefore reports to the political leadership that their environmental quality is excellent and that they are secure from environmental degradation. Then suppose that there is a severe energy crisis or a war which results in the cutoff of most of their oil supply, so that living off of international trade becomes prohibitively expensive. They are then forced to clear land for agriculture, timber, fuel wood, industry, etc., and are unable to export their wastes. The ecological integrity of the island's ecosystems deteriorates rapidly. Was their earlier judgment of ecological integrity, and sustainable use of their and others' resources, then justified?

The three papers in this section are unusual in that they deal directly with humans as *components* of ecosystems. More specifically, they deal with economic throughput, as defined by resource use and waste discharge, as the driver of human perturbations to the ecosystems of which they are a part. They thereby contribute in a significant way to understanding the conceptual basis for defining sustainability.

Daniel Campbell considers the human carrying capacity of the State of Maine, U.S., using a method of environmental accounting termed emergy analysis. Emergy (embodied energy, or energy memory) analysis is used to assess the relationships between the economy of a region and its supporting environment in equivalent terms. Using the emergy environmental accounting approach, Campbell concludes that Maine is capable of sustaining (i.e., continuously supporting) only 34% of its current population at the 1980 level of resource use, if it had to depend on Maine's renewable environmental resource bases alone. The fact that Maine appears to be doing quite well in terms of environmental quality is a result of its import of petroleum fuels, goods, and services.

Jae-Young Ko, Charles Hall, and Luis López Lemus examine the prospects for sustainability of five countries: Costa Rica, Korea, Mexico, the Netherlands, and the United States. These countries were chosen because (in addition to the author's direct experience in them) they exhibited very different characteristics of demography, economic development, and natural resource stocks. By examining trends of energy and agricultural efficiencies, environmental impacts of economic activities, and ecological footprints (the area required to provide consumption-related resource flows and waste sinks), Ko *et al.* conclude that none of these countries are sustainable, nor do they appear to be approaching sustainability over time. The United States is maintaining its unsustainability at a relatively constant level, in spite of its increasing population and per capita resource use rates, primarily by increases in the efficiency of energy use. The other four countries are becoming less sustainable. In searching for positive signs, the authors note some improvements in efficiency of energy and fertilizer use, but these are being canceled by growth of population and the economy. They conclude that when selecting indicators of sustainability, more than the traditional ecological indicators must be considered.

Mathis Wackernagel and David Yount discuss the conceptual basis for and outline the mechanics of ecological footprint analysis, an area-based indicator approach for quantifying the ratio between the human load on a defined region and the region's ability to sustain that load without depletion of natural capital. The region's ability to sustain the human load is termed the region's carrying capacity, and sustainability is quantified as the ratio of the carrying capacity to the load. (Note that this definition of carrying capacity is somewhat different than that employed by Campbell). The currency of this environmental accounting method is "ecologically productive area," and sustainability is a zero-sum game. In order for the total human impact on earth to be sustainable, regions whose environmental load is above their carrying capacity (such as the countries analyzed by Ko *et al.*) must be supported by regions which are living below their carrying capacity. As the authors demonstrate (with support from Campbell and from Ko *et al.*), sustainability does not exist for the biosphere as a whole. This overload of the biosphere's carrying capacity clearly constitutes an environmental risk, and the geographic location and the components of the overload, as measured by ecological footprint analysis, indicate opportunities for risk management.

Although the details differ, these three papers, taken collectively, present a strong case for an expansion of the concept of indicators of ecological integrity to include indicators pertaining to the total ecological-economic system.

References

USEPA: 1997, *Environmental Monitoring and Assessment Program (EMAP): Research Strategy 1997*, United States Environmental Protection Agency, Office of Research and Development, National Health and Environmental Effects Research Laboratory.

THE ECOLOGICAL FOOTPRINT: AN INDICATOR OF PROGRESS TOWARD REGIONAL SUSTAINABILITY

MATHIS WACKERNAGEL[1] and J. DAVID YOUNT[2]

[1] *Centro de Estudios para la Sustentabilidad, Universidad Anáhuac de Xalapa, Apdo. Postal 653, 91000 Xalapa, Ver., MEXICO, tel.: 52 (28) 14-96-11, fax: 52 (28) 19-04-53, e-mail: mathiswa@compuserve.com*
[2] *United States Environmental Protection Agency, National Health and Environmental Effects Research Laboratory, Mid-Continent Ecology Division, 6201 Congdon Blvd., Duluth, MN 55804, USA, tel.: 218-720-5752, fax: 218-720-5539, e-mail: yount.david@epamail.epa.gov*

Abstract. We define regional sustainability as the continuous support of human quality of life within a region's ecological carrying capacity. To achieve regional sustainability, one must first assess the current situation. That is, indicators of status and progress are required. The ecological footprint is an area-based indicator which quantifies the intensity of human resource use and waste discharge activity in relation to a region's ecological carrying capacity. If the ecological footprint of a human population is greater than the area which it occupies, the population must be doing at least one of the following: receiving resources from elsewhere, disposing of some of its waste outside of the area, or depleting the area's natural capital stocks. To achieve global sustainability, the sum of all regional footprints must not exceed the total area of the biosphere. This paper explains the mechanics of a footprint calculation method for nations and regions. As the method is standardized, the relative ecological load imposed by nations and regions can be compared. Further, a nation's or region's consumption can be contrasted with its local ecological production, providing an indicator of potential vulnerability and contribution to ecological decline.

1. Introduction: Why are Ecological Footprints needed in Environmental Monitoring and Assessment?

Environmental Monitoring and Assessment programs such as those presented at this EMAP conference attempt to monitor temporal changes in the environment, and to assess their possible causes. Almost without exception, however, the monitoring tools which are used focus on specific and isolated conditions of the physical, chemical, or biotic "environment," ignoring the activities of the dominant (or "keystone") species, *Homo sapiens*. Yet human population size and consumptive behavior are often the ultimate source of the stressors which result in a degradation of ecological integrity in a location.

Since the early 1970s, one report after another has warned that unlimited growth of human population and consumption is not sustainable. Among the most prominent of these reports are *The Limits to Growth* (Meadows *et al.*, 1972), the Brundtland Commission's *Our Common Future* (WCED, 1987), and the Worldwatch Institute's annual *State of the World* publications. In spite of these warnings the human economy continues to expand, with more people, more consumption, more waste and more poverty, along with less biodiversity, less forest area, less available fresh water, less soil, less fossil oil in the ground and less protective ozone in the stratosphere (World Resources Institute, 1994, 1996; United Nations Development Program, annual). We seem to be getting further and

Environmental Monitoring and Assessment **51:** 511–529, 1998.

further away from sustainability. But how far? Indicators of progress are needed. This paper presents one of them – the ecological footprint (Wackernagel and Rees, 1996), and shows how it can be applied as a planning and monitoring tool for sustainability.

Ecological footprint analysis is an area-based indicator (Rees, 1996) which quantifies the intensity of human resource use and waste discharge activity in a specified area in relation to the area's capacity to provide for that activity. Ecological footprint analysis is based on two assumptions. First, that it is possible to keep track of most of the resources that a human population consumes and most of the wastes that the population generates. Second, that these resource and waste flows can be converted to a biotically productive area necessary to provide the resources and to assimilate the wastes. The biotically productive area which performs these functions is termed the "ecological footprint" of the human population. Thus, ecological footprints quantify the biotically productive area that a population uses. Locations (nations, regions, states, watersheds, etc.) in which the ecological footprint of the resident human population is greater than the area which they occupy must be doing at least one of the following: receiving resources from elsewhere, disposing of some of its waste outside of the area, or depleting the area's natural capital stocks. To deplete natural capital stocks means to withdraw more ecological services than the biotic capacity of the defined area can regenerate; for example by harvesting timber faster than it can regrow or by discharging sewage at a rate faster than can be assimilated.

Attempts to estimate the biosphere's capacity to support human needs go back several centuries, and the debate continues (Cohen, 1995). These estimates are based on a variety of approaches. Many have assumed that human population size is limited by food, and have attempted to sum agricultural productivity in various regions of the earth to obtain total agriculturally productive area and capability. Others have considered limiting factors in addition to, or other than, food. Most estimates, however, attempt to determine human carrying capacity as the number of people that can be supported by a given area of the earth's surface. Ecological footprint analysis, on the other hand, inverts the process. Rather than asking how many people can live in an area, it estimates the area of the earth's surface required to support a given human population. For other terrestrial animal species the two approaches are equivalent and relatively invariant across subpopulations. For humans, however, the area required to support subpopulations is highly variable, because subpopulations differ greatly in their intensity of resource use and waste discharge. This variability has given rise to the erroneous assertion that human carrying capacity is meaningless.

The purpose of this paper is to explain a method for measuring how much ecological capacity humans use to sustain themselves, and to indicate how this method could be useful in an environmental monitoring and assessment context. To move toward sustainability – that is, to develop sustainability – a necessary step is to clarify what it means. Many confusing definitions and statements, including the one in the Brundtland report (WCED, 1987), have impeded progress. To sustain something means "to provide for its support or maintenance" (Webster's Third New International Dictionary, Unabridged, 1976). It also means "to continue without interruption or diminution." We therefore find it useful to define regional sustainability as "*the continuous support of human quality of life within a region's ecological carrying capacity.*" By "support of human quality of life" we mean that people's subjectively perceived well-being (that is their physical and psychological comfort, including their health, security, and friendly connections to other people) must be at least maintained (or possibly improved, in the case of the poor). Otherwise, people could feel worse off as society moves toward sustainability. In addition, not living decently and equitably may cause conflict and degrade the social fabric. This would make the necessary cooperation unworkable. By "a region's ecological carrying capacity" we mean the ecological or biotic capacity within a region to regenerate used resources and to assimilate waste.

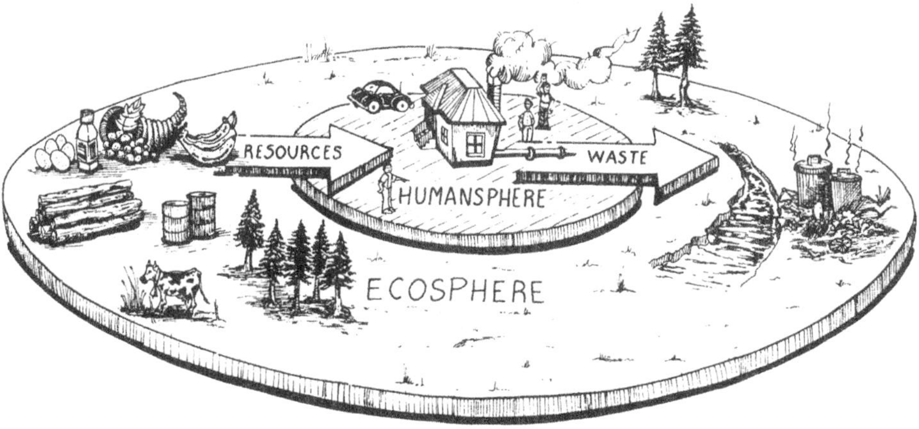

Fig. 1. People are part of nature. The humansphere is a dependent subsystem of nature. There are no activities of the human economy that fall outside of nature's economy. Nature supports humanity by dispensing resources, absorbing waste, and securing life-support services. (Illustration after Phil Testemale).

2. The conceptual basis of ecological footprints

People depend on the biosphere for a steady supply of the basic requirements for life: energy for warmth and mobility; wood for housing, furniture and paper products; fibers for clothing; quality food and water for healthy living; ecological sinks for waste absorption; and many non-consumptive life-support services. This human use of nature is termed the ecological footprint. Obviously, this footprint is not a continuous piece of land. Due to international trade, the land and water areas used by most global citizens are scattered all over the planet. It would take a great deal of research to determine where their exact locations are, assuming that an exact location could even be specified. To simplify comparisons among various regions of the earth, the occupied space is calculated by adding up the areas (using world average productivity) that are necessary to provide a human population with all the ecological services it consumes.

Every person, and every assemblage of people (e.g. a city or country), has an impact on the Earth. The ecological impact corresponds to the ecological footprint of the individual or assemblage. This use of nature includes the areas used for waste discharge assimilation, *and* the productive areas used for resource regeneration. The ecological footprint quantifies for any given population the mutually exclusive biotically productive area that must be in continuous use to provide its resource supplies and to assimilate its wastes. Area that is in continuous use to support one human population cannot simultaneously support another population without depleting natural capital stocks.

As mentioned above, ecological footprint studies build on a wide range of methods to assess nature's capacity to support human life. Apart from the early attempts (Cohen, 1995), much intellectual ground-work was laid in the 1960s and 1970s. Examples are Howard Odum's "emergy" analysis examining systems through embodied energy flows (Odum, 1994; Campbell, this volume), Jay Forrester's advancements on modeling world resource dynamics (Meadows *et al.*, 1972, 1992), John Holdren's and Paul Ehrlich's I=PAT formula (Holdren and Ehrlich, 1974), or, in the context of the International Biological Programme, Robert Whittaker's calculation of net primary productivity of the world's ecosystems (Whittaker, 1975, Lieth and Whittaker, 1975). The last ten years have witnessed exciting new developments: life cycle assessments (e.g., Abel *et al.*, 1990), lifestyle energy assessments (e.g., Hofstetter, 1991), environmental space calculations building on ideas of Johann Opshoor (Buitenkamp *et al.*, 1992), human appropriation of net primary productivity (Vitousek *et al.*, 1986), documentation of regional and industrial metabolisms (Ayres *et al.*, 1994), mass intensity measures such as Mass Intensity per Unit of Service MIPS (Schmidt-Bleek, 1994), measures of human processes such as the Sustainable Process Index SPI (Krotscheck and Narodoslawsky, 1996), national resource inventories (as performed by the Norwegians and the French), resource accounting input-output models (Duchin and Lange, 1994), computer based gradient models for analyzing land-use developments and ecological potentials (Hall, 1996), the "Polstar" scenario model (Gallopin *et al.*, 1997), and ecological footprint assessments (Wackernagel and Rees, 1996; Folke, 1996), to name just a few. Their applications and representations may vary, but their output is mostly the same: quantification of the human use of nature. As most of these approaches are compatible, results from one strengthens the others.

What differentiates the ecological footprint from other assessment methods is the way it interprets throughput analyses of human activities: it aggregates human impacts in an ecologically meaningful way, expressing them in mutually exclusive ecological spaces which are appropriated to provide the functions and services of nature. Therefore, we have also called the ecological footprint "appropriated carrying capacity." Of course, ecological functions that can be provided on the same space at the same time must only be counted once – otherwise the footprint overestimates the use of nature. This is why we refer to "mutually exclusive" biotically productive spaces. For example, in the case of double-cropping, photovoltaic use of roofs for energy supply, or water collection in a sufficiently humid timber plantation, only one utilization is added to the footprint. However, some forest uses are mutually exclusive. Biodiversity protection may depend on undisturbed ancient forests which cannot serve for timber-production without endangering biodiversity. On the other hand, recent research indicates that forests producing timber and agroforestry crops also may be credited with significant carbon dioxide (CO_2) sequestration in soils and long-lived forest commodities such as furniture or housing components (Moffat, 1997; Janzen, 1997).

This biogeophysical interpretation used by the ecological footprint concept has two advantages. First it makes the results more accessible. Everyone has experienced space, while many other quantities (like embodied energy content or erosion rates) may require more technical skills to interpret or to appreciate. Second, and more importantly, the human "demand" for ecological space can be compared easily to the earth's finite "supply" of space. The surface of the Earth is finite; therefore the available ecologically productive space must be finite. By providing the means of comparing human demand and nature's supply in the same units, the assessment results show clearly, at each geographical scale of analysis, the magnitude of the human load on the biosphere.

Fig. 2. The ecological footprint measures our use of nature. Every person, region or nation depends on ecological capacity to sustain itself. A population's ecological footprint corresponds to the aggregate land and water area in various ecosystem categories that is claimed by that population to produce all the resources it consumes, and to absorb all the waste it generates on a continuous basis, using prevailing technology. (Illustration: Phil Testemale).

3. Biotic productivity available on the Earth

Many human uses of nature compete for space. Land used for wheat production cannot be used for roads, forests or grazing, and vice versa. These mutually exclusive uses of nature are summed to assess the total ecological footprint. In this analysis, six main categories of ecologically productive area are distinguished: crop land, pasture, forest, ocean, built-up land and energy land:

Crop land is the land used to grow fruits, vegetables and grain for human consumption either directly, or indirectly by feeding it to livestock. Typically it is, from an ecological perspective, among the most productive land; it can grow the largest amount of human-consumable plant biomass per unit area. Today, there exists less than 0.25 hectares per capita worldwide of such highly productive land.

Pasture is grazing land for livestock, to produce dairy products and meat. Most of the 3.35 billion hectares of pasture, or 0.6 hectares per person, are significantly less productive than crop land. That is, its potential for accumulating biomass is much lower than that of crop land. In addition, conversion efficiencies from plant to animal reduce the available biochemical energy to humans by typically a factor of ten.

Forest refers to tree plantations or natural forests that can yield timber products. Of course, they may provide many other functions too, such as erosion prevention, climate stability, maintenance of hydrological cycles, and if they are managed properly, biodiversity protection. With 3.44 billion hectares covering the planet, there are 0.6 hectares per capita worldwide. Today, most of the remaining forests occupy ecologically less productive land.

Ocean covers 36.3 billion hectares of the planet, or a little over 6 hectares per person. Roughly 8 percent of this area, concentrated along the continental coasts, provides over 95% of the sea's ecological production. In per capita terms, there are 0.5 hectares of ecologically productive sea space out of these 6 hectares ocean. Measuring the ecological activity of the sea by its area (and not its volume as one might intuitively think) makes sense ecologically. It is surface which limits its productivity, as both the capturing of solar energy and gas exchanges with the atmosphere are proportional to surface area.

Built-up land refers to land used for human settlements and roads and consists of approximately 0.03 hectares per capita worldwide. As most human settlements are located in the most fertile areas of the world, *built-up land* often leads to the irrevocable loss of significant amounts of ecological capacity.

Energy land is the land that would be required for sequestration of CO_2 released by fossil fuel combustion. Alternately, it is the land area that would be required to accumulate an equivalent amount of usable energy via wood biomass. This latter approach would require a larger land area than for CO_2 absorption, because not all accumulated biomass would be usable for energy. Currently, no land is used exclusively to sequester CO_2 or to replenish the biochemical energy stock lost through fossil fuel burning (but see Moffat, 1997; Janzen, 1997).

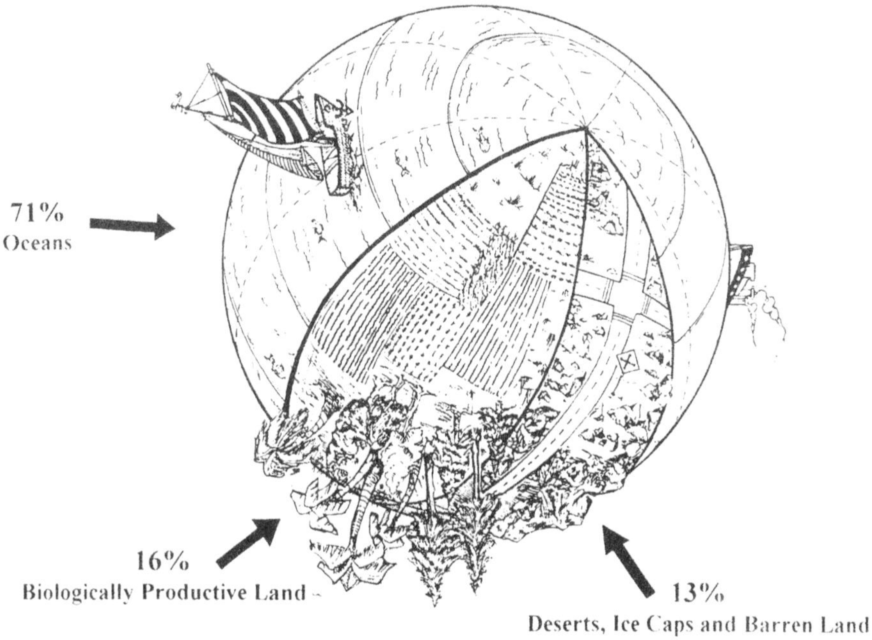

Fig. 3. The biotically productive areas on our planet. The Earth has a surface area of 51 billion hectares, of which 36.3 billion are sea and 14.7 billion are land. Only 8.3 billion hectares of the land area are biotically productive for human use. The remaining 6.4 billion hectares are marginally productive or unproductive for human use, as they are covered by ice, have unsuitable soil conditions or lack water. (Illustration after Phil Testemale).

4. A reference point for sustainability: ecological space per global citizen

Adding up the biotically productive land per capita worldwide; 0.25 hectares of arable crop land, 0.6 hectares of pasture, 0.6 hectares of forest and 0.03 hectares of built-up land; shows that there exist approximately 1.5 hectares per capita, or 2 hectares per capita including ecologically productive sea space. However not all that space is available for human use, as this area must also provide for the millions of species with whom humanity shares the planet. The World Commission on Environment and Development proposed to set aside for biodiversity protection at least 12 percent of the earth's ecological capacity, representing all ecosystem types. Although 12 percent may not be enough for securing biodiversity in the long term (Noss and Cooperrider, 1994), conserving more at this time may not be politically feasible.

If we accept 12 percent as a minimum number for biodiversity preservation, one can calculate that from the approximately 2 hectares per capita of biotically productive area that exist on our planet, **only 1.7 hectares per capita, at most, are available for human use**. These 1.7 hectares become the ecological benchmark for comparing ecological footprints.

It is the current ecological reality. Therefore, to achieve sustainability with current population numbers, the average footprint needs to be reduced to at least this size. If some people need or demand much more ecological capacity, then in a sustainable world economy others must use much less than the average amount available. Assuming no further ecological degradation, the amount of available biotically productive space will drop to approximately 1 hectare per capita if the world population reaches its predicted 10 billion. If current growth trends persist, this will happen in little more than 30 years.

5. Using the U.S. as an example: the calculation procedure for assessing national footprints

As indicated, footprints can be calculated at every scale: from global, national, regional, and municipal down to household size. In fact, footprint studies exist for each of the mentioned scales. To demonstrate the mechanics of these calculations, a national case is presented here.

Earlier national footprint calculations were much cruder and more simplified estimates, using an eclectic variety of data sources (Wackernagel *et al.*, 1993; Neumann, 1995; Wackernagel and Rees, 1996; Graszl, 1996). As they were still first attempts, they did not follow a consistent methodology. Subsequently, Wackernagel *et al.* (1997) have developed more consistent and complete national calculations in the form of spreadsheet-based yearly accounts of the resource flows of a nation. The presented example shows the U.S. footprint for 1993, the latest year with a complete United Nations data set available when the study was completed early in 1997 (see Table 1). Mainly United Nations data were used for this assessment to make countries comparable among themselves. These current estimates show larger footprints than previously, as consumption is documented more completely and as productivity data for forest and pasture are lower than assumed in earlier estimates. In fact, the presented calculations lead in the case of the U.S. to footprints that are about one quarter larger than the ones presented in Wackernagel and Rees (1996).

The full spreadsheet for the United States contains 120 lines and 14 columns. Table I presents a condensed version for illustrative purposes.

The spreadsheet is composed of three main areas. The upper part consists of a consumption analysis of over 20 main resources. The rows represent resources or product types. The columns specify the productivity, production, import, export and consumption of these resource or product types. Consumption is calculated by adding imports to production and subtracting exports. Using estimates from the Food and Agriculture Organization of the United Nations (FAO) of world average yield, consumption and waste absorption are translated into appropriated ecologically productive area. In other words, the consumption quantities are divided by their corresponding (world average) biotic productivity which gives us the land and sea areas necessary to sustain this consumption. These areas form a part of the total footprint.

Table I

Calculation of the American's average Ecological Footprint (1993 data)

population of the United States of America: 258,262,000

LAND AND SEA AREA ACCOUNTING *CATEGORIES* *units if not specified*	*Yield* [kg/ ha] (global average)	*Production* [t]	(biotic resources) *Import* [t]	*Export* [t]	*Consumption* [t]	***Footprint component*** [ha/cap]	
FOODS							
.meat (average animal units)	74	31,277,000	1,211,559	2,129,258	30,359,300		
..meat (fresh)			977,112	1,946,500			
..bovine, goat, mutton, buffalo	33	10,737,000	714,840	412,305	11,039,535	1.302 pasture	
..non-bovine,-goat,-mutton,-buffalo		20,540,000	496,719	1,716,953	19,319,765	(already in cereals)	
.dairy (milk equiv.)		68,303,000	1,521,211	7,127,226	62,696,985	0.483 pasture	
..milk	502	68,303,000	37,891	5,734,796			
.cheese	50		146,091	18,522			
..butter	50		2,241	120,721			
.marine fish	29		consumption in [kg/cap]		33	1.137 sea	
.cereals	2,744	258,952,000		100,659,208	158,292,792	0.223 arable land	
..wheat				35,666,000	(35,666,000)		
..maize				40,365,000	(40,365,000)		
.animal feed	2,744			18,758,000	(18,758,000)	-0.026 arable land	
.veg & fruit	18,000	63,040,000	11,259,373	8,073,933	66,225,440	0.014 arable land	
..veg etc			2,480,020	2,676,121			
..fresh fruit			5,367,738	3,430,552	1,937,186		
.roots and tubers	12,607	19,949,000	323,229	244,680	20,027,549	0.006 arable land	
.pulses	852	1,285,000	80,601	489,545	876,056	0.004 arable land	
.coffee & tea	566	500	2,187,899		2,188,399	0.015 arable land	
.cocoa	454				625,917	0.005 arable land	
.sugar	4,893	6,964,000	1,784,840	321,280	8,427,560	0.007 arable land	
.div.food	2,744			490,878	(490,878)	-0.001 arable land	
.oil seed (incl. soya)	1,856	63,340,000	505,696	2,388,813	61,456,883	0.128 arable land	
TIMBER[roundwood equivalent,m3]	1.48	449,150,000	103,565,960	72,943,407	479,772,553	1.255 forest	
.roundwood [m3/ha,m3]	*waste factors*	495,800,000	2,406,000	26,680,000	471,526,000	*final use*	
.fire wood	0.5	93,300,000	597,000	261,000	93,636,000	10 % of cons.	fire wood
.direct roundwood consumption [m3]	1				9,900,000	2 % of cons.	mines
.sawnwood [m3]	1.65	106,167,000	36,489,000	9,411,000	133,245,000	46 % of cons.	sawn wood
.wood based panels [m3]	2.48	31,568,000	5,446,000	3,359,000	33,655,000	17 % of cons.	panels
.wood pulp [t]	1.98	58,310,000	4,915,000	5,961,000	57,264,000	(not a final use)	
.paper and paper board [t]	1.47	77,250,000	11,885,000	7,146,000	81,989,000	25 % of cons.	paper
OTHER CROPS							
.tobacco	1,548	3,512,000		1,240,556	2,271,444	0.009 arable land	
.cotton	1,000				2,176,245	0.008 arable land	
.jute	1,500				76,686	0.000 arable land	
.rubber	1,000				842,954	0.003 arable land	
.wool	15				104,693	0.027 pasture	
.hide	33				459,663	0.054 pasture	

ENERGY BALANCE:			
	100 [Gj/ha/yr]	**fossil and nuclear energy consumption**	**313 [Gj/yr/cap]**
	1,000 [Gj/ha/yr]	**hydro-electricity consumption**	**4 [Gj/yr/cap]**
	100 [Gj/ha/yr]	**energy embodied in net import**	**6 [Gj/yr/cap]**

S U M M A R Y

DEMAND FOOTPRINT (per capita) Category	[ha/cap]	*SUPPLY* EXISTING CAPACITY WITHIN THE U.S.A. (per capita) Category	yield factor	local area [ha/cap]	yield adjusted area [ha/cap]
fossil energy	3.23	CO2 absorption land		0.00	0.00
arable land	0.43	arable land	1.56	0.73	1.14
pasture	1.84	pasture	1.92	0.93	1.78
forest	1.26	forest	2.72	1.11	3.02
built-up area	0.61	built-up area	1.56	0.39	0.61
sea	1.14	sea	1.00	1.24	1.24
TOTAL used	**8.49**	**TOTAL existing**		4.38	**7.77**

OTHER INDICATORS (all in [ha/capita] with world average productivity)

footprint on the land:	7.35	available capacity within the U.S.A. (incl. sea space):	6.84
existing land within the U.S.A.:	6.54	national ecological deficit:	1.65

The middle part of the table provides an energy balance of the traded goods. Such an analysis is necessary to adjust the energy directly consumed within the country by the amount of energy that was previously consumed in producing the exported and imported goods. This traded energy is calculated by multiplying, for each trade category, the amount of net import by the typical embodied energy of these commodities. Particularly for small countries, embodied energy in net imports can be a significant portion of the consumed energy. To keep the table simple, only the net result of this energy trade analysis is listed in Table I.

In the bottom part, the results are summarized in two boxes. Here all of the footprint components are added to obtain the total footprint. The left box itemizes the ecological footprint in six ecological categories and gives the total. The results are presented as *per capita* figures. Multiplying the per capita data by the country's population gives the total footprint of the nation. The right box shows how much biotically productive capacity exists within the country. Worldwide, there exist 2 hectares of ecologically productive space per person, as mentioned above, and with a 12% set-aside for biodiversity protection, only 1.7 hectares are available for human use. However, some countries are better endowed with ecological productivity by having either more space available and/or ecosystems and agroecosystems of higher productivity per unit area. Therefore, to document the ecological production available within a country, the number of physical hectares of biotically productive area that exist in each ecological category within the country (second column in the right box) is multiplied by the factor by which the country's ecosystems differ in productivity from the world average (first column in the right box). We call this factor the "yield factor." A yield factor of 1.5 would mean that the local productivity is 50 percent higher than world average – absorbing 50 percent more CO_2 or producing 50 percent more potatoes per hectare. A yield factor smaller than one indicates that the area is less productive than world average. Multiplying the yield factors by the number of physically existing hectares gives an equivalent area with world average productivity. This area we identify as the "yield adjusted area" (third column in the right box).

For example, the U.S. yield factor for arable land is assessed to be 1.56 based on the U.S. yield of cereals as compared to world average. The U.S. yield factor for forest is assumed to be 2.72. It is extrapolated from European yields as we have been unable so far to find a reliable estimate of sustainable timber yield in the U.S..

From the 7.77 hectares per person of existing yield-adjusted area, 12 percent is subtracted to get 6.84 ha/capita of locally available capacity. This area is a measure of the biotic capacity in the US. The number is listed at the bottom under "other indicators." With this adjustment, both footprints and ecologically productive spaces (or capacities) are expressed in the same units: in areas with world average productivity. Footprints and available capacity can now be compared among all nations of the world.

In this presented case study of the average U.S. citizen, the calculations show a footprint of 8.49 hectares. This means that over eight hectares of biotically productive space (based on world average productivity) must be in constant production to support the average United States citizen. This footprint occupies five times more space than the available 1.7 hectares per world citizen. Only countries with footprints lower than 1.7 hectares per person have a global impact that could sustainably be replicated by everybody; that is, without depleting the natural capital stock of the earth.

Fig. 4. National ecological deficits. The ecological footprint measures how much ecological capacity people occupy. Through the magnitude of their economic activity, some regions claim more ecological capacity than there is within their boundaries. This means that the region runs an ecological deficit. Consequently, the local population needs to import the missing ecological capacity -- or deplete their local natural capital stocks (on the left). Regions with footprints smaller than their capacity live within their nation's ecological means (on the right). Often, however, the remainder is used for producing export goods which partially cover the deficits incurred by other regions.

While the ecological footprint shows the global impact of local consumption, it may also be of interest to determine to what extent local ecological productivity could provide for local consumption. Therefore Wackernagel *et al.* (1997) compared the ecological footprints of 52 large countries in the world with the biotically productive space available within the country. If the footprint exceeds the available biotically productive area of the country, as in the U.S. example presented here, it runs a national ecological deficit. In that case, the country's area alone cannot sustainably provide sufficient ecological services to satisfy its population's current patterns of consumption. Consequently, as mentioned above, they need to either import services or deplete their natural capital stock. The United States is fortunate to have available 6.8 hectares of ecologically productive space per citizen. For the U.S., consequently, the national ecological deficit is 1.7 hectares per person (8.5 ha US footprint less 6.8 ha available in the U.S.). The U.S. global ecological deficit is even larger: 6.8 hectares per person (8.5 ha U.S. footprint less 1.7 ha available in the world). In comparison, the world as a whole with an average footprint of 2.3 hectares per capita and an available space of 1.7 runs a deficit of 0.6 hectares per capita (Wackernagel *et al.*, 1997). As the world cannot import ecological capacity from somewhere else, this global deficit corresponds to an unsustainable ecological overuse: more timber cut than can reproduce, more CO_2 released than can be absorbed, etc.

The above study by Wackernagel *et al.* (1997) showed that there are only ten countries whose citizens use less than the amount available on a worldwide per capita basis. In other words, if all people of the world adapted the lifestyle of the first 43 countries, the world's ecological assets would be rapidly depleted. Footprints beyond per capita available world capacity show local contribution to global ecological decline. The national ecological deficit shows that 41 of the countries examined consume beyond national ecological capacity. This national ecological deficit becomes an indicator of potential vulnerability to external instabilities.

6. Some Environmental Monitoring and Assessment applications

National calculations are just starting points for more comprehensive sustainability studies. Monitoring footprint assessments over time could reveal progress toward sustainability by tracking a country's or a region's ecological deficit. The paper by Ko *et al.* in this volume shows such a possible study (1997). Such monitoring analyses and assessments can indicate to what extent economic and demographic change have expanded or contracted a nation's or region's footprint. They thereby become indicators of countries' (in most cases increasing) potential vulnerability to economic dislocations and their contribution to global ecological decline. For most countries, these time series could be calculated with comparatively little effort as most of the necessary data can be found in already existing statistical collections of each nation. The appropriate data would only have to be fitted into the ecological footprint accounting framework explained above.

At the sub-national level, particularly for regions or watersheds that live directly from local resources, accounting for local natural capital becomes vital. For managing these resource-exporting or subsistence economies, decision-makers need to know how much natural capital is available for use within their region and how much, if any, is available for export. With the gradual extension of the global economy, a community's security may no longer be provided by government institutions. Further, for most countries the value of monetary savings is diminishing rapidly as their currencies lose purchasing power. For the majority of the people living in these countries, monetary savings are therefore also an unreliable option for securing their long-term well-being. In lack of institutional or market support, local natural capital therefore becomes the ultimate source of security and wealth. Thus nature preservation, such as erosion control, reforestation, or decontamination, becomes not merely an altruistic deed but a necessary investment strategy for the local community to ensure a better and more secure future.

Watershed or ecoregional (Omernick, 1987) assessments would start from an inventory of the ecologically productive spaces in the watershed or ecoregion. Ecoregional assessments, where ecoregion designations are available, have the advantage of relatively uniform ecological productivity throughout the region. In fact, this is one of the distinguishing characteristics of ecoregions. Geographic Information Systems (GIS) may be useful to capture the dimensions of all the ecologically productive space categories and to register their respective productivities and uses. Such a survey would provide an estimate of the local ecological capacity – the supply side. The demand side, that is the footprint of the local population, can be documented with various degrees of precision. Given an estimated per-capita demand for ecosystem services, and the productivity of the ecoregional categories in a watershed or other region, an estimate of the human carrying capacity deficit or excess of the region could, in principle, be obtained. If comparable watershed or ecoregional assessments were conducted over a gradient of ecological capacity and of human population footprint, it may even be possible to correlate carrying capacity deficits with more traditional indicators of ecological integrity.

A first estimate of per-capita demand may be extrapolated from the national footprint assessments. For more precision, the national per capita footprint can be adjusted according to the differential between national and local purchasing power. For even finer scaled analyses, consumption patterns of households representing the basic income categories need to be surveyed.

Fig. 5. Humanity's ecological impact. Every individual human has an impact. Is nature able to cope with humanity's cumulative impact? Our calculations show that the ecological footprint of humanity is already larger than the biotically productive space on the planet (Wackernagel et al., 1997). This overshoot results in costly degradation and erosion of natural capital. The ecological footprint offers a tool to measure this overuse and helps us plan for a sustainable future where people's quality of life can be supported within the carrying capacity of nature.

A "production footprint" can be determined, in addition. While the conventional ecological footprint answers the question of how much ecological capacity is necessary to support a population's consumption (with all its associated resource use and waste generation), the "production footprint" analyzes the ecological capacity necessary to keep the population's economic production running. This corresponds to the ecological functions and services required to generate the population's income so they can purchase their consumption. To illustrate, consider the example of an industrialized farm that sells all its products to the market. All the fields, plus the resources to work and harvest them, correspond to the farm family's production footprint – the natural capacity necessary to sustain their income. Their consumption footprint, however, would correspond to their private garden with their home grown vegetables plus the area required to generate all the food, furniture, medical bills and other consumption goods and services that they buy. Similarly, resource-exporting regions may give up significant amounts of ecological capacity while receiving little capacity in return via imported products. This becomes manifest in the discrepancy between a population's or region's consumption footprint and their production footprint. Such differences point to ecological leakages. These leakages represent potentials for improving the local standard of living by using the local natural capital more effectively.

Current ecological footprint assessments still omit some uses of ecosystems for resource production and waste absorption. Therefore, new developments in the footprint research will focus on making the calculation method more complete. Particularly in arid countries, fresh water supply becomes a critical resource that should be covered by footprint studies. There, human settlements, agriculture and other ecosystems compete for this resource (Pimentel *et al.* 1997). Water is appropriated for human use at high energy costs and often with significant ecological impacts. Some areas are used exclusively to catch water for domestic use (as is done in many areas to ensure the delivery of healthy water). Areas that are dedicated to absorb human waste water are additional ecological spaces that should be added to new footprint assessments. If water is withdrawn from surface freshwater ecosystems for human consumption, thereby reducing ecological productivity, an ecological area necessary to compensate for this loss needs to be included. If groundwater is used beyond its recharge rate, negative ecological effects may not be felt immediately. One possible way of calculating the footprint of groundwater use may be to assess the resource costs that would be associated with restoring the original groundwater level (e.g., by getting water from an area with surpluses). Further, the ecological capacity to provide the infrastructure and operational energy for water withdrawal, transport, distribution and cleaning represents additional ecological capacity to be added to the footprint of a human population.

Examples of ecologically productive area appropriated for human freshwater use include: the 0.27 to 0.37 hectares of land per capita set aside in typical Australian cities to collect domestic fresh water (Foran in Wackernagel and Rees 1996); the "Three Gorges Dam" hydroelectric power project on the Yangtze River in China which is trading off ecologically productive area in return for electricity (World Resources Institute 1994); the extraction of water from rivers supplying the Aral Sea, with loss of marine productive area and salinization of land irrigated by the extracted water (World Resources Institute 1996); reservoirs in the western United States accompanied by salinization problems (WRI 1996); and ground water depletion in Mexico City where pumping rate exceeds natural recharge rate by 50 to 80 percent (Postel 1996). In other cases, ecologically nonproductive desert areas may become productive by irrigation which avoids waterlogging and salinization (as in the Ladakh valley, or in some areas of Israel and California).

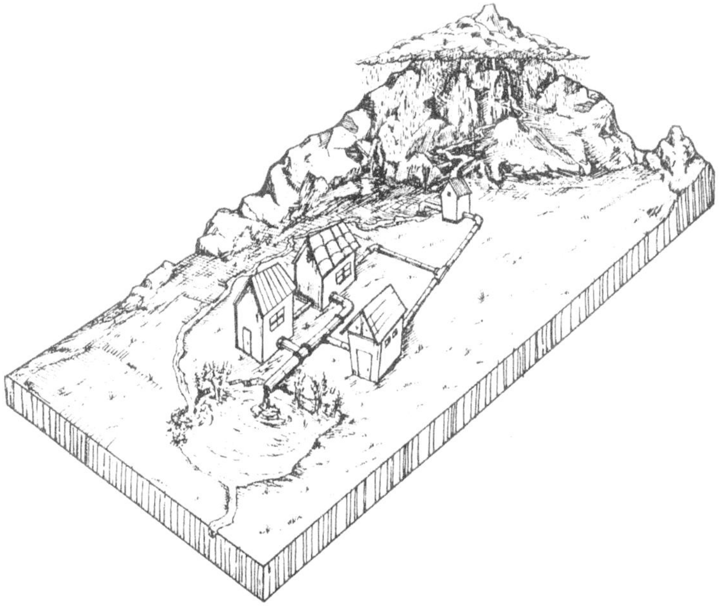

Fig. 6. The water footprint. Domestic water use occupies various ecological spaces: (a) exclusive areas to capture the water, or to compensate lost ecological productivity caused by excessive water withdrawal from an ecosystem, (b) embodied resources in the construction and operation of infrastructure to transport, distribute and dispose the water, and (c) ecological spaces (or human infrastructure) to assimilate and clean the waste water.

In addition to the area appropriated for fresh water or hydropower supply, the ecological impacts of contamination and waste streams are only partially included in current assessments. In current footprint calculations, the major waste stream included in current assessments is the land required to sequester CO_2 from fossil fuel burning. While marine ecosystems are also potential sinks for CO_2, here we focus on terrestrial ecosystems. However, considering the limited ways in which marine ecosystems can be successfully manipulated, the sea's potential as an additional CO_2 sink is questionable (see also Sarmiento and Le Quéré, 1996). Reforestation is therefore the most effective strategy for CO_2 absorption (Moffat, 1997).

Once the use of fresh water is added to the footprints, the appropriation of area by assimilation of waste water would show up as part of the occupied ecological space. Still, the ecological footprint of areas lost due to soil contamination, as manifested, for example, in industrial areas of the former Soviet Union, in radioactively contaminated areas such as Chernobyl, in soil salinization, or in the many cases of acid rain all over the world are still left out due to lack of data. Such contamination can reduce ecological productivity significantly or make products of nature unfit for human use. In the case of regional assessments, data may be available on locally contaminated areas as well as on the amounts of locally discharged contaminants (like SO_x or trophospheric ozone) leaving the region and impacting ecological productivity elsewhere. For example, ozone levels of 200 micrograms per cubic meter may reduce agricultural yields up to 15 percent, according to a Swiss government publication (Baudepartement Basel-Stadt, 1997). Also, the impacts of local solid waste on water, its potential for soil contamination, and the resources necessary for their management, for example in "Superfund" contaminated sites such as Love Canal in the USA (Rosenbaum, 1995) can be incorporated in local assessments if local information is available. These contaminated areas could be added to the "built-up" category as it is no longer useful to people for ecological services.

We could improve accuracy by analyzing fossil energy in finer categories. For example, compared to liquid fossil fuel the CO_2 release per energy unit for natural gas is approximately 25% lower and for coal 25% higher. Also, activities which release greenhouse gases should be included more systematically. Hydroelectricity could be analyzed more specifically to get a more accurate conversion figure (even though in most examples, hydroelectricity occupies only a small percentage of the total footprint). Furthermore, traded goods should be accounted for not only in terms of embodied energy but also according to their embodied material resources and waste discharges.

The merit of our current footprint calculation method is its easy replicability. It is sufficiently detailed to give a general indication of the magnitude of human impact globally. Also, by using the same assumptions for all assessments, the results of all countries are comparable in relative terms.

Current numbers probably underestimate ecological footprints. First, they use forest and agricultural productivities that are doubtlessly too high, at least for the long run. Industrial forestry and agriculture with its high yields may not be sustainable over long periods of time due to erosion and soil depletion. In poorer countries, UN statistics may underestimate production and hence footprint area as they may not capture adequately direct consumption and secondary crops in agricultural production. Second, present footprint calculations leave out various additional ecological functions such as water use and water-borne waste assimilation. These uses may add significant area to the footprint as it is now calculated. Furthermore, there are also clear limits to the accuracy of the presented footprint assessments. Within the current methodological approach, additional uncertainties arise from the lack of differentiation between the carbon intensity of the various fossil energy sources, and from the embodied energy figures and UN statistics which are not equally accurate for each nation. Even within the UN publications, Wackernagel *et al.* (1997) found discrepancies between the same data reported in different publications while preparing their "footprints of nations" study. However current estimates provide a first reference point. In this way, these national calculations offer an analytical framework that may be useful at the regional level and provide benchmark results with which to compare regional analyses.

7. Conclusions

The ecological footprint is designed to provide an area-based indicator of the extent of the human appropriation of nature's goods and services relative to what is available for appropriation. Quantification of the available ecological productivity which is appropriated by categories of human use provides information on where excess appropriation can be reduced. The figures should not lead merely to a more informed discussion of our challenges ahead, but more importantly, such assessments can help governments, businesses and NGOs shape sustainable development. The measure shows where we are, in which direction we need to go, and which projects and programs most effectively move us there. In more specific terms, these biophysical assessments can assist sustainability efforts on various levels. They:

- **Offer a measure of carrying capacity available for human use.** Many countries and other subdivisions of the earth live on footprints larger than what their own ecosystems can provide. This frames the sustainability challenge: if we wish to secure well-being to people for some generations to come and avoid human suffering caused by an ecological down-turn, we need to live again within our ecological means. Footprint assessments would give us an indication to what extent humanity's economic activities would have to become less resource consumptive and less contaminating. Also, it helps us to comprehend the ecological impact of humanity's growth trends.
- **Become an indicator of sustainability.** Not knowing what is sustainable, not knowing where we are or where we are going makes the future more risky. In contrast, understanding our ecological constraints and identifying future risks supports informed decision making. This reduces threatening uncertainties and points to new opportunities.
- **Integrate concerns about the relative importance of human population and consumption.** The numbers show the impact of both population level and per-capita consumption rates. Clearly, the high level of consumption in industrialized countries takes the biggest share of the planet's bounty. But with ever larger populations it becomes less likely that everyone's quality of life can be secured.
- **Provide a target for assessing progress.** Essentially, the sustainability debate reduces to the fact that there are on average only 1.7 biotically productive hectares available per person on this planet. Population growth and ecological deterioration are steadily reducing this area. The key question is therefore: can a high and attractive quality of life for everyone be obtained out of 1.7 hectares? Experiments and case studies to highlight this question and show how it might be possible to live within these limits would be helpful.

Ecological footprint assessments demonstrate that sustainability can be measured. The ecological footprint indicator shows clearly where we are and where we need to be. Ecological examinations as presented here can give direction for local, national and global efforts to close the sustainability gap. They become an effective planning tool and a guidepost for a more secure, equitable and sustainable future.

Acknowledgments

The study on which the data analysis of this paper is based was commissioned and funded by *The Earth Council* for the *Rio+5 Forum* held in Rio de Janeiro from March 13 to 19, 1997. This study would not have been possible without the collaboration of Larry Onisto, Alejandro Callejas Linares, Ina Susana López Falfán, Jesus Méndez García, Ana Isabel Suárez Guerrero and Ma. Guadalupe Suárez Guerrero. Iliana Pámanes produced the illustrations. We are particularly indebted to Charles Hall and Jae-Young Ko for their careful review and well thought-out suggestions which helped us significantly improve this paper.

References

Ayres, R. and Simmonis, U. (eds.): 1994, *Industrial Metabolism: Restructuring for Sustainable Development*. UN University Press, Tokyo and New York.

Abel, S., Braunschweig, A. and Müller-Wenk, R.: 1990, *Methodik für Ökobilanzen auf der Basis ökologischer Optimierung* (Methodology for life cycle assessment based on ecological optimization). Bern: Bundesamt für Umwelt, Wald und Landschaft. Schriftenreihe Umwelt, Vol.**133**.

Baudepartement Basel Stadt: 1997, 'Nur Abgasreduktion schützt dauerhaft vor Ozon' (only a reduction in emissions will durably reduce ozone). *Unser Lebensraum* 1/97, Baudepartement Basel Stadt, Switzerland.

Buitenkamp, M., Venner, H. and Wams, T. (editors): 1993, *1033 Action Plan Sustainable Netherlands*. Dutch Friends of the Earth. Amsterdam, the Netherlands.

Campbell, D. E.: (this volume), 'Emergy Analysis of Human Carrying Capacity and Regional Sustainability: An Example Using the State of Maine'.

Cohen, J. E.: 1995: *How Many People Can the Earth Support?* W. W. Norton & Co., New York.

Duchin, F. and Lange, G. M.: 1994: *The Future of the Environment: Ecological Economics and Technological Change*. Oxford University Press, Oxford.

Folke, C., *et al.*: 1996, 'Renewable Resource Appropriation by Cities.' in Costanza, R. *et al.*: 1996. *Getting Down to Earth*. Island Press, Washington DC.

Gallopin, G., Hammond, A., Raskin, P. and Swart, R.: 1997, *Branch Points: Global Scenarios and Human Choice*, PoleStar Series Report No.7, Stockholm Environment Institute.

Graszl, H.: 1996, *Der Fussabdruck Feldbachs* (The Footprint of Feldbach), Universität Graz, Austria.

Hall, C. A., Tian, H., Qi, Y., Pontius, G., Cornell, J. and Uhlig, J. 1995, *Spatially Explicit Models of Land Use Change and Their Application to the Tropics*, DOE Research Summary, CDIAC, Oak Ridge National Laboratory.

Hofstetter, P.: 1991. *Persönliche Energie - und CO_2-Bilanz*. (Personal Energy and CO_2 Balance). Second draft. Büro für Analyse und Ökologie, Zürich.

Holdren, J. and Ehrlich, P.: 1974. 'Human Population and the Global Environment.' *American Scientist* **62**, 282-292.

Janzen, D. H.: 1997. The Carbon Crop. *Science* **277**, 883.

Ko, J. Y., Hall, C. A. and L. G. L. Lemus: (this volume), 'Resource Use Rates and Efficiency as Indicators of Regional Sustainability: An Examination of Five Countries'.

Krotscheck, C. and Narodoslawsky, M. 1996, "The Sustainable Process Index: A New Dimension in Ecological Evaluation". *Ecological Engineering*, Vol.6 p241-258.

Lieth, H., and Whittaker, R. (eds.): 1975 *The Primary Productivity of the Biosphere*, Springer, New York

Meadows, D., Meadows, D. and Randers, J.: 1992, *Beyond the Limits*. Chelsea Green Publishing Co., Post Mills, Vermont, USA.

Meadows, D., Meadows, D., Randers, J. and Behrens, W.: 1972, *Limits to Growth*, Universe Books, New York.

Moffat, A. S., 1997, 'Resurgent Forests can be Greenhouse Gas Sponges', *Science* **277**, 315-316.

Neumann, I.: 1994, *Der ökologische Fussabdruck der Region Trier* (The Ecological Footprint of the Trier Region), Diplomarbeit, Universität Trier, Germany.

Noss, R. F. and Cooperrider, A. Y.: 1994, *Saving Nature's Legacy - Protecting and Restoring Biodiversity*, Island Press, Washington DC.

Odum, H. T.: 1994, *Ecological and General Systems*, revised edition. University of Colorado Press, Boulder.

Omernick, J. M.: 1987, 'Ecoregions of the conterminous United States', *Annals of the Association of American Geographers* **77**, 118-125.

Pimentel, D., Houser, J., Preiss, E., White, O., Fang, H., Mesnick, L., Barsky, T., Tariche, S., Schreck, J. and Alpert, S.: 1997. 'Water Resources: Agriculture, the Environment, and Society', *BioScience* **47**, 97-106.

Postel, S.: 1996, 'Forging a Sustainable Water Strategy', in Brown, L. *et al.*: 1996, *State of the World*, N.N. Norton, New York.

Rees, W. E.: 1996. 'Revisiting Carrying Capacity: Area-Based Indicators of Sustainability', *Population and Environment* **17**, 195-215.

Rosenbaum, W. A.: 1995, *Environmental Politics and Policy*. CQ Press, Congressional Quarterly Inc. Washington, DC, USA.

Schmidt-Bleek, F.: 1994, *Wieviel Umwelt braucht der Mensch: MIPS - das Mass für ökologisches Wirtschaften*. (How Much Environment Do People Need? MIPS: The Measure for Managing Ecological Economies). Birkhäuser, Basel, Boston. English edition forthcoming: "The Fossil Makers", New York.

Sarmiento, J. L. and Le Quéré, C.: 1996, 'Oceanic Carbon Dioxide Uptake in a Model of Century-Scale Global Warming', *Science* **274**, 1346-1350.

United Nations Development Program (UNDP) annual. *Human Development Report*. Oxford University Press, New York.

Vitousek, P. M., Ehrlich, P. R., Ehrlich, A. H. and Matson, P. A.: 1986. 'Human Appropriation of the Products of Photosynthesis', *BioScience* **34**, 368-373.

Wackernagel, M., Macintosh, J., Rees, W. E. and Willard, R.: 1993, *How Big Is Our Ecological Footprint? A Handbook for Estimating a Community's Appropriated Carrying Capacity*. Draft. The UBC Task Force on Healthy and Sustainable Communities, University of British Columbia, Vancouver, BC, Canada.

Wackernagel, M. and Rees, W. E.: 1996. *Our Ecological Footprint: Reducing Human Impact on the Earth*. New Society Publishers, Philadelphia, PA, USA

Wackernagel, M., Onisto, L., Linares, A. C., Falfán, I. S. L., García, J. M., Guerrero, A. I. S., Guerrero, M. G. S.: 1997, *Ecological Footprints of Nations: How Much Nature Do They Use? How Much Nature Do they Have?*. Commissioned for the Rio+5 Forum. International Council for Local Environmental Initiatives, Toronto. (available through ICLEI: iclei@iclei.org).

WCED: 1987, *Our Common Future*, World Commission on Environment and Development, (Gro Harlem Brundtland, chair), Oxford University Press, New York.

Whittaker, R. H.: 1975. *Communities and Ecosystems*, MacMillan Publishing New York.

World Resources Institute (WRI): 1994. *World Resources* 1994-95. Oxford University Press, New York.

World Resources Institute (WRI): 1996. *World Resources* 1996-97. Oxford University Press, New York.

EMERGY ANALYSIS OF HUMAN CARRYING CAPACITY AND REGIONAL SUSTAINABILITY: AN EXAMPLE USING THE STATE OF MAINE

DANIEL E. CAMPBELL

Atlantic Ecology Division, National Health and Environmental Effects Research Laboratory, United States Environmental Protection Agency, Narragansett, RI 02882

Abstract. The human carrying capacity for a region at a specified standard of living depends on the economic and environmental resources of the region and the exchange of resources across regional boundaries. The length of time that a human population living at a given standard can be sustained depends on the rates of use and renewal of the resource base. All environmental, economic, and social resources are produced as a result of energy transformations; therefore, the energy required for their production can be specified and evaluated in common terms by converting their energy values into emergy. Emergy is defined as the available energy of one kind, previously used up directly and indirectly to make a product or service. Its unit is the emjoule. Emergy values and indices are used to evaluate the resource base for Maine, a politically defined region, and to estimate its human carrying capacity at the 1980 standard of living and for possible future resource bases. Emergy indices for Maine are compared with similar indices for Florida, Texas, and the United States to demonstrate variations in human carrying capacity and sustainability among different regions. The 1980 standard of living for Maine, Florida, Texas, and the Nation as measured by emergy use per person fell within a relatively narrow range of 3.4E16 to 4.3E16 solar emjoules y^{-1}. The human carrying capacity for a region is considered within a pulsing paradigm for sustainabilty and within the constraints provided by a renewable resource base. For example, in the short-term the developed human carrying capacity for Maine is largely determined by the fuel emergy inflow relative to renewable emergy resources. If purchased emergy inflows relative to Maine's renewable emergy increase to the average ratio for a developed country around 1980, the population living in Maine at 1980 standards could increase to 2.9 million or 2.6 times Maine's 1980 population. In contrast, the human carrying capacity based on Maine's renewable resources alone was 0.37 million people at the 1980 standard of living or 33% of the 1980 population.

1. Introduction

The assessment of regional systems is complicated by the need to evaluate the network of interactions occurring between human beings and their environmental support systems. Traditionally, the disciplines that address human activities, principally economics and sociology, have been pursued separately from the disciplines of physics, chemistry, biology, and ecology that provide the context and constraints for human socio-economic activities (Hall 1992). Hall (1992) has documented the need for an alternative integrated approach to understand systems of humanity and nature with their environmental-economic interfaces. The alternative analysis system used in this paper to make regional assessments is the energy systems approach of H.T. Odum (1983, 1994) which integrates ecology and economics within the context of thermodynamics and general systems theory.

The assessment of regional systems defined on scales from tens to thousands of kilometers and tens to hundreds of years is addressed in this paper. A region is generally defined as a part of a larger system that may have naturally or arbitrarily determined

Environmental Monitoring and Assessment **51:** 531–569, 1998.

boundaries. For the purposes of environmental assessment, a region is usually considered to be a large continuous area within a larger defined surface. A state or ecological region within the United States of America, or a county or group of counties within a state, (e.g., the coastal counties of Maine or the potato growing region of northern Maine) are examples. Regional systems are complex networks composed of climatic, physiographic, biogeochemical, socio-economic, and cultural components and processes. Looking through a space-time window at these systems forces us to see the environmental and economic subsystems as parts of the same whole (Figure 1) because it views the scales of time and space over which human and environmental processes are co-dominant in their effects. Using a window of smaller scale usually leads to the view that human activities are forcing functions from the larger scale and gazing through a larger scale window often focuses our attention on human systems as they are constrained by long-term environmental patterns.

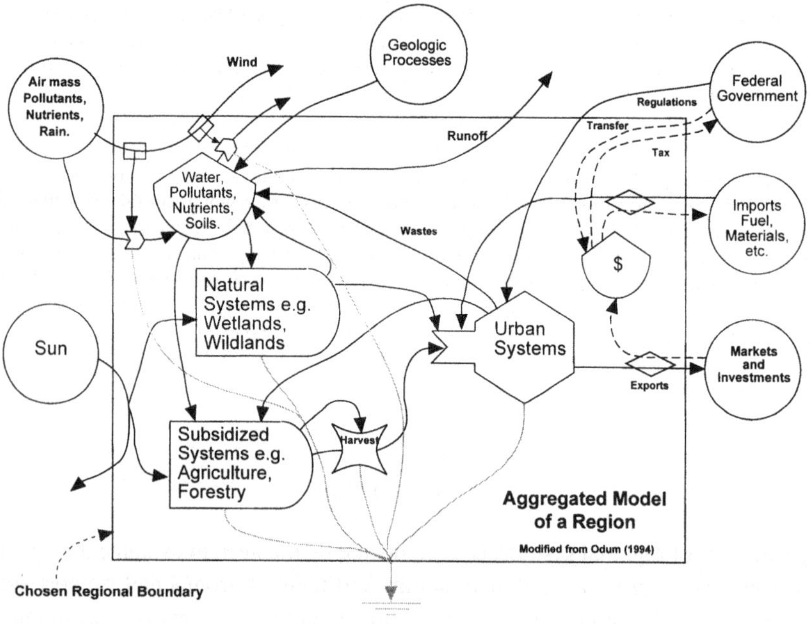

Fig. 1. An aggregated model of a regional system diagramed using Energy Systems Language (modified from H.T. Odum 1994).

An aggregated energy systems model of the components of a regional system at the environmental-economic interface is shown in Figure 1. Energy systems language (H.T. Odum, 1994) uses symbols (e.g., circles for external energy sources, bullets for producers, hexagons for consumers, rectangular arrowheads for interactions and diamonds for economic exchange, etc. (H.T. Odum, 1994) to represent the interactive network of a system diagrammatically. The regional system in Figure 1 is composed of components representing the human social and economic systems, the subsidized systems of agriculture, forestry, and aquaculture that feed and support the urban systems, and the natural ecosystems such as wetlands and wildlands that provide life support services to the human dominated and human subsidized subsystems (E.P. Odum, 1997). This regional system is constrained by the total available energy obtained from both the environment (circles on the left) and from the economy (circles on the right labeled fuel, etc.). Other aspects of a regional system shown in Figure 1 that determine its dynamic performance include (1) balance and recycle of materials, (e.g., water, waste etc.), (2) hierarchical relationships, (e.g., urban control of the subsidized systems), (3) control mechanisms from the next larger system, (e.g. inputs from the federal government and control of exchange by market prices), and (4) the thermodynamic limits on energy transformation efficiency, represented by the energy flows to the heat sink shown in gray as a fraction of the total energy transformed (H.T. Odum, 1987). These design characteristics determine the dynamic behavior of regional systems and they serve as a starting point for constructing simulation models to predict changes and trends in the environment and the economy of the region.

Identifying and assessing the ecological significance of risks to the human and environmental subsystems of a region requires that we evaluate both the economic and environmental components in equivalent terms. Traditional economic analyses are usually not broad enough in scope to adequately address the complex problems of regional systems which include environmental components (Pillet and Odum, 1984; Hall, 1992). The boundaries of most economic studies are fixed so that the creative and supportive work of the environment is external to the economy (Pillet and Odum, 1984). Environmental work is not valued by our monetary market system because money is paid to people for the human labor and capital investment in obtaining a product and not for the work of the environment which also contributes to the creation of the product. Since the products and services of nature are not given their true value by the economy, the ecological systems which produce them are like capital investments subject to depletion without provision being made for their eventual replacement or rehabilitation (Repetto, 1992). Emergy Analysis is an alternative means of determining worth which provides some unique insights into value not available by using monetary evaluation alone.

EMERGY ANALYSIS

Emergy Analysis (H.T. Odum, 1996) is a new method of environmental accounting which may be used to assess the complex relationships between the economy and its support environment because the work of both is expressed in equivalent terms. In this system of evaluation, emergy serves as a common denominator to express the value of environmental work as well as economic work in the manufacture, mining, growing, or creation of anything. Emergy is defined as the available energy of one kind previously used up directly and indirectly to make a product or service. Its unit is the emjoule (H.T. Odum, 1986). Available energy is potential energy capable of doing work (exergy). Emergy was originally called embodied energy but a new word was needed to distinguish it from other quantities also called embodied energy which were calculated in a different way (Odum 1996). The prefix em- comes from the words "energy" (e) "memory" (m) which captures the essential distinguishing characteristic of emergy which is that it is a physical quantity expressing the past use of energy upon which the form of present energy depends (Scienceman 1993). For example, a joule of sunlight, electricity, and human thinking have the same energy content but very different form and emergy content and thus different abilities to do work in a system.

The difference between emergy and energy can be illustrated by considering a small woodlot owner who fells a tree and cuts it into firewood for his wood stove. When the logs are burned over the course of the winter they yield a certain number of joules of heat which is the energy content of the logs. However, the emergy lost as the logs are burned includes the summation of all the solar emjoules (sej) of rain, fertilizer etc. without double counting, that supported the growth of the tree that produced the logs over the period that it stood in the wood lot and the solar equivalent joules required for harvest and processing the tree into firewood. The ratio of the solar emergy of a log to the heat energy it contains is the solar transformity of the log. For example, Doherty *et al.* (1995) calculated a solar transformity of 3846 sej per joule for spruce logs produced in Sweden. If the transformity and energy content of a product or service is known, its emergy can be immediately calculated.

Emergy, unlike dollars, is a true measure of relative importance because the total economic and environmental requirements for an item are accounted for in the same units by a scientific estimation process. The dollar value of a thing is receiver based and subjective because it depends on what individual humans are willing to pay for the thing. In contrast, emergy measures are donor based and objective because they are tied to measurements made on an efficient production process. Emergy expresses the true importance of a thing in the context of its system because it accounts for everything that was required for that thing to be a part of the system in which it occurs. Emergy is not a substitute for dollar values in market transactions; however, it is useful in determining the relative importance of things on the macroeconomic scale for public policy decision making. Both economic and environmental data and analyses are needed to make an emergy accounting. Therefore, emergy analysis is not a substitute for economic analysis, but a complement to it.

Emergy Analysis has been developed over the past 25 years by H.T. Odum and his collaborators. Since 1983 a great deal of research effort has been concentrated on this subject, and this work has culminated in the publication of a book describing the method (H.T. Odum, 1996). Emergy analysis has been used to characterize many regional systems including (1) nations such as, the United States (H.T. Odum and Alexander, 1977), Ecuador (H.T. Odum and Arding, 1991), Brazil (E.C. and H.T. Odum, 1984), Thailand (Brown and McClanahan, 1992), Switzerland (Pillet and H.T. Odum, 1984), and New Zealand (E.C. Odum, *et al.* 1982); (2) ecological regions such as the coastal region of Texas (H.T. Odum, *et al.* 1987a), the sea of Cortez (Brown *et al.*, 1991), Narayit, Mexico coastal region (Brown *et al.* 1992), south Florida (H.T. Odum and M.T. Brown, 1975), the Mississippi River region of the U.S. (H.T. Odum, *et al.* 1987b), and the Amazon Basin (H.T. Odum,. *et al.* 1986); and (3) the states of Florida (H.T. Odum *et al.*, 1986), Texas (H.T. Odum *et al.*, 1987a), Alaska (Brown *et al.*, 1993), and Maine (this paper).

This study applies Energy Systems Theory and Emergy Analysis to gain a better understanding of human carrying capacity and regional sustainability than can be obtained from using traditional environmental monitoring indicators alone. Emergy can be used to show the relative importance of environmental indicators and energy systems diagrams, and quantitatively capture the interrelationships between traditional indicators. Emergy indices calculated for a regional system network can be used to integrate information from different types of traditional environmental indicators. From an anthropocentric perspective the central question to be answered in a regional assessment is "What is the sustainable human carrying capacity for a region?" This paper first explores the ecological concepts of carrying capacity and sustainability and the special conditions that apply to human carrying capacity. Energy Systems Theory (H.T. Odum, 1994) is used to evaluate the concept of sustainability and to apply a new paradigm of pulsing (W.E. Odum *et al.*, 1995) to examine the patterns of development that may be sustainable in regions. The results of an emergy assessment of the environmental-economic system of the State of Maine, a politically defined region, are presented to examine the concepts of human carrying capacity and regional sustainability and to illustrate the Emergy Analysis method.

2. The Energy Basis for Human Carrying Capacity

The idea of a carrying capacity for animal populations derives from the logistic growth curve first investigated by Verhulst in 1838 (E.P. Odum, 1971). In the logistic growth model for animal populations, the population size, Q in Figure 2a, approaches an upper asymptote, K or the carrying capacity, because the negative effects of interactions among the individual population members increase with population size (Figure 2c and d). Many different premises can be used to derive the mathematical forms that have logistic growth dynamics as their solution (H.T. Odum, 1987). Two of these forms in which interactive unit effects limit growth are diagramed in Figure 2a and b using Energy Systems Language (H.T. Odum, 1983). Energy Systems Language is a visual mathematics which can be used for conceptual thinking, quantitative evaluation, and mathematical simulation. Energy systems diagrams and their mathematical translations are shown for the models simulated in the this paper (Figures 2a, 2b, 3a, 4a). Figure 2a shows the form of the logistic growth equation familiar in biology illustrating the intrinsic growth rate, r, and the carrying capacity, K. Energy resources are not explicitly considered in this form of the logistic equation; however, by comparison with Figure 2b, r is seen to be equivalent to a constant, k_1, times the energy source, E (Odum, 1987). This formulation results in exponential growth when the negative effects of population size are not density dependent. Theoretically, population growth can be constrained by limitations in the supply of energy resources as well as from interactive population unit effects (Odum, 1987). In the real world energy sources are never unlimited, so commonly observed logistic growth forms may result from limitation of the energy supply as well as from negative density dependent effects on growth e.g., crowding.

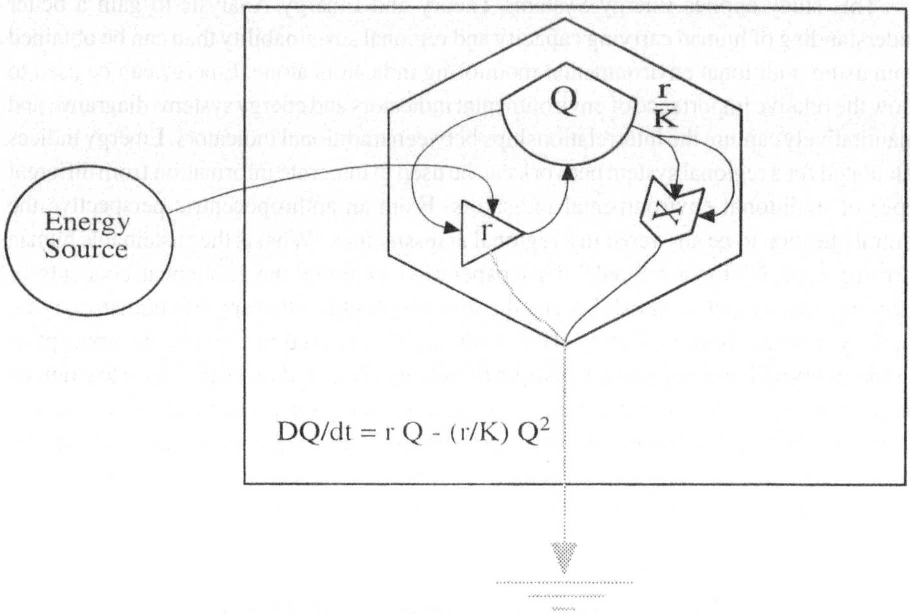

Fig. 2(a)-(d). **Energy systems diagrams and the equations for two versions of the logistic equation (a) the standard equation familiar in biology, (b) an equation showing the relation of r to the energy source (Odum 1987), (c) solutions for (b) when the available energy E is changed in increments of 25%, (d) solutions for (b) when k_2, the intensity of negative population unit interactions is changed in increments of 25%. The dotted line in (b) indicates that the fluxes on the two indicated arrows are taken as the net flux.**

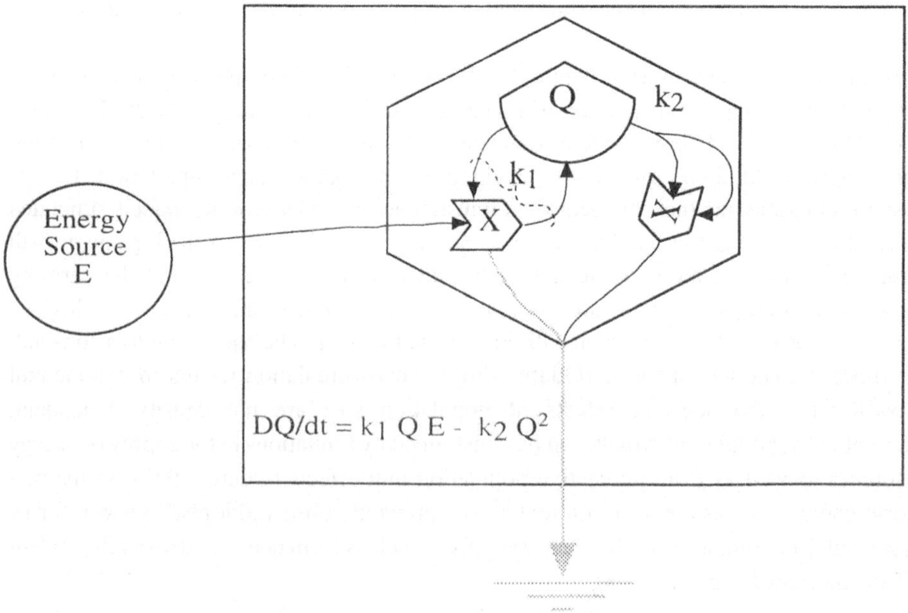

Fig. 2b

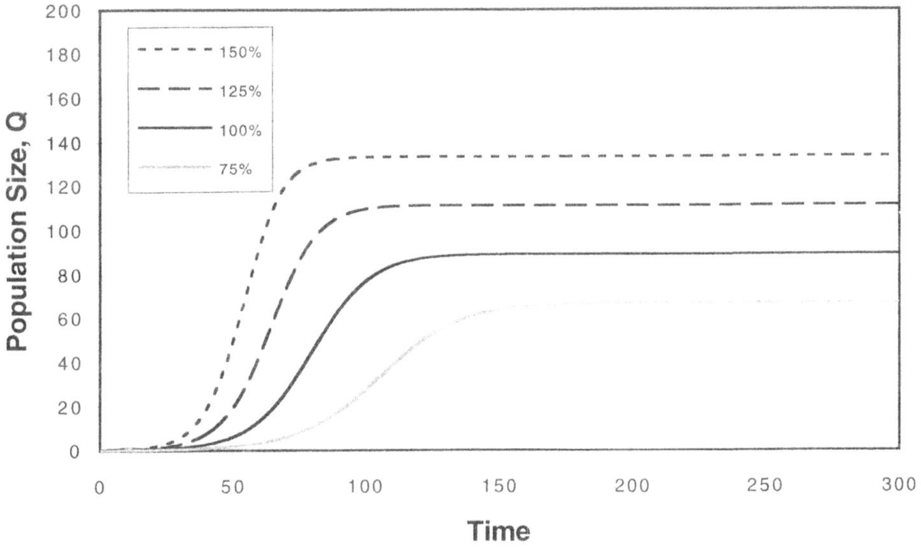

Fig. 2c

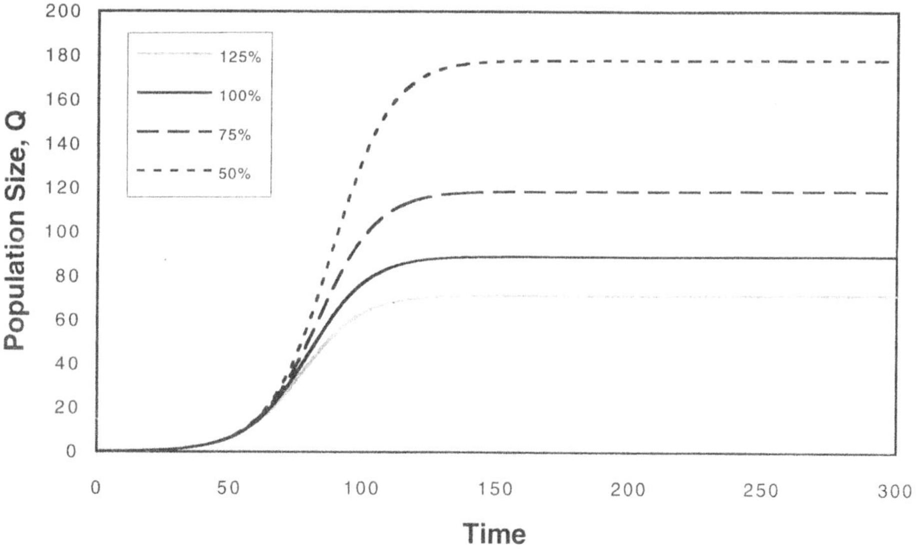

Fig. 2d

The model in Figure 2b was simulated using Extend™ , an advanced simulation tool, to illustrate some aspects of the solutions to the logistic equation that are helpful in thinking about human carrying capacity. Figure 2c shows that when k_1E or r is increased in percentage increments of 25% while all other factors are held constant, the population adjusts to progressively higher asymptotes for K. These new values for K represent progressively higher subsistence or maximum values for the population density, as contrasted with the safe or optimum density which is usually between 1/2 to 2/3 of the maximum (E.P. Odum, 1997). Some animal populations unregulated by predation tend to rapidly approach K and overshoot it, damaging their habitat, thereby diminishing the available energy resources and causing K to descend to a lower population density (McCollough, 1979). In contrast, animal populations under sufficient pressure from predators exist at densities closer to the optimum where their food supply is more secure and their resistance to the fluctuations of the environment is greater (E.P. Odum, 1997). The logistic model implies that the carrying capacity, whether the maximum or the optimum, is a level of population size that can be sustained indefinitely into the future.

Human populations can increase to the maximum or subsistence level, but this pattern of behavior is not obligatory as it is with other animal populations. Given sufficient understanding human societies can choose to put some of their energy resources into increasing the assets of the society, thereby, improving the quality of life rather than allowing all the additional energy to go into supporting additional people. Attaining an optimum or safe population density for humans depends on the societal and individual choices which determine how much of the available energy goes into subsistence versus how much goes into increasing some measure of societal assets per individual or the quality of life. The social choices of individual countries in the world represent a wide variety of solutions to the trade-off between more humans beings at a lower standard of living and fewer humans with more assets per person. Thus, carrying capacity for humans can not be defined unless a standard of living is also specified. In the social sciences, human carrying capacity is usually qualified in this manner (E.P. Odum, 1997).

Figure 2d shows the effect of decreasing the intensity of the negative density dependent interaction among individuals in a population. When all other factors are constant, the carrying capacity, K, increases exponentially as interactive effects are decreased. Humans can choose to apply the effective increase in energy resources gained through increased efficiency to improving their quality of life instead of increasing the subsistence carrying capacity of the population. The former kind of change in K is dependent on developing better or more efficient systems designs and it illustrates R. Buckminster Fuller's injunction to human society to learn how to get more for less (Fuller, 1981). Getting more for less is a good strategy for improving our standard of living but its efficacy is limited by the total energy available to support societal organization and by the thermodynamic optimum efficiency for maximum power (H.T. Odum, 1994) production in the system.

The intrinsic rate of increase, r, for a population has been shown to implicitly include ($r=k_1E$) the energy resources for that population (H.T. Odum, 1987). A better understanding of human carrying capacity can be gained by separating the energy resources for society into renewable and nonrenewable components and observing the growth patterns that each produces (Figure 3a). The distinction between renewable and nonrenewable resources is somewhat artificial because all resources on earth are renewed by the global web of ecological processes; however, those that are being renewed very slowly compared to their rate of use are said to be nonrenewable. The large space-time scale patterns in the development of human populations and societal assets for nations and hence regions within nations are almost always controlled by the dominant energy sources available to support development (Watt, 1992). Almost all present economic development is based on the use of these energy resources in a nonrenewable manner (Hall, 1992). This insight alters our view on the possibility of sustaining present population sizes and states of development.

The simple model shown in Figure 3a (Odum, 1987) may be used as the basis for an overview of the regional development process based on available energy. When this model is simulated using Extend™ the patterns shown in Figure 3b result. The nonrenewable resources, F, are depleted because they are being used at a rate that exceeds their rate of replacement. Their use builds a peak level of societal assets that is not sustainable. The human population depends on the assets produced to maintain its standard of living, and therefore, it must decline as assets decline to maintain the same standard of living. C.A.S. Hall (this volume) has presented examples of some inherently unsustainable agricultural, industrial and social activities practiced in the world today. The level of development supported by renewable resources can also be found in Figure 3b, as evidenced by the lower asymptote approached by the assets of society as nonrenewable resources are depleted. This asymptote is the K of the logistic curve produced by a renewable or flow limited energy source, e.g., sunlight is a flow limited resource because only a fixed quantity is available for use per unit area and time (the solar constant is approximately 2 g cal cm^{-2} min^{-1}).

From this analysis it is clear that a human carrying capacity at a specified standard of living is not sustainable unless it is based on the use of resources in a renewable way. Once the patterns dictated by energy use are recognized, some critical questions arise: "What is the total amount of nonrenewable resources available to a nation or a region and what are their rates of use?" For example, ancient ground water in many arid regions such as the southwest U.S. is being pumped at a rate exceeding its recharge to make the deserts bloom (Bowden, 1977). For a given set of technologies the quantity of the nonrenewable resource will determine the timing of the assets rise and fall and the duration of its peak. Technology determines the rate of use and the completeness of exploitation for a resource, but unless it can alter the rate at which the resource is produced that resource will continue to be nonrenewable if it is being used faster than it is being replenished.

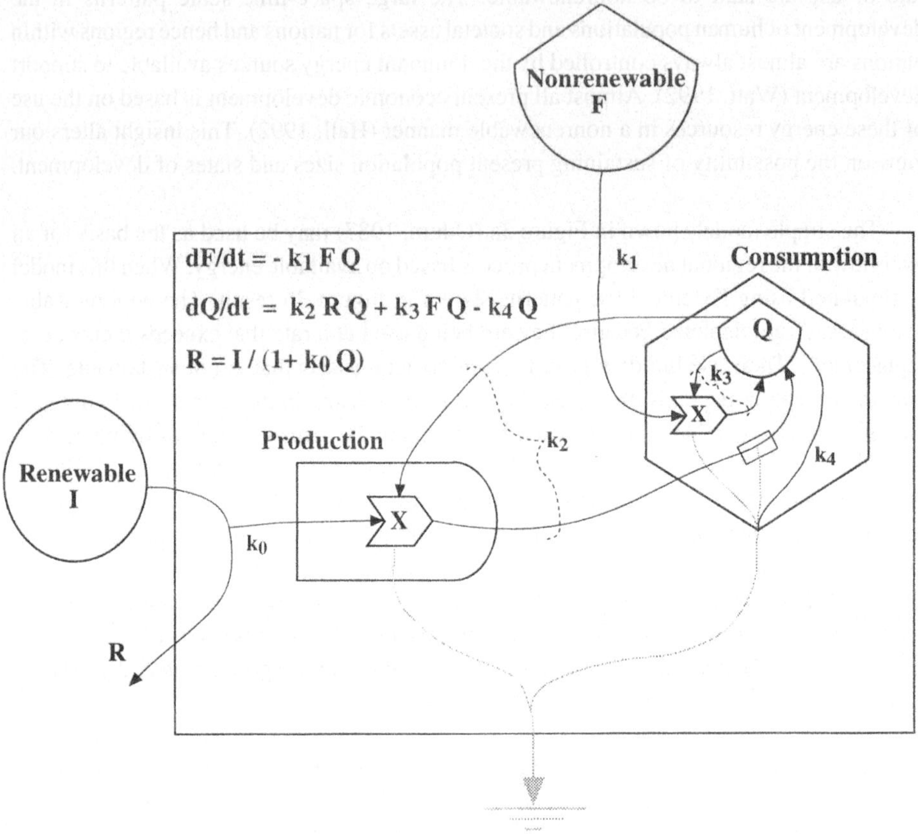

Fig. 3. (a) An energy systems diagram and equations showing the dependence of human populations and their assets on renewable and nonrenewable resources. (b) The solution to these equations showing the expected pattern of societal assets with time as nonrenewable resources are depleted. The dotted lines are net fluxes as in Figure 2.

3. Maximum Power and Sustainability

Models based on logistic growth such as the ones presented above reach a constant value for carrying capacity at steady state and contain the idea of sustainability in a familiar form, i.e., sustainability is the prolongation or maintenance of a state desired by humans. However, the word "sustainability" also means to keep an entity in existence and in this sense it is related to survival of a system. The prolongation of a certain state and the survival of the system that is in that state are two different things and each may require a different systems design. The survival of systems in evolutionary competition with others is hypothesized to depend on creating designs that maximize empower (emergy production and use per unit time) within the system network (Lotka, 1922; Odum, 1996).

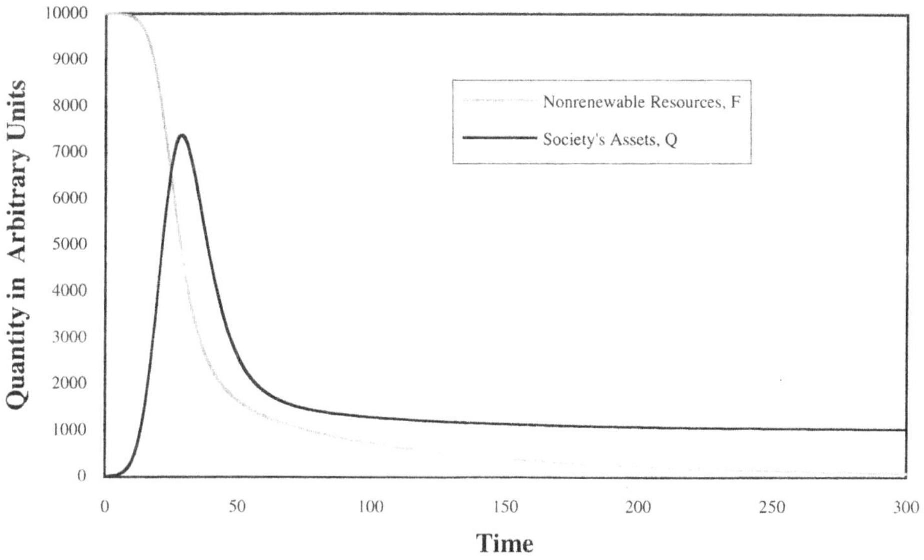

Fig. 3b

System designs that pulse are observed to be ubiquitous in nature occurring from the scale of biochemical reactions to the scale of the galaxies (Odum, 1994). W.E. Odum *et al.* (1995) and H.T. Odum (1996) present evidence and theory to support a new pulsing paradigm for understanding the patterns of humanity and nature that are required for survival in ecological systems. A working hypothesis to explain the broad occurrence of pulsing in nature is that systems which pulse will attain higher performance (i.e., develop greater empower) in the long run than those which maintain steady levels (W.E. Odum *et al.*, 1995).

Pulsing in regional systems can be investigated with the help of EMPULSE (H.T. Odum, 1996), a model illustrating a general pulsing mechanism, which is presented in Figure 4a. In this model Environmental Resources, Q, accumulate at a slow rate and are exploited both linearly (the box with k_2 entering and k_5 leaving), and autocatalytically (interactions symbols marked with an X where a feedback, k_7, from A, the economic assets of the region increases the use of environmental resources nonlinearly). This model includes both renewable, I, and "nonrenewable", Q, resources, but the space-time dimensions of the window of attention in Figure 3 have been widened to include scales which show the slow renewal of the formerly "nonrenewable" resource. Figure 4b shows the pulsing pattern of economic assets that occurs as a consequence of the rapid exploitation of slowly replaced environmental resources. The total material, T_m, in the system is a constant of which the dispersed fraction, M, is available for creating new environmental resources on the pathway indicated by k_1. The money circulating in the regional economy, G, is represented by the dotted line. The dashed box shows the

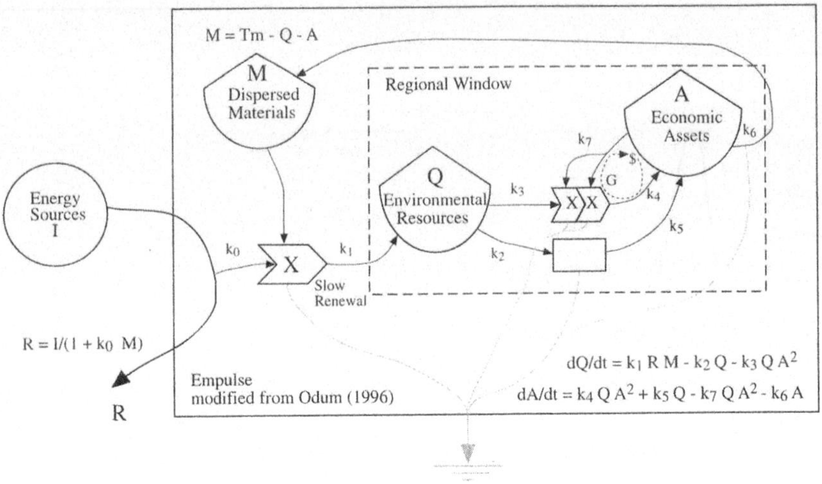

Fig. 4. (a) An energy systems diagram of a regional system emphasizing the environmental-economic interface. The Empulse model (H.T. Odum, 1996) was slightly modified and simulated using Extend to obtain the pulsing patterns for environmental resources and economic assets shown in (b).

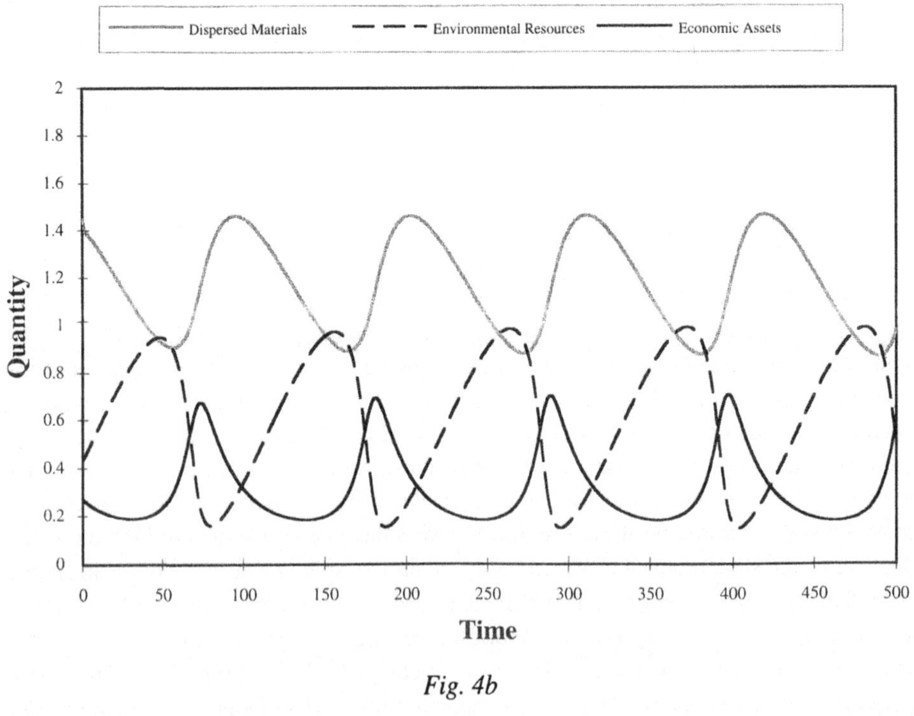

Fig. 4b

boundaries of a regional system which includes environmental resources, human assets and the processes by which humans utilize the environment (H.T. Odum, 1996). The region's environmental and economic assets (Figure 4b) go through a repeating oscillation that represents the only pattern that is sustainable in many cases. Processes of succession and climax that are part of the constant steady state idea of sustainability can be identified within the repeating cycles of environmental and economic assets (W.E. Odum *et al.*, 1995). However, the pulsing steady state paradigm requires that a period of descent or regression and a low energy period be added to the build-up period of succession and the peak period of climax to complete the cycle of change. The key to long-term sustainability of human populations and the economic assets determining their quality of life in a nation or a region may well lie in acquiring the knowledge and understanding to manage the pulsing of subsystems within the constraints of the cycle of change determined by the fundamental energy drivers from the larger system.

4. Human Carrying Capacity and Sustainability for Maine

The State of Maine can be considered as a politically defined region within the larger system of the United States of America (Figure 5). Other means of defining regional boundaries could have been used depending on the research questions to be answered (e.g., biogeographic zones, spheres of regulatory control or spheres of economic influence, etc.), but for the purpose of this paper Maine will serve to illustrate the method of regional environmental-economic assessment using emergy. The emergy analysis of Maine performed in this paper is a static analysis based on evaluating storages and flows in the regional system for the year 1980. Human carrying capacities, standards of living, and their sustainability will be considered in the context of the energy resource constraints illustrated in Figure 3b. Also, the role of pulsing (Figure 4b) in sustaining the patterns of human population and assets in the long-term will be considered as a conceptual basis for understanding what human carrying capacity may be sustainable in a region. Simulation models will be introduced in the methods section; however, they were not employed in this paper to dynamically consider the implications of pulsing to sustainability. Emergy indices were calculated for Maine and employed to provide insight into the nature of economic, environmental, and social interrelationships. These indices were compared to similar data from the states of Florida and Texas as well as from the nation as a whole. Finally, the assessment of human carrying capacity as an indicator of regional sustainability is discussed and recommendations for ensuring optimum patterns for human carrying capacity and standard of living in the future are given.

5. Methods

The method presented here has been used with some success investigating many complex problems which have both economic and environmental ramifications. For example, energy analysis has been used to: (1) develop a method of siting nuclear power plants (Odum *et al.*, 1983); (2) determine the economic and environmental feasibility of tree farming and paper production in the Amazon rain forest (Odum *et al.*, 1986); (3) determine appropriate environmental and economic policy for countries, states and regions as noted above and (4) to determine the net energy available from various energy sources including nuclear power, shale oil, solar technology, ocean thermal heat gradient, and ethanol from biomass production (summarized in H.T. Odum, 1996). Several predictions derived from the application of emergy analysis have proved to be correct (e.g., the lack of net energy in shale oil, H.T. Odum, 1996); however, only the future holds the answers to many of the larger questions that can be addressed now by this technique.

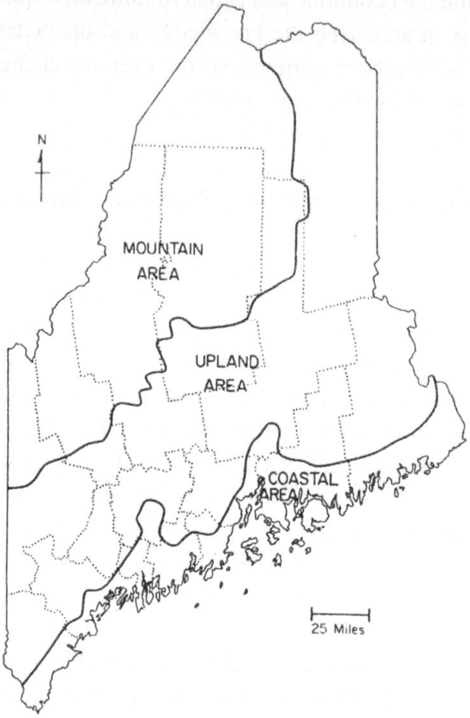

Fig. 5. A map of Maine, showing three regional areas delimited on the basis of their physiography.

The methods and techniques for performing a regional analysis using the energy systems perspective were first published by H.T. Odum *et al.* (1976). The method presented there has been modified and presented in this paper. Emergy analysis can be organized into eight principal steps with useful products produced at each step. The steps in the emergy analysis of a region are as follows:

1) Assemble the information and/or the individuals necessary to define the system boundaries, components, external causal influences, and processes. The product of this step is three lists, one for the main system components, one for the forcing functions or external causal factors, and one for the pathway flows that are generated by the processes through which the forcing functions and components interact. In the initial organizing meeting with knowledgeable individuals this step may be combined with step 3, construction of a detailed energy systems diagram. The diagraming process facilitates

discussion among participants and produces an understanding of what the important parts of the system are and how they work together.

2) Make or obtain land use maps to estimate the areas of the main system components. High transformity elements that create organization such as networks of water flow, transportation, and supply of fuels and electricity are identified and their patterns mapped. All important emergy flows whether of large extent or high concentration are included in the spatial classification.

3) Construct a detailed diagram of the regional system network categorizing the main components and showing the interactions among these components and the external forcing functions. Energy systems language uses mathematically defined symbols (H.T. Odum 1983; 1994) to represent the interactive network of a system. Producer, consumer and rectangle symbols can show a nested hierarchy by indicating intermediate levels of organization. A large rectangle shows the system boundaries and the hierarchy of organization within a system is represented from left to right in order of increasing transformity.

4) Simplify the detailed regional diagram by combining functionally similar components and processes to make one or more aggregated diagrams. This simplification is done not by cutting out pieces, but by combining them into aggregate variables. The aggregated diagram contains the variables that are important in describing major system trends or those which are germane to the analysis of specific problems and policy alternatives. To understand how a system works and to demonstrate these mechanisms clearly, the complex system must be simplified to a few main components and forcing functions. This simpler structure is easier to evaluate and simulate allowing specific research questions to be answered in the most straightforward way. The emergy flows that are large or controlling are guides in the simplification process. Identifying the fundamental external drivers that are changing and their interactions is also a key factor in simplification.

5) Evaluate the sources, storages, and main flows of energy, materials, and money on the aggregated regional systems diagrams.

A. For simulation models tables defining the storages, forcing functions, and flows are made showing their value, giving its source, and linking it to the aggregated diagram with an appropriate symbol.

B. For evaluating the emergy basis for a region a standard table with columns for (1) a footnote detailing the calculation, (2) the item, (3) the item's energy value, (4) the item's transformity, (5) the solar emergy value of the item, (6) the item's emdollar value (the emergy of the item divided by the emergy/dollar ratio for the economy). Emdollars express the value of an item in terms of dollars flowing in the gross state or gross national product, GSP or GNP, as if these dollars were distributed according to the emergy flows. The emergy analysis table provides a template for calculating solar emergy and emdollar value from data on the raw value of energy inputs and there respective transformities.

6) Analyze the emergy basis for the region using emergy indices, spatial plots of emergy use density, emergy power spectra, and other tools (H.T. Odum et al., 1976). Emergy indicators aid in understanding a region and in comparing it to other regions. Several emergy indicators are defined below.

7) Computer simulation of one or more aggregated models can be performed to predict future trends, evaluate policy alternatives, or test the model's sensitivity to critical variables. Step seven has a number of sub-steps related to the construction and analysis of simulation models. (a) Mathematical equations are written from the systems diagram, (see Figures 2a and b, 3a, and 4a). (b) The equations are programmed using a computer language such as Basic, Fortran or a simulation tool such as Extend™. (c) Each major aspect of the model is calibrated by comparing simulation results to the available data on that model output. (d) The model is verified by testing its ability to represent the entire calibration data set in simulation. (e) The verified model is validated using one or more independent data sets. (f) A sensitivity analysis of the effects of varying the forcing functions on predictions of the validated model can be used to evaluate management alternatives. Sensitivity analysis can be used to investigate other aspects of model behavior such as the robustness of model solutions as a parameter value is varied. The prediction of future trends can be enhanced by driving the regional model with a macroscopic minimodel (H.T. Odum, 1976) of the next larger system which dynamically describes the behavior of the fundamental forcing functions. Emergy magnitudes are used as a guide to identify these important forcing factors.

8) Combine all information from the emergy analysis to address issues of concern to public policy. For example, emergy values and their changes are indicators of the ecological significance of ecosystem components and processes. These indicators should be used with determinations of risk (probability of damage or harm) to evaluate the importance of an ecological change for use in the environmental decision making process. An emergy evaluation of environmental impacts and management alternatives allows us to predict those alternatives which have a higher probability of success because they result in a maximum contribution of empower to the regional and national system.

EMERGY INDICES

If maximizing empower production and use in regional systems determines the patterns of development that will survive and prosper in the long run, we may expect many emergy indices to be particularly meaningful for characterizing the condition of a region, for estimating human carrying capacity and its sustainability, and for determining the true relationship between the region and the next larger system. There follows a list and definition of the important emergy indices used in this study:

(1) The solar transformity of an object or resource is the energy in solar equivalent joules that it takes to create a unit of that product efficiently and quickly. For example, a large number of joules from solar heat and deep heat sources in the earth are necessary to warm the land and oceans and produce a joule of chemical potential energy in pure rain water (Odum, 1994). In fact one joule of chemical potential energy in rainfall requires 1.5 E4 solar energy joules for its creation on a global basis. The solar transformity of chemical potential energy in rain is therefore 1.5 E4 solar emjoules (sej) per joule of rain.

Solar transformities for a large number of objects and resources have been calculated by H.T. Odum and E.C. Odum (1983) and H.T. Odum (1996). These transformity calculations are based on the evaluation of subsystems or production processes that result in the creation of the object or resource. When a needed transformity is not available a subsystem analysis of the production process for that object must be performed. This involves summing all the energy inputs in equivalent units required to produce a unit of that type. Solar transformities provide a scale for value referenced to a planetary energy baseline. They are necessary factors for calculating the emergy value of resources from their energy or mass contents. The calculation of a revised solar transformity for tidal energy is presented in Appendix B. The older calculations of the emergy indices for Florida, Texas, and the nation have been modified to incorporate this new estimate of the transformity of tidal energy.

(2) Emergy exchange ratio is the ratio of the emergy received in trade for the emergy given. The trading partner that receives the greater emergy value in a trade will have its economy stimulated more by the exchange. In practice raw materials such as lumber, fish, furs, oil, agricultural products, and minerals have high emergy values relative to the emergy received in trade for them because payment is rendered for the human labor involved in obtaining these products and not for the work of the environment in creating them.

(3) The emergy to dollar ratio is the total emergy used by a country or state in a particular year divided by the gross national or gross state product for that year. The emergy used includes that from renewable environmental energies, sun, wind, rain etc. without double counting (H.T. Odum, 1996); nonrenewable emergy from fuel, soil, and water reserves; and the emergy in imported goods and services.

(4) Several emergy indices of an economy are useful in examining the human carrying capacity and sustainability of regional and national systems. The emergy flow per person is an index of the standard of living which includes environmental and economic contributions to the quality of life. The emergy flow per unit area is an indication of the spatial concentration of economic activities in a state or nation. The fraction of the total emergy inflow that comes from within the region or nation is an index of its self-sufficiency.

MODEL DEVELOPMENT

An energy systems model of the Maine economy at a medium scale of complexity is shown in Figure 6. This model could be evaluated in full but in this study the model is used as a conceptual guide for organizing the emergy analysis and thinking about the Maine regional system. Pathway flows indicated by the k_i's in Figure 6 are defined in Table I. The economic sectors that comprise the various aggregated compartments in the model are defined in Table II. The environmental energy sources that provide the basis for a productive economy are shown by the circles around the edge of the large rectangle which indicates the regional boundary. The regional boundary coincides with the State boundaries and the offshore area to the 100m isobath. The transformity of forcing functions increases from left to right around the outside of the rectangle. Emergy also enters the Maine economy as fuels, goods and services, and through the earth's geologic processes. Tourists, the federal government, and export markets supply money to the Maine economy in exchange for products and services of some perceived value. These external emergy sources for Maine are evaluated in Table III except for the glaciers which contributed to building the present Maine landforms over a time scale not evaluated in this analysis.

The model components include aggregated ecosystem variables for coastal ecosystems, forests and wildlands, and agricultural systems represented by the bullet shapes on the left (Figure 6). Soil, ground and surface water, and landform are storages of environmental resources used by the economy. Wastes are produced as a byproduct of human activities. These effluents some of which are toxic are released into the environment, often in partially treated form, where they impact aquatic and terrestrial ecosystems. The industrial sector is divided into resource industries and other export industries (Table II) according to the classification of Pease and Richards (1983). Commerce and service industries are lumped into a single component that accounts for a large share of the money circulating in the gross state product, GSP, represented by a storage variable in the upper right corner of the model. People and their households are shown by the hexagon symbol on the far right. State, local, and federal government installations are included within the rectangle designated, Government. Electrical power plants within the state are also shown by a multipurpose rectangular box.

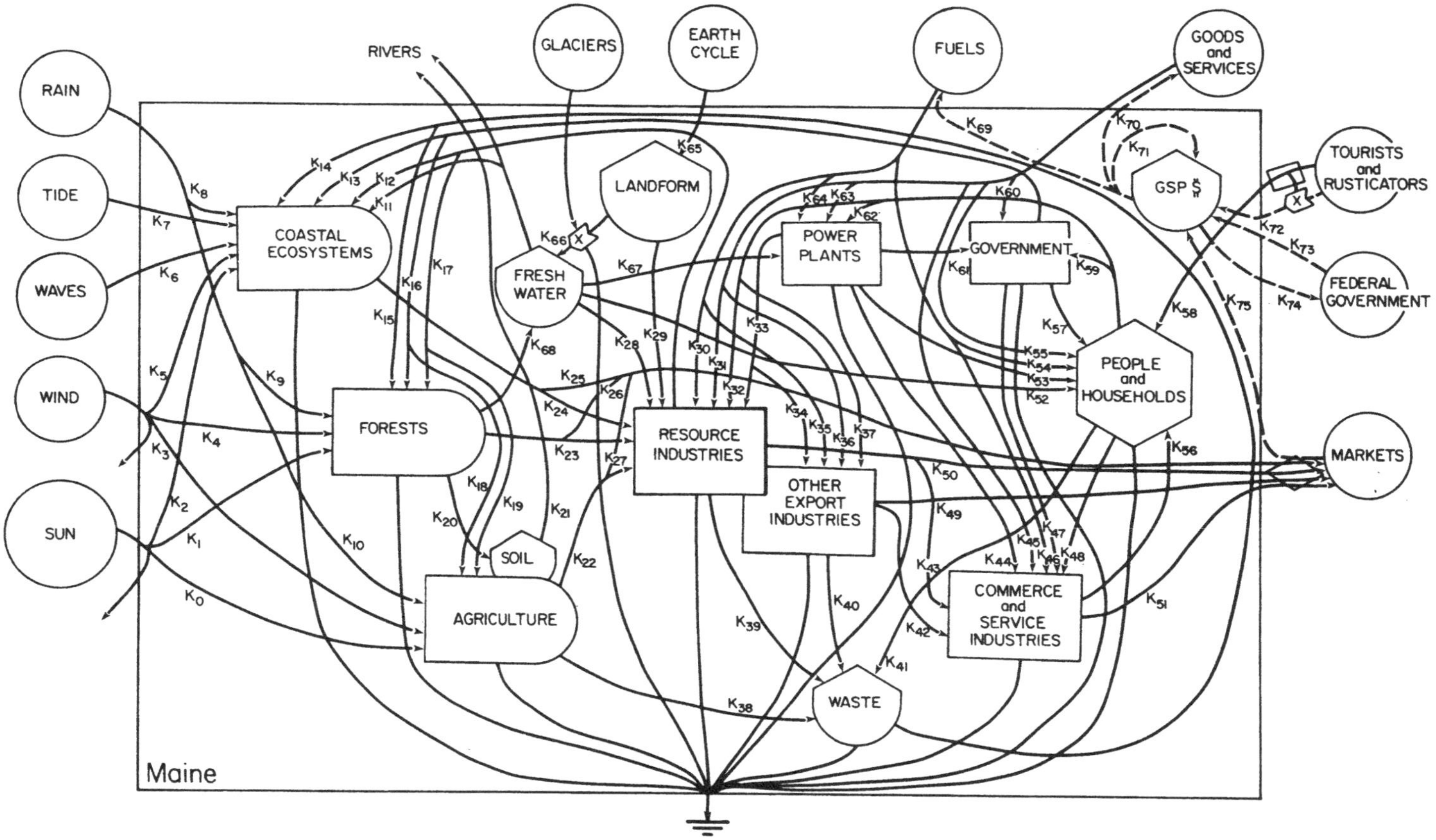

Fig. 6. A detailed conceptual energy systems model of Maine showing the environmental and economic interactions of system components and their emergy basis.

Table I
Definition of pathway flows in the conceptual energy systems model of Maine's environment and economy shown in Figure 6

Pathway Coefficient	Definition of Flow
k_0	Solar radiation absorbed by farm land.
k_1	Solar radiation absorbed by forest land.
k_2	Solar radiation absorbed by the coastal area.
k_3	Wind energy absorbed by farm land.
k_4	Wind energy absorbed by forest land.
k_5	Wind energy absorbed by the coastal area.
k_6	Wave energy absorbed along the shore.
k_7	Tidal energy absorbed by the coastal systems.
k_8	Chemical potential of rain falling on the coastal area.
k_9	Chemical potential of rain falling on the forest.
k_{10}	Chemical potential of rain falling on farm lands.
k_{11}	Chemical potential energy of river inflow to coastal ecosystems.
k_{12}	Environmental effects of commercial fishing and aquaculture.
k_{13}	Government expenditure on programs to help the coastal zone.
k_{14}	Waste discharged to coastal area.
k_{15}	Waste discharged to forests.
k_{16}	Government expenditure to improve forests.
k_{17}	Environmental effects of forest management practices.
k_{18}	Government expenditures for agriculture.
k_{19}	Environmental effects of agricultural activities.
k_{20}	Soil formed by forest land.
k_{21}	Soil lost through erosion
k_{22}	Agricultural products used by resource industries.
k_{23}	Forest products used by resource industries.
k_{24}	Fisheries products used by resource industries.
k_{25}	Fisheries products exported directly.
k_{26}	Forest products exported directly.
k_{27}	Farm products exported directly.
k_{28}	Fresh water used by resource industries.
k_{29}	Mined products used by resource industry.
k_{30}	Fuel used by resource industries.
k_{31}	Goods and services used by resource industries.
k_{32}	Labor used in resource industries.
k_{33}	Electric power used in resource industries.
k_{34}	Electric power used by other export industries.
k_{35}	Fuel used by other export industries.
k_{36}	Goods and services used by other export industries.
k_{37}	Labor used by other export industries.
k_{38}	Waste production by agriculture.
k_{39}	Waste production by resource industries.
k_{40}	Waste produced by other export industries.
k_{41}	Waste produced by people.
k_{42}	Other export industry products sold in the state.
k_{43}	Resource industry products sold in the state.
k_{44}	Electricity used by commerce and service industry.
k_{45}	Goods and services used by commerce and the service industry.
k_{46}	Fuel used by commerce and the service industry.
k_{47}	Government contributions to commerce and service industries
k_{48}	Labor used in commerce and service industry.
k_{49}	Products exported by other export industries.
k_{50}	Products exported by resource industries.
k_{51}	Commerce and service industry exports.
k_{52}	Fresh water used by people.
k_{53}	Electricity used by people.
k_{54}	Fuels used by people.
k_{55}	Imported goods and services purchased by people.
k_{56}	Goods and services purchased locally by people.
k_{57}	Government subsidies given directly to people.
k_{58}	Tourists, seasonal residents, net immigration.
k_{59}	Labor used by government.
k_{60}	Goods and services used by government.
k_{61}	Electrical power used by government.
k_{62}	Labor used by the power industry.
k_{63}	Goods and services purchased by the power industry.
k_{64}	Fuel purchased by the power industry.
k_{65}	Earth cycle energy flow driving land uplift.
k_{66}	Energy flow in glaciers creating landform.
k_{67}	Water used for hydroelectric power.
k_{68}	Fresh water recharge by forests.
k_{69}	Money spent for imported fuel.
k_{70}	Money spent for imported goods and services.
k_{71}	Money circulating in the State GSP.
k_{72}	Money brought into the state by tourists.
k_{73}	Federal subsidies to the state.
k_{74}	Federal taxes paid by the state.
k_{75}	Money received from the export trade.

Table II
Definition of the aggregated components in the energy systems model of Maine's environment and economy shown in Figure 6.

Component	Definition
Coastal Ecosystems	All marine, estuarine, and intertidal ecosystems <100 m in depth; including beaches, marshes, mudflats, estuaries, rocky coasts, coastal shelf.
Forests	All forest land both managed and unmanaged, including maple-beech, spruce-fir and pine forests, as well as, bogs, swamp, and marshes.
Agriculture	All crop, pasture and orchard land.
Soil	The storage topsoils in Maine.
Fresh water	The quantity of fresh water stored as ground-water and surface water.
Landform	The land and the minerals it contains.
Resource Industries	All primary and secondary fishing, farming, forest, and mining industries, including paper, lumber, furniture, and cord wood industries; the various fisheries, aquaculture, and seafood processing; potato, poultry, dairy, and fruit farms and food processing operations; sand, gravel, and limestone mining and cement manufacture.
Export Industries	All other manufacturers of durable and nondurable goods, including leather products, textiles, and apparel; engines, instruments and computers; large insurance carriers; and shipbuilding.
Power Plants	All fossil fuel, nuclear, and hydroelectric plants generating electricity in Maine.
Government	State, local, and federal government.
Commerce and Service Industries.	Retail and wholesale trade, hotels, restaurants, banking, real estate, insurance companies; the transportation industry; health, legal, social, personal, and repair services; waste treatment, schools other government services.
People	The population of Maine and their assets (households).
Waste	Waste products created by industry, people, and agriculture.
GSP	Gross State Product.

Table III

Emergy evaluation of the resource base for the Maine economy in 1980. The transformities are taken or modified from H.T. Odum (1996) except as noted.

Note †	Item	Energy Flow y^{-1}	Transformity SEJ $Unit^{-1}$	Solar Emergy E21 SEJ	Emdollars[#] E8 1980 $
Renewable Sources within Maine					
2	Sun	4.51E20 J	1 J^{-1}	0.45	1.7
3	Wind	5.47E17 J	1268 J^{-1}	0.69	2.7
4	Tides*	1.57E17 J	49383 J^{-1}	7.75	29.8
5	Waves	9.95E16 J	25890 J^{-1}	2.58	9.9
6	Rain, chemical	4.74E17 J	15423 J^{-1}	7.31	28.1
7	Rain, geo-potential	2.06E17 J	8888 J^{-1}	1.83	7.0
8	Rivers, chemical	2.99E17 J	41068 J^{-1}	12.30	47.2
9	Earth Cycle	1.42E17 J	34377 J^{-1}	4.88	18.8
10	Fuels, renewable from within the state				
	Hydropower	2.90E16 J	8.5E4 J^{-1}	2.46	9.5
	Wood	2.95E16 J	3.2E4 J^{-1}	0.94	3.6
Imports from Outside Maine					
10	Fuels, nonrenewable purchased out of state				
	Coal	2.00E15 J	4.0E4 J^{-1}	0.08	0.3
	Petroleum	2.54E17 J	5.4E4 J^{-1}	13.7	52.8
	Natural Gas.	2.32E15 J	4.8E4 J^{-1}	0.11	0.4
	Nuclear Elec	2.82E16 J	1.6E5 J^{-1}	4.51	17.4
	Canadian Elec	7.27E15 J	8.5E4 J^{-1}	0.62	2.4
11	Tourists	1.07E9 $	2.6E12 $\$^{-1}$	2.78	10.7
12	Goods and Services	2.98E9 $	2.6E12 $\$^{-1}$	7.80	29.8
13	Federal Government	0.86E9 $	2.6E12 $\$^{-1}$	2.24	8.6
18	Immigration	7500 people	3.4E16 $p.^{-1}$	0.26	0.9

Solar Emergy in column 5 divided by 2.6E12 sej $\$^{-1}$ for the U.S. in 1980.
* Calculated in Appendix B of this paper.
† Note 1 gives the area used in the calculations.

6. Results

Table III gives the emergy and emdollar values for the renewable and nonrenewable emergy base of Maine's economy. The numbers listed in column 1 of Table III and Table IVdirect the reader to a note in Appendix A where the value in column three is calculated and/or documented. The emergy sources generated by solar energy in order of decreasing magnitude are the chemical potential emergy in rain, the emergy in waves, and the geopotential emergy of rain. The emergy of the tides is the largest renewable emergy source, but it is only slightly greater than the chemical potential emergy of rain. The extremely large emergy value for the chemical potential energy in rivers is a further concentration of the large chemical potential emergy in rainfall and as such it is not considered to be a primary emergy source for Maine. The largest emergy input to Maine from all sources is in fuels. Petroleum is the most important fuel for Maine, although nuclear energy and hydroelectric power together supply half as much emergy as petroleum. The emergy contributed to the regional system by imported goods and services is second in magnitude behind petroleum fuels, and it is just slightly larger than the emergy supplied by the tides and in the chemical potential energy of rain.

The emergy and emdollar values in Maine's stored resources are shown in Table IV. The largest emergy value of the stored resources (natural capital) is found in Maine's extensive peatlands which are about 3% of the total state area (Hasbrouck 1979). The emdollar value stored in peat is worth about 50 times the total value of the 1980 gross state product. The second largest storage is in the accumulated talents and experience of the people of Maine (human capital) which has an emergy value that is 81% of the emergy stored in peat. Topsoil contains the third largest emergy storage. Even though the state as a whole is gaining topsoil, agricultural areas such as the potato growing region of Aroostock County are loosing this stored resource faster than it is being replaced. Economic assets, wood, and groundwater are also important stored resources in Maine. When all forest lands are taken into account, timber in Maine grows about twice as fast as it is harvested. However, this is not true for individual tree species which may be over harvested if they are of sufficient value. None of these stored resources were considered as part of the annual emergy basis for Maine's economy because over the whole state their stored values are not being used faster than they are being replaced. As noted above, this situation may change if a more detailed analysis of regions or industries within the state is performed.

Table IV
Evaluation of emergy storages in the environmental and economic resources of Maine in 1980.

Note	Item	Raw Units	Transformity sej unit^{-1}	Solar Emergy E21 sej	Emdollars[#] E8 1980 $
14	Peat*	6.8E19 J	1.9E4/J	1292	4969
15	Wood*	4.52E18 J	3.2E4/J	145	556
16	Groundwater†	1.71E18 J	1.1E5/J	188	723
17	Topsoil†	1.21E19 J	6.3E4/J	762	2932
18	Population	3.38E7 p-y	3.1E16/p-y	1048	4030
19	Assets	1.86E11 $	2.6E12/$	484	1861

Solar emergy in column 5 divided by 2.6E12 solar emjoules per dollar for the U.S. economy in 1980.
* H.T. Odum (1996)
† H.T. Odum *et al.* (1987)

Figure 7 is an aggregated model of Maine's economy suitable for gaining an overview of emergy flows across state boundaries. Table V summarizes the flows of emergy and dollars that form the basis of the Maine economy. Diffuse environmental resources (e.g., sunlight, rainfall, etc.) total 15.1E21 sej y^{-1} or 33% of the state's total emergy budget. Fuels account for 41% of the emergy use, and imported goods and services excluding fuels accounted for 17%. Renewable fuel resources such as wood and hydroelectric power found within the state comprise 7.3% of the state's total emergy use.

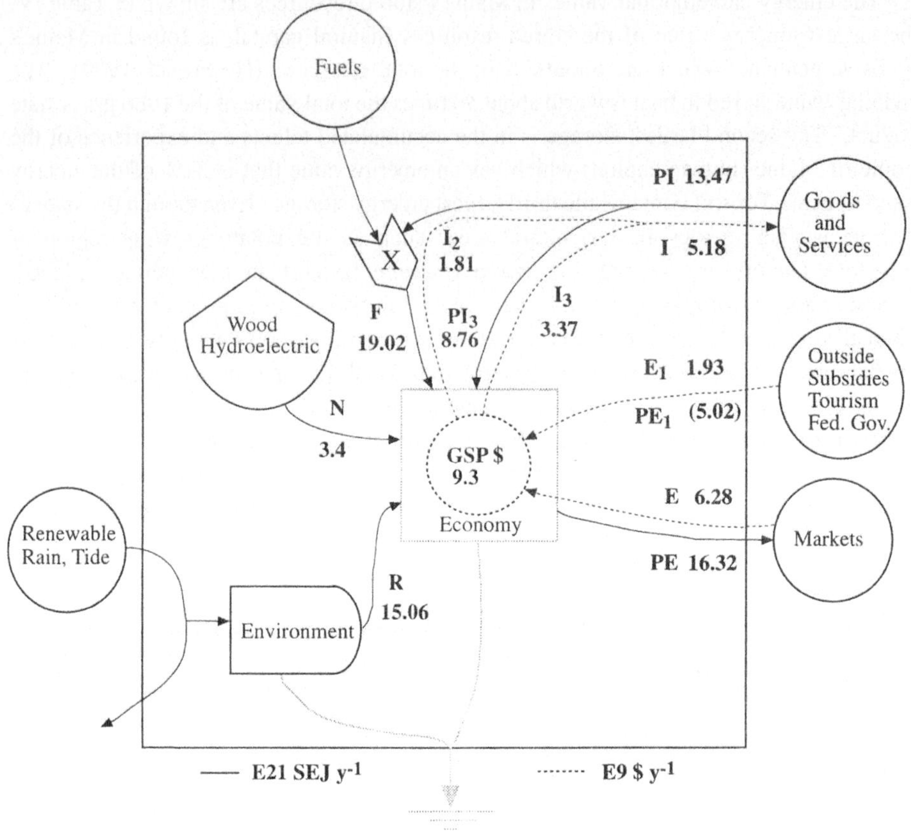

Fig. 7. An aggregated diagram of the Maine State economy and its emergy resource base for the calculation of emergy indices.

The gross state product, GSP, in 1980 was 9.3 E9 $ of which 56% was spent outside on goods and services. Fuels accounted for 35% of imports and 19.5% of the GSP on a dollar basis, whereas, on an emergy basis fuels account for 68% of imported emergy and 41% of the total emergy use by the economy. Twenty-one percent of the GSP or 1.93 E9 $ y^{-1} is attracted to Maine in tourist and net federal expenditures. There is no actual emergy flow which corresponds to all of the money spent; however, something of value is received for these expenditures (e.g., tourists consume resources obtain experiences which are taken with them) therefore, there is a virtual emergy outflow of 5.02 E21 sej y^{-1} corresponding to this monetary influx which is directly or indirectly dependent on the abundance of environmental resources in Maine and the image of attractiveness that they project. This virtual emergy flow is equivalent to 11% of the total actual emergy use, 23% of total exports (including the virtual flows), and 37% of imported goods and services in 1980.

Table V
Summary of emergy and dollar flows for Maine in 1980 (See Figure 7).

Symbol	Item	Solar Emergy E21 sej y^{-1}	Dollars E9 \$ y^{-1}
R	Renewable sources	15.06	
N	Energy sources within Maine	3.4	
F	Imported fuels	19.02	
I	\$ paid for imports		5.18
I_2	\$ paid for service in fuels		1.81
I_3	\$ paid for imports (minus fuel)		3.37
PI_3	Imported goods and services excluding services to fuel	8.76	
PI	Imported goods and services	13.47	
E	\$ paid for exports		6.28
PE	Exported goods and services	16.32	
E_1	\$ attracted by natural resources		1.93
PE_1	Virtual emergy flow representing the aesthetic value of the environment	5.02	
X	Gross State Product		9.3
P	U.S emergy to dollar ratio used for imports.	2.6E12 sej $\$^{-1}$	

Table VI defines a number of emergy indices that are useful in characterizing Maine and comparing its emergy profile to that of other states and to the nation as a whole. Locally renewable emergy (R/U = 0.33) is an indicator of a region's self sufficiency in the long run. At present 40% of Maine's emergy is derived from home sources which also indicates the potential for a region to be self-sufficient; however, 60% of the emergy use is purchased from outside the state which indicates a strong present dependence on national resources. Twenty nine percent of Maine's total emergy use comes from imported goods and services which compose 48% of the total imported emergy, the other 52% is mostly in the fuels themselves. In 1980 Maine had an emergy imbalance with the rest of the nation of 11.5 E21 sej (F+ PI3 - PE), equivalent to 41% of the emergy imported. Therefore, economic conditions in Maine are highly dependent on the conditions in the national economy which determine the availability of emergy for import. Actual exports are only 59% of imports, but if total exports which include the virtual flows are considered exports increase to 77% of imports. During 1980, 4.1 E16 sej per person were used in Maine which is an expression of the people's standard of living in that year. At the 1980 standard of living Maine's renewable resources could support a human carrying capacity of 0.37 E6 people. The developed human carrying capacity for Maine was 2.9 E6 people compared to the 1980 population of 1.13 E6 people.

Table VI
Emergy indices for an overview of Maine in 1980.

Expression	Name of Index	Quantity
R	Renewable emergy flow	15.1E21 sej y^{-1}
N	Flow from indigenous sources	3.4E21 sej y^{-1}
$F+PI_3$	Flow of imported emergy	27.8E21 sej y^{-1}
$R+N+F+PI_3$	Total emergy in flows	46.3E21 sej y^{-1}
$R+N+F+PI_3+N0$	Total emergy used*, U	46.3E21 sej y^{-1}
(R+N)/U	Fraction of emergy from Maine	0.40
R/U	Fraction of use locally renewable	0.33
(R+N0)/U	Fraction of use that is free†	0.33
$(F+PI_3)/U$	Fraction of use purchased outside	0.60
PI/U	Fraction of use in imported goods and services	0.29
PE	Exported emergy in goods and services	16.3E21 sej y^{-1}
PE_1	Virtual emergy export	5.0E21 sej y^{-1}
$PE+PE_1$	Total emergy export	21.3E21 sej y^{-1}
$(F+PI_3)-PE$	Imports - actual exports	11.5E21 sej y^{-1}
$PE/(F+PI_3)$	Ratio of actual exports to imports	0.59
$(PE+PE1)/(F+PI_3)$	Ratio of total exports to imports	0.77
$(F+N+PI_3)/R$	Ratio of concentrated to dispersed	2.00
U/area	Emergy use per unit area (9.4E10 m^{-2})	4.9E11 sej m^{-2}
U/population	Present emergy use per person (pop = 1.125E6)	4.1E16 sej $p.^{-1}$
R/(U/pop.)	Renewable carrying capacity at the present standard of living	0.37 E6 people
8R/(U/pop.)	Developed carrying capacity at the present standard of living.	2.9 E6 people
P1=U/GSP	Ratio of emergy use to GSP, or state emergy to dollar ratio	5.0 E12 sej $\$^{-1}$
el/U	Ratio of electricity to emergy use el = 10.26 E21 sej y^{-1}	0.22
fuel/pop.	Fuel use per person	2.2E16 sej $p.^{-1}$

* Total emergy use includes that from N0, the stored emergy resources, e.g. soil, groundwater, forest biomass, that are being used faster than their rate of replacement. For Maine initial calculations indicate that replacement rates exceed use of these resources in the state as a whole.

†Since N0 = 0 the fraction of locally renewable emergy is equal to the that which is free.

COMPARISON OF EMERGY INDICES FOR MAINE, FLORIDA, TEXAS, AND THE NATION

Table VII contains a comparison of various emergy indices for Maine, Florida, Texas, and the United States including Alaska. Maine is endowed with a large amount of renewable emergy relative to the average available in the nation. Maine accounts for 1% of the national area but receives 2% of the nation's renewable emergy. Florida is somewhat better endowed with renewable emergy since Maine contains 30% of Florida's land area but only 22% of her renewable emergy. In contrast, Maine receives 39% of the renewable emergy input to Texas over an area 13.4 % the size of Texas. Renewable emergy accounts for a larger fraction of present emergy use in Maine (0.33) than in the other states examined. Renewable emergy as a fraction of total use indicates the present degree of economic development in an area while the quantity of renewable emergy available per unit area indicates the potential for self-sufficiency in the long run.

Table VII
Comparison of emergy indices for Maine, Florida, Texas, and the United States circa 1980.

Index	Maine 1980	Florida# 1979	Texas† 1983	U.S.† 1983
Renewable emergy flow*	15.1	66.2	39	773
Indigenous emergy flow*	3.4	2.1	666	5346
Imported emergy flow*	27.8	284	307	1936
Total emergy inflow*	46.3	352	595	8055
Total emergy used*	46.3	380	628	7887
Fraction use from home	0.40	0.18	0.84	0.75
Actual emergy export*	16.3	95.7	501	870
Imports - exports *	11.5	188	-194	811
Ratio exports to imports	0.59	0.34	1.6	0.58
Renewable fraction of use	0.33	0.17	0.06	0.10
Purchased fraction of use	0.60	0.75	0.37	0.25
Fraction that is free	0.33	0.18	0.12	0.22
Ratio concentrated to dispersed	2.0	4.2	7.3	3.5
Area (m^{-2})	9.4E10	3.1E11	7E11	9.4E12
Population	1.13E6	8.8E6	15.7E6	234E6
Use per area sej m^{-2}	4.9E11	12E11	9E11	8.4E11
Use per person sej $p.^{-1}$	4.1E16	4.3E16	4.0E16	3.4E16
Renewable carrying capacity	0.37E6	1.53E6	0.98E6	23E6
Developed carrying capacity	2.9E6	12.3E6	7.8E6	183E6
Emergy to $ ratio sej $\$^{-1}$	5.0E12	4.3E12	2.6E12	2.4E12
Ratio electricity to use	0.22	0.23	0.18	0.17
Fuel use in sej $p.^{-1}$	2.2E16	2.3E16	2.9E16	1.5E16

* Flows in sej y^{-1} times E21.
# Data on Florida from H.T. Odum *et al.* (1986) with modified tidal input.
† Data on Texas and the United States are from H.T. Odum and E.C. Odum (1987) with modified emergy input from tides.

Maine, Florida, and the nation import more emergy than they export. Texas alone exports emergy, principally due to the sale of petroleum. Maine exports 59% of the emergy which it imports making it the state most similar to the national average. Florida is the state most dependent on the emergy available nationally because it's exports account for only 34% of the emergy imported. Texas is the state least dependent on the nation in the short run since it exports 1.6 times more emergy than it imports. Florida purchases 75% of its emergy use, while Maine buys 60% of the emergy it uses. Both of these fractions are large when compared to Texas which purchases only 37% of its emergy. The fraction of emergy purchased outside the state is an indicator of the dependence of a state on the availability of emergy in the national economy. The United States is less dependent (25% of its emergy is imported) on foreign emergy compared to the dependence of these individual states on the nation.

The ratio of emergy use in urban systems to emergy use in rural systems is an indicator of the degree of economic development. According to this criterion Maine is much less developed than Texas and less developed than Florida and the U.S. as a whole. This picture is mirrored in the index of emergy use per unit area, where the density of emergy use in Maine is 41% of that in Florida, 54% of that in Texas and 58% of the national emergy use density. Maine has an emergy to dollar ratio 1.9 times that of Texas, 2.1 times the national average, but only 1.2 times greater than Florida. This indicates that a dollar spent in Maine buys twice the emergy of an average dollar spent in the nation. This is true because a large fraction of the total resource base is provided by unpaid environmental work. The greater the emergy to dollar ratio the more competitive an area will be in attracting economic inflows, all other things being equal, because a dollar spent buys more free environmental service than in areas with a lower emergy to dollar ratio.

The emergy use per capita is an indicator of the standard of living which must be specified to describe human carrying capacity. Maine, Florida, and Texas all use more emergy per capita than the national average. Maine and Texas had a per capita emergy use which was 95% and 93%, respectively, of the highest use which was found in Florida. All the standards of living measured fell within a narrow range with Florida only 26% greater than the national average.

Maine can support 33% of her present population at their 1980 standard of living in a time when only renewable emergy is available for use. This estimate indicates that for a future similar to that predicted in Figure 3b, 370,000 people could be sustained indefinitely at the 1980 standard of living as defined by emergy use. This percentage is considerably larger than the national average of 9.8%. Florida and Texas could support 17.3% and 6.2% of their present populations, respectively, on their renewable emergy alone. The developed human carrying capacity of Maine and Florida exceeded their 1980 populations. The developed carrying capacity is defined as the number of people a state could support at their present standard of living if its emergy use was 8 times the renewable resource base. The factor of eight represents an average ratio for developed countries in the world circa 1980 (H.T. Odum et al., 1987a). Texas had the highest degree of development since it proved to be supporting twice as many people as expected for an average developed industrial country. Maine's population can be increased 2.6 times and Florida's 1.4 times before they support as many people as an average developed country at the 1980 standard of living.

7. Discussion

Aggregation of variables at the state level obscures regional differences and may lead to a somewhat distorted view of the actual situation in Maine. For example, the entire state area is used to determine the emergy use per unit area when in reality about two thirds of the state's area is sparsely settled forest land and the remaining third supports 85% of the population (Morris 1976). Therefore, an intrastate regional analysis may give different results than those found for the state as a whole. Similar regional differences in the emergy indices for the United States as a whole may exist because the vast hinterland of Alaska has been included in the calculation of national average values. The existence of regional differences within a nation or a state does not necessarily invalidate the whole analysis or the comparison of emergy indices among nations or states because these regional variations are present to a greater or lesser degree in most nations and states. However, these departures from uniformity do point out the need for emergy analysis to be pursued at the regional level within systems to address certain assessment questions.

The state of Maine was divided into three regions, which were used in averaging observations for subsequent analysis (Figure 5). The Coastal area is a zone rich in emergy inputs and popular with tourists, rusticators, and natives alike because of its great natural beauty. The Upland region is the industrial and farming center for Maine which is linked to the rest of New England by Interstate 95 (Barringer 1972), while the Mountain region is a hinterland containing vast forest tracts, recreational facilities, and few people (Barringer 1972). An emergy analysis of these three regions could determine the contribution each area makes to the support of economic activities within the state. The potato growing area in Aroostock County forms an additional subregion within the uplands that should be evaluated as a separate system because of its particular set of problems i.e., soil erosion. For similar reasons, the eastern coastal and upland area is logically combined into the economically isolated "Down East" region (Washington County) for separate analysis. Single sectors of the economy, such as the fishing, transportation, or forest products industries are also possible subjects for analysis using this method.

Several emergy indices show that Maine may have a high degree of self-sufficiency in a future with lower fossil fuel availability, whereas, in today's fossil fuel economy it has a low degree of self-sufficiency. In a low energy future it is probable that Maine would be even better off than indicated in Table VII. Several observations contribute to this opinion. Maine has large quantities of renewable energy which are not being fully exploited at present. For example, Maine forests are growing biomass at twice the present rate of harvest, the potential for using water power is not yet exhausted, and there is a great potential in tidal power that is unutilized. In addition, to these renewable resources, Maine has large quantities of peat stored in bogs that can be an important source of energy in the future whether it is exploited in a renewable or nonrenewable way. These extensive natural resources combined with the ingenuity that the Maine people have historically displayed in mastering the forest and the sea will probably make Maine a state with a standard of living close to the present high level despite lower energy in the future.

Despite some evidence of environmental degradation (Pollard 1973, Larsen 1989), emergy indices show that Maine is a state which at least for the present has both a high standard of living and a relatively unspoiled environment. Fuel, electricity, and emergy use per person are all far above the national average which indicate that the Maine standard of living is a good one. In addition, the fraction of emergy use in free and renewable sources is the highest of the four cases examined which reflects the relatively unspoiled and undeveloped state of Maine lands as a whole. Thus, many people in Maine enjoy the values of a developed economy along with the environmental resources of a state that is yet to become heavily developed. This view is altered somewhat if the regional distribution of development is considered. Southern Maine is within the expanding edge of the greater Boston metropolitan area, and at present is experiencing some of the environmental and social problems that development brings. In contrast, large areas of the state exist as unincorporated townships which lack the organization and infrastructure that we have come to expect as part of a developed country.

The challenge for Maine in the immediate future is to fulfill its development potential without compromising the environmental resources that support a high quality of life. What human carrying capacity will be sustainable in Maine? We have demonstrated using the conceptual model in Figure 3 that the answer to this question depends on the standard of living that is desired by the population and on the renewable energy basis for the region. The human carrying capacities for Maine, Florida, Texas, and the nation at the 1980 standard of living on their renewable environmental resource bases alone were 34%, 17%, 6%, and 10% of their present populations, respectively. The present human populations living in these states and in the United States as a whole indicate that our way of life is not sustainable using present system designs. In the long run human population and/or the standard of living must adjust to come within the range that the renewable environmental resource base can support. This can happen in two obvious ways. Human population size can be decreased given sufficient lead time (e.g., the one child policy instituted by the People's Republic of China) or the rate of resource consumption can decrease to a level which allows the existing population to survive on the renewable resource base. The latter course assumes that present populations can indeed subsist on the renewable environmental resource base. In either case, our standard of living will be ameliorated by testing and incorporating changes in our social, economic, and environmental system designs that improve efficiencies and help us obtain more for less from the existing environmental-economic interface. As mentioned earlier such design changes are not a panacea, but they may considerably soften the shock of a necessary decline in population size and/or standards of living to meet the constraints on the rates of resource use imposed by using our resources in a renewable way.

Pulsing may be nature's strategy for getting more for less. Developing a better understanding of how pulsing maximizes the empower production of environmental-economic systems may increase our ability to recognize system designs that will allow us to choose alternatives that optimize human populations and their standard of living at each stage in the cycle of change including a future time when we are more dependent on renewable environmental resources. For example, by developing presently unexploited renewable resources such as tidal power, and by exploiting renewable peatlands and forest biomass in phase coordinated pulses with other regions, Maine may be able to contribute to supporting a larger fraction of the nation's 1980 population in a future with low fossil fuel supplies than would otherwise be possible.

Optimizing human population and standards of living during the climax of fossil fuel use (in the short run) requires us to consider our present state of development in relation to the expectations for global development (see Wackernagel and Yount this volume). The present populations for Maine (39% of the average) and Florida (72% of the average) were less than the expected value for an average developed country in the world circa 1980, whereas, the populations of Texas (200% of the average) and the United States as a whole (128% of the average) exceeded the world average carrying capacity for a developed country. For the near term future Maine and Florida should focus on carefully managing their remaining growth potentials, whereas, Texas should consider strategies that will reduce its population size and focus research programs on developing system designs (perhaps centering on agriculture, H.T. Odum *et al.*, 1987a) that will maximize the quality of life for a smaller population in the future.

It may be unwise to develop regions more intensely than the average for a developed country at present because the most heavily developed regions may suffer the greatest hardship during a decline in fossil fuel availability. Alternatively, the less developed areas may suffer more because they will have to willingly or forcibly subsidize the developed areas during a period of decline. The pulsing paradigm for sustainability leads to the hypothesis that a climax state of emergy use may be prolonged by the coordinated out of phase pulsing of regions within or controlled by the system that is in climax. Design changes that reinforce this pattern may be successful in optimizing human carrying capacity at a chosen standard of living for regions, nations, and our planet during the portion of the pulsing cycle of change when our nonrenewable fossil fuel resources are peaking.

8. Conclusions

Several conclusions about human carrying capacity and regional sustainability can be derived from the discussion of the overview ideas presented in this paper. They are as follows:

(1) All regional development based on the nonrenewable use of resources is inherently unsustainable.

(2) Human carrying capacity at a specified standard of living represents the anthropogenic load on a given regional environmental resource base.

(3) Because the environmental resource base of a region is varying in a pulsing cycle of change (Figure 4b), human populations and/or standards of living must constantly be adjusted to maintain the same load on the resources.

(4) Pulsing appears to be the pattern that insures survival and maximum performance in natural systems. Therefore, a pulsing steady state with a cycle of change in carrying capacity may be the pattern that is sustainable for a region with a given environmental resource base rather than a steady state with a constant development level or carrying capacity.

These assumptions interpreted in the context of Energy Systems Theory lead to the following recommendations for optimizing the load on environmental resources which results from different human carrying capacity and level of development choices in a region through time.

(1) Identify the pulsing patterns in a region and the fundamental drivers (energy sources) so that we know where we are in the cycle of change, e.g. succession, climax, regression, or low energy steady state.

(2) Use an emergy perspective to assess environmental and economic problems while nonrenewable resources are still high and the range of possible responses to their future decline is correspondingly large.

(3) Search for and incorporate systems designs that will maximize empower production and use at each stage in the pulsing cycle of change.

(4) The key to prolonging a stage in the cycle and to insuring a smooth transition from one phase of the cycle to another may be in learning to recognize and manage the cycles of environmental-economic pulsing on multiple scales.

Acknowledgements

I thank Earthwatch for providing a format through which the general public can assist in performing scientific research. In particular, I acknowledge the excellent contributions of the Earthwatch Research Team members Bill Claypoole, Nancy Thwarp, Tim Tindol, Ellen Stienmetz, and Phyl Van Fleet. I thank the Maine Department of Marine Resources for supporting the original study. I thank David Yount of the USEPA, Mid-Continental Ecology Division for inviting me to join his symposium session. I thank the USEPA, Atlantic Ecology Division for providing the time to complete the final research and write-up of this paper. The opinions expressed in this paper are the author's and do not necessarily reflect those of the USEPA.

Appendix A - Evaluation of Maine Energy Resources

NOTE #1. AREAS

Land area of the state: 86156 km^2, (Maine State Development Office 1985 (our estimate from maps was 84261 km^2 which is 2.2% less than the area above).

Area by physiographic region in Figure 5.

Coastal region:	10596 km^2
Upland region:	42609 km^2
Mountain region:	31056 km^2
Continental shelf headlands to 100 m:	9822 km^2

NOTE #2. DIRECT SUNLIGHT

(area of state) (average insolation)

Area of Maine including the shelf area out to 100 m. = 94083 km^2

Northern Region: Mountains and Aroostook Co. 31056 km^2 + 10059 km^2 = 41115 km^2 = 4.1115 E10 m^2

Remaining land: Uplands and Coastal (land only) 32550 km^2 + 10596 km^2 = 43146 km^2 = 4.3146 E10 m^2

Shelf to 100 meter isobath: 9822 km^2 = 9.822 E9 m^2

Solar Energy Totals: The average daily solar insolation from 1961 - 1971 received at Caribou is used for the Northern region. The remainder of the land and the continental shelf are assumed to receive an average solar insolation similar to that received at Portland during the same time period (U.S. Dept. Commerce 1971).

(4.1115 E10 m^2) (4.643 E9 J m^{-2} y^{-1}) = 1.909 E20 J y^{-1}

(4.3146 E10 m^2) (4.917 E9 J m^{-2} y^{-1}) = 2.1215 E20 J y^{-1}

(9.8220 E9 m^2) (4.917 E9 J m^{-2} y^{-1}) = 4.8295 E19 J y^{-1}

Total Solar Energy = 4.5134 E20 J y^{-1}

NOTE #3 - KINETIC ENERGY IN WIND USED AT THE SURFACE

(height) (density) (diffusion coefficient) (wind gradient)^2 (area)]

Assume an annual average vertical eddy diffusion coefficient of 15 m^2 s^{-1} similar to Albany, NY. the closest station to Maine given in Odum et al. (1983). The annual average vertical velocity gradient was calculated for stations at Caribou and Portland (U.S. Dept. of Commerce 1980). Areas were the same as those used in the solar energy calculation.

Northern Region:

(1000 m) (1.23 kg m^{-3}) (15 m^3 m^{-1} s^{-1}) (3.154E7 s y^{-1}) (3.31E-3 m s^{-1} m^{-1})^2 (4.115E10 m^2) = 2.62E17 J y^{-1}

Uplands, Coast, and Shelf:

(1000 m) (1.23 kg m^{-3}) (15 m^3 m^{-1} s^{-1}) (3.154E7 s y^{-1}) (3.04E-3 m s^{-1} m^{-1})^2 (5.2968E10 m^2) = 2.84E17 J y^{-1}

Total wind energy = 5.47E17 J y^{-1}

NOTE #4 - TIDAL ENERGY ABSORBED

(area elevated) (tides per year) (height)^2 (density) (gravity)

Tidal energy absorbed on the shelf (9.822E9 m^2) assuming an average tidal height over the entire area of 1.5 m (Moody et al. 1984) and that 100% of the tidal energy is absorbed on the shelf or in nearshore waters.

(9.822E9 m^2) (706 y^{-1}) (1.5 m)^2 (1.0253E3 kg m^{-3}) (9.8 m s-^{2}) = 1.568E17 J y^{-1}

NOTE #5 - WAVE ENERGY ABSORBED

(Shore length) (1/8) (density) (gravity) (height)^2 (velocity)

The annual average wave height measured at NOAA's Portland buoy station was 1.0 m for the years 1982 - 1984 (National Climate Data Center 1986). Assume that d, the average water depth in the breaker zone along the Maine coast, is 5 m. The shore length at the headlands is our estimate.

Then wave speed, c= gd = 7 m s^{-1}.

(3.59E5 m) (1/8) (1.025E3 kg m^{-3}) (9.8 m s^2) (1.0 m)^2 (7 m s^{-1}) (3.145E7 s y^{-1}) = 9.95E16 J y^{-1}

NOTE #6 - CHEMICAL POTENTIAL ENERGY IN RAIN:

(Area including shelf) (Rainfall) (Gibbs Free Energy, G) =

The area weighted average annual rainfall is 102.04 cm y^{-1} based on the 1931 -55 average precipitation by area for 26 stations (U.S. Dept. Commerce 1972). G assumes 10 ppm dissolved solids concentration in rain. The spatial division averages were: Coastal = 114.44 cm y^{-1}, Upland = 101.73 cm y^{-1}, Mountain= 97.66 cm y^{-1}.

(9.41E10 m^2)(1.02 m y^{-1})(4.94 J g^{-1})(1E6 g m^{-3}) = 4.74E17 J y^{-1}

NOTE #7 - GEOPOTENTIAL ENERGY IN RAIN

(area) (mean elevation) (runoff) (density) (gravity)

The mean elevation for Maine is 244m from Odum et al. (1983).

(8.43E10 m^2)(244 m)(1.02 m y^{-1})(1.0E3 kg m^{-3})(9.8 m s^{-2})= 2.06E17 J y^{-1}

NOTE #8 - CHEMICAL POTENTIAL ENERGY IN RIVER

(Volume of Flow) (Density) (G), where G is the Gibbs free energy of river water relative to sea water. The volume of flow is a 30 year average from Bue (1970).

$$G = \left(\frac{(8.33\ J\ mole^{-1}\ degK^{-1})\ (\ 300\ °K)}{18\ g\ mole^{-1}} \right) \ln \left(\frac{1E6\ -\ S}{965{,}000} \right) J\ g^{-1}$$

where S = 50 ppm is the dissolved solids concentration in river water.

$$G = 138.8\ J\ g^{-1}\ \ln \left(\frac{999{,}950}{965{,}000} \right) = 4.94\ J\ g^{-1}$$

($6.05E10\ m^3\ y^{-1}$) ($1E6\ g\ m^3$) ($4.94\ J\ g^{-1}$) = $2.99E17\ J\ y^{-1}$

NOTE #9. EARTH CYCLE

(area) (heat flow per area)

53.54 mW m^{-2} average crustal heat flux in Maine estimated from Decker (1987).

Heat flux due to earth cycle = ($8.4261E10\ m^2$) ($1.689E6\ J\ m^{-2}\ y^{-1}$) = $1.422E17\ J\ y^{-1}$

NOTE #10. POWER SOURCES

Data in this section were extracted from Maine Office of Energy Resources (1985). The dollar value of Maine's 1980 fossil fuel use was $1.81E9. The use of energy in Maine during 1980 in J y^{-1} for the major sources is as follows: Coal, 2.003E15; Petroleum, 2.548E17; Natural Gas, 2.319E15; Wood, 2.951E16; Nuclear output, 2.825E16; Canadian Electric 7.273E15; Hydropower, 2.899E16.

NOTE #11 TOURISM

Rovelstad and Rovelstad (1987) estimate tourist expenditures in Maine to be $1.69E9 in 1985. A prior study by Arthur D. Little in 1973 (Pease and Richard 1983) estimated tourist expenditures in Maine to be $3.10E8 If tourism increased linearly between these two times, tourist expenditures in Maine for 1980 would have been $1.07E9.

NOTE #12 IMPORTED GOODS AND SERVICES

The U.S. Transportation Census estimated that Maine manufacturers received $3.2E9 worth of goods from all parts of the country in 1977 and shipped $4.2E9 worth of products in return. Intrastate shipments were valued at $0.8E9 which we assume are distributed equally between shipments and receipts. Maine foreign exports in 1977 were around $0.2E9 (Maine State Development Office, 1985). We estimate the total value of Maine manufacturers shipments in 1977 at $4.4E9. By 1981 this value was $7.8E9, if we assume a linear rate of increase, manufacturers shipments in 1980 are estimated to be $6.95E9. Exports to Canada in 1980 were estimated to be $0.22 E9 so about $6.73E9 were shipped domestically in 1980. If intrastate shipments were about 10% of the domestic shipments as they were in 1977, interstate shipments in 1980 were $6.06E9 and a total of $6.28 E9 goods and services were exported. If the trade balance between Maine and the rest of the nation in 1980 was the same as it was in 1977 (imports = 0.76 exports), she imported $4.6E9 of goods from the rest of the nation in 1980. Maine's imports from Canada in 1981 are estimated to be $0.58E9 and total imported goods and services are estimated to be $5.18E9. Data quoted here are from Maine Development Office (1985) and Pease and Richard (1983).

NOTE #13 FEDERAL GOVERNMENT

Total outlay of federal funds to Maine in 1980 was $2.56E9 of which $1.33E9 was direct transfer payments to individuals. Personal income taxes paid in Maine in 1980 were $9.64E8. If Maine's share of all taxes is similar to her share of personal income tax we estimate business taxes, social security taxes, and corporate income taxes paid in 1980 to be $8.25E7, $4.3E8, and $2.43E8 respectively. Therefore, Maine's total contribution to the federal government in 1980 was $1.72E9, leaving a surplus balance of $8.6E8 in government funds that was spent in Maine. Data in this footnote were taken from Pease and Richard (1983) and U.S. Bureau of Census (1985).

NOTE #14. POTENTIAL ENERGY IN STORED PEAT

(volume of material) (density) (organic fraction) (G)

Volume: Average depth of peat = 3.05m, Approximate area of peatlands = $2.53E9m^2$, 3.05 x 2.53E9 = 7.71E9 m Density: Peat density = $0.67g\ cm^{-3}$; % Organic: Ash content of peat = 10%; G: Assumed similar to lignite, 6300 Btu/lb, (13.88 Btu/g) (1,054 J/Btu) = 14,629 J/g

Energy stored in Maine peat = ($7.71E9\ m^3$) ($1E6\ cm^3\ m^{-3}$) (($0.67\ g\ cm^{-3}$) (.9) ($14{,}629\ J\ g^{-1}$) = 6.8E19 J

Presently peat is not mined extensively in Maine (about 5000 tons were mined in 1977). The information used to make the peat calculations is from Hasbrouck (1979).

NOTE #15 - POTENTIAL ENERGY STORED IN WOOD.

(Volume of material) (density) (organic fraction) (G) where G = (4.2 kcal g^{-1}) (4186 J $kcal^{-1}$)= 17581 J g^{-1}
Volume: 1.2E9 green tons = 6.97E8 m^3 with density = 0.64 g cm^3 (Maine State Development Office, 1985). Density: softwoods; 0.56 g cm^3, hardwoods; 0.785 g cm^3 , 35% of Maine timber stock is hardwood and 65% is softwood. Organic fraction: Ash content of wood is assumed to be 10%
Energy stored in wood biomass = (6.97E8 m^3)(1E 6 cm^3) m^{-3})(0.51g cm^{-3})(0.9)(17581 J g^{-1}) = 4.52E18 J
The annual rate of biomass growth for all species in 1984 was estimated at 30-40 million green tons per year, whereas, the annual harvest was 21 million green tons in that year (Maine Office of Energy Resources, 1985).

NOTE #16 - CHEMICAL POTENTIAL ENERGY OF WATER STORAGES

(water volume) (density) (G) where G is the Gibbs free energy of the water relative to sea water. The formula for calculating G is given in Footnote #8.
A) Groundwater. Groundwater volume for the state is estimated as 1.0E14 gallons by Hasbrouck (1985). This converts to a volume of 3.79E11 m^3. An S equal to 132 ppm for Maine groundwater was estimated as the average of five stations given in Haskell et al. (1984). This solute concentration gives a G of 4.467 J g^{-1}.
(3.79E11 m^3) (1.0E6 g m^{-3}) (4.47 J g^{-1}) = 1.69E18 J
B) Surface water. The average capacity for surface water in Maine was estimated as 75% of the maximum capacity calculated from Haskell et al. (1984) as 1.77E11 ft^3. This average capacity converts to 3.77E9 m^3. An S equal to 46 ppm for Maine surface waters was the average of 14 values from 3 stations given in Haskell et al. (1984). This solute concentration gives a G of 4.68 J g^{-1}.
(3.77E9 m^3) (1.0E6 g m^{-3}) (4.68 J g^{-1}) = 1.76E16 J and 1.71E18 J = stored emergy of water

NOTE #17 LAND USE PATTERNS AND TOPSOIL

Table A1 Erosion rates and soil loss from several land use types in Maine.

Land use	Area m^2	Rate of erosion $g\ m^{-2}\ y^{-1}$	Soil loss $g\ y^{-1}$
Cropland	3.67E9	673	2.47E12
Pasture land	1.0E9	67	0.07E12
Forest land	6.69E10	22.4	1.5E12

Data in the table above were found in U.S. Dept. Agriculture (1982).

Total topsoil lost to erosion is 4.04E12 g y^{-1}. Average soil formation rate for the forested area is 650 g m^{-2} y^{-1} assuming that the earth cycle is in steady state and net uplift is balanced by soil formation rate. Net uplift for upland and mountain regions of Maine is about 0.25 mm y^{-1} (estimated from Tyler & Ladd 1980), and the average density of rock is assumed to be 2.6 g cm^{-3}. Topsoil formation on forested land: (6.69E10 m^2) (650 g m^{-2} y^{-1}) = 4.35E13 g y^{-1} balances all erosion.
Estimate of energy storage in Maine topsoil.
(area of crop, pasture and forest land) (depth of soil)(soil density) (% organic) (energy per gram)
Energy storage in topsoil = (7.16E10 m^2) (0.5 m) (0.5E6 g m^{-3}) (0.03) (22604 J g^{-1}) = 1.21E19 J

NOTE #18 POPULATION

The 1980 census estimates the Maine population at 1.125E6 people. The median age of the U.S. population in 1980 was 30 y. Value stored in people of Maine = (population) (average age) 1.125E6 people x 30 years = 3.38E7 people-years

NOTE #19 ECONOMIC ASSETS

Economic assets are estimated assuming a depreciation rate for replacement of 5% per year. Therefore the dollar value of total assets is approximately 20 times the GSP. Economic Assets=20 (9.3E9 \$) =1.86E1 \$

Appendix B - Calculation of a Revised Solar Transformity for Tidal Energy Received and Tidal Energy Dissipated Globally

D.E. Campbell and H.T. Odum

The solar transformity of tidal energy can be calculated in a manner similar to that used by Odum (1996) to determine the solar transformity of the earth's heat using the following assumptions:
(1) The available geopotential energy of the elevated water in worlds oceans is similar for this purpose regardless of source.
(2) On the time scale of one year the available potential energy of the world's oceans is in steady state, thus all the potential energy that is created in a given year is dissipated in that year. If this assumption is not true on average, there would be an accumulation of potential energy in the global ocean which is not observed.
(3) The elevation of the ocean surface relative to a reference level is primarily caused by the solar heat engine including its effect in delivering fresh water streams or the gravitational pull of the sun and moon. Therefore, almost all the available potential energy of the oceans is created by one of these two sources.
(4) The dissipation of tidal energy in the deep oceans is less than 0.001 of that in shallow water (Miller 1966). This fraction does not take into account recent estimates of the importance of deep ocean internal waves generated by seamounts.

If the solar energy flux to earth is 3.93 E24 joules y^{-1} (Odum 1996), the gravitational energy transmitted to the earth is 8.515 E19 joules y^{-1} (Munk and MacDonald 1960), the tidal energy transmitted to shallow water is 5.2 E19 joules y^{-1} (Miller 1966), and the available potential energy in the top 1000 m of the global ocean is 21.4 E19 joules y^{-1} (Oort et al. 1989), the following calculation can be performed.

The fraction of the available potential energy of the oceans created by solar energy is equal to the total available potential energy minus the potential energy created by the tide. If almost all of the available potential energy produced by gravitational attraction is transmitted to shallow water and dissipated the amount of the global available potential energy produced by solar energy is 16.2 E 19 joules y^{-1}.

21.4 E19 joules y^{-1} - 5.2 E19 joules y^{-1} = 16.2 E 19 joules y^{-1}

If the deep heat energy input from the earth contributes to the formation of geopotential energy in the ocean by creating the continental land masses and coastal shelves, the non-tidal emergy input to this process should be 8.0 E24 sej y^{-1} (3.93 E24 sej y^{-1} from solar and 4.07 E24 sej y^{-1} from deep heat of the earth (Odum 1996). To see that there must be a geologic input to creating the potential energy of the oceans imagine an earth without continents and thus no geologic input to the upper zone. Would the oceanic geopotential energy created by the sun and tide be different? The solar transformity of the available potential energy in the oceans created by the solar heat engine is 8.0 E24 sej y^{-1} / 16.2 E19 joules y^{-1} = 49383 sej/j. Because the available potential energy created by the tides is the same "stuff" as the available potential energy created by solar energy it is logical to assume that it has a similar solar transformity.

The solar emergy used up globally in the dissipation of available potential energy produced annually by the tides is then 5.2 E19 joules y^{-1} * 49383 sej/j = 2.568 E24 sej y^{-1} and the solar transformity of the gravitational energy received by the earth is: 2.568 E24 sej / 8.515 E19joules y^{-1} = 30159 sej/j and the new planetary baseline is 3.93 + 4.07 + 2.57 = 10.57 E24 sej y^{-1}.

References

Barringer, R.:1972, *A Maine Manifest.*; Tower Publishing Co., Portland, p. 92.

Bowden, C.: 1977, *Killing the Hidden Waters;* University of Texas Press, Austin, p. 177.

Brown, M.T.: 1980, *Energy Basis for Hierarchies in Urban and Regional Landscapes;*Ph.D. Dissertation, University of Florida, Gainesville, p. 357.

Brown, M.T., McClanahan, T.R. : 1992, EMERGY Analysis Perspectives of Thailand and Mekong River Dam Proposals, Report to the Cousteau Society, Center for Wetlands and Water Resources, University of Florida, Gainesville, p. 60.

Brown, M.T., Tennenbaum, S., Odum, H.T.: 1991, EMERGY Analysis and Policy Perspectives for the Sea of Cortez, Mexico, Report to the Cousteau Society, Center for Wetlands (Pub. 88-04), University of Florida, Gainesville, p. 58.

Brown, M.T.,Green, P., Gonzalez, A. Venegas, J.: 1992, EMERGY Analysis Perspectives, Public Policy Options, and Development Guidelines for the Coastal Zone of Nayarit, Mexico, Report to the Cousteau Society, Center for Wetlands and Water Resources, University of Florida, Gainesville, Vol. 1, p. 255; Vol. 2, p. 145, and 31 map inserts.

Brown, M.T.,Woithe, R.D., Odum, H.T., Montague, C.L., Odum, E.C.: 1993, EMERGY Analysis Perspectives on the Exxon Valdez oil Spill in Prince William Sound, Alaska; Report to the Cousteau Society, Center for Wetlands and Water Resources, University of Florida, Gainesville, p. 122.

Bue, C.D. :1970, Streamflow from the United States into the Atlantic Ocean During 1931- 1960. *U.S. Geological Survey Water Supply Paper* **1899-I**, I1-I36.

Decker, E.R. :1987, Heat flow and basement radioactivity in Maine: First-Order results and preliminary interpretations, *Geophysical Research Letters.* **14**, 256-259.

Doherty, S. J., Odum, H.T., Nilsson, P.O.: 1995, *Systems Analysis for the Solar Emergy Basis for Forest Alternatives in Sweden,* Final Report to the Swedish State Power Board, College of Forestry, Garpenberg, Sweden, pp. 112.

Fuller, R. Buckminster: 1981, *Critical Path,* St. Martins Press, New York, 471 p.

Hall, C.A.S.: 1992, Economic Development or Developing Economics: What are our Priorities? pp. 101-126. In: M.K. Wali (ed.), *Ecosystem Rehabilitation, vol. 1: Policy Issues;* Academic Publishing, The Hague, The Netherlands.

Hall, C.A.S.: 1997, (this volume)

Hasbrouck, S.: 1979, *Maine's Land and Water Resources - Peat.*; Land and Water Resource Center, University of Maine, Orono, p. 12.

Hasbrouck, S.: 1985, *Resource Highlights - Maine's Groundwater.*; Land and Water Resource Center, University of Maine, Orono, p. 12.

Haskell, C.R., Bartlett ,W.P. Jr., Higgins, W.B,. Nichols, W.J. Jr.: 1984, *Water Resources Data Maine Water Year 1984.*, U.S. Geological Survey Water Data Report ME-84-1, 144 p.

Larsen, P.F.: 1989, An Overview of the Environmental quality of the Gulf of Maine, pp. 69-93. In: *The Gulf Of Maine*, NOAA Estuary-of-the-Month Seminar Series, 22, Washington, D.C.

Lotka, A.J.: 1922, Contribution to the Energetics of Evolution, *Proc. Natl. Acad. Sci.* **8**, 147-151.

Maine Office of Energy Resources: 1985, *Comprehensive Energy Resources Plan*, State of Maine, Office of Energy Resources, Augusta, p.123.

Maine State Development Office: 1985, *Maine, A Statistical Summary,* Maine State Development Office, Augusta, p. 60.

McCullough, D.R.: 1979, *The George Reserve Deer Herd: Population Ecology of a K Selected Species;* University of Michigan Press, Ann Arbor.

Miller, G.A.: 1966, The flux of tidal energy out of the deep oceans, *J. Geophysical Res.*, **71**, 2485-2489.

Moody, J.A., Butman, B., Beardsley, R.C., Brown, W.S., Daifuku, P., Irish, J,D., Mayer, D.A., Mofjeld, H.O., Petrie, B., Ramp. S., Smith, P., Wright, W.R.: 1984, *Atlas of Tidal Elevation and Current Observations on the Northeast American Continental Shelf and Slope*, U.S. Geological Survey Bulletin 1611, p. 122, pl. 24.

Morris, G.E. (ed.): 1976, *The Maine Bicentennial Atlas.*, The Maine Historical Society, Portland., p. 20, pl. 69.

Munk, W.H.,MacDonald, G.F.: 1960, *The Rotation of the Earth: A Geophysical Discussion*; Cambridge Univ. Press, London, p. 323.

National Climatic Data Center: 1986, *Climatic Summaries for NDBC Data Buoys*, U.S. Dept. Commerce, National Oceanic and Atmospheric Administration, National Weather Service, National Data Buoy Center.

Odum, E.C., Odum, H.T. 1984, System of ethanol production from sugarcane in Brazil, *Ciencia y Cultura* **37**, 1849-1855.

Odum, E.C., Scott, G., Odum, H.T.: 1982, *Energy and Environment in New Zealand,* University of Canterbury, Christchurch , New Zealand, p. 127.

Odum, E.P.: 1971, Fundamentals of Ecology (Third Edition); W.B. Saunders Co. Philadelphia, p. 184.

Odum, E.P.: 1997, *Ecology, A Bridge Between Science and Society;* Sinauer Associates Inc., p. 330.

Odum, H.T.: 1976, Macroscopic Minimodels of Man and Nature, pp. 249-280. In, B.Patten, (ed), *Systems Analysis and Simulation in Ecology,* Vol. 4. Academic Press, New York.

Odum, H.T.: 1983, *Systems Ecology* ; John Wiley and Sons, New York, p. 644.

Odum, H.T.: 1986, Emergy in Ecosystems, pp. 337-369. In: Polunin, N. (ed.) *Ecosystem Theory and Application.*; Environmental Monographs and Symposia, John Wiley and Sons, New York.

Odum, H.T.: 1987, Models for National, International, and Global Systems Policy, pp. 203-251. In: L.C. Bratt, W.F.J. van Litrop (eds.), *Economic-Ecologic Modeling*; Elsevier Sci. Pub., North Holland.

Odum, H.T.: 1994, *Ecological and General Systems: An Introduction to Systems Ecology;* University Press of Colorado, Niwot, CO, p. 644. (Rev. Ed. of Systems Ecology).

Odum, H.T.: 1996, *Environmental Accounting: Emergy and Environmental Decision Making*; John Wiley and Sons, NY, p. 370.

Odum, H.T., Alexander, J. (eds.): 1977, *Energy Analysis of Models of the United States*, Annual Report to the Dept. of Energy, Env. Eng. Sci., University of Florida, Gainesville, (contract EY-76-S-05-4398), p. 457.

Odum, H.T., Arding, J. E.: 1991, *EMERGY analysis of Shrimp Mariculture in Ecuador,* Report to Coastal Resource Center, University of Rhode Island, Center for Wetlands, Univ. of Florida, Gainesville, p. 87.

Odum, H.T.,Brown, M. (eds.): 1975, Carrying Capacity for Man and Nature in South Florida, Final Report to the National Park Service, U.S. Dept. Interior, and State of Florida, Division of State Planning, Center for Wetlands, University of Florida, Gainesville, p. 886.

Odum, H.T.,Odum, E.C. (eds.): 1983, *Energy Analysis Overview of Nations*, Working Paper WP-83-82, International Institute for Applied Systems Analysis,. Laxenburg, Austria, p. 469.

Odum, H.T., Odum, E.C., Blissett, M.: 1987a, *The Texas System, Emergy Analysis and Public Policy*. A Special Project Report, L.B. Johnson School of Public Affairs. University of Texas at Austin, and The Office of Natural Resources, Texas Department of Agriculture, Austin, p. 92.

Odum, H.T.,Brown, M., Costanza, R.: 1976, Developing a Steady State for Man and Land: Energy Procedures for Regional Planning, pp. 343-361. In: *Science for Better Environment*, Proceedings of the International Congress on Human Environment (HESC), Kyoto, Asahi Evening News, Tokyo, Japan.

Odum, H.T., Brown, M.T., Christianson,R.A.: 1986, *Energy Systems Overview of the Amazon Basin.*, University of Florida, Center for Wetlands, CFW Publication Number 86-1, p. 190.

Odum, H.T.,Diamond, C., Brown, M.: 1987b, *Energy Analysis Overview of the Mississippi Basin*, Rept. to the Cousteau Society, Center for Wetlands (Pub. 87-1), University of Florida, Gainesville, p. 107.

Odum, H.T., Odum, E.C., Brown, M.T. Scott, ,G.B., Lahart, D. , Bersok,C., Sendzimir,J.:1986. *Florida Systems and Environment.*; A supplement to the text Energy Systems and Environment. University of Florida, Center for Wetlands, p. 81.

Odum, H.T., Lavine, M.J.,Wang, F.C., Miller, M.A., Alexander, J. F., Butler, T.: 1983, *A Manual for Using Energy Analysis for Power Plant Citing*. U.S. Nuclear Regulatory Commission, NUREG/ CR-2443.

Odum, W.E., Odum, E.P., Odum, H.T.: 1995, Nature's pulsing paradigm, *Estuaries* **18**, 547-555.

Oort, A.H., Ascher, S.C., Levitus, S., Peixoto, J.P.: 1989, New estimates of the available potential energy in the oceans, *J. Geophysical Res.*, **94**, 3187-3200.

Pease, A., Richard,W. (eds.): 1983, *Maine, Fifty Years of Change.*; University of Maine, Maine State Planning Office, State Employment and Training Council, Maine Department of Labor. University of Maine Press, Orono, p. 194.

Pillet, G., Odum, H.T.: 1984, Energy externality and the economy of Switzerland, *Swiss Journal of Political Economy and Statistics.* **120**, 409-435.

Pollard, J.A.: 1973, *Polluted Paradise, The Story of the Maine Rape.*; Twin City Printery, Inc. Lewiston., p. 328.

Rappaport, R.A.: 1971, The flow of energy in an agricultural society, *Sci. Am.*, **224**, 116- 133.

Repetto, R.: 1992, Accounting for environmental assets, *Sci. Am.* (June), 94-100.

Rovelstad, M.E. Rovelstad, J.M.: 1987, *Maine Tourism Study 1985-1986 Economic Analysis,* Prepared for the Maine State Development Office by Center for Survey and Marketing Research, University of Wisconsin-Parkside, Kenosha, p. 69.

Scienceman, D.M.: 1993, The system of EMERGY units, pp. 214-223. In: Proceedings of the ISSS meeting in Western Sydney, Australia, R. Packham (ed.), International Society for Systems Sciences, Louisville, Ky.

Tyler,D.A.,Ladd, J.W.: 1980, *Vertical Crustal Movement in Maine*, Maine Geological Survey, Open-File No. 80-34, p.53.

U.S. Dept. Agriculture: 1982, *Basic Statistics 1977 National Resources Inventory.*, U.S.D.A., Soil Conservation Service, Iowa State University Statistical Laboratory, Statistical Bulletin Number 686, p. 267.

U.S. Bureau of Census: 1985, *Statistical Abstracts of the United States: 1986 (106th edition),* U.S. Dept. Commerce, Washington D.C.

U.S. Dept Commerce, NOAA, Environmental Data Service, National Climate Data Center:1971, *Climatological Data National Summary.*

U.S. Dept Commerce, NOAA, Environmental Data Service, National Climate Data Center:1980, *Climatological Data National Summary.*

U.S. Dept Commerce, NOAA, Environmental Data Service: 1972, *Climate of the States,* Climatography of the United States No. 60-17, p. 25.

Wackernagel, M., Yount, D.: 1997, (this volume)

Watt, K.E.F.: 1992, *Taming the Future, A Revolutionary Breakthrough in Scientific Forecasting,* The Contextured Web Press, Davis, California, p. 232.

RESOURCE USE RATES AND EFFICIENCY AS INDICATORS OF REGIONAL SUSTAINABILITY: AN EXAMINATION OF FIVE COUNTRIES

JAE-YOUNG KO, CHARLES A. S. HALL, and LUIS G. LÓPEZ LEMUS[1]

State University of New York, College of Environmental Science and Forestry. One Forestry Drive, Syracuse, New York, 13210

Abstract. We examine trends from 1970 to the mid 1990's of some variables related to development and sustainability for Costa Rica, Korea, Mexico, the Netherlands and the United States: first, by calculating energy and agricultural efficiencies over time, second, by examining the environmental impacts of economic activities, and third, by estimating ecological footprints. We find that many "optimistic" arguments about sustainability have been misleading, and that there is little or no indication that we are becoming any more sustainable or even efficient. Total quality-corrected energy consumption has increased for all five countries and the renewable energy portion is decreasing. The efficiency of turning energy into both agricultural production and GDP has declined for all countries except for the US. In general, there is a remarkable linearity between resource use and economic and agricultural production over all countries and all years, suggesting severe biophysical constraints to sustainable objectives. On the other hand, per capita ecological footprints have decreased somewhat in Costa Rica, Mexico, and the United States, while national ecological footprints have tended to remain constant except for Korea. While there has been a reduction of specific pollutants in the United States, some of this has been achieved by exporting heavy manufacturing industries. We conclude that continued population and economic growth in each country is likely to make the achievement of any kind of 'sustainability' increasingly unlikely. Sustainability, if that is desirable, requires a very different approach than what we have undertaken to date.

1. Introduction

The International Union for the Conservation of Nature and Natural Resources (IUCN) introduced the concept of sustainable development in 1980 (Lélé, 1991). Since then there has been a great deal of discussion about the meaning and feasibility of sustainable development within the context of an increasing recognition of both the vulnerability of ecosystems and the ever growing pressures for economic growth. The most popular definition of sustainable development was introduced by the World Commission on Environment and Development (WCED, often called the *Brundtland* Commission): "...development that meets the needs of the present without compromising the ability of future generations to meet their own needs (WCED, 1987: p.43)." The *Brundtland* Commission considered population control, food security, and energy supply as critical components of sustainability. The Commission also emphasized poverty as a major cause of declining world environmental quality. Consequently, the commission suggested that development, with simultaneous technological solutions, would reduce the overall environmental impact of human beings on Earth. In contrast, Ehrlich *et al.* (1977) argue

[1] Fulbright/García Robles Scholar.

Environmental Monitoring and Assessment **51:** 571–593, 1998.

that environmental impact = (population) * (affluence) * (technology), *i.e.* that more wealth generates more impact.

In practice, sustainable development is about sustainable *economic* development. The emphasis on *sustainable* development is on resource availability, *i.e.* the biosphere's ability to supply resources, its waste assimilation capacity, its environmental quality, and also about improving the *quality* of human life. *Conventional* economic development has focused on human material quality of life by increasing the *quantitative* outputs of economic activities. Therefore, the concept of sustainable development should be understood in terms of how long economic development can be maintained at some given level without exceeding the capacity of a region. Too frequently, the concepts "economic development" and "economic growth" are used interchangeably, but there are differences between the two. Economic *growth* is an one dimensional measurement of quantitative changes in economic capacity, while economic *development* pertains to multiple dimensions such as quality of life, social structure, industrialization, and also economic growth (Gillis *et al.*, 1996; Todaro, 1994). Thus, if a country achieves economic growth, it does not necessarily achieve economic development. For example, Gillis *et al.* (1996:p.8) argue that Libya and South Korea (here-in-after referred to as Korea) both achieved economic growth since 1960, but Korea has achieved economic development and Libya has not, because only Korea made *structural* changes in its economy. It is probably extremely difficult, however, to achieve economic development without at least some economic growth. Therefore, sustainable *development* probably *requires* at least some growth as a precondition. Such a statement may be consistent with the proposed relationship between per capita income and some measures of environmental degradation known as environmental Kuznets curves (*i.e.*, that pollution may be greater at an intermediate level of economic activity but decreases as nations become very wealthy). That is, without increasing economic capacity we may not have resources to compensate for the decline in environmental quality (Stern *et al.*, 1994; Selden *et al.*, 1994).

An "optimistic group" asserts that both goals, growth and sustainability, can be achieved through increased *efficiency* of resource use by technological development (*e.g.*, Simon and Kahn, 1984; Goldemberg *et al.*, 1988; Ausubel, 1996) and increased market economy-driven efficiency (Smith, 1994). A second group argues that without population and consumption control sustainable development is not attainable (*e.g.*, Jensen, 1978; Ehrlich and Ehrlich, 1990; Hall *et al.*, 1994). This view implies that any efficiency gains are insufficient to make up for increasing population and affluence.

The extreme "optimists" argue that population growth itself increases technological development, so that the global food base and pollution assimilation technology will be expanded indefinitely to meet the needs of a growing human population (Simon and Kahn, 1984). "Pessimists", ever since Malthus (1798), along with Ehrlich & Ehrlich (1990), and Pimentel & Pimentel (1996) emphasize the adverse consequences of population growth, including famine, due to the finite nature of the Earth's carrying and regenerative capacity.

Which side has the data in their favor? The answer is mixed. There have been many improvements in environmental quality in the United States in the 1980s. Levels of toxic materials such as in fish of the Great Lakes show significant declines. At the national level emissions of certain air pollutants such as carbon monoxide and sulfur-oxides have been reduced significantly (Council on Environmental Quality, 1992). One explanation is that new industrial devices are more environmentally sound and efficient than old models due to technological development (Ausubel, 1996). "Pessimists" argue that these analyses overlook the impact of change in industrial structure caused by changing trade patterns. Specifically, they argue that the environmental regulations in the United States since the 1970s have provided incentives to move manufacturing facilities to foreign countries which provide more favorable business environments due in part to less stringent environmental regulation or enforcement (Leonard, 1988).

Another significant issue related to sustainable development, perhaps the most critical, is supplying food to a growing population. Agricultural production has been energy-intensified under the name of the "green revolution." More lands are irrigated, more chemical fertilizers are applied, and more pesticides and machines are used (Brown *et al.*, 1997). Accordingly, the production of crops, such as cereals, has increased. Can we continue to increase agricultural production by using more and more energy inputs? Naylor (1996) argues that the agricultural production return from intensive inputs has declined everywhere. In their analysis of agricultural yield trends, Plucknett and Smith (1986) and Brown *et al.* (1997) say that it will be a real challenge to maintain even the current level of agricultural production. Pimentel *et al.* (1995) found that intensive agricultural production has degraded the global land resource base (*e.g.*, top soil) significantly. Hall and Hall (1993) concluded that if cheap fossil fuel is no longer available due to oil depletion, agricultural yield will fall to levels below what would have been achieved with a non-degraded resource base. But this is not seen in yields now because inputs are increasing.

Within this context our paper examines the possibility of sustainable development for five national economies: Costa Rica, Korea, Mexico, the Netherlands and the United States, representing very different levels and rates of economic development. We assess the impacts of economic growth during the last three decades for the five countries, and discuss the possibility of sustainability by analyzing trends in rates and efficiencies of resource use.

We define *economic development* as an increase in the generation of economic output as imperfectly measured in physical units (*e.g.*, tons) or inflation-corrected dollars of GDP. We define *sustainability* as not degrading the resource base that makes that economic process possible. We define *efficiency* as output over input. We examine some efficiencies and environmental impacts resulting from the link between energy consumption and both economic output and agricultural productivity from 1960 to 1994. We differentiate efficiencies of resource use into two categories: *micro-efficiency* and *macro-efficiency* (or aggregate efficiency). The micro-efficiency measures any energy or resource savings achieved by technological breakthroughs, such as increases in gas mileage of automobiles, while macro-efficiency is the efficiency of a regional unit such as a nation.

Discussion of carrying capacity must deal with efficiencies of resource consumption. The reduction of resource use rates increases the carrying capacity of a region, and indeed its status of sustainability. Most of the sustainability literature has operated only within the aegis of micro-efficiency, presumably from the belief that resource savings of a unit (*e.g.*, a car) is manifested as resource savings in a society. In fact, quite to the contrary, total consumption may be increased *because of* a unit's increased efficiency. This has been known for more than 100 years as "Jevons' paradox." Stanley Jevons, also known for his contribution to neoclassical economics, wrote in *The Coal Question* (1864) that all previous increases in efficiency of steam engines, done to stretch the supply of English coal, made steam power cheaper. Because it was cheaper, more was used and thus aggregate coal consumption increased. Thus, we believe it important to examine macro-efficiency, especially at the level of one nation.

Previous discussions about sustainable development have been limited mainly to a single region (*e.g.*, Knight *et al.*, 1997), one country (*e.g.*, Emmelin, 1973; Soussan *et al.*, 1991; López Lemus, 1997), or a global scale (e.*g.*, Goodland *et al.*, 1993; Schloss, 1993). Our comparison of time series data for five very different economies provides insights that cannot be found in a single case study either for a single country or at the global scale, making it possible to determine what patterns may exist independently of a given nation's development level. We also examine the response of these five countries to the oil shocks in the 1970s in terms of increasing or decreasing resource efficiency.

2. Methods

The five countries were chosen to represent very different characteristics of demography, economic development, and natural resource stocks (Table I). Costa Rica's population density is higher than two countries, and lower than the other two. Its per capita income and economic growth are relatively low, and has depended mainly on agriculture. Korea has a very high population density and has had high economic growth over the last three decades based on heavy industry. The population density of Mexico is lower than Costa Rica or Korea, but higher than the United States (US). The Mexican economy has had strong economic growth in recent years. The Netherlands, with a high population density, had a relatively stable 2.3% annual economic growth rate for the last decade. The United States has a low population density, a high material standard of living, and relatively low economic and population growth. For all elements of natural capital given in Table I (*i.e.*, pasture land, crop land, timber resources, non-timber resources, protected areas, and soil assets), international market prices were used and adjusted by an appropriate factor to represent the 'rent' portion of the traded price (World Bank, 1997).

TABLE I
A comparison of the five countries in relation to their economic output, demographic conditions, and natural capital assets

	Costa Rica	Korea	Mexico	Netherlands	USA
Per capita purchasing power parity GDP, 1992 (int$)	4,522	9,565	7,867	17,373	23,220
Population density,1995 (per 1,000 ha)	671	4,557	491	4,570	275
Average annual population growth (per cent), 1990-95	2.4	1.0	2.1	0.7	1.0
Per Capita Natural Capital (US$)	7,860	2,940	6,630	4,140	16,500

Source: World Resources Institute, 1996. *World Resources 1996-97: A Guide to the Global Environment*: pp.190-191 & pp.216-217; World Bank, 1997. *Expanding the Measure of Wealth: Indicators of Environmentally Sustainable Development*: pp.20-21.

Time series data for our study were obtained from several sources. The main source was the World Resources Institute's (WRI) 1996-97 data set, which compiles environmental information published by several international agencies, including the United Nations Food & Agriculture Organization (FAO) and the World Bank. The United Nations' *Energy Statistics Yearbooks* were used for additional information and for cross-checking the WRI energy data set. We downloaded agricultural data from the FAO Internet web site. We used trade data from the National Bureau of Economic Research and the US Department of Commerce to examine the impact of US trade on both its own economy and the ones of the other countries. All energy-related data were published originally as heat units (*i.e.*, gigajoules). Since each energy source has different quality in terms of its ability to do economic work (*i.e.*, electricity has higher energy quality and can do more work per joule than coal; Hall *et al.*, 1986), we adjusted the original values of all electricity (*i.e.*, fossil, nuclear and hydro-electric energies) for quality by applying a 38% thermal efficiency correction factor (*i.e.*, electricity values were multiplied by 2.63) (Nilsson, 1993). We also included unadjusted caloric estimates of traditional energy (*e.g.*, fuelwood, charcoal, plant, and animal waste).

The differences in traditional energy use among nations are significant. For example, 36% of Costa Rica's total energy consumption in 1993 was estimated to be traditional energy sources, but only 1% for the US (WRI, 1996: p.287). We believe, in agreement with Nilsson (1993), that the inclusion of traditional energy in these calculations is theoretically more sensible than using only commercial energy consumption (*i.e.*, fossil, nuclear and hydro-electric energies, sold in markets) as a measure of energy metabolism of human societies.

2.1. ECONOMIC ANALYSIS

We used gross domestic product (GDP) as an index of economic output for all countries. GDP figures presented in this study were corrected for inflation and "purchasing power parity" (PPP), thus resulting in *constant (1985) international dollars.* The inflation correction provides a more accurate comparison of economies over time. The international dollar-based PPP index is used to minimize the impact of price-level differentials of local currencies (WRI, 1996: p.171). The use of PPP allows a more accurate estimation of consumption power for citizens of each country, although for poorer countries it might overestimate purchasing power in international trade.

2.2. AGRICULTURAL ANALYSIS

We measured the *fertilizer efficiency* of cereal production as an index of agricultural efficiency. This was calculated by dividing the total cereal production by the NPK fertilizer used on all cereals (barley, corn, rice, sorghum, wheat, *etc.*). We also measured *fertilizer productivity* for cereal production, defined as the ratio of cereal yield per mass unit of fertilizer input on cereal crops. This was not easy as no year by year numbers are kept on fertilizer use on specific crops. To estimate the fertilizer input to cereals, we corrected total year by year national fertilizer use by relative area planted for grains, and for intensity of fertilizer use on grains, as follows:

We calculated the *ratio* of *fertilizer* for "total cereals" to "total crops" for the only year available from *Fertilizer Use by Crop* (FAO, 1992), which was derived from questionnaires sent to national governments who returned the data (for example, the fertilizer consumption ratio of total cereals over total crops was 9% for Costa Rica in 1991, which is the "index" year).

The *ratio* of harvested *area* for "total cereals" over "arable and permanent croplands" is available for each year from the agricultural statistics of FAO through the Internet web site of FAO (for example, that ratio for Costa Rica for 1991 was 17%). For the Netherlands, the single largest sector of fertilizer use is "permanent grasslands" (FAO, 1992). Consequently for this country only we calculated the denominator of the ratio as the sum of "arable and permanent croplands" and "permanent pasture." Fertilizer use on pastures was small for the other countries.

From I and II we estimated *relative fertilizer intensity* as "ratio of fertilizer" divided by "ratio of land" for the index year. This gives the relative per hectare intensity of fertilizer use on cereals for our index year (for example, this was (0.09)/(0.17) = 0.53 for Costa Rica in 1991).

The relative fertilizer intensity was then multiplied by the proportion of land in cereals for each year to get the *proportion* of fertilizer used on cereals (for example, (0.53)*(0.21) = 0.11 for Costa Rica in 1961, where 21% is that year's ratio of harvested area for total cereals to the arable and permanent crop lands). This factor corrects for more or less land area planted to cereals.

The proportion used each year then was multiplied by total fertilizer used each year from 1960 to 1994 to give an estimate of fertilizer used for cereal production [for example, (0.11)*(18,687 metric tons) = 2,055 metric tons for 1961, where 18,687 metric tons is the total fertilizer use for all crops for the year].

2.3. ECOLOGICAL FOOTPRINT ANALYSIS

We also calculated the ecological footprints of four countries of our sample for each year from 1970 to 1992. A country's ecological footprint measures the aggregate land and water area in various ecosystem categories that is used by that country —locally or elsewhere, to produce all the resources it consumes, and to absorb all the waste it generates on a continuous basis using prevailing technology (Wackernagel and Rees, 1996; see also Wackernagel and Yount, this volume).

The ecological footprint concept is based on the idea that for every unit of material or energy consumption, a certain amount of land in one or more ecosystem categories is required to provide the consumption-related resource flow and waste sinks. To measure the ecological footprints of the countries for the period, we followed a simplified approach considering only four important categories of domestic consumption: food, forest products, energy and "other" land. This avoided any significant double counting where we had no detailed information, yet was sufficient for our international comparisons of ecological footprints within the context of our discussion.

Estimates for energy land, farm land, and forest land were corrected for trade (*i.e.*, production + imports - exports), and national/local biotic productivity factors were used to correct estimates of both food land and forestry land. The energy footprint varies inversely as the productivity of the source: the higher the productivity (*e.g.*, high altitude hydro-electricity = 15,000 GJ/ha/yr), the smaller the footprint (= 0.0067 ha/100 GJ/yr). *Other* land here encompasses WRI's (1996) estimates of abandoned or uncultivated land, grassland not used for pasture, wetlands, wastelands, roads, and built-up areas. Marine area appropriated for human use is not included in present footprint calculations, thus our values are somewhat lower than those of the same countries as published elsewhere (Wackernagel *et al.*, 1997). We derived an [un]sustainability factor as the ratio of the national land area that is required to the land area that actually can be supplied using prevailing production methods (*i.e.*, hectares of ecological footprint/hectares of ecologically productive land such as croplands, permanent pastures, forests and woodlands). We also calculated a corresponding national ecological deficit estimate as the difference between the two. Both figures measure the actual amount by which each country's ecological footprints exceed their locally available national biological capacity (*i.e.*, their ecologically productive land area).

Finally we examined the relations between total energy consumption of a nation and both economic output and total population by using regression analyses for the five nations from 1970 to 1992. Our basic assumption is that true sustainability cannot be based on the depletion of non-renewable resources such as fossil fuel.

3. Results and Discussion

3.1. ENERGY USE

There was a large increase in energy used in each of the five countries and no indication that renewable energies were becoming more important except possibly in Costa Rica (Figures 1 and 2).

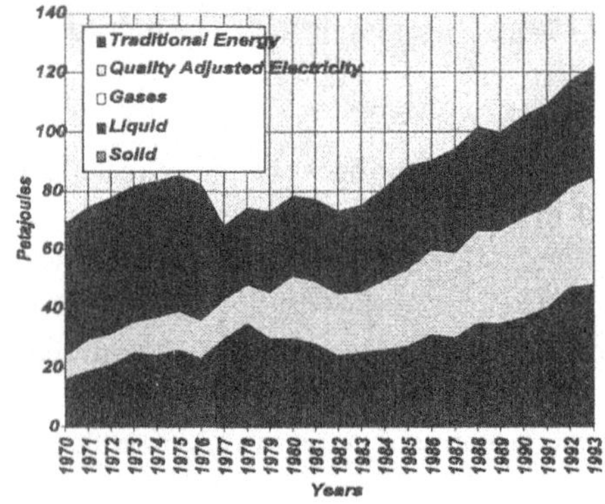

Fig. 1a. Trends of energy use by type in Costa Rica (1970-1993)

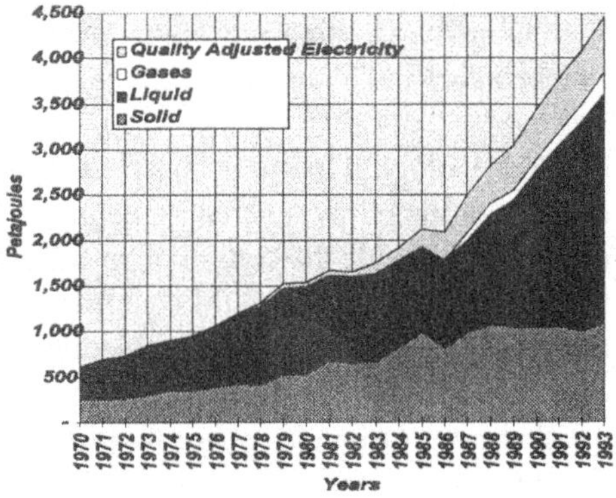

Fig. 1b. Trends of energy use by type in Korea (1970-1993)

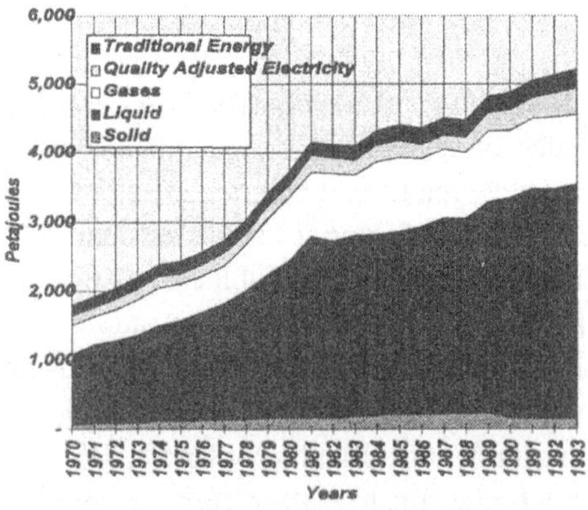

Fig. 1c. Trends of energy use by type in Mexico (1970-1993)

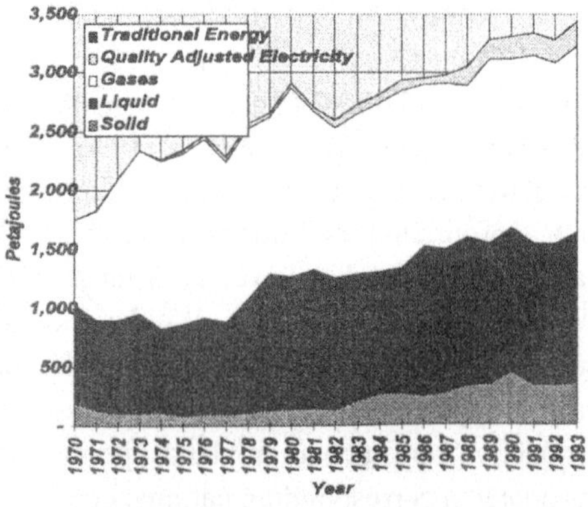

Fig. 1d. Trends of energy use by type in the Netherlands (1970-1993)

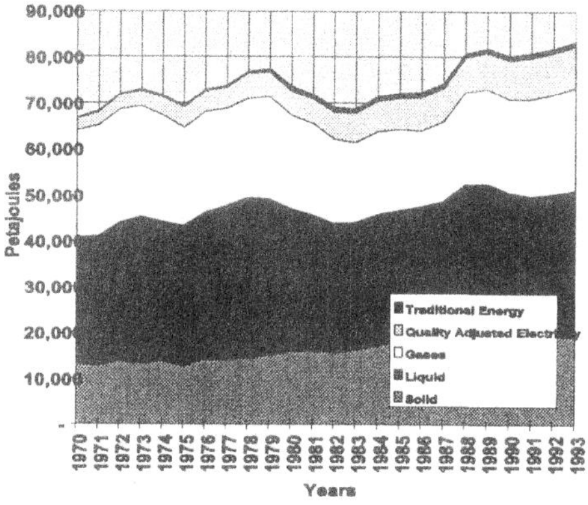

Fig. 1e. Trends of energy use by type in the United States(1970-1993)

3.1a. Energy use and economic output: Economic activity is correlated significantly with energy consumption for each of these five countries over 20 years (Table II). The level of correlation differs for each nation. Based on the raw UN energy data, Costa Rica had only a moderate correlation between economic growth and energy consumption. But the sudden drop of traditional energy use in Costa Rica in 1977, suggests a problem with our data source (Figure 1a). We assume that the traditional energy consumption level of 1976 is about the same as that of 1977, although the UN numbers differ by 54%. We multiplied the traditional values for years before 1976 by 0.54 to give a corrected traditional energy consumption. After this adjustment, the R-square value increased to 0.93. Korea and Mexico, which have had high economic growth in recent years, had higher R-square values than Costa Rica or the United States. The Netherlands, which has a slower growth rate than Korea and Mexico, has a somewhat lower R-square value (0.9397) than do those two countries (0.9897, and 0.9544 respectively). Thus, the higher the economic growth rate, the more tightly economic activity was tied to energy use.

TABLE II
The relation of GDP to energy consumption for the five countries,1970-92
Regression Model : GDP= a (energy consumption) + b

	Costa Rica	Korea	Mexico	Netherlands	USA
R-square	0.6418 (0.9308)*	0.9897	0.9544	0.9397	0.6152
P-values	0.0001 (0.0001)*	0.0001	0.0001	0.0001	0.0001

* Traditional energy consumption adjusted.
Source: United Nations. *Energy Statistics Yearbook*. (multiple years); WRI (1996). *World Resources 1996-97*, database diskette.

Another issue pertaining to sustainable development is liquid petroleum's portion of total energy use, since petroleum is probably the most depletable fuel. This proportion is increasing, particularly in Costa Rica, Korea and Mexico (Figures 1a, 1b, and 1c). The

Korean economy, which imports all of its oil from foreign countries, consumed about 7 times as much petroleum in 1993 as in 1970 (Figure 1b), and the Mexican economy used three times the oil in 1993 as in 1970 (Figure 1c). Oil consumption for the United States and the Netherlands have similar patterns of steady growth. In the United States oil consumption grew steadily from 1970 to 1993, except for a slight decrease in 1975 and 1980-1985. The importance of the large *Groningen* gas field to the Netherlands is obvious.

Total electricity use (*i.e.*, primary electricity plus that from fossil fuels) increased substantially in the five countries during the last two decades, although the patterns in the trends are different. Costa Rica and Korea had a relatively large increase (4.6 and 5.5 times, respectively, between 1970-93), while Mexico and the United States had lower increases (2.6 and 3.5 times, respectively, during the same period). Total energy consumption in Costa Rica and Korea has increased rapidly, while the United States had a smaller growth. In Mexico, the consumption of natural gas increased rapidly in these two decades, which may explain why Mexico had a slower growth rate in electricity consumption.

3.1b. Per capita relations: The per capita energy consumption of these five countries is shown in Figure 2. The use of energy is basically proportional to the economic level of the nation. An average person in Costa Rica consumed 37 gigajoules in 1993, while an average person in the United States consumed 323 gigajoules. Korea, where economic output increased rapidly during these two decades, had a proportional increase in energy consumption. The significant difference among the five countries in both the increase in energy consumption and GDP per capita during this 1970-92 period is illustrated in Figure 3. The per capita energy consumption of Costa Rica remained between 30 and 44 gigajoules, and its per capita GDP within a range of 2,904 to 3,821 (PPP) 1985 international dollars without any clear trend. There is a clear increase in both per capita economic level and energy consumption in Korea and Mexico. The Netherlands lies between the United States and the other three countries in both indices. Figure 3 illustrates the roughly positive linear proportional relationship between energy consumption and GDP per capita over time for four of the five nations.

Figure 3 also shows that the energy macro-efficiency of the United States, but no other nation, increased during this period. Thus in the 1990s a person in the United States attained the same economic output level as before the 1970s' oil-shock with less energy consumption. This has not been a smooth pattern. The energy macro-efficiency increased immediately after each oil shock, but subsequently the linear relation between GDP and energy use returned, although the level of energy efficiency remained higher than before. Thus while each person in the United States in 1992 consumed the same level of energy as in 1970, the GDP per capita increased from $12,963 to $17,945. This seems consistent with increasing macro efficiency, but not sustainability, since the per capita level of energy use remains much higher, and the efficiency lower, than in other countries, and total national energy use increased in proportion to population growth. Mexico, on the other hand, lost per capita economic production from 1981 to 1990 while per capita energy use remained constant.

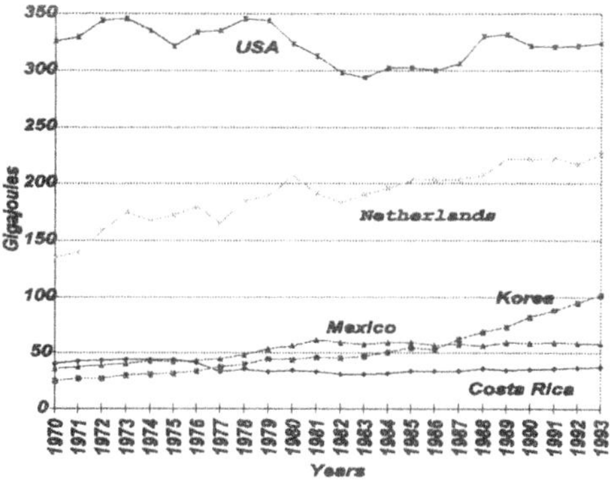

Fig. 2. Per capita total energy consumption of the five countries (1970-1993)

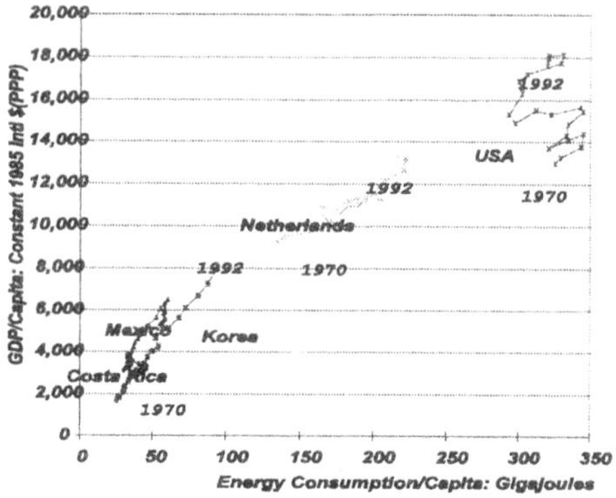

Fig. 3. Per capita energy consumption v. per capita GDP (1970-1992)

3.2. AGRICULTURAL EFFICIENCY

All five nations [except the Netherlands recently] used increasing quantities of fertilizer from 1961 to 1993 (Figure 4) and had decreasing fertilizer efficiencies (Figure 5). Costa Rica had a steady increase of fertilizer use for the last three decades, and Korea intensified fertilizer use rapidly (Figure 4). In the United States yields increased in proportion to increased fertilizer use, so that fertilizer efficiency remained steady after an initial decline (Figure 5). The Netherlands' fertilizer efficiency for cereal production has been increasing since 1979 (Figure 5), associated with a reduction in the intensity of fertilizer use (Figure 4).

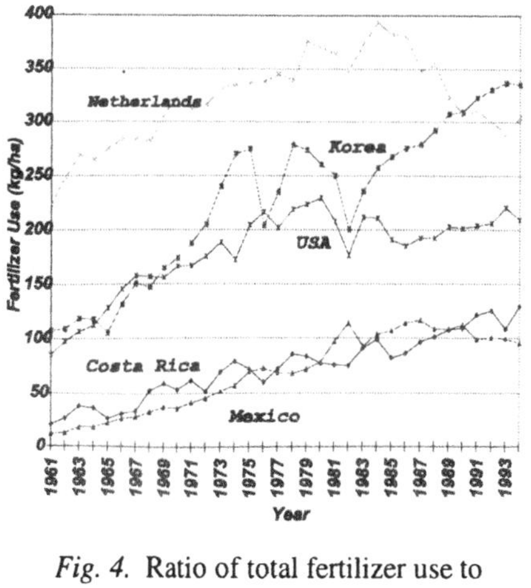

Fig. 4. Ratio of total fertilizer use to cultivated area for total cereals

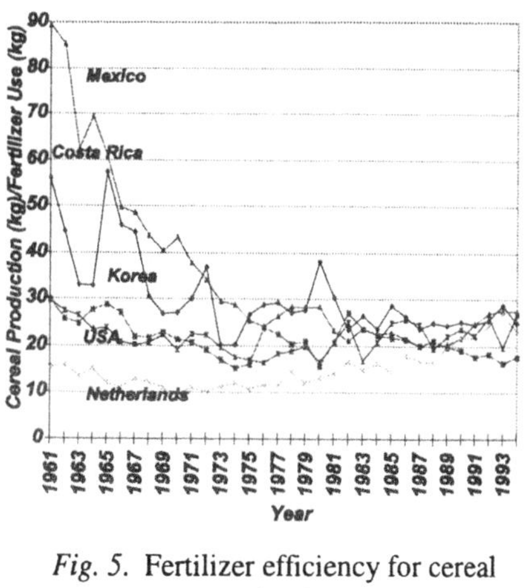

Fig. 5. Fertilizer efficiency for cereal production

There is a rough positive linear relation between fertilizer input and yield of cereals among all five nations taken together (Figure 6). But each nation independently seemed to have an asymptotic relation between fertilizer input and cereal yield, which implies that the more the fertilizer input, the less the yield per kg of fertilizer after reaching a threshold. The Netherlands appears to show increasing fertilizer efficiency for cereal production as fertilizer input decreases from 1985 to 1993. In other words, in the late 70s production was about 4 metric tons per ha at 300 kg of fertilizer/ha. In the 1990s, production was 7 to 8 metric tons per ha at the same fertilizer level. Does this show increasing efficiency due to technology? We do not think necessarily so. We think that another aspect of resource quality, site quality, explains much of increased efficiency of fertilizer, as David Ricardo (1817) argued. Due to the increasing trade of cereals between nations following the increased economic integration among European countries, the area harvested for total cereals in the Netherlands has decreased from 361,706 ha in 1970, to 237,843 ha in 1979, to 183,300 ha in 1993. The net import of barley, maize and wheat to the Netherlands was 2.3 million in 1961, but 4.6 million metric tons in 1995 (Figure 7, FAO, website). Presumably the land that was least productive for grain was turned into other purposes.

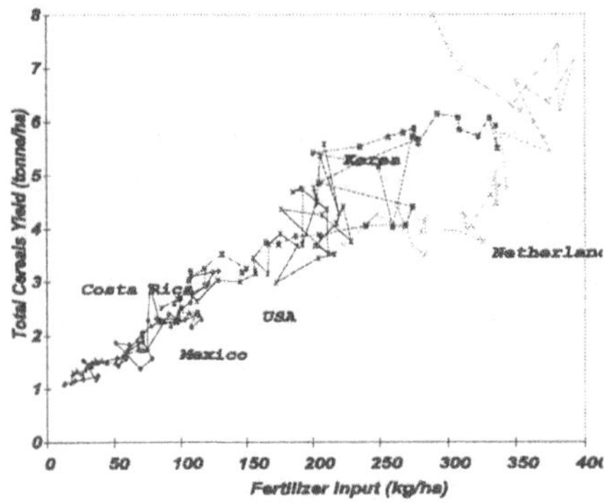

Fig. 6. Fertilizer input v. cereal yield (1961-94)

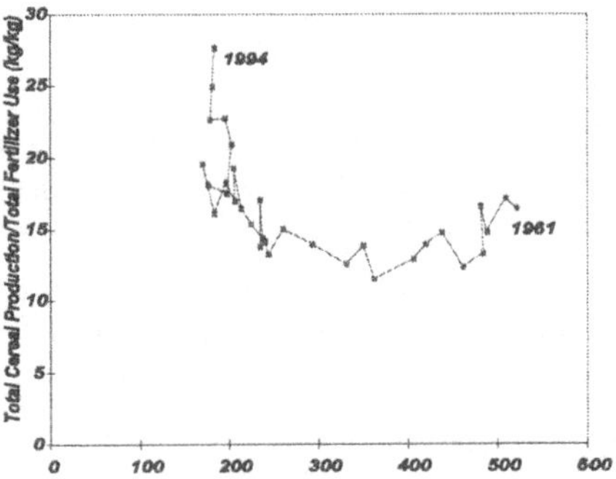

Fig. 7. Harvested area for total cereals v. fertilizer efficiency of total cereals in the Netherlands (1961-94)

The relation between fertilizer input and fertilizer efficiency from 1961 to 1994 shows a clear reverse relation : the higher the fertilizer input, the lower the fertilizer efficiency (Figure 8). The highest fertilizer efficiency lies in the area of low fertilizer input in the poor countries. The rich, industrialized nations show lower fertilizer efficiency. Thus one of our major conclusions is that *efficiency*, one important component of sustainability, is *inverse* to *intensity* for both fertilizer application and agricultural land use.

The sources of chemical fertilizers are globally limited (P) or energy intensive (N), and the residues of the fertilizers can trigger serious environmental contamination problems. Nitrogen fertilizer is made by the Haber-Busch process in which nitrogen in the air reacts with hydrogen from natural gas to make ammonia. The nitrogen fertilizer spread onto soils generally is oxidized to form soluble nitrate for absorption by plants. But the saturation level of nitrate in soils is low and unused nitrates are released into surface and ground waters. Thus heavy nitrogen fertilizer use becomes a main source of non-point water pollution which contaminates surface water and underground water (Kesler, 1994). One of the major sources of phosphorus is called *guano*, which is bird and bat excrement. Most of the *guano* type resources, which were abundant until the early twentieth century, are now almost depleted. For example, the deposits on Christmas Island, which was once the largest deposit in the Indian ocean, are depleted and production from the island ceased in 1991 (Kesler, 1994). Large but rapidly decreasing deposits remain in Florida and Morocco. Thus all commercial fertilizers are essentially non-renewable. We can increase the efficiency of their use by decreasing both intensity of application and land area used, but this is difficult in a world of increasing human populations and affluence.

3.3. ENVIRONMENTAL IMPACTS

3.3a. National scale: As populations grow , less land is available for natural areas and more fertilizers are added to surface waters, so that negative environmental impacts grow. But no data is summarized allowing a national comparison. A comparison of the five countries for some available indicators of environmental impact are shown in Table III.

Table III
A comparison of some indicators of environmental impact in the five countries

	Costa Rica	Korea	Mexico	Netherlands	USA
Cropland % change in 1991-93 since 1981-83	3.5	-4.9	0.2	12	-1.2
Forest and Woodland: % change in 1991-93 since 1981-83	-5.2	0	4.1	15.9	-2
Fertilizer use, 1993 (kg per ha of cropland)	208	474	71	560	108

Source: WRI (1996). *World Resources 1996-97: A Guide to the Global Environment:* pp.216-217, pp.240-241.

During the last decade, Costa Rica expanded its cropland and pastures at the expense of forests and woodlands. Korea lost cropland, kept the forest and woodland area almost unchanged, and used fertilizer intensively. Mexico has increased cropland slightly, and expanded forests and woodlands by using more lands from the *other* land category. The United States has lost both cropland and forest land. One of interesting things in Table III is that forest and woodlands, which have a capacity to assimilate carbon dioxide (CO_2) produced as a by-product of economic activities, have increased in Mexico and the Netherlands. Costa Rica and the United States have lost forest lands. Korea maintained its current small size of forest lands for the last several decades.

3.3b. Per capita base : An average person in Costa Rica uses one tenth the energy as his/her US citizen counterpart and is about 10% as wealthy (as measured by GDP) (Table IV), which supports the argument that wealthier people consume more energy. A US

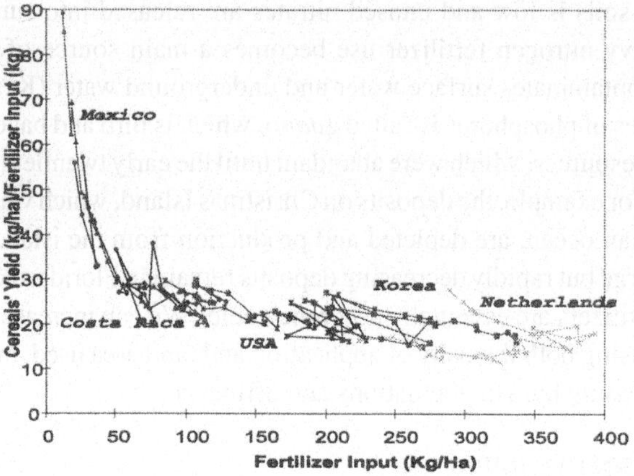

Fig. 8. Fertilizer input v. fertilizer efficiency for total cereals production (1961-94)

citizen consumes 60% more water than a Costa Rican. In contrast, people in densely-populated Korea and the Netherlands consume less water per capita than the less densely-populated US, Mexico or Costa Rica. The water use is not necessarily determined by affluence alone.

One of the environmental consequences of energy consumption is CO_2 emissions, which may contribute to global warming. Thus the per capita relationship between GDP and CO_2 emissions from industrial sources is an important issue with respect to sustaining the atmosphere. Costa Rica, Korea, and Mexico have a high linear proportional relationships between per capita GDP and CO_2 release (Figure 9). The Netherlands and the United States have higher output levels but improved efficiencies. The data for the United States shows the impact of oil shocks. In 1974-1975 and 1980-82 economic output dropped and CO_2 emissions decreased.

Table IV

A per capita comparison of the environmental impact in the five countries using some selected indicators

	Costa Rica	Korea	Mexico	Netherlands	USA
Total energy consumption per capita, 1991(GJ)	28	88	59	213	324
Per capita freshwater withdrawals (m^3)*	780	632	899	518	1870
Per capita CO_2 emissions for industrial processes, 1991 (metric tons)	1.21	6.56	3.77	9.16	19.13

Source: United Nations (1993). *Energy statistics yearbook 1991*.pp.95-101 ; WRI (1996). *World Resources 1996-97: A Guide to the Global Environment*, pp.306-307.

*The years of data are 1970 for Costa Rica, 1992 for Korea, 1991 for Mexico, 1991 for Netherlands, and 1990 for the United States.

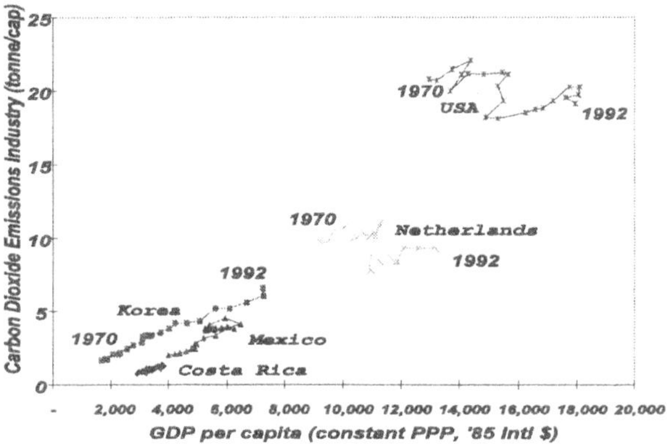

Fig. 9. CO2 emissions v. GDP per capita (1970-1992)

In the 1990s, CO_2 gas emissions declined, which was not driven by the oil shock, but apparently by the structural changes in the US economy, such as increasing foreign investment by US manufacturing industries or possibly genuine improvements in *macro*-efficiency. In addition, the production of low CO_2 release nuclear-generated energy doubled during the 1980's. The CO_2 released by the Netherlands, whose per capita income is less than the United States but higher than the other three countries, lies between the United States and the other three countries.

3.4. ECOLOGICAL FOOTPRINTS

Per capita ecological footprints have been decreasing slightly but continuously for Costa Rica, Mexico, and the United States. This decline represents a much smaller increase in agricultural area (due to the intensification of yield) relative to population growth, and in some cases increased imports from higher yielding regions. Korea's per capita ecological footprint has increased steadily (Figure 10a). However, national ecological footprints-calculated by multiplying these per capita estimates by the country's populations, have increased slightly in Mexico and Costa Rica, and radically in Korea, due to its high and increasing population density, and, particularly since 1986, to a heavy industrial tread on its scarce biological capacity. Costa Rica exceeded its ecological capacity by 15% in 1975 (*i.e.*, [un]sustainability factor = 1.15) and 26% in 1991; at the other extreme, Korea has increased this deficit tremendously from 130% up to 434% (*i.e.*, [un]sustainability factor = 5.34), during the same period (Figure 10b).

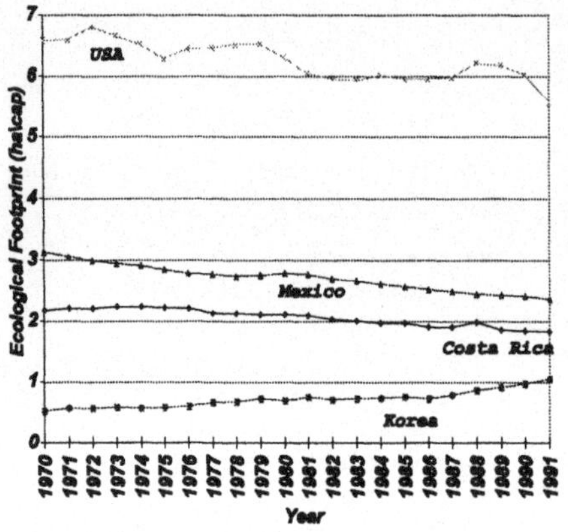

Fig. 10a. Per capita ecological footprints (1970-1991)

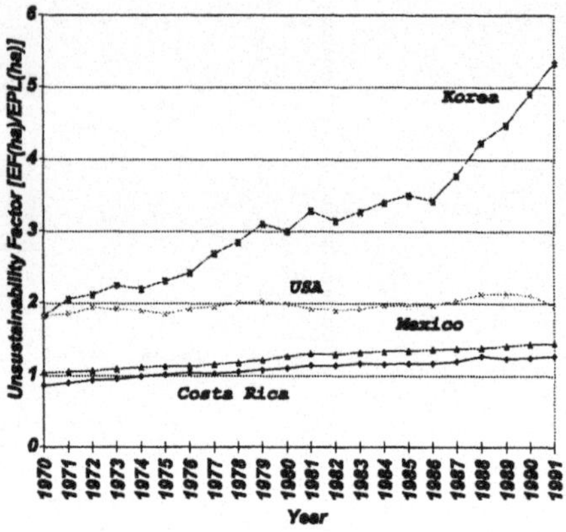

Fig. 10b. National Unsustainability Factor (1970-1991)
EF = National Ecological Footprint; EPL = Ecologically Productive Land

In measuring the ecological footprint of these nations, we wanted to ask whether these countries were living within their ecological means or at what rate they were using more than their share of productive biosphere. In short our basic question was: "How much nature do these countries use to sustain themselves?"

The US has maintained an almost constant [un]sustainability factor of 2 (Figure 10b), although its per capita ecological footprint estimates decreased slightly from 6.68 ha/cap in 1970 to 5.57 ha/cap 1991(Figure 10a). This is remarkable given that it is the country with the highest rates of energy use, CO_2 emissions, and material use per person. The reason is, of course, that the United States has much highly productive land and a low population density. The factor of 2 means that the US is "importing " a land use equal to its own actual ecological capacity, and has been doing that for almost 20 years! (Figure 10b). In contrast, in 1992 Korea's natural capital (i.e., ecological capacity) is barely 1.2% of the US' (*i.e.*, 8,716,000 ha v. 725,643,000 ha), while its population is about 17.1%. This difference in natural capital and the high population density accounts for Korea's huge ecological deficit, 952%, in contrast with the much lower figure for the US, 81% (Wackernagel and Rees, 1996).

This actual national biophysical capacity is also reflected in monetary value (Table I). We believe that the larger the ecological deficit of a nation, the larger its economic vulnerability, and the greater likelihood that it may be forced to overexploit its remaining stocks of natural capital. Or the gap could be compensated by its dependence on foreign countries. Current global trends aim to extend industrialization *fast* and at the *lowest* possible costs through monetary and trade agreements. From this perspective, Korea appears to be a good example of a country with scarce natural capital, growing the hard way by internal industrialization, through deficit-financed net imports of biophysical capacity. In monetary terms, Korea possesses only 37% of Costa Rica's natural capital assets per capita, and only 18% of the US', thus explaining the skyrocketing of its [un]sustainability factor to 5.5 in 1991. This represents the level of appropriation from other regions that is required to offset its own biophysical deficit and support its high population density and moderate standards of living.

Given their higher share of natural capital and their lower relative levels of population and industrialization, both Costa Rica and Mexico managed to live within their ecological means during the first half of the 70s. Since then, growth in these factors has increased somewhat their ecological deficits to slightly above their ecological capacities (*i.e.*, 1.2 and 1.3, respectively). This also represents these countries' increasing overshoot on their resource base and as such is an indicator of potential vulnerability.

3.5. TRADE IMPACT ON ENVIRONMENTAL CONSERVATION

There was a general slow-down of economic activity in the United States in the mid 1970s and especially the mid 1980s following the two oil shocks. The decrease in industrial activity appears to have contributed to the increase of energy efficiency, presumably because those factories that closed were the less efficient ones. Also, after the US economy recovered its vitality from the oil shocks of the 1970s, US manufacturing companies increased their investment in foreign countries, taking advantage of their low labor costs and/or lenient environmental regulations (Figure 11).

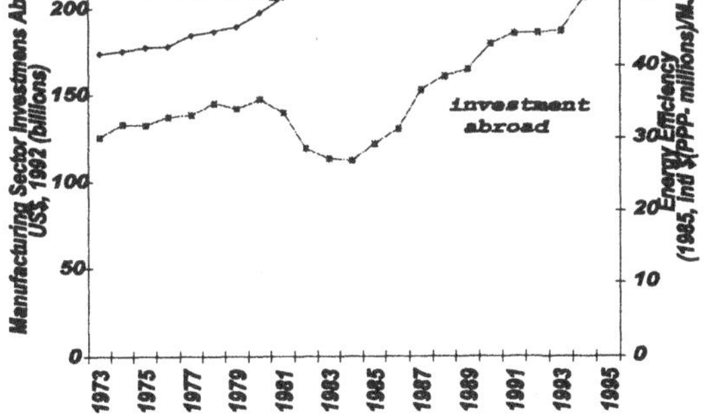

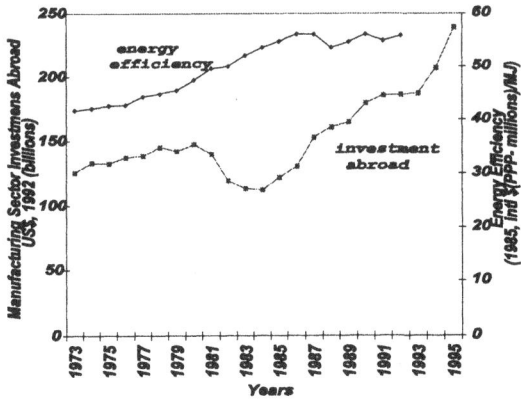

Fig. 11. Energy efficiency of the US and US's investment abroad

Thus, American automobile manufacturing companies now use large quantities of iron made in Korea (US Congress Office of Technology Assessment, 1990), and many chemical products are imported from US manufacturing branches in Mexico (US Department of Commerce, 1992). The effect of 1994 NAFTA agreement appears to have accelerated this trend.

Presumably this foreign investment has contributed to the increasing energy efficiency of the domestic economy of the United States, since the energy used in production was used elsewhere but much of the profits were added to the US economy. So, one could conclude that the United States has gained its environmental improvements partly at the cost of the environments of other countries where American companies have made investments. The net import of manufactured products to the United States from the other four countries illustrates one impact of increasing foreign investment by the American companies and an increasing U.S. dependence on foreign countries for manufactured goods (Figure 12). The United States has increased dramatically the volume of imported manufactured products from Korea. The United States reached its net exporting peak of manufactured products to Mexico in 1981. Since then, net exports to Mexico have decreased sharply, because the United States has increased its volume of imported manufactured products from Mexico. In other words, the United States has increased its use of Korea's and Mexico's manufacturing power, to meet its domestic demand for manufactured products, and this has contributed to its reduction of industrial pollution as measured by industrial CO_2 emissions in Figure 9.

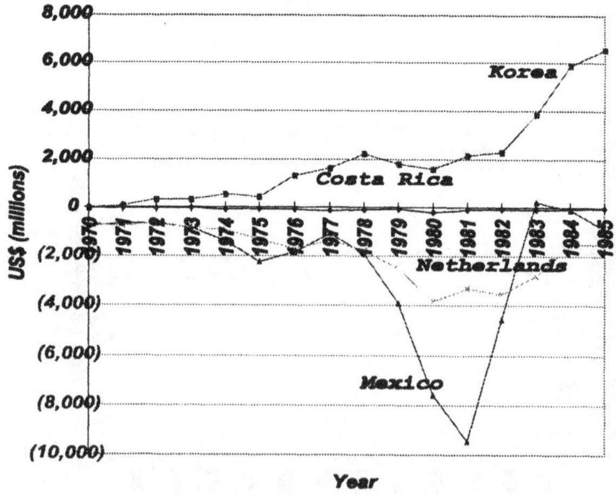

Fig. 12. Net imports of manufactured products of the US from the other four countries

4. Jevons' Paradox

The perverse link between increased efficiency and expanded resource exploitation was described first by Jevons (1864). He found that as the efficiency of steam engines increased, driven originally by the necessity to save coal, more coal was used because people found more uses for the cheaper engine. Jevons' paradox explains the gap between the micro-level efficiency of an item driven by technology and the macro-level efficiency observed for a region, and is one reason we see little reduction in national resource use.

Automobiles now are more energy efficient than those in the 1940s and emit less CO_2 per mile. These improvements are driven by engineering and technology. From evidence such as this, technological optimists project a decrease in pollution with the possibility of a joyful future including clean air, dematerialization, and recovered ecosystems (Ausubel, 1996). But despite the increase in efficiency, the gasoline consumed by a family in the 1990s is much higher than that of the 1940s. For example, Freund and Martin (1993) found that fuel consumption per car between 1970 and 1990 in the United States declined by 34%, but total auto fuel consumption increased by 7% for the same period, because now a family often has more cars, and drives each many more miles per year. Three factors are involved here. First, the "baby boomers" caused a significant growth in the driving age population. Second, per capita ownership has increased (Schipper *et.al.*, 1992). Finally, the life styles of today, including increased affluence, the development of suburban communities, the high mobility of the population, and mall shopping have been factors encouraging people to drive their cars more often (Freund and Martin, 1993). Therefore, an increase in energy micro-efficiency (e.g., an individual car's energy saving from technological innovations) does not necessarily lead to an increase in the energy macro-efficiency (*e.g.*, a reduction in energy use by a society). The sustainability of a region has a more direct relation to resource-use macro-efficiency than it does to the resource micro-efficiency of manufactured goods.

5. Conclusion

Figures 1, 2, 3, 6, 7, and 10b all imply limitations for the possibilities of sustainable development. While energy use in the United States increased only very slowly, that of the less wealthy world is increasing very rapidly. So far we have not seen a significant technological or social contribution to curb the increasing use of fossil fuel. Thus we conclude that the economies of these five countries are becoming less sustainable by our criteria relating to energy use. In addition, this fuel use is increasingly based on non-renewables which by definition are not sustainable. We also have demonstrated that economic development is dependent heavily on fossil fuel use, including petroleum, and that the last decade's effort to reduce energy use by using so-called state-of-art technology has not contributed significantly to energy efficiency. If another oil shortage hits the Korean or the United States economy, the economic output could not be sustained at today's level.

Second, as seen in Figures 6 and 8, data from four of five countries imply that there is a limit to feeding more people by increasing fertilizer input because yields have saturated and fertilizer efficiency has declined. Past increase in agricultural yield was driven principally by an ever-increasing use of fertilizer, whose supply is limited and whose negative impacts on the environment are serious (see also Pimentel and Pimentel, 1996). Now we may be reaching the limits of that approach implying limitations to feeding an increasing population. We also found that expanding the area cropped decreases the average cereal production.

Third, the recent increase in the United State's economy is accomplished in part at the cost of declining environmental quality in poorer countries.

Fourth, increasing the sustainability of development is more difficult when the population issue is considered. The results of multiple regressions of energy consumption with GDP and total population level for the five countries show the importance of both economic growth and population growth in decreasing sustainability of resource use, especially energy (Table V). Seventy-four per cent of the variation in energy consumption in the United States is correlated with these variables and the multiple regression explains more than the simple linear regression in Table II.

Table V
Determinants of energy consumption in the five countries, 1970-92
The proposed equation: Energy Consumption (ENGCONS) = a[GDP] + b[Total Population (POP)] + c

Country	Equations	Adjusted R-square	Durbin_ Watson	Significance
Costa Rica	ENGCONS = 3.68 GDP + 0.02 POP - 15.15	0.9745	1.359	0.0001
Korea	ENGCONS = 12.12 GDP - 0.01 POP + 486.94	0.9889	1.257	0.0001
Mexico	ENGCONS = 5.61GDP + 0.05 POP -11857.22	0.9713	0.313	0.0001
Netherlands	ENGCONS =8.24 GDP + 0.42 POP - 4572.89	0.9470	1.595	0.0001
USA	ENGCONS = 22.78GDP -0.68 POP+148101	0.7393	0.546	0.0001

Source: United Nations. *Energy Statistics Yearbook*. (multiple years); WRI (1996). *World Resources 1996-97*, database diskette.

The ecological footprint analysis for nations shows that these countries are using the ecological services of more lands than they have within their borders. To achieve sustainability, an alternative to current industrialization needs to be found, that provides decent and equitable living within the means of nature. Not living within these ecological limits leads to the depletion of each country's natural capital and/or increases each country's dependence on other's ecological capacity. Having insufficient natural resources, or not living decently and equitably may cause conflict and degrade social fabric, none of which leads to any sort of sustainability.

There are some positive signs. In the United States, micro-efficiency increases appear to have led to macro-efficiency in energy use, although it is not realized as a decrease in total energy use. The Netherlands has decreased intensities of both fertilizer and land use and has gained in the efficiencies of use for each. Per capita footprints are declining everywhere except Korea. This study at least shows that when developing or selecting indicators of sustainability, more than the traditional *ecological* indicators need to be considered.

In conclusion, the benefits of technological development are not quite clear. The argument for sustainable development may be an oxymoron (like 'jumbo shrimp'), which may meet our sentimental view only. Without a serious change in our idea of economic development and serious thoughts about population, economic growth and resource depletion, we may never be able to implement sustainable development in any meaningful way. On the other hand, perhaps we should not try to achieve the unachievable, and ask instead how we should use our remaining fossil fuels, soils and other resources as wisely as possible.

Acknowledgments:

We thank Sabine O'Hara, Mathis Wackernagel, David Yount and Timm Kroeger for critical reviews. Mario Giampietro brought Jevons' paradox to our attention. Luis G. López Lemus wishes to acknowledge the financial support of the Consejo Nacional de Ciencia y Tecnología (CONACyT, México), through the endorsement of the Fulbright Commission at the Institute for International Educacion (IIE, USA). Many of these ideas were first considered by Howard Odum.

References

Ausubel, J.: 1996, *Am. Scientist.* **84**, 166-178.

Brown, L. R, Flavin, C., French, H., Abramovitz, J., Bright, C., Garner, G., McGinn, A., Renner, M., Roodman, D., Starke, L.: 1997, *State of the World*, WorldWatch/Norton & Co., 229 pp.

Cleveland, C.J., Costanza, R., Hall, C, A., Kaufmann, R. K.: 1984, *Science*. **225**, 890-897.

Council on Environmental Quality: 1992, *Environmental Quality 1991*.

Daly, H. E.: 1991, *Steady-State Economics*. 2nd Ed., Island Press, 297 pp.

Ehrlich, P., Ehrlich, A.: 1990, *The Population Explosion*, Simon & Schuster, 320 pp.

Ehrlich, P., Ehrlich, A., Holdren, J. P.: 1977, *Ecoscience: Population, Resource, Environment*, W.H. Freeman, 1051 pp.

Emmelin, L.: 1973, *Ambio*. **2**, 26-36.

Food and Agricultural Organization of the United States (FAO) : 1992, *Fertilizer Use by Crop*, FAO, 67 pp.

FAO: 1997, Agricultural Statistics from the FAO *Internet Web site* : http:// www.fao.org.

Feensta, R.C., Lipsey, R.E., Bowen, H.P.: 1997, *World Trade Flows, 1970-1992, with Production and Tariff Data*. National Bureau of Economic Research. CD-ROM.

Freund, P., Martin, G.: 1993, *The Ecology of the Automobile*. Black Rose Books, 213 pp.

Gillis, M., Perkins, D. H., Roemer, M., Snodgrass, D. R.: 1996, *Economics of Development*. 4th ed. W.W. Norton & Co., 604 pp.

Goldemberg, J., Johansson, T. B., Reddy, A. K. N., Williams, R. H.: 1988, *Energy for a Sustainable Development*. Wiley Eastern Ltd.

Goodland, R. J. A., Daly, H. E., Serially, S. E.: 1993, *Environ Conservation*, **20**, 297-309.

Hall, C. A. S., Pontius, G., Coleman, L., Ko, J.: 1994, *Population and Environment* **15**, 505-524.

Hall, C. A. S., Hall, M. H. P.: 1993, *Agricultural Ecosystems and Environment* **46**, 1-30.

Hall, C. A. S., Cleveland, C., Kaufmann, R.: 1986, *Energy and Resource Quality: The Ecology of the Economic Process* . Wiley Interscience, 577pp.

International Union for Conservation of Nature and Natural Resources (IUCN).: 1980, *World Conservation Strategy: Living Resource Conservation for Sustainable Development*. United Nations Environmental Program and the World Wildlife Fund.

Jensen, N. F.: 1978, *Science* **201**, 317-320.

Jevons, W.S.: 1864, *The Coal Question: an Enquiry Concerning the Progress of the Nation, and the Probable Exhaustion of our Coal Mines*. Macmillian.

Kesler, S. E.: 1994, *Mineral Resources, Economics and the Environment*. Macmillan College Publishing Co., 391 pp.

Knight, D., Mitchell B., Wall, G.: 1997, *Ambio*. **26,** 97-100.

Lélé, S. M.: 1991, *World Development*. **19**, 607-621.

Leonard, H. J.:1988, *Pollution and the Struggle for the World Product: Multinational Corporations, Environment, and International Comparative Advantage*. Cambridge University Press. 254 pp.

López Lemus, L. G.: 1997, *Proceedings of the 20th Conference of the International Association of Energy Economics*. New Dehli.

Malthus, T. R.: 1798, *An Essay on the Principle of Population*. (J.M. Dent & Sons, London, 1961).

Naylor, R. L.: 1996, *Annu. Rev. Energy Environ.* **21,** 99-123.

Nilsson, L. J.: 1993, *Energy* **18**, 309-322.

Pimentel, D., Pimentel, M.: 1996, *Food, Energy and Society*. Rev. Ed., University Press of Colorado, 392 pp.

Pimentel, D., Harvey, C., Resosudamo, P., Sinclair, K., Kurz, D., McNair, M., Crist, S., Shpritz, L., Fitton, L., Saffouri, R., Blair, R.: 1995, *Science*. **267**, 1117-1123.

Plucknett, D. L., Smith, N. J .H.: 1986, *BioScience* **36**, 40-45.

Ricardo, D.: 1817, *The Principles of Political Economy and Taxation*. G. Bell and Sons.

Schipper, L., Meyers, S., Howarth, R. B., Steiner, R.: 1992, *Energy Efficiency and Human Activity: Past, Trends, Future Prospects*. Cambridge University Press, 385 pp.

Schloss, M.: 1993, *World Resource Rev.*, **5**, 214-27.

Selden, T. M., Song, D.:1995, *J. Environ. Econ. Management* **27,** 147-162.

Simon, J., Kahn, H. (Eds) :1984, *The Resourceful Earth*. Blackwell Scientific Publ., 585 pp.

Smith, F. L.:1994, *Boston College Environ. Affairs Law Rev.* **21,** 297-308.

Soussan, J., Gevers, E., Ghimire, K., O'Keefe, P.:1991, *World Development* **19**, 1299-1314.

Stern, D. I., Common, M. S., Barbier, E. D.: 1994, *Economic growth and environmental degradation: A critique of the environmental Kuznets curve*. Department of Environmental Economics and Environmental Management. University of York, York, U.K.

Todaro, M. P.: 1994, *Economic Development*. 5th Ed. Longman, 719 pp.

United Nations. *Energy Statistics Yearbook* (multiple years).

US Department of Commerce [Bureau of Economic Analysis]: 1992, *Business Statistics 1963-91*. US Government Printing Office, 194 pp.

US Congress Office of Technology Assessment.: 1990, *Energy Use and the US Economy*. US Government Printing Office, 65 pp.

Wackernagel, M., Onisto, L., Callejas Linares, A., López Falfán, I.S., Méndez García, J., Suárez Guerrero, .A. I., Suárez Guerrero, M. J.:1997, *Ecological Footprints of Nations: How Much Nature Do They Use? -- How Much Nature Do They Have?* A **Rio+5 Forum** study commissioned and financed by the Earth Council. Costa Rica, 32 pp.

Wackernagel, M., Rees, W.E. :1996, *Our Ecological Footprint: Reducing Human Impact on the Earth.* New Society Publishers, 160 pp.

World Bank. :1997, *Expanding the Measure of Wealth: Indicators of Environmentally Sustainable Development.* [Río +5 Edition — Draft for Discussion]. The World Bank.

World Commission on Environment and Development.:1987, *Our Common Future.* Oxford University Press, 400 pp.

World Resource Institute.:1996, *World Resources 1996-97: A Guide to Our Global Environment.* Oxford University Press, 365 pp.

CANADA's ECOLOGICAL MONITORING AND ASSESSMENT NETWORK: WHERE WE ARE AT AND WHERE WE ARE GOING

TOM BRYDGES and ASHOK LUMB

Ecological Monitoring Coordinating Office, Environment Canada, Canada Centre for Inland Waters, 867 Lakeshore Road, Burlington, Ontario, Canada, L7R 4A6

Abstract. Canada has established a National Ecological Monitoring and Assessment Network (EMAN). The Network's operating objective is to understand what changes are occurring in the ecosystems and why. Each site is designed to have long-term multidisciplinary monitoring programs in place with supporting research and manipulation experiments. About 85 sites have been incorporated into the network. A Directory of EMAN Sites is available and a list of the Goals, Objectives and Deliverables (GODs) for many sites is also available. Information can be obtained on the EMAN's website at http://www.cciw.ca/eman/. The network is operated in conjunction with a program of developing national environmental indicators, with increasing emphasis on indicators of sustainable development. A series of environmental assessments are being produced that are issue and/ or area focused. The assessment are designed as support for policy decisions. The national coordinating office supports the overall program of data gathering, reporting environmental indicators and produce assessments.

1. Introduction

Multidisciplinary environmental studies, particularly at the small watershed level, have been carried out in Canada for several decades. Studies were initiated by Governments and academic institutions, usually to deal with environmental problems of interest to the specific location. For example, in the 1960s, the Federal Government initiated studies on lake eutrophication at the Experimental Lakes Area near Kenora, Ontario (Hecky, Rosenburg and Campbell, 1994), and Laval University began the Centre for Arctic Studies at Kuujjaauapik which has focused on arctic and sub-arctic ecological processes. Also in the 1960s, studies at Kejimkujik National Park began to look at nutrient processes in surface waters. In the mid 1970s, the Ontario Government conducted a comprehensive study of the effects of cottage development on lakes in the Muskoka area (Hutchinson, Neary and Dillon, 1991). The Last Mountain Lake site was established as a National Wildlife Area. Many other sites have been established across the country to look at a variety of research questions and environmental factors. As new issues have emerged, other sites, for example, Turkey Lakes in Ontario and Duschenay in Quebec, were established in response to the need for more information on acid rain. These multi-year, interdisciplinary studies were very effective in resolving the site-specific scientific and policy questions set out by the supporting agencies.

Many urgent environmental problems confronting society, such as global warming (Intergovernmental Panel on Climate Change, 1995 Report), UV-B, depletion of the stratospheric ozone layer [Scientific Committee on Problems of the Environment (SCOPE), 1992 Report], acid rain, etc., are connected with man-made changes to the atmosphere.

Environmental Monitoring and Assessment **51:** 595–603, 1998.

These have an impact at the multinational regional level and are of global concern. The ecological effects of these stresses are subtle and show up over long periods of time. Equally, reversing the effects by pollution control measures will take a long time. The input of data collected for over 10 years at some of the 15 ecological study sites across eastern Canada and the United States provided enough information to establish the deposition targets in eastern North America. This represented a scientific basis for action and defined a solution that led to defining control measures needed to address the acid rain problem. Understanding the ecological consequences of global climate variability/change will require long-term ecological monitoring sites around the globe. These current environmental problems are scientifically much more complex in their ecological effects and they affect larger areas. Therefore, it has become necessary to further develop the concept of long-term (i.e., decades) multidisciplinary studies. Understanding how ecosystems are changing, and developing the scientific information required by decision-makers, are beyond the resources and abilities of any single department or agency. Consequently, it is necessary to develop partnerships within all components of the Canadian and international environmental science community. This is necessary to maximize the quality of the science and the efficiency of conducting the work at a time of economic restraint. These concepts led to the creation of Ecological Monitoring and Assessment Network.

2. Where we are at and how Does EMAN Operate?

In April 1994, Environment Canada established the Ecological Monitoring and Assessment Network (EMAN), with an overall goal of conducting long-term multi-disciplinary research and monitoring sufficient to provide answers to the questions of what is changing in ecosystems and why. To conduct this network's business, the Ecological Monitoring Coordinating Office (EMCO) was located at Canada Centre for Inland Waters, Burlington, Ontario, Canada. The EMAN has four overall objectives: 1) To provide a national perspective on how Canadian ecosystems are being affected by the multitude of stresses on the environment; 2) To provide scientifically defensible rationales for pollution control and resource management policies; 3) To evaluate and report to Canadians on the effectiveness of these policies; 4) To identify new environmental issues at the earliest possible stage.

The Ecological Monitoring and Assessment Network (EMAN) is a cooperative partnership of academic, governmental (local, provincial, and federal) and private sector scientists. EMCO's goal is to coordinate the ecological monitoring and research to meet national, regional, and local environmental needs for environmental information on ecosystem function and change.

The Ecological Monitoring Coordinating Office staff of five was given the responsibility of organizing the EMAN into a cohesive network of existing sites and also promoting the development of new sites where feasible. Some of these sites, mentioned above, have been established over the years for a number of reasons, and most of these are operated by Federal Departments, Provinces, Universities, Industries and NGOs. The

EMCO staff worked in conjunction with seven Regional Leaders in the five Environment Canada Regions. Pacific and Yukon, and Prairie and Northern Region each have assigned a leader for the southern and northern halves of these geographically large regions, while Atlantic, Quebec and Ontario Regions have one leader each. Site specific and program leadership are provided by staff of other Federal Departments, Provinces and Territorial agencies, universities, schools and the private sector.

As of January 1997, the EMCO included 85 sites (Figure 1) into the newly formed Ecological Monitoring and Assessment Network. Canada has been divided into fifteen land-based ecozones plus five marine ecozones, and it is the EMAN objective to have at least one monitoring site in each of these ecozones. These sites are organized into 17 terrestrial Ecological Science Cooperatives (ESCs) and are included in the National Directory (EMAN Occasional Paper Series, Report. 2, 1996a). Where there is more than one monitoring site, they will be loosely linked in an Ecological Science Cooperative, since sites in the same ecozone will have a number of common interests. For example, it will be important to compare all of the results from the Boreal Shield forest with regard to issues, such as climate change or UV-B radiation. Equally, it will be important to compare all sites within the Prairie ecozone regarding the response to issues such as increasing average temperature. For some issues, such as climate change or plant phenology, it may be relevant to compare results from all sites across the country. It is therefore anticipated that some new sites will be added and, in this time of restraint, some sites may cease to operate. Each of these sites have developed a statement of their Goals, Objectives and Deliverables (GODs) Declarations (EMAN Occasional Paper Series, Report. 3, 1996b) indicating the nature of the work being carried out at the site. The monitoring programs at the EMAN's sites are briefly described in Figure 2. There are over 100 agencies involved in conducting ecological monitoring and research, including the Federal government, Provinces, universities, private sector and NGO's. We anticipate that the issues covered will expand as new problems are found and new partners join the program.

A fundamental start-up procedure has been to conduct organizational workshops within each region or ecozone. These workshops bring together the interested parties to get to know each other, exchange information across disciplines and sectors, and compile a list of issues and sites within a given area. Out of a small EMCO budget, the "grease and glue" money is used to provide travel and organizational resources for these workshops. Such workshops have produced many comprehensive reports on environmental issues.

An annual national science meeting is held each January, rotating among the five Environment Canada Regions. This multidisciplinary meeting has representatives from Governments, EMAN sites, universities, NGO's and industry. The meeting is to help with the Network "construction" and to promote discussion on the scientific issues and results coming from the long-term multidisciplinary studies.

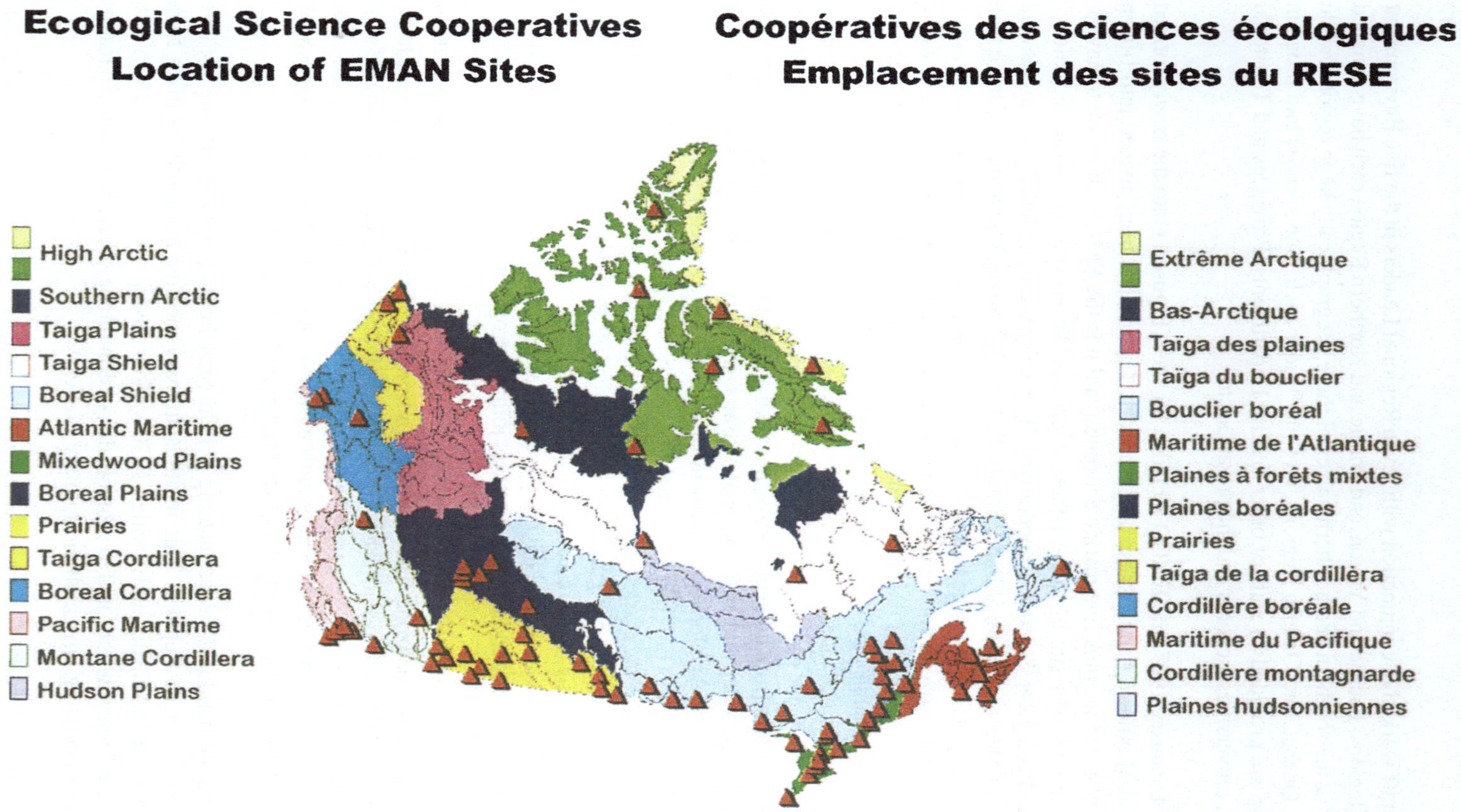
Ecological Science Cooperatives
Location of EMAN Sites
Coopératives des sciences écologiques
Emplacement des sites du RESE
High Arctic
Southern Arctic
Taiga Plains
Taiga Shield
Boreal Shield
Atlantic Maritime
Mixedwood Plains
Boreal Plains
Prairies
Taiga Cordillera
Boreal Cordillera
Pacific Maritime
Montane Cordillera
Hudson Plains
Extrême Arctique
Bas-Arctique
Taïga des plaines
Taïga du bouclier
Bouclier boréal
Maritime de l'Atlantique
Plaines à forêts mixtes
Plaines boréales
Prairies
Taïga de la cordillèra
Cordillère boréale
Maritime du Pacifique
Cordillère montagnarde
Plaines hudsonniennes

Fig. 1. Ecological Science Cooperatives Location of EMAN Sites.

Abiotic Monitoring

Hydrological
Water Quantity (surface water and ground water)
Water Quality (physico chemical, metals and organics):

Geophysical
(Geographic & geological research and monitoring land use changes, radar automated seismograph)

Atmospheric
Atmospheric Pollutants: (physico chemical, metals and organics):

Climate Change
Climate cycles -

Meteorological
(temperature, precipitation, evaporation, wind and radiation long and short wave)

Radiation
(Ozone and UV-B Radiation)

Pedological
(soil erosion: agricultural impacts on soil quality:)

Cryological
(ice cores: glaciology: sea ice: snow pillow studies: snow depth and moisture equivalence monitoring: surging glaciers: and permafrost)

Biotic Monitoring

Fresh Water Aquatic Biota
(Flora, invertebrate and vertebrate. Ecology/management... limnology, palaeo-limnology, aquatic wildlife, riparian ecology, lake carbon cycling, population dynamics, ecotoxicology, benthic communities, impacts of agricultural and industrial pollution on aquatic ecosystem.)

Terrestrial Biota
(Flora/vegetation, invertebrate, vertebrate and ecology/ management)

Marine
(Shellfish, beluga whale, intertidal zone, and aqua-culture impact)

Soil
(soil research - arthropod; micro-biota, soil dynamics)

Palaeo/History
(palaeo-limnology, palaeobiology palaeo-ecology especially fire history and dendrochronology)

Human Studies
(archaeology, heritage monitoring, traditional knowledge, oral history, human systems, recreation, cultural anthropology)

Biomass
(evaluation, above/below ground biomass)

Fig. 2. Monitoring Programs at various EMAN Sites.

3. Where we are Going?

In June 1996, the EMCO was combined with the Indicators group of the substantially downsized State of Environment Reporting Branch of Environment Canada. The newly formed Indicators, Monitoring and Assessment Branch (IMAB) was given a coordinating and facilitating role in the generation of data and the use of standard indicators, and assisted in the production of issue- or area-related assessments to provide a report to the Canadian people and decision-makers with information on the ecological condition of Canada. IMAB has two offices: Ecological Monitoring and Coordination Office (EMCO) in Burlington, Ontario is responsible for the coordination of EMAN; the second, the Indicators and Assessment Office in Ottawa, Ontario is responsible for developing and reporting the environmental indicators and assessments. The overall operating objective of IMAB (Figure 3) is to promote the gathering and use of scientific environmental information for the policy and management decision-making processes and to provide a better link between the policy requirements and the scientific community.

The most important function of the Network will be to serve as a major source of ecological and environmental information that is driven by a series of policy- or issue-related questions. The information will be assembled in the form of periodic issue- or area-related assessments. In addition, some components of the information will be used for the ongoing production of indicators, which will provide Canadians with the current status of various issues which will be dealt with in greater depth by the periodic assessments (Figure 3).

Substantial support activities for EMAN sites are being carried out by IMAB in the development of standard parameter lists, standard measurement protocols, data management systems, and Quality Assurance/Quality Control through the EMAN QA/QC Steering Committee and Biodiversity Science Advisory Board (See EMAN web site http://www.cciw.ca/eman/). This Committee and Board has the mandate to promote and develop protocols and standard procedures for environmental sampling, analysis, data collection, recording, and to organize the data into a structured system to allow for easy inputting of new data and data updates. The Committee and Board are also charged with ensuring that data are in most shareable form possible for any type of user using appropriate quality assurance and control measures. Indicators and Assessment Office of IMAB is committed to improving the practical usefulness of environmental information and optimizing its delivery in a manner that permits the easy integration of environmental, social and economic perspectives in support of sustainable development. The Biodiversity Science Advisory Board will also address the biodiversity issue, including Canadian commitments to the Convention on Biological Diversity.

EMCO has a major interest in and applying some resources to the development of extensive volunteer networks. The participants in the networks, such as weather networks, breeding bird surveys, plant phenology and amphibian surveys, have effectively gathered data for decades. These programs greatly assist various Departments in obtaining extensive

Fig. 3

Reporting on the State of the Environment: The Way Ahead Operations Flow Chart

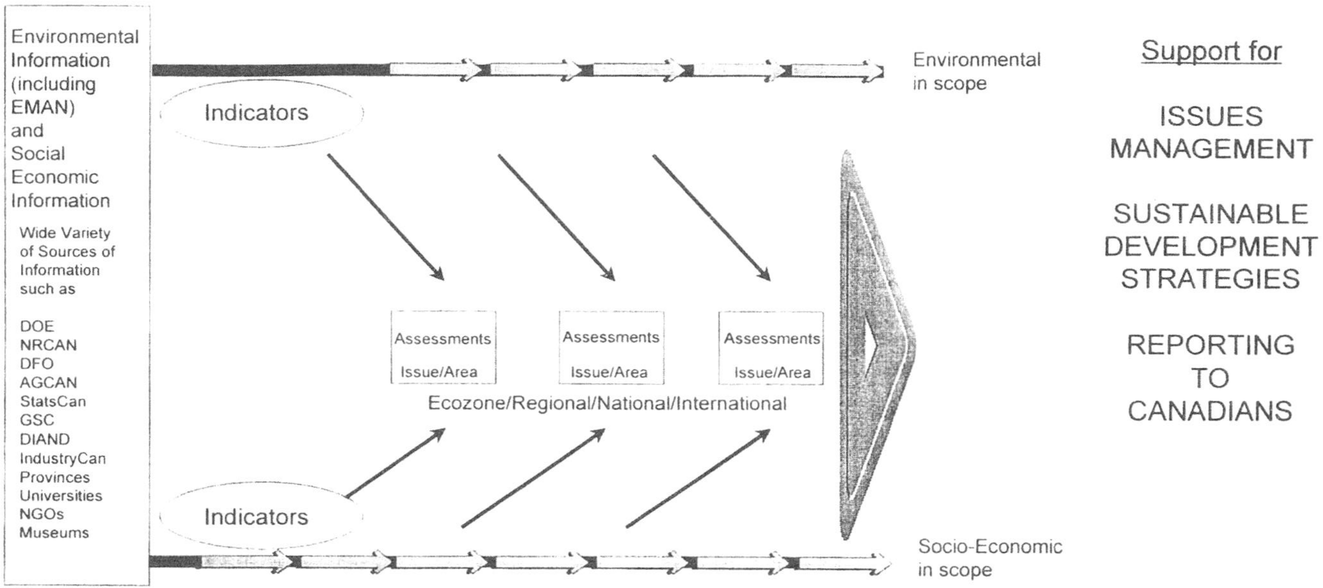

environmental information. It is the objective of EMCO to have as many as possible of these extensive networks collecting data from EMAN sites. This will provide additional opportunities for explaining any changes that are observed in these measurements, and in turn, being able to extrapolate the results from the EMAN sites to larger geographical areas covered by the extensive network. Within three years we hope that virtually all of the extensive volunteer networks, such as breeding bird survey, DAPCAN, plant phenology, ice phenology, frog watch, tree watch, etc., will have participants at every appropriate EMAN site. This, we hope, would promote the involvement of more professionals in the volunteer network activities and also provide some increased ability to interpret changes identified in the extensive volunteer network by using the detailed information available at the EMAN sites. In this way we can cover huge areas of the Canadian landscape with a coordinated monitoring network.

It is expected that the scientists and supporting agencies for all sites will become increasingly familiar with the nature of the policy issues and policy concerns at the local, regional, national, and global levels. Monitoring and research programs need to be oriented toward these issues. All supporting and funding agencies need to be aware of the overall organization and, during the resource allocation process, give priority to the EMAN activities. To this end, EMAN coordinators have already met with Natural Science and Engineering Research Council (NSERC) to explain the Network operations. There have been meetings with industries to explain the Network and the role of the private sector and this has resulted in funding for a number of sites and projects.

The Network will only be as effective as the scientists, governments, universities, and the concerned public make it. Many scientists have already seen the Network as an opportunity to develop joint projects over larger areas or with other disciplines. Pooling of data, and even resources, can result in an enhanced program and output from individual projects. The EMCO invites and encourages the scientists in all parts of the Network to take the initiative in organizing programs so that the scientific total is greater than the sum of the individual parts. The EMCO would like to see the day when more scientists become well known as experts on issues or components of issues. That does not mean that they do, or direct, all of the work, but that they serve as a focal point for speaking to the public and the media on a particular issue and for improving the communications among the scientific community dealing with their particular area of expertise. Overall, we see the Network as providing substantial opportunity for individual scientist development and recognition.

We would like to see teaching institutions, particularly at the high school and university level, incorporate EMAN concepts and activities in their curricula. The goal is to have the entire EMAN structure as a fully integrated "package" of policy questions, appropriate monitoring, and assessment activity, leading to policy answers followed up by routine reporting of indicators.

4. More Information About EMAN?

Point your Web browser at the EMAN Web site http://www.cciw.ca/eman/ or call Tom Brydges, Director, Ecological Monitoring Coordinating Office at 905 336-4410.

References

EMAN Occasional Paper Series, Report. 2, Ecological Science Cooperatives: Directory of EMAN Sites. Ecological Monitoring Coordinating Office (1996a).

EMAN Occasional Paper Series, Report. 3, EMAN's Goals, Objectives and Deliverables: 1996 Declarations. Ecological Monitoring Coordinating Office (1996b).

Hecky, R.E., Campbell, P., and Rosenburg, D.M.,: 1994, 'Introduction to Experimental Lakes and Natural Processes: 25 years of Observing Natural Ecosystems at the Experimental Lakes Area', *Cand. J. Fish. Aquat. Sci.* **Vol. 51,** 2721-2722, and the references cited there in.

Hutchinson, N.J., Neary, B.P., and Dillon, P.J.: 1991, 'Validation and Use of Ontario's Trophic Status Model for Establishing Lake Development Guidelines', *Lake and Reserv. Manage.* **7(1),** 13-23.

Intergovernmental Panel on Climate Change, 1995 Report. *'Climate Change 1995 - IPPC Second Assessment Report'*. CCP Office, 4th Floor, North Tower, Les Terrasses de la Chaudière, 10 Wellington Street, Hull, Quebec, K1A 0H3.

Scientific Committee on Problems of the Environment (SCOPE), 1992 Report, 'Effects of Increased Ultraviolet Radiation on Global Ecosystem', SCOPE Secretariat, 51 bd de Montmorency 75016 Paris, France.

The manufacturer's authorised representative in the EU is Springer Nature Customer Service Centre GmbH, Europaplatz 3, 69115 Heidelberg, Germany. If you have any concerns regarding our products, please contact ProductSafety@springernature.com

Printed and bound by CPI Group (UK) Ltd, Croydon, CR0 4YY
24/04/2026
02096316-0012